KT-465-377

How to access the supplemental online study guide

We are pleased to provide access to an online study guide that supplements your textbook, *Physiology of Sport and Exercise, Fourth Edition*. This study guide offers an interactive learning experience that will allow you to practice, review, and develop knowledge and skills about the physiology of sport and exercise. We are certain you will enjoy this unique online learning experience.

Accessing the online study guide is easy! Simply follow these steps:

1. Using your web browser, go to the **Physiology of Sport and Exercise** product Web site at **www.HumanKinetics.com/ PhysiologyofSportandExercise**.
2. Click on the **View Student Resources** button on the right side of the home page.
3. Click on the please register now link. You will create your personal profile and password at this time.
4. Write your e-mail and password down for future reference. Keep it in a safe place.
5. Once you are registered, enter the key code exactly as it is printed at the right, including all hyphens. Click **Submit**
6. Once the key code has been submitted, you will see a welcome screen. Click the **Continue** button to open your online study guide.
7. After you enter the key code the first time, you will not need to use it again to access the study guide. In the future, simply log in using your e-mail and the password you created.

For technical support, send an e-mail to:

support@hkusa.com . . . . U.S. and international customers
info@hkcanada.com . . . . Canadian customers
academic@hkeurope.com . . . . European customers
keycodesupport@hkaustralia.com . . . . Australian customers

Product: Physiology of Sport and Exercise, Fourth Edition online study guide

Key code: WILMORE-6Z464WQ6-0736062262

This unique code allows you access to the online study guide

Access is provided if you have purchased a new book.
Once submitted, the code may not be entered for any other user.

FOURTH EDITION

PHYSIOLOGY OF SPORT AND EXERCISE

Jack H. Wilmore, PhD
Margie Gurley Seay Centennial Professor Emeritus
Department of Kinesiology and Health Education
The University of Texas at Austin

David L. Costill, PhD
Emeritus John and Janice Fisher Chair in Exercise Science
Human Performance Laboratory
Ball State University

W. Larry Kenney, PhD
Professor of Physiology and Kinesiology
Department of Kinesiology
Penn State University

Human Kinetics

Library of Congress Cataloging-in-Publication Data

Wilmore, Jack H., 1938-
Physiology of sport and exercise / Jack H. Wilmore, David L. Costill, W. Larry Kenney.—4th ed.
p. ; cm.
Includes bibliographical references and index.
ISBN-13: 978-0-7360-5583-3 (hard cover)
ISBN-10: 0-7360-5583-5 (hard cover)
1. Exercise—Physiological aspects. 2. Sports—Physiological aspects. I. Costill, David L. II. Kenney, W. Larry. III. Title.
[DNLM: 1. Exercise—physiology. 2. Sports—physiology. 3. Physical Endurance. 4. Physical Fitness. QT 260.W744p 2008]
QP301.W6749 2008
612'.044—dc22

2007019594

ISBN-10: 0-7360-5583-5
ISBN-13: 978-0-7360-5583-3

Permission notices for photographs in this book from other sources can be found on page xvi.

The Web addresses cited in this text were current as of April 2007, unless otherwise noted.

For instructor and student resources, visit the *Physiology of Sport and Exercise* Web site at www.HumanKinetics.com/PhysiologyOfSportAndExercise.

Acquisitions Editor: Michael S. Bahrke, PhD; **Editor of Original Edition:** Lori Garrett; **Developmental Editor:** Maggie Schwarzentraub; **Assistant Editor:** Jillian Evans; **Copyeditor:** Joyce Sexton; **Proofreader:** Erin Cler; **Indexer:** Craig Brown; **Permission Manager:** Dalene Reeder; **Graphic Designer:** Fred Starbird; **Graphic Artist:** Dawn Sills; **Photo Asset Manager:** Laura Fitch; **Photo Office Assistant:** Jason Allen; **Cover Designer:** Keith Blomberg; **Photographer (cover):** Getty Images; **Photographer (interior):** see p. xvi for a full listing; **Art Manager:** Kelly Hendren; **Illustrators of Graphs and Computer-Generated Art:** Mic Greenberg, Chuck Nivens, and Tammy Page; **Medical Illustrators:** Mic Greenberg, Kristin Mount, Tammy Page, and Argosy; **Printer:** Quad Graphics

Printed in the United States of America 10 9 8 7 6 5 4 3 2 1

Human Kinetics
Web site: www.HumanKinetics.com

United States: Human Kinetics
P.O. Box 5076
Champaign, IL 61825-5076
800-747-4457
e-mail: humank@hkusa.com

Canada: Human Kinetics
475 Devonshire Road Unit 100
Windsor, ON N8Y 2L5
800-465-7301 (in Canada only)
e-mail: orders@hkcanada.com

Europe: Human Kinetics
107 Bradford Road
Stanningley
Leeds LS28 6AT, United Kingdom
+44 (0) 113 255 5665
e-mail: hk@hkeurope.com

Australia: Human Kinetics
57A Price Avenue
Lower Mitcham, South Australia 5062
08 8372 0999
e-mail: info@hkaustralia.com

New Zealand: Human Kinetics
Division of Sports Distributors NZ Ltd.
P.O. Box 300 226 Albany
North Shore City
Auckland
0064 9 448 1207
e-mail: info@humankinetics.co.nz

To those who have had the greatest impact on my life: to my loving wife, Dottie, and our three wonderful daughters, Wendy, Kristi, and Melissa, for their patience, understanding, and love; to my sons-in-law, Craig, Brian, and Randall, for being such great husbands and fathers; to my grandchildren, who are a constant source of joy and amazement; to Mom and Dad for their love, sacrifice, direction, and encouragement; to my present and former students, who are a continual source of joy and inspiration; and to my Lord, Jesus Christ, who provides for every one of my needs.

Jack H. Wilmore

To my grandchildren, Renee and David, who have added a new dimension to my life. To my wife, Judy, who gave me two loving daughters, Jill and Holly. To my college swimming coach, Bob Bartels, who "rescued my soul" on more than one occasion, and showed me the joys of research and teaching. To my former students, who taught me more than I taught them—their subsequent successes have been the highlight of my career.

David L. Costill

To my wife, Patti, a constant reminder and example of what's truly important in life, for all her love and support through the years. To Matt, as he maneuvers down his own educational and athletic path—enjoy the adventure. To Alex, my constant reminder that life is for living—be yourself and never lose that bounce. And to Lauren, the light of my life, you're the best—keep smiling. And to my parents, who gave me everything I really ever needed.

W. Larry Kenney

Contents

Event
Travel
VIESSMANN

Preface

The body is an amazingly complex machine. All of its various cells and tissues communicate with each other, and their activities are precisely coordinated. When you think of the numerous processes occurring within the body at any given time, it is truly remarkable that all the body systems function so well together. Even as you sit reading this, your heart pumps blood throughout your body, your intestines digest and absorb nutrients, your kidneys clear waste products, your lungs bring in oxygen, and your muscles hold this book while your brain concentrates on reading. Although you may feel at rest, your body is physiologically quite active. Imagine, then, how much more active all your body systems become when you engage in exercise. As your exercise intensity increases, so does your muscles' physiological activity. Exercising muscles require more nutrients, more oxygen, more metabolic activity, and thus more efficient clearance of waste products. How does your body respond to the increased physiological demands of exercise?

That is the key question when you study the physiology of sport and exercise. *Physiology of Sport and Exercise, Fourth Edition,* introduces you to the fields of sport and exercise physiology. Our goal is to build on the foundation of knowledge that you have developed through basic course work in human anatomy and physiology by applying the principles you've learned to how the body performs and responds to physical activity.

WHAT'S NEW IN THE FOURTH EDITION?

The fourth edition of *Physiology of Sport and Exercise* has changed substantially. Most importantly, we now have a third author, Dr. W. Larry Kenney, who is a professor in the Noll Laboratory within the Department of Kinesiology at Penn State University. As a former president of the American College of Sports Medicine and an acknowledged leader in the field of exercise physiology, Dr. Kenney has brought a fresh look to the organization and content of this book. His expertise, along with the expertise and advice of nine content experts who reviewed the third edition of this book, has helped us develop a better product. The names and university affiliations of these content experts are listed in the acknowledgments section of this book.

In this fourth edition, we have made a conscious effort to reduce the size of the book without reducing the content. Changes include a reduction in paper weight, along with a very small reduction in the font size and page margins; deletion of material in each chapter that was not central to the specific purpose of the chapter; and placement of the references and selected readings sections for each chapter at the end of the book. Further, new topics and material have been added, and the material included in the third edition has been reorganized. Several chapters were eliminated, but the material contained in those chapters has been combined with material from other chapters to provide for a more effective presentation. Also, new chapters have been added.

As with previous editions, with sport and exercise the focus is on muscle and how its needs are altered as someone goes from a resting to an active state. In chapters 1 through 7, we focus first on muscle and then look at how it is fueled through metabolism, controlled by the nervous system, and supplied with oxygen through the cardiovascular and respiratory systems, both at rest and during an acute bout of exercise. Chapters 8 through 10 are new chapters that concentrate on the principles of exercise training and the specific adaptations that occur in response to resistance training and aerobic and anaerobic training. In the final parts of the fourth edition, the only major change in organization was to combine the chapters on nutrition and body composition into a single chapter to reduce overlap of material.

BASIC ORGANIZATION OF THE FOURTH EDITION

We begin in the introduction with a historical overview of sport and exercise physiology as they have emerged from the parent disciplines of anatomy and physiology, and we explain basic concepts that will be used throughout the text. In parts I and II, we review the major physiological systems, focusing on their responses to acute bouts of exercise. In part I, we focus on how the muscular, metabolic, endocrine, and nervous systems interact to produce body movement. In part II, we look at how the cardiovascular and respiratory systems transport nutrients and oxygen to the active muscles and waste products away from them during physical activity. In part III, we consider how these systems adapt to long-term exposure to exercise—training.

We change perspective in part IV to examine the impact of the external environment on physical performance. We consider the body's response to heat and cold, and then we examine the impact of low atmospheric pressure experienced at altitude. In part V, we shift our attention to how athletes can optimize physical performance. We evaluate the effects of different amounts of training. We consider the importance of appropriate body composition for performance and examine athletes' special dietary needs and how nutrition can be used to enhance performance. Finally, we explore the use of ergogenic aids: substances purported to improve athletic ability.

In part VI, we examine unique considerations for specific populations of athletes. We look first at the processes of growth and development and how they affect the performance capabilities of young athletes. We evaluate changes that occur in physical performance as people age and explore the ways in which physical activity can prolong youthfulness. Finally, we examine issues and special physiological concerns of female athletes.

In the final part of the book, part VII, we turn our attention to the application of sport and exercise physiology to prevent and treat various diseases and the use of exercise for rehabilitation. We look at prescribing exercise to maintain health and fitness, and then we close the book with a discussion of cardiovascular disease, obesity, and diabetes.

SPECIAL FEATURES INCLUDED IN THE FOURTH EDITION

This fourth edition of *Physiology of Sport and Exercise* offers a novel approach to the study of sport and exercise physiology. It was designed for the student reader with the goal of making learning easy and enjoyable. This text is comprehensive, but we don't want readers to be overwhelmed by either its size or its scope. We have included special features to help the reader progress through the book.

Chapter outlines identify where material is located

Reminders within outlines indicate related **study guide activities**

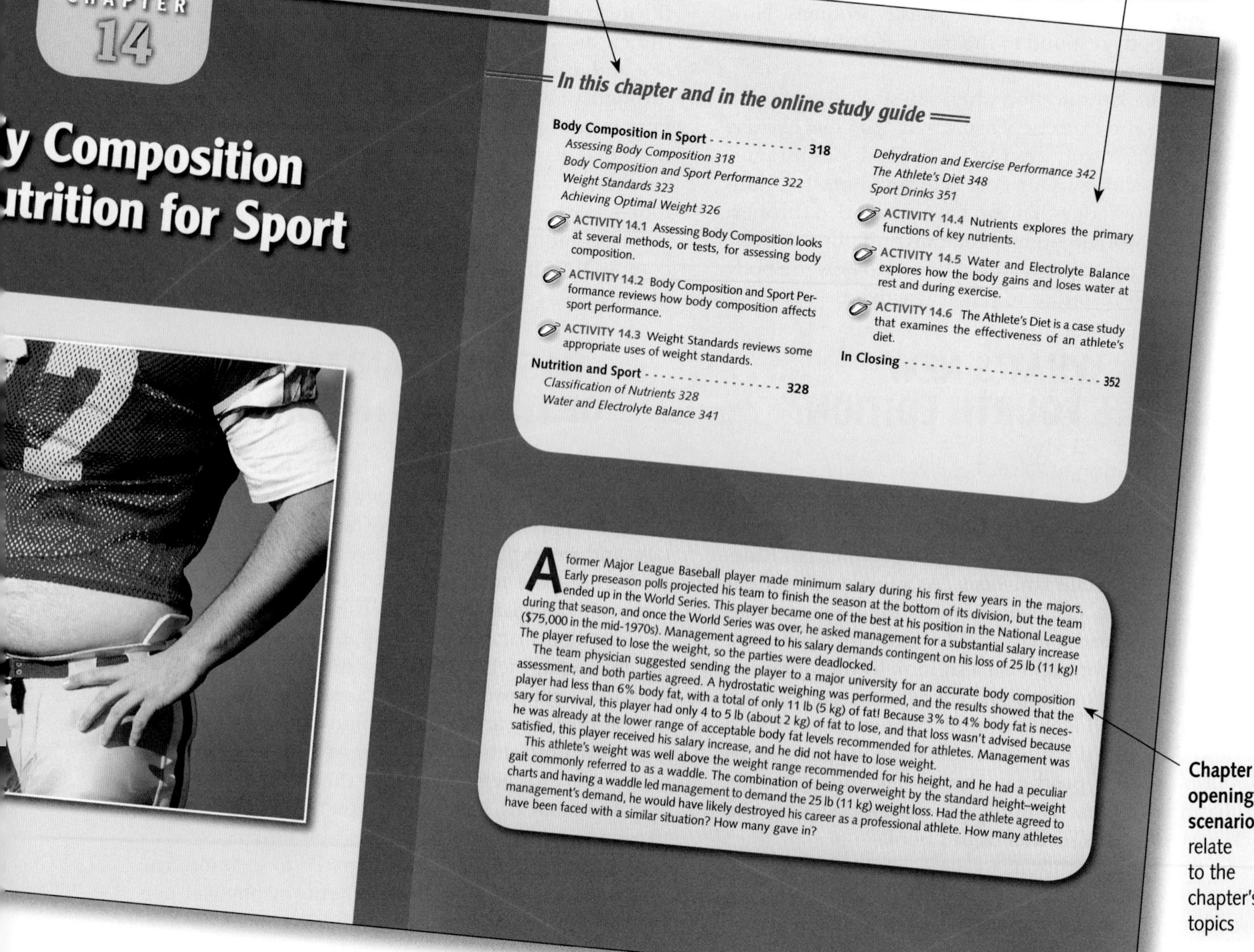

CHAPTER 14

y Composition
utrition for Sport

In this chapter and in the online study guide

A former Major League Baseball player made minimum salary during his first few years in the majors. Early preseason polls projected his team to finish the season at the bottom of its division, but the team ended up in the World Series. This player became one of the best at his position in the National League during that season, and once the World Series was over, he asked management for a substantial salary increase ($75,000 in the mid-1970s). Management agreed to his salary demands contingent on his loss of 25 lb (11 kg)! The player refused to lose the weight, so the parties were deadlocked.

The team physician suggested sending the player to a major university for an accurate body composition assessment, and both parties agreed. A hydrostatic weighing was performed, and the results showed that the player had less than 6% body fat, with a total of only 11 lb (5 kg) of fat! Because 3% to 4% body fat is necessary for survival, this player had only 4 to 5 lb (about 2 kg) of fat to lose, and that loss wasn't advised because he was already at the lower range of acceptable body fat levels recommended for athletes. Management was satisfied, this player received his salary increase, and he did not have to lose weight.

This athlete's weight was well above the weight range recommended for his height, and he had a peculiar gait commonly referred to as a waddle. The combination of being overweight by the standard height–weight charts and having a waddle led management to demand the 25 lb (11 kg) weight loss. Had the athlete agreed to management's demand, he would have likely destroyed his career as a professional athlete. How many athletes have been faced with a similar situation? How many gave in?

317

Chapter opening scenarios relate to the chapter's topics

Student and Instructor Resources

STUDENT RESOURCES

Students, visit the free **Online Study Guide** at www.HumanKinetics.com/PhysiologyOfSportAndExercise for dynamic and interactive learning activities, all of which can be conducted outside the lab or classroom. You'll be able to apply the key concepts learned by conducting self-made experiments and recording your own physiological responses to exercise. The guide also includes study questions and activities to test your knowledge as you prepare for quizzes or tests. You'll also have access to links to professional journals, organizations, and career information. The Web site format conveniently allows you to practice, review, and develop knowledge and skills about the physiology of sport and exercise via the Internet.

INSTRUCTOR RESOURCES

Instructor Guide

Specifically developed for instructors who have adopted *Physiology of Sport and Exercise, Fourth Edition,* the Instructor Guide includes sample lecture outlines, suggestions for class projects and student assignments, key points, laboratory experiences for case studies, and direct links to detailed sources on the Internet for every chapter in the text.

Test Package

The Test Package, created with Respondus 2.0, includes a bank of over 1,800 questions created especially for *Physiology of Sport and Exercise, Fourth Edition.* A variety of question types are included, such as true-false, fill-in-the-blank, essay/short answer, and multiple choice. With Respondus LE, a free version of the Respondus software, instructors can

- create print versions of their own tests by selecting from the question pool;
- create, store, and retrieve their own questions;
- select their own test forms and save them for later editing or printing; or
- export them into a word processing program.

Respondus also offers the capability to create and manage exams that can be published directly to Blackboard, eCollege, WebCT, and other course management systems. Instructions for downloading a free version of Respondus and for acquiring the Test Package are at www.HumanKinetics.com/PhysiologyOfSportAndExercise.

The Instructor Guide and Test Package are free to course adopters.

Presentation Package

The Presentation Package includes a comprehensive series of PowerPoint slides for each chapter. Learning-objective slides present the major topics covered in each chapter; text slides list key points; and illustration and photo slides contain the outstanding graphics found in the text. The Presentation Package has more than 700 slides that can be used directly with PowerPoint and used to print transparencies or slides or make copies for distribution to students. Instructors can easily add to, modify, or rearrange the order of the slides, as well as search for images based on key words. You may access the Presentation Package by visiting www.HumanKinetics.com/PhysiologyOfSportAndExercise.

Acknowledgments

We would like to thank Rainer Martens for his continued support of this fourth edition of *Physiology of Sport and Exercise.* Rainer and the staff at Human Kinetics have been supportive of our many demands and have been extremely dedicated to publishing a quality product. Special recognition must go to Lori Garrett, the editor of our first edition; Julie Rhoda, our developmental editor for both the second and third editions of this book; and Maggie Schwarzentraub, our developmental editor for this fourth edition, who has worked very hard to bring this edition to completion. Maggie has managed to keep the book on schedule while maintaining a quality product. She has also had to work with a "grumpy old man" and a much younger man who were not always considerate of her special needs and the conditions under which she had to work. A very sincere thank you to Maggie!

We especially appreciate the efforts of Lawrence Armstrong, Calvin Caplan, James Gale, Dan Halvorsen, Peter Iltis, Anthony Mahon, Donald Michielli, Robert Ruhling, Frans Verstappen, Lawrence Weiss, and Ben B. Yaspelkis, III, who provided us important feedback on the first edition, as well as the efforts of Scott Trappe, Howard G. Knuttgen, the anonymous colleagues who reviewed the second edition, and Jim Davis, who reviewed the third edition.

For this fourth edition, Human Kinetics allowed us to solicit the reviews of content experts in the field, who did an excellent job of reviewing their respective chapters. Special thanks go to Bob Armstrong (Texas A&M University), Larry Armstrong (University of Connecticut), Kirk Cureton (University of Georgia), Mark Hargreaves (The University of Melbourne), Bill Haskell (Stanford University), Donna Korzick (Penn State University), Jim Pawelczyk (Penn State University), Janet Walberg Rankin (Virginia Tech), and Chris Womack (Michigan State University) for their insightful and helpful comments and criticisms. Dr. Korzick also graciously gave us access to several figures that are new to the fourth edition.

Special recognition goes to Lacy Alexander Holowatz at Penn State University for her hard work in helping us reorganize and revise selected chapters of this fourth edition. Her insights and skill in editing have made this a much better book. Finally, we thank our families, who put up with our many long hours of isolation while we were writing, rewriting, editing, and finally proofing this book. Their patience and support can never be repaid.

Jack H. Wilmore
David L. Costill
W. Larry Kenney

Photo Credits

Chapter opener photos

Introduction (p. iv, xviii): © Human Kinetics
Chapter 1 (p. iv, 23, 24): © Visuals Unlimited/Corbis
Chapter 2 (p. iv, 23, 46): © PA Photos
Chapter 3 (p. iv, 23, 78): © Lester V. Bergman/CORBIS
Chapter 4 (p. iv, 23, 98): © 2003 Nigel Farrow
Chapter 5 (p. v, 121, 122): © Matthias Kulka/zefa/Corbis
Chapter 6 (p. v, 121, 142): © Getty Images
Chapter 7 (p. v, 121, 160): © StockByte
Chapter 8 (p. v, 185, 186): © Louie Psihoyos/CORBIS
Chapter 9 (p. v, 185, 202): © Human Kinetics
Chapter 10 (p. v, 185, 220): © Nigel Farrow
Chapter 11 (p. vi, 251, 252): © Getty Images
Chapter 12 (p. vi, 251, 278): © SportsChrome/Bongart
Chapter 13 (p. vi, 295, 296): © Nigel Farrow
Chapter 14 (p. vi, 295, 316): © Photodisc
Chapter 15 (p. vii, 295, 354): © Photodisc
Chapter 16 (p. vii, 381, 382): © Human Kinetics
Chapter 17 (p. vii, 381, 402): © Kevin Dodge/Corbis
Chapter 18 (p. vii, 381, 422): © PA Photos
Chapter 19 (p. vii, 447, 448): © Nigel Farrow
Chapter 20 (p. viii, 447, 470): © Getty Images
Chapter 21 (p. viii, 447, 492): © Getty Images

Photos courtesy of the authors

Figures 0.1*a-b,* 0.2*a-b,* 0.3*a-c,* 0.4*b-c,* 1.1*a-c,* 1.10, 1.11*a-c,* 4.9*a-b,* 17.6, 19.1, 19.2, 20.7, and 21.8*a*
Photos on p. 1, 11, 12, 17, 36, 38, 243, 457, and 573

Additional photos

Photo on p. 4: Photo courtesy of American College of Sports Medicine Archives. All rights reserved.
Photos on p. 9: Photo courtesy of American College of Sports Medicine Archives. All rights reserved.
Figure 0.4*a*: Photo courtesy of American College of Sports Medicine Archives. All rights reserved.
Figure 0.5: © Human Kinetics
Figure 0.6: © Bongarts/Getty Images
Figure 1.4: © Custom Medical Stock Photo
Figure 3.2*a*: © Custom Medical Stock Photo
Photo on p. 99: © Getty Images
Figure 4.2*a*: © Tom Roberts
Figure 4.2*b*: Photo courtesy of COSMED Engineering.
Figure 5.7*a*: © Lester Lefkowitz/CORBIS
Photo on p. 182: © Blair Seitz
Figure 8.1: © Tom Roberts
Figure 8.3: © Human Kinetics
Figure 8.5: © PA Photos
Figure 9.1: © BananaStock
Figure 9.3*a-b*: Photos courtesy of Dr. Michael Deschene's laboratory.
Figure 9.9: From R.C. Hagerman et al., 1984, "Muscle damage in marathon runners," *Physician and Sportsmedicine* 12: 39-48.

Figure 9.10*a-b*: From R.C. Hagerman et al., 1984, "Muscle damage in marathon runners," *Physician and Sportsmedicine* 12: 39-48.
Photos on p. 227: © Tom Roberts
Figure 11.3*a-b*: From Department of Health and Human Performance, Auburn University, Alabama. Courtesy of JohnEric Smith, Joe Molloy, and David D. Pascoe. By permission of David Pascoe.
Figure 14.2: Tom Pantages
Figure 14.3*a-b*: Photos courtesy of Hologic, Inc.
Figure 14.4: Photo courtesy of Life Measurement, Inc.
Figure 14.5: © Human Kinetics
Figure 14.6: © Human Kinetics
Photo on p. 355: © Gero Breloer/epa/Corbis
Photo on p. 383: © Associated Press
Figure 18.11*a-b*: Reproduced from *Journal of Bone and Mineral Research* 1986, 1:15-21 with permission of the American Society for Bone and Mineral Research. Photos provided courtesy of D.W. Dempster.
Figure 18.13: © SportsChrome
Figure 19.3: © Human Kinetics
Figure 19.5: © Getty Images
Figure 19.8: © Icon Sports Media
Figure 19.9: © Nigel Farrow
Figure 19.10: © Lee Cohen/Corbis
Photo on p. 493: © Getty Images
Figure 21.8*b-c*: From J.C. Seidell et al., 1987, "Obesity and fat distribution in relation to health - Current insights and recommendations," *World Review of Nutrition and Dietetics* 50: 57-91.

An Introduction to Exercise and Sport Physiology

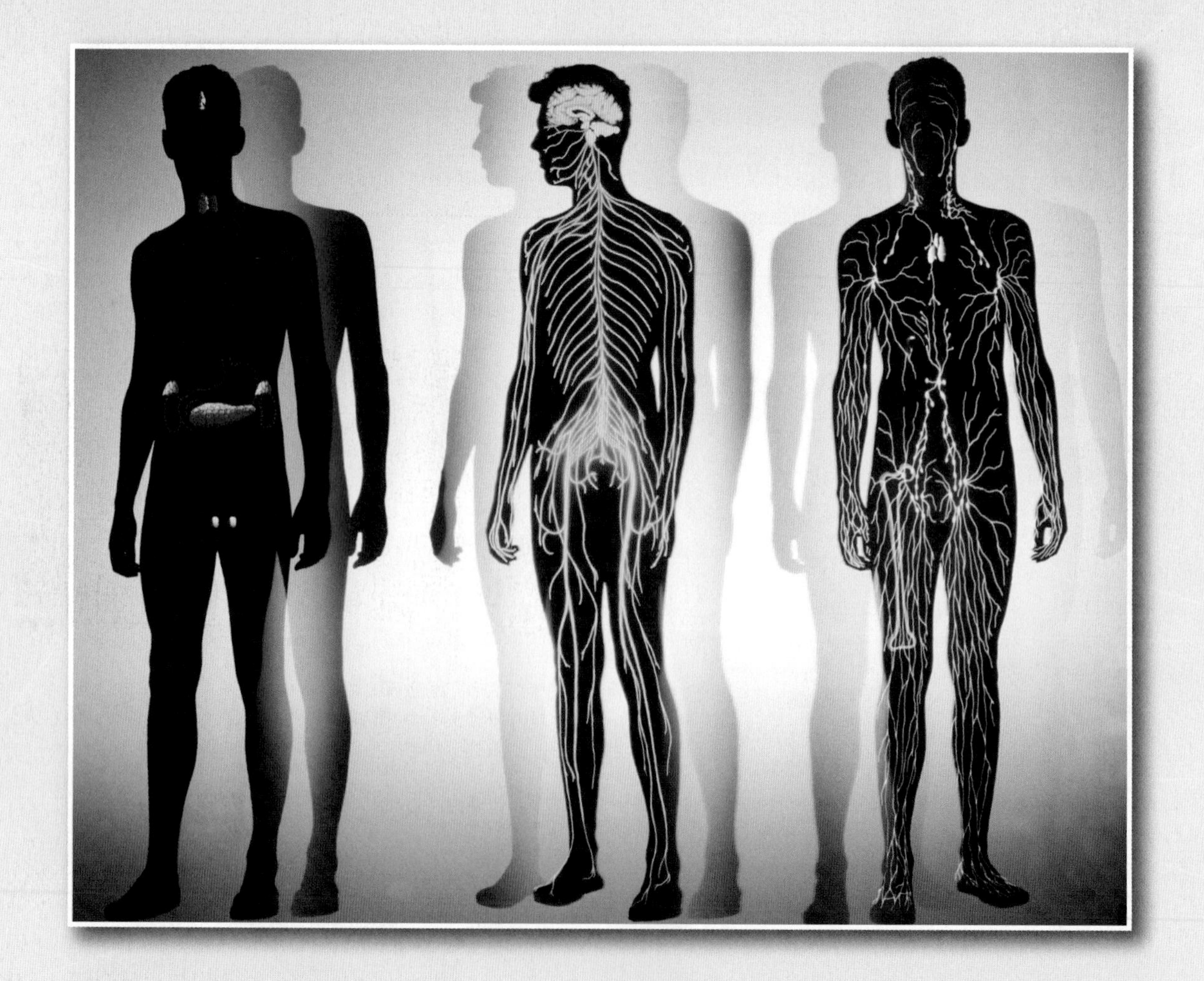

In this chapter and in the online study guide

Much of the history of exercise physiology in the United States can be traced to the effort of a Kansas farm boy, David Bruce (D.B.) Dill, whose interest in physiology first led him to study the composition of crocodile blood. Fortunately for us, this young scientist redirected his research to humans when he became the first director of the Harvard Fatigue Laboratory, established in 1927. Throughout his life he was intrigued by the physiology and adaptability of many animals that survive extreme environmental conditions; but he is best remembered for his research on human responses to exercise, heat, high altitude, and other environmental factors. Dr. Dill always served as one of the human "guinea pigs" in these studies. During the Harvard Fatigue Laboratory's 20-year existence, he and his coworkers produced approximately 350 scientific papers along with a classic book titled *Life, Heat, and Altitude.*[8]

After the Harvard Fatigue Laboratory closed its doors in 1947, Dr. Dill began a second career as deputy director of medical research for the Army Chemical Corps, a position he held until his retirement from that post in 1961. Dr. Dill was then 70 years old—an age he considered too young for retirement—so he moved his exercise research to Indiana University, where he served as a senior physiologist until 1966. In 1967 he obtained funding to establish the Desert Research Laboratory at the University of Nevada at Las Vegas. Dr. Dill used this laboratory as a base for his studies on human tolerance to exercise in the desert and at high altitude. He continued his research and writing until his final retirement at age 93, the same year he produced his last publication, a book titled *The Hot Life of Man and Beast.*[10] Dr. Dill once bragged that he was the only scientist to have retired four times.

(a) Dr. David Bruce (D.B.) Dill at the beginning of his career; *(b)* as director of the Harvard Fatigue Laboratory at age 42; and *(c)* Dr. Dill sitting in the same pose when retired at age 92.

The human body is an amazing machine. As you sit reading this introduction, countless perfectly coordinated events are occurring simultaneously in your body. These events allow complex functions, such as hearing, seeing, breathing, and information processing, to continue without your conscious effort. If you stand up, walk out the door, and jog around the block, almost all your body's systems will be called into action, enabling you to successfully shift from rest to exercise. If you continue this routine daily for weeks or months and gradually increase the duration and intensity of your jogging, your body will adapt so that you can perform better.

For example, as the busy executive goes out for his morning jog or the point guard directs her team down the basketball court on a fast break, their bodies must make many adjustments that require a series of complex interactions involving many body systems. Consider a few examples:

- The skeletal system provides the basic framework through which muscles act.
- The cardiovascular system delivers nutrients to the body's various cells and removes waste products.
- The cardiovascular and respiratory systems together provide oxygen to the cells and remove carbon dioxide.
- The integumentary system (skin) helps maintain body temperature by allowing the exchange of heat between the body and its surroundings.
- The urinary system helps maintain fluid and electrolyte balance and assists in the long-term regulation of blood pressure.
- The nervous and endocrine systems coordinate and direct all this activity to meet the body's needs.

Adjustments occur even at the cellular and molecular levels. For example, to enable the biceps muscle to contract to lift a 20 kg (44 lb) weight, nerve cells from the brain, referred to as motor neurons, conduct electrical impulses down the spinal cord to the arm. On reaching the biceps muscle, these neurons release chemical messengers that cross the gap between the nerve and muscle, each neuron exciting a number of individual muscle cells or fibers. Once the nerve impulses cross this gap, they spread along the length of each muscle fiber, entering into the muscle fiber through small pores. Once inside the muscle fiber, the impulse activates the muscle fiber's contraction processes, which involve specific protein molecules—actin and myosin—and an elaborate energy system to provide the fuel necessary to sustain a single contraction and subsequent contractions. It is at this level that other molecules, such as adenosine triphosphate (ATP) and phosphocreatine (PCr), become critical for providing the energy necessary to fuel contraction.

For centuries, scientists have studied **physiology**—how the human body functions. During the past 100 years or so, a growing group of physiologists have focused their studies on how the body functions during physical activity and sport. This introduction presents a historical overview of exercise and sport physiology and then explains some basic concepts that form the foundation for the chapters that follow.

FOCUS OF EXERCISE AND SPORT PHYSIOLOGY

Exercise and sport physiology have evolved from the fundamental disciplines of anatomy and physiology. Anatomy is the study of an organism's structure, or morphology. While anatomy focuses on the basic structure of various body parts and their interrelationships, physiology is the study of body *function.* In physiology, we study how our organ systems, tissues, cells, and the molecules within cells work and how their functions are integrated to regulate our internal environments, a process called **homeostasis.** Because physiology focuses on the functions of body structures, understanding anatomy is essential to learning physiology. Furthermore, both anatomy and physiology rely on a working knowledge of biology, chemistry, physics, and other basic sciences.

Exercise physiology is the study of how our bodies' structures and functions are altered when we are exposed to acute bouts of exercise, a challenge to homeostasis. Because the body adapts to repeated exercise bouts, study of the chronic adaptations to exercise forms the second cornerstone of exercise physiology. Finally, because the environment in which one performs exercise has a large impact, **environmental physiology** has emerged as a subdiscipline of exercise physiology. **Sport physiology** further applies the concepts of exercise

Exercise physiology has evolved from its parent discipline, physiology. It is concerned with the study of how the body adapts physiologically to the acute stress of exercise, or physical activity, and the chronic stress of physical training. Many exercise physiologists utilize exercise or environmental conditions (heat, cold, altitude, etc.) to stress the body in ways that uncover basic physiological mechanisms. Others examine exercise effects on health, disease, and well-being. Sport physiology applies the concepts of exercise physiology to athletes and sport performance.

physiology to training the athlete and enhancing the athlete's sport performance. Thus, sport physiology is derived from exercise physiology.

Let's consider an example to examine the relation between exercise and sport physiology. Through considerable research in exercise physiology, we have come to better understand how our bodies derive energy from the foods we eat to support muscle actions to initiate and sustain movement. Fat is our major energy source when we are at rest and during low-intensity exercise; but our bodies use proportionately more carbohydrate as exercise intensity increases, until carbohydrate becomes our primary energy source. Prolonged high-intensity exercise can substantially reduce our bodies' carbohydrate stores, which can contribute to exhaustion.

Recognizing that the body has limited carbohydrate energy stores, sport physiology uses this information about energy to find ways to

- increase the body's carbohydrate storage capacity (carbohydrate loading),
- decrease the rate at which the body uses carbohydrate during physical performance (carbohydrate sparing), and
- improve the athlete's diet both before and during competition to minimize the risk of depleting carbohydrate stores (the area of sport nutrition, a subdiscipline of sport physiology, is one of the most rapidly growing areas of research and practice).

Because exercise physiology and sport physiology are so closely related and integrated, it is often hard to clearly distinguish the two. For this reason, exercise and sport physiology are often considered together, as they are in this text.

HISTORICAL EVENTS

On initial study, contemporary exercise physiology may seem to present a plethora of new ideas never before studied with rigorous scientific scrutiny, but this is not the case. Rather, the information in this book represents the lifelong efforts of many outstanding scientists who have helped piece together the puzzle of human movement. The thoughts and theories of today's physiological detectives were often shaped by the efforts of scientists who may be long forgotten. What we consider original or new is most often an assimilation of previous findings or the application of basic science to problems in exercise physiology. There are, of course, a number of pivotal scientific findings and a number of investigators who made significant leaps in our knowledge of the physiological responses to physical activity. The following section briefly reflects on the history and on just a few of the people who shaped the field of exercise physiology. It is impossible in this short section to do justice to the hundreds of pioneering scientists who paved the way and laid the foundation for modern exercise physiologists.

Beginnings of Anatomy and Physiology

One of the earliest descriptions of human anatomy and physiology was Claudius Galen's Greek text *De fascius,* published in the first century. As a physician to the gladiators, Galen had ample opportunity to study and experiment on human anatomy. His theories of anatomy and physiology were so widely accepted that they remained unchallenged for nearly 1,400 years. Not until the 1500s were any truly significant contributions made to the understanding of both the structure and function of the human body. A landmark text by Andreas Vesalius, titled *Fabrica Humani Corporis [Structure of the Human Body],* presented his findings on human anatomy in 1543. Although Vesalius' book focused primarily on anatomical descriptions of various organs, he occasionally attempted to explain their functions as well. British historian Sir Michael Foster said, "This book is the beginning, not only of modern anatomy, but of modern physiology. It ended, for all time, the long reign of fourteen centuries of precedent and began in a true sense the renaissance of medicine" (p. 354).[13]

Most early attempts at explaining physiology were either incorrect or so vague that they could be considered no more than speculation. Attempts to explain how a muscle generates force, for example, were usually limited to a description of its change in size and shape during action because observations were limited to what could be seen with the eye. From such observations, Hieronymus Fabricius (ca. 1574) suggested that a muscle's contractile power resided in its fibrous tendons, not in its "flesh." Anatomists did not discover the existence of individual muscle fibers until Dutch scientist Anton van Leeuwenhoek introduced the microscope (ca. 1660). But how these fibers shortened and created force remained a mystery until the middle of the 20th century, when the intricate workings of muscle proteins could be studied by electron microscopy.

Historical Aspects of Exercise Physiology

Exercise physiology is a relative newcomer to the world of science, although muscular activity played an interesting role in a physiological study as early as 1793. A celebrated paper by Séguin and Lavoisier describes the oxygen consumption of a young man as measured in the resting state and while he lifted a weight of 7.3 kg (16 lb) numerous times to a total height of 200 m (219 yd) in 15 min.[19] At rest the man used 24 L of oxygen per hour (L/h), which increased to 63 L/h during exercise. Lavoisier believed that the site of oxygen utilization and carbon dioxide production was in the lungs. This belief was doubted by other physiologists of the time, but remained accepted doctrine until the middle of the 1800s, when several German physiologists demonstrated that combustion of oxygen occurred in tissues throughout the entire body.

Although advances in the understanding of circulation and respiration occurred during the 1800s, few efforts were made to focus on the physiology of physical activity. However, in 1888, an apparatus was described that enabled scientists to study subjects during mountain climbing, even though the subjects had to carry a 7 kg (15.4 lb) gasometer on their backs.[22]

Arguably the first published textbook on exercise physiology, *Physiology of Bodily Exercise,* was written in French by Fernand LaGrange in 1889.[15] Considering the small amount of research on exercise that had been conducted at that time, it is intriguing to read the author's accounts of such topics as "Muscular Work," "Fatigue," "Habituation to Work," and "The Office of the Brain in Exercise." This early attempt to explain the body's response to exercise was, in many ways, limited to speculation and theory, presenting little fact. Although some basic concepts of exercise biochemistry were emerging at that time, LaGrange was quick to admit that many details were still in the formative stages. For example, he stated that "vital combustion [energy metabolism] has become very complicated of late; we may say that it is somewhat perplexed, and that it is difficult to give in a few words a clear and concise summary of it. It is a chapter of physiology which is being rewritten, and we cannot at this moment formulate our conclusions" (p. 395).[15]

Since the early text by LaGrange offered only limited physiological insights regarding bodily functions during physical activity, it might be argued that the third edition of a text by F.A. Bainbridge titled *The Physiology of Muscular Exercise* should be considered the earliest scientific text on this subject.[3] Interestingly, that third edition was written by A.V. Bock and D.B. Dill, at the request of A.V. Hill, three key pioneers of exercise physiology discussed in this introductory chapter.

A.V. Hill

October 16, 1923, was a significant milestone in the history of exercise physiology. A.V. Hill was inaugurated that day as Joddrell Professor of Physiology at University College, London. In his inaugural address he stated the principles that subsequently shaped the field of exercise physiology:

"It is strange how often a physiological truth discovered on an animal may be developed and amplified, and its bearings more truly found, by attempting to work it out on man. Man has proved, for example, far the best subject for experiments on respiration and on the carriage of gases by the blood, and an excellent subject for the study of kidney, muscular, cardiac and metabolic function. . . . Experiment on man is a special craft requiring a special understanding and skill, and 'human physiology,' as it may be called, deserves an equal place in the list of those main roads which are leading to the physiology of the future. The methods, of course, are those of biochemistry, of biophysics, of experimental physiology; but there is a special kind of art and knowledge required of those who wish to make experiments on themselves and their friends, the kind of skill that the athlete and the mountaineer must possess in realizing the limits to which it is wise and expedient to go.

1921 Nobel Prize winner Archibald Hill (1927).

Quite apart from direct physiological research on man, the study of instruments and methods applicable to man, their standardization, their description, their reduction to routine, together with the setting up of standards of normality in man are bound to prove of great advantage to medicine; and not only to medicine but to all those activities and arts where normal man is the object of study. Athletics, physical training, flying, working, submarines, or coalmines, all require a knowledge of the physiology of man, as does also the study of conditions in factories. The observation of sick men in hospitals is not the best training for the study of normal man at work. It is necessary to build up a sound body of trained scientific opinion versed in the study of normal man, for such trained opinion is likely to prove of the greatest service, not merely to medicine, but in our ordinary social and industrial life. Haldane's unsurpassed knowledge of the human physiology of respiration has often rendered immeasurable service to the nation in such activities as coal mining or diving; and what is true of the human physiology of respiration is likely also to be true of many other normal human functions."

During the late 1800s, many theories were proposed to explain the source of energy for muscle contraction. Muscles were known to generate much heat during exercise, so some theories suggested that this heat was used directly or indirectly to cause muscle fibers to shorten. After the turn of the century, Walter Fletcher and Sir Frederick Gowland Hopkins observed a close relationship between muscle action and lactate formation.[11] This observation led to the realization that energy for muscle action is derived from the breakdown of muscle glycogen to lactic acid (see chapter 2), although the details of this reaction remained obscure.

Because of the high energy demands of exercising muscle, this tissue served as an ideal model to help unravel the mysteries of cellular metabolism. In 1921, Archibald V. (A.V.) Hill was awarded the Nobel Prize for his findings on energy metabolism. At that time, biochemistry was in its infancy, although it was rapidly gaining recognition through the research efforts of such other Nobel laureates as Albert Szent Gorgyi, Otto Meyerhof, August Krogh, and Hans Krebs, all of whom were actively studying how living cells generate and utilize energy.

Although much of Hill's research was conducted with isolated frog muscle, he also conducted some of the first physiological studies of runners. Such studies were made possible by the technical contributions of John S. Haldane, who developed the methods and equipment needed to measure oxygen use during exercise. These and other investigators provided the basic framework for our understanding of whole-body energy production, which became the focus of considerable research during the middle of the 20th century and is incorporated into the manual and computer-based systems that are used to measure oxygen uptake in exercise physiology laboratories throughout the world today.

Era of Scientific Exchange and Interaction

From the early 1900s through the 1930s, the medical and scientific environment in the United States was changing. This was an era of revolution in the education of medical students, led by changes at Johns Hopkins. More medical and graduate programs based their educational endeavors on the European model of experimentation and development of scientific insights. There were important advances in physiology in areas such as bioenergetics, gas exchange, and blood chemistry that served as the basis for advances in the physiology of exercise. Building on collaborations forged in the late 1800s, interactions among laboratories and scientists were promoted, and international meetings of organizations such as the International Union of Physiological Sciences created an atmosphere for free scientific exchange, discussion, and debate.

Harvard Fatigue Laboratory

No laboratory has had more impact on the field of exercise physiology than the Harvard Fatigue Laboratory (HFL). A visit by A.V. Hill to Harvard University in 1926 appeared to have a significant impact on the founding and early activities of the HFL, which was founded a year later in 1927. Interestingly, the early home of the HFL was the basement of Harvard's Business School, and its stated early mission was to conduct research on "fatigue" and other hazards in industry. Creation of this laboratory was due to the insightful planning of world-famous biochemist and Nobel laureate, Lawrence J. (L.J.) Henderson. Because Henderson was not interested in heading the laboratory's research program, he chose a young biochemist from Stanford University, David Bruce "D.B." Dill, as the first director of research, a title Dill held until the HFL closed in 1947.

As noted earlier, Dill had aided Arlen "Arlie" Bock in writing the third edition of Bainbridge's text on exercise physiology. Later in his career he credited the writing of that textbook with "shaping the program of the Fatigue Laboratory." Although he had little experience in applied human physiology, Dill's creative thinking and ability to surround himself with young, talented scientists created an environment that would lay the foundation for modern exercise and environmental physiology. For example, HFL personnel examined the physiology of endurance exercise and described the physical requirements for success in events such as distance running. Some of the most outstanding HFL investigations were conducted not in the laboratory but in the Nevada desert, on the Mississippi Delta, and on White Mountain in California (with an altitude of 3,962 m, or 13,000 ft). These and other studies provided the foundation for future investigations on the effects of the environment on physical performance and in exercise and sport physiology.

In its early years, the HFL focused primarily on general problems of exercise, nutrition, and health. For example, the first studies on exercise and aging were conducted in 1939 by Sid Robinson (see figure 0.1), a student at the HFL. On the basis of his studies of subjects ranging in age from 6 to 91 years, Robinson described the effect of aging on maximal heart rate and oxygen uptake.[18] But with the onset of World War II, Henderson and Dill realized the HFL's potential contribution to the war effort, and research at the HFL took a different direction. Harvard Fatigue Lab scientists and support personnel were instrumental in forming new laboratories for the Army, Navy, and Army Air Corps (now the Air Force). They also published the methodologies necessary for relevant military research, methods that are still in use throughout the world.

Today's exercise physiology students would be amazed at the methods and devices used in the early days of the HFL and at the time and energy committed to conducting research projects in those days. What is now accomplished in milliseconds with the aid of computers and automatic analyzers literally demanded days of effort by HFL personnel. Measurements of oxygen uptake during exercise, for example, required collecting expired air in Douglas bags and analyzing it for oxygen and carbon dioxide by using a manually oper-

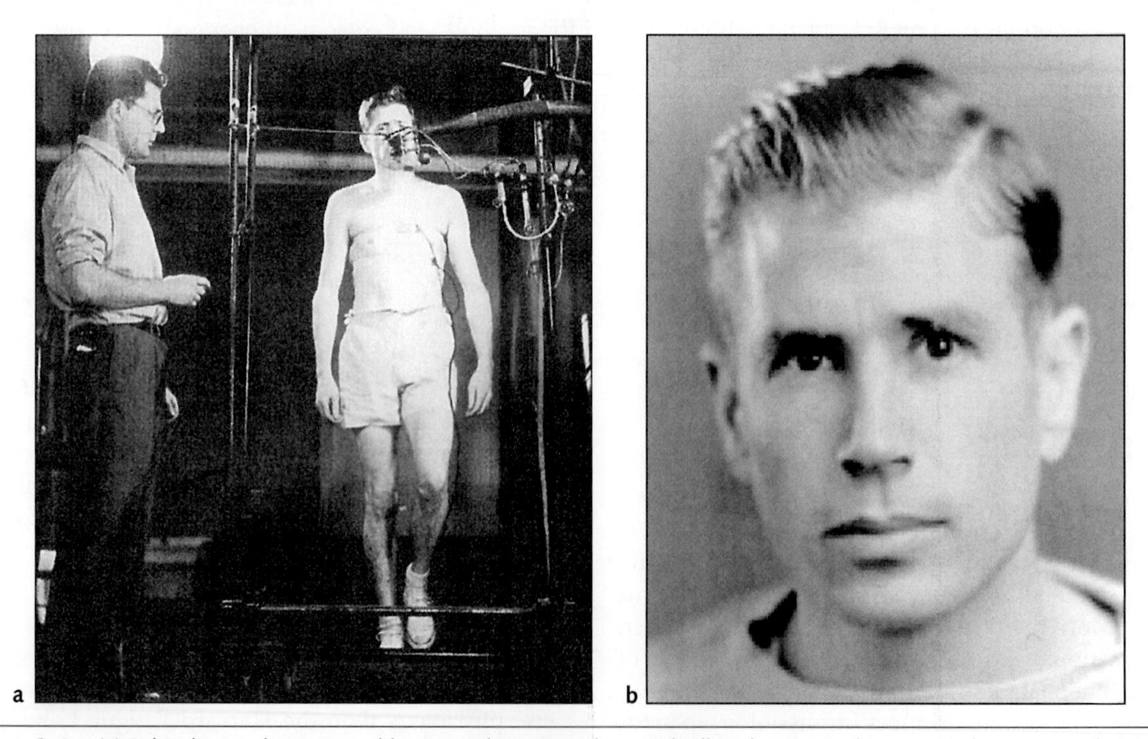

Figure 0.1 *(a)* Sid Robinson being tested by R.E. Johnson on the treadmill in the Harvard Fatigue Laboratory and *(b)* as a Harvard student and athlete in 1938.

ated chemical analyzer, without the help of a computer, of course (see figure 0.2). The analysis of a single 1 min sample of expired air required 20 to 30 min of effort by one or more laboratory workers. Today, scientists make such measurements almost instantaneously and with little physical effort. One must marvel at the dedication, diligence, and hard work of the HFL's exercise physiology pioneers. Using the equipment and methods available at the time, HFL scientists published approximately 350 research papers over a 20-year period.

The HFL was an intellectual environment that attracted young physiologists and physiology doctoral students from all over the globe. Scholars from 15 countries worked in the HFL between 1927 and its closure in 1947. Most went on to develop their own laboratories and become noteworthy figures in exercise physiology in the United States, including Sid Robinson, Henry Longstreet Taylor, Lawrence Morehouse, Robert E. Johnson, Ancel Keys, Steven Horvath, C. Frank Consolazio, and William H. Forbes. Notable international scientists who spent time at the HFL included August Krogh, Lucien Brouha, Edward Adolph, Walter B. Cannon, Peter Scholander, and Rudolfo Margaria, along with several other notable Scandinavian scientists discussed later. Thus, the HFL planted seeds of intellect at home and around the world that resulted in an explosion of knowledge and interest in this new field.

In focus

Founded by biochemist L.J. Henderson in 1927 and directed by D.B. Dill through its closure in 1947, the HFL trained most of those who became world leaders in exercise physiology during the 1950s and 1960s. Most contemporary exercise physiologists can trace their roots back to the HFL.

Scandinavian Influence

In 1909, Johannes Lindberg established a laboratory and fertile breeding ground for scientific endeavors at the University of Copenhagen in Denmark. Lindberg and 1920 Nobel Prize winner August Krogh teamed to conduct classic experiments and published many seminal papers on topics ranging from metabolic fuels for muscle to gas exchange in the lungs. This work was continued from the 1930s into the 1970s by three young Danes, Erik Hohwü-Christensen, Erling Asmussen, and Marius Nielsen.

As a result of contacts between D.B. Dill and August Krogh, these three Danish physiologists came to the HFL in the 1930s, where they studied exercise in the heat and at high altitude. After returning to Europe, each man established a separate line of research. Asmussen and Nielsen became professors at the University of Copenhagen, where Asmussen studied the mechanical properties of muscle and Nielsen conducted studies on control of body temperature. Both remained active at the University of Copenhagen's August Krogh Institute until their retirements.

In 1941, Hohwü-Christensen (see figure 0.3*a* on p. 8) moved to Stockholm to become the first physiology professor at the College of Physical Education at Gymnastik-och Idrottshögskolan (GIH). In the late 1930s, he teamed with Ole Hansen to conduct and publish a series of five studies of carbohydrate and fat metabolism during exercise. These studies still are cited frequently and are considered among the first and most important

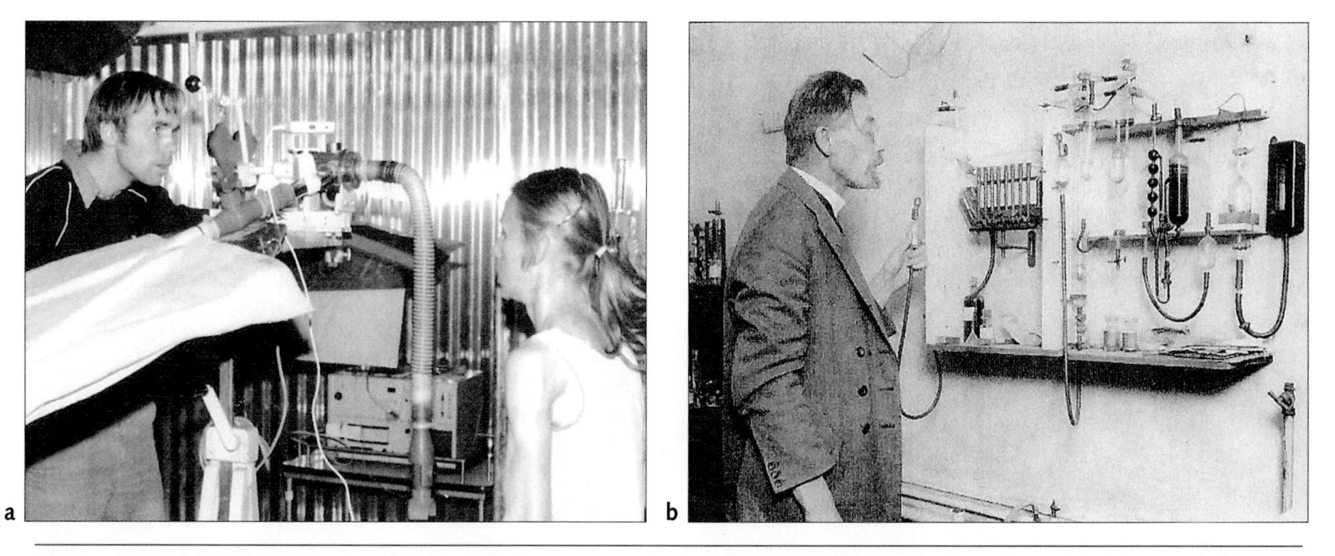

Figure 0.2 *(a)* Early measurements of metabolic responses to exercise required the collection of expired air in a sealed bag known as a Douglas bag. *(b)* A sample of that gas then was measured for oxygen and carbon dioxide using a chemical gas analyzer, as illustrated by this photo of Nobel laureate August Krogh.

sport nutrition studies. Hohwü-Christensen introduced Per-Olof Åstrand to the field of exercise physiology. Åstrand, who conducted numerous studies related to physical fitness and endurance capacity during the 1950s and 1960s, became the director of GIH after Hohwü-Christensen retired in 1960. During his tenure at GIH, Hohwü-Christensen mentored a number of outstanding scientists, including Bengt Saltin, who was the 2002 Olympic Prize winner for his many contributions to the field of exercise and clinical physiology (see figure 0.3*b*).

In addition to their work at GIH, both Hohwü-Christensen and Åstrand interacted with physiologists at the Karolinska Institute in Stockholm who studied clinical applications of exercise. It is hard to single out the most exceptional contributions from this institute, but Jonas Bergstrom's reintroduction of the biopsy needle (ca. 1966) to sample muscle tissue was a pivotal

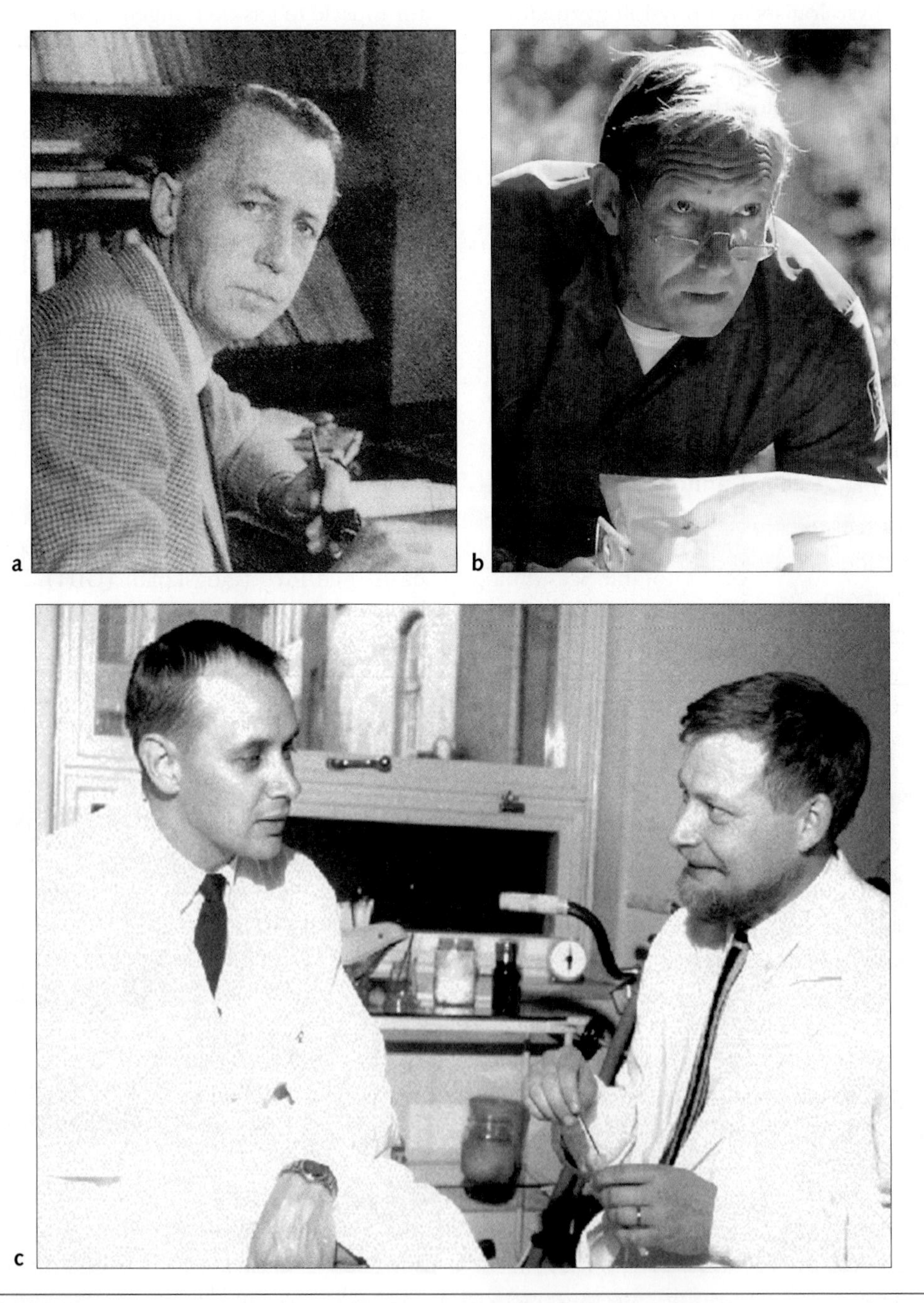

Figure 0.3 *(a)* Erik Hohwü-Christensen was the first physiology professor at the College of Physical Education at Gymnastik- och Idrottshögskolan in Stockholm, Sweden. *(b)* Bengt Saltin, winner of the 2002 Olympic Prize. *(c)* Jonas Bergstrom (left) and Eric Hultman (right) were the first to use the muscle biopsy to study muscle glycogen use and restoration before, during, and after exercise.

point in the study of human muscle biochemistry and muscle nutrition (figure 0.3*c*). This technique, which involves withdrawing a tiny sample of muscle tissue through a small incision, was first introduced in the early 1900s to study muscular dystrophy. The needle biopsy enabled physiologists to conduct histological and biochemical studies of human muscle before, during, and after exercise.

Other invasive studies of blood circulation were subsequently conducted by physiologists at GIH and at the Karolinska Institute. Just as the HFL had been the mecca of exercise physiology research between 1927 and 1947, the Scandinavian laboratories have been equally noteworthy since the late 1940s. Many leading investigations during the past 35 years were collaborations between American and Scandinavian exercise physiologists. Norwegian Per Scholander introduced a gas analyzer in 1947. Finn Martii Karvonen published a formula for calculating exercise heart rate that is still widely used today. (For a more detailed listing of the Scandinavian contributions to exercise physiology, consult Åstrand's review.[1])

Exercise Physiology and Other Fields

Physiology has always been the basis for clinical medicine. In the same manner, exercise physiology has provided essential knowledge for many other areas, such as physical education, physical fitness, physical therapy, and health promotion. In the late 1800s and early 1900s, physicians such as Amherst College's Edward Hitchcock, Jr., and Harvard's Dudley Sargent studied body proportions (anthropometry) and the effects of physical training on strength and endurance. Although a number of physical educators introduced science to the undergraduate physical education curriculum, Peter Karpovich, a Russian immigrant who had been briefly associated with the HFL, played a major role in introducing physiology to physical education. Karpovich established his own research facility and taught physiology at Springfield College (Massachusetts) from 1927 until his death in 1968. Although he made numerous contributions to physical education and exercise physiology research, he is best remembered for the outstanding students he advised, including Charles Tipton and Loring Rowell, both recipients of the American College of Sports Medicine Honor and Citation Awards.

(a) Peter Karpovich introduced the field of exercise physiology during his tenure at Springfield College. *(b)* Thomas K. Cureton directed the exercise physiology laboratory at the University of Illinois at Urbana-Champaign from 1941 to 1971.

Another Springfield faculty member, swim coach T.K. Cureton, created an exercise physiology laboratory at the University of Illinois in 1941. He continued his research and taught many of today's leaders in physical fitness and exercise physiology until his retirement in 1971. Physical fitness programs developed by Cureton and his students, as well as Kenneth Cooper's 1968 book, *Aerobics*, established a physiological rationale for using exercise to promote a healthy lifestyle.[7]

Although there was some awareness as early as the mid-1800s of a need for regular physical activity to maintain optimal health, this idea did not gain popular acceptance until the late 1960s. Subsequent research has continued to support the importance of exercise in resisting the physical decline associated with aging.

Awareness of the need for physical activity has alerted the public to the importance of preventive medicine and the establishment of wellness programs. Although exercise physiology cannot be given credit for the current wellness movement, it provided the basic knowledge and justification for including exercise as an integral component of a healthy lifestyle and laid the foundation for the science of exercise prescription in both sickness and health.

Contemporary Exercise and Sport Physiology

Much advancement in exercise physiology must be credited to improvements in technology. In the late 1950s, Henry L. Taylor and Elsworth R. Buskirk published two seminal papers [6, 20] describing the criteria for measuring maximal oxygen uptake, and establishing this measure as the "gold standard" for cardiorespiratory fitness. In the 1960s, development of electronic analyzers to measure respiratory gases made studying energy metabolism much easier and more productive than before. This technology and radiotelemetry (which uses radio-transmitted signals), used to monitor heart rate and body temperature during exercise, were developed as a result of the U.S. space program. Although such instruments took much of the labor out of research, they did not alter the direction of scientific inquiry. Until the late 1960s, most exercise physiology studies focused on the whole body's response to exercise. The majority of investigations involved measurements of such variables as oxygen uptake, heart rate, body temperature, and sweat rate. Cellular responses to exercise received little attention.

In the mid-1960s, three biochemists emerged who were to have a major impact on the field of exercise physiology. John Holloszy (figure 0.4*a*) at Washington University (St. Louis), Charles "Tip" Tipton (figure 0.4*b*) at the University of Iowa, and Phil Gollnick (figure 0.4*c*) at Washington State University first used rats and mice to study muscle metabolism and to examine factors related to fatigue. Their publications and their training of graduate and postdoctoral students have resulted in a more biochemical approach to exercise physiology research. Holloszy was ultimately awarded the 2000 Olympic Prize for his contributions to exercise physiology and health.

At about the time that Bergstrom reintroduced the needle biopsy procedure, exercise physiologists who were well trained as biochemists emerged. In Stockholm, Bengt Saltin realized the value of this procedure for studying human muscle structure and biochemistry. He first collaborated with Bergstrom in the late 1960s to study the effects of diet on muscle endurance and muscle nutrition. About the same time, Reggie Edgerton (University of California at Los Angeles) and Phil Gollnick were using rats to study the characteristics of individual muscle fibers and their responses to training. Saltin subsequently combined his knowledge of the biopsy procedure with Gollnick's biochemical talents. These researchers were responsible for many early studies on human muscle fiber characteristics and use during exercise. Although many biochemists have used exercise to study metabolism, few have had more impact on the current direction of human exercise physiology than Bergstrom, Saltin, Tipton, Holloszy, and Gollnick.

For more than 100 years, athletes have served as subjects for study of the upper limits of human endurance. Perhaps the first physiological studies on athletes occurred in 1871. Austin Flint studied one of the most celebrated athletes of that era, Edward Payson Weston, an endurance runner/walker. Flint's investigation involved measuring Weston's energy balance (i.e., food intake vs. energy expenditure) during Weston's attempt to walk 400 mi (644 km) in five days. Although the study resolved few questions about muscle metabolism during

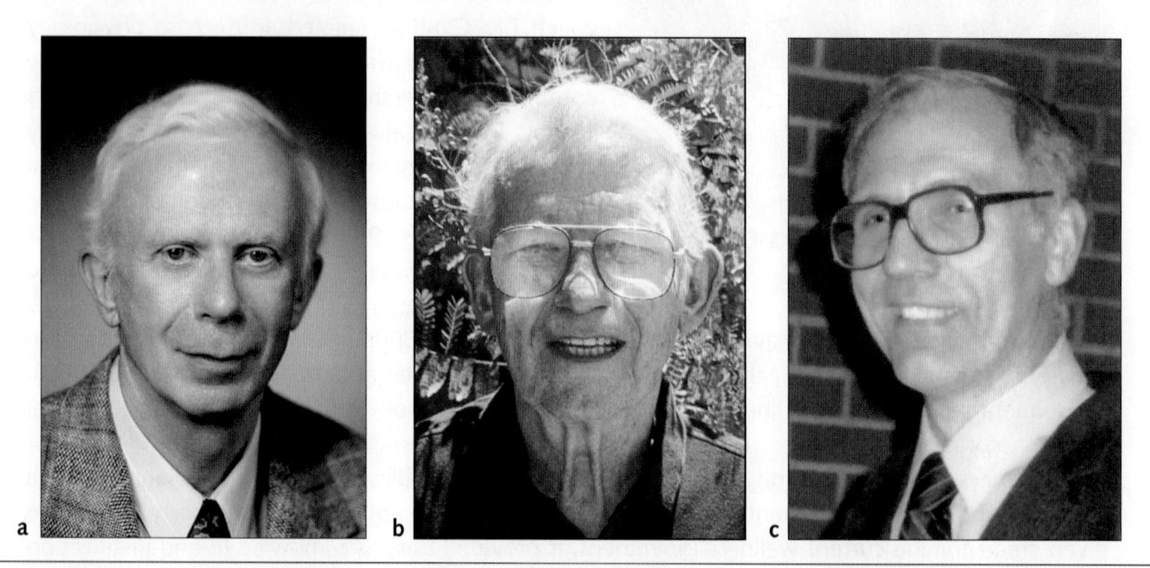

Figure 0.4 *(a)* John Holloszy was the winner of the 2000 Olympic Prize for scientific contributions in the field of exercise science. *(b)* Charles Tipton was a professor at the University of Iowa and the University of Arizona, and a mentor to many students who have become the leaders in molecular biology and genomics. *(c)* Phil Gollnick conducted muscle and biochemical research at Washington State University.

exercise, it did demonstrate that some body protein is lost during prolonged heavy exercise.[12]

Throughout the 20th century, athletes were used repeatedly to assess the physiological capabilities of human strength and endurance and to ascertain the characteristics needed for record-setting performances. Some attempts have been made to use the technology and knowledge derived from exercise physiology to

Exercise Physiology Beyond 2010

Dr. James A. Pawelczyk.

An important segment of exercise physiology concerns the response and adaptation of people to extremes of heat, cold, depth, and altitude. Understanding and controlling the physiological stresses and adaptations that occur at these environmental limits have contributed directly to notable societal achievements such as construction of the Brooklyn Bridge, the Hoover Dam, pressurized aircraft, and underwater habitats for the commercial diving industry.

Will the next environmental challenges require such physiological expertise? Absolutely! In January 2004, President George Bush announced the Vision for Space Exploration, a strategy to first return humans to the moon, then send explorers to the planet Mars, over the next 30 years. This ambitious plan to construct permanent human outposts on the moon beginning in 2017, followed by 2.5-year missions to the planet Mars, will require effective countermeasures to minimize the physiological changes that put space explorers at risk.

The continuous pull of gravity contributes to the growth and adaptation of postural skeletal muscles; loads bones, which increases their size and density; and requires the cardiovascular system to maintain blood pressure and brain blood flow. In a microgravity environment (free fall around the earth or constant-velocity conditions in deep space), the reduction in loading leads to dramatic losses in muscle mass and strength, osteoporosis, and exercise intolerance at rates that mimic those seen in spinal cord–injured patients.

A series of dedicated space shuttle flights have addressed these problems in detail. In 1983, the National Aeronautics and Space Administration (NASA) began flying the European Space Agency–developed Spacelab module, ushering a new era of internationally sponsored scientific research into low-earth orbit. The Spacelab Life Sciences (SLS-1, SLS-2) missions (STS-40 and STS-58) emphasized the study of cardiorespiratory, vestibular, and musculoskeletal adaptations to microgravity. Subsequently the Federal German Aerospace Research Establishment (DLR) sponsored two missions (STS-61A and STS-68), perfecting a model of multidisciplinary, international investigation that was emulated by the Life and Microgravity Sciences Spacelab mission (STS-78), which concentrated on neuromuscular adaptation. The 1998 Neurolab Spacelab mission (STS-90), with an exclusive neuroscience theme, concluded flights of the Spacelab module. Dr. James A. Pawelczyk, a Penn State exercise physiologist and mission specialist on that flight, cotaught the first exercise physiology class from space! Even now, 250 mi (402 km) overhead, an active biomedical research program continues on the International Space Station.

For the exercise physiologist, the question is what combination of resistance and "aerobic" exercise training can prevent or diminish the changes that occur in space. At this time, the answer remains unclear. Furthermore, if physical conditioning is required before and during space exploration and as part of postflight rehabilitation, how should exercise prescriptions be individualized, evaluated, and updated? Without doubt, further research in exercise and environmental physiology will be essential to complete what is destined to be the largest exploration feat of the 21st century.

predict performance, to prescribe training, or to identify athletes with exceptional potential. In most cases, however, these applications of physiological testing are of little more than academic interest because few laboratory or field tests can accurately assess all the qualities required to become a champion.

The intent of this section has been to provide readers with an overview of the personalities and technologies

Women in Exercise Physiology

As in many areas of science, the contributions of female exercise physiologists have been slow to gain recognition. In 1954, Irma Rhyming collaborated with her future husband, P.-O. Åstrand, to publish a classic study that provided a means to predict aerobic capacity from submaximal heart rate.[2] Although this indirect method of assessing physical fitness has been challenged over the years, its basic concept is still in use today.

In the 1970s, two Swedish women, Birgitta Essen and Karen Piehl, gained international attention for their research on human muscle fiber composition and function. Essen, who collaborated with Bengt Saltin, was instrumental in adapting microbiochemical methods to study the small amounts of tissue obtained with the needle biopsy procedure. Her efforts enabled others to conduct studies on the muscle's use of carbohydrates and fats and to identify different muscle fiber types. Piehl published a number of studies that illustrated which muscle fiber types were activated during both aerobic and anaerobic exercise.

In the 1970s and 1980s, a third Scandinavian female physiologist, Bodil Nielsen, daughter of Marius Nielsen, actively conducted studies on human responses to environmental heat stress and dehydration. Her studies even encompassed measurements of body temperature during immersion in water. Interestingly, at about the same time an American female exercise physiologist, Barbara Drinkwater, was doing similar work at the University of California at Santa Barbara. Her studies were often conducted in collaboration with Steven Horvath, D.B. Dill's son-in-law and director of UCSB's Environmental Physiology Laboratory. Drinkwater's contributions to environmental physiology and the physiological problems confronting the female athlete gained international recognition. In addition to their scientific contributions, the legacy of these and other women in physiology includes the credibility they earned and the roles they played in attracting other young women to the field of exercise physiology and medicine.

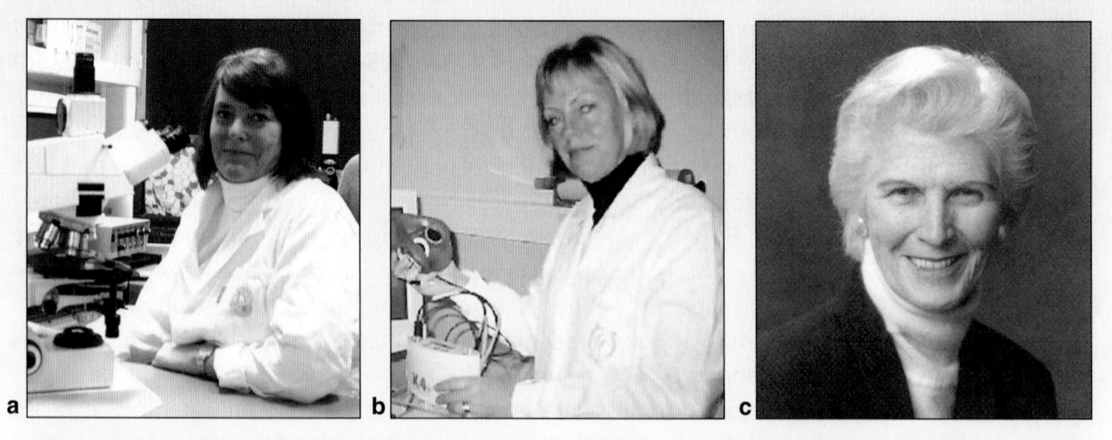

(a) Birgitta Essen collaborated with Bengt Saltin and Phil Gollnick in publishing the earliest studies on muscle fiber types in human muscle. *(b)* Karen Piehl was among the first physiologists to demonstrate that the nervous system selectively recruits type 1 (slow-twitch) and type 2 (fast-twitch) fibers during exercise of differing intensities. *(c)* Barbara Drinkwater was among the first to conduct studies on female athletes and to address issues specifically related to the female athlete.

that have helped to shape the field of exercise physiology. Naturally, a comprehensive review of all the scientists and research associated with this field is not possible in a text intended as an introduction to exercise physiology; but for those students who wish to take an in-depth look at the historical background in exercise physiology, there are several good sources.

In 1968, D.B. Dill wrote a chapter, "History of the Physiology of Exercise," that detailed many of the events and scientists who contributed to this field before the founding of the HFL.[9] In that same year, Roscoe Brown, Jr., the first African American exercise physiologist, coauthored *Classical Studies on Physical Activity*.[4] Although the authors subjectively selected those scientific studies they considered worthy of publication, the edited book provides an excellent sampling of important exercise physiological research from the early 1900s.

In the early 1970s, D.B. Dill's son-in-law and daughter (Steven and Betty Horvath) published a detailed history of the HFL, including the laboratory and field studies conducted by the key scientists of that era.[14] Although others have written different versions of exercise physiology history,[5, 21] most tend to provide the authors' views of important scientists and events, perhaps as we have done here. Finally, McArdle, Katch, and Katch[16] published one of the most comprehensive reviews of the evolution of exercise physiology. Their description of the early anatomists, physiologists, and exercise physiologists clearly illustrates the complexity and diversity of this field of science.

Now that we understand the historical basis for the discipline of exercise physiology, from which sport physiology emerged, we can explore the scope of exercise and sport physiology.

ACUTE AND CHRONIC RESPONSES TO EXERCISE

The study of exercise and sport physiology involves learning the concepts associated with two distinct exercise patterns. First, exercise physiologists are often concerned with how the body responds to an individual bout of exercise, such as running on a treadmill for an hour or undergoing a strength training session. An individual bout of exercise is called **acute exercise,** and the responses to that exercise bout are referred to as acute responses. When examining the acute response to exercise, we are concerned with the body's immediate response to a single exercise bout.

The other major area of interest in exercise and sport physiology is how the body responds over time to the stress of repeated exercise bouts or **chronic adaptation** to exercise, sometimes referred to as **training effects.** When one performs regular exercise over a period of weeks, the body adapts. The physiological adaptations that occur with chronic exposure to exercise or training improve both exercise capacity and efficiency. With resistance training, the muscles become stronger. With aerobic training, the heart and lungs become more efficient, and endurance capacity increases. As discussed in this introductory chapter, these adaptations are highly specific to the type of training the person does.

RESEARCH: THE FOUNDATION FOR UNDERSTANDING

Exercise and sport scientists actively engage in research to better understand the mechanisms that regulate the body's physiological responses to acute bouts of exercise as well as its adaptations to training and detraining. Most of this research is conducted at major research universities, medical centers, and specialized institutes using standardized research approaches and select tools of the exercise physiologist.

Use of Ergometers

When physiological responses to exercise are assessed in a laboratory setting, the participant's physical effort must be controlled to provide a constant and measurable pace of exercise. This is generally accomplished through use of ergometers. An **ergometer** (*ergo* = work; *meter* = measure) is an exercise device that allows the intensity of exercise to be controlled (standardized) and measured.

Treadmills

Treadmills are the ergometers of choice for most researchers and clinicians, particularly in the United States. With these devices, a motor-and-pulley system drives a large belt on which a subject can either walk or run; thus these ergometers are often called motor-driven treadmills (see figure 0.5). Belt length and width must accommodate the individual's body size and stride length. It is nearly impossible to test elite athletes on treadmills that are too narrow or too short.

> **In focus**
>
> Treadmills generally produce higher peak values than other ergometers for almost all assessed physiological variables, such as heart rate, ventilation, and oxygen uptake.

Treadmills offer a number of advantages. Walking is a very natural activity for almost everyone, so individuals normally adjust to the skill required for walking on a treadmill within a few minutes. Also, most people almost always

Figure 0.5 A motor-driven treadmill.

achieve their peak physiological values on the treadmill, although some athletes (e.g., competitive cyclists) achieve higher values on ergometers that more closely match their mode of training or competition.

Treadmills do have some disadvantages. They are generally more expensive than simpler ergometers, like the cycle ergometers discussed next. They are also bulky, require electrical power, and are not very portable. Accurate measurement of blood pressure during treadmill exercise can be difficult because both the noise associated with normal treadmill operation and subject movement can make hearing through a stethoscope difficult.

Cycle Ergometers

For many years, the **cycle ergometer** was the primary testing device in use, and it is still used extensively in both research and clinical settings. Cycle ergometers can be designed to allow subjects to pedal either in the normal upright position (see figure 0.6) or in reclining or semireclining positions.

Cycle ergometers in a research setting generally use either mechanical friction or electrical resistance. With mechanical friction devices, a belt encompassing a flywheel is tightened or loosened to adjust the resistance against which the cyclist pedals. The power output depends on the combination of the resistance and the pedaling rate—the faster one pedals, the greater the power output. To maintain the same power output throughout the test, one must maintain the same pedaling rate, so pedal rate must be constantly monitored. This is sometimes done with use of a metronome.

With electrically braked cycle ergometers, the resistance to pedaling is provided by an electrical conductor that moves through a magnetic or electromagnetic field. The strength of the magnetic field determines the resistance to pedaling. These ergometers can be controlled so that the resistance increases automatically as pedal rate decreases, and decreases as pedal rate increases, to provide a constant power output.

Similar to treadmills, cycle ergometers offer some advantages and disadvantages compared to other ergometers. The upper body can remain relatively stable on a cycle ergometer, allowing for more accurate determination of blood pressure and easier blood sampling during exercise. Furthermore, the exercise intensity on a cycle ergometer does not depend on the subject's body weight. This is important when one is investigating physiological responses to a standard rate of work (power output). As an example, if someone lost 5 kg (11 lb), data derived from treadmill testing could not be compared with data obtained before the weight loss because physiological responses to a set speed and grade on the treadmill vary with body weight. After the weight loss, the rate of work at the same speed and grade would be less than before. With the cycle ergometer, weight loss does not have as great an effect on physiological response to a standardized power output. Thus, walking/running

In focus

Cycle ergometers are the most appropriate devices for evaluating changes in submaximal physiological function before and after training in people whose weights have changed. Unlike the situation with treadmill exercise, cycle ergometer intensity is largely independent of body weight.

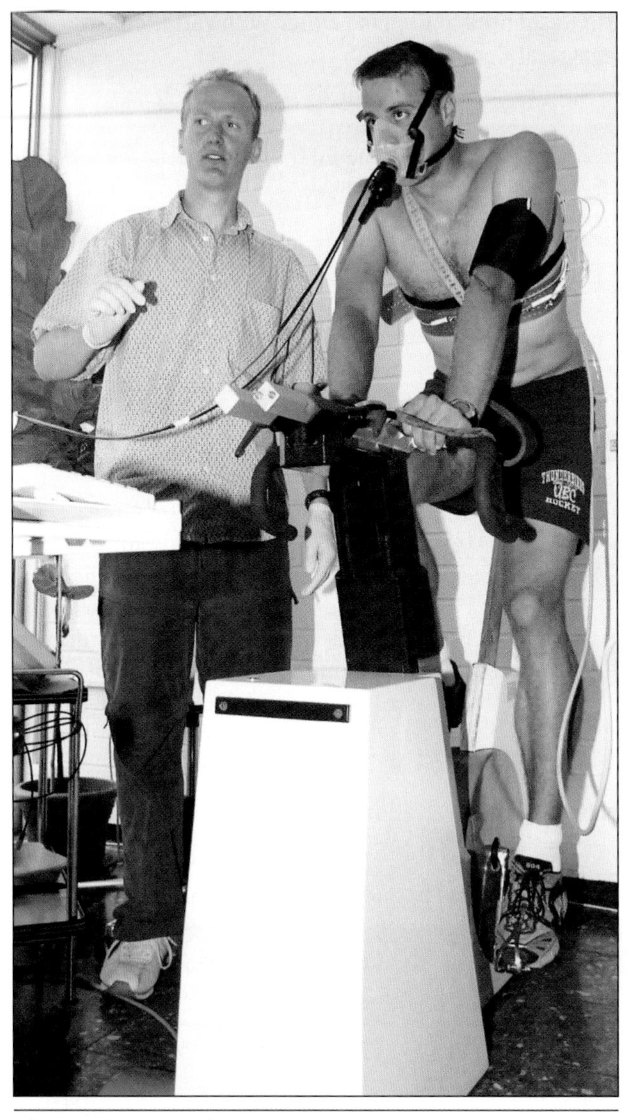

Figure 0.6 A cycle ergometer.

is often referred to as weight-dependent exercise, while cycling is weight independent.

Cycle ergometers also have disadvantages. If the subject does not regularly engage in that form of exercise, the leg muscles will likely fatigue early in the exercise bout. This may prevent a subject from attaining his or her peak exercise intensity. In addition, peak (maximal) values for some physiological variables obtained on a cycle ergometer are frequently lower than comparable maximal values obtained on a treadmill. This may be attributable to local leg fatigue, blood pooling in the legs (less blood returns to the heart), or the use of a smaller muscle mass during cycling than during treadmill exercise. Trained cyclists, however, tend to achieve their highest peak values on the cycle ergometer.

Other Ergometers

Other ergometers allow athletes who compete in specific sports or events to be tested in a manner that more closely approximates their training and competition. For example, an arm ergometer may be used to test athletes or nonathletes who use primarily their arms and shoulders in physical activity. Arm ergometry has also been used extensively to test and train athletes paralyzed below arm level. The rowing ergometer was devised to test competitive rowers.

Valuable research data have been obtained by instrumenting swimmers and monitoring them during swimming in a pool. However, the problems associated with turns and constant movement led to the use of two devices—tethered swimming and swimming flumes. In tethered swimming, the swimmer is attached to a harness connected to a rope, a series of pulleys, and counterbalancing weights. The swimmer swims at an effort that maintains a constant body position in the pool. As weights are added, the swimmer must swim with greater effort to maintain position.

A swimming flume allows swimmers to more closely simulate their natural swimming strokes. The swimming flume operates by pumps that circulate water past the swimmer, who attempts to maintain body position in the flume. The pump circulation can be increased or decreased to vary the speed at which the swimmer must swim. The swimming flume, which unfortunately is very expensive, has at least partially resolved the problems with tethered swimming and has created new opportunities to investigate the sport of swimming.

When one is choosing an ergometer, the concept of specificity is particularly important with highly trained athletes. The more specific the ergometer is to the actual pattern of movement used by the athlete in his or her sport, the more meaningful will be the test results.

Factors to Consider During Monitoring

Many factors can alter the body's acute response to a bout of exercise. For example, environmental conditions must be carefully controlled. Factors such as the temperature and humidity of the laboratory and the amount of light and noise in the test area can markedly affect physiological responses, both at rest and during exercise. Even the timing, volume, and content of the last meal and the quantity and quality of sleep the night before must be carefully controlled in research studies.

To illustrate this, table 0.1 shows how varying environmental and behavioral factors can alter heart rate at rest and during running on a treadmill at 14 km/h (9 mph). The subject's heart rate response during exercise differed by 25 beats/min when the air temperature was increased from 21 °C (70 °F) to 35 °C (95 °F). Most physiological variables that are normally measured during exercise are similarly influenced by environmental fluctuations. Whether one is comparing a person's exercise results from one day to another, or comparing the responses of two different subjects, all of these factors must be controlled as carefully as possible.

Physiological responses, both at rest and during exercise, also vary throughout the day. The term **diurnal variation** refers to fluctuations that occur during a 24 h day. Table 0.2 illustrates the diurnal variation in heart rate at rest, during various levels of exercise, and during recovery. Body temperature shows similar fluctuations throughout the day. As seen in table 0.2, testing the same person in the morning on one day and in the afternoon on the next can and will produce different results. Test times must be standardized to control for this diurnal effect.

At least one other physiological cycle must also be considered. The normal 28-day menstrual cycle often involves considerable variations in

- body weight,
- total body water and blood volume,
- body temperature,
- metabolic rate, and
- heart rate and stroke volume (the amount of blood leaving the heart with each contraction).

Exercise scientists must control for menstrual cycle phase or the use of oral contraceptives (which similarly alter hormonal status), or both, when testing women. When older women are being tested, testing strategies must take into account menopause and hormone replacement therapies.

TABLE 0.1 Heart Rate Responses to Running Differ With Variations in Environmental and Behavioral Conditions

	Heart rate (beats/min)	
Environmental and behavioral factors	Rest	Exercise
Temperature (50% humidity)		
21 °C (70 °F)	60	165
35 °C (95 °F)	70	190
Humidity (21 °C)		
50%	60	165
90%	65	175
Noise level (21 °C, 50% humidity)		
Low	60	165
High	70	165
Food intake (21 °C, 50% humidity)		
Small meal 3 h before exercising	60	165
Large meal 30 min before exercising	70	175
Sleep (21 °C, 50% humidity)		
8 h or more	60	165
6 h or less	65	175

TABLE 0.2 An Example of Diurnal Variations in Heart Rate at Rest and During Exercise

	Time of day					
	2 a.m.	6 a.m.	10 a.m.	2 p.m.	6 p.m.	10 p.m.
Condition	Heart rate (beats/min)					
Resting	65	69	73	74	72	69
Light exercise	100	103	109	109	105	104
Moderate exercise	130	131	138	139	135	135
Maximal exercise	179	179	183	184	181	181
Recovery, 3 min	118	122	129	128	128	125

Data from T. Reilly and G.A. Brooks (1990), "Selective persistence of circadian rhythms in physiological responses to exercise," *Chronobiology International*, 7: 59-67.

In focus

Conditions under which research participants are monitored, at rest and during exercise, must be carefully controlled. Environmental factors, such as temperature, humidity, altitude, and noise, can affect the magnitude of response of all basic physiological systems, as can behavioral factors such as eating patterns and sleep. Likewise, physiological measurements must be well controlled for diurnal and menstrual cycle variations.

Evolution of Exercise Physiology Tools and Techniques

The history of exercise physiology has, in some ways, been driven by advancements in technologies adapted from basic sciences. The early studies of energy metabolism during exercise were made possible by the invention of gas-collecting equipment and chemical analysis of oxygen and carbon dioxide. Chemical determination of blood lactic acid seemed to provide some insights regarding the aerobic and anaerobic aspects of muscular activity, but these data told us little regarding the production and removal of this by-product of exercise. Likewise, blood glucose measurements taken before, during, and after exhaustive exercise proved to be interesting data but were of limited value for understanding the energy exchange at the cellular level.

Prior to the 1960s, there were few biochemical studies on the adaptations of muscle to training. Although the field of biochemistry can be traced to the early part of the 20th century, this special area of chemistry was not applied to human muscle until Bergstrom and Hultman reintroduced and popularized the needle biopsy procedure in 1966. Initially, this procedure was used to examine glycogen depletion during exhaustive exercise and its resynthesis during recovery. In the early 1970s, as noted earlier, a number of exercise physiologists used the muscle biopsy method, histological staining, and the light microscope to determine human muscle fiber types.

Over the last 30 years, muscle physiologists have used various chemical procedures to understand how muscles generate energy and adapt to training. Test tube experiments (in vitro) with muscle biopsy samples have been used to measure muscle proteins (enzymes) and to determine the muscle fiber's capacity to use oxygen. Although these studies provided a snapshot of the fiber's potential to generate energy, they often left more questions than answers. It was natural, therefore, for the sciences of cell biology to move to an even deeper level. It was apparent that the answers to those questions must lie within the fiber's molecular makeup.

(a) Frank Booth and (b) Ken Baldwin.

Although not a new science, molecular biology has become a useful tool for exercise physiologists who wish to delve deeper into the cellular regulation of metabolism and adaptations to the stress of exercise. Physiologists like Frank Booth and Ken Baldwin have dedicated their careers to understanding the molecular regulation of muscle fiber characteristics and function and have laid the groundwork for our current understanding of the genetic controls of muscle growth and atrophy. The use of molecular biological techniques to study the contractile characteristics of single muscle fibers is discussed in chapter 1.

Well before James Watson and Francis Crick unraveled the structure of DNA (1953), scientists appreciated the importance of genetics in predetermining the structure and function of all living organisms. The newest frontier in exercise physiology combines the study of molecular biology and genetics. Since the early 1990s, scientists have attempted to explain how exercise emits signals that affect the expression of genes within skeletal muscle.

In retrospect, it is apparent that since the beginning of the 20th century, the field of exercise physiology has evolved from measuring whole-body function (i.e., oxygen consumption, respiration, and heart rate) to molecular studies of muscle fiber genetic expression. There is little doubt that exercise physiologists of the future will need to be well grounded in biochemistry, molecular biology, and genetics.

Reading and Interpreting Tables and Graphs

This book contains references to specific research studies that have made a major impact on our understanding within a given area. Once a group of scientists completes a research project, they submit the results of their research to one of the many research journals in sport and exercise physiology. Some of the more widely used research journals appear on the list of selected readings and references at the back of this book as well as in the study guide on the Web site www.HumanKinetics.com/PhysiologyOfSportAndExercise.

Most of the quantitative research results published in these journals are presented in the form of tables and graphs. Thus, this textbook has included tables and graphs from selected research studies to help readers become familiar with research data and how they are interpreted.

Tables and graphs provide an efficient way for researchers to communicate the results of their studies to other scientists. For the student in exercise and sport physiology, a working knowledge of how to read and interpret tables and graphs is critical.

With respect to tables, let's use table 0.2 as an example. It is important to first look at the title of the table, which identifies what information is being presented. In this case, the table is designed to illustrate how heart rate varies throughout a 24 h day, at rest and during exercise and recovery from exercise. The left-hand column specifies the conditions under which the heart rate was measured, and the times across the top of the table are the specific times these measurements were taken. In every good table and graph, the units for each variable are clearly presented; in this table, heart rate is expressed in "beats/min," or beats per minute. Pay careful attention to the units of measure used when interpreting a table or graph. From this table we see that heart rate is at its lowest early in the morning and highest midafternoon, both at rest and during exercise. These data effectively illustrate the importance of standardizing the time of day at which measurement is taken for a given variable, either when the same person is being compared across different days or when two or more individuals are being compared on the same day.

Graphs can provide a better view of trends in data, response patterns, and comparisons of data collected from two or more groups of subjects. For some students, graphs can be more difficult to read and interpret. First, every graph has a horizontal or *x*-axis for the **independent variable** and one vertical or *y*-axis (sometimes two) for the **dependent variable** or variables. Independent variables are those factors that are manipulated or controlled by the researcher, while dependent variables are those that change with—that is, depend on—the independent variables.

In table 0.2, time of day is the independent variable and would therefore be placed along the *x*-axis of a graph, while heart rate is the dependent variable (since heart rate *depends on* the time of day) and would therefore be plotted on the *y*-axis, as illustrated in figure 0.7*a*. The units of measure for each variable are also displayed on the graph. While figure 0.7*a* is in the form of a line graph, these same data can also be plotted in the format of a bar graph (figure 0.7*b*).

If we were interested in comparing the diurnal pattern of responses for both heart rate and the metabolic rate (oxygen uptake) during light exercise, two separate *y*-axes would be needed. For example, the left-hand *y*-axis could show heart rate while the right-hand *y*-axis could show oxygen uptake, both during light exercise (figure 0.7*c*). From figure 0.7*c*, we can see that oxygen consumption and heart rate change in the same pattern over the course of the day.

Research Settings

Research can be conducted either in the laboratory or in the field. Laboratory tests are usually more accurate because more specialized equipment can be used and conditions can be carefully controlled. As an example, the direct laboratory measurement of maximal oxygen uptake ($\dot{V}O_{2max}$) is considered the most accurate estimate of cardiorespiratory endurance capacity. However, some field tests, such as the 1.5 mi (2.4 km) run, are used to predict or estimate $\dot{V}O_{2max}$. The field test is not totally accurate; but it provides a reasonable estimate of $\dot{V}O_{2max}$ and is inexpensive to conduct, and many people can be tested in a short time. To measure $\dot{V}O_{2max}$ directly, one would need to go to a university or clinical laboratory, but one could estimate $\dot{V}O_{2max}$ from the time it takes to run 1.5 mi on a flat surface. Field tests can be conducted in the workplace, on a running track or in a swimming pool, or during athletic competitions.

Research Designs

In exercise physiology research, there are two basic types of research design: cross-sectional and longitudinal. With a **cross-sectional research design,** a cross section of the population of interest (that is, a representative sample) is tested at one specific time, and the differences between subgroups from that population are compared. With a **longitudinal research design,** research subjects are retested periodically after initial testing to measure changes over time in variables of interest.

The differences between these two approaches are best understood through an example. The objective of a hypothetical study might be to determine whether a regular program of distance running increases the

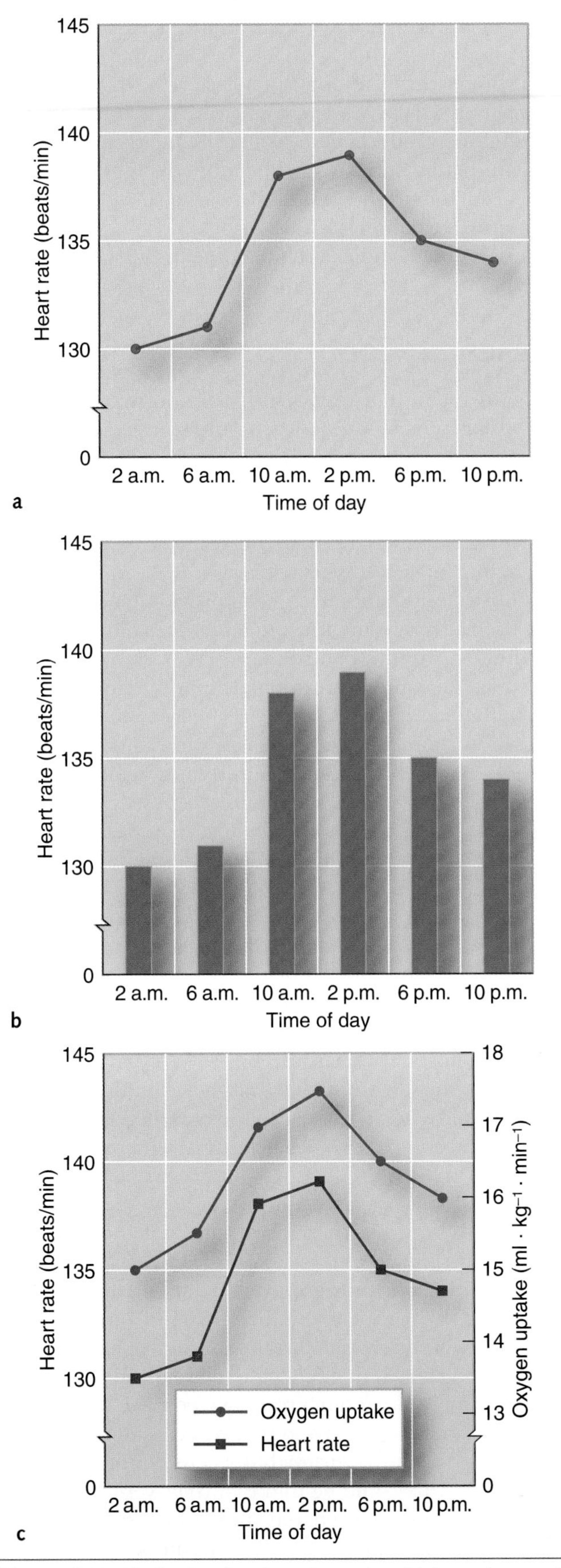

Figure 0.7 Understanding how to read and interpret a graph. *(a)* Line graph; *(b)* bar graph; *(c)* line graph with two *y*-axes.

concentration of high-density lipoprotein cholesterol (HDL-C) in the blood. High-density lipoprotein cholesterol is the desirable form of cholesterol; increased concentrations are associated with reduced risk for heart disease. Using the cross-sectional approach, one could, for example, test a large number of people who fall into the following categories:

- A group of subjects who do no training (the "control group")
- A group of subjects who run 24 km (15 mi) per week
- A group of subjects who run 48 km (30 mi) per week
- A group of subjects who run 72 km (45 mi) per week
- A group of subjects who run 96 km (60 mi) per week

One would then compare the results from each group, basing one's conclusions on how much running was done. Using this approach, exercise scientists found that weekly running results in elevated HDL-C levels, suggesting a positive health benefit related to running distance. Furthermore, as illustrated in figure 0.8, there was a **dose–response relationship** between these variables—the higher the "dose" of exercise training, the higher the resulting concentration of HDL-C. It is important to remember, however, that with a cross-sectional design, these are different groups of runners, not the same runners at different training volumes.

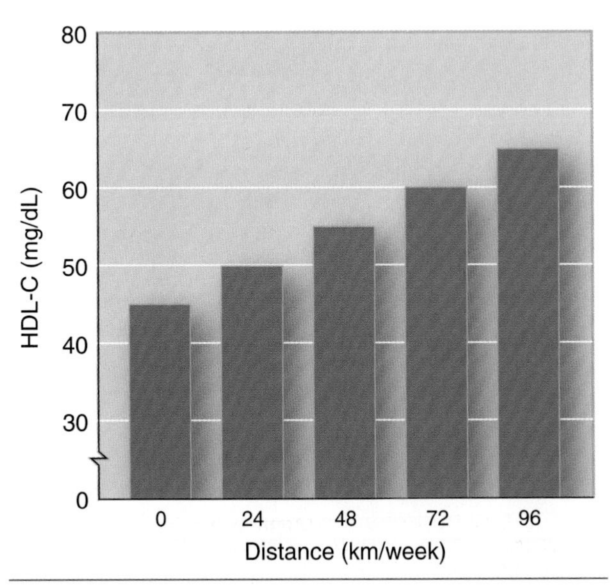

Figure 0.8 The relationship between distance run per week and average high-density lipoprotein cholesterol (HDL-C) concentrations across five groups: nontraining control (0 km/week), 24 km/week, 48 km/week, 72 km/week, and 96 km/week. This illustrates a cross-sectional study design.

Using the longitudinal approach to test the same question, one could design a study in which untrained people would be recruited to participate in a 12-month distance-running program. One could, for example, recruit 40 people willing to begin running and then randomly assign 20 to a training group and the remaining 20 to a control group. Both groups would be followed for 12 months. Blood samples would be tested at the beginning of the study, then at three-month intervals, concluding at 12 months when the program ended. With this design, both the running group and the control group would be followed over the entire period of the study, and changes in their HDL-C levels could be determined across each period. Actual studies have been conducted using this longitudinal design to examine changes in HDL-C with training, but their results have not been as clear as the results of the cross-sectional studies. See figure 0.9 as an example. Note that in this figure, in contrast to figure 0.8, there is only a small increase in HDL-C in the subjects who are training. The control group stays relatively stable, with only minor fluctuations in their HDL-C from one three-month period to the next.

In focus

Longitudinal research studies are generally the most accurate for studying changes in physiological variables over time. However, it is not always feasible to use a longitudinal design, and valuable information can be derived from cross-sectional studies.

A longitudinal research design is usually best suited to studying changes in variables over time. Too many factors that may taint results can influence cross-sectional designs. For example, genetic factors might interact so that those who run long distances are also those who have high HDL-C levels. Also, different populations might follow different diets; but in a longitudinal study, diet and other variables can be more easily controlled. However, longitudinal studies are time-consuming, are expensive to conduct, and are not always possible; and cross-sectional studies provide some insight into these questions.

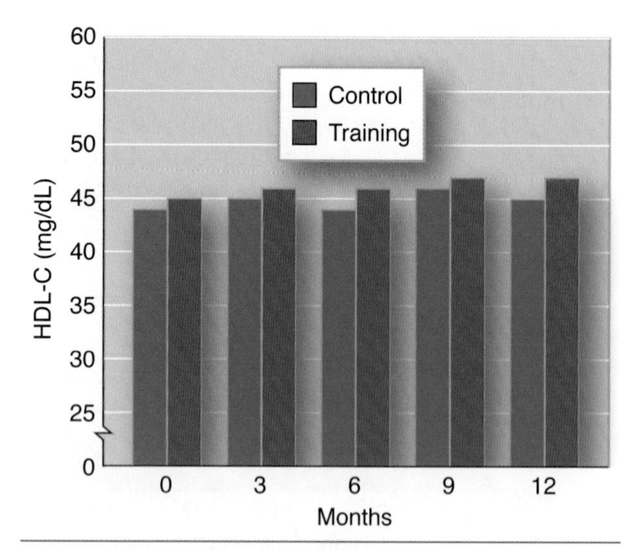

Figure 0.9 The relationship between months of distance-running training and average high-density lipoprotein cholesterol (HDL-C) concentrations in an experimental group (20 subjects, distance training) and a sedentary (20 subjects) control group. This illustrates a longitudinal study design.

Research Controls

When we conduct research, it is important to be as careful as possible in designing the study and collecting the data. We saw from figure 0.9 that changes in a variable over time resulting from an intervention such as exercise can be very small. Yet, even small changes in a variable such as HDL-C can mean a substantial reduction in risk for heart disease. Recognizing this, scientists design studies aimed at providing results that are both accurate and reproducible. This requires that studies be carefully controlled.

Research controls are applied at various levels. Starting with the design of the research project, the scientist must determine how to control for variation in the subjects used in the study. The scientist must determine if it is important to control for the subjects' sex, age, or body size. To use age as an example, for certain variables, the response to an exercise training program might be different for a child or an aged person compared with a young or middle-aged adult. Is it important to control for the subject's smoking or dietary status? Considerable thought and discussion are needed to make sure that the subjects used in a study are appropriate for the specific research question being asked.

For almost all studies, it is critical to have a control group. In the longitudinal research design for the cholesterol study described earlier, the **control group** acts as a comparison group to make certain that any changes observed in the running group are attributable solely to the training program and not to any other factors, such as the time of the year or aging of the subjects during the course of the study. Experimental designs often employ a **placebo group.** Thus, in a study in which a subject might expect to have a benefit from the proposed intervention, such as the use of a specific food or drug, a scientist might decide to use three groups of subjects: an intervention group that receives the actual substance, a placebo group that receives an inert substance, and a control group that receives nothing. (The placebo group would be told that they were receiving the specific intervention, e.g., food or drug, yet would be given an inert substance that has no known physiological effects.) If the intervention and placebo groups improve their performance to the same level and the control group does not improve performance, then the improvement is likely the result of the "placebo effect," or the expectation that the substance will improve performance. If the intervention group improves performance and the placebo and control groups do not, then we can conclude that the intervention does improve performance.

One other way of controlling for the placebo effect is to conduct a study that uses a **crossover design.** In this case, each group undergoes both treatment and control trials at different times. For example, one group is administered the intervention for the first half of the study (e.g., 6 months of a 12-month study) and serves as a control the last half of the study. The second group serves as a control during the first half of the study and receives the intervention during the second half. In some cases, a placebo can be used in the control phase of the study. Chapter 15, "Ergogenic Aids and Sport," provides further discussion of placebo groups.

It is equally important to control data collection. The equipment must be calibrated so the researcher knows that the values generated by a given piece of equipment are accurate, and the procedures used in collecting data must be standardized. For example, when using a scale to measure the weight of subjects, researchers need to calibrate that scale by using a set of calibrated weights (e.g., 10 kg, 20 kg, 30 kg, and 40 kg) that have been measured on a precision scale. These weights are placed on the weighing scale to be used in the study, individually and in combination, at least once a week to provide certainty that the scale is measuring the weights accurately. As another example, electronic analyzers used to measure respiratory gases need to be calibrated frequently with gases of known concentration to ensure the accuracy of these analyses.

Finally, it is important to know that all test results are reproducible. In the example illustrated in figure 0.9, the HDL-C of an individual is measured every three months. If that person is tested five days in a row before he or she starts the training program, one would expect the HDL-C results to be similar across all five days, providing diet, exercise, sleep, and time of day for testing remained the same. In figure 0.9, the values for the control group across 12 months varied from about 44 to 45 mg/dL, whereas the exercise group increased from 45 to 47 mg/dL. Over five consecutive days, the measurements should not vary by more than 1 mg/dL for any one person if the researcher is going to pick up this small change over time. To control for reproducibility of results, scientists generally take several measurements, sometimes on different days, and then average the results, before, during, and at the end of an intervention.

In Closing

In this introduction we highlighted the historical roots and scientific underpinnings of exercise and sport physiology. We learned that the current state of knowledge in these fields builds on the past and is merely a bridge to the future—many questions remain unanswered. We briefly defined the acute responses to exercise bouts and chronic adaptations to long-term training. We concluded with an overview of the principles used in sport and exercise physiology research.

In part I, we begin examining physical activity the way exercise physiologists do as we explore the essentials of movement. In the next chapter, we examine the structure and function of skeletal muscle, how it produces movement, and how it responds during exercise.

KEY TERMS

acute exercise
chronic adaptation
control group
crossover design
cross-sectional research design
cycle ergometer
dependent variable
diurnal variation
dose–response relationship
environmental physiology
ergometer
exercise physiology
homeostasis
independent variable
longitudinal research design
physiology
placebo group
sport physiology
training effect
treadmill

STUDY QUESTIONS

1. What is exercise physiology? How does sport physiology differ?
2. Describe the evolution of exercise physiology from the early studies of anatomy. Who were some of the key figures in the development of this field?
3. Describe the founding and the key areas of research emphasized by the Harvard Fatigue Laboratory. Who was the first research director of this laboratory?
4. Name the three Scandinavian physiologists who conducted research in the Harvard Fatigue Laboratory.
5. Provide an example of what is meant by studying acute responses to a single bout of exercise.
6. Describe what is meant by studying chronic adaptations to exercise training.
7. List several environmental conditions that could affect one's response to an acute bout of exercise.
8. What is meant by diurnal variation?
9. Why is it important when testing to control for environmental and diurnal conditions?
10. What is an ergometer? Name the two most commonly used ergometers and explain their advantages and disadvantages.
11. What factors must researchers consider when designing a research study to ensure that they get accurate and reproducible results?

STUDY GUIDE ACTIVITIES

In addition to the activities listed in the chapter opening outline on page 1, two other activities are available in the online study guide, located at

www.HumanKinetics.com/PhysiologyOfSportAndExercise.

The chapter **SUMMARY** reviews the key concepts covered in the chapter, and the end-of-chapter **QUIZ** tests your understanding of the material covered in the chapter.

PART

I

Exercising Muscle

In the introduction, we explored the foundations of exercise and sport physiology. We defined these two fields of study, gained a historical perspective of their development, and established basic concepts that underlie the remainder of this book. With this foundation, we can begin our quest to understand how the human body performs physical activity. We start our journey in chapter 1, "Structure and Function of Exercising Muscle," where we focus on skeletal muscle, examining the structure and function of skeletal muscles and muscle fibers and how they produce body movement. We will learn how muscle fiber types differ and why these differences are important to specific types of activity. In chapter 2, "Fuel for Exercising Muscle," we study basic principles of metabolism, focusing on the primary source of energy, adenosine triphosphate (ATP), and how it is provided through three energy systems by hormonal control. In chapter 3, "Neural Control of Exercising Muscle," we discuss how the nervous system coordinates muscle action by integrating sensory information coming from all parts of the body and then signaling the appropriate muscles to act. Finally, in chapter 4, "Energy Expenditure and Fatigue," we learn how the body's energy expenditure varies from resting conditions up through maximal rates of exercise, and how fatigue can result when energy demands exceed energy supply.

CHAPTER

1

Structure and Function of Exercising Muscle

In this chapter and in the online study guide

According to a UPI newspaper report, 9-year-old Jeremy Schill weighed only 29.5 kg (65 lb), but that didn't stop him from lifting the rear of the family's 1860 kg (4100 lb) car off his father's chest! The car had slipped off the jack while Rique Schill was working underneath it, pinning him under the rear axle. When Jeremy realized that his father was slowly suffocating, the third-grader lifted the car enough to enable his father to breathe and to allow his mother to place another jack under the rear bumper. Could a small boy lift such a great weight? In this chapter we consider how his muscles might have generated the force that enabled him to save his father's life if this is, in fact, a true story.

When our hearts beat, when a meal we've eaten moves through our intestines, and when we move any part of our bodies, muscle is involved. The myriad functions of the muscular system are performed by only three types of muscle (see figure 1.1): smooth, cardiac, and skeletal.

Smooth muscle is called involuntary muscle because it is not directly under conscious control. It is found in the walls of most blood vessels, allowing them to constrict or dilate to regulate blood flow. It is also found in the walls of most internal organs, allowing them to contract and relax, perhaps to move food along the digestive tract, to expel urine, or to give birth.

Cardiac muscle is found only in the heart, composing most of the heart's structure. It shares some characteristics with skeletal muscle but, like smooth muscle, is not under conscious control. Cardiac muscle controls itself, with some fine-tuning by the nervous and endocrine systems. Cardiac muscle is discussed fully in chapter 5.

Skeletal muscles are under conscious control and are so named because most attach to and move the skeleton. We know many of these muscles by their names—such as deltoid, pectorals, and biceps—but the human body contains more than 600 skeletal muscles. The thumb alone is controlled by nine separate muscles!

Exercise requires movement of the body, which is accomplished through the action of skeletal muscles. Because this is an exercise and sport physiology book, our primary interest is in the structure and function of skeletal muscle. Although the anatomical structures of smooth, cardiac, and skeletal muscle differ somewhat, their control mechanisms and principles of action are similar.

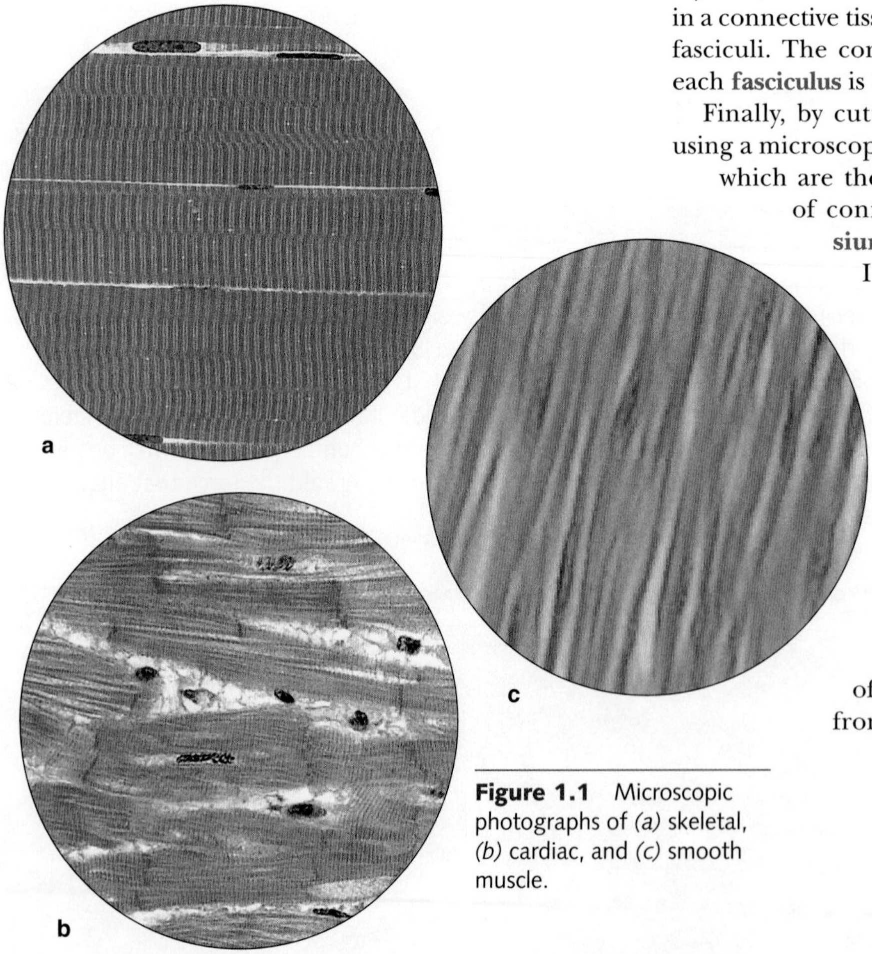

a

b

c

Figure 1.1 Microscopic photographs of *(a)* skeletal, *(b)* cardiac, and *(c)* smooth muscle.

FUNCTIONAL ANATOMY OF SKELETAL MUSCLE

When we think of muscles, we visualize each muscle as a whole, that is, as a single unit. This is natural because a skeletal muscle seems to act as a single entity. But skeletal muscles are far more complex than that (see figure 1.2).

If a person were to dissect a muscle, he or she would first cut through the outer connective tissue covering. This is the **epimysium.** It surrounds the entire muscle, holding it together. Once one had cut through the epimysium, one would see small bundles of fibers wrapped in a connective tissue sheath. These bundles are called fasciculi. The connective tissue sheath surrounding each **fasciculus** is the **perimysium.**

Finally, by cutting through the perimysium and using a microscope, one would see the **muscle fibers,** which are the individual muscle cells. A sheath of connective tissue, called the **endomysium,** also covers each muscle fiber. It is generally thought that muscle fibers extend from one end of the muscle to the other; but under the microscope, muscle bellies often divide into compartments or more transverse fibrous bands (inscriptions). Because of this compartmentalization, the longest human muscle fibers are about 12 cm (4.7 in.), which corresponds to about 500,000 sarcomeres, the basic functional unit of the myofibril. The number of fibers in different muscles ranges from several hundred (e.g., tensor tym-

In focus

A single muscle cell is known as a muscle fiber.

pani, attached to the eardrum) to more than a million (e.g., medial gastrocnemius muscle).[6]

Muscle Fiber

Muscle fibers range in diameter from 10 to 120 μm, so they are nearly invisible to the naked eye. The following sections describe the structure of the individual muscle fiber.

Plasmalemma

If one looked closely at an individual muscle fiber, one would see that it is surrounded by a plasma membrane, called the **plasmalemma** (figure 1.3 on page 28). The plasmalemma is part of a larger unit referred to as the **sarcolemma.** The sarcolemma is composed of the plasmalemma and the basement membrane. Some textbooks use the term sarcolemma to describe just the plasmalemma.[6] At the end of each muscle fiber, its plasmalemma fuses with the tendon, which inserts into the bone. Tendons are made of fibrous cords of connective tissue that transmit the force generated by muscle fibers to the bones, thereby creating motion. So typically, individual muscle fibers are ultimately attached to bone via the tendon.

The plasmalemma has several unique features that are important to muscle fiber function. It appears as a series of shallow folds along the surface of the fiber when the fiber is contracted or in a resting state, but these folds disappear when the fiber is stretched. This folding allows stretching of the muscle fiber without disrupting the plasmalemma. The plasmalemma also has junctional folds in the innervation zone at the motor end plate, which assists in the transmission of the action potential from the motor neuron to the muscle fiber as discussed later in this chapter. Finally, the plasmalemma helps to maintain acid–base balance and transports metabolites from the capillary blood into the muscle fiber.[6]

Satellite cells are located between the plasmalemma and the basement membrane. These cells are involved in the growth and development of skeletal muscle and in muscle's adaptation to injury, immobilization, and training. This will be discussed in greater detail in subsequent chapters.

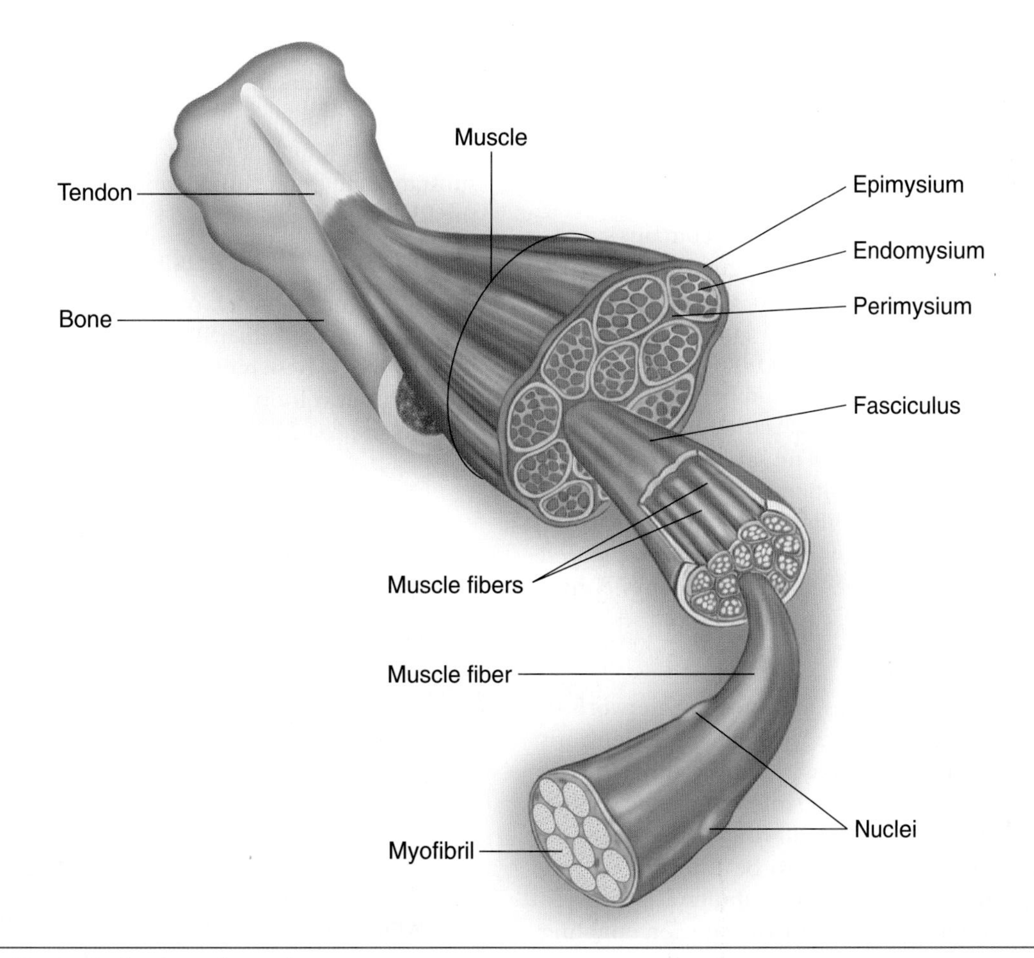

Figure 1.2 The basic structure of muscle.

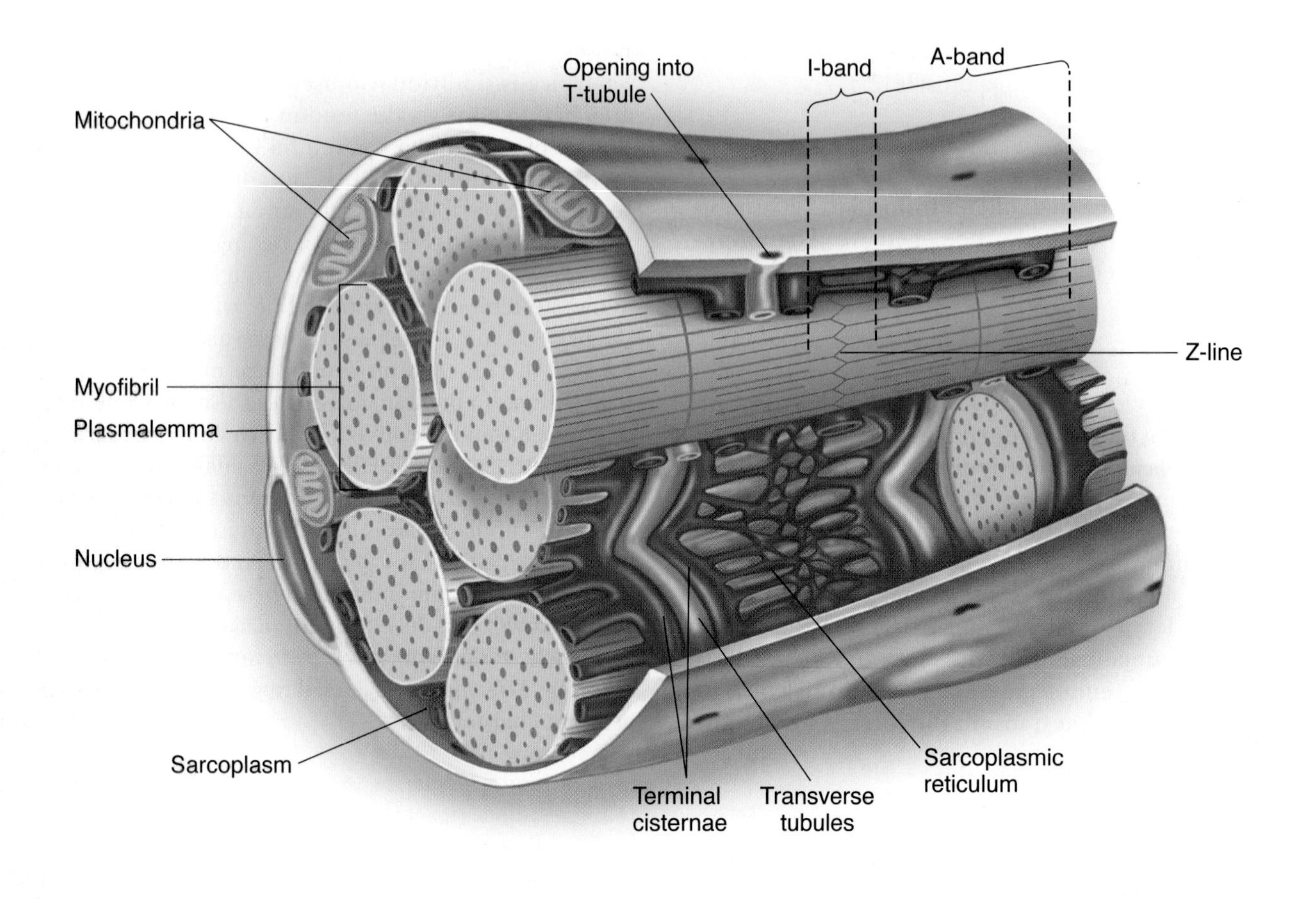

Figure 1.3 The structure of a single muscle fiber.

Sarcoplasm

Inside the plasmalemma, a muscle fiber contains successively smaller subunits, as shown in figure 1.3. The largest of these are myofibrils, which we discuss separately. For now, consider myofibrils to be rodlike structures running the length of the muscle fibers. A gelatin-like substance fills the spaces within and between the myofibrils. This is the **sarcoplasm.** It is the fluid part of the muscle fiber—its cytoplasm. The sarcoplasm contains mainly dissolved proteins, minerals, glycogen, fats, and the necessary organelles. It differs from the cytoplasm of most cells because it contains a large quantity of stored glycogen as well as the oxygen-binding compound myoglobin, which is quite similar to hemoglobin.

The Transverse Tubules

The sarcoplasm also houses an extensive network of **transverse tubules (T-tubules),** which are extensions of the plasmalemma that pass laterally through the muscle fiber. These tubules are interconnected as they pass among the myofibrils, allowing nerve impulses received by the plasmalemma to be transmitted rapidly to individual myofibrils. The tubules also provide pathways from outside the fiber to its interior, enabling substances to enter the cell and waste products to leave the fibers.

The Sarcoplasmic Reticulum

A longitudinal network of tubules, known as the **sarcoplasmic reticulum (SR),** is also found within the muscle fiber. These membranous channels parallel the myofibrils and loop around them. The SR serves as a storage site for calcium, which is essential for muscle

In review

- An individual muscle cell is called a muscle fiber.
- A muscle fiber is enclosed by a plasma membrane called the plasmalemma.
- The cytoplasm of a muscle fiber is called the sarcoplasm.
- The extensive tubule network found in the sarcoplasm includes T-tubules, which allow communication and transport of substances throughout the muscle fiber, and the SR, which stores calcium.

contraction. Figure 1.3 depicts the T-tubules and the SR. Their functions are discussed in more detail later in this chapter when we describe the process of muscle contraction.

Myofibril

Each muscle fiber contains several hundred to several thousand **myofibrils.** These are the contractile elements of skeletal muscle. Myofibrils appear as long strands of still smaller subunits—the sarcomeres.

Striations and the Sarcomere

Under a light microscope, skeletal muscle fibers have a distinctive striped appearance. Because of these markings, or striations, skeletal muscle is also called striated muscle. This striation also is seen in cardiac muscle, so it too can be considered striated muscle.

Refer to figure 1.4, showing myofibrils within a single muscle fiber, and note the striations. Note that dark regions, known as A-bands, alternate with light regions, known as I-bands. Each dark A-band has a lighter region in its center, the H-zone, which is visible only when the myofibril is relaxed. There is a dark line in the middle of the H-zone called the M-line. The light I-bands are interrupted by a dark stripe referred to as the Z-disk, also known as the Z-line.

A **sarcomere** is the basic functional unit of a myofibril and the basic contractile unit of muscle. Each myofibril is composed of numerous sarcomeres joined end to end at the Z-disks. Each sarcomere includes what is found between each pair of Z-disks, in this sequence:

The sarcomere is the basic functional unit of a muscle.

- An I-band (light zone)
- An A-band (dark zone)
- An H-zone (in the middle of the A-band)
- An M-line in the middle of the H-zone
- The rest of the A-band
- A second I-band

Looking at individual myofibrils through an electron microscope, one can differentiate two types of small protein filaments that are responsible for muscle contraction. The thinner filaments are composed primarily of **actin,** and the thicker filaments are primarily **myosin.** The striations seen in muscle fibers result from alignment of these filaments, as illustrated in figure 1.4. The light I-band indicates the region of the sarcomere where there are only thin filaments. The dark A-band represents the regions that contain both thick and thin filaments. The H-zone is the central portion of the A-band and contains only thick filaments. The absence of thin filaments causes the H-zone to appear lighter than the adjacent A-band. In the center of the H-zone is the M-line, which is composed of proteins that serve as the attachment site for the thick filaments and assist in stabilizing the structure of the sarcomere. Z-disks, composed of proteins, appear at each end of the sarcomere. Along with two additional proteins, titin and nebulin, they provide points of attachment and stability for the thin filaments.

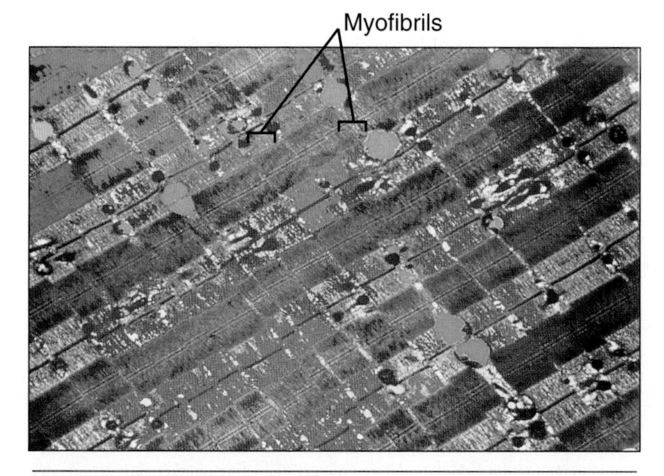

Figure 1.4 An electron micrograph of myofibrils. Note the presence of striations. The blue regions are the A-bands and the pink regions are the I-bands.

Thick Filaments

About two-thirds of all skeletal muscle protein is myosin, the principal protein of the thick filament. Each myosin filament typically is formed by about 200 myosin molecules.

Each myosin molecule is composed of two protein strands twisted together (see figure 1.5). One end of each strand is folded into a globular head, called the myosin head. Each thick filament contains many such heads, which protrude from the thick filament to form cross-bridges that interact during muscle contraction with specialized active sites on the thin filaments. There is an array of fine filaments, composed of **titin,** that stabilizes the myosin filaments along their longitudinal axis (see figure 1.5). Titin filaments extend from the Z-disk to the M-line.

Thin Filaments

Each thin filament, although often referred to simply as an actin filament, is actually composed of three different protein molecules—actin, **tropomyosin,** and **troponin.** Each thin filament has one end inserted into a Z-disk,

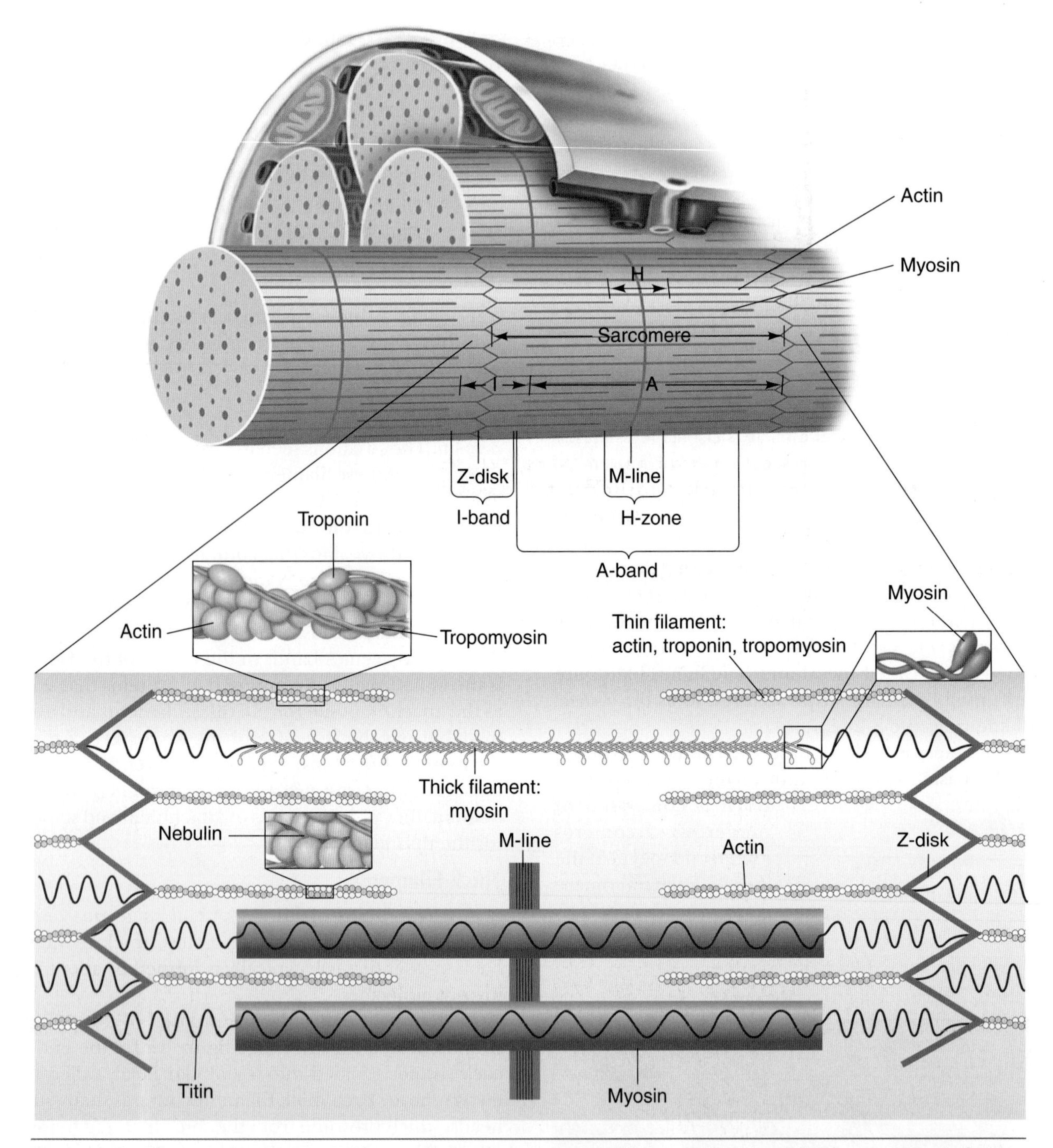

Figure 1.5 The basic functional unit of a myofibril is the sarcomere, which contains a specialized arrangement of actin and myosin filaments. The role of titin is to position the myosin filament to maintain equal spacing between the actin filaments. Nebulin is often referred to as an "anchoring protein" because it provides a framework that helps stabilize the position of actin.

with the opposite end extending toward the center of the sarcomere, lying in the space between the thick filaments. **Nebulin,** an anchoring protein for actin, coextends with actin and appears to play a regulatory role in mediating actin and myosin interactions (figure 1.5). Each thin filament contains active sites to which myosin heads can bind.

Actin forms the backbone of the filament. Individual actin molecules are globular proteins (G-actin) and join together to form strands of actin molecules. Two

In review

- Myofibrils are composed of sarcomeres, the smallest functional units of a muscle.
- A sarcomere is composed of two different-sized filaments, thick and thin filaments, which are responsible for muscle contraction.
- Myosin, the primary protein of the thick filament, is composed of two protein strands, each folded into a globular head at one end.
- The thin filament is composed of actin, tropomyosin, and troponin. One end of each thin filament is attached to a Z-disk.

strands then twist into a helical pattern, much like two strands of pearls twisted together.

Tropomyosin is a tube-shaped protein that twists around the actin strands. Troponin is a more complex protein that is attached at regular intervals to both the actin strands and the tropomyosin. This arrangement is depicted in figure 1.5. Tropomyosin and troponin work together in an intricate manner along with calcium ions to maintain relaxation or initiate contraction of the myofibril, which we discuss later in this chapter.

Muscle Fiber Contraction

An **α-motor neuron** is a neuron that connects with and innervates many muscle fibers. A single α-motor neuron and all the muscle fibers it supplies are collectively termed a **motor unit** (see figure 1.6). The synapse or gap between the α-motor neuron and a muscle fiber is referred to as a neuromuscular junction. This is where communication between the nervous and muscular systems occurs.

In focus

A motor unit consists of a single α-motor neuron and all the muscle fibers it supplies.

Action Potential

The events that trigger a muscle fiber to contract are complex. The process, depicted in figure 1.7 on page 32, is initiated by an electrical signal, or **action potential,** from the brain or spinal cord to an α-motor neuron. The action potential arrives at the α-motor neuron's dendrites, specialized receptors on the neuron's cell body. From here, the action potential travels down the axon to the axon terminals, which are located very close to the plasmalemma. When the action potential arrives at the axon terminals, these nerve endings secrete a neurotransmitter substance called acetylcholine (ACh), which binds to receptors on the plasmalemma (see figure 1.7*a*). If enough ACh binds to the receptors, the action potential will be transmitted the full length of the muscle fiber as ion gates open in the muscle cell membrane and allow sodium to enter. This process is referred to as depolarization. An action potential must be generated in the muscle cell before the muscle cell can act. These neural events are discussed more fully in chapter 3.

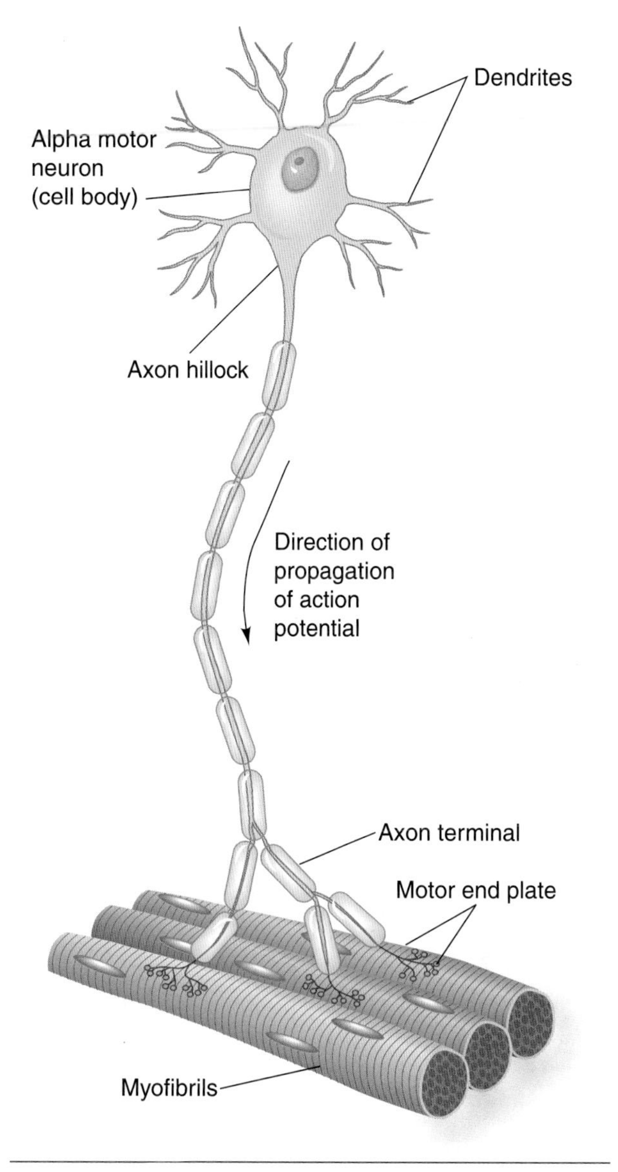

Figure 1.6 Motor units include the α-motor neurons and the muscle fibers they innervate.

Role of Calcium in the Muscle Fiber

In addition to depolarizing the fiber membrane, the action potential travels over the fiber's network of tubules (T-tubules) to the interior of the cell. The arrival

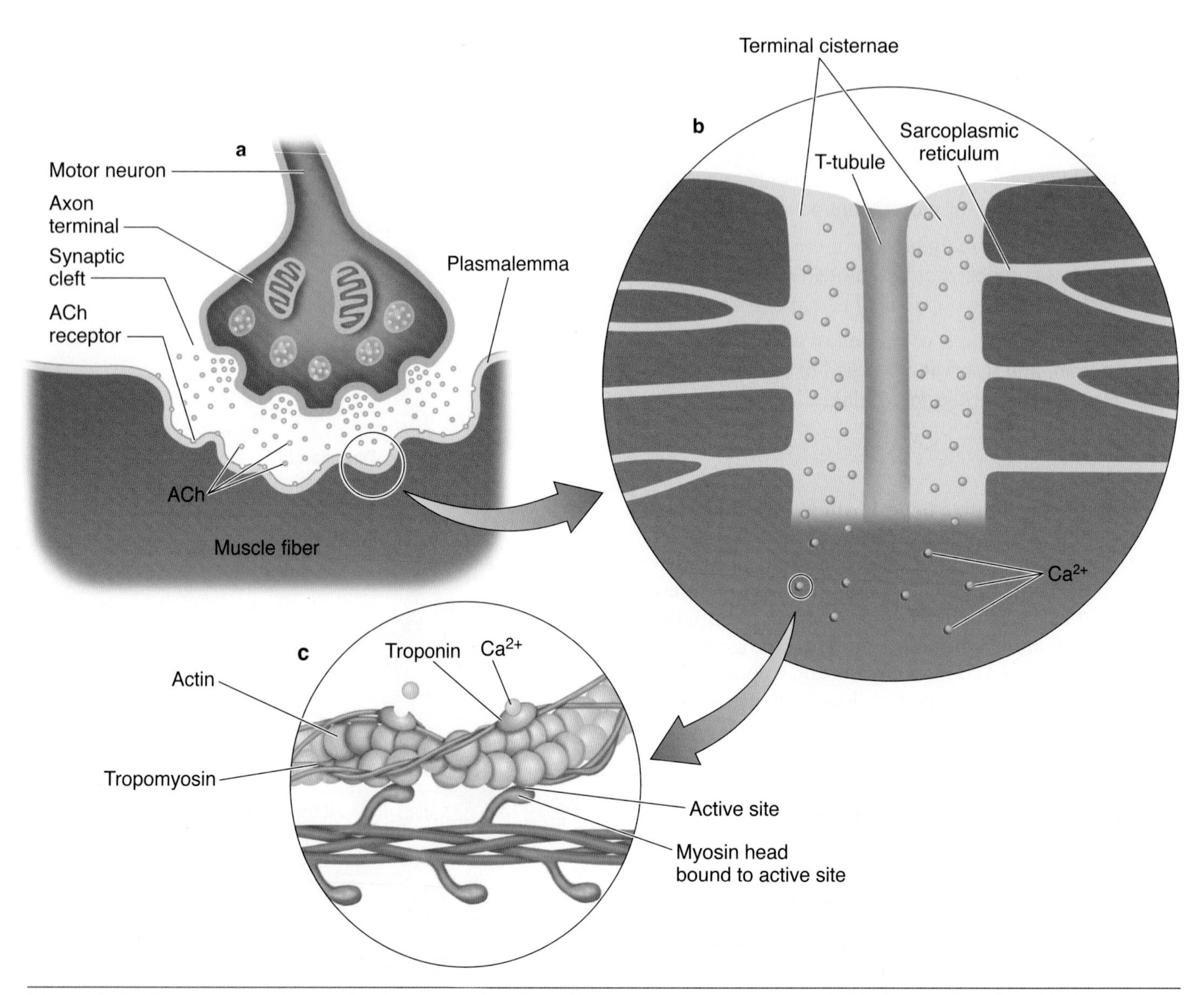

Figure 1.7 The sequence of events leading to muscle action. *(a)* A motor neuron releases acetylcholine (ACh), which binds to receptors on the plasmalemma. If enough ACh binds, an action potential is generated in the muscle fiber. *(b)* The action potential triggers release of calcium ions (Ca^{2+}) from the terminal cisternae of the sarcoplasmic reticulum into the sarcoplasm. *(c)* The Ca^{2+} binds to troponin on the actin filament, and the troponin pulls tropomyosin off the active sites, allowing myosin heads to attach to the actin filament.

of an electrical charge causes the adjacent SR to release large quantities of stored calcium ions (Ca^{2+}) into the sarcoplasm (see figure 1.7*b*).

In the resting state, tropomyosin molecules cover the myosin-binding sites on the actin molecules, preventing the binding of the myosin heads. Once calcium ions are released from the SR, they bind to the troponin on the actin molecules. Troponin, with its strong affinity for calcium ions, is believed to then initiate the contraction process by moving the tropomyosin molecules off the myosin-binding sites on the actin molecules. This is shown in figure 1.7*c*. Because tropomyosin normally covers the myosin-binding sites, it blocks the attraction between the **myosin cross-bridges** and actin molecules. However, once the tropomyosin has been lifted off the binding sites by troponin and calcium, the myosin heads can attach to the binding sites on the actin molecules.

Sliding Filament Theory: How Muscle Creates Movement

When muscle contracts, muscle fibers shorten. How do they shorten? The explanation for this phenomenon is

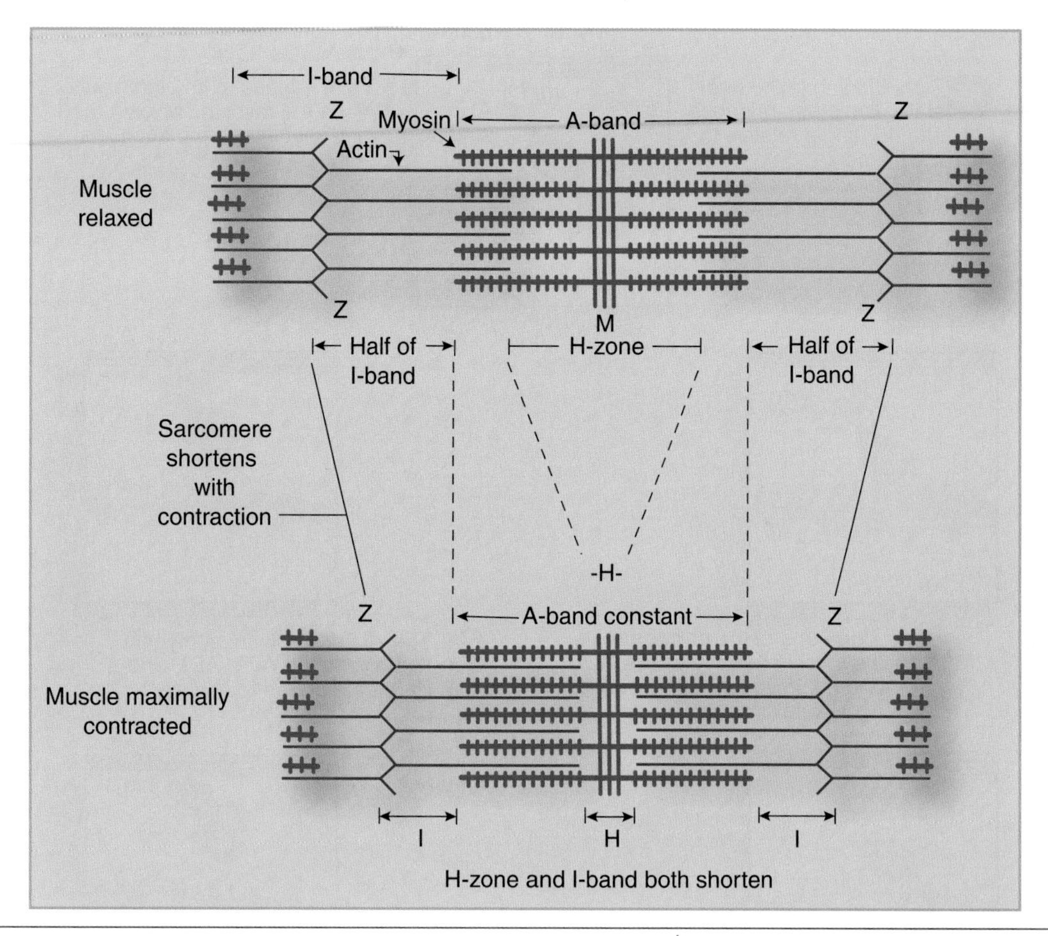

Figure 1.8 A sarcomere in its relaxed and contracted state, illustrating the sliding of the actin and myosin filaments with contraction.

termed the **sliding filament theory.** When the myosin cross-bridges are activated, they bind with actin, resulting in a conformational change in the cross-bridge, which causes the myosin head to tilt and to drag the thin filament toward the center of the sarcomere (see figures 1.8 and 1.9). This tilting of the head is referred to as the **power stroke.** The pulling of the thin filament past the thick filament shortens the sarcomere and generates force. When the fibers are not contracting, the myosin head remains in contact with the actin molecule, but the molecular bonding at the site is weakened or blocked by tropomyosin.

Immediately after the myosin head tilts, it breaks away from the active site, rotates back to its original position, and attaches to a new active site farther along the actin filament. Repeated attachments and power strokes cause the filaments to slide past one another, giving rise to the term *sliding filament theory*. This process continues until the ends of the myosin filaments reach the Z-disks, or until the Ca^{2+} is pumped back into the sarcoplasmic reticulum. During this sliding (contraction), the thin filaments move toward the center of the sarcomere and protrude into the H-zone, ultimately overlapping. When this occurs, the H-zone is no longer visible.

Energy for Muscle Contraction

Muscle contraction is an active process requiring energy. In addition to the binding site for actin, a myosin head contains a binding site for **adenosine triphosphate (ATP).** The myosin molecule must bind with ATP for muscle contraction to occur because ATP supplies the needed energy.

The enzyme **adenosine triphosphatase (ATPase),** which is located on the myosin head, splits the ATP to yield adenosine diphosphate (ADP), inorganic phosphate (P_i), and energy. The energy released from this breakdown of ATP is used to power the tilting of the myosin head. Thus, ATP is the chemical source of energy for muscle contraction. We discuss this in much more detail in chapter 2.

End of Muscle Contraction

Muscle contraction continues as long as calcium is available in the sarcoplasm. At the end of a muscle contraction, calcium is pumped back into the SR,

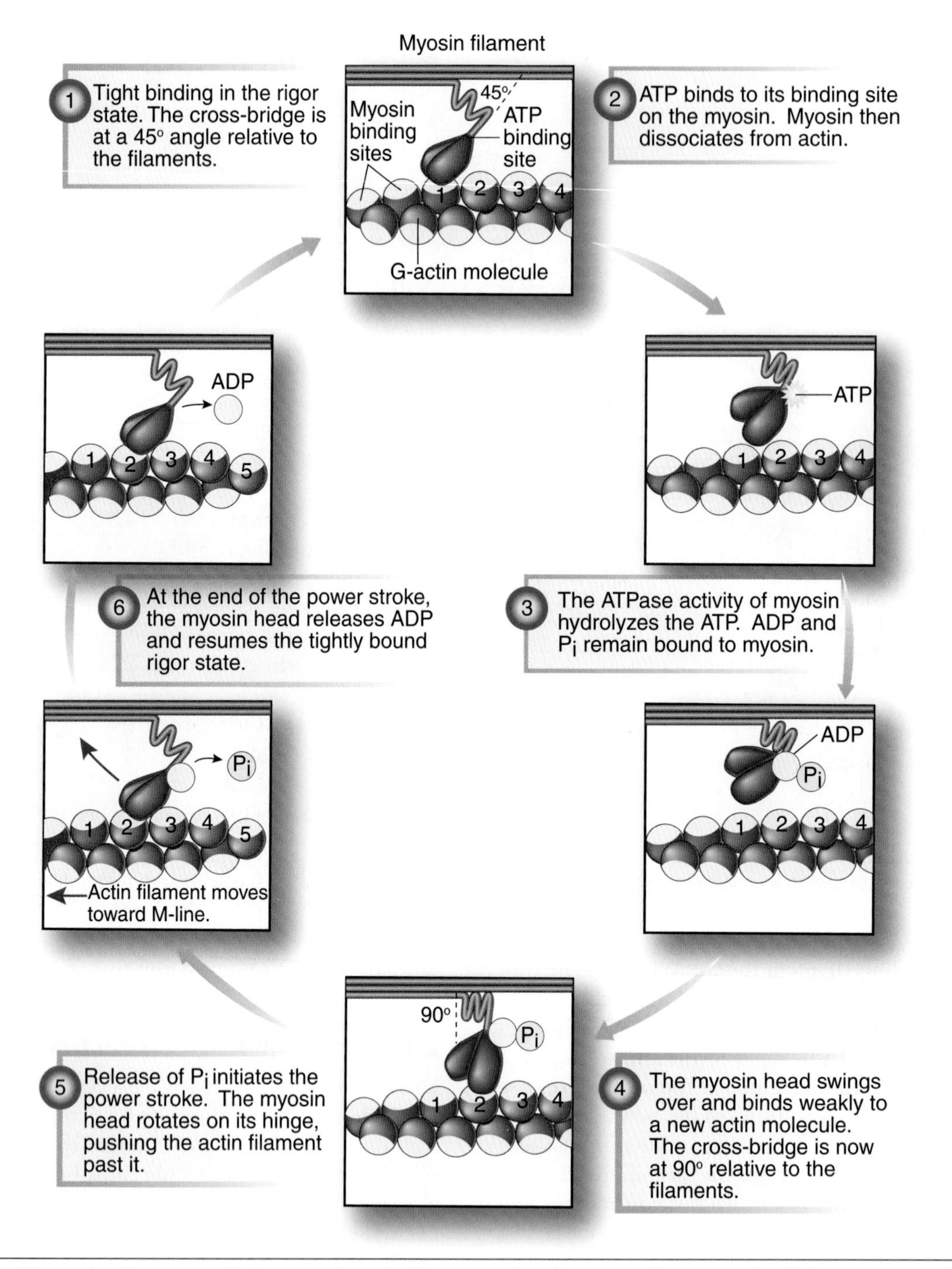

Figure 1.9 The molecular events of a contractile cycle illustrating the changes in the myosin head during various phases of the power stroke.

Fig. 12.9, p. 405 from HUMAN PHYSIOLOGY, 4th ed. By Dee Unglaub Silverthorn. Copyright © 2007 by Pearson Education, Inc. Adapted by permission.

where it is stored until a new action potential arrives at the muscle fiber membrane. Calcium is returned to the SR by an active calcium-pumping system. This is another energy-demanding process that also relies on ATP. Thus, energy is required for both the contraction and relaxation phases.

When the calcium is pumped back into the SR, troponin and tropomyosin return to the resting conformation. This blocks the linking of the myosin cross-bridges and actin molecules and stops the use of ATP. As a result, the thick and thin filaments return to their original relaxed state.

In review

- Muscle contraction is initiated by an α-motor neuron impulse or action potential. The motor neuron releases ACh, which opens up ion gates in the muscle cell membrane, allowing sodium to enter the muscle cell (depolarization). If the cell is sufficiently depolarized, an action potential is fired and muscle contraction occurs.
- The action potential travels along the plasmalemma, then moves through the T-tubule system, causing stored calcium ions to be released from the SR.
- Calcium ions bind with troponin. Then troponin moves the tropomyosin molecules off of the myosin-binding sites on the actin molecules, opening these sites to allow the myosin heads to bind with them.
- Once a strong binding state is established with actin, the myosin head tilts, pulling the thin filament past the thick filament. The tilting of the myosin head is the power stroke.
- Energy is required for muscle contraction to occur. The myosin head binds to ATP, and ATPase found on the head splits ATP into ADP and P_i, releasing energy to fuel the contraction.
- Muscle contraction ends when calcium is actively pumped out of the sarcoplasm back into the SR for storage. This process, leading to relaxation between the myosin heads and the binding sites, also requires energy supplied by ATP.

SKELETAL MUSCLE AND EXERCISE

Having reviewed the overall structure of muscle and the process by which it develops force, we now look more specifically at muscle function during exercise. Strength, endurance, and speed during exercise depend largely on the muscle's ability to produce energy and force. This section examines how muscle accomplishes this task.

Muscle Fiber Types

Not all muscle fibers are alike. A single skeletal muscle contains fibers having different speeds of shortening and strength: slow-twitch, or type I, fibers, and fast-twitch, or type II, fibers. **Slow-twitch fibers** take approximately 110 ms to reach peak tension when stimulated. **Fast-twitch fibers,** on the other hand, can reach peak tension in about 50 ms. While the terms "slow twitch" and "fast twitch" continue to be used, scientists now prefer to use the terminology type I and type II, as is the case in this textbook.

Although only one form of type I fiber has been identified, type II fibers can be further classified. The two major forms of type II fibers are fast-twitch type a (type IIa) and fast-twitch type x (type IIx). Type IIx fibers in humans are approximately the equivalent of type IIb fibers in animals. Figure 1.10 is a micrograph of human muscle in which thinly sliced (10 μm) cross sections of a muscle sample have been chemically stained to differentiate the fiber types. The type I fibers are stained black; type IIa fibers are unstained and appear white; and type IIx fibers appear gray. Although not apparent in this figure, a third subtype of fast-twitch fibers has also been identified: type IIc.

The differences among the type IIa, type IIx, and type IIc fibers are not fully understood, but type IIa fibers are believed to be the most frequently recruited. Only type I fibers are recruited more frequently than type IIa fibers. Type IIc fibers are the least often used. On the average, most muscles are composed of roughly 50% type I fibers and 25% type IIa fibers. The remaining 25% are mostly type IIx, with type IIc fibers making up only 1% to 3% of the muscle. Because knowledge about type IIc fibers is limited, we will not discuss them further. The exact percentage of each of these fiber types varies greatly in various muscles and among individuals, so the numbers listed here are only averages. This extreme variation is most evident in athletes, as we will see later in this chapter when we compare fiber types in athletes across sports and events within sports.

Characteristics of Type I and Type II Fibers

Different muscle fiber types play different roles in physical activity. This is largely due to differences in their characteristics.

ATPase

The type I and type II fibers differ in their speed of contraction. This difference results primarily from different forms of myosin ATPase. Recall that myosin ATPase is the enzyme that splits ATP to release energy

The Muscle Biopsy Needle

It was once difficult to examine human muscle tissue from a living person. Most early (pre-1900) muscle research used muscle from laboratory animals or muscle from humans obtained by open-incision surgery. In the early 1900s, a needle biopsy procedure was developed to study muscular dystrophy. In the 1960s, this technique was adapted to sample muscle for studies in exercise physiology.

Samples are removed by muscle biopsy, which involves removing a very small piece of muscle from the muscle belly for analysis. The area from which the sample is taken is first deadened with a local anesthetic, and then a small incision (approximately 1 cm, or 0.4 in.) is made with a scalpel through the skin, subcutaneous tissue, and connective tissue. A hollow needle is then inserted to the appropriate depth into the belly of the muscle. A small plunger is pushed through the center of the needle to snip off a very small sample of muscle.

The biopsy needle is withdrawn, and the sample, weighing 10 to 100 mg, is removed, cleaned of blood, mounted, and quickly frozen. It is then thinly sliced, stained, and examined under a microscope.

This method allows us to study muscle fibers and gauge the effects of acute exercise and chronic training on fiber composition. Microscopic and biochemical analyses of the samples aid our understanding of the muscles' machinery for energy production.

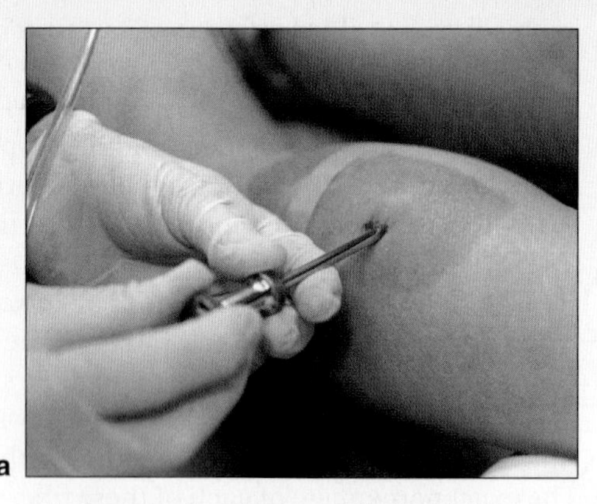
a

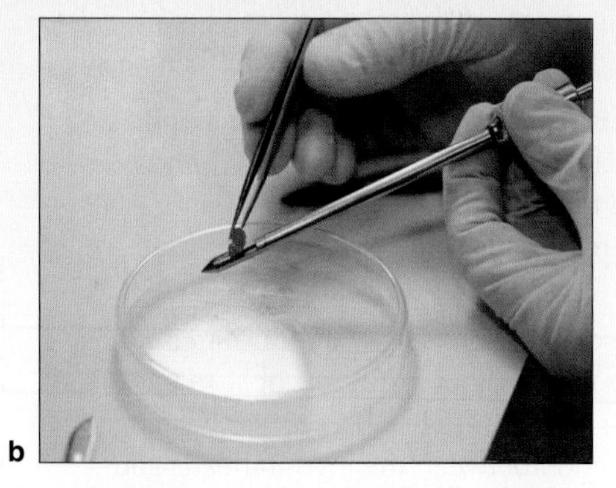
b

(a) The use of a biopsy needle to obtain a sample from the leg muscle of an elite female runner.
(b) A close-up view of a muscle biopsy needle and a small piece of muscle tissue.

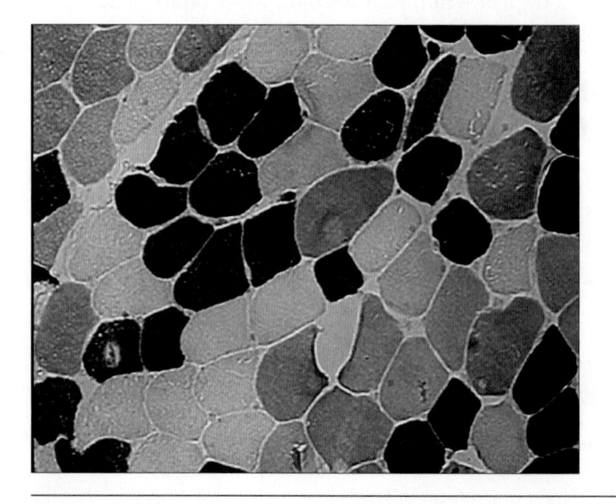

Figure 1.10 A photomicrograph showing type I (black), type IIa (white), and type IIx (gray) muscle fibers.

to drive contraction. Type I fibers have a slow form of myosin ATPase, whereas type II fibers have a fast form. In response to neural stimulation, ATP is split more rapidly in type II fibers than in type I fibers. As a result, cross-bridges cycle more rapidly in type II fibers.

One of the methods used to classify muscle fibers is a chemical staining procedure applied to a thin slice of tissue. This staining technique measures the ATPase activity in the fibers. Thus, the type I, type IIa, and type IIx fibers stain differently, as we saw in figure 1.10. This technique makes it appear that each muscle fiber has only one type of ATPase, but fibers can have a mixture of ATPase types. Some have a predominance of type I-ATPase, but others have mostly type II-ATPase. Their appearance in a stained slide preparation should be viewed as a continuum rather than as absolutely distinct types.

A newer method for identifying fiber types is to chemically separate the different types of myosin molecules (isoforms) by using a process called gel electrophoresis. As shown in figure 1.11, the isoforms are separated and stained to show the bands of protein (i.e., myosin) that characterize type I, type IIa, and type IIx fibers. Although our discussion here categorizes fiber types simply as slow twitch (type I) and fast twitch (type IIa and type IIx), scientists have further subdivided these fiber types. The use of electrophoresis has led to the detection of myosin hybrids or fibers that possess two or more forms of myosin. With this method of analysis, the fibers are classified as I; Ic (I/IIa); IIc (IIa/I); IIa; IIax; IIxa; and IIx.[6] In this book, we will use the histochemical method of identifying fibers by their primary isoforms, types I, IIa, and IIx.

Table 1.1 on page 38 summarizes the characteristics of the different muscle fiber types. The table also includes alternative names that are used in other classification systems to refer to the various muscle fiber types.

Sarcoplasmic Reticulum

Type II fibers have a more highly developed SR than do type I fibers. Thus, type II fibers are more adept at delivering calcium into the muscle cell when stimulated. This ability is thought to contribute to the faster speed of contraction (V_o) of type II fibers. On average, human type II fibers have a V_o that is five to six times faster than that of type I fibers. Although the amount of force (P_o) generated by type II and type I fibers having the same diameter is about the same, the calculated power ($\mu N \cdot$ fiber length$^{-1} \cdot s^{-1}$) of a type II fiber is three to five times greater than that of a type I fiber because of a faster shortening velocity. This may explain in part why individuals who have a predominance of type II fibers in their leg muscles tend to be better sprinters than individuals who have a high percentage of type I fibers.

Motor Units

Recall that a motor unit is composed of a single α-motor neuron and the muscle fibers it innervates. The α-motor neuron appears to determine whether the fibers are type I or type II. The α-motor neuron in a type I motor unit has a smaller cell body and typically innervates a cluster of ≤300 muscle fibers. In contrast, the α-motor neuron in a type II motor unit has a larger cell body and innervates ≥300 muscle fibers. This difference in the size of motor units means that when a single type I α-motor neuron stimulates its fibers, far fewer muscle fibers contract than when a single type II α-motor neuron stimulates its fibers. Consequently, type II muscle fibers reach peak tension faster and collectively generate more force than type I fibers.[2]

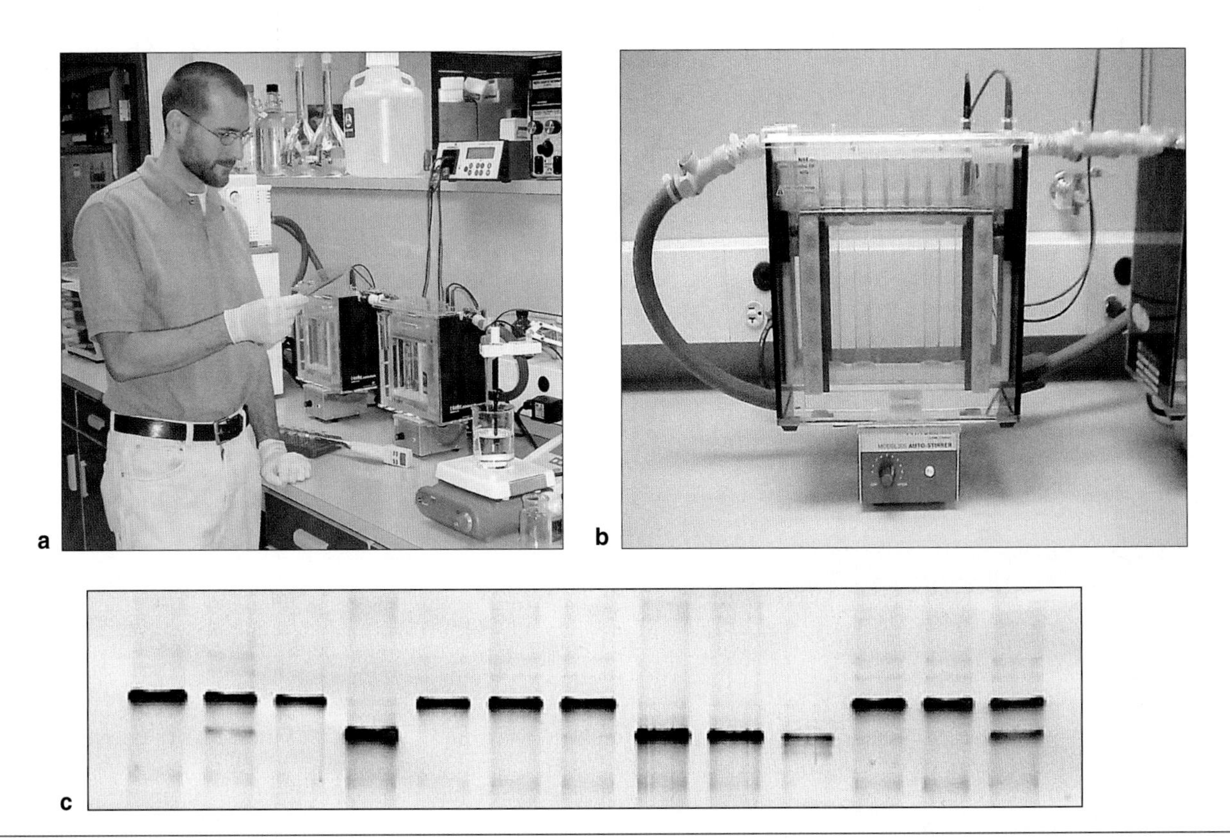

Figure 1.11 Electrophoretic separation of myosin isoforms to identify type I, type IIa, and type IIx fibers. *(a)* Single fibers are isolated under a dissecting microscope. *(b)* The myosin isoforms are separated for each fiber using electrophoretic techniques. *(c)* The isoforms are then stained to show the myosin that indicates the fiber type.

TABLE 1.1 Classification of Muscle Fiber Types

	Fiber classification		
System 1	Type I	Type IIa	Type IIx
System 2	Slow twitch (ST)	Fast twitch a (FTa)	Fast twitch x (FTx)
System 3	Slow oxidative (SO)	Fast oxidative/glycolytic (FOG)	Fast glycolytic (FG)
	Characteristics of fiber types		
Oxidative capacity	High	Moderately high	Low
Glycolytic capacity	Low	High	Highest
Contractile speed	Slow	Fast	Fast
Fatigue resistance	High	Moderate	Low
Motor unit strength	Low	High	High

Note. In this text we use system 1 to classify muscle fiber types. Other terminologies (systems 2 and 3) have also been used to identify the three primary fiber types.

Single Muscle Fiber Physiology

One of the most advanced methods for the study of human muscle fibers is to dissect fibers out of a muscle biopsy sample, suspend a single fiber between force transducers, and measure its strength and **single-fiber contractile velocity (V_o)**.

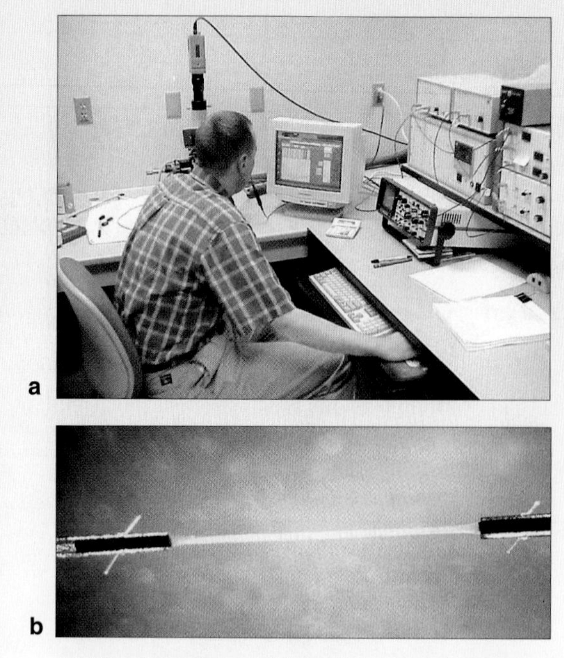

(a) The dissection and *(b)* suspension of a single muscle fiber to study the physiology of different fiber types.

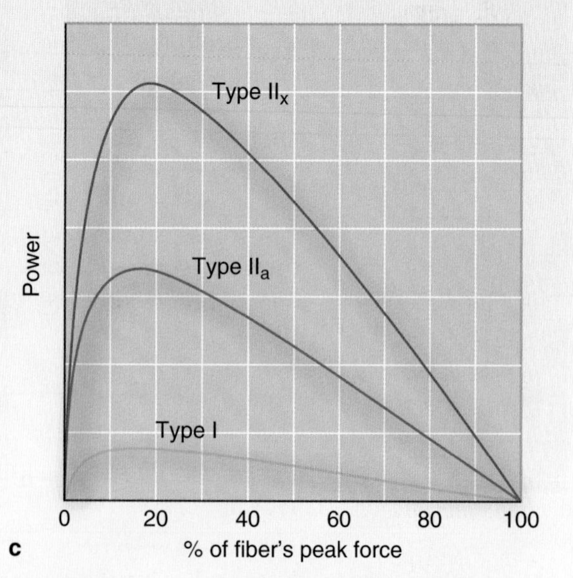

Differences in peak power generated by each fiber type at various percentages of maximal force. Note that all of the fibers tend to reach their peak power when the fibers are generating only about 20% of their peak force. It is quite clear that the peak power of the type II fibers is considerably higher than that of the type I fibers.

Distribution of Fiber Types

As mentioned earlier, the percentages of type I and type II fibers are not the same in all the muscles of the body. Generally, a person's arm and leg muscles have similar fiber compositions. Studies have shown that people with a predominance of type I fibers in their leg muscles will likely have a high percentage of type I fibers in their arm muscles as well. A similar relationship exists for type II fibers. There are some exceptions, however. The soleus muscle (beneath the gastrocnemius in the calf), for example, is composed of a very high percentage of type I fibers in everyone.

Fiber Type and Exercise

Because of these differences in type I and type II fibers, one might expect that these fiber types would also have different functions when people are physically active. Indeed, this is the case.

Type I Fibers

In general, type I muscle fibers have a high level of aerobic endurance. Aerobic means "in the presence of oxygen," so oxidation is an aerobic process. Type I fibers are very efficient at producing ATP from the oxidation of carbohydrate and fat, which will be discussed in chapter 2.

Recall that ATP is required to provide the energy needed for muscle fiber contraction and relaxation. As long as oxidation occurs, type I fibers continue producing ATP, allowing the fibers to remain active. The ability to maintain muscular activity for a prolonged period is known as muscular endurance, so type I fibers have high aerobic endurance. Because of this, they are recruited most often during low-intensity endurance events (e.g., marathon running) and during most daily activities for which the muscle force requirements are low (e.g., walking).

In focus

The difference in maximal isometric force development between type II and type I motor units is attributable to the number of muscle fibers per motor unit and the difference in size of type II and type I fibers. Type I and type II fibers of the same diameter generate about the same force. On average, however, type II fibers tend to be larger than type I fibers. So, when stimulated, type II motor units generate more force because they have larger fibers and more muscle fibers per motor unit than do the type I motor units.

Type II Fibers

Type II muscle fibers, on the other hand, have relatively poor aerobic endurance when compared to type I fibers. They are better suited to perform anaerobically (without oxygen). This means that in the absence of adequate oxygen, ATP is formed through anaerobic pathways, not oxidative pathways. (We discuss these pathways in detail in chapter 2.)

Type IIa motor units generate considerably more force than do type I motor units, but type IIa motor units also fatigue more easily because of their limited endurance. Thus, type IIa fibers appear to be the primary fiber type used during shorter, higher-intensity endurance events, such as the mile run or the 400 m swim.

Although the significance of the type IIx fibers is not fully understood, they apparently are not easily activated by the nervous system. Thus they are used rather infrequently in normal, low-intensity activity but are predominantly used in highly explosive events such as the 100 m dash and the 50 m sprint swim. Characteristics of the various fiber types are summarized in table 1.2.

Determination of Fiber Type

The characteristics of muscle fibers appear to be determined early in life, perhaps within the first few years. Studies with identical twins have shown that muscle fiber type, for the most part, is genetically determined, changing little from childhood to middle age. These studies reveal that identical twins have nearly identical fiber types, whereas fraternal twins differ in their fiber type profiles. The genes we inherit from our parents likely determine which α-motor neurons innervate our individual muscle fibers. After innervation is established, muscle fibers differentiate (become specialized) according to the type of α-motor neuron that stimulates

TABLE 1.2 Structural and Functional Characteristics of Muscle Fiber Types

	Fiber type		
Characteristic	Type I	Type IIa	Type IIx
Fibers per motor neuron	≤ 300	≥ 300	≥ 300
Motor neuron size	Smaller	Larger	Larger
Nerve conduction velocity	Slower	Faster	Faster
Contraction speed (ms)	110	50	50
Type of myosin ATPase	Slow	Fast	Fast
Sarcoplasmic reticulum development	Low	High	High

Adapted, by permission, from R. Close, 1967, "Properties of motor units in fast and slow skeletal muscles of the rat," *Journal of Physiology* 193: 45-55.

In review

- Most skeletal muscles contain both type I and type II fibers.
- The different fiber types have different myosin ATPase activities. The ATPase in the type II fibers acts faster, providing energy for muscle contraction more quickly than the ATPase in type I fibers.
- Type II fibers have a more highly developed SR, enhancing the delivery of calcium needed for muscle contraction.
- α-Motor neurons innervating type II motor units are larger and innervate more fibers than do α-motor neurons for type I motor units. Thus, type II motor units have more fibers to contract and can produce more force than type I motor units.
- The proportions of type I and type II fibers in an individual's arm and leg muscles are usually similar.
- Type I fibers have higher aerobic endurance and are well suited to low-intensity endurance activities.
- Type II fibers are better suited for anaerobic activity. Type IIa fibers play a major role in high-intensity exercise. Type IIx fibers are activated when the force demanded of the muscle is high.

them. Some recent evidence, however, suggests that endurance training, strength training, and muscular inactivity may cause a shift in the myosin isoforms. Consequently, training may induce a small change, perhaps less than 10%, in the percentage of type I and type II fibers. Further, both endurance and resistance training have been shown to reduce the percentage of type IIx fibers while increasing the fraction of type IIa fibers.

Studies of older men and women have shown that aging may alter the distribution of type I and type II fibers. As we grow older, muscles tend to lose type II motor units, which increases the percentage of type I fibers.

Muscle Fiber Recruitment

When an α-motor neuron carries an action potential to the muscle fibers in the motor unit, all fibers in the unit develop force. Activating more motor units produces more force. When little force is needed, only a few motor units are stimulated to act. Recall from our earlier discussion that type IIa and type IIx motor units contain more muscle fibers than type I motor units do. Skeletal muscle contraction involves a progressive recruitment of type I and then type II motor units, depending on the requirements of the activity being performed. As the intensity of the activity increases, the number of fibers recruited increases in the following order, in an additive manner: type I → type IIa → type IIx.

Orderly Recruitment of Muscle Fibers and the Size Principle

Most researchers agree that motor units are generally activated on the basis of a fixed order of fiber recruitment. This is known as the **principle of orderly recruitment,** in which the motor units within a given muscle appear to be ranked. Let's use the biceps brachii as an example: Assume a total of 200 motor units, which are ranked on a scale from 1 to 200. For an extremely fine muscle contraction requiring very little force production, the motor unit ranked number 1 would be recruited. As the requirements for force production increase, numbers 2, 3, 4, and so on would be recruited, up to a maximal muscle contraction that would activate most, if not all, of the motor units. For the production of a given force, the same motor units are usually recruited each time and in the same order.

A mechanism that may partially explain the principle of orderly recruitment is the **size principle,** which states that the order of recruitment of motor units is directly related to their motor neuron size. Motor units with smaller motor neurons will be recruited first. Because the type I motor units have smaller motor neurons, they are the first units recruited in graded movement (going from very low to very high rates of force production). The type II motor units then are recruited as the force needed to perform the movement increases. It is unclear at this time how the size principle relates to complex athletic movements.

During events that last several hours, exercise is performed at a submaximal pace, and the tension in the muscles is relatively low. As a result, the nervous system tends to recruit those muscle fibers best adapted to endurance activity: the type I and some type IIa fibers. As the exercise continues, these fibers become depleted of their primary fuel supply (glycogen), and the nervous system must recruit more type IIa fibers to maintain muscle tension. Finally, when the type I and type IIa fibers become exhausted, the type IIx fibers may be recruited to continue exercising.

This may explain why fatigue seems to come in stages during events such as the marathon, a 42 km (26.1 mi) run. It also may explain why it takes great conscious effort to maintain a given pace near the finish of the event. This conscious effort results in the activation of muscle fibers that are not easily recruited. Such information is of practical importance to our understanding of the specific requirements of training and performance.

In review

- Motor units give all-or-none responses. When an α-motor neuron carries an action potential to its muscle fibers, all muscle fibers in that motor unit are activated.
- Activating more motor units and thus more muscle fibers produces more force.
- In low-intensity activity, most muscle force is generated by type I fibers. As the intensity increases, type IIa fibers are recruited, and at the higher intensities, the type IIx fibers are activated. The same pattern of recruitment is followed during events of long duration.

Fiber Type and Athletic Success

From what we have just discussed, it appears that athletes who have a high percentage of type I fibers might have an advantage in prolonged endurance events, whereas those with a predominance of type II fibers could be better suited for high-intensity, short-term, and explosive activities. Can it be that the proportions of an athlete's various muscle fiber types determine athletic success?

The muscle fiber makeup of successful athletes from a variety of athletic events and of nonathletes is shown in table 1.3. As anticipated, the leg muscles of distance runners, who rely on endurance, have a predominance of type I fibers.[3] Studies of elite male and female distance runners revealed that many of these athletes' gastrocnemius (calf) muscles contain more than 90% type I fibers. Also, although muscle fiber cross-sectional area varies markedly among elite distance runners, type I fibers in their leg muscles average about 22% more cross-sectional area than type II fibers.

TABLE 1.3 Percentages and Cross-Sectional Areas of Type I and Type II Fibers in Selected Muscles of Male and Female Athletes

					Cross-sectional area (μm²)	
Athlete	**Sex**	**Muscle**	**% type I**	**% type II**	**Type I**	**Type II**
Sprint runners	M	Gastrocnemius	24	76	5,878	6,034
	F	Gastrocnemius	27	73	3,752	3,930
Distance runners	M	Gastrocnemius	79	21	8,342	6,485
	F	Gastrocnemius	69	31	4,441	4,128
Cyclists	M	Vastus lateralis	57	43	6,333	6,116
	F	Vastus lateralis	51	49	5,487	5,216
Swimmers	M	Posterior deltoid	67	33	–	–
Weightlifters	M	Gastrocnemius	44	56	5,060	8,910
	M	Deltoid	53	47	5,010	8,450
Triathletes	M	Posterior deltoid	60	40	–	–
	M	Vastus lateralis	63	37	–	–
	M	Gastrocnemius	59	41	–	–
Canoeists	M	Posterior deltoid	71	29	4,920	7,040
Shot-putters	M	Gastrocnemius	38	62	6,367	6,441
Nonathletes	M	Vastus lateralis	47	53	4,722	4,709
	F	Gastrocnemius	52	48	3,501	3,141

In focus

World champions in the marathon are reported to possess 93% to 99% type I fibers in their gastrocnemius muscles. World-class sprinters, on the other hand, have only about 25% type I fibers in this muscle.

In contrast, the gastrocnemius muscles are composed principally of type II fibers in sprint runners, who rely on speed and strength. Although swimmers tend to have higher percentages of type I fibers (60-65%) in their arm muscles than untrained subjects (45-55%), fiber type differences between good and elite swimmers are not apparent.[4, 5]

The fiber composition of muscles in distance runners and sprinters is markedly different. However, it may be a bit risky to think we can select champion distance runners and sprinters solely on the basis of predominant muscle fiber type. Other factors, such as cardiovascular function, motivation, training, and muscle size, also contribute to success in such events of endurance, speed, and strength. Thus, fiber composition alone is not a reliable predictor of athletic success.

Use of Muscles

We have examined the different muscle fiber types. We understand that all fibers in a motor unit, when stimulated, act at the same time and that different fiber types are recruited in stages, depending on the force required to perform an activity. Now we can move back to the gross level, turning our attention to how whole muscles work to produce movement.

Types of Muscle Contraction

Muscle movement generally can be categorized into three types of contractions—concentric, static, and eccentric. In many activities, such as running and jumping, all three types of contraction may occur in the execution of a smooth, coordinated movement. For the sake of clarity, though, we will examine each type separately.

A muscle's principal action, shortening, is referred to as a **concentric contraction,** the most familiar type of contraction. To understand muscle shortening, recall our earlier discussion of how the thin and thick filaments slide across each other. In a concentric contraction, the thin filaments are pulled toward the center of the sarcomere. Because joint movement is produced, concentric contractions are considered **dynamic contractions.**

Muscles can also act without moving. When this happens, the muscle generates force, but its length remains static (unchanged). This is called a **static,** or **isometric, muscle contraction,** because the joint angle does not change. A static contraction occurs, for example, when one tries to lift an object that is heavier than the force generated by the muscle, or when one supports the weight of an object by holding it steady with the elbow flexed. In both cases, the person feels the muscles tense, but there is no joint movement. In a static contraction, the myosin cross-bridges form and are recycled, producing force, but the external force is too great for the thin filaments to be moved. They remain in their normal position, so shortening can't occur. If enough motor units can be recruited to produce sufficient force to overcome the resistance, a static contraction can become a dynamic one.

Muscles can exert force even while lengthening. This movement is an **eccentric contraction.** Because joint movement occurs, this is also a dynamic contraction. An example of an eccentric contraction is the action of the biceps brachii when one extends the elbow to lower a heavy weight. In this case, the thin filaments are pulled farther away from the center of the sarcomere, essentially stretching it.

Generation of Force

Whenever muscles contract, whether the contraction is concentric, static, or eccentric, the force developed must be graded to meet the needs of the task or activity. Using golf as an example, the force needed to tap in a 1 m (~40 in.) putt is far less than that needed to drive the ball 100 m (109 yd) from the tee to the middle of the fairway. The amount of muscle force developed is dependent on the number and type of motor units activated, the frequency of stimulation of each motor unit, the size of the muscle, the muscle fiber and sarcomere length, and the muscle's speed of contraction.

Motor Units and Muscle Size

More force can be generated when more motor units are activated. Type II motor units generate more force than type I motor units because a type II motor unit contains more muscle fibers than a type I motor unit. In a similar manner, larger muscles, having more muscle fibers, can produce more force than smaller muscles.

Frequency of Stimulation of the Motor Units: Rate Coding

A single motor unit can exert varying levels of force dependent on the frequency at which it is stimulated. This is illustrated in figure 1.12.[1] The smallest contractile response of a muscle fiber or a motor unit to a single electrical stimulus is termed a **twitch.** A series of three stimuli in rapid sequence, prior to complete relaxation from the first stimulus, can elicit an even greater increase in force or tension. This is termed **summation.** Continued stimulation at higher frequencies can lead to the state of **tetanus,** resulting in the peak force or tension of the muscle fiber or motor unit. **Rate coding** is the term used to describe the process by which the tension of a given motor unit can vary from that of a twitch to that of tetanus by increasing the frequency of stimulation of that motor unit.

Muscle Fiber and Sarcomere Length

There is an optimal length of each muscle fiber relative to its ability to generate force. Recall that a given muscle fiber is composed of sarcomeres connected end to end and that these sarcomeres are composed of both thick and thin filaments. The optimal sarcomere length is defined as that length where there is optimal overlap of the thick and thin filaments, thus maximizing cross-bridge interaction. This is illustrated in figure 1.13.[6] When a sarcomere is fully stretched (A) or shortened (E), little or no force can be developed since there is little cross-bridge interaction.

Speed of Contraction

The ability to develop force also depends on the speed of muscle contraction. During concentric (shortening) contractions, maximal force development decreases progressively at higher speeds. When people try to lift a very heavy object, they tend to do it slowly, maximizing the force they can apply to it. If they grab it and quickly try to lift it, they will likely fail, if not injure themselves. However, with eccentric (lengthening) contractions, the opposite is true. Fast eccentric contractions allow maximal application of force. These relationships are depicted in figure 1.14. Eccentric contractions are shown on the left and concentric on the right.

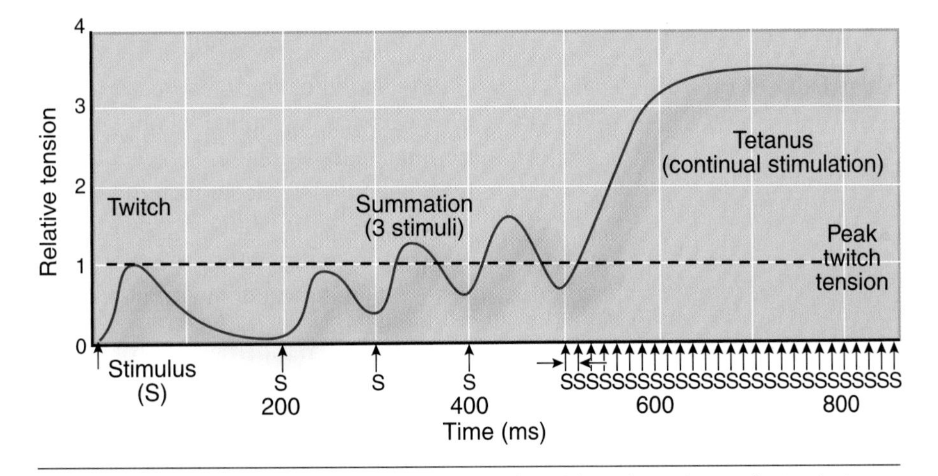

Figure 1.12 Variation in force or tension produced based on electrical stimulation frequency, illustrating the concept of a twitch, summation, and tetanus.

Adapted, by permission, from G.A. Brooks, et al., 2005, *Exercise Physiology: human bioenergetics and its applications,* 4th ed. (New York: McGraw-Hill), 388. With permission of the McGraw-Hill Companies.

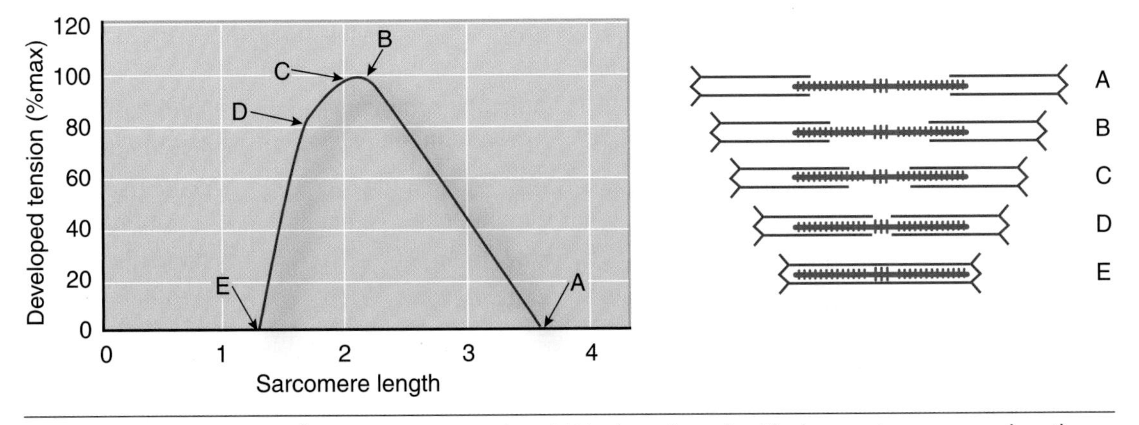

Figure 1.13 Variation in force or tension produced (% of maximum) with changes in sarcomere length, illustrating the concept of optimal length for force production.

Adapted, by permission, from B.R. MacIntosh, P.F. Gardiner, and A.J. McComas, 2006, *Skeletal muscle: Form and function,* 2nd ed. (Champaign, IL: Human Kinetics), 156.

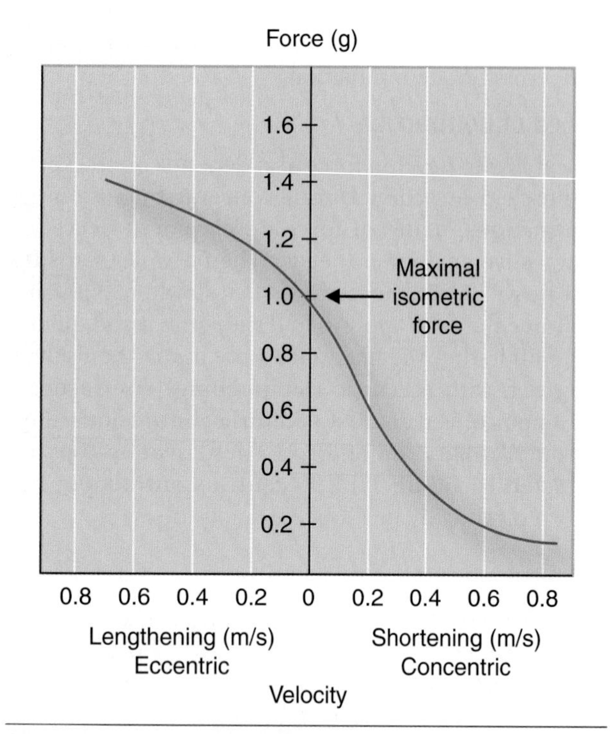

Figure 1.14 The relationship between muscle lengthening and shortening velocity and force production. Note that the capacity for the muscle to generate force is greater during eccentric (lengthening) actions than during concentric (shortening) actions.

In review

- Muscle fiber type composition differs in athletes by sport and event, with speed and strength events characterized by higher percentages of type II fibers and endurance events by higher percentages of type I fibers.
- The three main types of muscle contraction are concentric, in which the muscle shortens; static or isometric, in which the muscle acts but the joint angle is unchanged; and eccentric, in which the muscle lengthens.
- Force production is increased both through the recruitment of more motor units and through increase in the frequency of stimulation (rate coding) of the motor units.
- Force production is maximized at the muscle's optimal length. At this length, the amount of energy stored and the number of linked actin-myosin cross-bridges are optimal.
- Speed of contraction also affects the amount of force produced. For concentric contraction, maximal force is achieved with slower contractions. The closer to zero velocity (static), the more force can be generated. With eccentric contractions, however, faster movement allows more force production.

In Closing

In this chapter, we reviewed the components of skeletal muscle. We considered the differences in fiber types and their impact on physical performance. We learned how muscles generate force and produce movement. Now that we understand how movement is produced, it is time to turn our attention to how movement is fueled. In the next chapter, we focus on metabolism and energy production.

KEY TERMS

actin
action potential
adenosine triphosphatase (ATPase)
adenosine triphosphate (ATP)
α-motor neuron
concentric contraction
dynamic contraction
eccentric contraction
endomysium
epimysium
fasciculus
fast-twitch (type II) fiber
motor unit
muscle fiber
myofibril
myosin
myosin cross-bridge
nebulin
perimysium
plasmalemma
power stroke
principle of orderly recruitment
rate coding
sarcolemma
sarcomere
sarcoplasm
sarcoplasmic reticulum (SR)
satellite cells
single-fiber contractile velocity (V_o)
size principle
sliding filament theory
slow-twitch (type I) fiber
static (isometric) contraction
summation
tetanus
titin
transverse tubules (T-tubules)
tropomyosin
troponin
twitch

STUDY QUESTIONS

1. List and define the components of a muscle fiber.
2. List the components of a motor unit.
3. What is the role of calcium in muscle contraction?
4. Describe the sliding filament theory. How do muscle fibers shorten?
5. What are the basic characteristics of type I and type II muscle fibers?
6. What is the role of genetics in determining the proportions of muscle fiber types and the potential for success in selected activities?
7. Describe the relationship between muscle force development and the recruitment of type I and type II motor units.
8. Differentiate and give examples of concentric, static, and eccentric contractions.
9. What two mechanisms are used by the body to increase force production in a single muscle?
10. What is the optimal length of a muscle for maximal force development?
11. What is the relationship between maximal force development and the speed of shortening (concentric) and lengthening (eccentric) contractions?

STUDY GUIDE ACTIVITIES

In addition to the activities listed in the chapter opening outline on page 25, two other activities are available in the online study guide, located at

www.HumanKinetics.com/PhysiologyOfSportAndExercise.

The chapter **SUMMARY** reviews the key concepts covered in the chapter, and the end-of-chapter **QUIZ** tests your understanding of the material covered in the chapter.

Fuel for Exercising Muscle

Metabolism and Hormonal Control

In this chapter and in the online study guide

In 1972, two of this book's coauthors ran 16 km (10 mi) per day for five successive days. To add to the physical stress, these runs were performed in the bright summer sun and heat (approximately 30-35 °C, or 86-95 °F) of Davis, California. Each lost 3 to 4 kg (6.6-8.8 lb) of sweat daily. Blood samples taken each morning during this period revealed that their hemoglobin concentrations and hematocrits (percentage of blood composed of red blood cells) were declining. Yet isotope studies showed that they were not losing hemoglobin or blood cells. Rather, the plasma, the fluid part of the blood, was increasing day by day. Their bodies were trying to offset the detrimental effects of daily dehydration by retaining water to minimize the loss of plasma volume that accompanied such heavy sweating. At least three hormones—aldosterone, renin, and antidiuretic hormone—functioned in concert to help increase plasma volume and minimize dehydration on each successive day of exercise in the heat.

During exercise, the body is faced with tremendous demands that require a multitude of physiological adjustments. The rate of energy production must increase, and metabolic by-products must be cleared. While the body's internal environment is in a constant state of flux even at rest, during exercise these well-orchestrated changes must occur rapidly and frequently. Cells must convert foodstuffs, or substrates, such as carbohydrates, fats, and proteins, to a usable form of energy. The first part of this chapter describes those intricate processes.

The more rigorous the exercise, the more difficult it is to maintain homeostasis. Much of the regulation required during exercise is accomplished by the nervous system (chapter 3). But another system affects virtually every cell in the body. It constantly monitors the body's internal environment, noting all changes that occur and responding quickly to ensure that homeostasis is not drastically disrupted. It is the endocrine system that exerts this control through the hormones it releases. In the latter part of this chapter, we focus on the importance of hormones in making adjustments and maintaining homeostasis amid all the internal processes that support physical activity. Because we cannot cover all aspects of endocrine control during exercise, the focus will be on hormonal control of metabolism and body fluids.

METABOLISM AND BIOENERGETICS

All energy originates from the sun as light energy. Chemical reactions in plants (photosynthesis) convert light into stored chemical energy. In turn, we obtain energy by eating plants or animals that feed on plants. Nutrients from ingested foods are provided and stored as carbohydrates, fats, and proteins. These three basic fuels, or energy **substrates,** can be broken down in our cells to release the stored energy. Each cell contains chemical pathways that convert these substrates to energy that can then be used by that cell and other cells of the body, a process called **bioenergetics.** All of the chemical reactions in the body are collectively called **metabolism.**

Because all energy eventually degrades to heat, the amount of energy released in a biological reaction can be calculated from the amount of heat produced. Energy in biological systems is measured in calories. By definition, 1 calorie (cal) equals the amount of heat energy needed to raise 1 g of water 1 °C, from 14.5 °C to 15.5 °C. In humans, energy is expressed in **kilocalories (kcal),** where 1 kcal is the equivalent of 1,000 cal. Sometimes the term *Calorie* (with a capital C) is used synonymously with kilocalorie, but kilocalorie is more technically and scientifically correct. Thus, when one reads that someone eats or expends 3,000 Cal per day, it really means that person is ingesting or expending 3,000 *kcal* per day.

Some free energy in the cells is used for growth and repair throughout the body. Such processes build muscle mass during training and repair muscle damage after exercise or injury. Energy also is needed for active transport of many substances, such as sodium, potassium, and calcium ions, across cell membranes. Active transport is critical to the survival of cells and the maintenance of homeostasis. Myofibrils also use some of the energy released in our bodies to cause sliding of the actin and myosin filaments, resulting in muscle action and force generation, as we saw in chapter 1.

Energy Sources

Energy is released when chemical bonds—the bonds that hold elements together to form molecules—are broken. Foods are composed primarily of carbon, hydrogen, oxygen, and (in the case of protein) nitrogen. The molecular bonds that hold these elements together are relatively weak and therefore provide little energy when broken. Consequently, food is not used directly for cellular operations. Rather, the energy in food molecular bonds is chemically released within our cells and then stored in the form of the high-energy compound introduced in chapter 1, adenosine triphosphate (ATP).

The formation of ATP provides cells with a high-energy compound for storing and—when broken down, releasing—energy. It serves as the immediate source of energy for most body functions including muscle contraction.

At rest, the energy that the body needs is derived almost equally from the breakdown of carbohydrates and fats. Proteins serve important functions as enzymes that aid chemical reactions and as structural building blocks, but usually provide little energy for metabolism. During intense, short-duration muscular effort, more carbohydrate is used, with less reliance on fat to generate ATP. Longer, less intense exercise utilizes carbohydrate and fat for sustained energy production.

Carbohydrate

The amount of **carbohydrate** utilized during exercise is related to both the carbohydrate availability and the muscles' well-developed system for carbohydrate metabolism. All carbohydrates are ultimately converted to **glucose,** a monosaccharide (one-unit, or simple, sugar) that is transported through the blood to all body tissues. Under resting conditions, ingested carbohydrate is stored in muscles and liver in the form of a more complex sugar molecule, **glycogen.** Glycogen is stored in the cytoplasm of muscle cells until those cells use it to form ATP. The glycogen stored in the liver is converted back to glucose as needed and then transported by the blood to active tissues, where it is metabolized.

Liver and muscle glycogen stores are limited and can be depleted during prolonged, intense exercise unless the diet contains a reasonable amount of carbohydrate. Thus, we rely heavily on dietary sources of starches and sugars to continually replenish our carbohydrate reserves. Without adequate carbohydrate intake, muscles can be deprived of their primary energy source.

Fat

Fat provides a large portion of the energy during prolonged, less intense exercise. Body stores of potential energy in the form of fat are substantially larger than the reserves of carbohydrate, in terms of both weight and potential energy. Table 2.1 provides an indication of the total body stores of these two energy sources in a lean person (12% body fat). For the average middle-aged adult with more body fat (adipose tissue), the fat stores would be approximately twice as large, whereas the carbohydrate stores would be about the same. But fat is less readily available for cellular metabolism because it must first be reduced from its complex form, **triglyceride,** to its basic components, glycerol and **free fatty acids (FFAs).** Only FFAs are used to form ATP.

In focus

Carbohydrate stores in the liver and skeletal muscle are limited to about 2,500 to 2,600 kcal of energy, or the equivalent of the energy needed for about 40 km (25 mi) of running. Fat stores can provide more than 70,000 kcal of energy.

TABLE 2.1 Body Stores of Fuels and Energy

	g	kcal
Carbohydrates		
Liver glycogen	110	451
Muscle glycogen	500	2,050
Glucose in body fluids	15	62
Fat		
Subcutaneous and visceral	7,800	73,320
Intramuscular	161	1,513
Total	7,961	74,833

Note. These estimates are based on a body weight of 65 kg (143 lb) with 12% body fat.

Substantially more energy is derived from breaking down a gram of fat (9.4 kcal/g) than from the same amount of carbohydrate (4.1 kcal/g). Nonetheless, the rate of energy release from fat is too slow to meet all of the energy demands of intense muscular activity.

Other types of fats found in the body serve non-energy-producing functions. Phospholipids are a key structural component of all cell membranes and form protective sheaths around some large nerves. Steroids are also found in cell membranes and function as hormones and building blocks of hormones such as estrogen and testosterone.

Protein

Protein also can be used as a minor energy source, but it must first be converted into glucose. In the case of severe energy depletion or starvation, protein may even be used to generate FFAs for cellular energy. The process by which protein or fat is converted into glucose is called **gluconeogenesis.** The process of converting protein into fatty acids is termed **lipogenesis.** Protein can supply up to 5% or 10% of the energy needed to sustain prolonged exercise. Only the most basic units of protein—the amino acids—can be used for energy. A gram of protein yields about 4.1 kcal.

Rate of Energy Release

To be useful, free energy must be released from chemical compounds at a controlled rate. This rate is partially determined by the choice of the primary fuel source. Large amounts of one particular fuel can cause cells to rely more on that source than on alternatives. This influence of energy availability is termed the *mass action effect.*

Specific protein molecules called enzymes control the rate of free-energy release. Many of these enzymes facilitate the breakdown **(catabolism)** of chemical compounds. The way these enzymes speed catabolism has been characterized as a "lock-and-key" mechanism, as illustrated in figure 2.1. However, many enzymes also become altered in structure after binding to the chemical compound. Thus, the structure and function of enzymes may be more complex than we have shown in figure 2.1, but the concept of the lock and key provides a useful model of the interactions between energy compounds (e.g., glucose) and enzymes important to energy transfer within the cell. Although the enzyme names are quite complex, most end with the suffix *-ase.* For example, an important enzyme that acts to break down ATP and release stored energy is adenosine triphosphatase (ATPase).

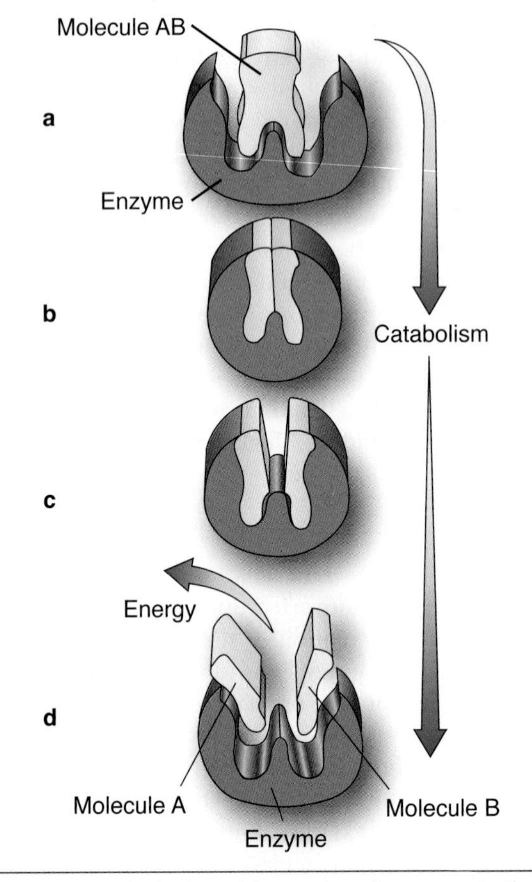

Figure 2.1 The lock-and-key action of enzymes in the catabolism (breakdown) of compounds. *(a)* The compound AB is bound into the enzyme molecule, *(b)* undergoes modification while in combination with the enzyme, and *(c)* is broken down, thereby *(d)* releasing the energy that originally held the compound together.

In review

- We derive our energy from three substrates in foods: carbohydrate, fat, and protein. Proteins provide very little of the energy used for metabolism under most normal conditions.
- The energy we derive from food is stored in cells in the form of a high-energy compound, ATP.
- Carbohydrate and protein each provide about 4.1 kcal of energy per gram, compared with about 9.4 kcal/g for fat.
- Carbohydrate, stored as glycogen in muscle and the liver, is more quickly accessible than either protein or fat. Glucose, directly from food or broken down from glycogen, is the usable form of carbohydrate.
- Fat, stored as triglycerides in adipose tissue, is an ideal storage form of energy. Free fatty acids from the breakdown of triglycerides are converted to energy.

Bioenergetics: The Basic Energy Systems

An ATP molecule (figure 2.2*a*) consists of adenosine (a molecule of adenine joined to a molecule of ribose) combined with three inorganic phosphate (P_i) groups.

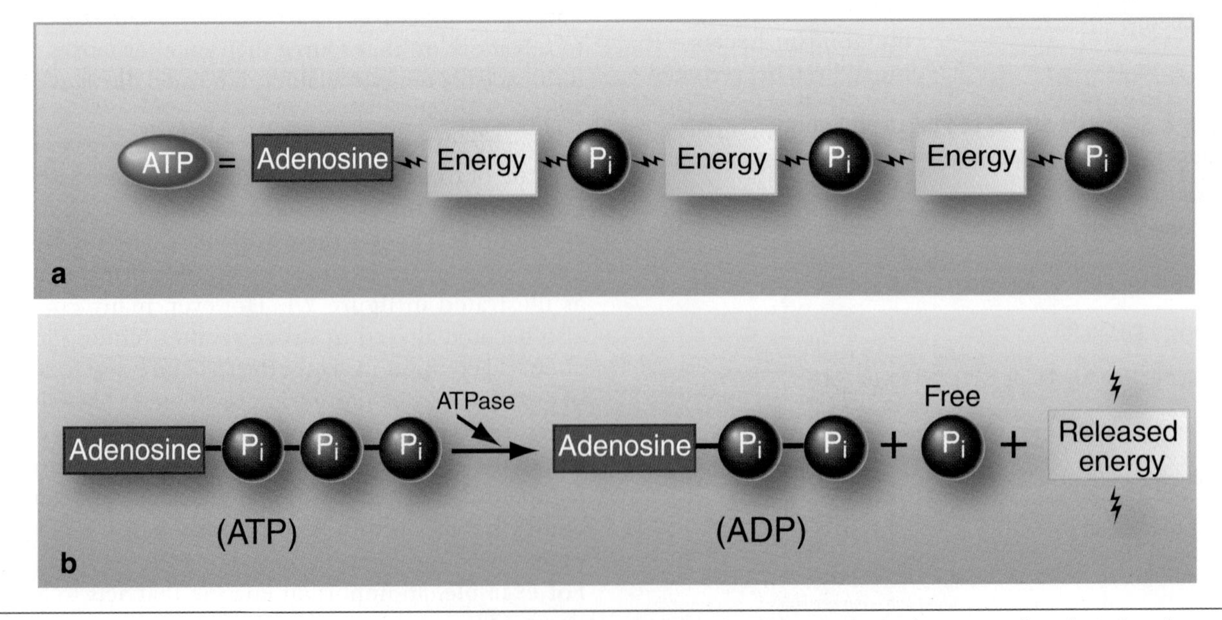

Figure 2.2 *(a)* The structure of an adenosine triphosphate (ATP) molecule, showing the high-energy phosphate bonds. *(b)* When the third phosphate on the ATP molecule is separated from adenosine by the action of adenosine triphosphatase (ATPase), energy is released.

Adenine is a nitrogen-containing base, and ribose is a five-carbon sugar. When an ATP molecule is combined with water (hydrolysis) and acted on by the enzyme ATPase, the last phosphate group splits away, rapidly releasing a large amount of free energy (approximately 7.3 kcal per mole of ATP under standard conditions, but possibly up to 10 kcal per mole of ATP or greater within the cell). This reduces the ATP to **adenosine diphosphate (ADP)** and P_i (figure 2.2*b*). But how was that energy originally stored?

To generate ATP, a phosphate group is added to the relatively low-energy compound, ADP, a process called *phosphorylation.* Some ATP is generated independent of oxygen availability, and such metabolism is called substrate-level phosphorylation. Other ATP-producing reactions occur without oxygen, a process called **anaerobic metabolism.** When these reactions occur with the aid of oxygen, the overall process is called **aerobic metabolism,** and the aerobic conversion of ADP to ATP is *oxidative phosphorylation.*

Cells generate ATP through three different processes or systems:

1. The ATP-PCr system
2. The glycolytic system (glycolysis)
3. The oxidative system (oxidative phosphorylation)

ATP-PCr System

The simplest of the energy systems is the **ATP-PCr system,** shown in figure 2.3. In addition to storing a very small amount of ATP directly, cells contain another high-energy phosphate molecule that stores energy. This molecule is called **phosphocreatine,** or **PCr** (sometimes called creatine phosphate). Unlike freely available ATP, energy released by the breakdown of PCr is not directly used for cellular work. Instead, it regenerates ATP to maintain a relatively constant supply.

The release of energy from PCr is facilitated by the enzyme creatine kinase, which acts on PCr to separate P_i from creatine. The energy released can then be used to add a P_i molecule to an ADP molecule, forming ATP. As energy is released from ATP by the splitting of a phosphate group, cells can prevent ATP depletion by breaking down PCr, providing energy and P_i to re-form ATP from ADP.

This process is rapid and can be accomplished without any special structures within the cell. The ATP-PCr system is classified as substrate-level metabolism. Although it can occur in the presence of oxygen, this process does not require oxygen.

During the first few seconds of intense muscular activity, such as sprinting, ATP is maintained at a relatively constant level, but PCr declines steadily as it is used to replenish the depleted ATP (see figure 2.4). At exhaustion, however, both ATP and PCr levels are low and are unable to provide energy for further muscle contraction and relaxation. Thus, the capacity to maintain ATP levels with the energy from PCr is limited. The combination of ATP and PCr stores can sustain the muscles' energy needs for only 3 to 15 s during an all-out sprint. Beyond that time, muscles must rely on other processes for ATP formation: glycolytic and oxidative combustion of fuels.

Glycolytic System

Another method of ATP production involves the liberation of energy through the breakdown (lysis) of glucose. This system is called the glycolytic system because it entails **glycolysis,** which is the breakdown of glucose through a pathway that involves a sequence of glycolytic enzymes. An overview of this process is depicted in figure 2.5.

Glucose accounts for about 99% of all sugars circulating in the blood. Blood glucose comes from the digestion of carbohydrate and the breakdown of liver glycogen. Glycogen is synthesized from glucose by a process

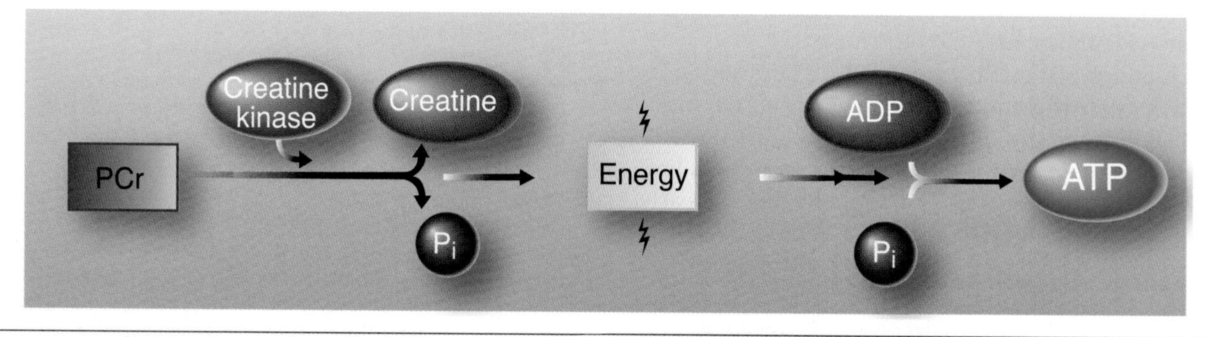

Figure 2.3 Adenosine triphosphate (ATP) can be recreated via the binding of an inorganic phosphate (P_i) to adenosine diphosphate (ADP, or adenosine plus two phosphates) with the energy derived from phosphocreatine (PCr).

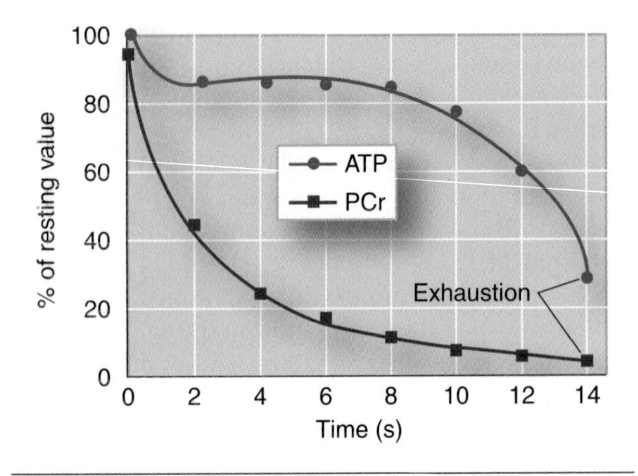

Figure 2.4 Changes in type II (fast-twitch) skeletal muscle adenosine triphosphate (ATP) and phosphocreatine (PCr) during 14 s of maximal muscular effort (sprinting). Although ATP is being used at a very high rate, the energy from PCr is used to synthesize ATP, preventing the ATP level from decreasing. However, at exhaustion, both ATP and PCr levels are low.

called glycogenesis. Glycogen is stored in the liver or in muscle until needed. At that time, the glycogen is broken down to glucose-1-phosphate, which enters the glycolysis pathway, a process termed **glycogenolysis.**

Before either glucose or glycogen can be used to generate energy, it must be converted to a compound called glucose-6-phosphate. Even though the goal of glycolysis is to release ATP, the conversion of a molecule of glucose to glucose-6-phosphate requires one ATP molecule. In the conversion of glycogen, glucose-6-phosphate is formed from glucose-1-phosphate without this energy expenditure. Glycolysis technically begins once the glucose-6-phosphate is formed.

Glycolysis, which is far more complex than the ATP-PCr system, requires 10 to 12 enzymatic reactions for the breakdown of glycogen to lactic acid. All these enzymes operate within the cell cytoplasm. The net gain from this process is 3 moles (mol) of ATP formed for each mole of glycogen broken down. If glucose is used instead of glycogen, the gain is only 2 mol of ATP because 1 mol was used for the conversion of glucose to glucose-6-phosphate.

This energy system does not produce large amounts of ATP. Despite this limitation, the combined actions of the ATP-PCr and glycolytic systems allow the muscles to generate force even when the oxygen supply is limited. These two systems predominate during the early minutes of high-intensity exercise.

Another major limitation of anaerobic glycolysis is that it causes an accumulation of lactic acid in the muscles and body fluids. Glycolysis produces pyruvic acid. This process does not require oxygen, but the presence of oxygen determines the fate of the pyruvic acid. Anaerobically, the pyruvic acid is converted

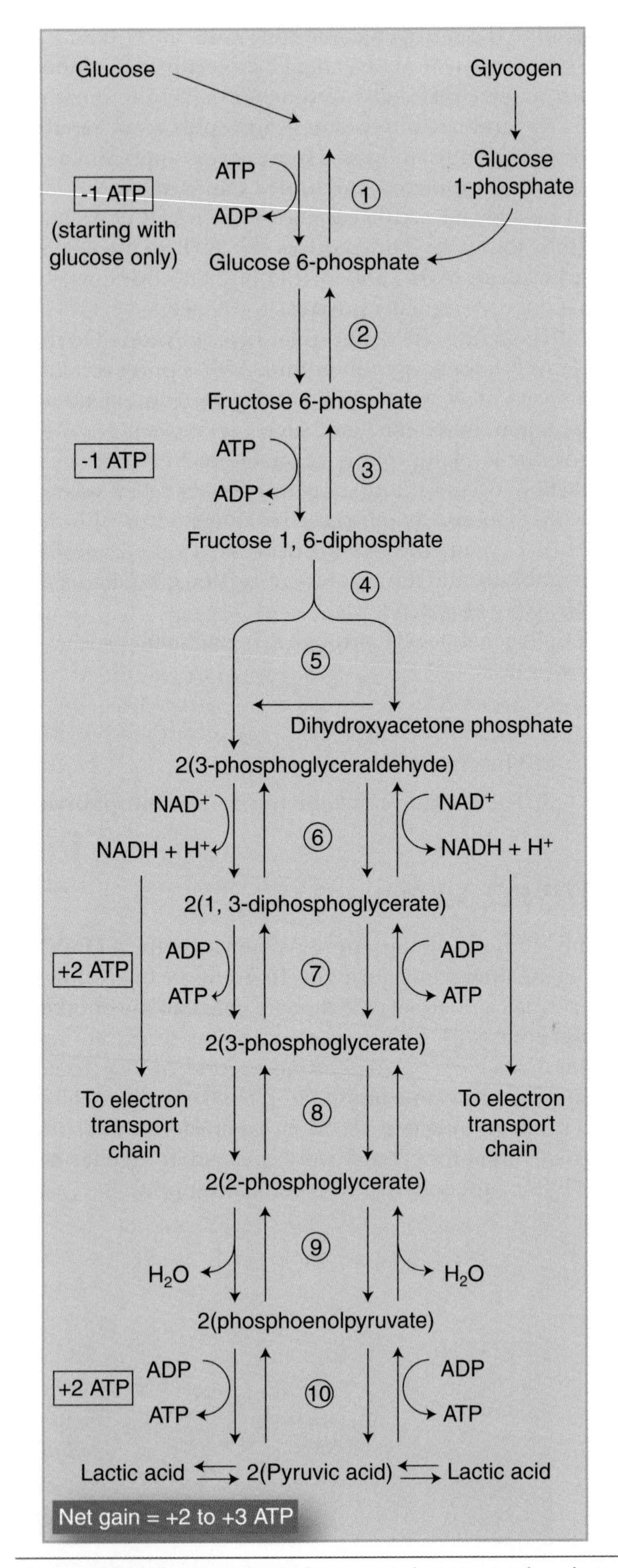

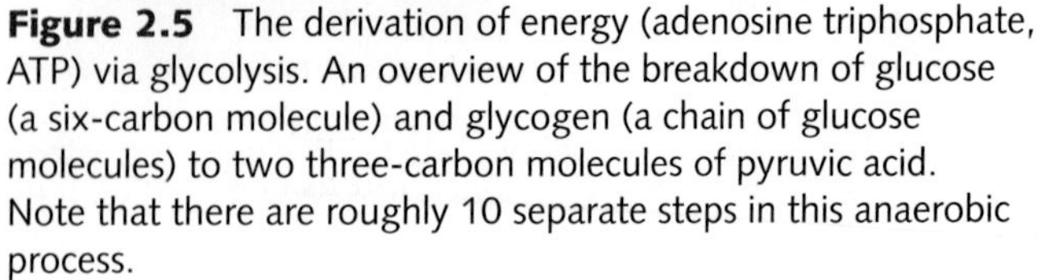

Figure 2.5 The derivation of energy (adenosine triphosphate, ATP) via glycolysis. An overview of the breakdown of glucose (a six-carbon molecule) and glycogen (a chain of glucose molecules) to two three-carbon molecules of pyruvic acid. Note that there are roughly 10 separate steps in this anaerobic process.

directly to lactic acid, an acid with the chemical formula $C_3H_6O_3$. When lactic acid releases hydrogen ions (H^+), the remaining compound joins with sodium ions (Na^+) or potassium ions (K^+) to form a salt, called lactate. Anaerobic glycolysis produces lactic acid, but it quickly dissociates, and lactate is formed. For this reason, the terms often are used interchangeably, although they do not refer to the same molecule.

In all-out sprint events lasting 1 or 2 min, the demands on the glycolytic system are high, and muscle lactic acid concentrations can increase from a resting value of about 1 mmol/kg of muscle to more than 25 mmol/kg. This acidification of muscle fibers inhibits further glycogen breakdown because it impairs glycolytic enzyme function. In addition, the acid decreases the fibers' calcium-binding capacity and thus may impede muscle contraction.

A muscle fiber's rate of energy use during exercise can be 200 times greater than at rest. The ATP-PCr and glycolytic systems alone cannot supply all the needed energy. Furthermore, these two systems are not capable of supplying all of the energy needs for all-out activity lasting more than 2 min or so. Prolonged exercise relies on the third energy system, the oxidative system.

In review

- Adenosine triphosphate is generated through three energy systems:
  1. The ATP-PCr system
  2. The glycolytic system
  3. The oxidative system
- In the ATP-PCr system, P_i is separated from PCr through the action of creatine kinase. The P_i can then combine with ADP to form ATP using the energy released from the breakdown of PCr. This system is anaerobic, and its main function is to maintain ATP levels. The energy yield is 1 mol of ATP per 1 mol of PCr.
- The glycolytic system involves the process of glycolysis, through which glucose or glycogen is broken down to pyruvic acid. When glycolysis occurs without oxygen, the pyruvic acid is converted to lactic acid. One mole of glucose yields 2 mol of ATP, but 1 mol of glycogen yields 3 mol of ATP.
- The ATP-PCr and glycolytic systems are major contributors of energy during short-burst activities lasting up to 2 min and during the early minutes of longer high-intensity exercise.

Oxidative System

The final system of cellular energy production is the **oxidative system.** This is the most complex of the three energy systems, and only a brief overview is provided here. The process by which the body breaks down substrates with the aid of oxygen to generate energy is called cellular respiration. Because oxygen is used, this is an aerobic process. This oxidative production of ATP occurs within special cell organelles called mitochondria. In muscles, these are adjacent to the myofibrils and are also scattered throughout the sarcoplasm. (See figure 1.3, chapter 1.)

Muscles need a steady supply of energy to continuously produce the force needed during long-term activity. Unlike anaerobic ATP production, the oxidative system is slow to turn on; but it has a tremendous energy-yielding capacity, so aerobic metabolism is the primary method of energy production during endurance events. This places considerable demands on the cardiovascular and respiratory systems to deliver oxygen to the active muscles.

Oxidation of Carbohydrate

As shown in figure 2.6, oxidative production of ATP involves three processes:

- Aerobic glycolysis (figure 2.6*a*)
- The Krebs cycle (figure 2.6*b*)
- The electron transport chain (figure 2.6*c*)

Aerobic Glycolysis In carbohydrate metabolism, glycolysis plays a role in both anaerobic and aerobic ATP production. The process of glycolysis is the same regardless of whether oxygen is present. The presence of oxygen determines only the fate of the end product, pyruvic acid. Recall that anaerobic glycolysis produces lactic acid and only 3 mol of ATP per mole of glycogen, or 2 mol of ATP per mole of glucose. In the presence of oxygen, however, the pyruvic acid is converted into a compound called **acetyl coenzyme A (acetyl CoA).**

Krebs Cycle Once formed, acetyl CoA enters the **Krebs cycle** (also called the citric acid cycle), a complex series of chemical reactions that permit the complete oxidation of acetyl CoA. At the end of the Krebs cycle, two additional moles of ATP have been formed directly, and the substrate (the original carbohydrate) has been broken down into carbon dioxide and hydrogen.

Electron Transport Chain During glycolysis, hydrogen ion is released when glucose is metabolized to pyruvic acid. Additional hydrogen ion is released during the Krebs cycle. If it remained in the system, the inside of the cell would become too acidic. What happens to this hydrogen?

The Krebs cycle is coupled to a series of reactions known as the **electron transport chain.** The hydrogen released during glycolysis and during the Krebs cycle combines with two coenzymes: nicotinamide adenine dinucleotide (NAD) and flavin adenine dinucleotide (FAD). These carry the hydrogen atoms to the electron transport chain, where they are split into protons and electrons. At the end of the chain, the H^+ combines with oxygen to form water, thus preventing acidification. The electrons that were split from the hydrogen pass through a chain of reactions (hence the name electron transport chain) and ultimately provide energy for the phosphorylation of ADP, thus forming ATP. Because this process relies on oxygen, it is referred to as oxidative phosphorylation. This is illustrated in figure 2.7.

Energy Yield From Oxidation of Carbohydrate The complete oxidation of carbohydrate can generate 37 to 39 molecules of ATP from one molecule of muscle glycogen. If the process begins with glucose, the maximal net gain is 38 ATP molecules (recall that one ATP molecule is used for conversion to glucose-6-phosphate before glycolysis begins). The sites of the ATP produced are summarized in table 2.2.

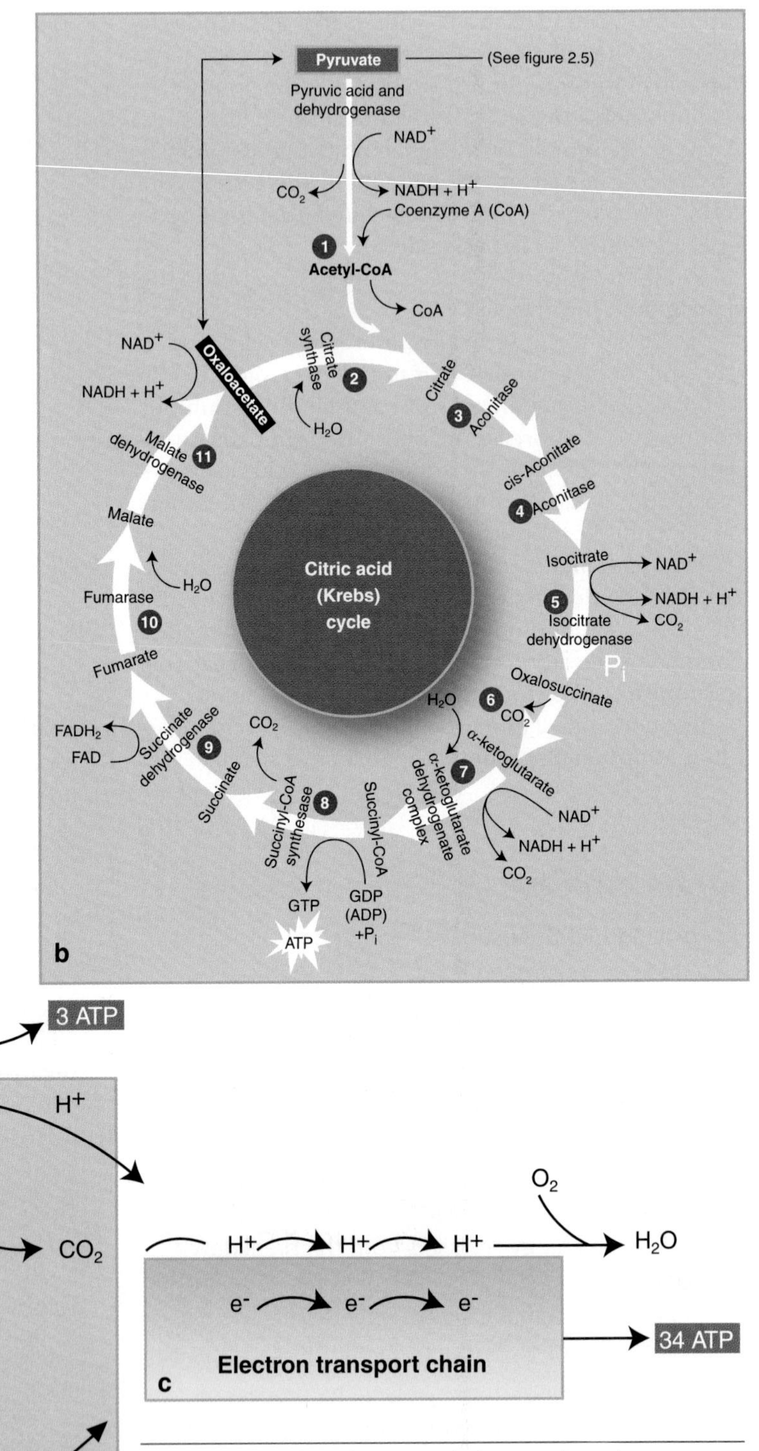

Figure 2.6 After glucose and glycogen have been reduced to pyruvate, *(a)* pyruvate is catalyzed to acetyl coenzyme A (acetyl CoA), which can enter *(b)* the Krebs cycle, where oxidative phosphorylation occurs. Hydrogen released during the Krebs cycle then combines with two coenzymes that carry the hydrogen atoms to *(c)* the electron transport chain.

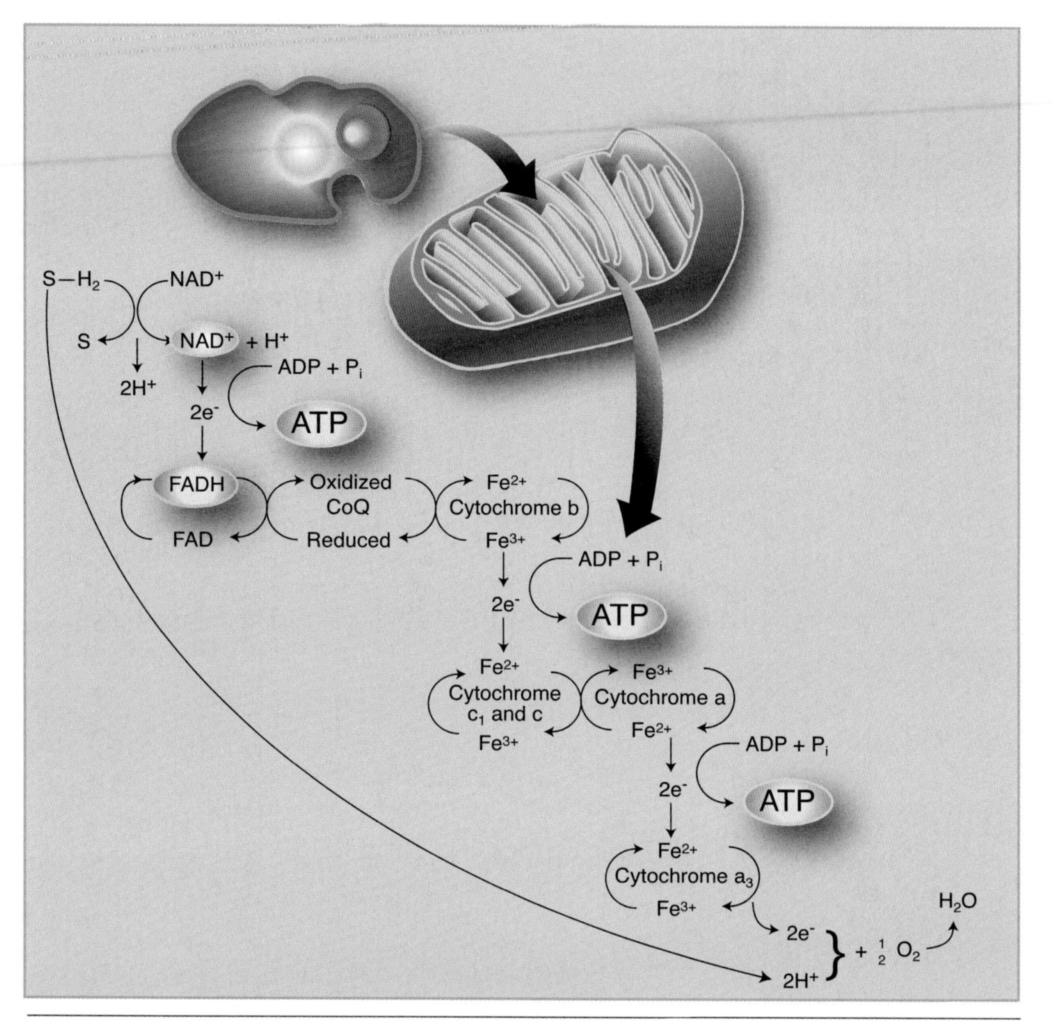

Figure 2.7 Within the mitochondrion, adenosine triphosphate (ATP) is formed at three sites along the electron transport chain. This process is known as oxidative phosphorylation.

TABLE 2.2 Energy Production From the Oxidation of Muscle Glycogen

Stage of process	Direct	By oxidative phosphorylation[a]
Glycolysis (glucose to pyruvic acid)	3	4-6[b]
Pyruvic acid to acetyl coenzyme A	0	6
Krebs cycle	2	22
Subtotal	5	32-34
Total	37-39	

[a]Refers to adenosine triphosphate (ATP) produced by transferring H^+ and electrons to the electron transport chain.

[b]The energy yield differs depending on whether reduced nicotinamide adenine dinucleotide (NADH) or reduced flavin adenine dinucleotide (FADH) is the carrier molecule to transport the electron through the mitochondrial membrane and the electron transport chain, with NADH yielding up to 39 molecules of ATP and FADH yielding 37 molecules of ATP.

It should be noted that the molecules of reduced NAD (termed NADH) formed in the cytoplasm cannot directly enter the mitochondria. They must donate their electrons to either NADH or reduced FAD (FADH) carrier molecules in the electron transport chain. Two cytoplasmic NADH molecules donating their electrons to mitochondrial NADH yield six ATP molecules, as opposed to only four ATP molecules when their electrons are donated to mitochondrial FADH. Thus, when FADH is the carrier, only up to 36 ATP molecules can be generated from glucose and 37 ATP molecules from glycogen.

Oxidation of Fat

As noted earlier, fat also contributes importantly to muscles' energy needs. Muscle and liver glycogen stores can provide only ~2,500 kcal of energy, but the fat stored inside muscle fibers and in fat cells can supply at least 70,000 to 75,000 kcal, even in a lean adult.

Although many chemical compounds (such as triglycerides, phospholipids, and cholesterol) are

In focus

Although fat provides more kilocalories of energy per gram than carbohydrate, fat oxidation requires more oxygen than carbohydrate oxidation. The energy yield from fat is 5.6 ATP molecules per oxygen molecule used, compared with carbohydrate's yield of 6.3 ATP per oxygen molecule. Oxygen delivery is limited by the oxygen transport system, so carbohydrate is the preferred fuel during high-intensity exercise.

classified as fats, only triglycerides are major energy sources. Triglycerides are stored in fat cells and between and within skeletal muscle fibers. To be used for energy, a triglyceride must be broken down to its basic units: one molecule of glycerol and three FFA molecules. This process is called **lipolysis,** and it is carried out by enzymes known as lipases.

Free fatty acids are the primary energy source. Once liberated from glycerol, FFAs can enter the blood and be transported throughout the body, entering muscle fibers by simple diffusion or by transporter-mediated (facilitated) diffusion. Their rate of entry into the muscle fibers depends on the concentration gradient. Increasing the concentration of FFAs in the blood increases the rate of their transport into muscle fibers.

β-Oxidation Although the various FFAs in the body differ structurally, their metabolism is essentially the same, as shown in the left half of figure 2.8. On entering the muscle fiber, FFAs are enzymatically activated with energy from ATP, preparing them for catabolism (breakdown) within the mitochondria. This enzymatic catabolism of fat by the mitochondria is termed **β-oxidation.**

In this process, the carbon chain of an FFA is cleaved into separate two-carbon units of acetic acid. For example, if an FFA originally has a 16-carbon chain, β-oxidation yields eight molecules of acetyl CoA.

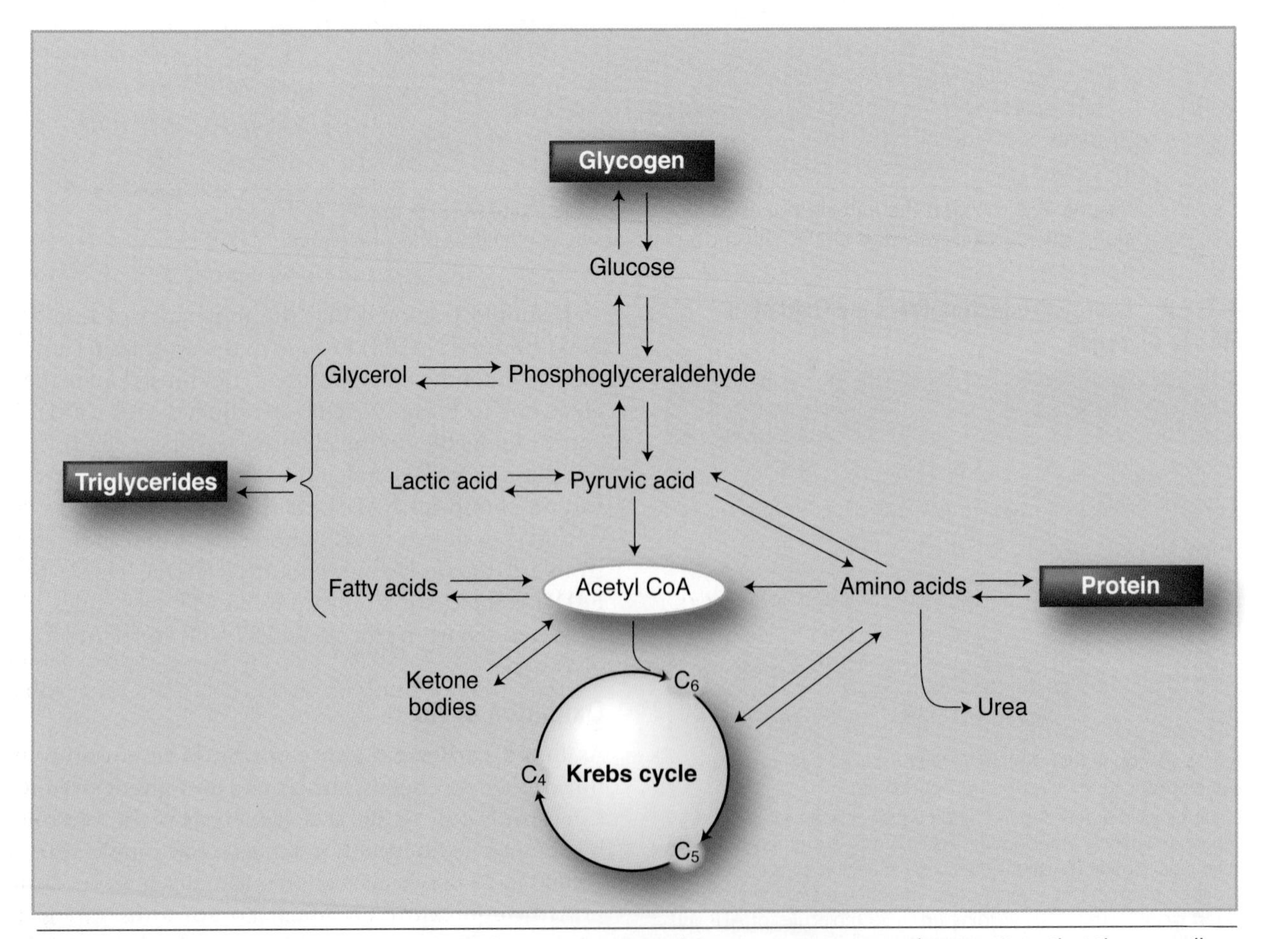

Figure 2.8 The metabolism of fat, carbohydrate, and protein share some common pathways. Note that they are all reduced to acetyl coenzyme A (CoA) and enter the Krebs cycle.

In focus

The maximum rate of high-energy phosphate formation from lipid oxidation is too low to match the rate of utilization of high-energy phosphate during higher-intensity exercise. This explains the reduction in an athlete's race pace when carbohydrate stores are depleted and fat, by default, becomes the predominant fuel source.

Krebs Cycle and the Electron Transport Chain From this point on, fat metabolism follows the same path as oxidative carbohydrate metabolism. Acetyl CoA formed by β-oxidation enters the Krebs cycle. The Krebs cycle generates hydrogen, which is transported to the electron transport chain along with the hydrogen generated during β-oxidation to undergo oxidative phosphorylation (see figure 2.7). As in glucose metabolism, the by-products of FFA oxidation are ATP, H_2O, and carbon dioxide (CO_2). However, the complete combustion of an FFA molecule requires more oxygen because an FFA molecule contains considerably more carbon than a glucose molecule.

The advantage of having more carbon in FFAs than in glucose is that more acetyl CoA is formed from the metabolism of a given amount of fat, so more acetyl CoA enters the Krebs cycle and more electrons are sent to the electron transport chain. This is why fat metabolism can generate so much more energy than glucose metabolism. Unlike glycogen, fats are heterogeneous, and the amount of ATP produced depends on the specific fat oxidized.

Consider the example of palmitic acid, a rather abundant 16-carbon FFA. The combined reactions of oxidation, the Krebs cycle, and the electron transport chain produce 129 molecules of ATP from one molecule of palmitic acid (as shown in table 2.3), compared with only 38 molecules of ATP from glucose or 39 from glycogen.

TABLE 2.3 Energy Production From the Oxidation of Palmitic Acid

	Adenosine triphosphate produced from one molecule of $C_{16}H_{32}O_2$	
Stage of process	Direct	By oxidative phosphorylation
Fatty acid activation	0	−2
β-oxidation	0	35
Krebs cycle	8	88
Subtotal	8	121
Total	129	

Oxidation of Protein

As noted earlier, carbohydrates and fatty acids are the preferred fuels. But proteins, or rather the amino acids that compose proteins, are also used. Some amino acids can be converted into glucose (by gluconeogenesis). Alternatively, some can be converted into various intermediates of oxidative metabolism (such as pyruvate or acetyl CoA) to enter the oxidative process.

Protein's energy yield is not as easily determined as that of carbohydrate or fat because protein also contains nitrogen. When amino acids are catabolized, some of the released nitrogen is used to form new amino acids, but the remaining nitrogen cannot be oxidized by the body. Instead it is converted into urea and then excreted, primarily in the urine. This conversion requires the use of ATP, so some energy is spent in this process.

When protein is broken down through combustion in the laboratory, the energy yield is 5.65 kcal/g. However, because of the energy expended in converting nitrogen to urea, when protein is metabolized in the body, the energy yield is only about 4.1 kcal/g, 27.4% less than the laboratory value.

To accurately assess the rate of protein metabolism, the amount of nitrogen being eliminated from the body must be determined. These measurements require urine collection for 12 to 24 h periods, a time-consuming process. Because the healthy body uses little protein during rest and exercise (usually not more than 5% of total energy expended), estimates of total energy expenditure generally ignore protein metabolism.

Interaction of the Three Energy Systems

The three energy systems do not work independently of one another. When a person is exercising at the highest intensity possible, from the shortest sprints (less than 10 s) to endurance events (greater than 30 min), each of the energy systems is contributing to the total energy needs of the body. Generally one energy system dominates, except when there is a transition from the predominance of one energy system to another. As an example, in a 10 s, 100 m sprint, the ATP-PCr system is the predominant energy system, but both the anaerobic glycolytic and oxidative systems provide a small portion of the energy needed. At the other extreme, in a 30 min, 10,000 m (10,936 yd) run, the oxidative system is predominant, but both the ATP-PCr and anaerobic glycolytic systems contribute some energy as well.

Figure 2.9 shows the reciprocal relationship among the energy systems with respect to power and capacity. The PCR energy system can provide energy at a fast rate but has a low capacity for energy production. Thus it supports exercise that is intense but of very short duration. By contrast, fat oxidation takes longer to gear up and produces energy at a slower rate; however, the amount of energy it can produce is unlimited.

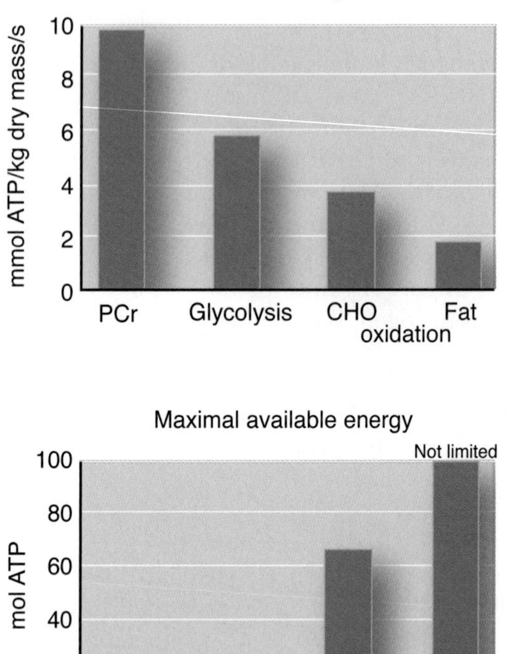

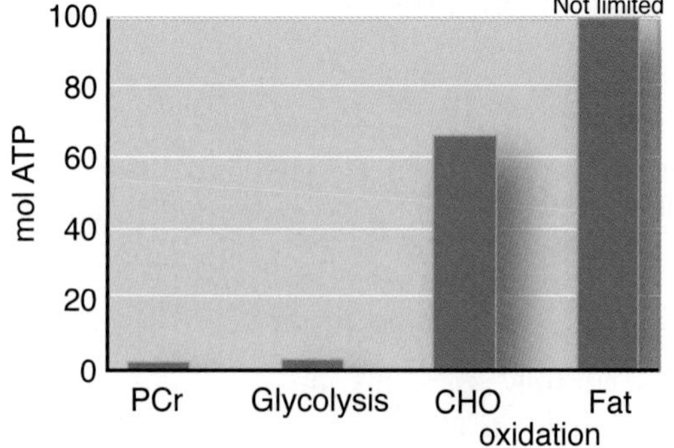

Figure 2.9 There is a reciprocal relationship among the various energy systems with respect to the rate at which energy can be produced (power) and the capacity to produce that energy.

Oxidative Capacity of Muscle

We have seen that the processes of oxidative metabolism have the highest energy yields. It would be ideal if these processes always functioned at peak capacity. But, as with all physiological systems, they operate within certain constraints. The oxidative capacity of muscle ($\dot{Q}O_2$) is a measure of its maximal capacity to use oxygen. This measurement is made in the laboratory where a small amount of muscle tissue can be tested to determine its capacity to consume oxygen when chemically stimulated to generate ATP.

Enzyme Activity

Muscle fibers' capacity to oxidize carbohydrate and fat is difficult to determine. Numerous studies have shown a close relationship between a muscle's ability to perform prolonged aerobic exercise and the activity of its oxidative enzymes. Because many enzymes are required for oxidation, the enzyme activity of the muscle fibers provides a reasonable indication of their oxidative potential.

Measuring all the enzymes in muscles is impossible, so a few representative enzymes have been selected to reflect the aerobic capacity of the fibers. The enzymes most frequently measured include succinate dehydrogenase and citrate synthase, mitochondrial enzymes involved in the Krebs cycle (see figure 2.6). Figure 2.10 illustrates the close relationship between succinate dehydrogenase activity in the vastus lateralis muscle and the muscle's oxidative capacity. Endurance athletes' muscles have oxidative enzyme activities nearly two to four times greater than those of untrained men and women.

Fiber Type Composition and Endurance Training

A muscle's fiber type composition primarily determines its oxidative capacity. As noted in chapter 1, slow-twitch, or type I, fibers have a greater capacity for aerobic activity than the fast-twitch, or type II, fibers because type I fibers have more mitochondria and higher concentrations of oxidative enzymes. Type II fibers are better suited for glycolytic energy production. Thus, in general, the more type I fibers in one's muscles, the greater the oxidative capacity of those muscles. Elite distance runners, for example, have been reported to possess more type I fibers, more mitochondria, and higher muscle oxidative enzyme activities than do untrained individuals.

Endurance training enhances the oxidative capacity of all fibers, especially type II fibers. Training that places demands on oxidative phosphorylation stimulates the muscle fibers to develop more mitochondria that also are larger and contain more oxidative enzymes. By increasing the fibers' enzymes for β-oxidation, this training also enables the muscle to rely more heavily on fat for ATP production. Thus, with endurance training, even people with large percentages of type II fibers can increase their muscles' aerobic capacities. But it is generally agreed that an endurance-trained type II fiber

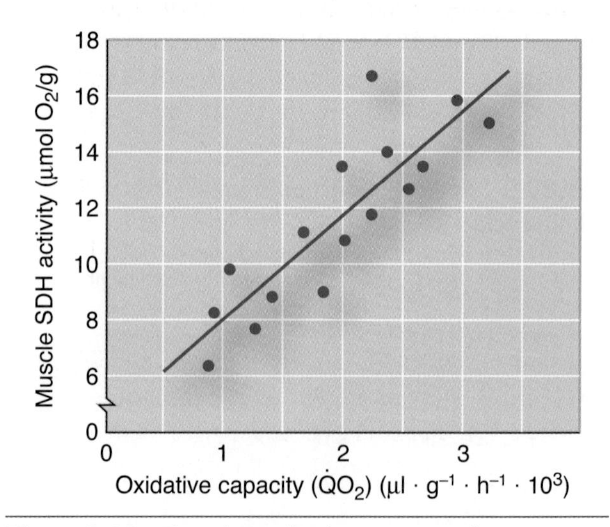

Figure 2.10 The relationship between muscle succinate dehydrogenase (SDH) activity and its oxidative capacity ($\dot{Q}O_2$), measured in a biopsy sample taken from the vastus lateralis.

will not develop the same high endurance capacity as a similarly trained type I fiber.

Oxygen Needs

Although the oxidative capacity of a muscle is determined by the number of mitochondria and the amount of oxidative enzymes present, oxidative metabolism ultimately depends on an adequate supply of oxygen. At rest, the need for ATP is relatively small, requiring minimal oxygen delivery. As exercise intensity increases, so do energy demands. To meet them, the rate of oxidative ATP production increases. In an effort to meet the muscles' need for oxygen, the rate and depth of respiration increase, improving gas exchange in the lungs, and the heart beats faster and more forcefully, pumping more oxygenated blood to the muscles. Arterioles dilate to facilitate delivery of arterial blood to muscle capillaries.

In review

- The oxidative system involves breakdown of substrates in the presence of oxygen. This system yields more energy than the ATP-PCr or the glycolytic system.
- Oxidation of carbohydrate involves glycolysis, the Krebs cycle, and the electron transport chain. The end result is H_2O, CO_2, and 38 or 39 ATP molecules per carbohydrate molecule.
- Fat oxidation begins with β-oxidation of FFAs and then follows the same path as carbohydrate oxidation: the Krebs cycle and the electron transport chain. The energy yield for fat oxidation is much higher than for carbohydrate oxidation, and it varies with the FFA being oxidized. However, the maximum rate of high-energy phosphate formation from lipid oxidation is too low to match the rate of utilization of high-energy phosphate during higher-intensity exercise, and the energy yield of fat per oxygen molecule used is much less than that for carbohydrate.
- Protein oxidation is more complex because protein (amino acids) contains nitrogen, which cannot be oxidized. Protein contributes relatively little to energy production, generally less than 5%, so its metabolism is often considered negligible.
- A muscle's oxidative capacity depends on its oxidative enzyme concentrations, fiber type composition, and oxygen availability.

The human body stores little oxygen. Therefore, the amount of oxygen entering the blood as it passes through the lungs is directly proportional to the amount used by the tissues for oxidative metabolism. Consequently, a reasonably accurate estimate of aerobic energy production can be made by measuring the amount of oxygen consumed at the lungs (see chapter 4).

HORMONAL CONTROL

As the body transitions from a resting to an active state, the rate of metabolism must increase to provide the necessary energy. This requires coordinated integration of many physiological and biochemical systems. Such integration is possible only if the multiple tissues and systems can efficiently communicate. Although the nervous system is responsible for much of this communication, fine-tuning the physiological responses to any disturbance of homeostasis is primarily the responsibility of the endocrine system. The endocrine and nervous systems work in concert to initiate and control movement and all physiological processes that movement involves. The nervous system functions quickly, having short-lived, localized effects, whereas the endocrine system responds more slowly but has longer-lasting effects.

The endocrine system includes all tissues or glands that secrete **hormones.** The major endocrine glands are illustrated in figure 2.11. Endocrine glands secrete their hormones directly into the blood where they act as chemical signals throughout the body. When secreted by the specialized endocrine cells, hormones are transported via the blood to specific **target cells**—cells that possess specific hormone receptors. On reaching their destinations, hormones can control the activity of the target tissue. A unique feature of hormones is that they travel away from the cells that secrete them and specifically affect the activities of other cells and organs. Some hormones affect many body tissues, whereas others target specific cells of the body.

Hormones

Hormones are involved in most physiological processes, so their actions are relevant to many aspects of exercise and sport performance. A discussion of hormonal control is included in this chapter because hormones significantly affect metabolism. However, hormones play key roles in almost every system of the body, and total coverage of that topic is well beyond the scope of this book. In the following sections, general mechanisms by which hormones act and their chemical nature are discussed.

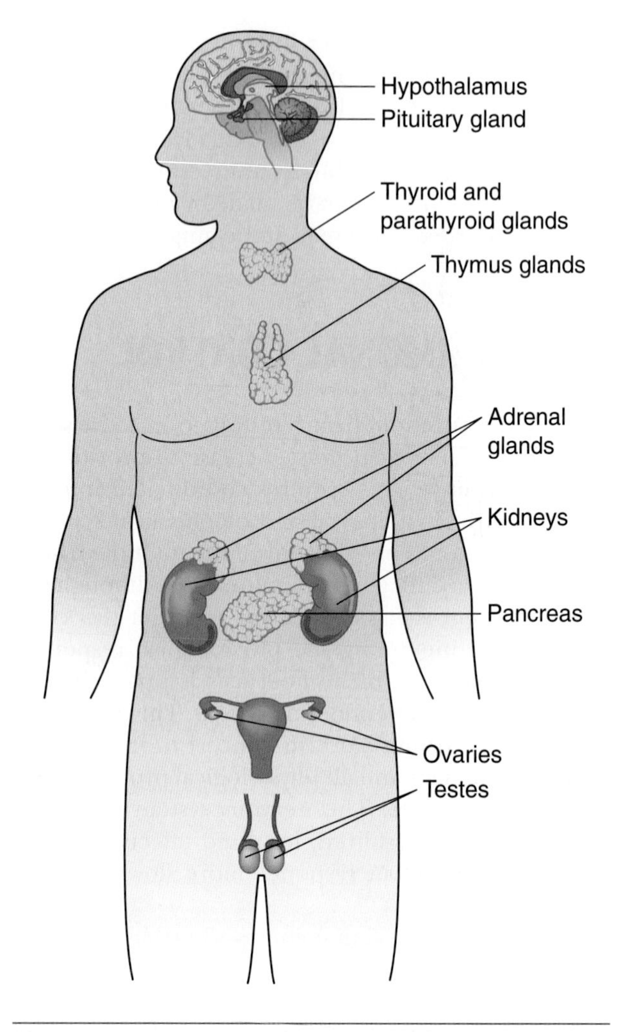

Figure 2.11 Locations of the major endocrine organs.

Chemical Classification of Hormones

Hormones can be categorized as two basic types: steroid hormones and nonsteroid hormones. **Steroid hormones** have a chemical structure similar to cholesterol, since most are derived from cholesterol. For this reason, they are soluble in lipids and diffuse rather easily through cell membranes. This group includes the hormones secreted by

- the adrenal cortex (such as cortisol and aldosterone),
- the ovaries (estrogen and progesterone),
- the testes (testosterone), and
- the placenta (estrogen and progesterone).

Nonsteroid hormones are not lipid soluble, so they cannot easily cross cell membranes. The nonsteroid hormone group can be subdivided into two groups: protein or peptide hormones and amino acid–derived hormones. The two hormones from the thyroid gland (thyroxine and triiodothyronine) and the two from the adrenal medulla (epinephrine and norepinephrine) are amino acid hormones. All other nonsteroid hormones are protein or peptide hormones.

Hormone Actions

Because hormones travel in the blood, they contact virtually all body tissues. How, then, do they limit their effects to specific targets? This ability is attributable to the specific hormone receptors possessed by the target tissues. The interaction between the hormone and its specific receptor has been compared with a lock (receptor) and key (hormone) arrangement, in which only the correct key can unlock a given action within the cells (see figures 2.12 and 2.13). The combination of a hormone bound to its receptor is referred to as a hormone–receptor complex.

Each cell typically has from 2,000 to 10,000 receptors. Receptors for nonsteroid hormones are located on the cell membrane, whereas those for steroid hormones are found either in the cell's cytoplasm or in its nucleus. Each hormone is usually highly specific for a single type of receptor and binds only with its specific receptors, thus affecting only tissues that contain those specific receptors. Numerous mechanisms allow hormones to control the actions of cells.

As mentioned earlier, steroid hormones are lipid soluble and thus pass easily through the cell membrane. Their mechanism of action is illustrated in figure 2.12. Once inside the cell, a steroid hormone binds to its specific receptors. The hormone–receptor complex then enters the nucleus, binds to part of the cell's DNA, and activates certain genes. This process is referred to as **direct gene activation.** In response to this activation, mRNA is synthesized within the nucleus. The mRNA then enters the cytoplasm and promotes protein synthesis. These proteins may be

- enzymes that can have numerous effects on cellular processes,
- structural proteins to be used for tissue growth and repair, or
- regulatory proteins that can alter enzyme function.

Because nonsteroid hormones cannot cross the cell membrane, they react with specific receptors outside the cell, on the cell membrane. A nonsteroid hormone molecule binds to its receptor and triggers a series of enzymatic reactions that lead to the formation of an intracellular **second messenger.** A widely distributed second messenger that mediates a specific hormone–receptor response is **cyclic adenosine monophosphate** (cyclic AMP, or **cAMP**). This mechanism of action is depicted in figure 2.13. In this case, attachment of the

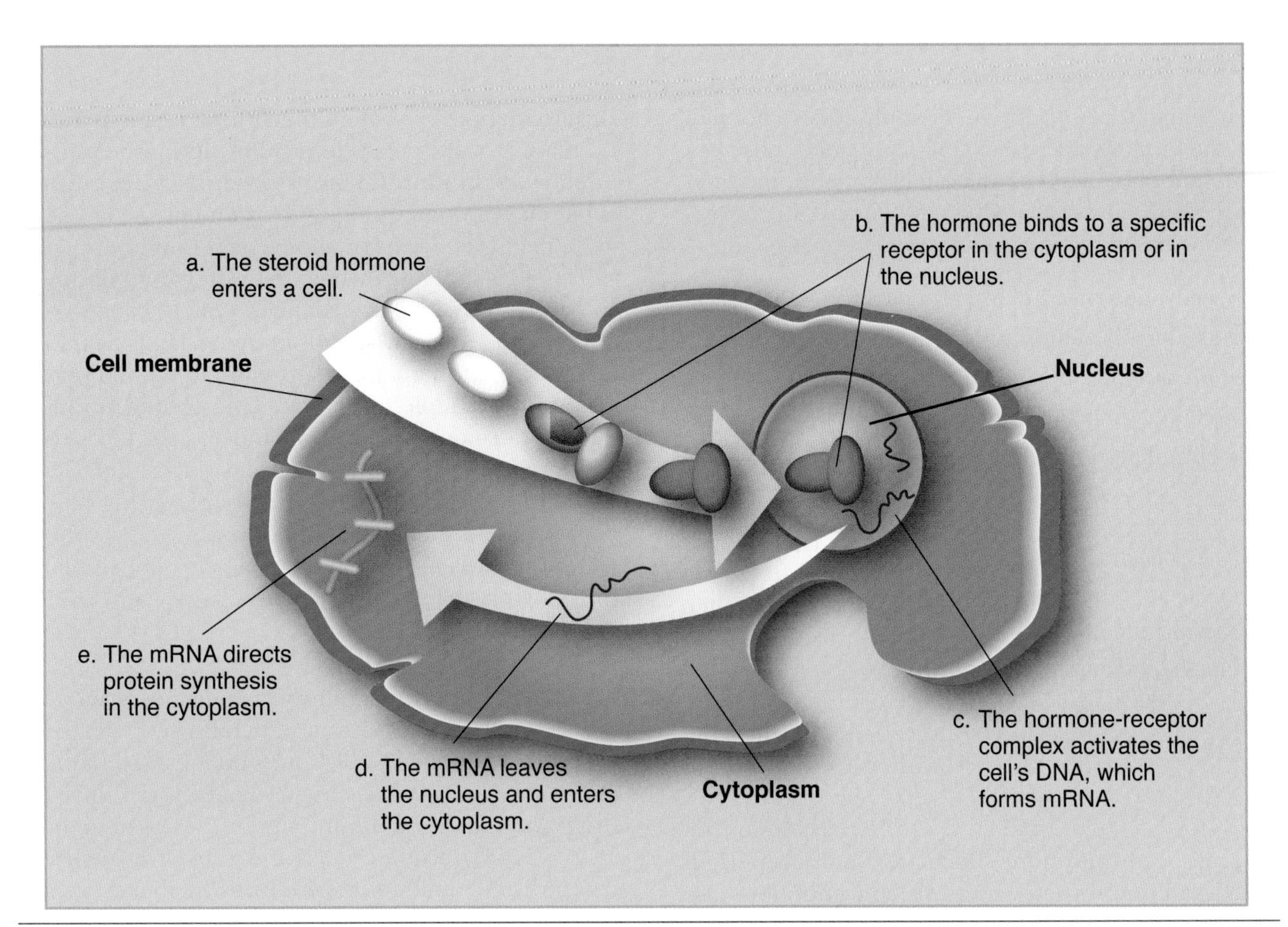

Figure 2.12 The mechanism of action of a steroid hormone, leading to direct gene activation.

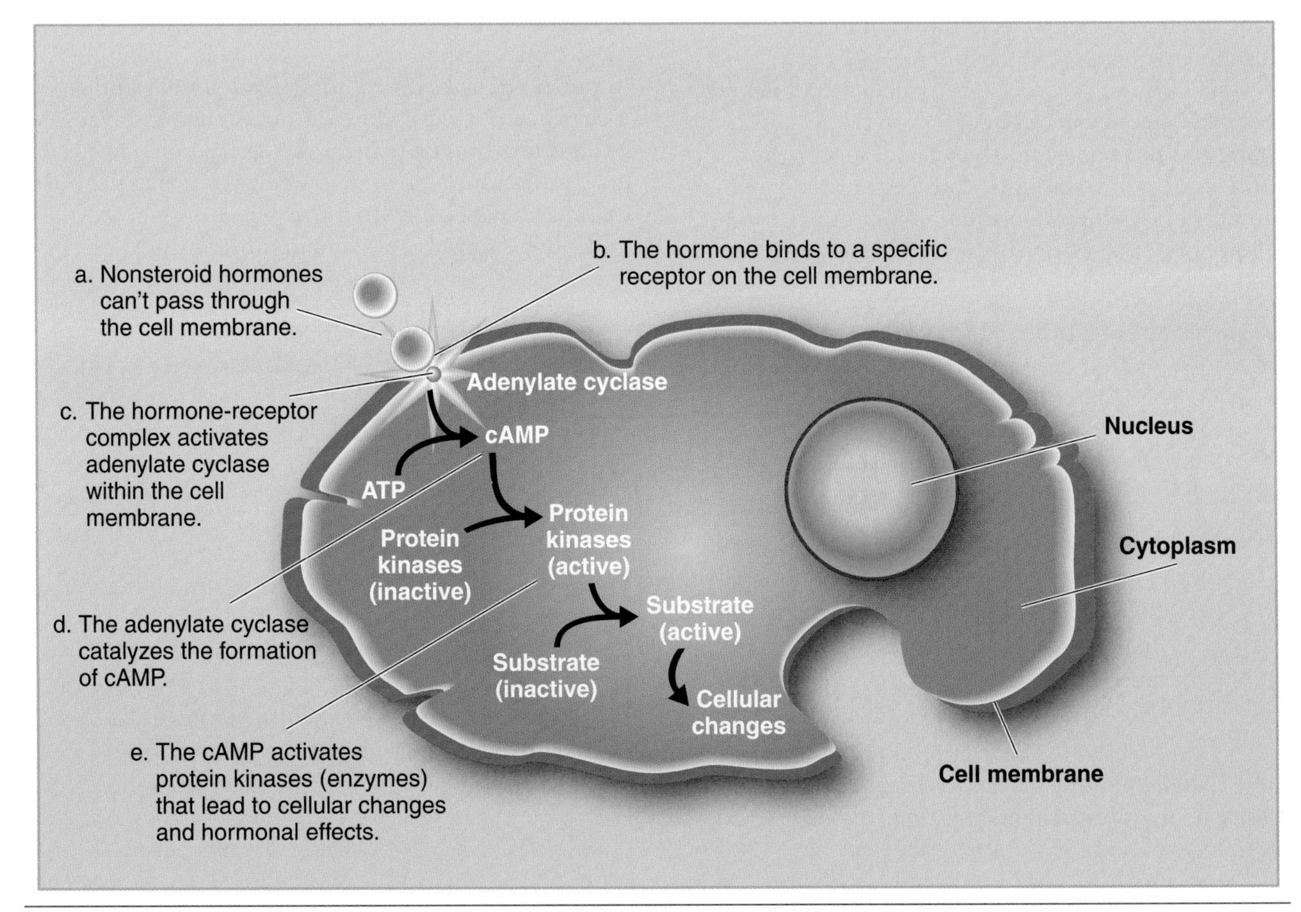

Figure 2.13 The mechanism of action of a nonsteroid hormone, using a second messenger (cyclic adenosine monophosphate, or cAMP) within the cell.

hormone to the appropriate membrane receptor activates an enzyme, adenylate cyclase, situated within the cell membrane. This enzyme catalyzes the formation of cAMP from cellular ATP. Cyclic AMP then can produce specific physiological responses, which can include

- activation of cellular enzymes,
- change in membrane permeability,
- promotion of protein synthesis,
- change in cellular metabolism, or
- stimulation of cellular secretions.

> **In focus**
>
> Hormones influence specific target tissues or cells through a unique interaction between the hormone and the specific receptors for that hormone on the cell membrane or within the cell.

Thus, nonsteroid hormones typically activate the cAMP system of the cell, which then alters intracellular functions.

Hormones are not secreted uniformly, but rather are released in relatively brief bursts, so plasma concentrations of specific hormones fluctuate over short periods such as an hour or less. But these concentrations also fluctuate over longer periods of time, showing daily or even monthly cycles (such as monthly menstrual cycles). How do endocrine glands know when to release their hormones?

Most hormone secretion is regulated by a **negative feedback** system. Secretion of a hormone causes some change in the body, and this change in turn inhibits further hormone secretion. Consider how a home thermostat works. When the room temperature decreases below some preset level, the thermostat signals the furnace to produce heat. When the room temperature increases to the preset level, the thermostat's signal ends, and the furnace stops producing heat. When the temperature again falls below the preset level, the cycle begins anew. In the body, secretion of a specific hormone is similarly turned on or off (or up or down) by specific physiological changes.

Negative feedback is the primary mechanism through which the endocrine system maintains homeostasis. Using the example of plasma glucose concentrations and the hormone insulin, when the plasma glucose concentration is high, the pancreas releases insulin. Insulin increases cellular uptake of glucose, lowering plasma concentration of glucose. When plasma glucose concentration returns to normal, insulin release is inhibited until the plasma glucose level increases again.

Hormone Receptors

The plasma concentration of a specific hormone is not always the best indicator of that hormone's activity because the number of receptors on target cells can be altered to increase or decrease that cell's sensitivity to the hormone. Most commonly, an increased amount of a specific hormone decreases the number of cell receptors available to it. When this happens, the cell becomes less sensitive to that hormone, because with fewer receptors, fewer hormone molecules can bind. This is referred to as **downregulation,** or desensitization. In some people with obesity, for example, the number of insulin receptors on their cells appears to be reduced. Their bodies respond by increasing insulin secretion from the pancreas, so their plasma insulin concentrations increase. To obtain the same degree of plasma glucose control as normal, healthy people, these individuals must release much more insulin.

Prostaglandins

Prostaglandins, although technically not hormones, are often considered to be a third class of hormones. These substances are derived from a fatty acid, arachidonic acid, and they are associated with the plasma membranes of almost all body cells. Prostaglandins typically act as *local* hormones, exerting their effects in the immediate area where they are produced. But some also survive long enough to circulate through the blood to affect distant tissues. Prostaglandin release can be triggered by many stimuli, such as other hormones or a local injury. Their functions are quite numerous because there are several different types of prostaglandins. They often mediate the effects of other hormones. They are also known to act directly on blood vessels, increasing vascular permeability (which promotes swelling) and vasodilation. In this capacity, they are important mediators of the inflammatory response. They also sensitize the nerve endings of pain fibers; thus, they promote both inflammation and pain.

In review

- Hormones can be classified as either steroid or nonsteroid. Steroid hormones are lipid soluble, and most are formed from cholesterol. Nonsteroid hormones are formed from proteins, peptides, or amino acids.
- Hormones generally are secreted into the blood and then circulate to target cells. They act by binding in a lock-and-key manner with specific receptors found only in the target tissues.
- Steroid hormones pass through cell membranes and bind to receptors inside the cell. They use a mechanism called direct gene activation to cause protein synthesis.
- Nonsteroid hormones cannot easily enter the cells, so they bind to receptors on the cell membrane. This activates a second messenger within the cell, which in turn can trigger numerous cellular processes.
- A negative feedback system regulates secretion of most hormones.
- The number of receptors for a specific hormone can be altered to meet the body's demands. Upregulation refers to an increase in available receptors, and downregulation refers to a decrease. These two processes change cell sensitivity to hormones.

In a few instances, a cell may respond to the prolonged presence of large amounts of a hormone by increasing its number of available receptors. When this happens, the cell becomes more sensitive to that hormone because more can be bound at one time. This is referred to as **upregulation.** In addition, one hormone occasionally can regulate the receptors for another hormone.

Endocrine Glands and Their Hormones: An Overview

The endocrine glands and their respective hormones are listed in table 2.4. This table also lists each hormone's target and actions. Because the endocrine system is extremely complex, the presentation here has been greatly simplified to focus on those endocrine glands and hormones of greatest importance to sport and physical activity.

Because hormones play such an important role in regulation of many physiological variables during exercise, it is not surprising that hormone release changes during acute bouts of activity. The hormonal responses to an acute bout of exercise and exercise training are summarized in table 2.5. This table is limited to those hormones suspected of playing major roles in sport and physical activity. Further details of these exercise-induced hormonal responses are provided in the following discussion of specific endocrine glands and their hormones.

Anterior Pituitary

The pituitary gland is a marble-sized gland at the base of the brain. The secretory action of the pituitary is controlled by either neural mechanisms or hormones secreted by the hypothalamus. Therefore, the pituitary gland can be thought of as the relay between central nervous system control centers and peripheral endocrine glands.

The pituitary gland is composed of three lobes: anterior, intermediate, and posterior. The intermediate lobe is very small and is thought to play little or no role in humans, but both the posterior and anterior lobes have major endocrine functions. The anterior pituitary has a major role in fluid and electrolyte balance as is discussed later in this chapter.

The anterior pituitary, also called the adenohypophysis, secretes six hormones in response to **releasing factors** and **inhibiting factors** (hormones) secreted by the hypothalamus. Communication between the hypothalamus and the anterior lobe of the pituitary occurs through a specialized circulatory system that transports the releasing and inhibiting factors from the hypothalamus to the anterior pituitary. The major functions of each of the anterior pituitary hormones, along with their releasing and inhibiting factors, are listed in table 2.4. Exercise appears to be a strong stimulant to the hypothalamus because exercise increases the release rate of all anterior pituitary hormones (see table 2.5 on p. 66).

Of the six anterior pituitary hormones, four are tropic hormones, meaning they affect the functioning of other endocrine glands. The exceptions are growth hormone and prolactin. **Growth hormone** is a potent anabolic agent (a substance that promotes constructive metabolism). It promotes muscle growth and hypertrophy by facilitating amino acid transport into the cells. In addition, growth hormone directly stimulates fat metabolism (lipolysis) by increasing the synthesis of enzymes involved in this process. Growth hormone concentrations are elevated during aerobic exercise, apparently in proportion to the exercise intensity, and typically remain elevated for some time after exercise.

TABLE 2.4 The Endocrine Glands, Their Hormones, Target Organs, and Controlling Factors, and Their Functions

Endocrine gland	Hormone	Target organ	Controlling factor	Major functions
Anterior pituitary	Growth hormone (GH)	All cells in the body	Hypothalamic GH-releasing hormone; GH-inhibiting hormone (somatostatin)	Promotes development and enlargement of all body tissues until maturation; increases rate of protein synthesis; increases mobilization of fats and use of fat as an energy source; decreases rate of carbohydrate use
	Thyrotropin (TSH)	Thyroid gland	Hypothalamic TSH-releasing hormone	Controls the amount of thyroxin and triiodothyronine produced and released by the thyroid gland
	Adrenocorticotropin (ACTH)	Adrenal cortex	Hypothalamic ACTH-releasing hormone	Controls the secretion of hormones from the adrenal cortex
	Prolactin	Breasts	Prolactin-releasing and -inhibiting hormones	Stimulates milk production by the breasts
	Follicle-stimulating hormone (FSH)	Ovaries, testes	Hypothalamic FSH-releasing hormone	Initiates growth of follicles in the ovaries and promotes secretion of estrogen from the ovaries; promotes development of the sperm in the testes
	Luteinizing hormone (LH)	Ovaries, testes	Hypothalamic FSH-releasing hormone	Promotes secretion of estrogen and progesterone and causes the follicle to rupture, releasing the ovum; causes testes to secrete testosterone
Posterior pituitary	Antidiuretic hormone (ADH or vasopressin)	Kidneys	Hypothalamic secretory neurons	Assists in controlling water excretion by the kidneys; elevates blood pressure by constricting blood vessels
	Oxytocin	Uterus, breasts	Hypothalamic secretory neurons	Controls contraction of uterus; milk secretion
Thyroid	Thyroxine (T_4) and triiodothyronine (T_3)	All cells in the body	TSH and T_3 and T_4 concentrations	Increase the rate of cellular metabolism; increase rate and contractility of the heart
	Calcitonin	Bones	Plasma calcium concentrations	Controls calcium ion concentration in the blood
Parathyroid	Parathyroid hormone (PTH or parathormone)	Bones, intestines, and kidneys	Plasma calcium concentrations	Controls calcium ion concentration in the extracellular fluid through its influence on bones, intestines, and kidneys
Adrenal medulla	Epinephrine	Most cells in the body	Baroreceptors, glucose receptors, brain and spinal centers	Stimulates breakdown of glycogen in liver and muscle and lipolysis in adipose tissue and muscle; increases skeletal muscle blood flow; increases heart rate and contractility; increases oxygen consumption
	Norepinephrine	Most cells in the body	Baroreceptors, glucose receptors, brain and spinal centers	Stimulates lipolysis in adipose tissue and in muscle to a lesser extent; constricts arterioles and venules, thereby elevating blood pressure

Adrenal cortex	Mineralocorticoids (aldosterone)	Kidneys	Angiotensin and plasma potassium concentrations; renin	Increase sodium retention and potassium excretion through the kidneys
	Glucocorticoids (cortisol)	Most cells in the body	ACTH	Control metabolism of carbohydrates, fats, and proteins; exert an anti-inflammatory action
	Androgens and estrogens	Ovaries, breasts, and testes	ACTH	Assist in the development of female and male sex characteristics
Pancreas	Insulin	All cells in the body	Plasma glucose and amino acid concentrations	Controls blood glucose levels by lowering glucose levels; increases use of glucose and synthesis of fat
	Glucagon	All cells in the body	Plasma glucose and amino acid concentrations	Increases blood glucose; stimulates the breakdown of protein and fat
	Somatostatin	Islets of Langerhans and intestines	Plasma glucose, insulin, and glucagon concentrations	Depresses the secretion of both insulin and glucagon
Kidney	Renin	Adrenal cortex	Plasma sodium concentrations	Assists in blood pressure control
	Erythropoietin (EPO)	Bone marrow	Low tissue oxygen concentrations	Stimulates erythrocyte production
Testes	Testosterone	Sex organs, muscle	FSH and LH	Promotes development of male sex characteristics, including growth of testes, scrotum, and penis, facial hair, and change in voice; promotes muscle growth
Ovaries	Estrogens and progesterone	Sex organs and adipose tissue	FSH and LH	Promote development of female sex organs and characteristics; increase storage of fat; assist in regulating the menstrual cycle

TABLE 2.5 Hormone Response to Acute Exercise and Change in Response With Exercise Training

Endocrine gland	Hormone	Response to acute exercise (untrained)	Effect of exercise training
Anterior pituitary	Growth hormone (GH)	Increases with increasing rates of work	Attenuated response at same rate of work
	Thyrotropin (TSH)	Increases with increasing rates of work	No known effect
	Adrenocorticotropin (ACTH)	Increases with increasing rates of work and duration	Attenuated response at same rate of work
	Prolactin	Increases with exercise	No known effect
	Follicle-stimulating hormone (FSH)	Small or no change	No known effect
	Luteinizing hormone (LH)	Small or no change	No known effect
Posterior pituitary	Antidiuretic hormone (ADH or vasopressin)	Increases with increasing rates of work	Attenuated response at same rate of work
	Oxytocin	Unknown	Unknown
Thyroid	Thyroxine (T_4) and triiodothyronine (T_3)	Free T_3 and T_4 increase with increasing rates of work	Increased turnover of T_3 and T_4 at same rate of work
	Calcitonin	Unknown	Unknown
Parathyroid	Parathyroid hormone (PTH or parathormone)	Increases with prolonged exercise	Unknown
Adrenal medulla	Epinephrine	Increases with increasing rates of work, starting at about 75% of $\dot{V}O_{2max}$	Attenuated response at same rate of work
	Norepinephrine	Increases with increasing rates of work, starting at about 50% of $\dot{V}O_{2max}$	Attenuated response at same rate of work
Adrenal cortex	Aldosterone	Increases with increasing rates of work	Unchanged
	Cortisol	Increases only at high rates of work	Slightly higher values
Pancreas	Insulin	Decreases with increasing rates of work	Attenuated response at same rate of work
	Glucagon	Increases with increasing rates of work	Attenuated response at same rate of work
Kidney	Renin	Increases with increasing rates of work	Unchanged
	Erythropoietin (EPO)	Unknown	Unchanged
Testes	Testosterone	Small increases with exercise	Resting levels decreased in male runners
Ovaries	Estrogens and progesterone	Small increases with exercise	Resting levels might be decreased in highly trained women

Thyroid Gland

The thyroid gland is located along the midline of the neck, immediately below the larynx. It secretes two important nonsteroid hormones, **triiodothyronine** (T_3) and **thyroxine** (T_4), which regulate metabolism in general, and an additional hormone, calcitonin, which assists in regulating calcium metabolism.

The two metabolic thyroid hormones share similar functions. Triiodothyronine and thyroxine increase the metabolic rate of almost all tissues and can increase the body's basal metabolic rate by as much as 60% to 100%. These hormones also

- increase protein synthesis (thus also enzyme synthesis);
- increase the size and number of mitochondria in most cells;
- promote rapid cellular uptake of glucose;
- enhance glycolysis and gluconeogenesis; and
- enhance lipid mobilization, increasing FFA availability for oxidation.

Release of **thyrotropin** (thyroid-stimulating hormone, or **TSH**) from the anterior pituitary increases during exercise. Thyroid-stimulating hormone controls the release of triiodothyronine and thyroxine, so the exercise-induced increase in TSH would be expected to stimulate the thyroid gland. Exercise does increase plasma thyroxine concentrations, but a delay occurs between the increase in TSH concentrations during exercise and the increase in plasma thyroxine concentration. Furthermore, during prolonged submaximal exercise, thyroxine concentrations remain relatively constant after a sharp initial increase as exercise begins, and triiodothyronine concentrations tend to decrease.

Adrenal Glands

The adrenal glands are situated directly atop each kidney and are composed of the inner adrenal medulla and the outer adrenal cortex. The hormones secreted by these two parts are quite different, so we consider them separately. The adrenal medulla produces and releases two hormones, **epinephrine** and **norepinephrine,** which are collectively referred to as **catecholamines.** Because of its origin in the adrenal gland, a synonym for epinephrine is **adrenaline.** When the adrenal medulla is stimulated by the sympathetic nervous system, approximately 80% of its secretion is epinephrine and 20% is norepinephrine, although these percentages vary with different physiological conditions. The catecholamines have powerful effects similar to those of the sympathetic nervous system. Recall that these same catecholamines function as neurotransmitters in the sympathetic nervous system; however, the hormones' effects last longer because these substances are removed from the blood relatively slowly compared to the quick reuptake and degradation of the neurotransmitters. These two hormones prepare a person for immediate action, often called the "fight-or-flight response."

Although some of the specific actions of these two hormones differ, the two work together. Their combined effects include

- increased heart rate and force of contraction,
- increased metabolic rate,
- increased glycogenolysis (breakdown of glycogen to glucose) in the liver and muscle,
- increased release of glucose and FFAs into the blood,
- redistribution of blood to the skeletal muscles (through vasodilation of vessels supplying skeletal muscles and vasoconstriction of vessels to the skin and viscera),
- increased blood pressure, and
- increased respiration.

Release of epinephrine and norepinephrine is affected by a wide variety of factors, including changes in body position, psychological stress, and exercise. Plasma concentrations of these hormones increase as individuals gradually increase their exercise intensity. Plasma norepinephrine concentrations increase markedly at work rates above 50% of $\dot{V}O_{2max}$, but epinephrine concentrations do not increase significantly until the exercise intensity exceeds 60% to 70% of $\dot{V}O_{2max}$. During long-duration steady-state activity of moderate intensity, blood concentrations of both hormones increase. When the exercise bout ends, epinephrine returns to resting concentrations within only a few minutes of recovery, but norepinephrine can remain elevated for several hours.

The adrenal cortex secretes more than 30 different steroid hormones, referred to as corticosteroids. These generally are classified into three major types: mineralocorticoids (discussed later in the chapter), glucocorticoids, and gonadocorticoids (sex hormones).

The **glucocorticoids** are essential components in the ability to adapt to external changes and stress. They also help maintain fairly consistent plasma glucose concentrations even when we go for long periods without ingesting food. **Cortisol,** also known as hydrocortisone, is the major corticosteroid. It is responsible for about 95% of all glucocorticoid activity in the body. Cortisol

- stimulates gluconeogenesis to ensure an adequate fuel supply;
- increases mobilization of FFAs, making them more available as an energy source;

- decreases glucose utilization, sparing it for the brain;
- stimulates protein catabolism to release amino acids for use in repair, enzyme synthesis, and energy production;
- acts as an anti-inflammatory agent;
- depresses immune reactions; and
- increases the vasoconstriction caused by epinephrine.

We discuss cortisol's important role in exercise later in this chapter when we consider the regulation of glucose and fat metabolism.

Pancreas

The pancreas is located behind and slightly below the stomach. Its two major hormones are insulin and glucagon. The balance of these two opposing hormones provides the major control of plasma glucose concentrations. When plasma glucose is elevated **(hyperglycemia)**, as after a meal, the pancreas receives signals to release insulin into the blood.

Among its actions, **insulin**

- facilitates glucose transport into the cells, especially those in muscle;
- promotes glycogenesis; and
- inhibits gluconeogenesis.

Insulin's main function is to reduce the amount of glucose circulating in the blood. But it is also involved in protein and fat metabolism, promoting cellular uptake of amino acids and enhancing synthesis of protein and fat.

The pancreas secretes **glucagon** when the plasma glucose concentration falls below normal concentrations **(hypoglycemia)**. Its effects generally oppose those of insulin. Glucagon promotes increased breakdown of liver glycogen to glucose (glycogenolysis) and increased gluconeogenesis, both of which increase plasma glucose levels.

During exercise lasting 30 min or longer, the body attempts to maintain plasma glucose concentrations; however, insulin concentrations tend to decline. Research has shown that the ability of insulin to bind to its receptors on muscle cells increases during exercise, due in large part to increased blood flow to muscle. This increases the body's sensitivity to insulin and reduces the need to maintain high plasma insulin concentrations for transporting glucose into the muscle cells. Plasma glucagon, on the other hand, shows a gradual increase throughout exercise. Glucagon primarily maintains plasma glucose concentrations by stimulating liver glycogenolysis. This increases glucose availability to the cells, maintaining adequate plasma glucose concentrations to meet increased metabolic demands. The responses of these hormones are usually blunted in trained individuals, and those who are well trained are better able to maintain plasma glucose concentrations.

Regulation of Metabolism During Exercise

As noted earlier in the chapter, carbohydrate and fat metabolism are responsible for maintaining muscle ATP levels during prolonged exercise. Various hormones work to ensure glucose and FFA availability for muscle energy metabolism. In the next two sections we examine how the metabolism of glucose and fat are affected by these hormones during exercise. Because carbohydrate is the primary fuel used during both brief and prolonged exhaustive exercise, we first consider the hormones that regulate its availability.

Regulation of Glucose Metabolism During Exercise

As we saw earlier in the chapter, the heightened energy demands of exercise require that more glucose be made available to the muscles. Recall that glucose is stored in the body as glycogen, primarily in the muscles and the liver. Glucose must be freed from its storage form of glycogen, so glycogenolysis must increase. Glucose freed from the liver enters the blood to circulate throughout the body, allowing it access to active tissues. Plasma glucose concentration also can be increased through gluconeogenesis.

Regulation of Plasma Glucose Concentration

Four hormones work to increase the amount of circulating plasma glucose:

- Glucagon
- Epinephrine
- Norepinephrine
- Cortisol

The plasma glucose concentration during exercise depends on a balance between glucose uptake by exercising muscles and its release by the liver. At rest, glucose release from the liver is facilitated by glucagon, which promotes both liver glycogen breakdown and glucose formation from amino acids. During exercise, glucagon secretion increases. Muscular activity also increases the rate of catecholamine release from the adrenal medulla, and these hormones (epinephrine and norepinephrine) work with glucagon to further increase glycogenolysis. Cortisol concentrations also increase during exercise. Cortisol increases protein catabolism, freeing amino acids to be used within the liver for gluconeogenesis. Thus, all four of these hormones can increase plasma glucose by enhancing the processes of glycogenolysis (breakdown of glycogen) and gluconeogenesis (making glucose from other substrates). In addition to the effects of the four major glucose-controlling hormones, growth hormone increases mobilization of FFAs and decreases cellular uptake of glucose, so less glucose is used by the cells (more remains in circulation); and the thyroid hormones promote glucose catabolism and fat metabolism.

The amount of glucose released by the liver depends on exercise intensity and the duration. As intensity increases, so does the rate of catecholamine release. This can cause the liver to release more glucose than is being taken up by the active muscles. Consequently, during or shortly after an explosive, short-term sprint bout, blood glucose concentrations may be 40% to 50% above the resting level, illustrating that the glucose release by the liver is greater than the uptake by the muscles.

The greater the exercise intensity, the greater the catecholamine release, and thus the glycogenolysis rate is significantly increased. This process occurs not only in the liver but also in the muscle. The glucose released from the liver enters the blood to become available to the muscle. But the muscle has a more readily available source of glucose: its own glycogen. The muscle uses its own glycogen stores before using the plasma glucose during explosive, short-term exercise. Glucose released from the liver is not used as readily, so it remains in the circulation, elevating the plasma glucose. Following exercise, plasma glucose concentrations decrease as the glucose enters the muscle to replenish the depleted muscle glycogen stores.

During exercise bouts that last for several hours, however, the rate of liver glucose release more closely matches the muscles' needs, keeping plasma glucose at or only slightly above the resting concentrations. As muscle uptake of glucose increases, the liver's rate of glucose release also increases. In most cases, plasma glucose does not begin to decline until late in the activity as liver glycogen stores become depleted, at which time glucagon concentrations increase significantly. Glucagon and cortisol together enhance gluconeogenesis, providing more fuel.

Figure 2.14 illustrates the changes in plasma concentrations of epinephrine, norepinephrine, glucagon, cortisol, and glucose during 3 h of cycling. Although the hormonal regulation of glucose remains intact throughout such long-term activities, the liver's glycogen supply may become critically low. As a result, the liver's rate of glucose release may be unable to keep pace with the muscles' rate of glucose uptake. Under this condition, plasma glucose may decline despite strong hormonal stimulation. Glucose ingestion during the activity can play a major role in maintaining plasma glucose concentrations.

In focus

Plasma glucose concentrations are increased by glucagon, epinephrine, norepinephrine, and cortisol. This is important during exercise, particularly long-duration or high-intensity exercise, during which blood glucose concentrations might otherwise decline. Glucose ingestion during exercise also helps maintain plasma glucose concentrations.

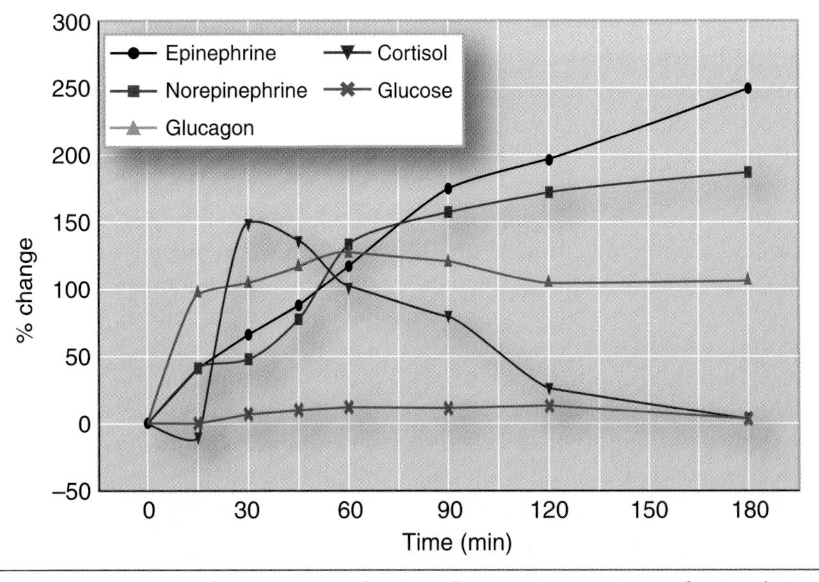

Figure 2.14 Changes (as a percentage of preexercise values) in plasma concentrations of epinephrine, norepinephrine, glucagon, cortisol, and glucose during 3 h of cycling at 65% $\dot{V}O_{2max}$.

Glucose Uptake by Muscle

Merely releasing sufficient amounts of glucose into the blood does not ensure that the muscle cells will have enough glucose to meet their energy demands. Not only must the glucose be released and delivered to these cells; it also must be taken up by the cells. Transport of glucose through the cell membranes and into muscle cells is controlled by insulin. Once glucose is delivered to the muscle, insulin facilitates its transport into the fibers.

Surprisingly, as seen in figure 2.15, plasma insulin concentration tends to decrease during prolonged submaximal exercise, despite a slight increase in plasma glucose concentration and glucose uptake by muscle. This apparent contradiction between the plasma insulin concentrations and the muscles' need for glucose serves as a reminder that a hormone's activity is determined not only by its concentration in the blood but also by a cell's sensitivity to that hormone. In this case, the cell's sensitivity to insulin is at least as important as the amount of circulating hormone. Exercise may enhance insulin's binding to receptors on the muscle fiber, thereby reducing the need for high concentrations of plasma insulin to transport glucose across the muscle cell membrane into the cell. This is important, because during exercise four hormones are trying to release glucose from its storage sites and create new glucose. High insulin concentrations would oppose their action, preventing this needed increase in plasma glucose supply.

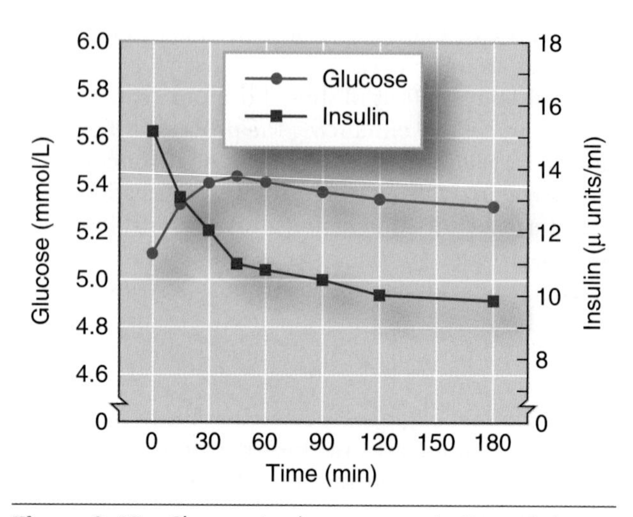

Figure 2.15 Changes in plasma concentrations of glucose and insulin during prolonged cycling at 65% to 70% of $\dot{V}O_{2max}$. Note the gradual decline in insulin throughout the exercise, suggesting an increased sensitivity to insulin during prolonged effort.

Regulation of Fat Metabolism During Exercise

Although fat generally contributes less than carbohydrate to muscles' energy needs during exercise, mobilization and oxidation of FFAs are critical to performance in endurance exercise. During such prolonged activity, carbohydrate reserves become depleted, and muscle must rely more heavily on the oxidation of fat for energy production. When carbohydrate reserves are low (low plasma glucose and low muscle glycogen), the endocrine system can accelerate the oxidation of fats (lipolysis), thus ensuring that muscles' energy needs can be met.

Free fatty acids are stored as triglycerides in fat cells and inside muscle fibers. Adipose tissue triglycerides, however, must be broken down to release the FFAs, which are then transported to the muscle fibers. The rate of FFA uptake by active muscle correlates highly with the plasma FFA concentration. Increasing this concentration would increase cellular uptake of the FFA. The rate of triglyceride breakdown may determine, in part, the rate at which muscles use fat as a fuel source during exercise.

The rate of lipolysis is controlled by at least five hormones:

- (Decreased) Insulin
- Epinephrine
- Norepinephrine
- Cortisol
- Growth hormone

The major factor responsible for adipose tissue lipolysis during exercise is a fall in circulating insulin. Lipolysis is also enhanced through the elevation of epinephrine and norepinephrine. In addition to having a role in gluconeogenesis, cortisol accelerates the mobilization and use of FFAs for energy during exercise. Plasma cortisol concentration peaks after 30 to 45 min of exercise and then decreases to near-normal levels. But the plasma FFA concentration continues to increase throughout the activity, meaning that lipase continues to be activated by other hormones. The hormones that continue this process are the catecholamines and growth hormone. The thyroid hormones also contribute to the mobilization and metabolism of FFAs, but to a much lesser degree.

Free fatty acids are a primary source of energy at rest and during exercise. They are derived from triglycerides through the action of the enzyme lipase, which breaks down triglycerides into FFA and glycerol.

Thus, the endocrine system plays a critical role in regulating ATP production during exercise as well as con-

In review

- Plasma glucose concentration is increased by the combined actions of glucagon, epinephrine, norepinephrine, and cortisol. These hormones promote glycogenolysis and gluconeogenesis, thus increasing the amount of glucose available for use as a fuel source.
- Insulin helps the released glucose enter the cells, where it can be used for energy production. But insulin concentrations decline during prolonged exercise, indicating that exercise facilitates the action of insulin so that less of the hormone is required during exercise than at rest.
- When carbohydrate reserves are low, the body turns more to fat oxidation for energy, and lipolysis increases. This process is facilitated by a decreased insulin concentration and increased concentrations of epinephrine, norepinephrine, cortisol, and growth hormone.

trolling the balance between carbohydrate and fat metabolism.

Hormonal Regulation of Fluid and Electrolyte Balance During Exercise

Fluid balance during exercise is critical for optimal metabolic, cardiovascular, and thermoregulatory function. At the onset of exercise, water is shifted from the plasma volume to the interstitial and intracellular spaces. This water shift is specific to the amount of muscle that is active and the intensity of effort. Metabolic by-products begin to accumulate in and around the muscle fibers, increasing the osmotic pressure there. Water is then drawn into these areas by diffusion. Also, increased muscular activity increases blood pressure, which in turn drives water out of the blood (hydrostatic forces). In addition, sweating increases during exercise. The combined effect of these actions is that the muscles and sweat glands gain water at the expense of plasma volume. For example, running at approximately 75% of $\dot{V}O_{2max}$ decreases plasma volume by 5% to 10%. Reduced plasma volume decreases blood pressure and the amount of blood flow to the skin and muscles. Both of these effects can impede athletic performance.

The endocrine system plays a major role in monitoring fluid levels and correcting imbalances, along with regulating electrolyte balance, especially that of sodium. The two major hormones involved in this regulation are antidiuretic hormone released from the posterior pituitary and aldosterone, a mineralocorticoid released from the adrenal cortex. The kidneys are the primary target organ for both of these hormones.

Posterior Pituitary

The pituitary's posterior lobe is an outgrowth of neural tissue from the hypothalamus. For this reason, it is also referred to as the neurohypophysis. It secretes two hormones: **antidiuretic hormone (ADH;** also called vasopressin or arginine vasopressin) and oxytocin. Both of these hormones are actually produced in the hypothalamus. They travel through the neural tissue and are stored in vesicles within nerve endings in the posterior pituitary. These hormones are released into capillaries as needed in response to neural impulses from the hypothalamus.

Of the two posterior pituitary hormones, only ADH is known to play an important role during exercise. Antidiuretic hormone promotes water conservation by increasing the water permeability of the kidneys' collecting ducts. As a result, less water is excreted in the urine, creating an "antidiuresis."

Muscular activity and sweating cause electrolytes to become concentrated in the blood plasma. This is called **hemoconcentration,** and it increases the plasma **osmolality** (the ionic concentration of dissolved substances in the plasma). This is the primary physiological stimulus for ADH release. The increased osmolality is sensed by osmoreceptors in the hypothalamus. A second and related stimulus for ADH release is a low plasma volume. In response to either stimuli, the hypothalamus sends neural impulses to the posterior pituitary, stimulating ADH release. The ADH enters the blood, travels to the kidneys, and promotes water retention in an effort to dilute the plasma electrolyte concentration back to normal levels. This hormone's role in conserving body water minimizes the extent of water loss and therefore the risk of severe dehydration during periods of heavy sweating and hard exercise. Figure 2.16 illustrates this process.

Loss of fluid (plasma) from the blood results in a concentration of the constituents of the blood, a phenomenon referred to as hemoconcentration. Conversely, a gain of fluid in the blood results in a dilution of the constituents of the blood, which is referred to as hemodilution.

Adrenal Cortex Revisited

The **mineralocorticoids,** secreted from the adrenal cortex, maintain electrolyte balance in the extracellular

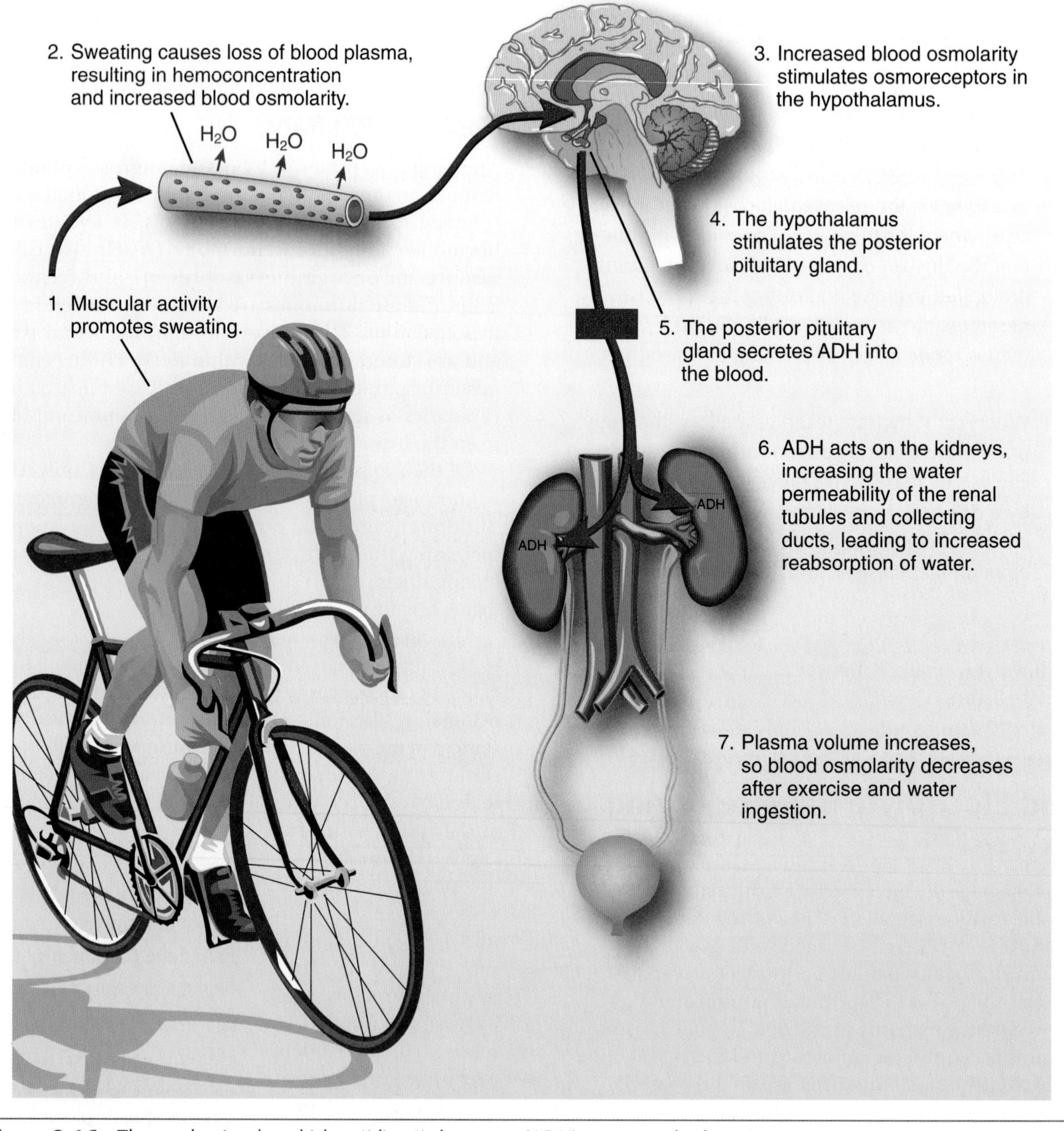

Figure 2.16 The mechanism by which antidiuretic hormone (ADH) conserves body water.

fluids, especially that of sodium (Na^+) and potassium (K^+). **Aldosterone** is the major mineralocorticoid, responsible for at least 95% of all mineralocorticoid activity. It works primarily by promoting renal reabsorption of sodium, thus causing the body to retain sodium. When sodium is retained, so is water; thus, aldosterone, like ADH, results in water retention. Sodium retention also enhances potassium excretion, so aldosterone plays a role in potassium balance as well. For these reasons, aldosterone secretion is stimulated by many factors, including decreased plasma sodium, decreased blood volume, decreased blood pressure, and increased plasma potassium concentration.

Kidneys

Although the kidneys are not typically considered major endocrine organs, they release a hormone called erythropoietin. **Erythropoietin (EPO)** regulates red blood cell (erythrocyte) production by stimulating

Osmolality

Body fluids contain many dissolved molecules and minerals. The presence of these particles in various body fluid compartments (i.e., intracellular, plasma, and interstitial spaces) generates an osmotic pressure or attraction to retain water within that compartment. The amount of osmotic pressure exerted by a body fluid is proportional to the number of molecular particles (osmoles, or Osm) in solution. A solution that has 1 Osm of solute dissolved in each kilogram (or liter) of water is said to have an osmolality of 1 osmole per kilogram (1 Osm/kg), whereas a solution that has 0.001 Osm/kg has an osmolality of 1 milliosmole per kilogram (1 mOsm/kg). Normally, body fluids have an osmolality of 300 mOsm/kg. Increasing the osmolality of the solutions in one body compartment generally causes water to be drawn away from adjacent compartments that have a lower osmolality (i.e., more water).

bone marrow cells. The red blood cells are essential for transporting oxygen to the tissues and removing carbon dioxide, so this hormone is extremely important in our adaptation to training and altitude. The kidneys also release renin, a hormone and enzyme involved in blood pressure control and fluid and electrolyte balance.

The kidneys have a strong regulatory influence on blood pressure that also allows them to regulate fluid balance. Plasma volume is a major determinant of blood pressure: When plasma volume decreases, so does blood pressure. Blood pressure is monitored by specialized cells within the kidneys. During exercise, these cells can be stimulated by decreased blood pressure, decreased blood flow to the kidneys through increased sympathetic nervous activity accompanying exercise, or direct stimulation from the sympathetic nerves.

Figure 2.17 shows the mechanism involved in renal control of blood pressure, the **renin–angiotensin-aldosterone mechanism.** The kidneys respond to decreased blood pressure or blood flow by forming an enzyme and hormone called **renin.** Renin, in turn, converts a plasma protein called angiotensinogen into an active form called angiotensin I. In the blood, angiotensin I is converted to angiotensin II when it encounters the enzyme **angiotensin converting enzyme (ACE)** in the lungs. Angiotensin converting enzyme inhibitors are a class of drugs used in the treatment of high blood pressure. They lower blood pressure by blocking, or inhibiting, the conversion of angiotensin I to angiotensin II. Angiotensin II acts in two ways. First, it is a potent constrictor of blood vessels. Through this action, peripheral resistance increases, which raises the blood pressure. The second job of angiotensin II is to trigger aldosterone release from the adrenal cortex.

Recall that aldosterone's primary action is to promote sodium reabsorption in the kidneys. Because water follows sodium, this renal conservation of sodium causes the kidneys to also retain water. The net effect is to conserve the body's fluid content, thereby minimizing the loss of plasma volume while keeping blood pressure near normal. Figure 2.18 illustrates the changes in plasma volume and aldosterone concentrations during 2 h of exercise.

The hormonal influences of ADH and aldosterone persist for up to 12 to 48 h after exercise, reducing urine production and protecting the body from further dehydration. In fact, aldosterone's prolonged enhancement of Na^+ reabsorption may cause the body's Na^+ concentration to increase above normal following an exercise bout. In an effort to compensate for this elevation in Na^+ concentrations, more of the water ingested shifts into the extracellular compartment.

As shown in figure 2.19, individuals who are subjected to three repeated days of exercise and dehydration show a significant increase in plasma volume that continues to increase throughout the period of activity. This increase in plasma volume appears to parallel the body's retention of dietary Na^+. When the daily bouts of activity are terminated, the excess Na^+ and water are excreted in urine.

In focus

Hemoglobin is one of the substances diluted by plasma expansion. For this reason, some athletes who actually have normal hemoglobin concentrations may appear to be anemic as a consequence of Na^+-induced hemodilution. This condition, not to be confused with true anemia, can be remedied with a few days of rest, allowing time for aldosterone concentrations to return to normal and for the kidneys to unload the extra Na^+ and water.

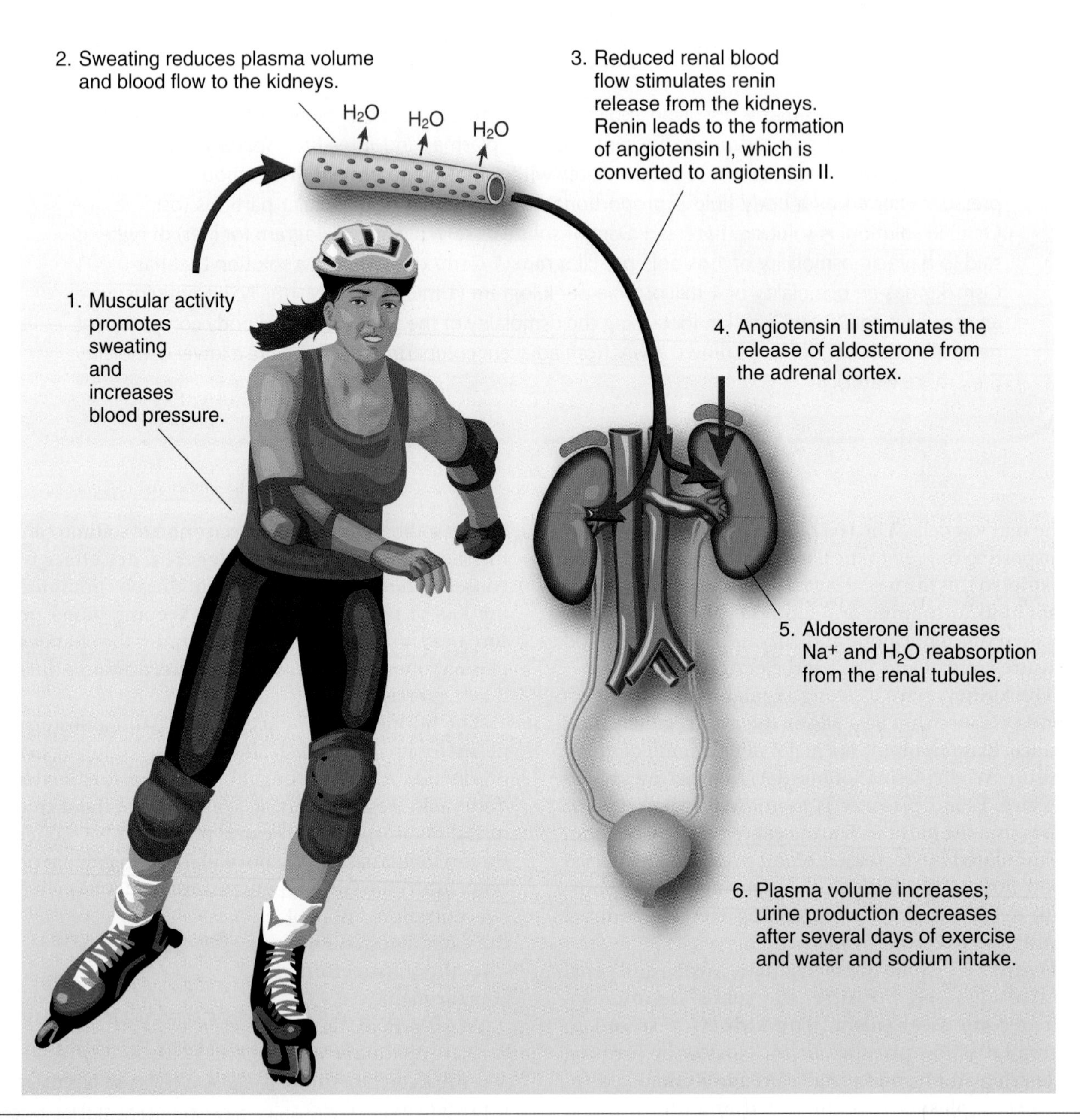

Figure 2.17 The influence of water loss from plasma during exercise leads to a sequence of events that promotes sodium (Na^+) and water reabsorption from the renal tubules, thereby reducing urine production. In the hours after exercise when fluids are consumed, the elevated aldosterone concentrations cause an increase in the extracellular volume and an expansion of plasma volume.

Most athletes involved in heavy training have an expanded plasma volume, which dilutes various blood constituents. The actual amount of proteins and electrolyte (solutes) within the blood remains unaltered, but the substances are dispersed throughout a greater volume of water (plasma), so they are diluted and their concentration decreases. This phenomenon is called **hemodilution.**

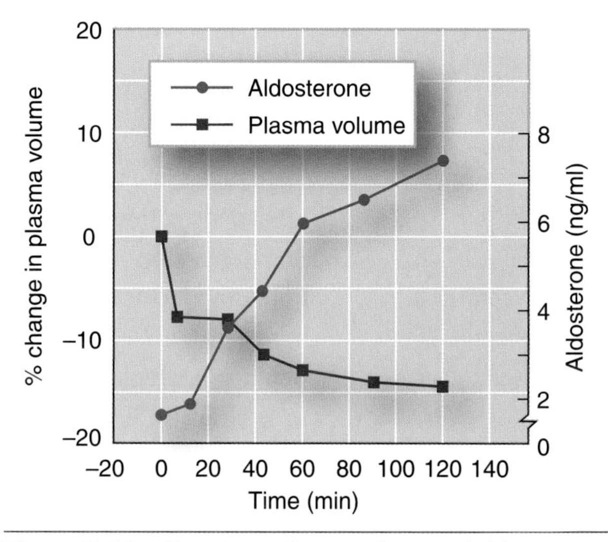

Figure 2.18 Changes in plasma volume and aldosterone concentrations during 2 h of cycling exercise. Note that plasma volume declines rapidly during the first few minutes of exercise and then shows a smaller rate of decline despite large sweat losses. Plasma aldosterone concentration, on the other hand, increases rather steadily throughout the exercise.

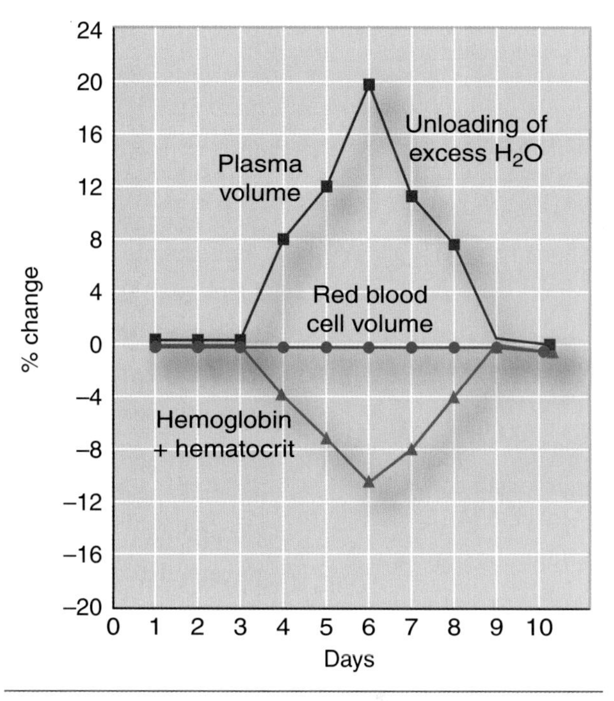

Figure 2.19 Changes in plasma volume during three days of repeated exercise and dehydration. Subjects exercised in heat on days 3 to 6. Note the sudden decline in plasma volume when the subjects stopped training (6th day). The changes in hemoglobin and hematocrit reflect the expansion and contraction of plasma volume during and after the three-day training period.

In review

- The two primary hormones involved in the regulation of fluid balance are ADH and aldosterone.
- Antidiuretic hormone is released in response to increased plasma osmolality. When osmoreceptors in the hypothalamus sense this increase, the hypothalamus triggers ADH release from the posterior pituitary. Low blood volume is a second stimulus for ADH release.
- Antidiuretic hormone acts on the kidneys, promoting water conservation. Blood osmolality decreases.
- When plasma volume or blood pressure decreases, the kidneys form an enzyme called renin that converts angiotensinogen into angiotensin I, which later becomes angiotensin II. Angiotensin II increases peripheral resistance, increasing the blood pressure.
- Angiotensin II also triggers the release of aldosterone from the adrenal cortex. Aldosterone promotes sodium reabsorption in the kidneys, which in turn causes water retention, thus minimizing the loss of plasma volume.

In Closing

In this chapter, we focused on metabolism and the role of the endocrine system in regulating metabolism and many other physiological processes that accompany exercise. We also discussed the role of hormones in regulating the metabolism of glucose and fat and maintaining fluid balance. We next look at the neural control of exercising muscle.

KEY TERMS

acetyl coenzyme A (acetyl CoA)
adenosine diphosphate (ADP)
adrenaline
aerobic metabolism
aldosterone
anaerobic metabolism
angiotensin converting enzyme (ACE)
antidiuretic hormone (ADH)
ATP-PCr system
β-oxidation
bioenergetics
carbohydrate
catabolism
catecholamines
cortisol
cyclic adenosine monophosphate (cAMP)
direct gene activation
downregulation
electron transport chain
epinephrine
erythropoietin (EPO)
free fatty acids (FFAs)
glucagon
glucocorticoids
gluconeogenesis
glucose
glycogen
glycogenolysis
glycolysis
growth hormone
hemoconcentration
hemodilution
hormone
hyperglycemia
hypoglycemia
inhibiting factors
insulin
kilocalories
Krebs cycle
lipogenesis
lipolysis
metabolism
mineralocorticoids
negative feedback
nonsteroid hormones
norepinephrine
osmolality
oxidative system
phosphocreatine (PCr)
prostaglandins
releasing factors
renin
renin–angiotensin-aldosterone mechanism
second messenger
steroid hormones
substrate
target cells
thyrotropin (TSH)
thyroxine (T_4)
triglycerides
triiodothyronine (T_3)
upregulation

STUDY QUESTIONS

1. What is ATP and how is it of importance in metabolism?
2. What is the primary substrate used to provide energy at rest? During high-intensity exercise?
3. What is the role of PCr in energy production? Describe the relationship between muscle ATP and PCr during sprint exercise.
4. Describe the three energy systems.
5. Why are the ATP-PCr and glycolytic energy systems considered anaerobic?
6. What role does oxygen play in the process of aerobic metabolism?
7. Describe the by-products of energy production from ATP-PCr, glycolysis, and oxidation.
8. What is lactic acid and why is it important?
9. What is an endocrine gland, and what are the functions of hormones?
10. Explain the difference between steroid hormones and nonsteroid hormones.
11. How can hormones have very specific functions when they reach nearly all parts of the body through the blood?
12. How are plasma concentrations of specific hormones controlled?
13. Briefly outline the major endocrine glands, their hormones, and the specific action of these hormones.
14. Which of the hormones outlined in question 13 are of major significance during exercise?
15. What is hemoconcentration, and how does the endocrine system relate to it?

16. Describe the hormonal regulation of metabolism during exercise. What hormones are involved, and how do they influence the availability of carbohydrates and fats for energy during exercise lasting for several hours?
17. Describe the hormonal regulation of fluid balance during exercise.
18. What is hemodilution, and how does the endocrine system relate to it?

STUDY GUIDE ACTIVITIES

In addition to the activities listed in the chapter opening outline on page 47, two other activities are available in the online study guide, located at

www.HumanKinetics.com/PhysiologyOfSportAndExercise.

The chapter **SUMMARY** reviews the key concepts covered in the chapter, and the end-of-chapter **QUIZ** tests your understanding of the material covered in the chapter.

CHAPTER 3

Neural Control of Exercising Muscle

In this chapter and in the online study guide

In 1959, at age 15, Jimmie Heuga was the youngest male ever to make the U.S. Ski Team. He raced internationally for 10 years, competing on the 1964 and 1968 Olympic teams and on the 1962 and 1966 World Championship teams. In 1964, Heuga and teammate Billy Kidd made history by winning the first U.S. Olympic medals in men's alpine skiing. In 1967, Heuga won a World Cup in the giant slalom, finishing third in the world for the entire season. He also won the Arlberg-Kandahar at Garmisch, Germany, one of the oldest and most prestigious alpine ski races.

After competing in the 1968 Olympics, troubled by unknown physical ailments, Heuga retired from the U.S. Ski Team. In 1970, he was diagnosed with multiple sclerosis (MS), a neurological disorder. At that time, people with MS were told that physical activity would exacerbate their condition, so he was advised to live a quiet and tranquil life. Heuga followed that advice and began feeling unhealthy, unmotivated, and less energetic. He began to deteriorate physically and mentally.

Six years later, Heuga decided to defy medical convention. He developed a cardiovascular endurance exercise program and began stretching and strengthening exercises. He established realistic goals for his personal wellness program. With this program, Heuga regained his health within the constraints of MS. In 1984, inspired by his own success, he created the Jimmie Heuga Center, a nonprofit organization based in Edwards, Colorado. Since then, thousands of people with MS have gone through the center's medical program. Furthermore, the center has been able to fund research studies and provide financial support for both doctoral and postdoctoral students. The center's most important contribution to research on MS, published in the *Annals of Neurology* in 1996,[4] demonstrated that a physical activity training program enhances physiological and psychological function and general quality of life in MS patients, countering the medical wisdom at that time, which dictated a life of restricted activity.

Source: Dr. Richard W. Hicks, Executive Director; Jimmie Heuga Center; Edwards, Colorado; June 1997.

All physiological activity in the human body is or can be influenced by the nervous system. Nerves provide the wiring through which electrical impulses are sent to and received from virtually all parts of the body. The brain acts as a computer, integrating all incoming information, selecting an appropriate response, and then instructing the involved body parts to take appropriate action. Thus, the nervous system forms a vital network, allowing communication and coordination of interaction among the various tissues in the body as well as with the outside world.

The nervous system is one of the body's most complex systems. Many of its functions are not yet fully understood. For these reasons, and because this book is concerned only with neural control of voluntary movement, we will limit our coverage of this complex system. We first review the structure and function of the nervous system and then focus on specific topics relevant to sport and exercise.

OVERVIEW OF THE NERVOUS SYSTEM

Before we examine the intricate details of the nervous system, it is important to step back and look at the big picture—how the nervous system is organized and how that organization functions to integrate the body's movement. First, the nervous system as a whole is composed of two components: the **central nervous system (CNS)** and the **peripheral nervous system (PNS).** The CNS is composed of the brain and spinal cord, while the PNS is composed of two major divisions, the **sensory division** (or **afferent division**) and the **motor division** (or **efferent division**). The sensory division is responsible for informing the CNS about what is going on within and outside the body. The motor division is responsible for sending information from the CNS to the various parts of the body in response to the signals coming in from the sensory division. The motor division is composed of two parts, the autonomic nervous system and the somatic nervous system. Figure 3.1 provides a schematic of these relationships. Much more detail concerning each of these individual units of the nervous system is presented later in this chapter.

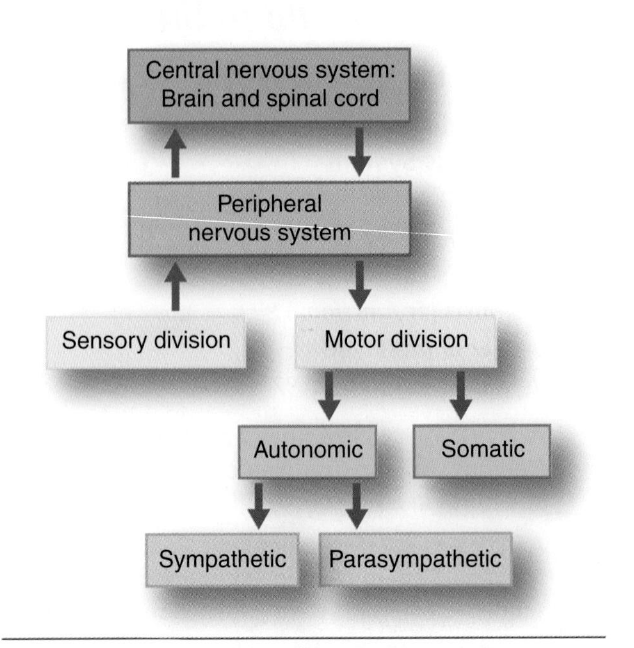

Figure 3.1 Organization of the nervous system.

STRUCTURE AND FUNCTION OF THE NERVOUS SYSTEM

The **neuron** is the structural unit of the nervous system. We first review the anatomy of the neuron and then look at how it functions—allowing electrical impulses to be transmitted throughout the body.

Neuron

Individual nerve fibers (nerve cells), depicted in figure 3.2, are called neurons. A typical neuron is composed of three regions:

- The cell body, or soma
- The dendrites
- The axon

The cell body contains the nucleus. Radiating out from the cell body are the cell processes: the dendrites and the axon. On the side toward the axon, the cell body tapers into a cone-shaped region known as the **axon hillock.** The axon hillock has an important role in impulse conduction, as discussed later.

Most neurons contain many dendrites. These are the neuron's receivers. Most impulses, or action potentials, coming into the neuron from sensory stimuli or from adjacent neurons typically enter the neuron via the dendrites. These processes then carry the impulses toward the cell body.

In contrast, most neurons have only one axon. The axon is the neuron's transmitter and conducts impulses away from the cell body. Near its end, an axon splits into numerous **end branches.** The tips of these branches are dilated into tiny bulbs known as **axon terminals** or synaptic knobs. These terminals or knobs house numerous vesicles (sacs) filled with chemicals known as **neurotransmitters** that are used for communication between a neuron and another cell. (This is discussed later in this chapter in more detail.) The structure of the neuron allows nerve impulses to enter the neuron through the dendrites, and to a lesser extent through

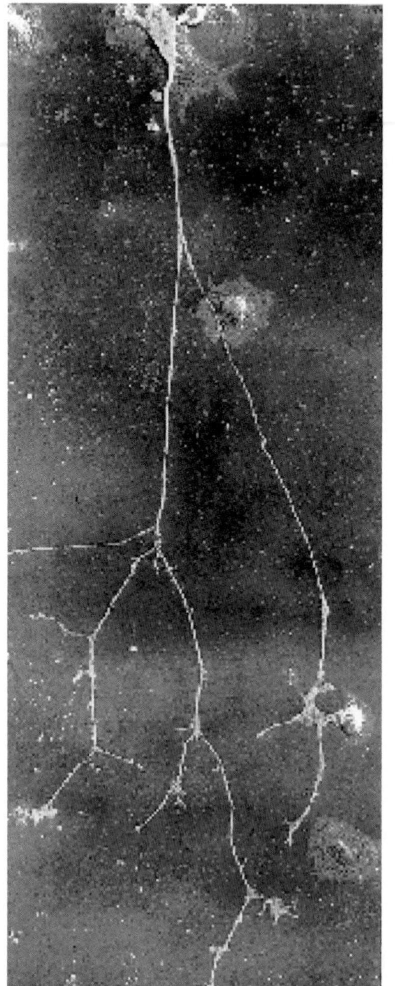

the cell body, and to travel through the cell body and axon hillock, down the axon, and out through the end branches to the axon terminals. We next explain in more detail how this happens, including how these impulses travel from one neuron to another and from a motor neuron to muscle fibers.

Nerve Impulse

A **nerve impulse**—an electrical charge—is the signal that passes from one neuron to the next and finally to an end organ, such as a group of muscle fibers, or back to the CNS. For simplicity, think of the nerve impulse traveling through a neuron much as electricity travels through the electrical wires in a home. This section describes how the electrical impulse is generated and how it travels through a neuron.

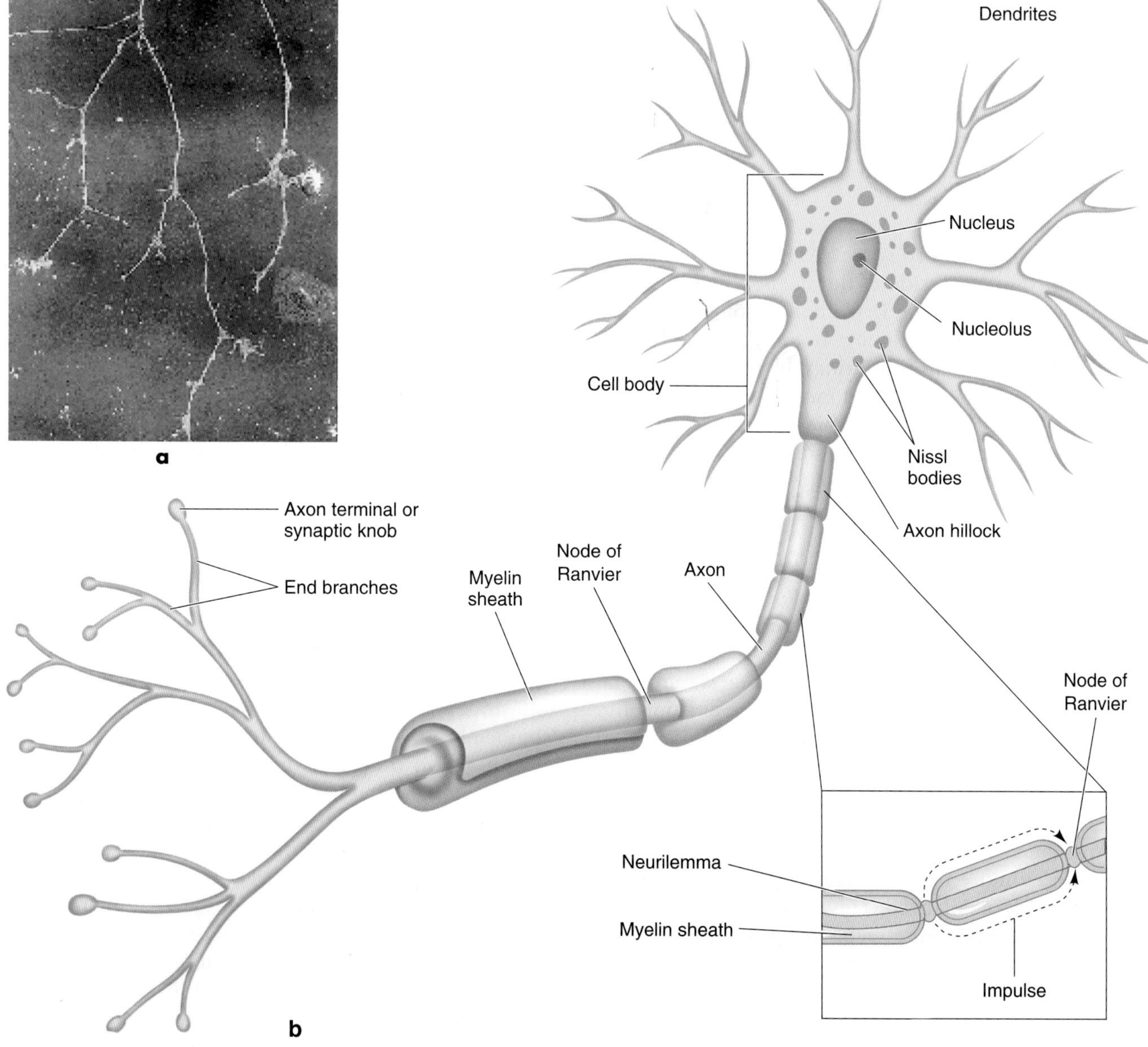

Figure 3.2 *(a)* A photomicrograph of a neuron and *(b)* its structure.

Resting Membrane Potential

The cell membrane of a neuron at rest has a negative electrical potential of about –70 mV. This means that if one were to insert a voltmeter probe inside the cell, the electrical charges found there and the charges found outside the cell would differ by 70 mV, and the inside would be negative relative to the outside. This electrical potential difference is known as the **resting membrane potential (RMP).** It is caused by a separation of charges across the membrane. When the charges across the membrane differ, the membrane is said to be polarized.

The neuron has a high concentration of potassium ions (K^+) on the inside of the membrane and a high concentration of sodium ions (Na^+) on the outside. The imbalance in the number of ions inside and outside the cell causes the RMP. This imbalance is maintained in two ways. First, the cell membrane is much more permeable to K^+ than to Na^+, so the K^+ can move more freely. Because ions tend to move to establish equilibrium, some of the K^+ will move to an area where it is less concentrated, outside the cell. The Na^+ cannot move to the inside as easily. Second, **sodium-potassium pumps** in the neuron membrane, which contain Na^+-K^+ adenosine triphosphatase (ATPase), maintain the imbalance on each side of the membrane by actively transporting potassium ions in and sodium ions out. The sodium-potassium pump moves three Na^+ out of the cell for each two K^+ it brings in. The end result is that more positively charged ions are outside the cell than inside, creating the potential difference across the membrane. Maintenance of a constant RMP of about –70 mV is primarily a function of the sodium-potassium pump.

Depolarization and Hyperpolarization

If the inside of the cell becomes less negative relative to the outside, the potential difference across the membrane decreases. The membrane will be less polarized. When this happens, the membrane is said to be depolarized. Thus, **depolarization** occurs any time the charge difference becomes less than the RMP of –70 mV, moving closer to zero. This typically results from a change in the membrane's Na^+ permeability.

The opposite can also occur. If the charge difference across the membrane increases, moving from the RMP to an even more negative value, then the membrane becomes more polarized. This is known as **hyperpolarization.** Changes in the membrane potential are actually signals used to receive, transmit, and integrate information within and between cells. These signals are of two types, graded potentials and action potentials. Both are electrical currents created by the movement of ions.

Graded Potentials

Graded potentials are localized changes in the membrane potential. These changes can be either depolarizations or hyperpolarizations. The membrane contains ion channels with ion gates that act as doorways into and out of the neuron. These gates are usually closed, preventing ion flow; but they open with stimulation, allowing ions to move from the outside to the inside or vice versa. This ion flow alters the charge separation, changing the polarization of the membrane.

Graded potentials are triggered by a change in the neuron's local environment. Depending on the location and type of neuron involved, the ion gates may open in response to the transmission of an impulse from another neuron or in response to sensory stimuli such as changes in chemical concentrations, temperature, or pressure.

Recall that most neuron receptors are located on the dendrites (although some are on the cell body), yet the impulse is always transmitted from the axon terminals at the opposite end of the cell. For a neuron to transmit an impulse, the impulse must travel almost the entire length of the neuron. Although a graded potential may result in depolarization of the entire cell membrane, it is usually just a local event such that the depolarization does not spread very far along the neuron. To travel the full distance, an impulse must generate an action potential.

Nerve impulses typically pass from the dendrites to the cell body and from the cell body along the length of the axon to its axon terminals.

Action Potentials

An **action potential** is a rapid and substantial depolarization of the neuron's membrane. It usually lasts only about 1 ms. Typically, the membrane potential changes from the RMP of about –70 mV to a value of about +30 mV and then rapidly returns to its resting value. This is illustrated in figure 3.3. How does this marked change in membrane potential occur?

All action potentials begin as graded potentials. When enough stimulation occurs to cause a depolarization of at least 15 to 20 mV, an action potential results. In other words, if the membrane depolarizes from the RMP of –70 mV to a value of –50 to –55 mV, the cell will experience an action potential. The membrane voltage at which a graded potential becomes an action potential is called the depolarization **threshold.** Any depolarization that does not attain the threshold will not result in an action potential. For example, if the membrane potential changes from the RMP of –70

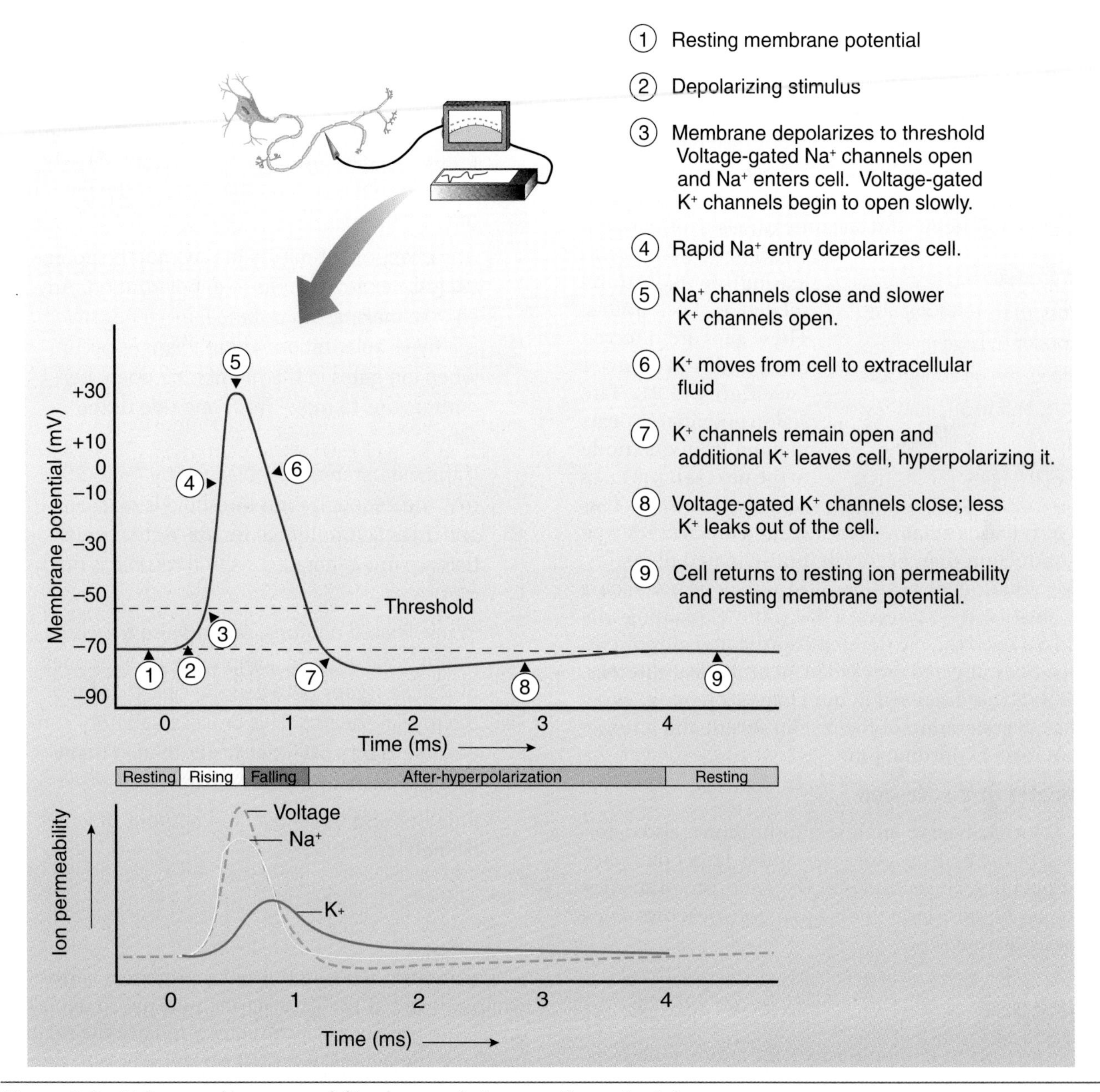

Figure 3.3 Voltage and ion permeability changes during an action potential.

mV to –60 mV, the change is only 10 mV and does not reach the threshold; thus, no action potential occurs. But any time depolarization reaches or exceeds the threshold, an action potential will result. This is the all-or-none principle.

When a given segment of an axon is generating an action potential and its sodium gates are open, it is unable to respond to another stimulus. This is referred to as the *absolute refractory period.* When the sodium gates are closed, the potassium gates are open, and repolarization is occurring, the segment of the axon can then respond to a new stimulus, but the stimulus must be of substantially greater magnitude to evoke an action potential. This is referred to as the *relative refractory period.*

Propagation of the Action Potential

Now that we understand how a neural impulse, in the form of an action potential, is generated, we can look at how the impulse is propagated, or how it travels through the neuron. Two characteristics of the neuron become particularly important when we consider how quickly an impulse can pass through the axon: myelination and diameter.

Myelin Sheath

The axons of many neurons, especially large neurons, are myelinated, meaning they are covered with a sheath formed by myelin, a fatty substance that insulates the cell membrane. This **myelin sheath** (see figure 3.2) is formed by specialized cells called Schwann cells.

The sheath is not continuous. As it spans the length of the axon, the myelin sheath exhibits gaps between adjacent Schwann cells, leaving the axon uninsulated at those points. These gaps are referred to as nodes of Ranvier (see figure 3.2). The action potential appears to jump from one node to the next as it traverses a myelinated fiber. This is referred to as **saltatory conduction,** a much faster type of conduction than occurs in unmyelinated fibers.

In focus

The velocity of nerve impulse transmission in large myelinated fibers can be as high as 100 m/s, or 5 to 50 times faster than that in unmyelinated fibers of the same size.

Myelination of peripheral motor neurons occurs over the first several years of life, partly explaining why children need time to develop coordinated movement. Individuals affected by certain neurological diseases, such as MS as discussed in our chapter opening, experience degeneration of the myelin sheath and a subsequent loss of coordination.

Diameter of the Neuron

The velocity of nerve impulse transmission is also determined by the neuron's size. Neurons of larger diameter conduct nerve impulses faster than neurons of smaller diameter because larger neurons present less resistance to local current flow.

In review

- A neuron's RMP of –70 mV results from the separation of sodium and potassium ions maintained primarily by the sodium-potassium pump, coupled with low sodium permeability and high potassium permeability of the neuron membrane.
- Any change that makes the membrane potential less negative results in depolarization. Any change making this potential more negative is a hyperpolarization. These changes occur when ion gates in the membrane open, permitting ions to move from one side to the other.
- If the membrane is depolarized by 15 to 20 mV, the depolarization threshold is reached and an action potential results. Action potentials are not generated if the threshold is not met.
- In myelinated neurons, the impulse travels through the axon by jumping between nodes of Ranvier (gaps between the cells that form the myelin sheath). This process, saltatory conduction, is 5 to 50 times faster than in unmyelinated fibers of the same size.
- Impulses also travel faster in neurons of larger diameters.

Synapse

For a neuron to communicate with another neuron, first an action potential must occur. Once the action potential occurs, it travels the full length of the axon, ultimately reaching the axon terminals. How does the action potential move from the neuron in which it starts to another neuron?

Neurons communicate with each other across synapses. A **synapse** is the site of action potential transmission from one neuron to another. There are both chemical and mechanical synapses, but the most common type is the chemical synapse, which is our focus.

As seen in figure 3.4, a synapse between two neurons includes

- the axon terminals of the neuron sending the action potential,
- receptors on the neuron receiving the action potential, and
- the space between these structures.

The neuron sending the action potential across the synapse is called the presynaptic neuron, so axon terminals are presynaptic terminals. Similarly, the neuron receiving the action potential on the opposite side of the synapse is called the postsynaptic neuron, and it has postsynaptic receptors. The axon terminals and postsynaptic receptors are not physically in contact with each other. A narrow gap, the synaptic cleft, separates them.

The action potential can be transmitted across a synapse in only one direction: from the axon terminals of the presynaptic neuron to the postsynaptic receptors, usually on the dendrites, of the postsynaptic neuron. Impulses also can go directly to receptors on the cell body: About 5% to 20% of the axon terminals are adjacent to the cell body instead of the dendrites.[2] Why can the action potential go in only one direction?

The presynaptic terminals of the axon contain a large number of saclike structures, called synaptic vesicles. These sacs contain neurotransmitter chemicals. When the impulse reaches the presynaptic axon terminals, the synaptic vesicles respond by dumping their chemicals

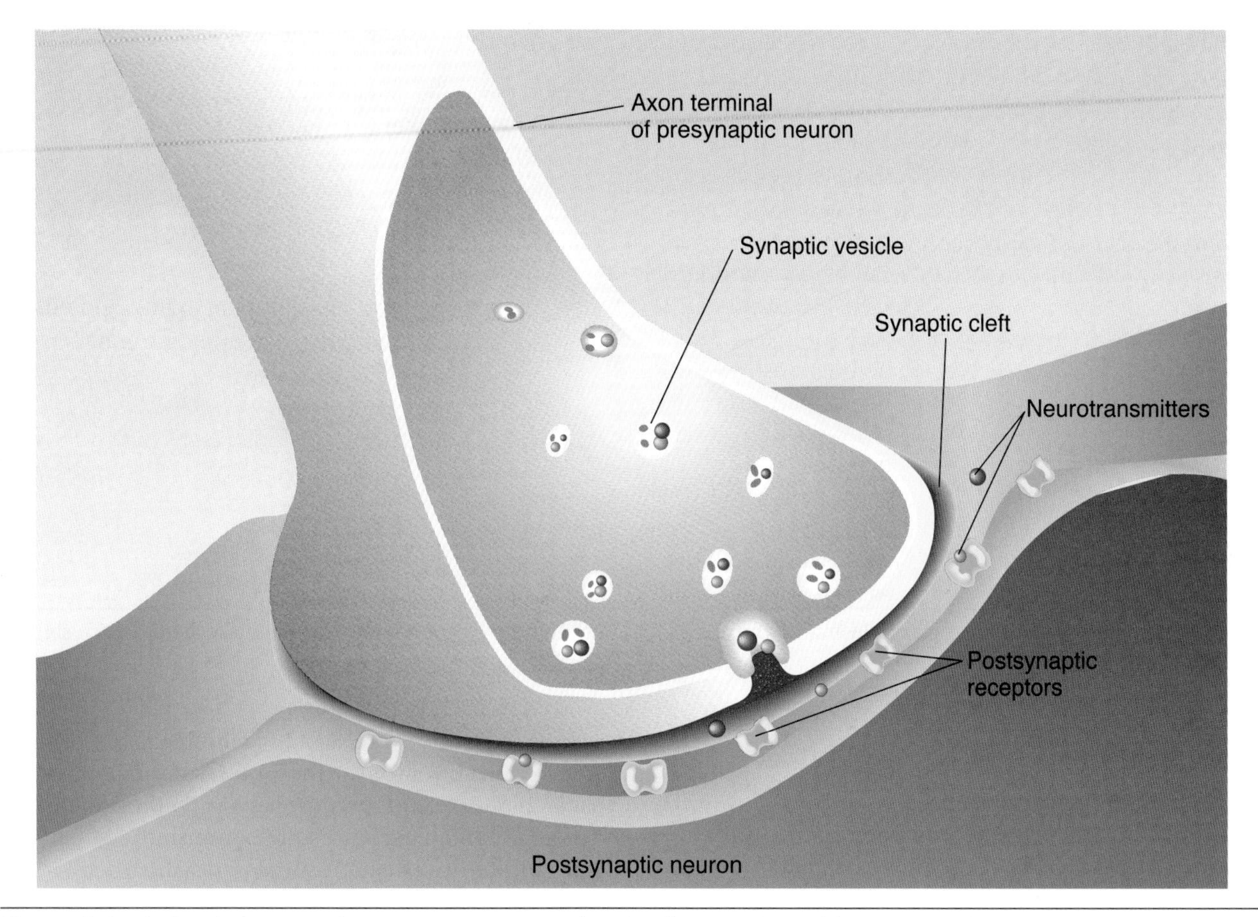

Figure 3.4 A chemical synapse between two neurons, showing the synaptic vesicles.

> ***In focus***
>
> Neurons communicate with one another across synapses by the action of neurotransmitters.

into the synaptic cleft. These neurotransmitters then diffuse across the synaptic cleft to the postsynaptic neuron's receptors. The postsynaptic receptors bind the neurotransmitter once it diffuses across the synaptic cleft. When this binding occurs, the impulse has been transmitted successfully to the next neuron and can be transmitted onward.

Neuromuscular Junction

Whereas neurons communicate with other neurons at synapses, an α-motor neuron communicates with muscle fibers at a site known as a **neuromuscular junction.** The function of the neuromuscular junction is essentially the same as that of a synapse. In fact, the proximal part of the neuromuscular junction is the same: It starts with the axon terminals of the motor neuron, which release neurotransmitters into the space between the motor nerve and the muscle fiber in response to an action potential. However, in the neuromuscular junction, the axon terminals protrude into motor end plates, which are troughlike segments on the plasmalemma (see figure 3.5).

The motor end plate is invaginated (folded to form cavities). The cavity thus formed is called the synaptic gutter. As with synapses, the space between the neuron and the muscle fiber is the synaptic cleft.

Neurotransmitters released from the α-motor neuron axon terminals diffuse across the synaptic cleft and bind to receptors on the muscle fiber's plasmalemma. This binding typically causes depolarization by opening sodium ion channels, allowing more sodium to enter the muscle fiber. As always, if the depolarization reaches the threshold, an action potential is formed. It spreads across the plasmalemma into the T-tubules, initiating muscle fiber contraction. As in the neuron, the plasmalemma, once depolarized, must undergo repolarization. During the period of repolarization, the sodium gates are closed and the potassium gates are open; thus, like the neuron, the muscle fiber is unable to respond to any further stimulation. This period is referred to as the refractory period. Once the electrical conditions of the muscle fiber are restored to resting levels, the fiber can respond to another stimulus. Thus, the refractory period limits the motor unit's firing frequency.

Now we know how the impulse is transmitted between two cells. But to understand what happens once the impulse is transmitted, we must first examine the chemical signals that accomplish transmission.

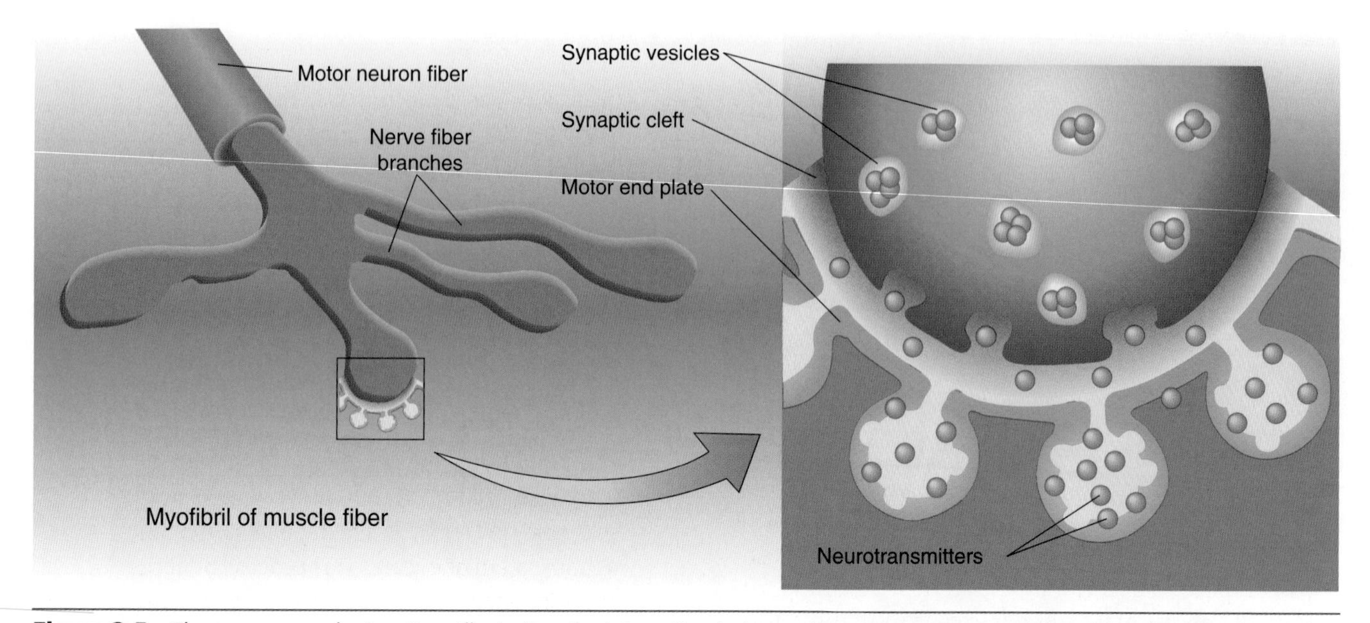

Figure 3.5 The neuromuscular junction, illustrating the interaction between the motor neuron and the plasmalemma of a single muscle fiber.

Neurotransmitters

More than 50 neurotransmitters have been positively identified or are suspected as potential candidates. These can be categorized as either (a) small-molecule, rapid-acting neurotransmitters or (b) neuropeptide, slow-acting neurotransmitters. The small-molecule, rapid-acting transmitters, which are responsible for most neural transmissions, are our main concern.

Acetylcholine and norepinephrine are the two major neurotransmitters involved in regulating our physiological responses to exercise. **Acetylcholine** is the primary neurotransmitter for the motor neurons that innervate skeletal muscle and for most parasympathetic neurons. It is generally an excitatory neurotransmitter, but it can have inhibitory effects at some parasympathetic nerve endings, such as in the heart. **Norepinephrine** is the neurotransmitter for most sympathetic neurons, and it too can be either excitatory or inhibitory, depending on the receptors involved. The sympathetic and parasympathetic nervous systems are discussed later in this chapter.

In review

- Neurons communicate with each other across synapses.
- A synapse involves
    1. the axon terminals of the presynaptic neuron,
    2. the postsynaptic receptors on the dendrite or cell body of the postsynaptic neuron, and
    3. the space (the synaptic cleft) between the two neurons.
- A nerve impulse causes chemicals called neurotransmitters to be released from the presynaptic axon terminals into the synaptic cleft.
- Neurotransmitters diffuse across the cleft and are bound to the postsynaptic receptors.
- Once sufficient neurotransmitters are bound, the impulse is successfully transmitted and the neurotransmitter is then destroyed by enzymes, removed by reuptake into the presynaptic terminal for future use, or diffused away from the synapse.
- Neurotransmitter binding at the postsynaptic receptors opens ion gates in that membrane and can cause depolarization (excitation) or hyperpolarization (inhibition), depending on the specific neurotransmitter and the receptors to which it binds.
- Neurons communicate with muscle cells at neuromuscular junctions. A neuromuscular junction involves presynaptic axon terminals, the synaptic cleft, and motor end-plate receptors on the plasmalemma of the muscle fiber. The neuromuscular junction functions much like a neural synapse.
- The neurotransmitters most important in regulating exercise are acetylcholine and norepinephrine.

In focus

Acetylcholine and norepinephrine are the two major neurotransmitters—chemical substances that transmit nerve impulses across synapses and neuromuscular junctions.

Once the neurotransmitter binds to the postsynaptic receptor, the nerve impulse has been successfully transmitted. The neurotransmitter is then either degraded by enzymes, actively transported back into the presynaptic terminals for reuse, or diffused away from the synapse.

Postsynaptic Response

Once the neurotransmitter binds to the receptors, the chemical signal that traversed the synaptic cleft once again becomes an electrical signal. The binding causes a graded potential in the postsynaptic membrane. An incoming impulse may be either excitatory or inhibitory. An excitatory impulse causes depolarization, known as an **excitatory postsynaptic potential (EPSP).** An inhibitory impulse causes a hyperpolarization, known as an **inhibitory postsynaptic potential (IPSP).**

The discharge of a single presynaptic terminal generally changes the postsynaptic potential less than 1 mV. Clearly this is not sufficient to generate an action potential, because reaching the threshold requires a change of at least 15 to 20 mV. But when a neuron transmits an impulse, several presynaptic terminals typically release their neurotransmitters so that they can diffuse to the postsynaptic receptors. Also, presynaptic terminals from numerous axons can converge on the dendrites and cell body of a single neuron. When multiple presynaptic terminals discharge at the same time, or when only a few fire in rapid succession, more neurotransmitter is released. With an excitatory neurotransmitter, the more that is bound, the greater the EPSP will be.

Triggering an action potential at the postsynaptic neuron depends on the combined effects of all incoming impulses from these various presynaptic terminals. A number of impulses are needed to cause sufficient depolarization to generate an action potential. Specifically, the sum of all changes in the membrane potential must equal or exceed the threshold. This summing of the individual impulses' effects is called **summation.**

For summation, the postsynaptic cell must keep a running total of the neuron's responses, both EPSPs and IPSPs, to all incoming impulses. This task is done at the axon hillock, which lies on the axon just past the cell body. Only when the sum of all individual graded potentials meets or exceeds the threshold can an action potential occur.

Summation refers to the cumulative effect of all individual graded potentials as processed by the axon hillock.

In review

- Excitatory postsynaptic potentials are depolarizations of the postsynaptic membrane; IPSPs are hyperpolarizations of that membrane.
- A single presynaptic terminal cannot generate enough of a depolarization to fire an action potential. Multiple signals are needed. These may come from numerous neurons or from a single neuron when numerous axon terminals release neurotransmitters repeatedly and rapidly.
- The axon hillock keeps a running total of all EPSPs and IPSPs. When their sum meets or exceeds the threshold for depolarization, an action potential occurs. This process of accumulating incoming signals is known as summation.

Individual neurons are grouped together into bundles. In the CNS (brain and spinal cord), these bundles are referred to as tracts, or pathways. Neuron bundles in the PNS are referred to as nerves.

CENTRAL NERVOUS SYSTEM

To comprehend how even the most basic stimulus can cause muscle activity, we must next consider the complexity of the CNS. In this section, we present an overview of the components of the CNS and their functions.

The CNS houses more than 100 billion neurons.

Brain

The brain is composed of numerous parts. For our purposes, we subdivide it into the four major regions illustrated in figure 3.6: the cerebrum, diencephalon, cerebellum, and brain stem.

Cerebrum

The cerebrum is composed of the right and left cerebral hemispheres. These are connected to each other by fiber bundles (tracts) referred to as the *corpus callosum,* which allows the two hemispheres to communicate with each other. The cerebral cortex forms the outer portion of the cerebral hemispheres and has been referred to as the site of the mind and intellect. It is also called the gray matter, which simply reflects its distinctive color resulting from lack of myelin on the cell bodies located in this area. The cerebral cortex is the conscious brain.

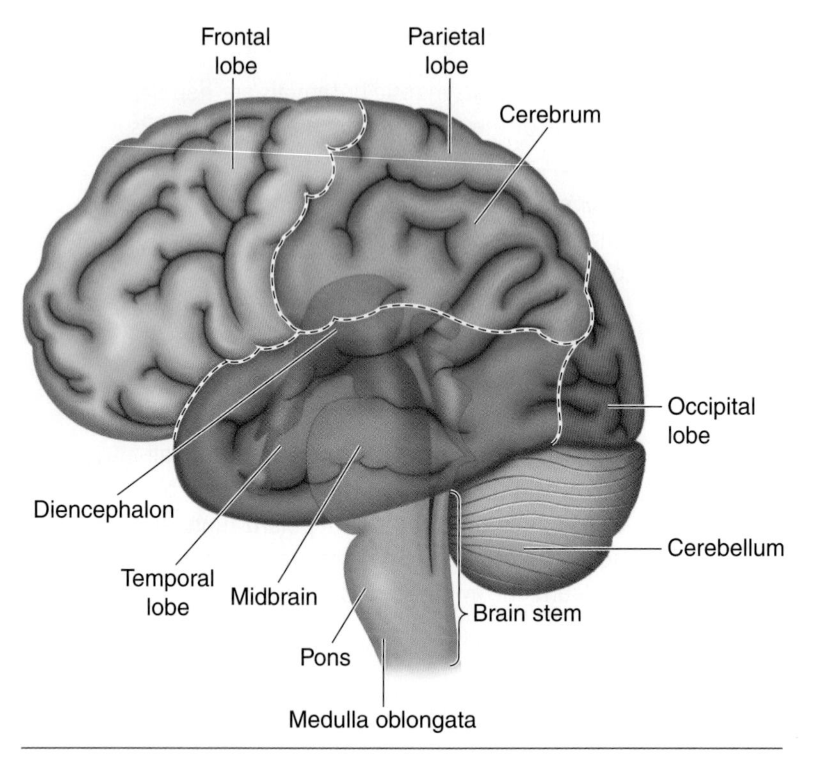

Figure 3.6 Four major regions of the brain and four outer lobes of the cerebrum.

It allows people to think, to be aware of sensory stimuli, and to voluntarily control their movements.

The cerebrum consists of five lobes—four outer lobes and the central insula, which we will not discuss. Its four outer lobes (see figure 3.6) have the following general functions:

- Frontal lobe: general intellect and motor control
- Temporal lobe: auditory input and its interpretation
- Parietal lobe: general sensory input and its interpretation
- Occipital lobe: visual input and its interpretation

The three areas in the cerebrum that are of primary concern to our discussion and that we discuss later in this chapter are the primary motor cortex, in the frontal lobe; the basal ganglia, in the white matter below the cerebral cortex; and the primary sensory cortex, in the parietal lobe.

Diencephalon

The region of the brain known as the diencephalon is composed mostly of the thalamus and the hypothalamus. The thalamus is an important sensory integration center. All sensory input (except smell) enters the thalamus and is relayed to the appropriate area of the cortex. The thalamus regulates what sensory input reaches the conscious brain and thus is very important for motor control.

The hypothalamus, directly below the thalamus, is responsible for maintaining homeostasis by regulating almost all processes that affect the body's internal environment. Neural centers here assist in the regulation of

- blood pressure, heart rate and contractility, respiration, and digestion;
- body temperature;
- fluid balance;
- neuroendocrine control;
- emotions;
- thirst;
- food intake; and
- sleep–wake cycles.

Cerebellum

The cerebellum is located behind the brain stem. It is connected to numerous parts of the brain and has a crucial role in coordinating movement, as we see later in this chapter.

Brain Stem

The brain stem, composed of the midbrain, the pons, and the medulla oblongata (see figure 3.6), is the stalk of the brain, connecting the brain and the spinal cord. Sensory and motor neurons pass through the brain stem as they relay information between the brain and the spinal cord. This is the site of origin for 10 of the 12 pairs of cranial nerves. The brain stem also contains the major autonomic regulatory centers that control the respiratory and cardiovascular systems.

A specialized collection of neurons in the brain stem, known as the *reticular formation,* is influenced by, and has an influence on, nearly all areas of the CNS. These neurons help

- coordinate skeletal muscle function,
- maintain muscle tone,
- control cardiovascular and respiratory functions, and
- determine our state of consciousness (both arousal and sleep).

The brain has a pain control system, called an analgesia system. The enkephalins and β-endorphin are important opiate substances that act on the opiate receptors in the analgesia system to help reduce pain. Research has demonstrated that exercise of long duration increases the natural levels of these opiate substances.

In review

- The CNS is composed of the brain and the spinal cord.
- The four major divisions of the brain are the cerebrum, the diencephalon, the cerebellum, and the brain stem.
- The cerebral cortex is the conscious brain.
- The diencephalon includes the thalamus, which receives all sensory input entering the brain, and the hypothalamus, which is a major control center for homeostasis.
- The cerebellum, which is connected to numerous parts of the brain, is critical for coordinating movement.
- The brain stem is composed of the midbrain, the pons, and the medulla oblongata.
- The spinal cord contains both sensory and motor fibers that transmit action potentials between the brain and the periphery.

Spinal Cord

The lowest part of the brain stem, the medulla oblongata, is continuous with the spinal cord below. The spinal cord is composed of tracts of nerve fibers that allow two-way conduction of nerve impulses. The sensory (afferent) fibers carry neural signals from sensory receptors, such as those in the skin, muscles, and joints, to the upper levels of the CNS. Motor (efferent) fibers from the brain and upper spinal cord transmit action potentials to end organs (e.g., muscles, glands).

PERIPHERAL NERVOUS SYSTEM

The PNS contains 43 pairs of nerves: 12 pairs of cranial nerves that connect with the brain and 31 pairs of spinal nerves that connect with the spinal cord. Cranial and spinal nerves directly supply the skeletal muscles. Functionally, the PNS has two major divisions: the sensory division and the motor division. Let's examine each briefly.

Sensory Division

The sensory division of the PNS carries sensory information toward the CNS. Sensory (afferent) neurons originate in such areas as

- blood and lymph vessels,
- internal organs,
- special sense organs (taste, touch, smell, hearing, vision),
- the skin, and
- muscles and tendons.

Sensory neurons in the PNS end either in the spinal cord or in the brain, and they continuously convey information to the CNS concerning the body's constantly changing status. By relaying this information, these neurons allow the brain to sense what is going on in all parts of the body and in the immediate environment. Sensory neurons within the CNS carry the sensory input to appropriate areas, where the information can be processed and integrated with other incoming information.

The sensory division receives information from five primary types of receptors:

1. Mechanoreceptors that respond to mechanical forces such as pressure, touch, vibrations, or stretch
2. Thermoreceptors that respond to changes in temperature
3. Nociceptors that respond to painful stimuli
4. Photoreceptors that respond to electromagnetic radiation (light) to allow vision
5. Chemoreceptors that respond to chemical stimuli, such as from foods, odors, or changes in blood or tissue concentrations of substances such as oxygen, carbon dioxide, glucose, and electrolytes

Several of these receptors are important in exercise and sport. Let's consider just a few. Free nerve endings detect crude touch, pressure, pain, heat, and cold. Thus, they function as mechanoreceptors, nociceptors, and thermoreceptors. These nerve endings are important for preventing injury during athletic performance. Special muscle and joint nerve endings are of many types and functions, and each type is sensitive to a specific stimulus. Here are some important examples:

- Joint kinesthetic receptors located in the joint capsules are sensitive to joint angles and rates of change in these angles. Thus, they sense the position and any movement of the joints.
- Muscle spindles sense muscle length and rate of change in length.
- Golgi tendon organs detect the tension applied by a muscle to its tendon, providing information about the strength of muscle contraction.

Muscle spindles and Golgi tendon organs are discussed later in this chapter.

Motor Division

The CNS transmits information to various parts of the body through the motor, or efferent, division of the PNS. Once the CNS has processed the information it receives from the sensory division, it decides how the body should respond to that input. From the brain and spinal cord, intricate networks of neurons go out to all parts of the body, providing detailed instructions to the target areas—for our purposes, muscles.

Autonomic Nervous System

The autonomic nervous system, often considered part of the motor division of the PNS, controls the body's involuntary internal functions. Some of these functions that are important to sport and activity include heart rate, blood pressure, blood distribution, and lung function.

The autonomic nervous system has two major divisions: the sympathetic nervous system and the parasympathetic nervous system. These originate from different sections of the spinal cord and from the base of the brain. The effects of the two systems are often antagonistic, but the systems always function together.

Sympathetic Nervous System

The sympathetic nervous system is sometimes called the fight-or-flight system: It prepares the body to face a crisis and sustains its function during that crisis. When excited, the sympathetic nervous system produces a massive discharge throughout the body, preparing it for action. A sudden loud noise, a life-threatening situation, or those last few seconds before the start of an athletic competition are examples of circumstances in which this massive sympathetic discharge occurs. The effects of sympathetic stimulation are important to the athlete:

- Heart rate and strength of cardiac contraction increase.
- Coronary vessels dilate, increasing the blood supply to the heart muscle to meet its increased demands.
- Peripheral vasodilation allows more blood to enter the active skeletal muscles.
- Vasoconstriction in most other tissues diverts blood away from them and to the active muscles.
- Blood pressure increases, allowing better perfusion of the muscles and improving the return of venous blood to the heart.
- Bronchodilation improves gas exchange.
- Metabolic rate increases, reflecting the body's effort to meet the increased demands of physical activity.
- Mental activity increases, allowing better perception of sensory stimuli and more concentration on performance.
- Glucose is released from the liver into the blood as an energy source.
- Functions not directly needed are slowed (e.g., renal function, digestion), conserving energy so that it can be used for action.

These basic alterations in bodily function facilitate motor responses, demonstrating the importance of the autonomic nervous system in preparing the body for and sustaining it during acute stress or physical activity.

Parasympathetic Nervous System

The parasympathetic nervous system is the body's housekeeping system. It has a major role in carrying out such processes as digestion, urination, glandular secretion, and conservation of energy. This system is more active when one is calm and at rest. Its effects tend to oppose those of the sympathetic system. The parasympathetic division causes decreased heart rate, constriction of coronary vessels, and bronchoconstriction.

The various effects of the sympathetic and parasympathetic divisions of the autonomic nervous system are summarized in table 3.1.

In review

- The PNS contains 43 pairs of nerves: 12 cranial and 31 spinal.
- The PNS can be subdivided into the sensory and motor divisions. The motor division also includes the autonomic nervous system.
- The sensory division carries information from sensory receptors to the CNS so that the CNS is constantly aware of the person's current status and environment.
- The motor division carries motor impulses from the CNS to the muscles, organs, and other tissues.
- The autonomic nervous system includes the sympathetic nervous system (the fight-or-flight system) and the parasympathetic system (the housekeeping system). Although these systems often oppose each other, they always function together to create an appropriately balanced response.

TABLE 3.1 Effects of the Sympathetic and Parasympathetic Nervous Systems on Various Organs

Target organ or system	Sympathetic effects	Parasympathetic effects
Heart muscle	Increases rate and force of contraction	Decreases rate of contraction
Heart: coronary blood vessels	Cause vasodilation	Cause vasoconstriction
Lungs	Cause bronchodilation; mildly constrict blood vessels	Cause bronchoconstriction
Blood vessels	Increase blood pressure; cause vasoconstriction in abdominal viscera and skin to divert blood when necessary; cause vasodilation in the skeletal muscles and heart during exercise	Little or no effect
Liver	Stimulates glucose release	No effect
Cellular metabolism	Increases metabolic rate	No effect
Adipose tissue	Stimulates lipolysis[a]	No effect
Sweat glands	Increase sweating	No effect
Adrenal glands	Stimulate secretion of epinephrine and norepinephrine	No effect
Digestive system	Decreases activity of glands and muscles; constricts sphincters	Increases peristalsis and glandular secretion; relaxes sphincters
Kidney	Causes vasoconstriction; decreases urine formation	No effect

[a]Lipolysis is the process of breaking down triglyceride to its basic units to be used for energy.

SENSORY-MOTOR INTEGRATION

Having discussed the components and divisions of the nervous system, we now discuss how a sensory stimulus gives rise to a motor response. How, for example, do the muscles in the hand know to pull one's finger away from a hot stove? When someone decides to run, how do the muscles in the legs coordinate while supporting weight and propelling the person forward? To accomplish these tasks, the sensory and motor systems must communicate with each other.

This process is called **sensory-motor integration,** and it is depicted in figure 3.7 on page 92. For the body to respond to sensory stimuli, the sensory and motor divisions of the nervous system must function together in the following sequence of events:

1. A sensory stimulus is received by sensory receptors (e.g., pinprick).
2. The sensory action potential is transmitted along sensory neurons to the CNS.
3. The CNS interprets the incoming sensory information and determines which response is most appropriate, or reflexively initiates a motor response.
4. The action potentials for the response are transmitted from the CNS along α-motor neurons.
5. The motor action potential is transmitted to a muscle, and the response occurs.

Sensory Input

Recall that sensations and physiological status are detected by sensory receptors throughout the body. The action potentials resulting from sensory stimulation are transmitted via the sensory nerves to the spinal cord. When they reach the spinal cord, they can trigger a local reflex at that level, or they can travel to the upper regions of the spinal cord or to the brain. Sensory pathways to the brain can terminate in sensory areas of the brain stem, the cerebellum, the thalamus, or the cerebral cortex. An area in which the sensory impulses terminate is referred to as an integration center. This is where the sensory input is interpreted and linked to the motor system. Figure 3.8 on page 93 illustrates various sensory receptors and their nerve pathways back to the spinal cord and up into various areas of the brain. The integration centers vary in function:

Neurons here, known as *pyramidal cells,* let us consciously control movement of our skeletal muscles. Think of the primary motor cortex as the part of the brain that decides what movement one wants to make. For example, in baseball, if a player is in the batter's box waiting for the next pitch, the decision to swing the bat is made in the primary motor cortex, where the entire body is carefully mapped out. The areas that require the finest motor control have a greater representation in the motor cortex; thus, more neural control is provided to them.

The cell bodies of the pyramidal cells are housed in the primary motor cortex, and their axons form the extrapyramidal tracts. These are also known as the corticospinal tracts because the nerve processes extend from the cerebral cortex down to the spinal cord. These tracts provide the major voluntary control of skeletal muscles.

In addition to the primary motor cortex, there is a premotor cortex just anterior to the precentral gyrus in the frontal lobe. Learned motor skills of a repetitious or patterned nature are stored here. This region can be thought of as the memory bank for skilled motor activities.[3]

Basal Ganglia

The basal ganglia (nuclei) are not part of the cerebral cortex. Rather, they are in the cerebral white matter, deep in the cortex. These ganglia are clusters of nerve cell bodies. The complex functions of the basal ganglia are not well understood, but the ganglia are known to be important in initiating movements of a sustained and repetitive nature (such as arm swinging during walking), and thus they control complex movements such as walking and running. These cells also are involved in maintaining posture and muscle tone.

Cerebellum

The cerebellum is crucial to the control of all rapid and complex muscular activities. It helps coordinate the timing of motor activities and the rapid progression from one movement to the next by monitoring and making corrective adjustments in the motor activities that are elicited by other parts of the brain. The cerebellum assists the functions of both the primary motor cortex and the basal ganglia. It facilitates movement patterns by smoothing out the movement, which would otherwise be jerky and uncontrolled.

The cerebellum acts as an integration system, comparing the programmed or intended activity with the actual changes occurring in the body and then initiating corrective adjustments through the motor system. It receives information from the cerebrum and other parts of the brain and also from sensory receptors (proprioceptors) in the muscles and joints that keep the cerebellum informed about the body's current position. The cerebellum also receives visual and equilibrium input. Thus, it notes all incoming information about the exact tension and position of all muscles, joints, and tendons and the body's current position relative to its surroundings; then it determines the best plan of action to produce the desired movement.

The primary motor cortex is the part of the brain that makes the decision to perform a movement. This decision is relayed to the cerebellum. The cerebellum notes the desired action and then compares the intended movement with the actual movement based on sensory feedback from the muscles and joints. If the action is different than planned, the cerebellum informs the higher centers of the discrepancy so corrective action can be initiated.

In review

- Sensory-motor integration is the process by which the PNS relays sensory input to the CNS and the CNS interprets this information and then sends out the appropriate motor signal to elicit the desired motor response.
- Sensory input can terminate at various levels of the CNS. Not all of this information reaches the brain.
- Reflexes are the simplest form of motor control. These are not conscious responses. For a given sensory stimulus, the motor response is always identical and instantaneous.
- Muscle spindles trigger reflexive muscle action when the muscle spindle is stretched.
- Golgi tendon organs trigger a reflex that inhibits contraction if the tendon fibers are stretched from high muscle tension.
- The primary motor cortex, located in the frontal lobe, is the center of conscious motor control.
- The basal ganglia, in the cerebral white matter, help initiate some movements (sustained and repetitive ones) and help control posture and muscle tone.
- The cerebellum is involved in all rapid and complex movement processes and assists the primary motor cortex and the basal ganglia in coordinating the response. It is an integration center that decides how to best execute the desired movement, given the body's current position and the muscles' current status.

MOTOR RESPONSE

Now that we have discussed how sensory input is integrated to determine the appropriate motor response, the last step in the process is how muscles respond to motor action potentials once they reach the muscle fibers.

Once an action potential reaches an α-motor neuron, it travels the length of the neuron to the neuromuscular junction. From there, the action potential spreads to all muscle fibers innervated by that particular α-motor neuron. Recall that the α-motor neuron and all muscle fibers it innervates form a single motor unit. Each muscle fiber is innervated by only one α-motor neuron, but each α-motor neuron innervates up to several thousand muscle fibers, depending on the function of the muscle. Muscles controlling fine movements, such as those controlling the eyes, have only a small number of muscle fibers per α-motor neuron. Muscles with more general functions have many fibers per α-motor neuron.

The muscles that control eye movements (the extraocular muscles) have an innervation ratio of 1:15, meaning that one α-motor neuron controls only 15 muscle fibers. In contrast, the gastrocnemius and tibialis anterior muscles of the lower leg have innervation ratios of almost 1:2,000.

The muscle fibers in a specific motor unit are homogeneous with respect to fiber type. Thus, one will not find a motor unit that has both type II and type I fibers. In fact, as mentioned in chapter 1, it is generally believed that the characteristics of the α-motor neuron actually determine the fiber type in that motor unit.[1, 5]

In Closing

We have seen how muscles respond to neural stimulation, whether through reflexes or under complex control of the higher brain centers. We discussed how the individual motor units respond and how they are recruited in an orderly manner depending on the required force. Thus, we have learned how the body functions to allow people to move. In the next chapter, we examine the energy needs of the body at rest and during exercise.

KEY TERMS

- acetylcholine
- action potential
- afferent division
- axon hillock
- axon terminal
- central nervous system (CNS)
- depolarization
- efferent division
- end branches
- excitatory postsynaptic potential (EPSP)
- Golgi tendon organ
- graded potential
- hyperpolarization
- inhibitory postsynaptic potential (IPSP)
- motor division
- motor reflex
- muscle spindle
- myelin sheath
- nerve impulse
- neuromuscular junction
- neuron
- neurotransmitter
- norepinephrine
- peripheral nervous system (PNS)
- resting membrane potential (RMP)
- saltatory conduction
- sensory division
- sensory-motor integration
- sodium-potassium pump
- summation
- synapse
- threshold

STUDY QUESTIONS

1. What are the major divisions of the nervous system? What are their major functions?
2. Name the different parts of a neuron.
3. Explain the resting membrane potential. What causes it? How is it maintained?
4. Describe an action potential. What is required before an action potential is activated?
5. Explain how an action potential is transmitted from a presynaptic neuron to a postsynaptic neuron. Describe a synapse and a neuromuscular junction.
6. What brain centers have major roles in controlling movement, and what are these roles?
7. How do the sympathetic and parasympathetic systems differ? What is their significance in performing physical activity?
8. Explain how reflex movement occurs in response to touching a hot object.
9. Describe the role of the muscle spindle in controlling muscle contraction.
10. Describe the role of the Golgi tendon organ in controlling muscle contraction.

STUDY GUIDE ACTIVITIES

In addition to the activities listed in the chapter opening outline on page 79, two other activities are available in the online study guide, located at

www.HumanKinetics.com/PhysiologyOfSportAndExercise.

The chapter **SUMMARY** reviews the key concepts covered in the chapter, and the end-of-chapter **QUIZ** tests your understanding of the material covered in the chapter.

CHAPTER 4

Energy Expenditure and Fatigue

In this chapter and in the online study guide

They call it the greatest football game ever played. On January 2, 1982, the Miami Dolphins and the San Diego Chargers battled in the hot, sticky night for over 4 h. Players were carted off the field time and time again, only to return to the field each time. Hall of Fame tight end Kellen Winslow overcame severe fatigue and excruciating back spasms and became one of many heroes of this epic test of wills. As Rick Reilly's *Sports Illustrated* (October 25, 1999) story noted, "No player on either team would ever take himself that far or that high again." One player commented, "You hear coaches say, Leave everything on the field. Well that actually happened that day. Both teams." Another player quipped, "Guys would refuse to come out of the game . . . just so they didn't have to run all the way to the sideline!" Perhaps no account so vividly highlights the concepts of energy and fatigue, the topics discussed in this chapter.

Energy expenditure during prolonged exercise diminishes performance and can lead to fatigue.

One cannot understand exercise physiology without understanding some key concepts about energy expenditure at rest and during exercise. In chapter 2, we discussed the formation of adenosine triphosphate (ATP), the major form of chemical energy stored within our cells. We produce ATP from substrates by processes that are known collectively as metabolism. In the first half of this chapter we discuss various techniques for measuring the body's rate of metabolism or energy expenditure; then we describe how energy expenditure varies from basal or total resting conditions up to maximal exercise intensities. If exercise is sustained for some time, eventually muscular contraction cannot be sustained and performance will diminish. This inability to maintain muscle contraction is broadly called "fatigue." Fatigue is a complex, multidimensional phenomenon that may or may not result from an inability to maintain metabolism and expend energy. Because fatigue sometimes has a metabolic component, it is discussed in the present chapter along with energy expenditure.

MEASURING ENERGY EXPENDITURE

Energy utilization by contracting muscle fibers during exercise cannot be directly measured. But numerous indirect laboratory methods can be used to calculate whole-body energy expenditure at rest and during exercise. Several of these methods have been in use since the early 1900s. Others are new and have only recently been used in exercise physiology.

Direct Calorimetry

Only about 40% of the energy liberated during the metabolism of glucose and fats is used to produce ATP. The remaining 60% is converted to heat, so one way to gauge the rate and quantity of energy production is to measure the body's heat production. This technique is called **direct calorimetry,** since the basic unit of heat is the **calorie (cal).**

This approach was first described by Zuntz and Hagemann in the late 1800s.[10] They developed the **calorimeter** (illustrated in figure 4.1), which is an insulated, airtight chamber. The walls of the chamber contain copper tubing through which water is passed. In the chamber, the heat produced by the body radiates to the walls and warms the water. The water temperature change is recorded, as are temperature changes of the air entering and leaving the chamber. These changes are caused by the heat the body generates. One's metabolism can be calculated from the resulting values.

Calorimeters are expensive to construct and to use and are slow to generate results. Their only real advantage is that they measure heat directly, but they have several disadvantages for exercise physiology. Although a calorimeter can provide an accurate measure of total-body energy expenditure, it cannot follow rapid changes in energy expenditure. Therefore, while direct calorimetry is useful for measuring resting metabolism, energy metabolism during most exercise situations cannot be adequately studied with a direct calorimeter. Second, exercise equipment such as a motor-driven treadmill also gives off heat of its own. Third, not all heat is liberated from the body; some is stored and causes body temperature to rise. And finally, sweating affects the measurements and the constants used in the calculations of heat produced. Consequently, this method is seldom used today because it is easier and less expensive to measure energy expenditure by assessing the exchange of oxygen and carbon dioxide that occurs during oxidative phosphorylation.

Indirect Calorimetry

As discussed in chapter 2, oxidative metabolism of glucose and fat—the main substrates for exercise—depends on O_2 availability and produces CO_2 and water. The rate of O_2 and CO_2 exchanged in the lungs normally equals the rate of usage and release by the body tissues. With this knowledge, we can measure caloric expenditure by measuring respiratory gases. This method of estimating total-body energy expenditure is called **indirect calorimetry** because heat production is not measured directly. Rather, energy expenditure is calculated from the respiratory exchange of O_2 and CO_2.

In order for oxygen consumption to reflect energy metabolism accurately, energy production must be almost completely oxidative. If a large portion of energy is being produced anaerobically, respiratory gas measurements will not reflect all metabolic processes. Therefore, this technique is limited to steady-state activities lasting about 60 s or longer, which fortunately takes into account most daily activities including exercise.

Respiratory gas exchange is determined through measurement of the volume of O_2 and CO_2 that enters and leaves the lungs during a given period of time. Because O_2 is removed from the inspired air in the alveoli and CO_2 is added to the alveolar air, the expired O_2 concentration is less than the inspired, whereas the CO_2 concentration is higher in expired air than in inspired air. Consequently, the difference between inspired and expired air tells us how much O_2 is being taken up and how much CO_2 is being produced. Because the body has only limited O_2 storage, the amount taken up at the lungs accurately reflects the body's use of O_2. Although a number of sophisticated and expensive methods are available for measuring the respiratory exchange of O_2

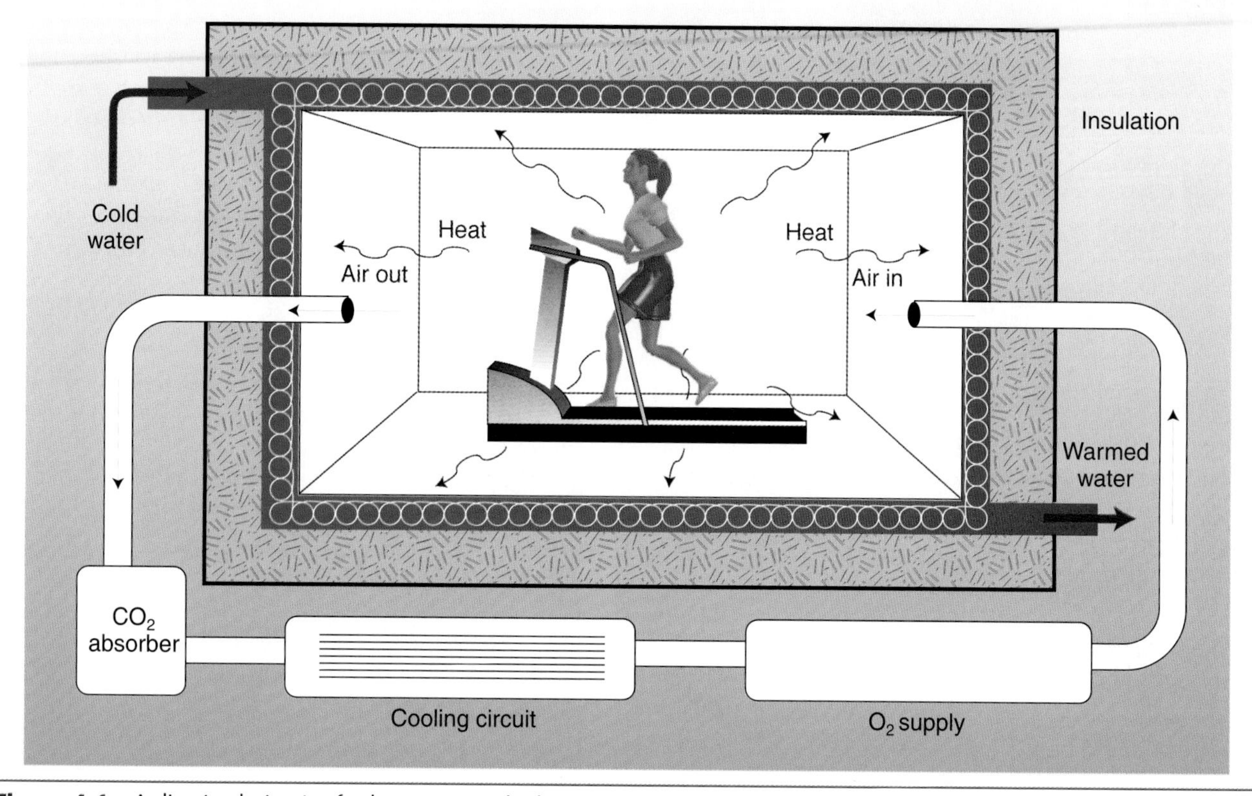

Figure 4.1 A direct calorimeter for human use. The heat generated by the subject's body is transferred to the air and walls of the chamber (through conduction, convection, and evaporation). This heat produced by the subject is then measured via recording of the temperature change in the air and water flowing through and around the chamber, respectively. This heat change is a measure of the subject's metabolic rate.

and CO_2, the simplest and oldest methods (i.e., Douglas bag and chemical gas analysis) are probably the most accurate, but they are relatively slow and permit only a few measurements during each session. Modern electronic computer systems for respiratory gas exchange measurements offer a large time savings and multiple measurements.

Notice in figure 4.2 that the gas expired by the subject passes through a hose into a mixing chamber, where samples are pumped to electronic oxygen and carbon dioxide analyzers. In this setup, a computer uses the measurements of expired gas (air) volume and the fraction (percentage) of oxygen and carbon dioxide in a sample of that expired air to calculate O_2 uptake and CO_2 production. Sophisticated equipment can do these calculations breath by breath, but calculations are more typically done over discrete time periods lasting one to several minutes.

Calculating Oxygen Consumption and Carbon Dioxide Production

Using equipment like that illustrated in figure 4.2, exercise physiologists can measure the three variables needed to calculate the actual volume of oxygen consumed ($\dot{V}O_2$) and volume of CO_2 produced ($\dot{V}CO_2$). Generally, values are presented as oxygen consumed per minute ($\dot{V}O_2$) and CO_2 produced per minute ($\dot{V}CO_2$). The dot over the V ($\dot{V}$) is used to indicate a rate of O_2 consumption or CO_2 production, for example, liters per minute.

In its simplest form, $\dot{V}O_2$ is equal to the volume of O_2 inspired minus the volume of O_2 expired. To calculate the volume of O_2 inspired, we multiply the volume of air inspired by the fraction of that air that is composed of O_2; the volume of O_2 expired is equal to the volume of air expired multiplied by the fraction of the expired air that is composed of O_2. The same holds true for CO_2.

Thus, calculation of $\dot{V}O_2$ and $\dot{V}CO_2$ requires the following information:

- Volume of air inspired ($\dot{V}_I$)
- Volume of air expired ($\dot{V}_E$)
- Fraction of oxygen in the inspired air (F_IO_2)
- Fraction of CO_2 in the inspired air (F_ICO_2)
- Fraction of oxygen in the expired air (F_EO_2)
- Fraction of CO_2 in the expired air (F_ECO_2)

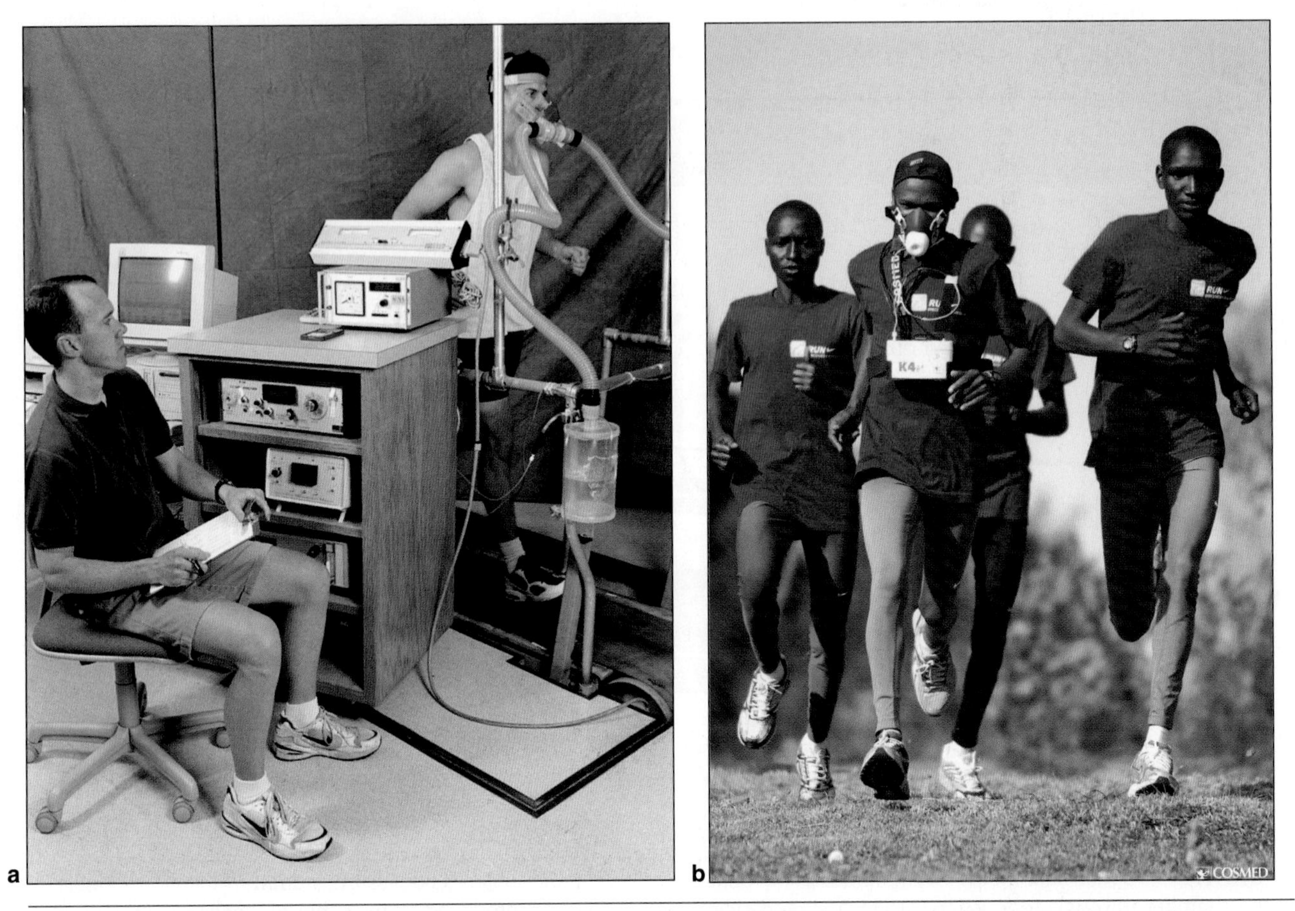

Figure 4.2 *(a)* Typical equipment that is routinely used by exercise physiologists to measure O_2 consumption and CO_2 production. These values can be used to calculate $\dot{V}O_2$ and the respiratory exchange ratio, and therefore energy expenditure. Although this equipment is cumbersome and limits movement, smaller versions have recently been adapted for use under a variety of conditions in the laboratory, on the playing field, and elsewhere. *(b)* Illustration of the Cosmed K4 portable metabolic gas analyzer being used to monitor the O_2 consumption of a subject performing aerobic exercise.

The oxygen consumption, in liters of oxygen consumed per minute, can then be calculated as follows:

$$\dot{V}O_2 = (\dot{V}_I \times F_IO_2) - (\dot{V}_E \times F_EO_2).$$

The CO_2 production is similarly calculated as

$$\dot{V}CO_2 = (\dot{V}_E \times F_ECO_2) - (\dot{V}_I \times F_ICO_2).$$

Haldane Transformation

Over the years, scientists have attempted to simplify the actual calculation of oxygen consumption and CO_2 production. Several of the measurements needed in the preceding equations are known and do not change. The gas concentrations of the three gases that make up inspired air are known: oxygen accounts for 20.93%, CO_2 accounts for 0.04%, and nitrogen accounts for 79.03% of the inspired air. What about the volume of inspired and expired air? Aren't they the same, such that we would need to measure only one of the two?

Inspired air volume equals expired air volume only when the volume of oxygen consumed equals the volume of CO_2 produced. When the volume of oxygen consumed is greater than the volume of CO_2 produced, $\dot{V}_I$ is greater than $\dot{V}_E$. Likewise, $\dot{V}_E$ is greater than $\dot{V}_I$ when the volume of CO_2 produced is greater than the volume of oxygen consumed. However, the one thing that is constant is that the volume of nitrogen inspired ($\dot{V}_IN_2$) is equal to the volume of nitrogen expired ($\dot{V}_EN_2$). Because $\dot{V}_IN_2 = \dot{V}_I \times F_IN_2$ and $\dot{V}_EN_2 = \dot{V}_E \times F_EN_2$, we can calculate $\dot{V}_I$ from $\dot{V}_E$ by using the following equation, which has been referred to as the **Haldane transformation:**

$$(1)\ \dot{V}_I \times F_IN_2 = \dot{V}_E \times F_EN_2,$$

which can be rewritten as

$$(2)\ \dot{V}_I = (\dot{V}_E \times F_EN_2) / F_IN_2.$$

Furthermore, because we are actually measuring the concentrations of O_2 and CO_2 in the expired gases, we can calculate F_EN_2 from the sum of F_EO_2 and F_ECO_2, or

$$(3)\ F_EN_2 = 1 - (F_EO_2 + F_ECO_2).$$

So, in pulling all of this information together, we can rewrite the equation for calculating $\dot{V}O_2$ as follows:

$$\dot{V}O_2 = (\dot{V}_I \times F_IO_2) - (\dot{V}_E \times F_EO_2).$$

By substituting equation 2, we get the following:

$$\dot{V}O_2 = [(\dot{V}_E \times F_EN_2) / (F_IN_2 \times F_IO_2)] - [(\dot{V}_E) \times (F_EO_2)].$$

By substituting known values for (F_IO_2) of 0.2093 and for F_IN_2 of 0.7903, we get the following:

$$\dot{V}O_2 = [(\dot{V}_E \times F_EN_2) / 0.7903) \times 0.2093] - [(\dot{V}_E) \times (F_EO_2)].$$

By substituting equation 3, we get the following:

$$\dot{V}O_2 = [(\dot{V}_E) \times (1 - (F_EO_2 + F_ECO_2)) \times (0.2093 / 0.7903)] - [(\dot{V}_E) \times (F_EO_2)].$$

or, simplified,

$$\dot{V}O_2 = (\dot{V}_E) \times [(1 - (F_EO_2 + F_ECO_2)) \times (0.265)] - (\dot{V}_E) \times (F_EO_2).$$

or, further simplified,

$$\dot{V}O_2 = (\dot{V}_E) \times \{[(1 - (F_EO_2 + F_ECO_2)) \times (0.265)] - (F_EO_2)\}.$$

This final equation is the one actually used in practice by exercise physiologists, although computers now do the calculating automatically in most laboratories. One final correction is necessary. When air is expired, it is at body temperature (BT), is at the prevailing atmospheric or ambient pressure (P), and is saturated (S) with water vapor, or what are referred to as BTPS conditions. Each of these influences would not only add error to the measurement of $\dot{V}O_2$ and $\dot{V}CO_2$, but would also make it difficult to compare measurements made in laboratories at different altitudes, for example. For that reason, every gas volume is routinely converted to its standard temperature (ST: 0 °C) and pressure (P: 760 mmHg), dry equivalent (D), or STPD. This is accomplished by a series of correction equations.

Respiratory Exchange Ratio

To estimate the amount of energy used by the body, it is necessary to know the type of food substrate (combination of carbohydrate, fat, protein) being oxidized. The carbon and oxygen contents of glucose, free fatty acids (FFAs), and amino acids differ dramatically. As a result, the amount of oxygen used during metabolism depends on the type of fuel being oxidized. Indirect calorimetry measures the amount of CO_2 released ($\dot{V}CO_2$) and oxygen consumed ($\dot{V}O_2$). The ratio between these two values is termed the **respiratory exchange ratio (RER).**

$$RER = \dot{V}CO_2 / \dot{V}O_2.$$

In general, the amount of oxygen needed to completely oxidize a molecule of carbohydrate or fat is proportional to the amount of carbon in that fuel. For example, glucose ($C_6H_{12}O_6$) contains six carbon atoms. During glucose combustion, six molecules of oxygen are used to produce six CO_2 molecules, six H_2O molecules, and 38 ATP molecules:

$$6\ O_2 + C_6H_{12}O_6 \rightarrow 6\ CO_2 + 6\ H_2O + 38\ ATP.$$

By evaluating how much CO_2 is released compared with the amount of O_2 consumed, we find that the RER is 1.0:

$$RER = \dot{V}CO_2 / \dot{V}O_2 = 6\ CO_2 / 6\ O_2 = 1.0.$$

As shown in table 4.1, the RER value varies with the type of fuels being used for energy. Free fatty acids have considerably more carbon and hydrogen but less oxygen than glucose. Consider palmitic acid, $C_{16}H_{32}O_2$. To completely oxidize this molecule to CO_2 and H_2O requires 23 molecules of oxygen:

$$16\ C + 16\ O_2 \rightarrow 16\ CO_2$$
$$32\ H + 8\ O_2 \rightarrow 16\ H_2O$$

Total = 24 O_2 needed
−1 O_2 provided by the palmitic acid
23 O_2 must be added

TABLE 4.1 Caloric Equivalence of the Respiratory Exchange Ratio (RER) and % kcal From Carbohydrates and Fats

	Energy	% kcal	
RER	kcal/L O_2	Carbohydrates	Fats
0.71	4.69	0	100
0.75	4.74	16	84
0.80	4.80	33	67
0.85	4.86	51	49
0.90	4.92	68	32
0.95	4.99	84	16
1.00	5.05	100	0

Ultimately, this oxidation results in 16 molecules of CO_2, 16 molecules of H_2O, and 129 molecules of ATP:

$$C_{16}H_{32}O_2 + 23\ O_2 \rightarrow 16\ CO_2 + 16\ H_2O + 129\ ATP.$$

Combustion of this fat molecule requires significantly more oxygen than combustion of a carbohydrate molecule. During carbohydrate oxidation, approximately 6.3 molecules of ATP are produced for each molecule of O_2 used (38 ATP per 6 O_2), compared with 5.6 molecules of ATP per molecule of O_2 during palmitic acid metabolism (129 ATP per 23 O_2).

Although fat provides more energy than carbohydrate, more oxygen is needed to oxidize fat than carbohydrate. This means that the RER value for fat is substantially lower than for carbohydrate. For palmitic acid, the RER value is 0.70:

$$RER = \dot{V}CO_2 / \dot{V}O_2 = 16 / 23 = 0.70.$$

Once the RER value is determined from the calculated respiratory gas volumes, the value can be compared with a table (table 4.1) to determine the food mixture being oxidized. If, for example, the RER value is 1.0, the cells are using only glucose or glycogen, and each liter of oxygen consumed would generate 5.05 kcal. The oxidation of only fat would yield 4.69 kcal/L of O_2, and the oxidation of protein would yield 4.46 kcal/L of O_2 consumed. Thus, if the muscles were using only glucose and the body were consuming 2 L of O_2/min, then the rate of heat energy production would be 10.1 kcal/min (2 L/min · 5.05 kcal/L).

Limitations of Indirect Calorimetry

While indirect calorimetry is a common and important tool of exercise physiologists, it has limitations. Calculations of gas exchange assume that the body's O_2 content remains constant and that CO_2 exchange in the lung is proportional to its release from the cells. Arterial blood remains almost completely oxygen saturated (about 98%), even during intense effort. We can accurately assume that the oxygen being removed from the air we breathe is in proportion to its cellular uptake. Carbon dioxide exchange, however, is less constant. Body CO_2 pools are quite large and can be altered simply by deep breathing or by performance of highly intense exercise. Under these conditions, the amount of CO_2 released in the lung may not represent that being produced in the tissues, so calculations of carbohydrate and fat used based on gas measurements appear to be valid only at rest or during steady-state exercise.

Use of the RER also can lead to inaccuracies. Recall that protein is not completely oxidized in the body because nitrogen is not oxidizable. This makes it impossible to calculate the body's protein use from the RER. As a result, the RER is sometimes referred to as nonprotein RER because it simply ignores protein oxidation.

Traditionally, protein was thought to contribute little to the energy used during exercise, so exercise physiologists felt justified in using the nonprotein RER when making calculations. But more recent evidence suggests that in exercise lasting for several hours, protein may contribute up to 5% of the total energy expended under certain circumstances.

The body normally uses a combination of fuels. Respiratory exchange ratio values vary depending on the specific mixture being oxidized. At rest, the RER value is typically in the range of 0.78 to 0.80. During exercise, though, muscles rely increasingly on carbohydrate for energy, resulting in a higher RER. As exercise intensity increases, the muscles' carbohydrate demand also increases. As more carbohydrate is used, the RER value approaches 1.0.

This increase in the RER value to 1.0 reflects the demands on blood glucose and muscle glycogen, but it also may indicate that more CO_2 is being unloaded from the blood than is being produced by the muscles. At or near exhaustion, lactate accumulates in the blood. The body tries to reverse this acidification by releasing more CO_2. Lactate accumulation increases CO_2 production because excess acid causes carbonic acid in the blood to be converted to CO_2. As a consequence, the excess CO_2 diffuses out of the blood and into the lungs for exhalation, increasing the amount of CO_2 released. For this reason, RER values approaching 1.0 may not accurately estimate the type of fuel being used by the muscles.

Another complication is that glucose production from the catabolism of amino acids and fats in the liver produces an RER below 0.70. Thus, calculations of carbohydrate oxidation from the RER value will be underestimated if energy is derived from this process.

Despite its shortcomings, indirect calorimetry still provides the best estimate of energy expenditure at rest and during aerobic exercise.

Isotopic Measurements of Energy Metabolism

In the past, determining an individual's total daily energy expenditure depended on recording food intake over several days and measuring body composition changes during that period. This method, although widely used, is limited by the individual's ability to keep accurate records and by the ability to match the individual's activities to accurate energy costs.

Fortunately, the use of isotopes has expanded our ability to investigate energy metabolism. Isotopes are elements with an atypical atomic weight. They can be either radioactive (radioisotopes) or nonradioactive (stable isotopes). As an example, carbon-12 (^{12}C) has a molecular weight of 12, is the most common natural form of carbon, and is nonradioactive. In contrast, carbon-14 (^{14}C) has two more neutrons than ^{12}C, giving it an atomic weight of 14. ^{14}C is created in the laboratory and is radioactive.

Carbon-13 (^{13}C) constitutes about 1% of the carbon in nature and is used frequently for studying energy metabolism. Because ^{13}C is nonradioactive, it is less easily traced within the body than ^{14}C. But although radioactive isotopes are easily detected in the body, they pose a hazard to body tissues and thus are used infrequently in human research.

^{13}C and other isotopes such as hydrogen 2 (deuterium, or ^2H) are used as tracers, meaning that they can be selectively followed in the body. Tracer techniques involve infusing isotopes into an individual and then following their distribution and movement.

Although the method was first described in the 1940s, studies that used doubly labeled water for monitoring energy expenditure during normal daily living in humans were not conducted until the 1980s. The subject ingests a known amount of water labeled with two isotopes ($^2H_2{}^{18}O$), hence the term doubly labeled water. The deuterium (^2H) diffuses throughout the body's water, and the oxygen-18 (^{18}O) diffuses throughout both the water and the bicarbonate stores (where much of the CO_2 derived from metabolism is stored). The rate at which the two isotopes leave the body can be determined by analysis of their presence in a series of urine, saliva, or blood samples. These turnover rates then can be used to calculate how much CO_2 is produced, and that value can be converted to energy expenditure through the use of calorimetric equations.

Because isotope turnover is relatively slow, energy metabolism must be measured for several weeks. Thus, this method is not well suited for measurements of acute exercise metabolism. However, its accuracy (more than 98%) and low risk make it well suited for determining day-to-day energy expenditure. Nutritionists have hailed the doubly labeled water method as the most significant technical advance of the past century in the field of energy metabolism.

In review

- Direct calorimetry involves using a large chamber to directly measure heat produced by the body; it is not a useful tool for exercise measurements.
- Indirect calorimetry involves measuring O_2 consumption and CO_2 production. By calculating the RER value (the ratio of these two gas volumes) and comparing the RER value with standard values to determine the metabolic substrates being oxidized, we can calculate the energy expended per liter of oxygen consumed in kilocalories.
- The RER value at rest is usually 0.78 to 0.80. The RER value for the oxidation of fat is 0.70 and is 1.00 for carbohydrates.
- Isotopes can be used to determine metabolic rate over long periods of time. They are injected or ingested into the body. The rates at which they are cleared can be used to calculate CO_2 production and then caloric expenditure.

ENERGY EXPENDITURE AT REST AND DURING EXERCISE

With the techniques described in the previous section, exercise scientists can measure the amount of energy a person expends at rest as well as during and following exercise. This section deals with the body's rates of energy expenditure, or metabolic rate, under resting conditions, during submaximal and maximal exercise intensities, and during the period of recovery following an exercise bout.

Basal and Resting Metabolic Rates

The rate at which the body uses energy is termed the metabolic rate. Estimates of energy expenditure at rest and during exercise are often based on measurement of whole-body oxygen consumption ($\dot{V}O_2$) and its caloric equivalent. At rest, an average person consumes about 0.3 L of O_2/min. This equals 18 L of O_2/h or 432 L of O_2/day.

Knowing an individual's $\dot{V}O_2$ allows us to calculate that person's caloric expenditure. Recall that at rest, the body usually burns a mixture of carbohydrate and fat. An RER value of 0.80 is fairly common for most resting individuals on a mixed diet. The caloric equivalence of an RER value of 0.80 is 4.80 kcal per liter of O_2 consumed (from table 4.1). Using these common values, we can calculate this individual's caloric expenditure as follows:

$$\begin{aligned}\text{kcal/day} &= \text{liters of } O_2 \text{ consumed per day} \times \text{kcal used per liter of } O_2 \\ &= 432 \text{ L } O_2\text{/day} \times 4.80 \text{ kcal/L } O_2 \\ &= 2{,}074 \text{ kcal/day.}\end{aligned}$$

This value closely agrees with the average resting energy expenditure expected for a 70 kg (154 lb) man. Of course, it does not include the extra energy needed for normal daily activity.

One standardized measure of energy expenditure at rest is the **basal metabolic rate (BMR).** The BMR is the rate of energy expenditure for an individual at rest in a supine position, measured immediately after at least 8 h of sleep and at least 12 h of fasting. This value reflects the minimum amount of energy required to carry on essential physiological functions.

Because muscle has high metabolic activity, the BMR is directly related to an individual's fat-free mass and is generally reported in kilocalories per kilogram of fat-free mass per minute ($kcal \cdot kg\ FFM^{-1} \cdot min^{-1}$). The more fat-free mass, the more total calories expended in a day. Because women tend to have a lower fat-free mass and a greater fat mass than men, women tend to have lower BMRs than men of similar weight.

Body surface area also affects BMR. The higher the surface area, the more heat loss occurs from the skin, which raises the BMR because more energy is needed to maintain body temperature. For this reason, the BMR is also often reported in kilocalories per square meter of body surface area per hour ($kcal \cdot m^{-2} \cdot h^{-1}$). Because we are discussing daily energy expenditure, we've opted for a simpler unit: kcal/day.

Many other factors affect BMR, including these:

- Age: BMR gradually decreases with increasing age, generally because of a decrease in fat-free mass.
- Body temperature: BMR increases with increasing temperature.
- Psychological stress: Stress increases activity of the sympathetic nervous system, which increases the BMR.
- Hormones: For example, thyroxine from the thyroid gland and epinephrine from the adrenal medulla both increase the BMR.

Instead of BMR, most researchers now use the term **resting metabolic rate (RMR),** because most measurements follow the same conditions required for measuring BMR but do not require the individual to sleep over in a hospital or research laboratory. Basal metabolic rate and RMR values are essentially identical, and they range from 1,200 to 2,400 kcal/day. But the average total metabolic rate of an individual engaged in normal daily activity ranges from 1,800 to 3,000 kcal.

The energy expenditure for very large athletes engaged in intense daily training can exceed 10,000 kcal/day!

Metabolic Rate During Submaximal Exercise

Exercise increases the energy requirement well in excess of RMR. Metabolism increases in direct proportion to the increase in exercise intensity, as shown in figure 4.3*a*. The subject exercised on a cycle ergometer for 5 min at 50 watts (W); oxygen consumption ($\dot{V}O_2$) increased from its resting value to a steady-state value within 1 to 2 min. The same subject then cycled on the same or the next day for 5 min at 100 W, and again a steady-state $\dot{V}O_2$ was reached in 1 or 2 min. In a similar manner, the subject cycled for 5 min at 150 W, 200 W, 250 W, and 300 W, respectively, and steady-state values were achieved at each power output. The steady-state $\dot{V}O_2$ values were plotted against their respective power outputs (right portion of figure 4.3*a*), showing clearly that there is a linear increase in the $\dot{V}O_2$ with increases in power output. The steady-state $\dot{V}O_2$ value represents the energy cost for that specific power output.

From more recent studies, it is clear that the $\dot{V}O_2$ response at higher rates of work does not follow the steady-state response pattern shown in figure 4.3*a* but rather follows that illustrated in figure 4.3*b*. It appears that at power outputs above the lactate threshold (the lactate response is indicated by the dashed line in the right half of figure 4.3, *a* & *b*), the oxygen consumption continues to increase beyond the typical 1 to 2 min needed to reach a steady-state value. This increase has been called the slow component of oxygen uptake kinetics.[4] The most likely mechanism for this slow component is an alteration in muscle fiber recruitment patterns, with the recruitment of more type II muscle fibers, which are less efficient (i.e., they require a higher $\dot{V}O_2$ to achieve the same power output).[2, 4]

A similar, but unrelated, phenomenon is referred to as $\dot{V}O_2$ drift. **$\dot{V}O_2$ drift** is defined as a slow increase in $\dot{V}O_2$ during prolonged, submaximal, constant power output exercise. Unlike the slow component, $\dot{V}O_2$ drift is observed at power outputs well below lactate threshold, and the magnitude of the increase in $\dot{V}O_2$ drift is much less. Although not understood completely, $\dot{V}O_2$ drift is likely attributable to an increase in ventilation and effects of increased circulating catecholamines.

Maximal Capacity for Aerobic Exercise

In figure 4.3*a*, it is clear that when the subject cycled at 300 W, the $\dot{V}O_2$ response was not different from that achieved at 250 W. This indicates that the subject had reached a maximal limit of his ability to increase his $\dot{V}O_2$. This peak value is referred to as aerobic capacity,

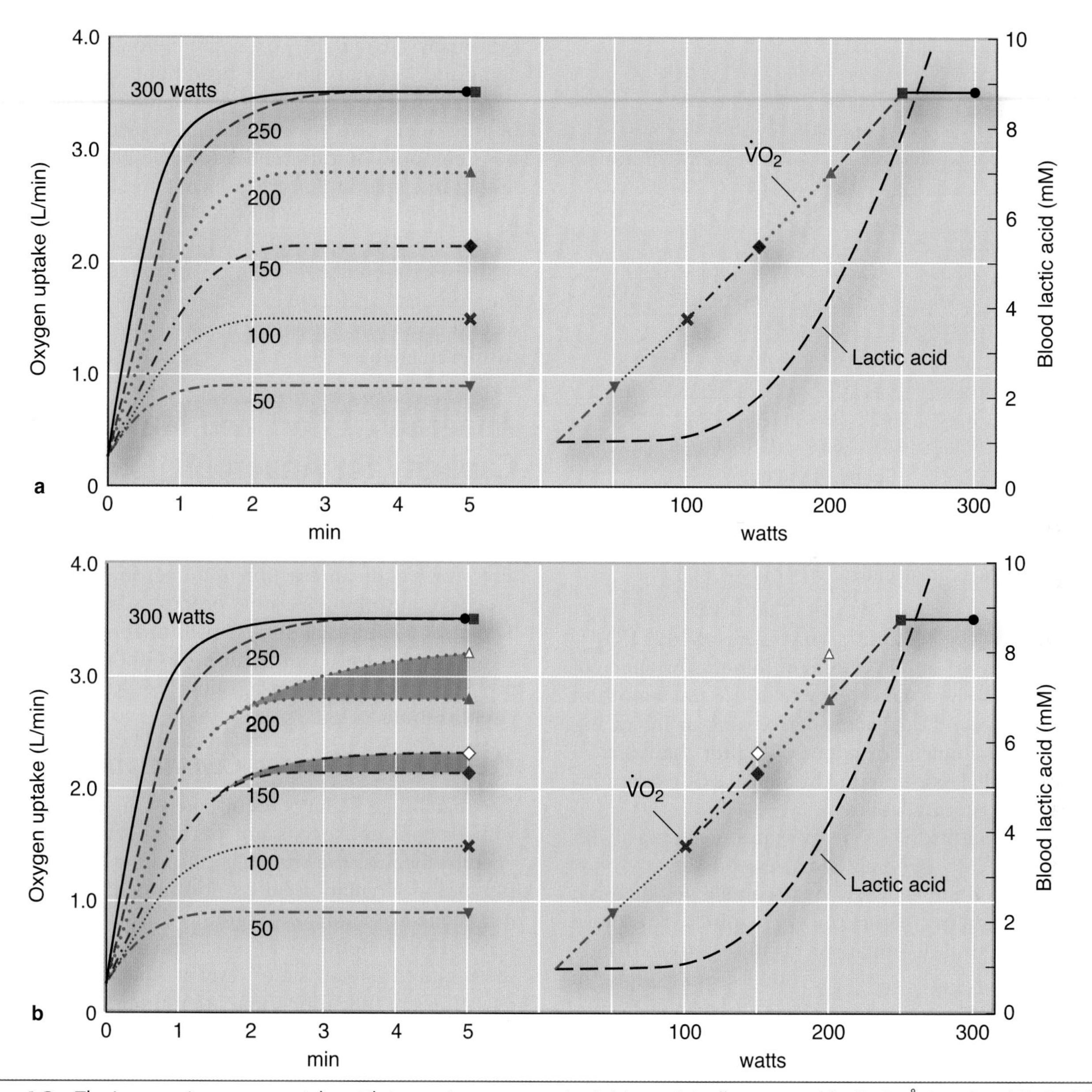

Figure 4.3 The increase in oxygen uptake with increasing power output *(a)* as originally proposed by P.-O. Åstrand and K. Rodahl (1986), *Textbook of work physiology: Physiological bases of exercise,* 3rd ed. (New York: McGraw-Hill), p. 300; and *(b)* as redrawn by Gaesser and Poole (1996, p. 36). See the text for a detailed explanation of the significance of this figure.

Reprinted, by permission, from G.A. Gaesser and D.C. Poole, 1996, "The slow component of oxygen uptake kinetics in humans," *Exercise and Sport Sciences Reviews* 24: 36.

maximal oxygen uptake, or $\dot{V}O_{2max}$. $\dot{V}O_{2max}$ is regarded by most as the best single measurement of cardiorespiratory endurance and aerobic fitness. This concept is further illustrated in figure 4.4, which compares the $\dot{V}O_{2max}$ of a trained and an untrained man.

Although some sport scientists have suggested that $\dot{V}O_{2max}$ is a good predictor of success in endurance events, the winner of a marathon race cannot be predicted from the runner's laboratory-measured $\dot{V}O_{2max}$. Likewise, an endurance running performance test is only a modest predictor of one's $\dot{V}O_{2max}$. This suggests that a good performance requires more than a high $\dot{V}O_{2max}$, a concept discussed in chapters 10 and 13.

Also, research has documented that $\dot{V}O_{2max}$ increases with physical training for only 8 to 12 weeks and that this value then plateaus despite continued higher-intensity training. Although $\dot{V}O_{2max}$ does not continue to increase, the participants continue to improve their endurance performance. It appears that these individuals develop the ability to perform at a higher percentage of their $\dot{V}O_{2max}$. Most runners, for example, can complete a 42 km (26.1 mi) marathon at an average pace that requires them to use approximately 75% to 80% of their $\dot{V}O_{2max}$.

Consider the case of Alberto Salazar, a near world-record holder in the marathon. His measured $\dot{V}O_{2max}$

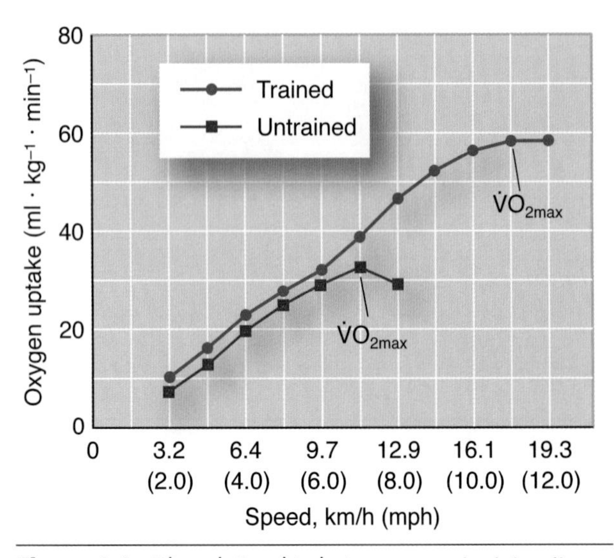

Figure 4.4 The relationship between exercise intensity (running speed) and oxygen uptake, illustrating $\dot{V}O_{2max}$ in a trained and an untrained man.

was 70 $ml \cdot kg^{-1} \cdot min^{-1}$. That is below the $\dot{V}O_{2max}$ expected based on his best marathon performance of 2 h 8 min. He was, however, able to run the marathon at 86% of his $\dot{V}O_{2max}$ when performing at his racing pace, a percentage considerably higher than that of other world-class runners. This may partly explain his world-class running ability.

Because individuals' energy requirements vary with body size, $\dot{V}O_{2max}$ generally is expressed relative to body weight, in milliliters of oxygen consumed per kilogram of body weight per minute ($ml \cdot kg^{-1} \cdot min^{-1}$). This allows a more accurate comparison of different-sized individuals who exercise in weight-bearing events, such as running. In non-weight-bearing activities, such as swimming and cycling, endurance performance is more closely related to $\dot{V}O_{2max}$ measured in liters per minute.

> **In focus**
>
> Aerobic capacities of 80 to 84 $ml \cdot kg^{-1} \cdot min^{-1}$ have been observed among elite male long-distance runners and cross-country skiers. The highest $\dot{V}O_{2max}$ value recorded for a man is that of a champion Norwegian cross-country skier who had a $\dot{V}O_{2max}$ of 94 $ml \cdot kg^{-1} \cdot min^{-1}$. The highest value recorded for a woman is 77 $ml \cdot kg^{-1} \cdot min^{-1}$ for a Russian cross-country skier. In contrast, poorly conditioned adults may have values below 20 $ml \cdot kg^{-1} \cdot min^{-1}$.

Normally active, but untrained, 18- to 22-year-old college students have average $\dot{V}O_{2max}$ values of 38 to 42 $ml \cdot kg^{-1} \cdot min^{-1}$ for women and 44 to 50 $ml \cdot kg^{-1} \cdot min^{-1}$ for men. After the age of 25 to 30 years, inactive people's $\dot{V}O_{2max}$ values decrease about 1% per year. This is probably attributable to a combination of biological aging and sedentary lifestyle. In addition, adult women generally have $\dot{V}O_{2max}$ values considerably below those of adult men. Two reasons for this sex difference are body composition differences (women generally have less fat-free mass and more fat mass) and blood hemoglobin content (women have less, thus they have less oxygen-carrying capacity). But it is unclear how much of the sex difference in $\dot{V}O_{2max}$ is attributable to actual physiological differences and how much might be caused by a more sedentary lifestyle. This is discussed further in chapter 18.

Anaerobic Effort and Maximal Capacity for Anaerobic Exercise

As noted earlier, the methods we have discussed ignore the anaerobic processes that accompany aerobic exercise. How can the interaction of the aerobic (oxidative) processes and the anaerobic processes be evaluated? The most common methods for estimating anaerobic effort involve the examination of either the excess postexercise oxygen consumption (EPOC) or the lactate threshold.

Postexercise Oxygen Consumption

The matching of energy requirements during exercise with oxygen delivery is not perfect. When aerobic exercise begins, the oxygen transport system (respiration and circulation) does not immediately supply the needed quantity of oxygen to the active muscles. Oxygen consumption requires several minutes to reach the required (steady state) level at which the aerobic processes are fully functional, even though the body's oxygen requirements increase the moment exercise begins.

Because oxygen needs and oxygen supply differ during the transition from rest to exercise, the body incurs an oxygen deficit, as shown in figure 4.5. This deficit accrues even at low exercise intensities. The oxygen deficit is calculated simply as the difference between the oxygen required for a given exercise intensity (steady state) and the actual oxygen consumption. Despite the insufficient oxygen delivery at the onset of exercise, the active muscles are able to generate the ATP needed through anaerobic pathways.

During the initial minutes of recovery, even though muscle activity has stopped, oxygen consumption does not immediately decrease. Rather, oxygen consumption remains elevated temporarily (figure 4.5). This consumption, which exceeds that usually required at rest, traditionally has been referred to as the "oxygen debt." The more common term today is **excess postexercise oxygen consumption (EPOC)**. The EPOC is the volume

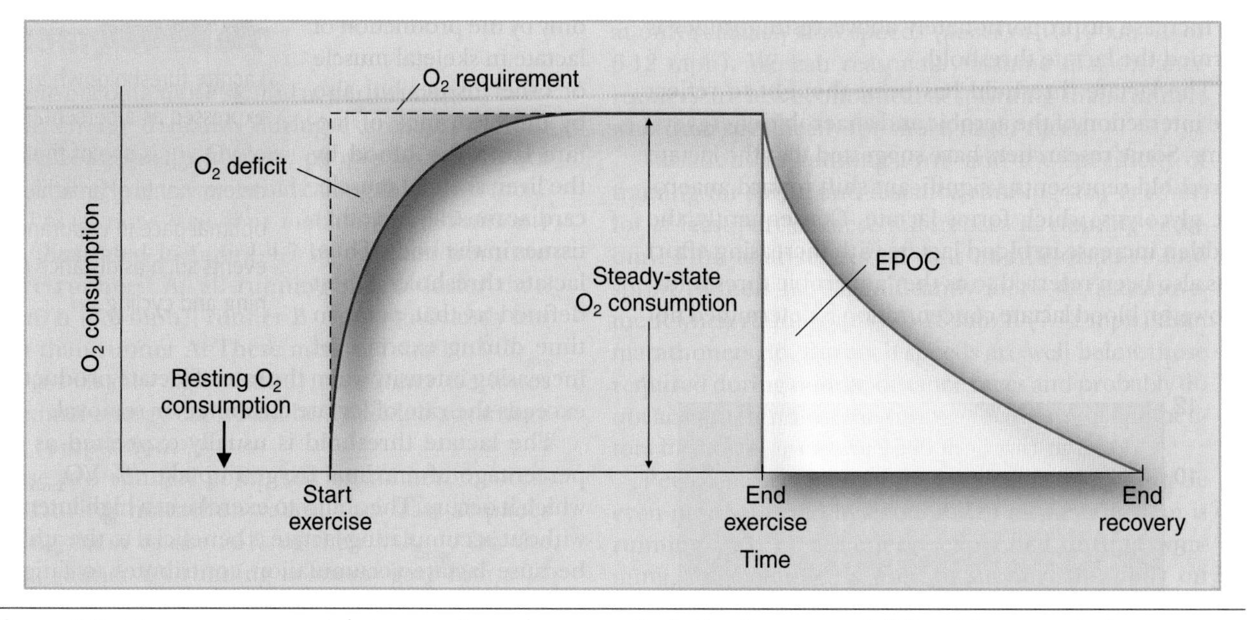

Figure 4.5 Oxygen requirement during exercise and recovery, illustrating the oxygen deficit and the concept of excess postexercise oxygen consumption (EPOC).

of oxygen consumed above that normally consumed at rest. Everyone has experienced this phenomenon at the end of an intense exercise bout: A fast climb up several flights of stairs leaves one with a rapid pulse and breathing hard. These physiological adjustments are serving to support the EPOC. After several minutes of recovery, the pulse and breathing return to resting rates.

For many years, the EPOC curve was described as having two distinct components: an initial fast component and a secondary slow component. According to classical theory, the fast component of the curve represented the oxygen required to rebuild the ATP and phosphocreatine (PCr) used during exercise, especially in its initial stages. Without sufficient oxygen, the high-energy phosphate bonds in these compounds were broken to supply the required energy. During recovery, these bonds would need to be re-formed, via oxidative processes, to replenish the energy stores, or repay the debt. The slow component of the curve was thought to result from removal of accumulated lactate from the tissues, by either conversion to glycogen or oxidation to CO_2 and H_2O, thus providing the energy needed to restore glycogen stores.

According to this theory, both the fast and slow components of the curve reflected the anaerobic activity that had occurred during exercise. The belief was that by examining the postexercise oxygen consumption, one could estimate the amount of anaerobic activity that had occurred.

However, more recently researchers have concluded that the classical explanation of EPOC is too simplistic. For example, during the initial phase of exercise, some oxygen is borrowed from the oxygen stores (hemoglobin and myoglobin). That oxygen must be replenished during recovery. Also, respiration remains temporarily elevated following exercise partly in an effort to clear CO_2 that has accumulated in the tissues as a by-product of metabolism. Body temperature also is elevated, which keeps the metabolic and respiratory rates high, thus requiring more oxygen; and elevated levels of norepinephrine and epinephrine during exercise have similar effects.

Thus, the EPOC depends on many factors other than merely the rebuilding of ATP and PCr and the clearing of lactate produced by anaerobic metabolism. The physiological mechanisms responsible for the EPOC are not yet clearly defined.

Lactate Threshold

Many investigators consider the lactate threshold to be a good indicator of an athlete's potential for endurance exercise. The **lactate threshold** is defined as the point at which blood lactate begins to accumulate substantially above resting concentrations during exercise of increasing intensity. For example, a runner might be required to run on the treadmill at different speeds with a rest between each speed. After each run, a blood sample is taken from his or her fingertip, or from a catheter in one of the arm veins, from which blood lactate is measured. As illustrated in figure 4.6, the results of such testing can be used to plot the relationship between blood lactate and running velocity. At low running velocities, blood lactate concentrations remain at or near resting levels. But as running speed increases, the blood lactate concentration increases rapidly beyond some threshold velocity. The point at which blood lactate appears

increases during intense short-term exercise because of the breakdown of PCr, is a potential cause of fatigue in this type of exercise.[9]

To delay fatigue, the athlete must control the rate of effort through proper pacing to ensure that PCr and ATP are not prematurely exhausted. This holds true even in endurance-type events. If the beginning pace is too rapid, available ATP and PCr concentrations will quickly decrease, leading to early fatigue and an inability to maintain the pace in the event's final stages. Training and experience allow the athlete to judge the optimal pace that permits the most efficient use of ATP and PCr for the entire event.

Glycogen Depletion

Muscle ATP concentrations are also maintained by the aerobic and anaerobic breakdown of muscle glycogen. In events lasting longer than a few seconds, muscle glycogen becomes the primary energy source for ATP synthesis. Unfortunately, glycogen reserves are limited and are depleted quickly. Since the muscle biopsy technique was first established, studies have shown a correlation between muscle glycogen depletion and fatigue during prolonged exercise.

As with PCr use, the rate of muscle glycogen depletion is controlled by the intensity of the activity. Increasing the intensity results in a disproportionate decrease in muscle glycogen. During sprint running, for example, muscle glycogen may be used 35 to 40 times faster than during walking. Muscle glycogen can be a limiting factor even during mild effort. The muscle depends on a constant supply of glycogen to meet the high energy demands of exercise.

Muscle glycogen is used more rapidly during the first few minutes of exercise than in the later stages, as seen in figure 4.8.[3] The illustration shows the change in muscle glycogen content in the subject's gastrocnemius (calf) muscle during the test. Although the subject ran the test at a steady pace, the rate of muscle glycogen metabolized from the gastrocnemius was greatest during the first 75 min.

The subject also reported his perceived exertion (how difficult his effort seemed to be) at various times during the test. He felt only moderately stressed early in the run, when his glycogen stores were still high, even though he was using glycogen at a high rate. He did not perceive severe fatigue until his muscle glycogen levels were nearly depleted. Thus, the sensation of fatigue in long-term exercise coincides with a decreased concentration of muscle glycogen, but not with its rate of depletion. Marathon runners commonly refer to the sudden onset of fatigue that they experience at 29 to 35 km (18-22 mi) as "hitting the wall." At least part of this sensation can be attributed to muscle glycogen depletion.

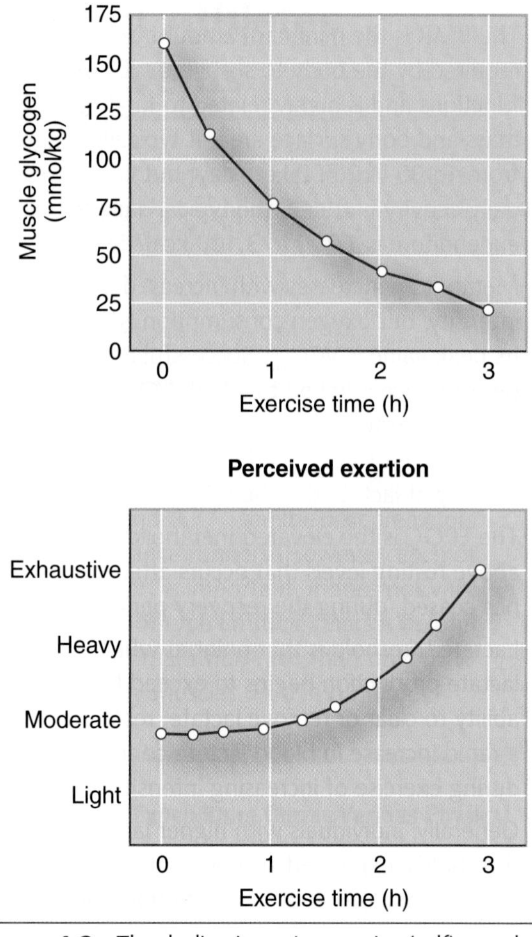

Figure 4.8 The decline in gastrocnemius (calf) muscle glycogen (upper panel) during 3 h of treadmill running at 70% of $\dot{V}O_{2max}$. The subject's subjective rating of the effort is recorded in the lower panel. Note that the effort was rated as moderate for nearly 1.5 h of the run, although glycogen was decreasing steadily. Not until the muscle glycogen became quite low (less than 50 mmol/kg) did the rating of effort increase.

Adapted, by permission, from D.L. Costill, 1986, *Inside running: Basics of sports physiology* (Indianapolis: Benchmark Press). Copyright 1986 Cooper Publishing Group, Carmel, In.

Glycogen Depletion in Different Fiber Types

Muscle fibers are recruited and deplete their energy reserves in selected patterns. The individual fibers most frequently recruited during exercise may become depleted of glycogen. This reduces the number of fibers capable of producing the muscular force needed for exercise.

This glycogen depletion is illustrated in figure 4.9, which shows a micrograph of muscle fibers taken from a runner after a 30 km (18.6 mi) run. Figure 4.9*a* has been stained to differentiate type I and type II fibers. One of the type II fibers is circled. Figure 4.9*b* shows a second sample from the same muscle, stained to show glycogen. The redder (darker) the stain, the more gly-

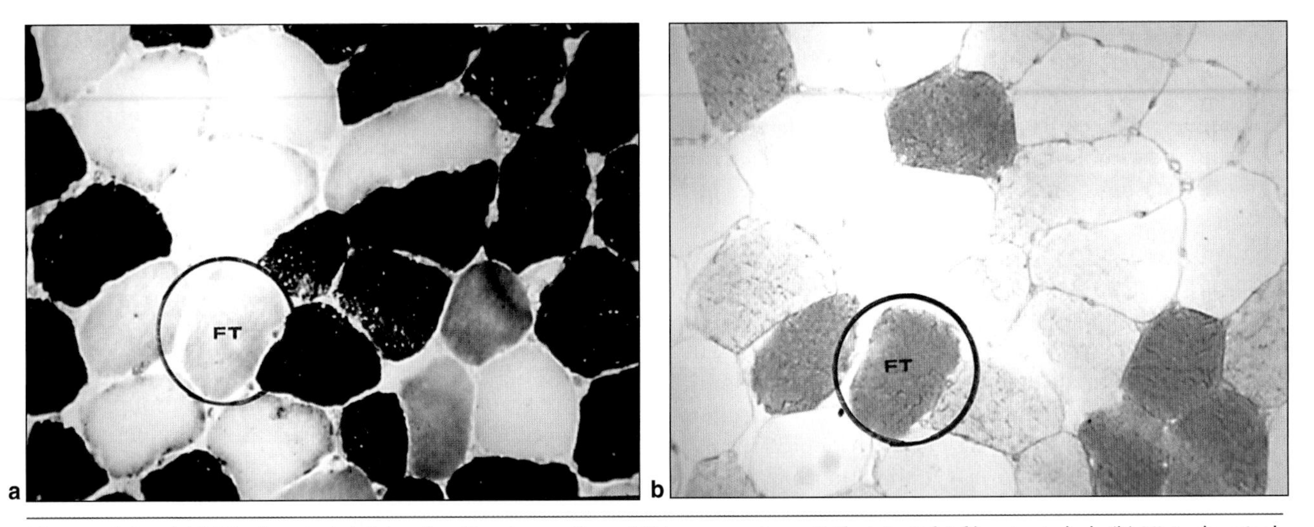

Figure 4.9 *(a)* Histochemical staining for fiber type after a 30 km run; a type II (fast-twitch) fiber is circled. *(b)* Histochemical staining for muscle glycogen after the run. Note that a number of type II fibers still have glycogen, as noted by their darker stain, whereas most of the type I (slow-twitch) fibers are empty of glycogen.

cogen is present. Before the run, all fibers were full of glycogen and appeared red (not depicted). In figure 4.9*b* (after the run), the lighter type I fibers are almost completely depleted of glycogen. This suggests that type I fibers are used more heavily during endurance exercise that requires only moderate force development, such as the 30 km run.

The pattern of glycogen depletion from type I and type II fibers depends on the exercise intensity. Recall that type I fibers are the first fibers to be recruited during light exercise. As muscle tension requirements increase, type IIa fibers are added to the workforce. In exercise approaching maximal intensities, the type IIx fibers are added to the pool of recruited fibers.

Depletion in Different Muscle Groups

In addition to selectively depleting glycogen from type I or type II fibers, exercise may place unusually heavy demands on select muscle groups. In one study, subjects ran on a treadmill positioned for uphill, downhill, and level running for 2 h at 70% of $\dot{V}O_{2max}$. Figure 4.10 compares the resultant glycogen depletion in three muscles of the lower extremity: the vastus lateralis (knee extensor), the gastrocnemius (ankle extensor), and the soleus (another knee extensor).

The results show that whether one runs uphill, downhill, or on a level surface, the gastrocnemius uses more glycogen than does the vastus lateralis or the soleus. This suggests that the ankle extensor muscles are more likely to become depleted during distance running than are the thigh muscles, isolating the site of fatigue to the lower leg muscles.

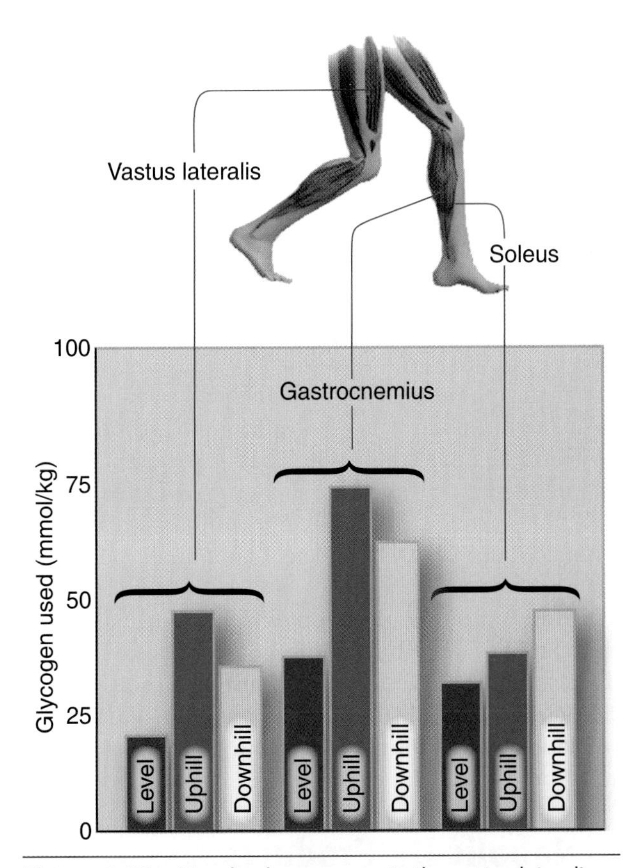

Figure 4.10 Muscle glycogen use in the vastus lateralis, gastrocnemius, and soleus muscles during 2 h of level, uphill, and downhill running on a treadmill at 70% of $\dot{V}O_{2max}$. Note that the greatest glycogen use is in the gastrocnemius during uphill and downhill running.

Glycogen Depletion and Blood Glucose

Muscle glycogen alone cannot provide enough carbohydrate for exercise lasting several hours. Glucose delivered by the blood to the muscles contributes a lot of energy during endurance exercise. The liver breaks down its stored glycogen to provide a constant supply of blood glucose. In the early stages of exercise, energy production requires relatively little blood glucose; but in the later stages of an endurance event, blood glucose may make a large contribution. To keep pace with the muscles' glucose uptake, the liver must break down increasingly more glycogen as exercise duration increases.

Liver glycogen stores are limited, and the liver cannot produce glucose rapidly from other substrates. Consequently, blood glucose levels can decrease when muscle uptake exceeds the liver's glucose output. Unable to obtain sufficient glucose from the blood, the muscles must rely more heavily on their glycogen reserves, accelerating muscle glycogen depletion and leading to earlier exhaustion. On the other hand, most studies have shown no effect of carbohydrate ingestion on net muscle glycogen utilization during prolonged, strenuous exercise.

Not surprisingly, endurance performances improve when the muscle glycogen supply is elevated before the start of activity. The importance of muscle glycogen storage for endurance performance is discussed in chapter 14. For now, note that glycogen depletion and hypoglycemia (low blood sugar) limit performance in activities lasting longer than 60 to 90 min.[6]

Mechanisms of Fatigue With Glycogen Depletion

It does not appear likely that glycogen depletion directly causes fatigue during endurance exercise performance. Rather, the depletion of muscle glycogen may be the first step in a series of events that leads to fatigue. A certain level of muscle glycogen metabolism is necessary to maintain oxidative metabolism of both carbohydrates and fats using the Krebs cycle. That is, we now know that a certain rate of glycogen breakdown is needed for the optimal production of reduced nicotinamide adenine dinucleotide (NADH) and to maintain the electron transport system.

Additionally, as glycogen is depleted, exercising muscle relies more heavily on the metabolism of FFAs. To accomplish this, more FFAs must be moved into the mitochondria, and the rate of transfer may limit FFA oxidation to the point where it can no longer keep up with the need for fat oxidation.

Metabolic By-Products and Fatigue

Various by-products of metabolism have been implicated as factors causing, or contributing to, fatigue. One example is P_i, which increases during intense short-term exercise as PCr and ATP are being broken down.[9] Additional metabolic by-products that have received the most attention in discussing fatigue are heat, lactate, and hydrogen ions.

Heat, Muscle Temperature, and Fatigue

Recall that energy expenditure results in a relatively large heat production, some of which is retained in the body, causing core temperature to rise. Exercise in the heat can increase the rate of carbohydrate utilization and hasten glycogen depletion, effects that may be stimulated by the increased secretion of epinephrine. It is hypothesized that high muscle temperatures impair both skeletal muscle function and muscle metabolism.

The ability to continue moderate- to high-intensity cycle performance is affected by ambient temperature. Galloway and Maughan[5] studied performance time to exhaustion of male cyclists at four different air temperatures: 4 °C (38 °F), 11 °C (51 °F), 21 °C (70 °F), and 31 °C (87 °F). Results of that study are shown in figure 4.11. Time to exhaustion was longest when subjects exercised at an air temperature of 11 °C, but lower at colder and warmer temperatures. Fatigue set in earliest at 31 °C. Precooling of muscles similarly prolongs exercise, while preheating causes earlier fatigue. Heat acclimation, discussed in chapter 11, spares glycogen and reduces lactate accumulation.

Lactic Acid, Hydrogen Ions, and Fatigue

Recall that lactic acid is a by-product of anaerobic glycolysis. Although most people believe that lactic acid is responsible for fatigue in all types of exercise,

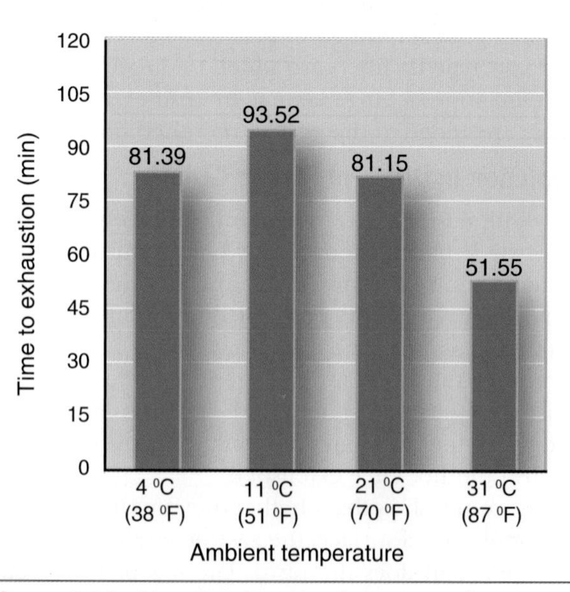

Figure 4.11 Time to exhaustion for a group of men performing cycle exercise at about 70% $\dot{V}O_{2max}$. The subjects were able to perform longer (delay fatigue longer) in a cool environment of 11 °C. Exercising in colder or warmer conditions hastened fatigue.

Adapted, by permission, from S.D.R. Galloway and R.J. Maughan, 1997, "Effects of ambient temperature on the capacity to perform prolonged cycle exercise in man," *Medicine and Science in Sports and Exercise*, 29: 1240-1249.

lactic acid accumulates within the muscle fiber only during relatively brief, highly intense muscular effort. Marathon runners, for example, may have near-resting lactic acid levels at the end of the race, despite their fatigue. As noted in the previous section, their fatigue is likely caused by inadequate energy supply, not excess lactic acid.

Short sprints in running, cycling, and swimming all lead to large accumulations of lactic acid. But the presence of lactic acid should not be blamed for the feeling of fatigue in itself. When not cleared, the lactic acid dissociates, converting to lactate and causing an accumulation of hydrogen ions. This H^+ accumulation causes muscle acidification, resulting in a condition known as acidosis.

Activities of short duration and high intensity, such as sprint running and sprint swimming, depend heavily on anaerobic glycolysis and produce large amounts of lactate and H^+ within the muscles. Fortunately, the cells and body fluids possess buffers, such as bicarbonate (HCO_3), that minimize the disrupting influence of the H^+. Without these buffers, H^+ would lower the pH to about 1.5, killing the cells. Because of the body's buffering capacity, the H^+ concentration remains low even during the most severe exercise, allowing muscle pH to decrease from a resting value of 7.1 to no lower than 6.6 to 6.4 at exhaustion.

However, pH changes of this magnitude adversely affect energy production and muscle contraction. An intracellular pH below 6.9 inhibits the action of phosphofructokinase, an important glycolytic enzyme,

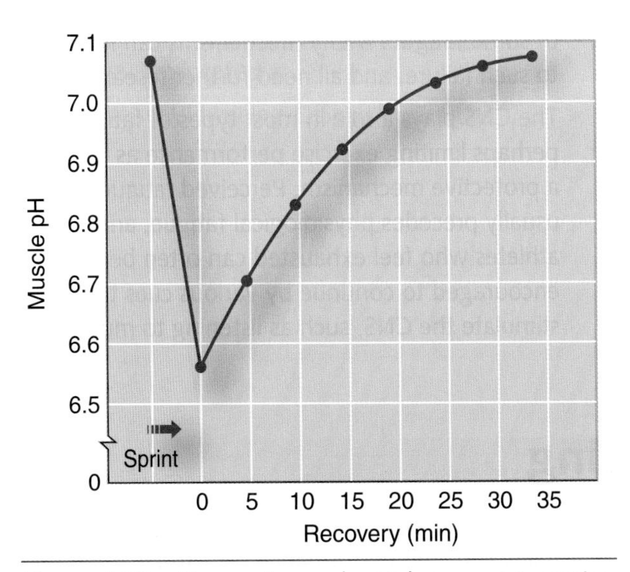

Figure 4.12 Changes in muscle pH during sprint exercise and recovery. Note the drastic decrease in muscle pH during the sprint and the gradual recovery to normal after the effort. Note that it took more than 30 min for pH to return to its preexercise level.

slowing the rate of glycolysis and ATP production. At a pH of 6.4, the influence of H^+ stops any further glycogen breakdown, causing a rapid decrease in ATP and ultimately exhaustion. In addition, H^+ may displace calcium within the fiber, interfering with the coupling of the actin-myosin cross-bridges and decreasing the muscle's contractile force. Most researchers agree that low muscle pH is the major limiter of performance and the primary cause of fatigue during maximal, all-out exercise lasting more than 20 to 30 s.

As seen in figure 4.12, reestablishing the preexercise muscle pH after an exhaustive sprint bout requires about 30 to 35 min of recovery. Even when normal pH is restored, blood and muscle lactate levels can remain quite elevated. However, experience has shown that an athlete can continue to exercise at relatively high intensities even with a muscle pH below 7.0 and a blood lactate level above 6 or 7 mmol/L, four to five times the resting value.

Some coaches and sport physiologists have attempted to use blood lactate measurements to gauge the intensity and volume of training needed to produce an optimal training stimulus. Such measurements provide an index of training intensity, but they might not reflect the anaerobic processes or the state of acidosis in the muscles. Because lactate and H^+ are generated in the muscles, both diffuse out of the cells. They then are diluted in the body fluids and transported to other areas of the body to be metabolized. Consequently, blood lactate concentrations depend on the rates of production, diffusion, oxidation, and clearance. A variety of factors can influence these processes, so measuring blood lactate is of questionable value for fine-tuning training.

Neuromuscular Fatigue

Thus far we have considered only factors within the muscle that might be responsible for fatigue. Evidence also suggests that under some circumstances, fatigue may result from an inability to activate the muscle fibers, a function of the nervous system. As noted in chapter 2, the nerve impulse is transmitted across the neuromuscular junction to activate the fiber's membrane, and it causes the fiber's sarcoplasmic reticulum to release calcium. The calcium, in turn, binds with troponin to initiate muscle contraction. Two of several possible neural mechanisms that could disrupt this process and possibly contribute to fatigue are discussed next.

Neural Transmission

Fatigue may occur at the neuromuscular junction, preventing nerve impulse transmission to the muscle fiber membrane. Studies in the early 1900s clearly established such a failure of nerve impulse transmission in fatigued

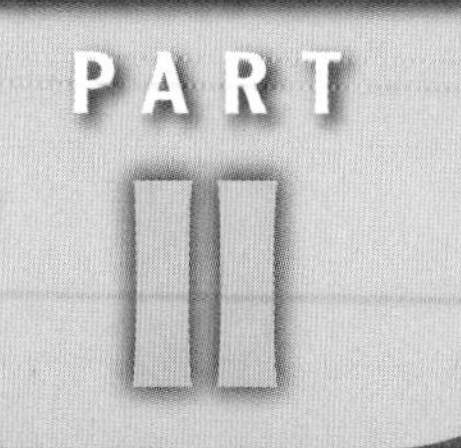

PART II

Cardiovascular and Respiratory Function

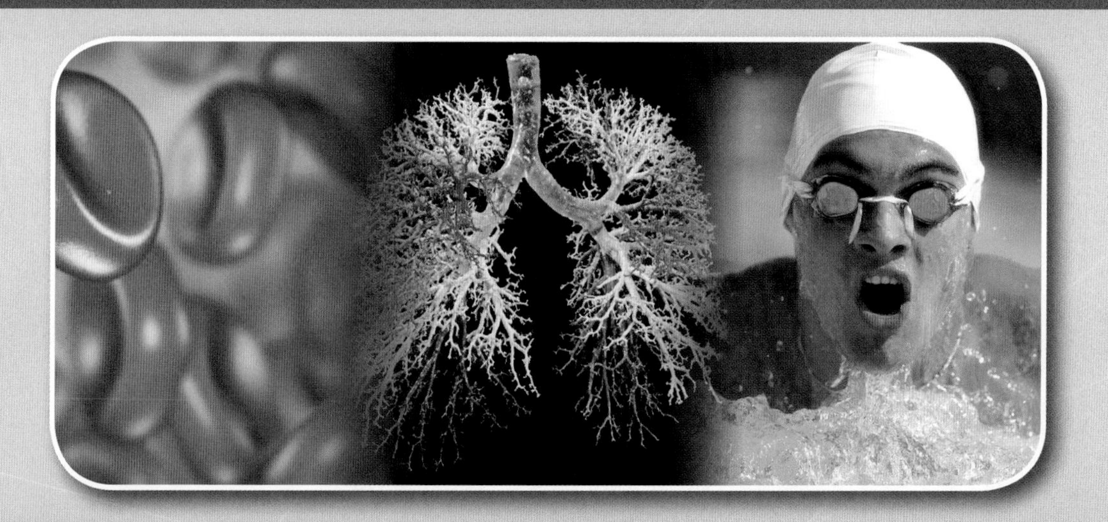

In the previous part, we learned how skeletal muscle contracts and how the body produces energy through metabolism to fuel its movement. In part II, we focus on how the cardiovascular and respiratory systems provide oxygen and fuel to the active muscles, how they rid the body of carbon dioxide and metabolic wastes, and how these systems respond to exercise. In chapter 5, "The Cardiovascular System and Its Control," we look at the structure and function of the cardiovascular system: the heart, the blood vessels, and the blood. Our primary focus is on how this system provides all parts of the body with an adequate blood supply to meet their needs under all conditions. In chapter 6, "The Respiratory System and Its Regulation," we examine the mechanics and regulation of breathing, the process of gas exchange in the lungs and at the muscles, and the transportation of oxygen and carbon dioxide in the blood. We also see how this system regulates the body's pH within a very narrow range. In chapter 7, "Cardiorespiratory Responses to Acute Exercise," we concentrate on the cardiovascular and respiratory changes that occur in response to an acute bout of exercise.

The Cardiovascular System and Its Control

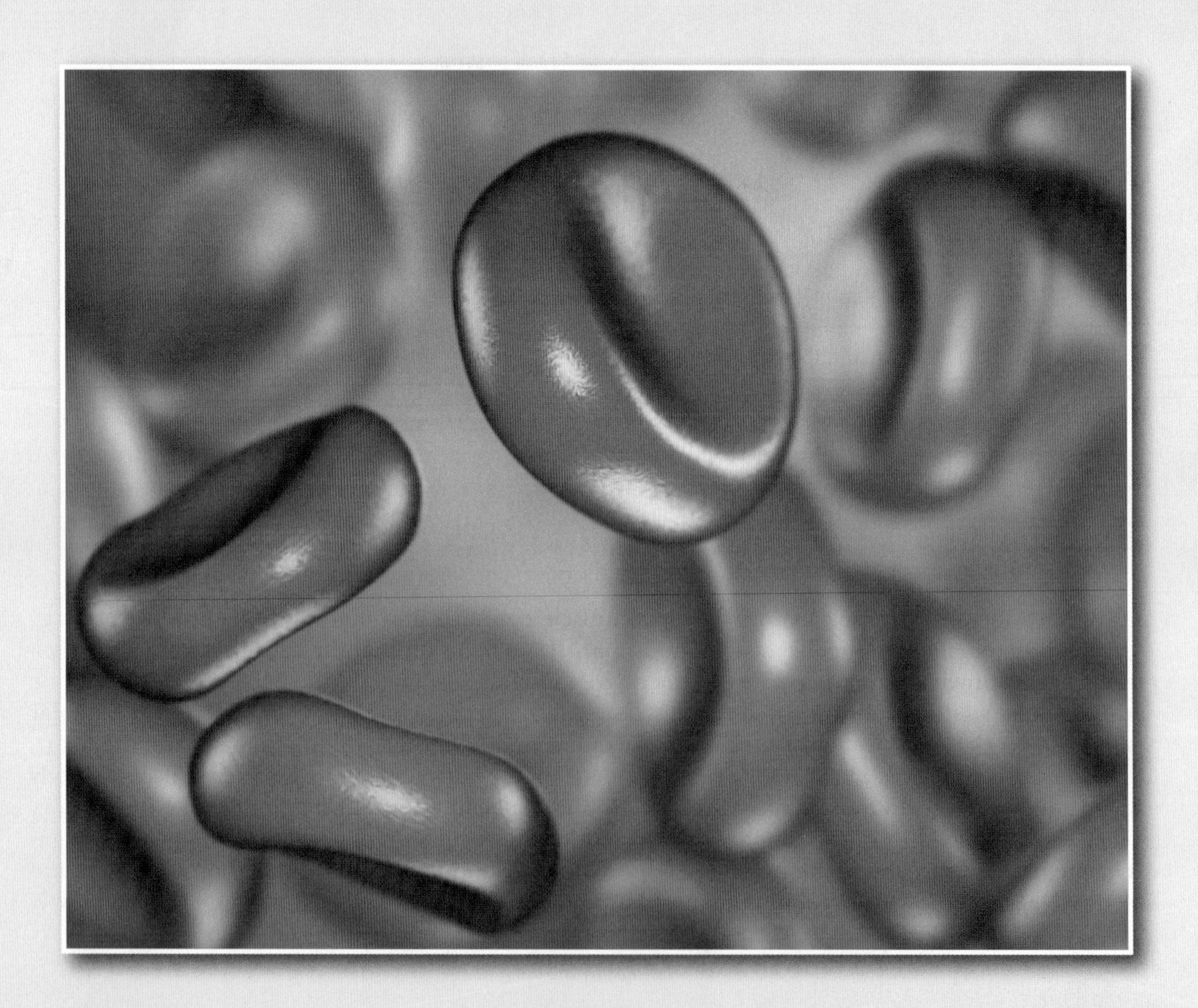

In this chapter and in the online study guide

On January 5, 1988, the sport world lost one of its greatest athletes. "Pistol Pete" Maravich, former National Basketball Association star, collapsed and died of cardiac arrest at 40 years of age during a pickup basketball game. His death came as a shock, and its cause surprised the medical experts. Maravich's heart was abnormally enlarged, primarily because he was born with only a single coronary artery on the right side of his heart—he was missing the two coronary arteries that supply the left side of the heart! The medical community was amazed that this single right coronary artery had taken over supplying the left side of Maravich's heart and that this adaptation had allowed him to compete for many years as one of the top players in the history of basketball. Although Maravich's death was a tragedy and shocked the sport world, he was able to perform at the highest level of competition for 10 years in one of the most physically demanding sports. More recently, a number of highly promising high school and college athletes have had their lives cut short by sudden cardiac death. Most of these deaths are attributable to hypertrophic cardiomyopathy, a disease characterized by an abnormally enlarged heart muscle mass, generally involving the left ventricle. In about half of the cases, the disease is inherited. Although this remains the major cause of sudden cardiac death in adolescent and young adult athletes (~36%), it is relatively rare, with an estimated occurrence of between 1 and 2 cases per 1 million athletes annually.

The cardiovascular system serves a number of important functions in the body, most of which support other physiological systems. The major cardiovascular functions fall into six categories:

- Delivery of oxygen and other nutrients
- Removal of carbon dioxide and other metabolic waste
- Transport of hormones
- Thermoregulation
- Maintenance of acid–base balance and overall body fluid balance
- Immune function

The cardiovascular system delivers oxygen and nutrients to, and removes carbon dioxide and metabolic waste products from, every cell in the body. It transports hormones (chapter 2) from endocrine glands to their target receptors. The cardiovascular system supports body temperature regulation (chapter 11), and the blood's buffering capabilities help control the body's pH. The cardiovascular system maintains appropriate fluid balance in the body and helps prevent infection from invading organisms. Although this is just an abbreviated list of roles, the cardiovascular functions listed here are important for understanding the physiological bases of physical activity. Obviously these roles change with the challenges imposed during exercise.

All physiological functions and virtually every cell in the body depend in some way on the cardiovascular system. Any system of circulation requires three components:

- A pump (the heart)
- A system of channels or tubes (the blood vessels)
- A fluid medium (the blood)

In order to keep blood circulating continually, the heart must generate sufficient pressure to drive the blood through the continuous network of blood vessels in this closed-loop system. Thus, the primary goal of the cardiovascular system is to ensure that there is adequate blood flow throughout the circulation to meet the metabolic demand of the tissues. We look first at the heart.

HEART

About the size of a fist and located in the center of the thoracic cavity, the heart is the primary pump that circulates blood through the entire vascular system. As shown in figure 5.1, the heart has two atria that act as receiving chambers and two ventricles acting as pumping units. It is enclosed in a tough membranous sac called the **pericardium.** The thin cavity between the pericardium and the heart is filled with pericardial fluid, which is necessary to reduce friction between the sac and the beating heart.

Blood Flow Through the Heart

The heart is sometimes considered to be two separate pumps, with the right side of the heart pumping deoxygenated blood to the lungs through the pulmonary circulation and the left side of the heart pumping oxygenated blood to all other tissues in the body through the systemic circulation. Blood that has circulated through the body, delivering oxygen and nutrients and picking up waste products, returns to the heart through the great veins—the superior vena cava and inferior vena cava—to the right atrium. This chamber receives all the deoxygenated blood from the systemic circulation.

From the right atrium, blood passes through the tricuspid valve into the right ventricle. This chamber pumps the blood through the pulmonary valve into the pulmonary artery, which carries the blood to the lungs. Thus, the right side of the heart is known as the pulmonary side, sending the blood that has circulated throughout the body into the lungs for reoxygenation.

After blood is oxygenated in the lungs, it is transported back to the heart through the pulmonary veins. All freshly oxygenated blood is received from the pulmonary veins by the left atrium. From the left atrium, the blood passes through the mitral valve into the left ventricle. Blood leaves the left ventricle by passing through the aortic valve into the aorta and is distributed to the systemic circulation. The left side of the heart is known as the systemic side. It receives the oxygenated blood from the lungs and then sends it out to supply all other body tissues.

Myocardium

Cardiac muscle is collectively called the **myocardium** or myocardial muscle. Myocardial thickness at various locations in the heart varies according to the amount of stress placed on it. The left ventricle is the most powerful of the four heart chambers. This chamber must contract to generate sufficient pressure to pump blood through the entire body. When a person is sitting or standing, the left ventricle must contract with enough force to overcome the effect of gravity, which tends to pool blood in the lower extremities.

The left ventricle must generate a considerable amount of force to pump blood to the systemic circulation, and this is reflected by the greater thickness of its muscular wall compared with that of the other heart

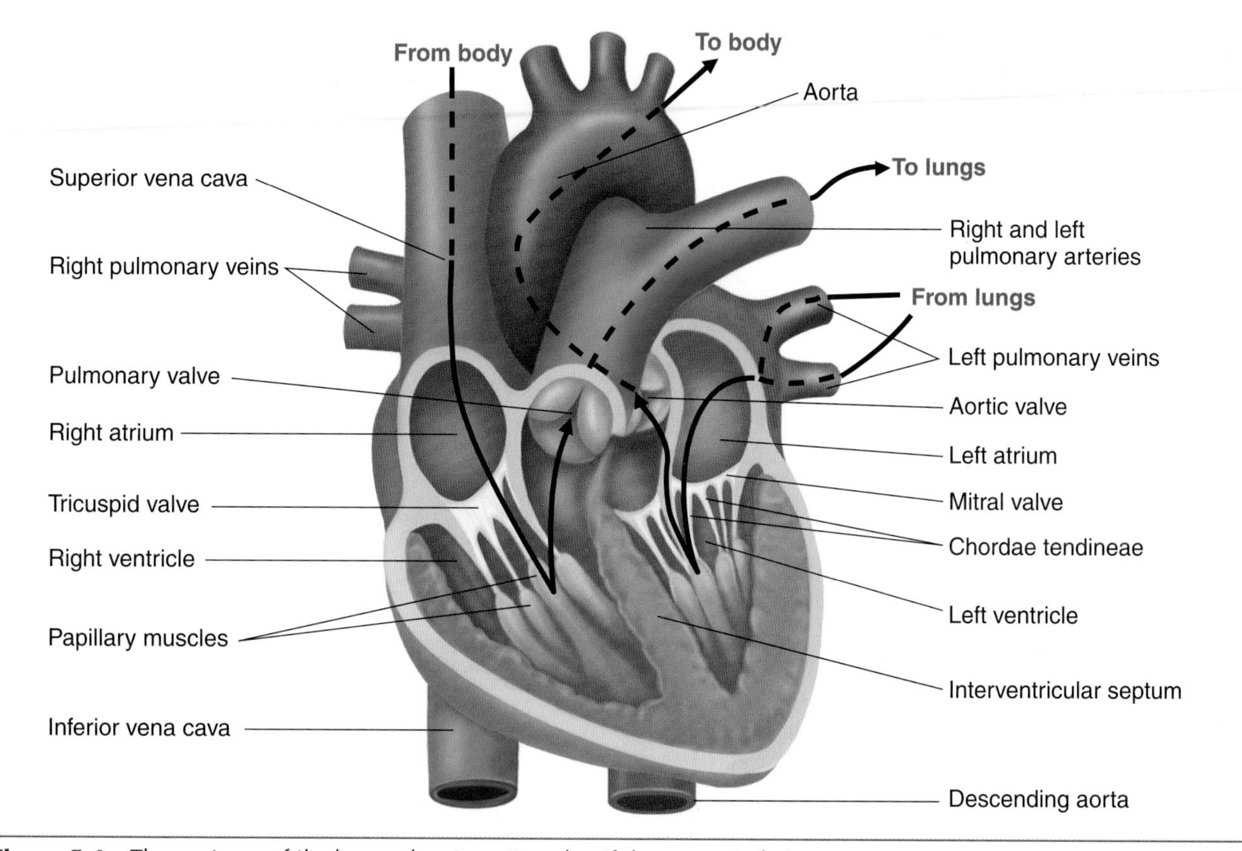

Figure 5.1 The anatomy of the human heart, sectioned as if the person is facing you.

chambers. This hypertrophy is the result of the pressure placed on the left ventricle at rest or under normal conditions of moderate activity. With more vigorous exercise—particularly intense aerobic activity, during which the working muscles' need for blood increases considerably—the demand on the left ventricle to deliver blood to exercising muscles is high. In response to both intense aerobic and resistance training, the left ventricle will hypertrophy. In contrast to the positive adaptations that occur as a result of physical training, cardiac muscle also hypertrophies as a result of diseases, such as high blood pressure or valvular heart disease. In response to either training or disease, over time the left ventricle adapts by increasing its size and pumping capacity, similar to the way skeletal muscle adapts to physical training. However, the mechanisms for adaptation and cardiac performance with disease are different from those observed with aerobic training.

Although striated in appearance, the myocardium differs from skeletal muscle in several important ways. First, cardiac muscle fibers are anatomically interconnected end to end by dark-staining regions called intercalated disks. These disks have desmosomes, which are structures that anchor the individual cells together so that they do not pull apart during contraction, and gap junctions, which allow rapid transmission of the action potentials that signal the heart to contract as one unit. Secondly, the myocardial fibers are rather homogeneous in contrast to the mosaic of fiber types in skeletal muscle. The myocardium contains only one fiber type, thought to be similar to type I fibers in skeletal muscle in that it is highly oxidative, is highly capillarized, and has a large number of mitochondria.

In addition to these differences, the mechanism of muscle contraction also differs between skeletal and cardiac muscle. Cardiac muscle contraction occurs by "calcium-induced calcium release" (figure 5.2). The action potential spreads rapidly along the myocardial sarcolemma from cell to cell via gap junctions, and also to the inside of the cell through the T-tubules. Upon stimulation, calcium enters the cell by the dihydropyridine receptor in the T-tubules. Unlike what happens in skeletal muscle, the amount of calcium that enters the cell is not sufficient to directly cause the cardiac muscle to contract; but it serves as a trigger to another type of receptor, called the ryanodine receptor, to release calcium from the sarcoplasmic reticulum. Figure 5.3 summarizes some of the similarities and differences between cardiac and skeletal muscle.

The myocardium, just like skeletal muscle, must have a blood supply to deliver oxygen and nutrients and remove waste products. Although blood courses

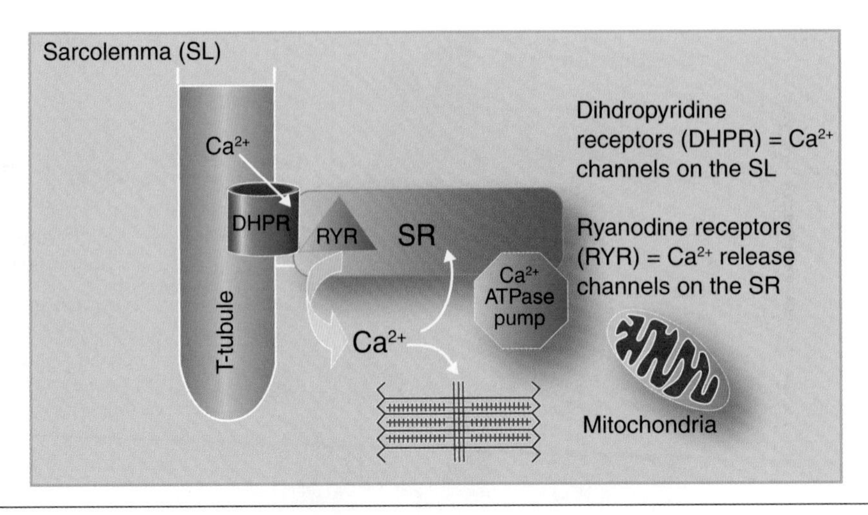

Figure 5.2 Mechanism of cardiac muscle contraction.

Courtesy of Dr. Donna H. Korzick, Pennsylvania State University.

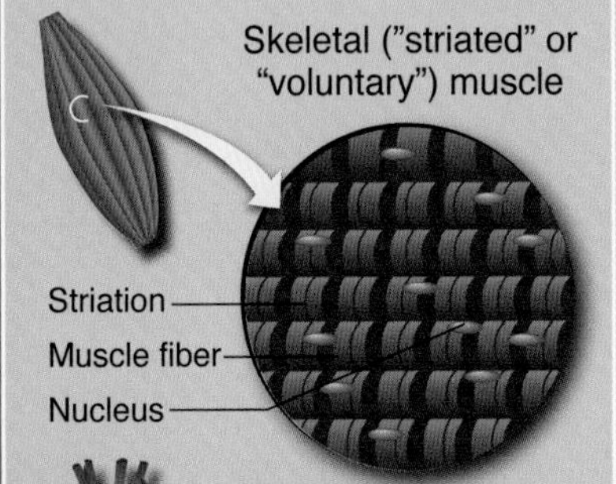

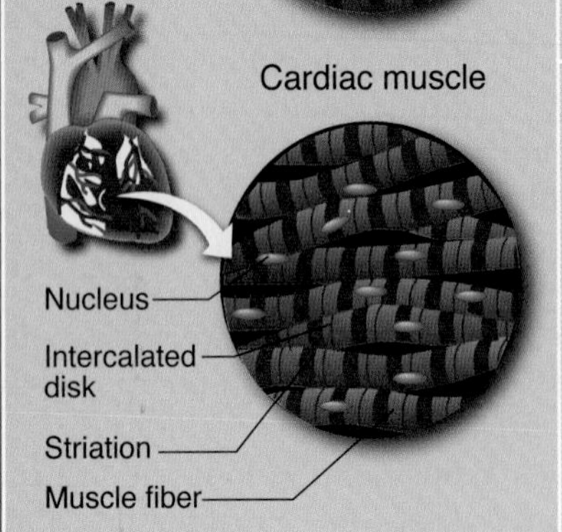

Muscle type	Location	Appearance	Type of activity	Stimulation
Skeletal ("striated" or "voluntary") muscle Striation Muscle fiber Nucleus	Named muscle (e.g., the biceps of the arm) attached to the skeleton and fascia of limbs, body wall, and head/neck	Large, long, unbranched, cylindrical fibers with transverse striations (stripes) arranged in parallel bundles; multiple, peripherally located nuclei	Strong, quick intermittent (phasic) contraction above a baseline tonus; acts primarily to produce movement or resist gravity	Voluntary (or reflexive) by the somatic nervous system
Cardiac muscle Nucleus Intercalated disk Striation Muscle fiber	Muscle of heart (myocardium) and adjacent portions of the great vessels (aorta, vena cava)	Branching and anastomosing shorter fibers with transverse striations (stripes) running parallel and connected end to end by complex junctions (intercalated disks); single, central nucleus	Strong, quick continuous rhythmic contraction; pumps blood from the heart	Involuntary; intrinsically (myogenically) stimulated and propagated; rate and strength of contraction modified by the autonomic nervous system

Figure 5.3 Functional and structural characteristics of skeletal and cardiac muscle.

Adapted, by permission, from K.L. Moore, and A.F. Dalley, 1999, *Clinically oriented anatomy*, 4th ed. (Baltimore, MD: Lippincott, Williams, and Wilkins), 27.

through each chamber of the heart, little nourishment comes from this blood supply. The primary blood supply to the heart is provided by the right and left coronary arteries, which arise from the base of the aorta and encircle the outside of the myocardium (figure 5.4). The right coronary artery supplies the right side of the heart, dividing into two primary branches, the marginal artery and the posterior interventricular artery. The left coronary artery, also referred to as the left main coronary artery, also divides into two major branches, the circumflex artery and the anterior descending artery. The posterior interventricular artery and the

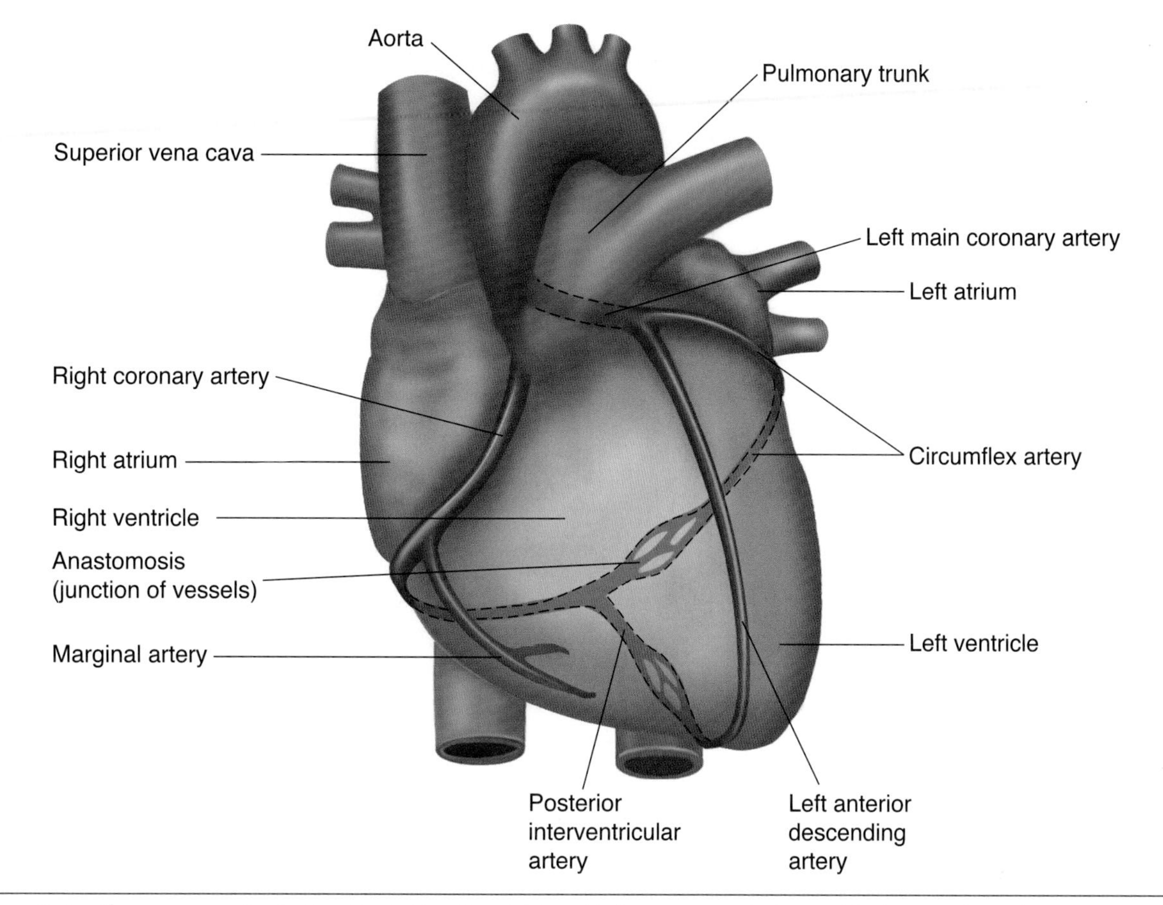

Figure 5.4 The coronary circulation, illustrating the right and left coronary arteries and their major branches.

anterior descending artery merge, or anastomose, in the lower posterior area of the heart, as does the circumflex. These arteries are very susceptible to atherosclerosis, or narrowing by the accumulation of plaque and inflammation, leading to coronary artery disease. This disease is discussed in greater detail in chapter 20. Anomalies—shortenings, blockages, or misdirections—sometimes occur in the coronary arteries, and such congenital abnormalities are a common cause of sudden death in athletes.

The ability of the myocardium to contract as a single unit depends on initiation and propagation of an electrical signal through the heart, the cardiac conduction system.

Cardiac Conduction System

Cardiac muscle has the unique ability to generate its own electrical signal, called spontaneous rhythmicity, which allows it to contract without any external stimulation. The contraction is rhythmical, in part because of the anatomical coupling of the conduction cells through gap junctions. With neither neural nor hormonal stimulation, the intrinsic heart rate (HR) averages ~100 beats (contractions) per minute. This resting heart rate of about 100 beats/min can be observed in patients who have undergone cardiac transplant surgery, because their transplanted hearts lack neural innervation.

Figure 5.5 illustrates the four main components of the cardiac conduction system:

- Sinoatrial (SA) node
- Atrioventricular (AV) node
- AV bundle (bundle of His)
- Purkinje fibers

The impulse for normal heart contractions is initiated in the **sinoatrial (SA) node,** a group of specialized cardiac muscle fibers located in the upper posterior wall of the right atrium. These specialized cells spontaneously depolarize at a faster rate than other myocardial muscle cells because they are especially leaky to sodium. Because this tissue generates the electrical impulse, typically at a frequency of about 100 beats/min—the fastest intrinsic firing rate—the SA node is known as the heart's pacemaker, and the rhythm it establishes is called the sinus rhythm. The electrical impulse generated by the SA node spreads through both atria and

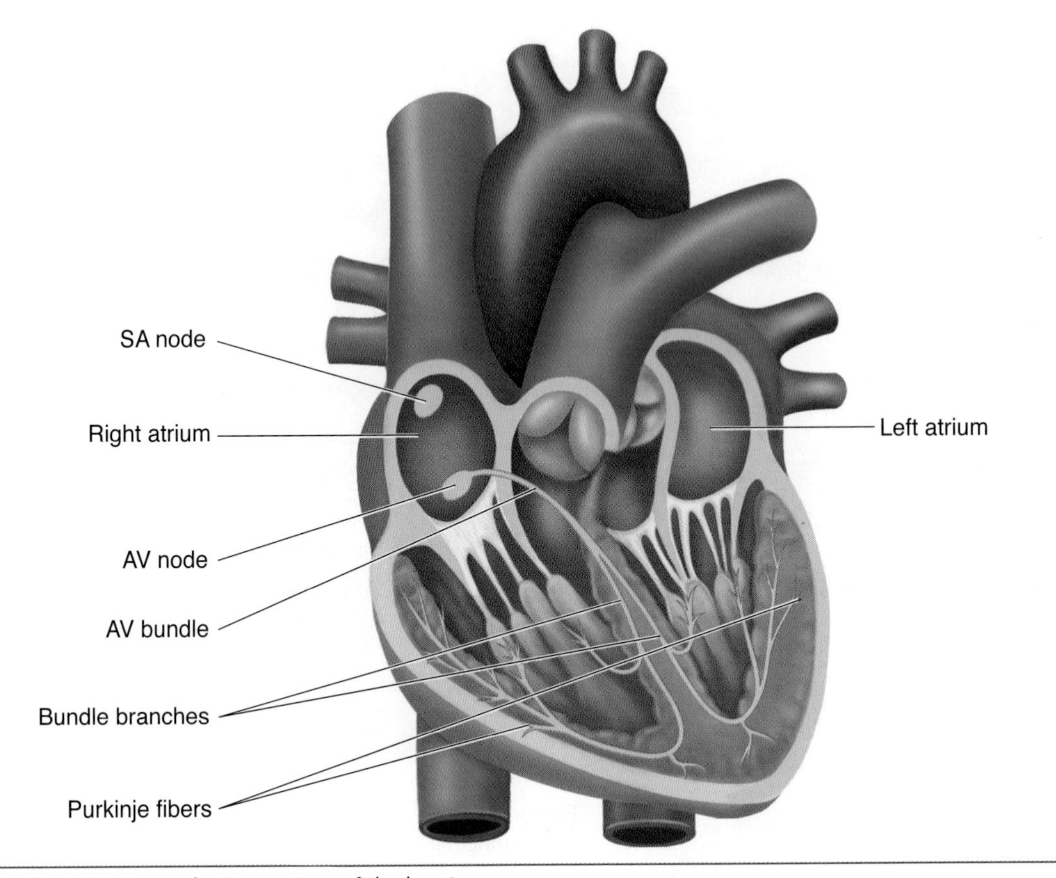

Figure 5.5 The intrinsic conduction system of the heart.

reaches the **atrioventricular (AV) node,** located in the right atrial wall near the center of the heart. As the electrical impulse spreads through the atria, they are signaled to contract.

The AV node conducts the electrical impulse from the atria into the ventricles. The impulse is delayed by about 0.13 s as it passes through the AV node, and then it enters the AV bundle. This delay allows blood from the atria to completely empty into the ventricles to maximize ventricular filling before the ventricles contract. While most blood moves passively from the atria to the ventricles, active contraction of the atria (sometimes called the "atrial kick") completes the process. The AV bundle travels along the ventricular septum and then sends right and left bundle branches into both ventricles. These branches send the impulse toward the apex of the heart and then outward. Each bundle branch subdivides into many smaller ones that spread throughout the entire ventricular wall. These terminal branches of the AV bundle are the Purkinje fibers. They transmit the impulse through the ventricles approximately six times faster than through the rest of the cardiac conduction system. This rapid conduction allows all parts of the ventricle to contract at virtually the same time.

Occasionally, chronic problems develop within the cardiac conduction system, hampering its ability to maintain appropriate sinus rhythm throughout the heart. In such cases, an artificial pacemaker can be surgically installed. This small, battery-operated electrical stimulator, usually implanted under the skin, has tiny electrodes attached to the right ventricle. An electrical stimulator is useful, for example, to treat a condition called AV block. With this disorder, the SA node creates an impulse, but the impulse is blocked at the AV node and cannot reach the ventricles, resulting in the heart rate's being controlled by the intrinsic firing rate of the pacemaker cells in the ventricles (closer to 40 beats/min). The artificial pacemaker takes over the role of the disabled AV node, supplying the needed impulse and thus controlling ventricular contraction.

Extrinsic Control of Heart Activity

Although the heart initiates its own electrical impulses (intrinsic control), both the rate and effect can be altered. Under normal conditions, this is accomplished primarily through three extrinsic systems:

- The parasympathetic nervous system
- The sympathetic nervous system
- The endocrine system (hormones)

Heart Murmur

The four heart valves prevent backflow of blood, ensuring one-way flow through the heart. These valves maximize the amount of blood pumped out of the heart during contraction. A heart murmur is a condition in which abnormal heart sounds are detected with the aid of a stethoscope. Normally, a heart valve makes a distinct clicking sound when it snaps shut. With a murmur, the click is replaced by a sound similar to blowing. This abnormal sound can indicate the turbulent flow of blood through a narrowed valve or backward (retrograde) flow back toward the atria through a leaky valve. It could also indicate errant blood flow through a hole in the wall (septum) separating the right and left sides of the heart (septal defect).

Mild heart murmurs are fairly common in growing children and adolescents. During growth periods, valve development does not always keep up with enlargement of the heart openings. Valves can leak as a result of disease, such as stenosis, in which the valve is narrowed and often thickened and rigid. This condition can require surgical replacement of the valve. With mitral valve prolapse, the mitral valve allows some blood to flow back into the left atrium during ventricular contraction. This disorder, relatively common in adults (6-17% of the population including athletes), usually has little clinical significance unless there is significant backflow.

Most murmurs in athletes are benign, affecting neither the heart's pumping nor the athlete's performance. Only when there is a functional consequence, such as light-headedness or dizziness, are murmurs a cause for immediate concern.

Although an overview of these systems' effects is offered here, they are discussed in more detail in chapters 2 and 3.

The parasympathetic system, a branch of the autonomic nervous system, originates centrally in a region of the brain stem called the medulla oblongata and reaches the heart through the vagus nerve (cranial nerve X). The vagus nerve carries impulses to the SA and AV nodes, and when stimulated releases acetylcholine, which causes hyperpolarization of the conduction cells. The result is a decrease in heart rate. At rest, parasympathetic system activity predominates and the heart is said to have "vagal tone." Recall that, in the absence of vagal tone, intrinsic heart rate would be approximately 100 beats/min. The vagus nerve has a depressant effect on the heart: It slows impulse generation and conduction and thus decreases the heart rate. Maximal vagal stimulation can decrease the heart rate to as low as 20 to 30 beats/min. The vagus nerve also decreases the force of cardiac muscle contraction.

The sympathetic nervous system, the other branch of the autonomic system, has opposite effects. Sympathetic stimulation increases the rate of impulse generation and conduction speed, and thus heart rate. Maximal sympathetic stimulation allows the heart rate to increase up to 250 beats/min. Sympathetic input also increases the contraction force of the ventricles. The sympathetic system predominates during times of physical or emotional stress, when the heart rate is greater than 100 beats/min. The parasympathetic system dominates when heart rate is less than 100. Thus, when exercise begins, or if exercise is at a low intensity, heart rate first increases due to withdrawal of vagal tone, with further increases if necessary due to sympathetic activation, as shown in figure 5.6.

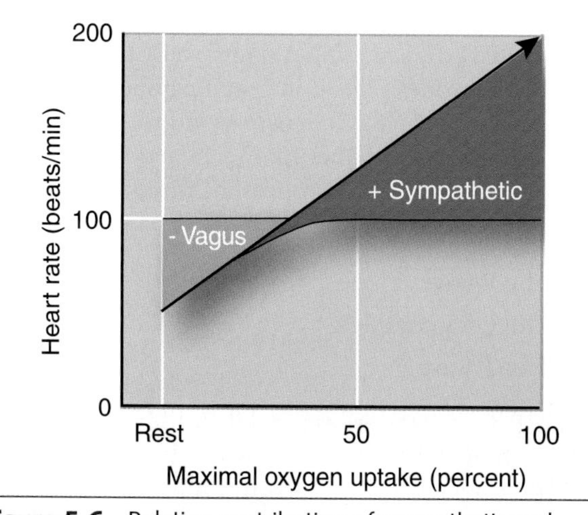

Figure 5.6 Relative contribution of sympathetic and parasympathetic nervous systems to the rise in heart rate during exercise.

Adapted from L.B. Rowell. *Human Cardiovascular Control.* Oxford University Press, 1993.

The third extrinsic influence, the endocrine system, exerts its effect through two hormones released by the adrenal medulla: norepinephrine and epinephrine (see chapter 2). These hormones are also known as catecholamines. Like norepinephrine that serves as the major neurotransmitter in the sympathetic nervous system, norepinephrine and epinephrine stimulate the heart, increasing its rate and contractility. In fact, release of these hormones from the adrenal medulla is triggered by sympathetic stimulation during times of stress, and their actions prolong the sympathetic response.

Normal resting heart rate (RHR) typically varies between 60 and 100 beats/min. With extended periods of endurance training (months to years), the RHR can decrease to 35 beats/min or less. A RHR as low as 28 beats/min has been observed in a world-class, long-distance runner. These lower training-induced RHRs are postulated to result from increased parasympathetic stimulation (vagal tone), with reduced sympathetic activity playing a lesser role.

In focus

Heart rate is established by the SA node, the pacemaker, but can be altered by the sympathetic and parasympathetic nervous systems as well as endocrine hormones.

Electrocardiogram (ECG)

The electrical activity of the heart can be recorded to monitor cardiac changes or diagnose potential cardiac problems. Because body fluids contain electrolytes, they are good electrical conductors. Electrical impulses generated in the heart are conducted through body fluids to the skin, where they can be amplified, detected, and printed out by a machine called an **electrocardiograph** (figure 5.7*a*). This printout is called an **electrocardiogram,** or **ECG** (figure 5.7*b*). A standard ECG is recorded from 10 electrodes placed in specific anatomical locations. These 10 electrodes correspond to 12 leads that represent different views of the heart. Three basic components of the ECG represent important aspects of cardiac function:

- The P wave
- The QRS complex
- The T wave

The P wave represents atrial depolarization and occurs when the electrical impulse travels from the SA node through the atria to the AV node. The QRS complex represents ventricular depolarization and occurs as the impulse spreads from the AV bundle to the **Purkinje fibers** and through the ventricles. The T wave represents ventricular repolarization. Atrial repolarization cannot be seen, because it occurs during ventricular depolarization (QRS complex).

Electrocardiograms are often obtained during exercise as clinical diagnostic tests of cardiac function. As exercise intensity increases, the heart must beat faster and work harder to deliver more blood to active muscles. Indications of coronary artery disease, not evident at rest, may show up on the ECG as the heart increases its rate of work. Exercise ECGs are also invaluable tools for research in exercise physiology because they provide a convenient method for tracking heart rate and rhythm changes during acute exercise.

An ECG provides a graphic record of the electrical activity of the heart and can be used to aid clinical diagnoses, for example in someone who has had a myocardial infarction in the past or is at risk for one in the future. It is important to remember that the ECG provides no information about the pumping capacity of the heart, only its electrical activity.

Cardiac Arrhythmias

Occasionally, disturbances in the normal sequence of cardiac events can lead to an irregular heart rhythm, called an arrhythmia. These disturbances vary in degree

In review

- The atria receive blood from the veins; the ventricles eject blood from the heart.
- Because the left ventricle must produce more force than other chambers to pump blood throughout the systemic circulation, its myocardial wall is thicker.
- Cardiac tissue is capable of spontaneous rhythmicity and has its own conduction system. It initiates its own depolarization without neural or hormonal control.
- The SA node is normally the heart's pacemaker, establishing its rate.
- Heart rate and contraction strength can be altered by the autonomic nervous system (sympathetic and parasympathetic) and the endocrine system through catecholamines (epinephrine and norepinephrine).
- The ECG is a recording of the heart's electrical activity. An exercise ECG can sometimes be used to detect underlying cardiac disorders.

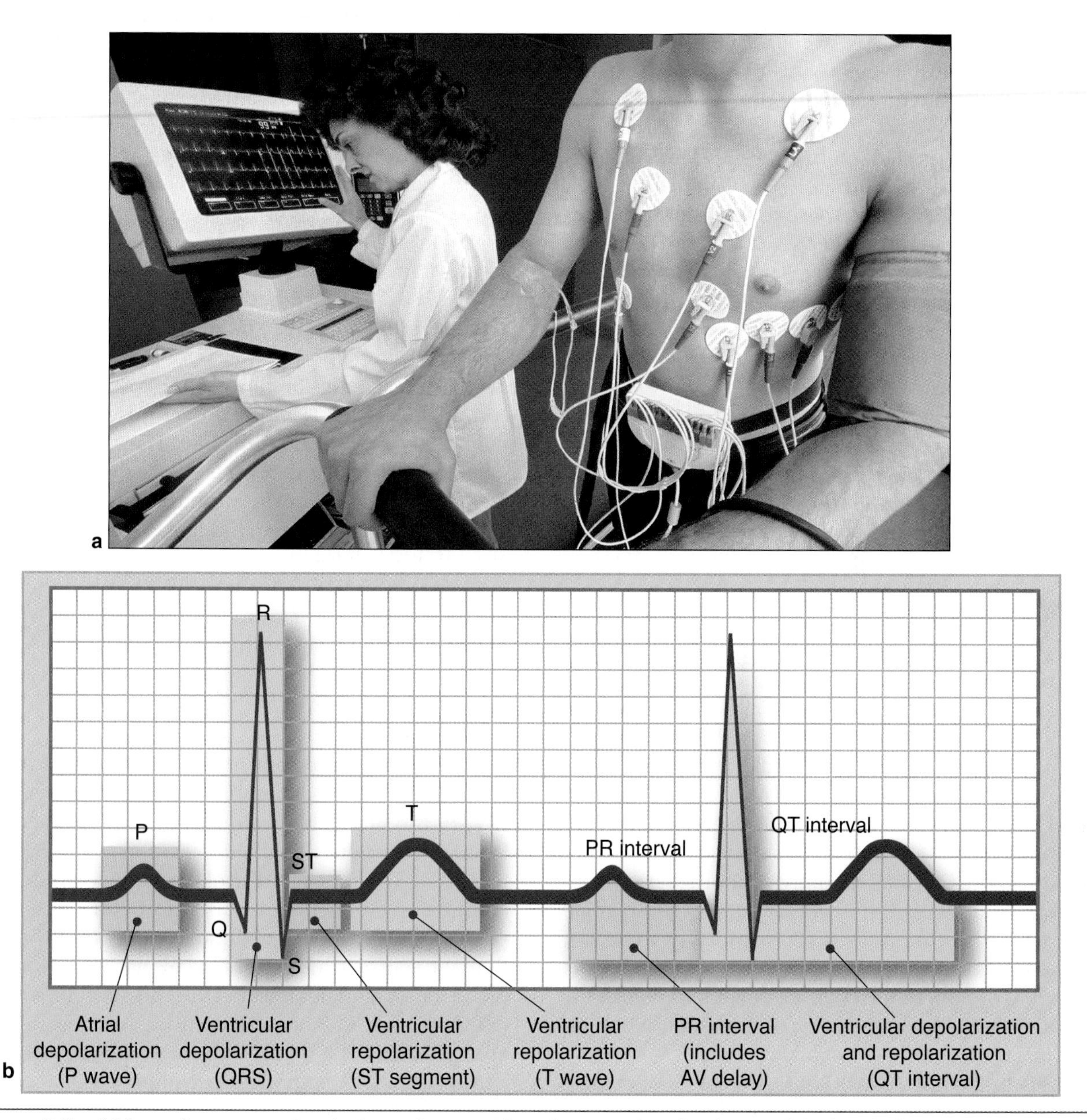

Figure 5.7 *(a)* Recording an exercise electrocardiogram. The subject is advised to let go of the handrail during testing (after he steadies himself). *(b)* A graphic illustration of the various phases of the resting electrocardiogram.

of seriousness. Bradycardia and tachycardia are two types of arrhythmias. **Bradycardia** is defined as a RHR lower than 60 beats/min, whereas **tachycardia** is defined as a resting rate greater than 100 beats/min. With these arrhythmias, the sinus rhythm is normal, but the rate is altered. In extreme cases, bradycardia or tachycardia can affect maintenance of sufficient blood pressure. Symptoms of both arrhythmias include fatigue, dizziness, light-headedness, and fainting. Tachycardia can sometimes be sensed as palpitations or a "racing" pulse.

Other arrhythmias also occur. For example, **premature ventricular contractions (PVCs),** which result in the feeling of skipped or extra beats, are relatively common and result from impulses originating outside the SA node. Atrial flutter, in which the atria contract at rates of 200 to 400 beats/min, and atrial fibrillation, in which the atria contract in a rapid and uncoordinated manner, are more serious arrhythmias that cause ventricular filling problems. **Ventricular tachycardia,** defined as three or more consecutive premature ventricular contractions, is a very serious arrhythmia that can lead to **ventricular fibrillation,** in which contraction of the ventricular tissue is uncoordinated. When this happens, the heart is extremely inefficient, with the result that little or no blood is pumped out of the heart. Most cardiac deaths result from ventricular fibrillation. Use of a defibrillator to shock the heart back into a normal sinus rhythm must occur within minutes if the victim is to survive. Chances of survival are greater if emergency treatment, including defibrillation, is provided quickly.

Interestingly, most highly trained endurance athletes develop low RHRs, an advantageous adaptation, as a result of training. Also, the heart rate naturally increases during physical activity to meet the increased demands of exercising muscle for blood flow. These adaptations should not be confused with pathological bradycardia or tachycardia, which are abnormal alterations in the RHR that usually indicate a pathological problem.

Terminology of Cardiac Function

The following terms are essential to understanding the work done by the heart and to our later discussions of cardiac responses to exercise: cardiac cycle, stroke volume, ejection fraction, and cardiac output ($\dot{Q}$).

Cardiac Cycle

The **cardiac cycle** includes all the mechanical and electrical events that occur during one heartbeat. In mechanical terms, it consists of all heart chambers undergoing a relaxation phase (diastole) and a contraction phase (systole). During diastole, the chambers fill with blood. During systole, the ventricles contract and expel blood into the aorta and pulmonary arteries. The diastolic phase is approximately twice as long as the systolic phase. Consider an individual with a heart rate of 74 beats/min. At this heart rate, the entire cardiac cycle takes 0.81 s to complete (60 s divided by 74 beats). Of the total cardiac cycle at this rate, diastole accounts for 0.50 s, or 62% of the cycle, and systole accounts for 0.31 s, or 38%. As the heart rate increases, these time intervals shorten proportionately.

Refer to the normal ECG in figure 5.7. One cardiac cycle spans the time between one systole and the next. Ventricular contraction (systole) begins during the QRS complex and ends in the T wave. Ventricular relaxation (diastole) occurs during the T wave and continues until the next contraction. Although the heart is continually working, it spends slightly more time in the diastole (~2/3 of the cardiac cycle) than in systole (~1/3 of the cardiac cycle).

The pressure inside the heart chambers rises and falls during each cardiac cycle. When the atria are relaxed, blood from the venous circulation fills the atria. About 70% of the blood filling the atria during this time passively flows directly through the mitral and tricuspid valves into the ventricles. When the atria contract, the atria push the remaining 30% of their volume into the ventricles.

During ventricular diastole, the pressure inside the ventricles is low, allowing the ventricles to passively fill with blood. After atrial contraction, pressure inside the ventricles increases slightly due to the increase in blood volume delivered from the atria. As the ventricles contract, pressure inside the ventricles rises sharply. This increase in ventricular pressure forces the atrioventricular valves (i.e., tricuspid and mitral valves) closed, preventing any backflow of blood from the ventricles to the atria. The closing of the atrioventricular valves results in the first heart sound. Furthermore, when ventricular pressure exceeds the pressure in the pulmonary artery and the aorta, the pulmonary and aortic valves open, allowing blood to flow into the pulmonary and systemic circulations, respectively. Following ventricular contraction, pressure inside the ventricles falls and the pulmonary and aortic valves close. The closing of these valves corresponds to the second heart sound.

The interactions of the various events of the heart are illustrated in figure 5.8, called a Wiggers diagram after the physiologist who created it. The diagram integrates information from the electrical conduction signals (ECG), heart sounds from the heart valves, pressure changes within the heart chambers, and left ventricular volume.

Stroke Volume

During systole, most, but not all, of the blood in the ventricles is ejected. This amount is the **stroke volume (SV)** of the heart—the volume of blood pumped per beat (contraction). This is depicted in figure 5.9*a* on page 134. To understand stroke volume, consider the amount of blood in the ventricle before and after contraction. At the end of diastole, just before contraction, the ventricle has completed filling. The volume of blood it now contains is called the **end-diastolic volume (EDV)**. At rest in a normally active adult, this value is approximately 100 ml. At the end of systole, just after contraction, the ventricle has completed its ejection phase, but not all the blood is pumped out of the heart. The volume of blood remaining in the ventricle is called the **end-systolic volume (ESV)** and is approximately 40 ml under resting conditions. Stroke volume is the volume of blood that was ejected and is merely the difference between the volume originally there and the amount remaining in the ventricle after contraction. So stroke volume is simply the difference between EDV and ESV; that is, SV = EDV – ESV (example: SV = 100 ml – 40 ml = 60 ml).

Ejection Fraction

The fraction of the blood pumped out of the left ventricle in relation to the amount of blood that was in the ventricle before contraction is called the **ejection fraction (EF)**. We determine this value, as seen in figure 5.9*b*, by dividing the stroke volume by EDV (60 ml / 100 ml = 60%). The EF, generally expressed as a percentage,

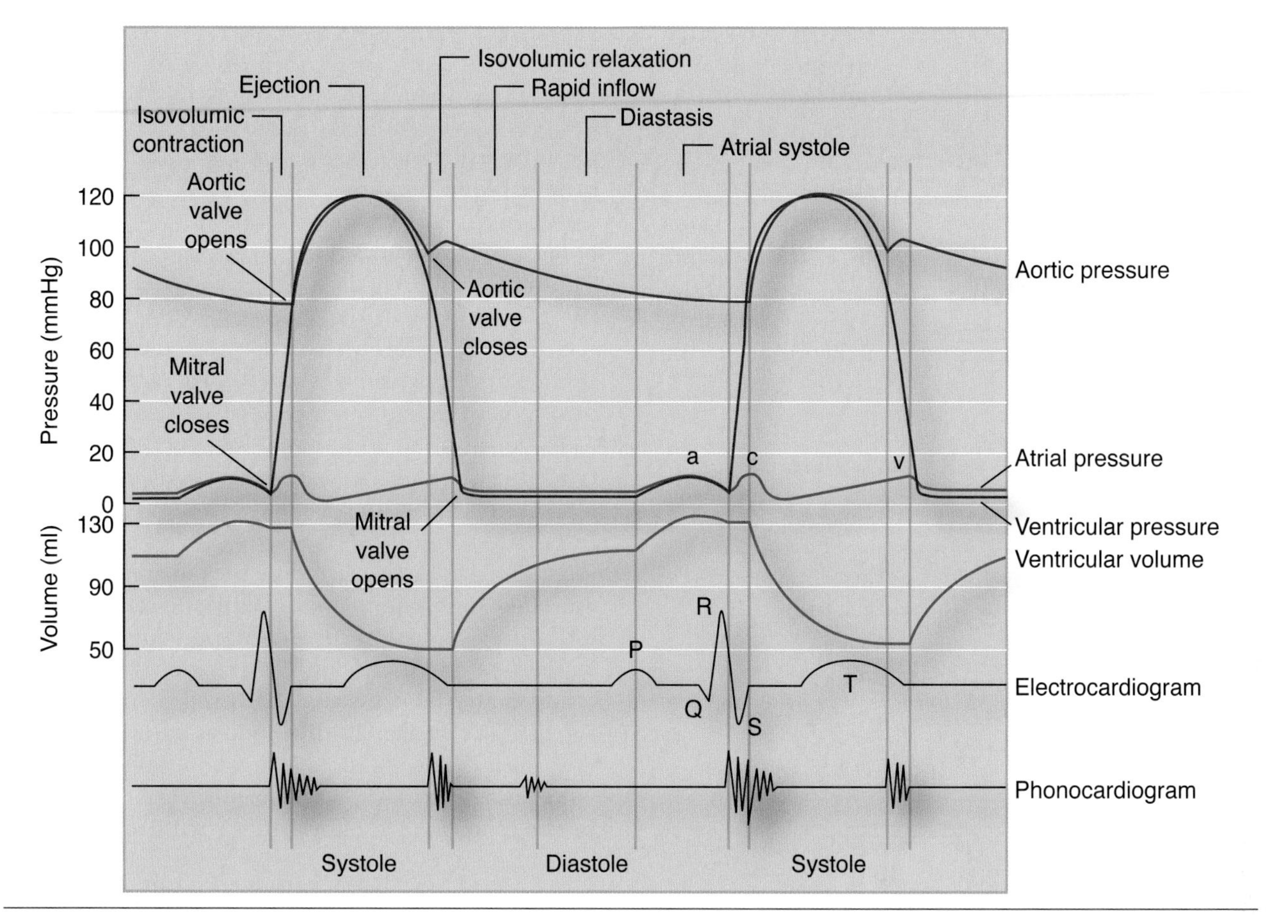

Figure 5.8 The Wiggers diagram, illustrating the events of the cardiac cycle for left ventricular function. Integrated into this diagram are the changes in left atrial and ventricular pressure, aortic pressure, ventricular volume, electrical activity (electrocardiogram), and heart sounds.

averages about 60% at rest. Thus, 60% of the blood in the ventricle at the end of diastole is ejected with the next contraction, and 40% remains in the ventricle. Ejection fraction is often used clinically as an index of the pumping ability of the heart.

Cardiac Output

Cardiac output ($\dot{Q}$), as shown in figure 5.9*c*, is the total volume of blood pumped by the ventricle per minute, or simply the product of HR and SV. The SV at rest in the standing posture averages between 60 and 80 ml of blood in most adults. Thus, at an RHR of 70 beats/min, the resting cardiac output will vary between 4.2 and 5.6 L/min. The average adult body contains about 5 L of blood, so this means that the equivalent of our total blood volume is pumped through our hearts about once every minute.

To calculate stroke volume, ejection fraction, and cardiac output:

$SV = EDV - ESV$ (ml).

$EF = (SV/EDV) \times 100$ (%).

$\dot{Q} = HR \times SV$ (L/min).

Understanding the electrical and mechanical activity of the heart provides a basis for understanding the cardiovascular system, but the heart is only one part of this system. In addition to the heart's serving as the primary pump, the system contains an intricate network of tubes that serve as a delivery system carrying the blood to all body tissues.

VASCULAR SYSTEM

The vascular system contains a series of vessels that transport blood from the heart to the tissues and back: the arteries, arterioles, capillaries, venules, and veins.

Arteries are large muscular, elastic, conduit vessels for transporting blood away from the heart to the arterioles. The aorta is the major artery transporting blood from the left ventricle to all regions of the body, and it

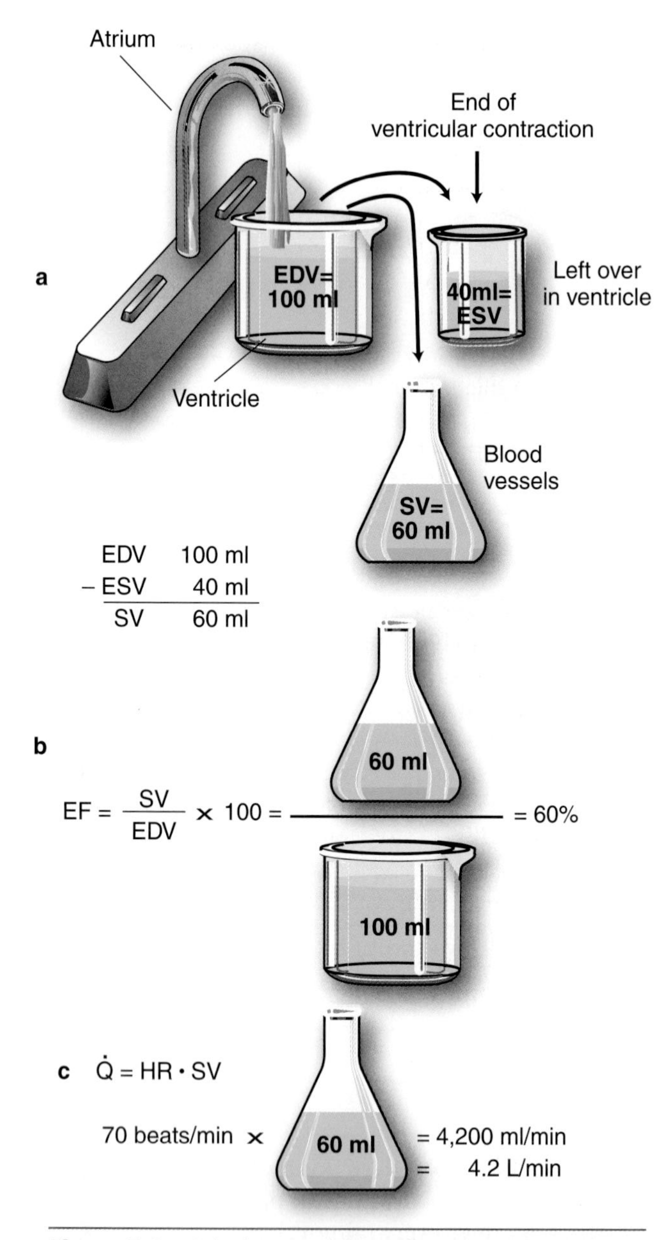

Figure 5.9 Calculations of *(a)* stroke volume (SV), which is the difference between end-diastolic volume (EDV) and end-systolic volume (ESV); *(b)* ejection fraction (EF); and *(c)* cardiac output ($\dot{Q}$).

branches into smaller arteries that become progressively smaller, finally branching into arterioles. The arterioles are the site of greatest control of the circulation by the sympathetic nervous system, so arterioles are sometimes called resistance vessels.

From the **arterioles,** blood enters the **capillaries,** the narrowest vessels, with walls only one cell thick. Virtually all exchange between the blood and the tissues occurs at the capillaries. Blood leaves the capillaries to begin the return trip to the heart in the **venules,** and the venules form larger vessels—the **veins.** The vena cava is the great vein transporting blood back to the right atrium from all regions of the body above (superior vena cava) and below (inferior vena cava) the heart.

Blood Pressure

Blood pressure is the pressure exerted by the blood on the vessel walls, and the term usually refers to arterial blood pressure. It is expressed by two numbers: the **systolic blood pressure (SBP)** and the **diastolic blood pressure (DBP).** The higher number is the SBP; it represents the highest pressure in the artery that occurs during ventricular systole. Ventricular contraction pushes the blood through the arteries with tremendous force, and that force exerts high pressure on the arterial walls. The lower number is the DBP and represents the lowest pressure in the artery, corresponding to ventricular diastole when the ventricle is filling.

Systolic blood pressure is the highest pressure within the vascular system, whereas diastolic blood pressure is the lowest pressure. Mean arterial pressure is the average pressure on the vessel walls during a cardiac cycle.

Mean arterial pressure (MAP) represents the average pressure exerted by the blood as it travels through the arteries. Since diastole takes about twice as long as systole in a normal cardiac cycle, mean arterial pressure can be estimated from the DBP and SBP as follows:

$$MAP = 2/3\ DBP + 1/3\ SBP.$$

Alternately,

$$MAP = DBP + [0.333 \times (SBP - DBP)].$$

(SBP – DBP) is also called pulse pressure.

To illustrate, with a normal resting blood pressure of 120 mmHg over 80 mmHg, the MAP = 80 + [0.333 × (120 – 80)] = 93 mmHg.

General Hemodynamics

Recall that the cardiovascular system is a continuous closed-loop system. Blood flows in this closed-loop system because of the pressure gradient that exists between the arterial and venous sides of the circulation. To understand regulation of blood flow to the tissues it is necessary to understand the intricate relationship between pressure, flow, and resistance.

In order for blood to flow in a vessel there must be a pressure difference from one end of the vessel to the other end. Blood will flow from the region of the vessel with high pressure to the region of the vessel with low pressure. Alternatively, if there is no pressure difference across the vessel, there is no driving force and therefore no blood flow. In the circulatory system, the mean arterial pressure in the aorta is approximately 100 mmHg at rest, and the pressure in the right atrium is very close to 0 mmHg. Therefore, the pressure difference across

Blood Flow to the Heart: Coronary Artery Blood Flow

The mechanism of blood flow to and through the coronary arteries is quite different from that of blood flow to the rest of the body. During contraction, when blood is forced out of the left ventricle under high pressure, the aortic semilunar valve is forced open. When this valve is open, its flaps block the entrances to the coronary arteries. As the pressure in the aorta decreases, the semilunar valve closes, and these entrances are exposed so that blood can then enter the coronary arteries. This design ensures that the coronary arteries are spared the very high blood pressure created by contraction of the left ventricle, thus protecting these vessels from damage.

the entire circulatory system is 100 mmHg – 0 mmHg = 100 mmHg.

The reason for the pressure differential from the arterial to the venous circulation is that the blood vessels themselves provide resistance or impedance to blood flow. The resistance that the vessel provides is largely dictated by the properties of the blood vessels and the blood itself. These properties include the length and radius of the blood vessel and the viscosity or thickness of the blood flowing through the vessel. Resistance to flow can be calculated as

$$\text{resistance} = [\eta L/r^4].$$

where η is the viscosity of the blood, L is the length of the vessel, and r is the radius of the vessel, which is raised to the fourth power.

Blood flow is proportional to the pressure difference across the system and is inversely proportional to resistance. This relationship can be illustrated by the following equation:

$$\text{blood flow} = \Delta\text{pressure}/\text{resistance}.$$

Notice that blood flow can increase by either an increase in the pressure difference (Δpressure), a decrease in resistance, or a combination of the two. Altering resistance to control blood flow is much more advantageous because very small changes in blood vessel radius equate to large changes in resistance. This is due to the fourth-power mathematical relationship between vascular resistance and vessel radius.

Changes in vascular resistance are largely due to changes in blood vessel radius or diameter, as the viscosity of the blood and the length of the vessels do not change significantly under normal conditions. Therefore, regulation of blood flow to organs is accomplished by small changes in blood vessel radius through **vasoconstriction** and **vasodilation.** This allows the cardiovascular system to divert blood flow to the areas where it is needed most.

As mentioned earlier, most resistance to blood flow occurs in the arterioles. Figure 5.10 shows the blood pressure changes across the entire vascular system. The arterioles are responsible for ~70% to 80% of the drop in mean arterial pressure across the entire cardiovascular system. This is important because small changes in arteriole radius can greatly affect the regulation of mean arteriole pressure and the local control of blood flow. At the capillary level, changes due to systole and diastole are no longer evident and the flow is smooth (laminar) rather than turbulent.

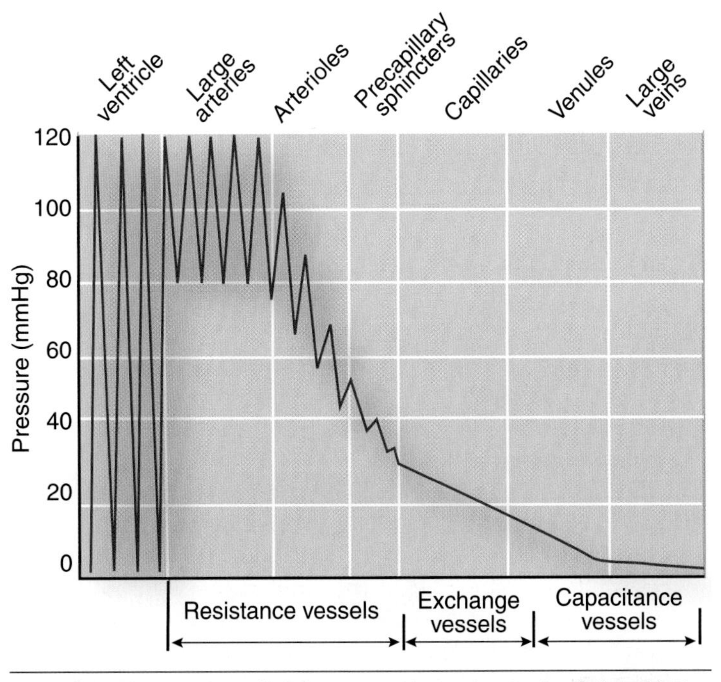

Figure 5.10 Pressure changes across the systemic circulation. Notice the large pressure drop that occurs across the arteriole portion of the system.

In focus

In terms of the entire cardiovascular system, cardiac output is the amount of blood flow to the entire system; the Δ pressure is the difference between aortic pressure when blood leaves the heart and venous pressure when blood returns to the heart; and resistance is the impedance to blood flow from the blood vessels. Blood flow is mainly controlled by small changes in blood vessel (arteriole) radius that greatly affect resistance.

Distribution of Blood

Distribution of blood to the various body tissues varies tremendously depending on the immediate needs of a specific tissue compared with that of other areas of the body. At rest under normal conditions, the most metabolically active tissues receive the greatest blood supply. The liver and kidneys combine to receive almost half the blood being circulated, and resting skeletal muscles receive only about 15% to 20%.

During exercise, blood is redirected to the areas where it is needed most. During heavy endurance exercise, muscles receive up to 80% or more of the available blood. This redistribution, along with increases in cardiac output (to be discussed in chapter 7), allows up to 25 times more blood flow to active muscles (see figure 5.11).

Similarly, after one eats a big meal, the digestive system receives more of the available cardiac output than when the digestive system is empty. Along the same lines, during increasing environmental heat stress, skin blood flow increases to a greater extent as the body attempts to maintain normal temperature. The cardiovascular system responds accordingly to redistribute blood, whether it is to the exercising muscle to match metabolism, for digestion, or to facilitate thermoregulation. These changes in the distribution of cardiac output are controlled by the sympathetic nervous system, primarily by increasing or decreasing

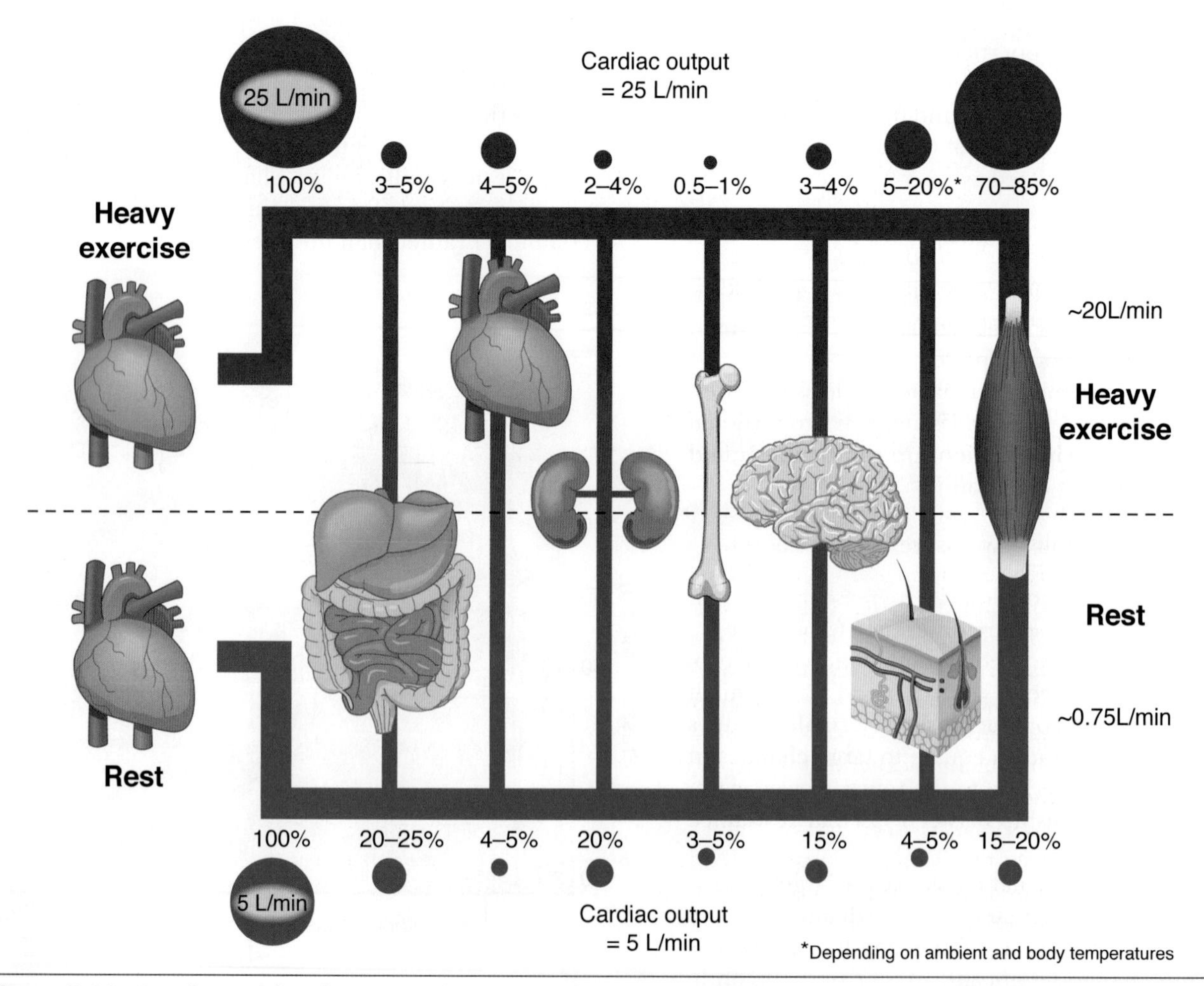

Figure 5.11 Distribution of cardiac output during rest and maximal exercise.

Reprinted, by permission, from P.O. Åstrand et al., 2003, *Textbook of work physiology: Physiological bases of exercise*, 4th ed. (Champaign, IL: Human Kinetics), 143.

In focus

The body has tremendous capacity to redistribute blood away from areas where the need is low to areas where there is increased need. Skeletal muscle normally receives about 15% of the total blood flow, or cardiac output, at rest. This can increase to 80% or more during heavy endurance exercise. Distribution of blood to various areas is controlled primarily at the level of the arterioles.

arteriolar diameter. These vessels have a strong muscular wall that can significantly alter vessel diameter, are highly innervated by sympathetic nerves, and have the capacity to respond to local control mechanisms.

Intrinsic Control of Blood Flow

Intrinsic control of blood distribution refers to the ability of the local tissues to vasodilate or vasoconstrict the arterioles that serve them and alter regional blood flow depending on the immediate needs of those tissues. With exercise and the increased metabolic demand of the exercising skeletal muscles, the arterioles undergo locally mediated vasodilation, opening up to allow more blood to enter that highly active tissue.

There are essentially three types of intrinsic control of blood flow. The strongest stimulus for the release of local vasodilating chemicals is metabolic, in particular an increased oxygen demand. As the tissue's oxygen use increases, available oxygen is diminished. Local arterioles vasodilate to allow more blood, and thus more oxygen, to perfuse that area. Other chemical changes that can stimulate increased blood flow are decreases in other nutrients and increases in by-products (carbon dioxide, K^+, H^+, lactic acid) or inflammatory chemicals. Second, several vasodilating substances can be produced in the endothelium (inner lining) of arterioles and initiate vasodilation in the vascular smooth muscle of the arterioles. These substances include nitric oxide (NO), prostaglandins, and endothelium-derived hyperpolarizing factor (EDHF). These endothelium-derived vasodilators are important to the regulation of blood flow at rest and during exercise in humans, although the precise mechanisms and interplay between these vasodilators are still being studied. Finally, pressure changes within the vessels themselves can also cause vasodilation and vasoconstriction. This is referred to as the myogenic response. The vascular smooth muscle contracts in response to an increase in pressure across the vessel wall and relaxes in response to a decrease in pressure across the vessel wall. Additionally, acetylcholine and adenosine also have been proposed as potential vasodilators for the increase in muscle blood flow during exercise. Increased blood flow can either bring in needed substances such as oxygen, or clear out metabolic waste such as carbon dioxide, or (usually) both. Figure 5.12 illustrates the three types of intrinsic control of vascular tone.

Extrinsic Neural Control

The concept of intrinsic local control explains redistribution of blood within an organ or tissue mass; however, the cardiovascular system must divert blood flow to where it is needed, beginning at a site upstream of the local environment. Redistribution at the system or body level is controlled by neural mechanisms. This is known as **extrinsic neural control** of blood flow, because the control comes from outside the specific area (extrinsic) instead of from locally inside the tissues (intrinsic).

Blood flow to all body parts is regulated largely by the sympathetic nervous system. The circular layers of smooth muscle within the artery and arteriole walls are supplied by sympathetic nerves. In most vessels, an increase in sympathetic nerve activity causes these muscle cells to contract, constricting blood vessels and thereby decreasing blood flow.

Under normal conditions, the sympathetic nerves transmit impulses continuously to the blood vessels, keeping the vessels moderately constricted to maintain adequate blood pressure. This state of tonic vasoconstriction is referred to as vasomotor tone. When sympathetic stimulation increases, further constriction of the blood vessels in a specific area decreases blood flow into that area and allows more blood to be dis-

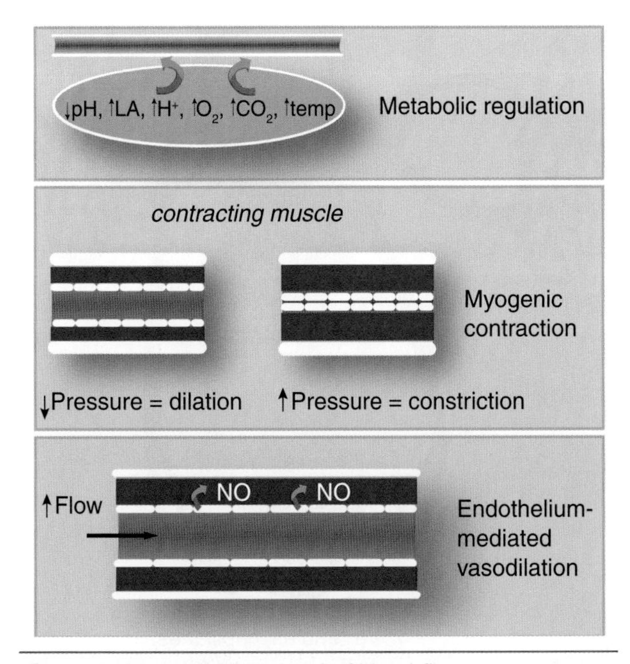

Figure 5.12 Intrinsic control of blood flow.

Figure courtesy of Dr. Donna H. Korzick, Pennsylvania State University.

In focus

Blood flow can be controlled at the local tissue level by the release of locally acting metabolic dilators, endothelium-dependent vasodilators (NO, prostaglandins, EDHF), and the myogenic response as well as by extrinsic neural control. The sympathetic nervous system plays a major role in redirecting blood flow from areas of low need to areas of high need.

tributed elsewhere. But if sympathetic stimulation decreases below the level needed to maintain tone, constriction of vessels in the area is lessened, so the vessels passively vasodilate, increasing blood flow into that area. Therefore, sympathetic stimulation will cause vasoconstriction in most vessels, but blood flow is altered by either increasing or decreasing the amount of vasoconstriction relative to normal vasomotor tone.

Distribution of Venous Blood

While flow to tissues is controlled by changes on the arterial side of the system, most of the blood volume normally resides in the venous side of the system. At rest,

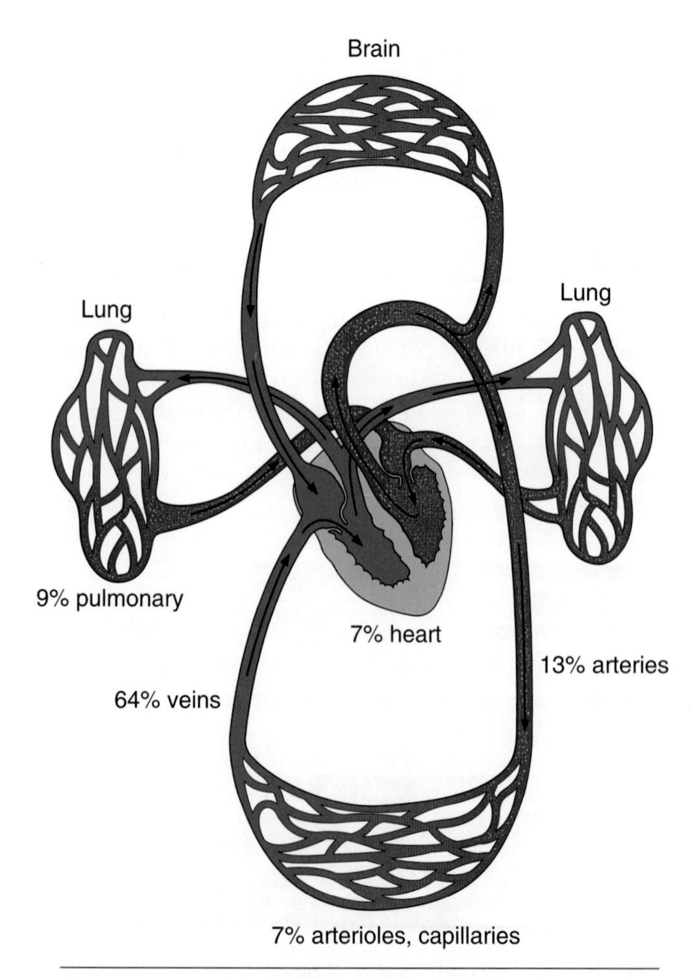

Figure 5.13 Blood volume distribution within the vasculature when the body is at rest.

the blood volume is distributed among the vasculature as shown in figure 5.13. The venous system has a great capacity to hold blood volume. There is little vascular smooth muscle in the veins, and they are very elastic and "balloon-like." Thus, the venous system provides a large reservoir of blood available to be rapidly distributed back to the heart (venous return) and to the arterial circulation. This is accomplished through sympathetic stimulation of the venules and veins, which causes the vessels to constrict.

Integrative Control of Blood Pressure

Blood pressure is normally maintained by reflexes from the autonomic nervous system. Specialized pressure sensors located in the aortic arch and the carotid arteries, called **baroreceptors,** are sensitive to changes in arterial pressure. They send information about the current blood pressure to the cardiovascular control centers in the brain where autonomic reflexes are initiated to respond to changes in blood pressure. For example, when blood pressure is elevated, the baroreceptors are stimulated by an increase in stretch. They relay this information to the cardiovascular control center in the brain. In response to the increased pressure there is a reflex increase in vagal tone, to decrease heart rate, and a decrease in sympathetic activity, which serves to normalize blood pressure. In response to a decrease in blood pressure, less stretch is sensed by the baroreceptors, and the response is to increase heart rate by vagal withdrawal and to increase sympathetic nervous activity, thus correcting the low-pressure signal.

There are also other specialized receptors, called **chemoreceptors** and **mechanoreceptors,** that send information about the chemical environment in the muscle and the length and tension of the muscle to the cardiovascular control centers. These receptors can also modify the blood pressure response and are especially important during exercise.

Return of Blood to the Heart

Because we spend so much time in an upright position, the cardiovascular system requires mechanical assistance to overcome the force of gravity when blood returns from the lower parts of the body to the heart. Three basic mechanisms assist in this process:

- Valves in the veins
- The muscle pump
- The respiratory pump

The veins contain valves that allow blood to flow in only one direction, thus preventing backflow and pooling of blood in the lower body. These venous valves also complement the action of the skeletal muscle pump, mechanical compression of the veins from rhythmic skeletal muscle contraction (figure 5.14). This pushes blood volume in the veins back toward the heart. Finally, the changes in pressure in the abdominal and thoracic cavities during breathing assist blood return to the heart.

BLOOD

Blood serves many useful purposes in regulating normal body function. The three functions of primary importance to exercise and sport are

- transportation,
- temperature regulation, and
- acid–base (pH) balance.

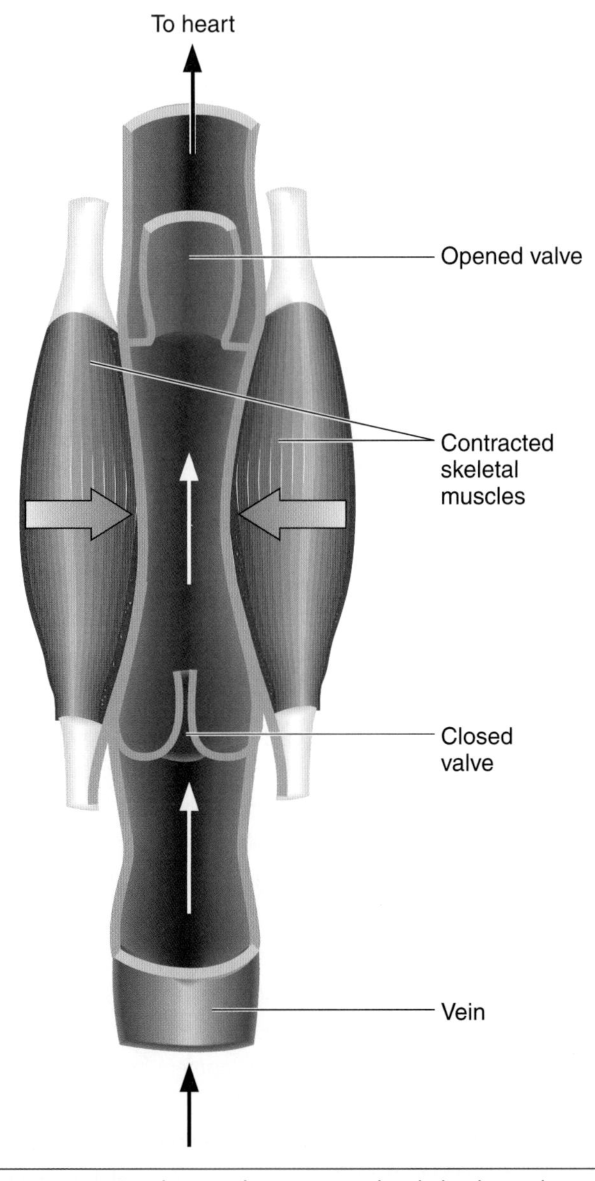

Figure 5.14 The muscle pump. As the skeletal muscles contract, they squeeze the veins in the legs and assist in the return of blood to the heart. Valves within the veins ensure the unidirectional flow of blood back to the heart.

In review

- Blood is distributed throughout the body based on the needs of the individual tissues. The most active tissues receive the most blood.
- Redistribution of blood is controlled locally by the release of chemical substances, which causes vasodilation, thus increasing blood flow to the area.
- Extrinsic neural control of blood flow distribution is accomplished by the sympathetic nervous system, primarily through vasoconstriction. The arterioles are the site of major control.
- Blood returns to the heart through the veins, assisted by valves within the veins, the muscle pump, and respiratory pressure changes.

We are most familiar with blood's transporting functions. In addition, blood is critical in temperature regulation during physical activity; it picks up heat from the exercising muscle and delivers it to the skin where it can be dissipated to the environment (see chapter 11). Blood also buffers the acids produced by anaerobic metabolism and maintains proper pH for metabolic processes (see chapters 2 and 6).

Blood Volume and Composition

The total volume of blood in the body varies considerably with an individual's size and state of training. Larger blood volumes are associated with greater lean body mass and higher levels of endurance training. The blood volume of people of average body size and normal physical activity generally ranges from 5 to 6 L in men and 4 to 5 L in women.

Blood is composed of plasma and formed elements (see figure 5.15). Plasma normally constitutes about 55% to 60% of total blood volume but can decrease by 10% of its normal amount or more with intense exercise in heat, or can increase by 10% or more with endurance training or acclimation to heat. Approximately 90% of the plasma volume is water; 7% consists of plasma proteins; and the remaining 3% includes cellular nutrients, electrolytes, enzymes, hormones, antibodies, and wastes.

The formed elements, which normally constitute the other 40% to 45% of total blood volume, are the red blood cells (erythrocytes), white blood cells (leukocytes), and platelets (thrombocytes). Red blood cells constitute more than 99% of the formed-element volume; white blood cells and platelets together account

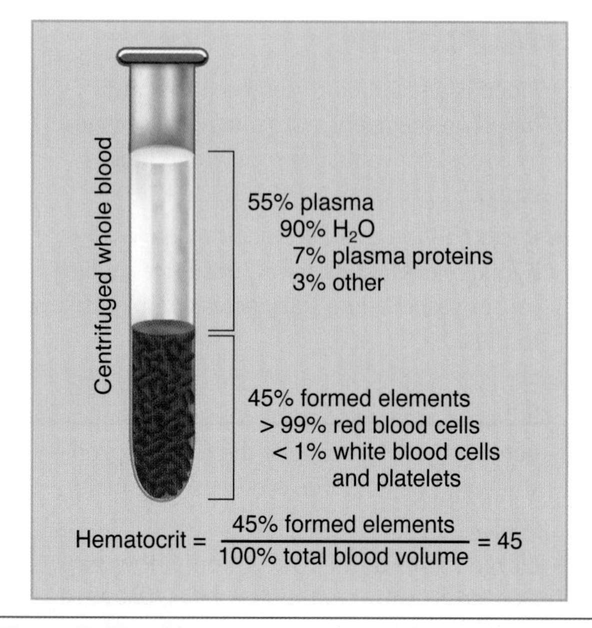

Figure 5.15 The composition of whole blood, illustrating the plasma volume (fluid portion) and the cellular volume (red cells, white cells, and platelets) after the blood sample has been centrifuged.

> **In focus**
>
> The hematocrit is the ratio of the formed elements in the blood (red cells, white cells, and platelets) to the total blood volume.

for less than 1%. The percentage of the total blood volume composed of cells or formed elements is referred to as the **hematocrit.**

White blood cells protect the body from infection either by directly destroying the invading agents through phagocytosis (ingestion) or by forming antibodies to destroy them. Adults have about 7,000 white blood cells per cubic millimeter of blood.

The remaining formed elements are the blood platelets. These are cell fragments that are required for blood coagulation (clotting), which prevents excessive blood loss. Exercise physiologists are most concerned with red blood cells.

Red Blood Cells

Mature red blood cells (erythrocytes) have no nucleus, so they cannot reproduce as other cells can. They must be replaced with new cells on a reoccurring basis, a process called **hematopoiesis.** The normal life span of a red blood cell is about four months. Thus, these cells are continuously produced and destroyed at about equal rates. This balance is very important, because adequate oxygen delivery to tissues depends on having a sufficient number of red blood cells to transport oxygen. Decreases in their number or function can hinder oxygen delivery and thus affect exercise performance.

Red blood cells transport oxygen, which is primarily bound to hemoglobin. **Hemoglobin** is composed of a protein (globin) and a pigment (heme). Heme contains iron, which binds oxygen. Each red blood cell contains approximately 250 million hemoglobin molecules, each able to bind four oxygen molecules—so each red blood cell can bind up to a billion molecules of oxygen! There is an average of 15 g of hemoglobin per 100 ml of whole blood. Each gram of hemoglobin can combine with 1.33 ml of oxygen, so as much as 20 ml of oxygen can be bound for each 100 ml of blood. Therefore, when arterial blood is saturated with oxygen, it has an oxygen-carrying capacity of 20 ml of oxygen per 100 ml of blood.

Blood Viscosity

Viscosity refers to the thickness of the blood. Recall from our discussion of vascular resistance that the more viscous a fluid, the more resistant it is to flow. The viscosity of blood is normally about twice that of water. Blood viscosity, and thus resistance to flow, increases with higher hematocrits.

Because of oxygen transport by the red blood cells, an increase in their number would be expected to maximize oxygen transport. But if an increase in red blood cell count is not accompanied by a similar increase in plasma volume, blood viscosity and vascular resistance will increase, which could result in reduced blood flow. This generally is not a problem unless the hematocrit reaches 60% or more.

Conversely, the combination of a low hematocrit with a high plasma volume, which decreases the blood's viscosity, appears to have certain benefits for the blood's transport function because the blood can flow more easily. Unfortunately, a low hematocrit frequently results from a reduced

> **In focus**
>
> When we donate blood, the removal of one "unit," or nearly 500 ml, represents approximately an 8% to 10% reduction in both the total blood volume and the number of circulating red blood cells. Donors are advised to drink plenty of fluids. Because plasma is primarily water, simple fluid replacement returns plasma volume to normal within 24 to 48 h. However, it takes at least six weeks to reconstitute the red blood cells because they must go through full development before they are functional. Blood loss greatly compromises the performance of endurance athletes by reducing oxygen delivery capacity.

red blood cell count, as in diseases such as anemia. Under these circumstances, the blood can flow easily, but it contains fewer carriers, so oxygen transport is impeded. For optimal physical performance, a low hematocrit with a normal or slightly elevated number of red blood cells is desirable. This combination facilitates oxygen transport. Many endurance athletes achieve this combination as part of their cardiovascular system's normal adaptation to training. This adaptation is discussed in chapter 10.

In review

- Blood is about 55% to 60% plasma and 40% to 45% formed elements.
- Oxygen is transported primarily by binding to the hemoglobin in red blood cells.
- As blood viscosity increases, so does resistance to flow.

In Closing

In this chapter, we reviewed the structure and function of the cardiovascular system. We learned how blood flow and blood pressure are regulated to meet the body's needs, and explored the role of the cardiovascular system in transporting and delivering oxygen and nutrients to the body's cells while clearing away metabolic wastes, including carbon dioxide. Knowing how substances are moved within the body, we now look more closely at the movement of oxygen and carbon dioxide. In the next chapter, we explore the respiratory system, considering how oxygen is moved into the body, how oxygen is delivered to the body's cells, and how carbon dioxide is removed.

KEY TERMS

arteries
arterioles
atrioventricular (AV) node
baroreceptor
bradycardia
capillaries
cardiac cycle
cardiac output ($\dot{Q}$)
chemoreceptor
diastolic blood pressure (DBP)
ejection fraction (EF)
electrocardiogram (ECG)
electrocardiograph
end-diastolic volume (EDV)
end-systolic volume (ESV)
extrinsic neural control
hematocrit
hematopoiesis
hemoglobin
mean arterial pressure (MAP)
mechanoreceptors
myocardium
pericardium
premature ventricular contraction (PVC)
Purkinje fibers
sinoatrial (SA) node
stroke volume (SV)
systolic blood pressure (SBP)
tachycardia
vasoconstriction
vasodilation
veins
ventricular fibrillation
ventricular tachycardia
venules

STUDY QUESTIONS

1. Describe the structure of the heart, the pattern of blood flow through the valves and chambers of the heart, how the heart as a muscle is supplied with blood, and what happens when the resting heart must suddenly supply an exercising body.
2. What events take place that allow the heart to contract, and how is heart rate controlled?
3. What is the difference between systole and diastole, and how do they relate to SBP and DBP?
4. What is the relationship between pressure, flow, and resistance?
5. How is blood flow to the various regions of the body controlled?
6. Describe the three important mechanisms for returning blood back to the heart when someone is exercising in an upright position.
7. Describe the primary functions of blood.

STUDY GUIDE ACTIVITIES

In addition to the activities listed in the chapter opening outline on page 123, two other activities are available in the online study guide, located at

www.HumanKinetics.com/PhysiologyOfSportAndExercise.

The chapter **SUMMARY** reviews the key concepts covered in the chapter, and the end-of-chapter **QUIZ** tests your understanding of the material covered in the chapter.

The Respiratory System and Its Regulation

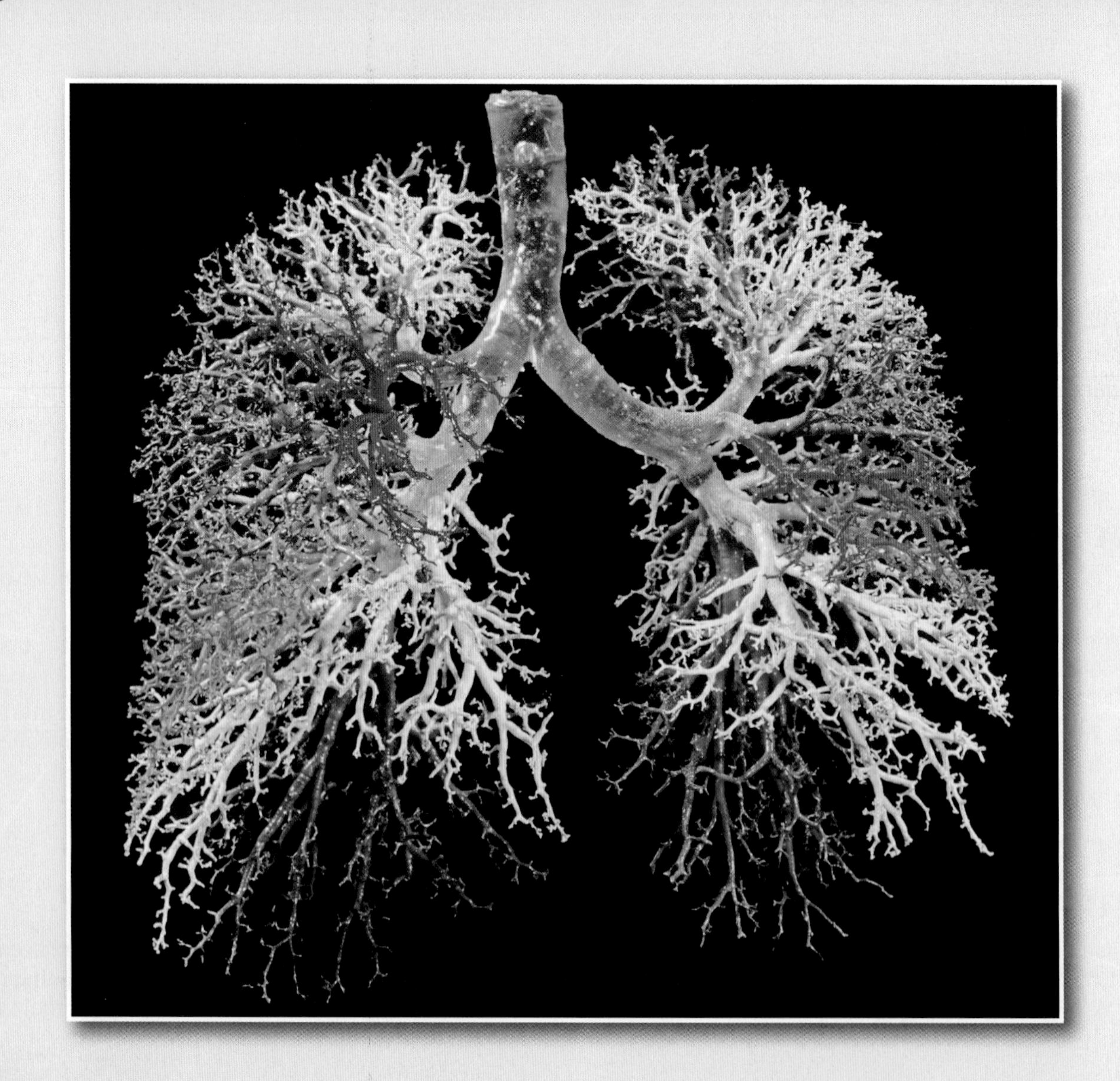

In this chapter and in the online study guide

During the 1972 Summer Olympic Games in Munich, a 16-year-old swimmer, Rick DeMont, won the 400 m freestyle event, but his gold medal was taken away when officials discovered traces of ephedrine, a banned substance, in his blood. DeMont, who had asthma, had taken drugs for allergies and also was receiving weekly shots for his allergies. Ironically, DeMont had declared these drugs on his medical statement, but the information was not provided to the appropriate authorities at the International Olympic Committee. The United States Olympic Committee, in December 2001, admitted that it had not acted properly in its handling of DeMont's medical information. His name has now been cleared with the USOC, and now the IOC must decide if he should get his medal back. As a final note, it is still not clear if the more popular allergy and asthma medications used today have ergogenic properties when administered to individuals who do not have asthma.

The respiratory and cardiovascular systems combine to provide an effective delivery system that carries oxygen to and removes carbon dioxide from tissues. This transportation involves four separate processes:

- Pulmonary ventilation (breathing): movement of air into and out of the lungs
- Pulmonary diffusion: the exchange of oxygen and carbon dioxide between the lungs and the blood
- Transport of oxygen and carbon dioxide via the blood
- Capillary diffusion: the exchange of oxygen and carbon dioxide between the capillary blood and the metabolically active tissues

The first two processes are referred to as **external respiration** because they involve moving gases from outside the body into the lungs and then the blood. Once the gases are in the blood, they must be transported to the tissues. When blood arrives at the tissues, the fourth step of respiration occurs. This gas exchange between the blood and the tissues is called **internal respiration.** Thus, external and internal respiration are linked by the circulatory system. The following sections examine all four components of respiration.

PULMONARY VENTILATION

Pulmonary ventilation, commonly referred to as breathing, is the process by which we move air into and out of the lungs. The anatomy of the respiratory system is illustrated in figure 6.1. Air is typically drawn into the lungs through the nose, although the mouth must also be used when the demand for air exceeds the amount

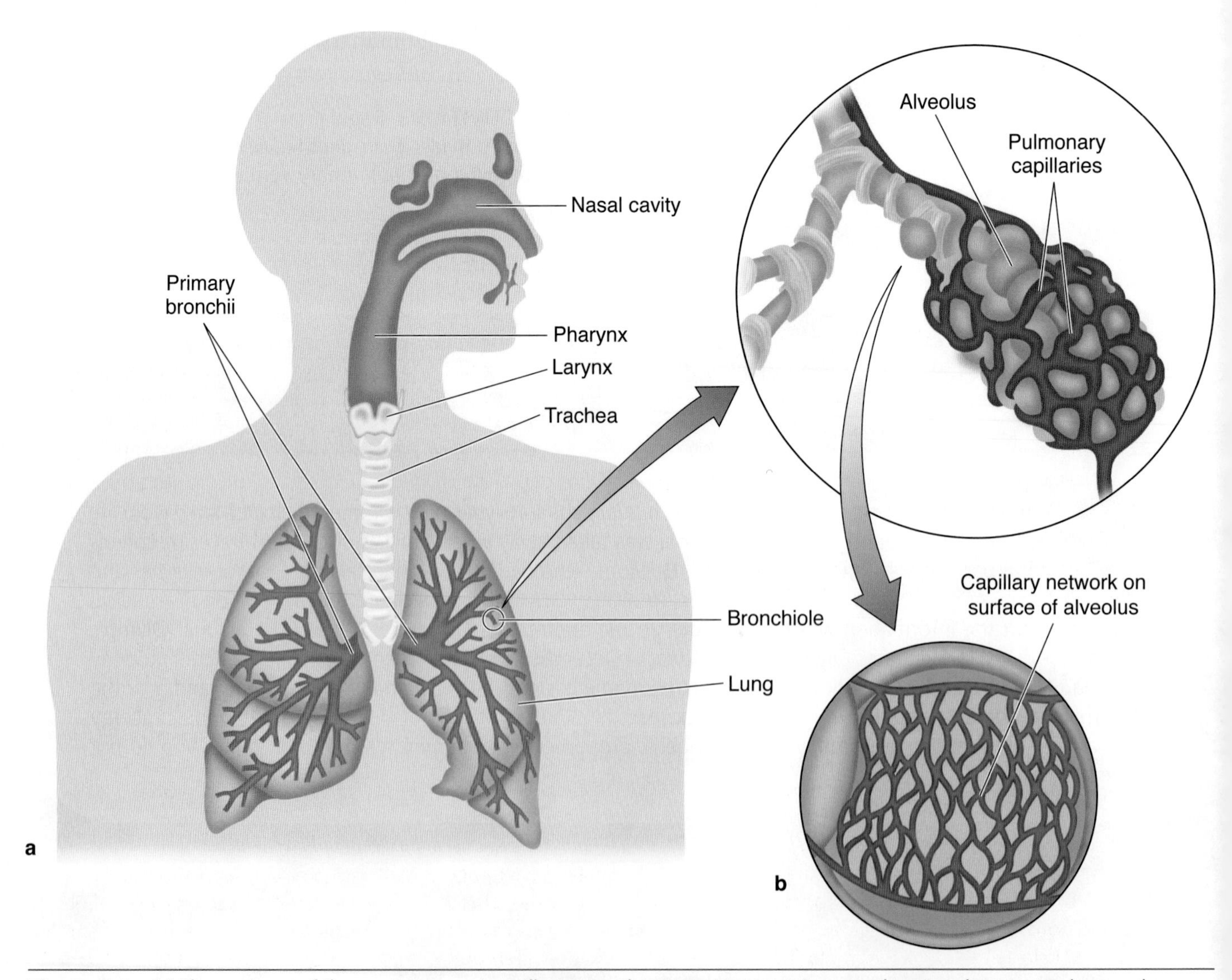

Figure 6.1 *(a)* The anatomy of the respiratory system, illustrating the respiratory tract (i.e., nasal cavity, pharynx, trachea, and bronchi). *(b)* The enlarged view of the alveolus shows the regions of gas exchange between the alveolus and pulmonary blood in the capillaries.

that can comfortably be brought in through the nose. Nasal breathing is advantageous because the air is warmed and humidified as it swirls through the bony irregular sinus surfaces (turbinates or conchae). Of equal importance, the turbinates churn the inhaled air, causing dust and other particles to contact and adhere to the nasal mucosa. This filters out all but the tiniest particles, minimizing irritation and the threat of respiratory infections. From the nose and mouth, the air travels through the pharynx, larynx, trachea, and bronchial tree. These anatomical structures serve as the transport zone of the lungs because gas exchange does not occur in these structures. Exchange of oxygen and carbon dioxide occurs when air finally reaches the smallest respiratory units: the respiratory bronchioles and the alveoli. The respiratory bronchioles are primarily transport tubes also, but are included in this region because they contain clusters of alveoli. This is known as the respiratory zone because it is the site of gas exchange in the lungs.

In focus

Breathing through the nose helps humidify and warm the air during inhalation and filters out foreign particles from the air.

The lungs are not directly attached to the ribs. Rather, they are suspended by the pleural sacs. The pleural sacs have a double wall: the parietal pleura, which lines the thoracic wall, and the visceral or pulmonary pleura, which lines the outer aspects of the lung. These pleural walls envelop the lungs and have a thin film of fluid between them that reduces friction during respiratory movements. In addition, these sacs are connected to the lungs and to the inner surface of the thoracic cage, causing the lungs to take the shape and size of the rib or thoracic cage as the chest expands and contracts.

The anatomy of the lungs, the pleural sacs, and the thoracic cage determines airflow into and out of the lungs, that is, inspiration and expiration.

Inspiration

Inspiration is an active process involving the diaphragm and the external intercostal muscles. Figure 6.2*a* shows the resting positions of the diaphragm and the thoracic cage, or thorax. With inspiration, the ribs and sternum are moved by the external intercostal muscles. The ribs swing up and out and the sternum swings up and forward. At the same time, the diaphragm contracts, flattening down toward the abdomen.

These actions, illustrated in figure 6.2*b*, expand all three dimensions of the thoracic cage, increasing the volume inside the lungs. When the lungs are expanded they have a greater volume, so the air within them has more space to fill. According to **Boyle's gas law,** which states that pressure × volume is constant (at a constant temperature), the pressure within the lungs decreases. As a result, the pressure in the lungs (intrapulmonary pressure) is less than the air pressure outside the body. Because the respiratory tract is open to the outside, air rushes into the lungs to reduce this pressure difference. This is how air moves into the lungs during inspiration.

During forced or labored breathing, as during heavy exercise, inspiration is further assisted by the action of other muscles, such as the scaleni (anterior, middle, and posterior) and sternocleidomastoid in the neck and the pectorals in the chest. These muscles help raise the ribs even more than during regular breathing.

In focus

The pressure changes required for adequate ventilation at rest are really quite small. For example, at standard atmospheric pressure (760 mmHg), inspiration may decrease the pressure in the lungs (intrapulmonary pressure) by only about 2 to 3 mmHg. However, during maximal respiratory effort, such as during exhaustive exercise, the intrapulmonary pressure can decrease by 80 to 100 mmHg.

Expiration

At rest, **expiration** is usually a passive process involving relaxation of the inspiratory muscles and elastic recoil of the lung tissue. As the diaphragm relaxes, it returns to its normal upward, arched position. As the external intercostal muscles relax, the ribs and sternum move back into their resting positions (figure 6.2*c*). While this happens, the elastic nature of the lung tissue causes it to recoil to its resting size. This increases the pressure in the lungs and causes a proportional decrease in volume in the thorax, and therefore air is forced out of the lungs.

During forced breathing, expiration becomes a more active process. The internal intercostal muscles actively pull the ribs down. This action can be assisted by the latissimus dorsi and quadratus lumborum muscles. Contracting the abdominal muscles increases the intra-abdominal pressure, forcing the abdominal viscera upward against the diaphragm and accelerating its return to the domed position. These muscles also pull the rib cage down and inward.

The changes in intra-abdominal and intrathoracic pressure that accompany forced breathing also help return venous blood back to the heart; this is similar to the action of the muscle pump in the legs in assisting the return of venous volume. As intra-abdominal and intrathoracic pressure increases, it is transmitted to the great veins—the pulmonary veins and superior and

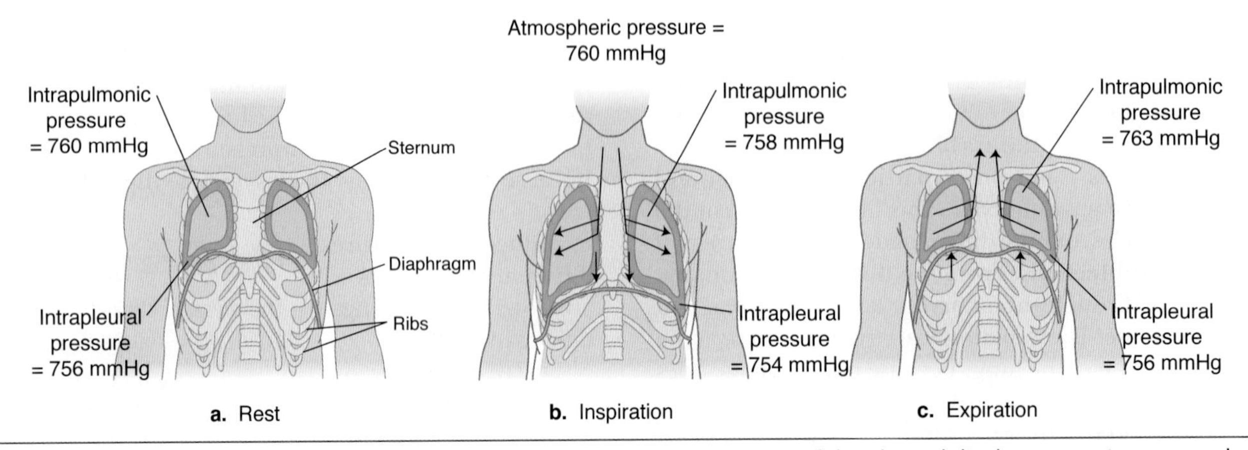

Figure 6.2 The process of inspiration and expiration, showing how movement of the ribs and diaphragm can increase and decrease the size of the thorax. *(a)* Note the size of the rib cage at rest. *(b)* The dimensions of the lungs and the thoracic cage increase during inspiration, forming a negative pressure that draws air into the lungs. *(c)* During expiration, the lung volume decreases, thereby forcing air out of the lungs.

inferior venae cavae—that transport blood back to the heart. When the pressure decreases, the veins return to their original size and fill with blood. The changing pressures within the abdomen and thorax squeeze the blood in the veins, assisting its return through a milking action. This phenomenon is known as the **respiratory pump** and is essential in maintaining adequate venous return.

PULMONARY VOLUMES

The volume of air in the lungs can be measured with a technique called **spirometry.** A spirometer measures the volumes of air inspired and expired and therefore changes in lung volume. Although more sophisticated spirometers are used today, a simple spirometer contains a bell filled with air that is partially submerged in water. A tube runs from the person's mouth under the water and emerges inside the bell, just above the water level. As the person exhales, air flows down the tube and into the bell, causing the bell to rise. The bell is attached to a pen, and its movement is recorded on a simple rotating drum (figure 6.3).

This technique is used clinically to measure lung volumes, capacities, and flow rates as an aid in diagnosing such respiratory diseases as asthma, as mentioned in the opening story, and emphysema.

The amount of air entering and leaving the lungs with each breath is known as the **tidal volume.** The **vital capacity (VC)** is the greatest amount of air that can be expired after a maximal inspiration. The amount of air remaining in the lungs after a maximal expiration is the **residual volume (RV).** The residual volume cannot

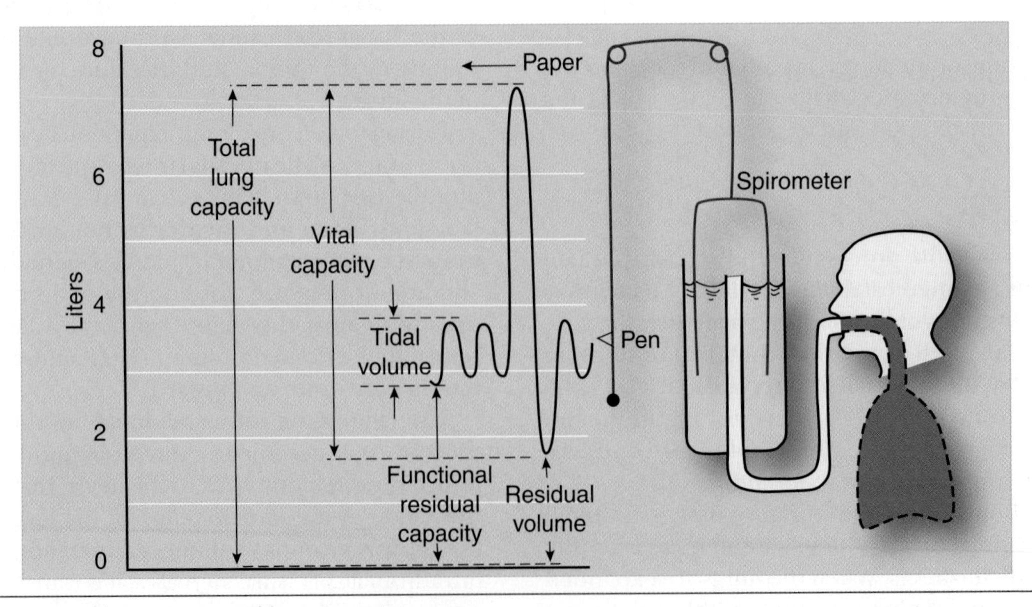

Figure 6.3 Lung volumes measured by spirometry.

Reprinted, by permission, from J. West, 2000, *Respiratory physiology: The essentials* (Baltimore, MD: Lippincott, Williams, and Wilkins), 14.

In review

- Pulmonary ventilation (breathing) is the process by which air is moved into and out of the lungs. It has two phases: inspiration and expiration.
- Inspiration is an active process in which the diaphragm and the external intercostal muscles contract, increasing the dimensions, and thus the volume, of the thoracic cage. This decreases the pressure in the lungs, causing air to flow in.
- Expiration at rest is normally a passive process. The inspiratory muscles and diaphragm relax and the elastic tissue of the lungs recoils, returning the thoracic cage to its smaller, normal dimensions. This increases the pressure in the lungs and forces air out.
- Forced or labored inspiration and expiration are active processes, dependent on accessory muscle actions.
- Lung volumes and capacities, along with rates of airflow into and out of the lungs, are measured by spirometry.

be measured with spirometry. The **total lung capacity (TLC)** is the sum of the vital capacity and the residual volume.

PULMONARY DIFFUSION

Gas exchange in the lungs, called **pulmonary diffusion,** serves two major functions:

- It replenishes the blood's oxygen supply, which is depleted at the tissue level as it is used for oxidative energy production.
- It removes carbon dioxide from returning systemic venous blood.

Air is brought into the lungs during pulmonary ventilation, enabling gas exchange to occur between this air and the blood through pulmonary diffusion. Oxygen from the air diffuses from the alveoli into the blood in the pulmonary capillaries, and carbon dioxide diffuses from the blood into the alveoli in the lungs. The **alveoli** are grapelike clusters, or air sacs, at the ends of the terminal bronchioles.

Blood from most of the body returns through the vena cavae to the right side of the heart. From the right ventricle, this blood is pumped through the pulmonary artery to the lungs, ultimately working its way into the pulmonary capillaries. These capillaries form a dense network around the alveolar sacs. These vessels are so small that the red blood cells must pass through them in single file, such that each cell is exposed to the surrounding lung tissue. This is where pulmonary diffusion occurs.

Blood Flow to the Lungs at Rest

At rest the lungs receive approximately 4 to 6 L/min of blood flow, depending on body size. Because cardiac output from the right side of the heart approximates cardiac output from the left side of the heart, blood flow to the lungs matches blood flow to the systemic circulation. However, pressure and vascular resistance in the blood vessels in the lungs are different than in the system circulation. The mean pressure in the pulmonary artery is ~15 mmHg (systolic pressure is ~25 mmHg and diastolic pressure is ~8 mmHg) compared to the mean pressure in the aorta of ~95 mmHg. The pressure in the left atrium where blood is returning to the heart from the lungs is ~5 mmHg; thus there is not a great pressure difference across the pulmonary circulation (15 – 5 mmHg). Figure 6.4 illustrates the differences in pressures between the pulmonary and systemic circulation.

Recalling the discussion of blood flow in the cardiovascular system from chapter 5, pressure = flow × resistance. Since blood flow to the lungs is equal to that of the systemic circulation, and there is a substantially lower change in pressure across the pulmonary vascular system, resistance is proportionally lower compared to that in the systemic circulation. This is reflected in differences in the anatomy of the vessels in the pulmonary versus systemic circulation: The pulmonary blood vessels are thin walled, with relatively little smooth muscle.

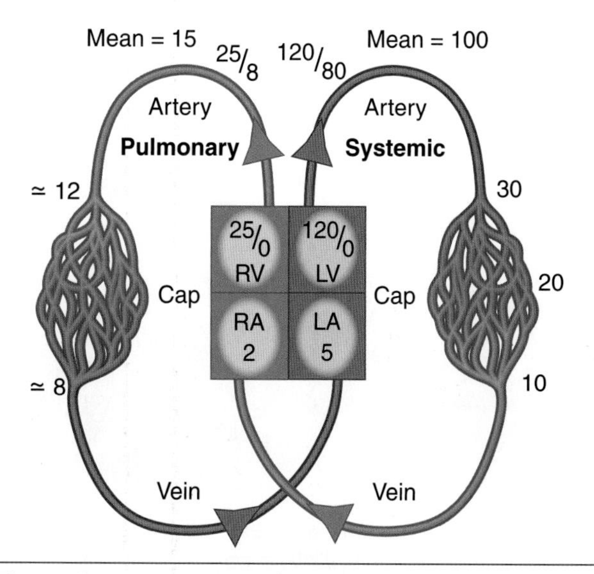

Figure 6.4 Comparison of pressure (mmHg) in the pulmonary and systemic circulations.

Reprinted, by permission, from J. West, 2000, *Respiratory physiology: The essentials* (Baltimore, MD: Lippincott, Williams, and Wilkins), 36.

Respiratory Membrane

Gas exchange between the air in the alveoli and the blood in the pulmonary capillaries occurs across the **respiratory membrane** (also called the alveolar-capillary membrane). This membrane, depicted in figure 6.5, is composed of

- the alveolar wall,
- the capillary wall, and
- their basement membranes.

In focus

Gas exchange occurs only at the alveoli.

The primary function of these membranous surfaces is for gas exchange. The respiratory membrane is very thin, measuring only 0.5 to 4.0 µm. As a result, the gases in the nearly 300 million alveoli are in close proximity to the blood circulating through the capillaries.

Partial Pressures of Gases

The air we breathe is a mixture of gases. Each exerts a pressure in proportion to its concentration in the gas mixture. The individual pressures from each gas in a mixture are referred to as **partial pressures.** According to Dalton's law, the total pressure of a mixture of gases equals the sum of the partial pressures of the individual gases in that mixture.

Consider the air we breathe. It is composed of 79.04% nitrogen (N_2), 20.93% oxygen (O_2), and 0.03% carbon dioxide (CO_2). These percentages remain constant regardless of altitude. At sea level, the atmospheric (or barometric) pressure is approximately 760 mmHg, which is also referred to as standard atmospheric pressure. Thus, if the total atmospheric pressure is 760 mmHg, then the partial pressure of nitrogen (PN_2) in air is 600.7 mmHg (79.04% of the total 760 mmHg pressure). Oxygen's partial pressure (PO_2) is 159.1 mmHg (20.93% of 760 mmHg), and carbon dioxide's partial pressure (PCO_2) is 0.2 mmHg (0.03% of 760 mmHg).

In focus

The total pressure of a mixture of gases equals the sum of the partial pressures of the individual gases in that mixture.

In the human body, gases are usually dissolved in fluids, such as blood plasma. According to **Henry's law,** gases dissolve in liquids in proportion to their partial pressures, depending also on their solubilities in the specific fluids and on the temperature. A gas's solubility in blood is a constant, and blood temperature also remains relatively constant at rest. Thus, the most critical factor for gas exchange between the alveoli and the blood is the pressure gradient between the gases in the two areas.

Gas Exchange in the Alveoli

Differences in the partial pressures of the gases in the alveoli and the gases in the blood create a pressure gradient across the respiratory membrane. This forms

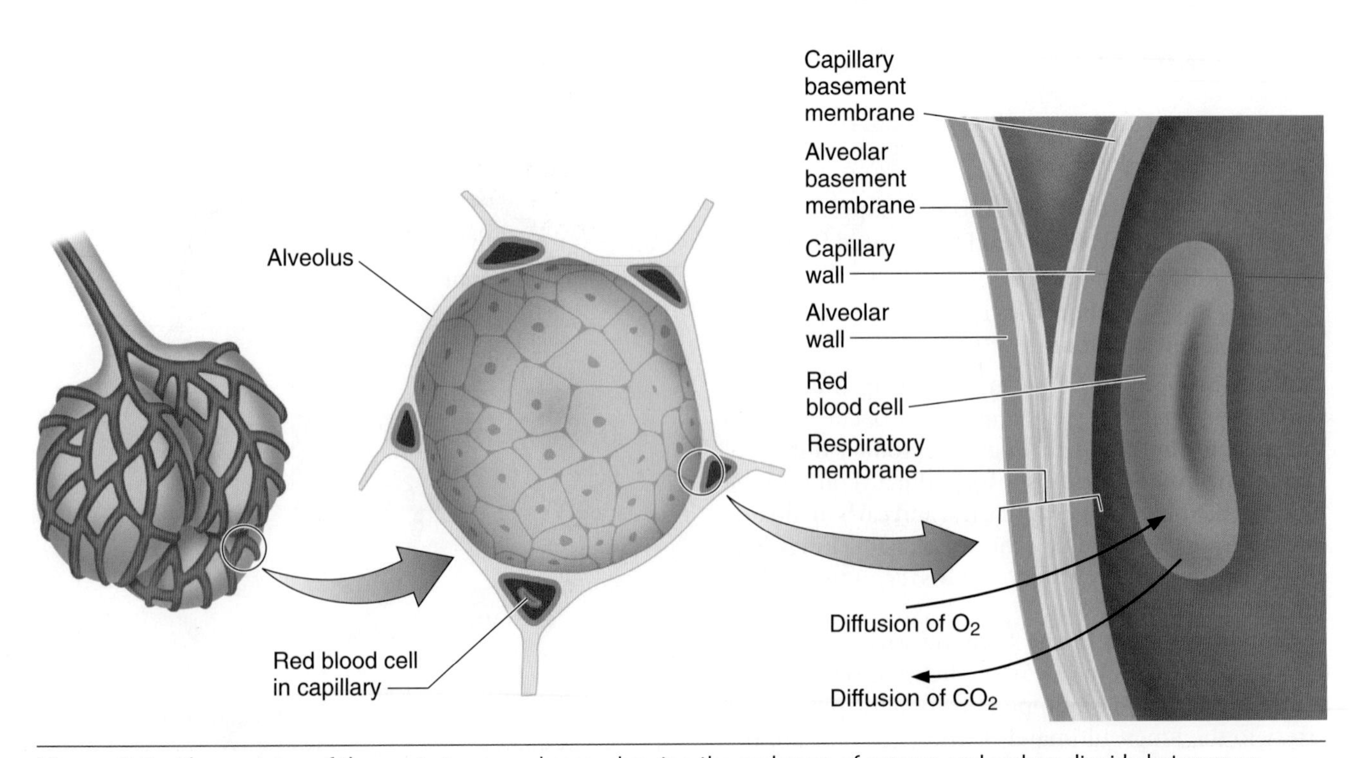

Figure 6.5 The anatomy of the respiratory membrane, showing the exchange of oxygen and carbon dioxide between an alveolus and pulmonary capillary blood.

the basis of gas exchange during pulmonary diffusion. If the pressures on each side of the membrane were equal, the gases would be at equilibrium and would not move. But the pressures are not equal, so gases move according to partial pressure gradients.

Oxygen Exchange

The PO_2 of air outside the body at standard atmospheric pressure is 159 mmHg. But this pressure decreases to about 105 mmHg when air is inhaled and enters the alveoli, where it is moistened and mixes with the air in the alveoli. The alveolar air is saturated with water vapor (which has its own partial pressure) and contains more carbon dioxide than the inspired air. Both the increased water vapor pressure and increased partial pressure of carbon dioxide contribute to the total pressure in the alveoli. Fresh air that ventilates the lungs is constantly mixed with the air in the alveoli while some of the alveolar gases are exhaled to the environment. As a result, alveolar gas concentrations remain relatively stable.

The blood, stripped of much of its oxygen by the tissues, typically enters the pulmonary capillaries with a PO_2 of about 40 mmHg (see figure 6.6). This is about 60 to 65 mmHg less than the PO_2 in the alveoli. In other words, the pressure gradient for oxygen across the respiratory membrane is typically about 65 mmHg. As noted earlier, this pressure gradient drives the oxygen from the alveoli into the blood to equilibrate the pressure of the oxygen on each side of the membrane.

The PO_2 in the alveoli stays relatively constant at about 105 mmHg. As the deoxygenated blood enters the pulmonary artery, the PO_2 in the blood is only about 40 mmHg. But as the blood moves along the pulmonary capillaries, gas exchange occurs. By the time the pulmonary blood reaches the venous end of these capillaries, the PO_2 in the blood equals that in the alveoli (approximately 105 mmHg), and the blood is now considered to be saturated with oxygen at its full carrying capacity. The blood leaving the lungs through the pulmonary veins and subsequently returning to the systemic (left) side of the heart has a rich supply of oxygen to deliver to the tissues. Notice, however, that the PO_2 in the pulmonary vein is 100 mmHg, not the 105 mmHg found in the alveolar air and pulmonary capillaries. This difference is attributable to the fact that about 2% of the blood is shunted from the aorta directly to the lung to meet the oxygen needs of the lung itself. This blood has a lower PO_2 and reenters the pulmonary vein along with fully saturated blood returning to the left atrium that has just completed gas exchange. This blood mixes and thus decreases the PO_2 of the blood returning to the heart.

In focus

The greater the pressure gradient across the respiratory membrane, the more rapidly oxygen diffuses across it.

Diffusion through tissues is described by **Fick's law** (figure 6.7). Fick's law states that the rate of diffusion

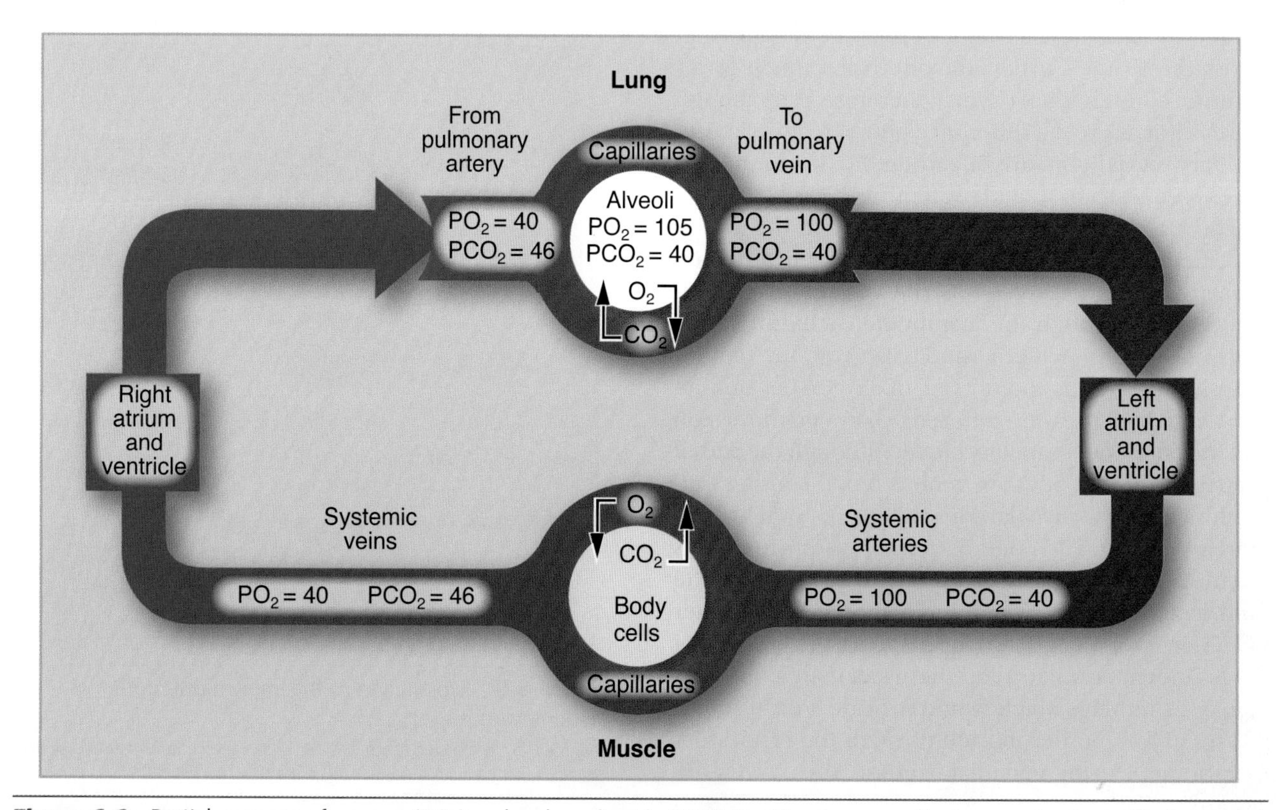

Figure 6.6 Partial pressure of oxygen (PO_2) and carbon dioxide (PCO_2) in blood as a result of gas exchange in the lungs and gas exchange between the capillary blood and tissues.

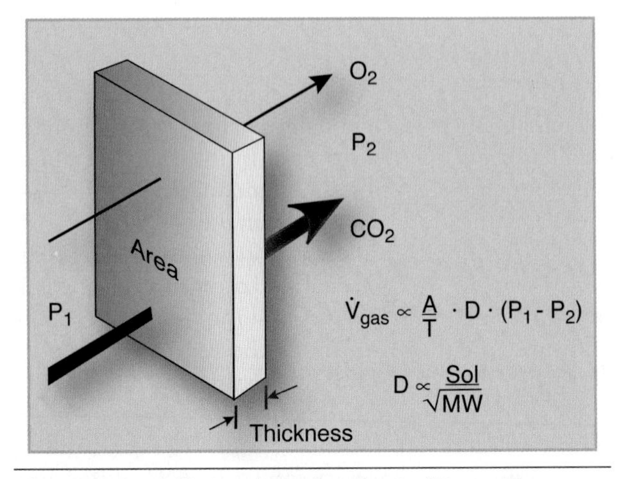

Figure 6.7 Diffusion through a sheet of tissue. The amount of gas ($\dot{V}_{gas}$) transferred is proportional to the area (A), a diffusion constant (D), and the difference in partial pressure ($P_1 - P_2$) and is inversely proportional to the thickness (T). The constant is proportional to the gas solubility (Sol) but inversely proportional to the square root of its molecular weight (MW).

Reprinted, by permission, from J. West, 2000, *Respiratory physiology: The essentials* (Baltimore, MD: Lippincott, Williams, and Wilkins), 26.

through a tissue such as the respiratory membrane is proportional to the surface area and the difference in the partial pressure of gas between the two sides of the tissue. The rate of diffusion is also inversely proportional to the thickness of the tissue in which the gas must diffuse. Additionally, the diffusion constant, which is unique to each gas, influences the rate of diffusion across the tissue. Carbon dioxide has a much lower diffusion constant than oxygen; therefore, even though there is not as great a difference between alveolar and capillary partial pressure of carbon dioxide as there is for oxygen, carbon dioxide still diffuses easily.

The rate at which oxygen diffuses from the alveoli into the blood is referred to as the **oxygen diffusion capacity** and is expressed as the volume of oxygen that diffuses through the membrane each minute for a pressure difference of 1 mmHg. At rest, the oxygen diffusion capacity is about 21 ml of oxygen per minute per 1 mmHg of pressure difference between the alveoli and the pulmonary capillary blood. Although the partial pressure gradient between venous blood coming into the lung and the alveolar air is about 65 mmHg (105 mmHg – 40 mmHg), the oxygen diffusion capacity is calculated on the basis of the mean pressure in the pulmonary capillary, which has a substantially higher PO_2. The gradient between the mean partial pressure of the pulmonary capillary and the alveolar air is approximately 11 mmHg, which would provide a diffusion of 231 ml of oxygen per minute through the respiratory membrane. During maximal exercise, the oxygen diffusion capacity may increase by up to three times the resting rate, because blood is returning to the lungs severely desaturated and thus there is a greater partial pressure gradient from the alveoli to the blood. In fact, rates of more than 80 ml/min have been observed among highly trained athletes.

The increase in oxygen diffusion capacity from rest to exercise is caused by a relatively inefficient, sluggish circulation through the lungs at rest, which results primarily from limited perfusion of the upper regions of the lungs attributable to gravity. If the lung is divided into three zones as depicted in figure 6.8, at rest only the bottom third (zone 3) of the lung is perfused with blood. During exercise, however, blood flow through the lungs is greater, primarily as a result of elevated blood pressure, which increases lung perfusion.

Carbon Dioxide Exchange

Carbon dioxide, like oxygen, moves along a pressure gradient. As shown in figure 6.6, the blood passing from the right side of the heart through the alveoli has a PCO_2 of about 46 mmHg. Air in the alveoli has a PCO_2 of about 40 mmHg. Although this results in a relatively small pressure gradient of only about 6 mmHg, it is more than adequate to allow for exchange of CO_2. Carbon dioxide's diffusion coefficient is 20 times greater than that of oxygen, so CO_2 can diffuse across the respiratory membrane much more rapidly.

The partial pressures of gases involved in pulmonary diffusion are summarized in table 6.1. Note that the

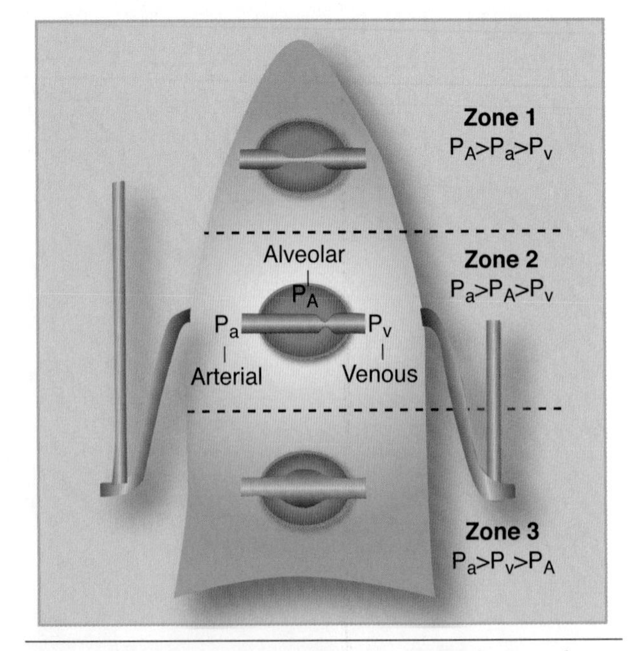

Figure 6.8 Explanation of the uneven distribution of blood flow in the lung.

Reprinted, by permission, from J. West, 2000, *Respiratory physiology: The essentials* (Baltimore, MD: Lippincott, Williams, and Wilkins), 44.

TABLE 6.1 Partial Pressures of Respiratory Gases at Sea Level

		Partial pressure (mmHg)				
Gas	% in dry air	Dry air	Alveolar air	Arterial blood	Venous blood	Diffusion gradient
H_2O	0.00	0	47	47	47	0
O_2	20.93	159.1	105	100	40	60
CO_2	0.03	0.2	40	40	46	6
N_2	79.04	600.7	568	573	573	0
Total	100.00	760	760	760	706[a]	0

[a]See text for an explanation of the decrease in total pressure.

In review

- Pulmonary diffusion is the process by which gases are exchanged across the respiratory membrane in the alveoli.
- The amount and rate of gas exchange that occurs across the membrane depend primarily on the partial pressure of each gas, although other factors are also important, as shown by Fick's law. Gases diffuse along a pressure gradient, moving from an area of higher pressure to one of lower pressure. Thus, oxygen enters the blood and carbon dioxide leaves it.
- Oxygen diffusion capacity increases as one moves from rest to exercise. When exercising muscles require more oxygen to be used in the metabolic processes, venous oxygen is depleted and oxygen exchange at the alveoli is facilitated.
- The pressure gradient for carbon dioxide exchange is less than for oxygen exchange, but carbon dioxide's diffusion coefficient is 20 times greater than that of oxygen, so carbon dioxide crosses the membrane readily without a large pressure gradient.

total pressure in the venous blood is only 706 mmHg, 54 mmHg lower than the total pressure in dry air and alveolar air. This is the result of a much greater decrease in PO_2 compared with the increase in PCO_2 as the blood goes through the body's tissues.

TRANSPORT OF OXYGEN AND CARBON DIOXIDE IN THE BLOOD

We have considered how air moves into and out of the lungs via pulmonary ventilation and how gas exchange occurs via pulmonary diffusion. Next we consider how gases are transported in the blood to deliver oxygen to the tissues and to remove the carbon dioxide that the tissues produce.

Oxygen Transport

Oxygen is transported by the blood either combined with hemoglobin in the red blood cells (greater than 98%) or dissolved in the blood plasma (less than 2%). Only about 3 ml of oxygen are dissolved in each liter of plasma. Assuming a total plasma volume of 3 to 5 L, only about 9 to 15 ml of oxygen can be carried in the dissolved state. This limited amount of oxygen cannot adequately meet the needs of even resting body tissues, which generally require more than 250 ml of oxygen per minute (depending on body size). However, **hemoglobin,** a protein contained within each of the body's 4 to 6 billion red blood cells, allows the blood to transport nearly 70 times more oxygen than can be dissolved in plasma.

Hemoglobin Saturation

As just noted, over 98% of oxygen is transported in the blood bound to hemoglobin. Each molecule of hemoglobin can carry four molecules of oxygen. When oxygen binds to hemoglobin, it forms oxyhemoglobin; hemoglobin that is not bound to oxygen is referred to as deoxyhemoglobin. The binding of oxygen to hemoglobin depends on the PO_2 in the blood and the bonding strength, or affinity, between hemoglobin and oxygen. The curve in figure 6.9 is an oxygen–hemoglobin dissociation curve, which shows the amount of hemoglobin saturated with oxygen at different PO_2 values. The shape of the curve is extremely important for its function in the body. The relatively flat upper portion means that, at high PO_2s such as in the lungs, large drops in PO_2 result in only small changes in hemoglobin saturation. This is called the "loading" portion of the curve. A high blood PO_2 results in almost complete hemoglobin saturation, which means that the maximal amount of oxygen is bound. But as the PO_2 decreases, so does hemoglobin saturation.

The steep portion of the curve coincides with PO_2 values typically found in the tissues of the body. Here, relatively small changes in PO_2 result in large changes in saturation. This is also advantageous because this is the "unloading" portion of the curve where hemoglobin loses its oxygen to the tissues.

In focus

Increased temperature and hydrogen ion (H^+) concentration (lowered pH) in an exercising muscle shift the oxygen dissociation curve rightward, allowing more oxygen to be unloaded to supply the active muscle. Because of the sigmoid shape of the curve, loading of hemoglobin with oxygen in the lungs is only minimally affected by the shift.

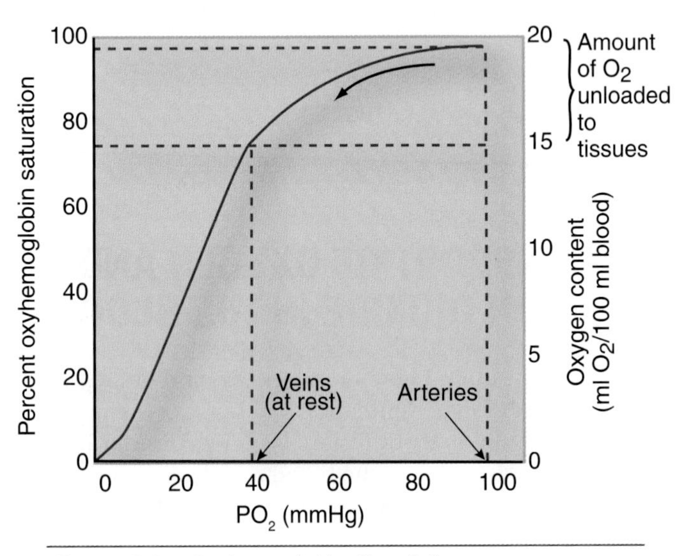

Figure 6.9 Oxyhemoglobin dissociation curve.

Reprinted, by permission, from S.K. Powers and E.T. Howley, 2004, *Exercise physiology: Theory and application to fitness and performance*, 5th ed. (New York: McGraw-Hill Companies), 205. With permission from The McGraw-Hill Companies.

Many factors determine the hemoglobin saturation. If, for example, the blood becomes more acidic, the dissociation curve shifts to the right. This indicates that more oxygen is being unloaded from the hemoglobin at the tissue level. This rightward shift of the curve (see figure 6.10*a*), attributable to a decline in pH, is referred to as the Bohr effect. The pH in the lungs is generally high, so hemoglobin passing through the lungs has a strong affinity for oxygen, encouraging high saturation.

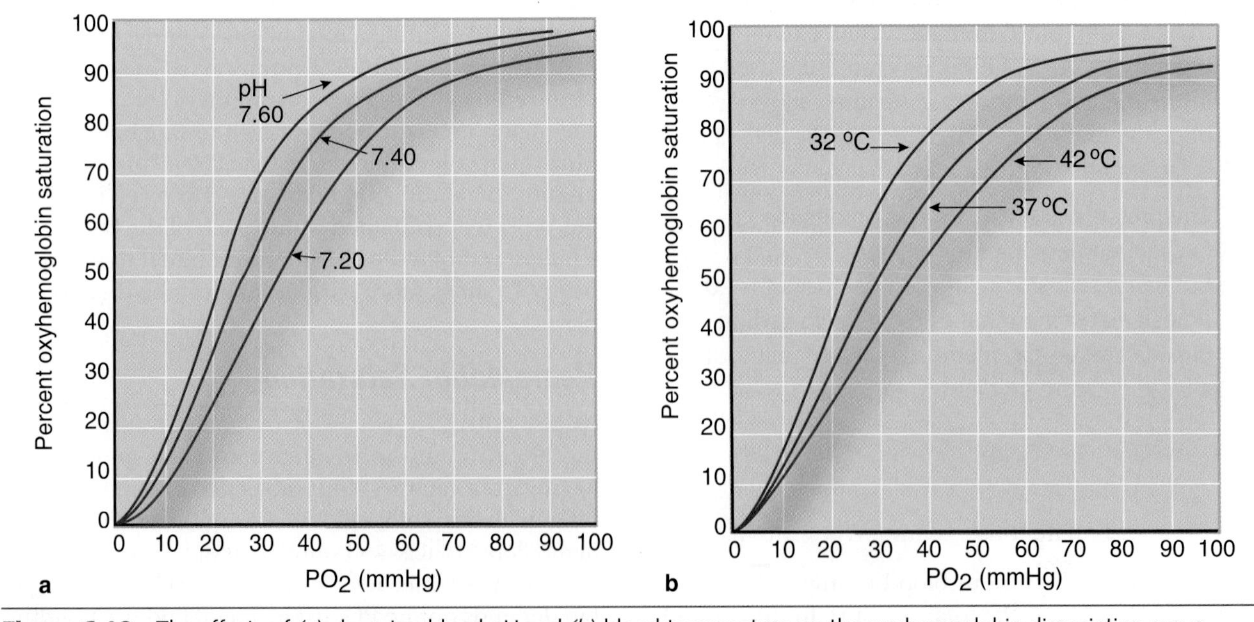

Figure 6.10 The effects of *(a)* changing blood pH and *(b)* blood temperature on the oxyhemoglobin dissociation curve.

Reprinted, by permission, from S.K. Powers and E.T. Howley, 2004, *Exercise physiology: Theory and application to fitness and performance*, 5th ed. (New York: McGraw-Hill Companies), 206. With permission from The McGraw-Hill Companies.

At the tissue level, especially during exercise, the pH is lower, causing oxygen to dissociate from hemoglobin, thereby supplying oxygen to the tissues. With exercise, the ability to unload oxygen to the muscles increases as the muscle pH decreases.

Blood temperature also affects oxygen dissociation. As shown in figure 6.10*b*, increased blood temperature shifts the dissociation curve to the right, indicating that oxygen is unloaded from hemoglobin more readily at higher temperatures. Because of this, the hemoglobin unloads more oxygen when blood circulates through the metabolically heated active muscles.

Blood Oxygen-Carrying Capacity

The oxygen-carrying capacity of blood is the maximal amount of oxygen the blood can transport. It depends primarily on the blood hemoglobin content. Each 100 ml of blood contains an average of 14 to 18 g of hemoglobin in men and 12 to 16 g in women. Each gram of hemoglobin can combine with about 1.34 ml of oxygen, so the oxygen-carrying capacity of blood is approximately 16 to 24 ml per 100 ml of blood when blood is fully saturated with oxygen. At rest, as the blood passes through the lungs, it is in contact with the alveolar air for approximately 0.75 s. This is sufficient time for hemoglobin to become 98% to 99% saturated. At high intensities of exercise, the contact time is greatly reduced, which can reduce the binding of hemoglobin to oxygen and slightly decrease the saturation, although the unique S shape of the curve guards against large drops.

People with low hemoglobin concentrations, such as those with anemia, have reduced oxygen-carrying capacities. Depending on the severity of the condition, these people might feel few effects of anemia while they are at rest because their cardiovascular system can compensate for reduced blood oxygen content by increasing cardiac output. However, during activities in which oxygen delivery can become a limitation, such as highly intense aerobic effort, reduced blood oxygen content limits performance.

Carbon Dioxide Transport

Carbon dioxide also relies on the blood for transportation. Once carbon dioxide is released from the cells, it is carried in the blood primarily in three forms:

- As bicarbonate ions resulting from the dissociation of carbonic acid
- Dissolved in plasma
- Bound to hemoglobin (called carbaminohemoglobin)

Bicarbonate Ion

The majority of carbon dioxide is carried in the form of bicarbonate ion. Bicarbonate accounts for the transport of 60% to 70% of the carbon dioxide in the blood. Carbon dioxide and water molecules combine to form carbonic acid (H_2CO_3). This reaction is catalyzed by the enzyme carbonic anhydrase, which is found in red blood cells. Carbonic acid is unstable and quickly dissociates, freeing a hydrogen ion (H^+) and forming a bicarbonate ion (HCO_3^-):

$$CO_2 + H_2O \rightarrow H_2CO_3 \rightarrow H^+ + HCO_3^-$$

The H^+ subsequently binds to hemoglobin, and this binding triggers the Bohr effect, mentioned previously, which shifts the oxygen–hemoglobin dissociation curve to the right. The bicarbonate ion diffuses out of the red blood cell and into the plasma. In order

In focus

The majority of carbon dioxide produced by the active muscle is transported back to the lungs in the form of bicarbonate ions.

In review

- Oxygen is transported in the blood primarily bound to hemoglobin (as oxyhemoglobin), although a small part of it is dissolved in blood plasma.
- Hemoglobin unloading of oxygen (desaturation) is enhanced when
  1. PO_2 decreases,
  2. pH decreases, or
  3. temperature increases.

  Each of these conditions can reflect increased local oxygen demand. They increase oxygen unloading in the metabolically active tissue.
- Hemoglobin is usually about 98% saturated with oxygen. This is a much higher oxygen content than our bodies require, so the blood's oxygen-carrying capacity seldom limits performance in healthy individuals.
- Carbon dioxide is transported in the blood primarily as bicarbonate ion. This prevents the formation of carbonic acid, which can cause H^+ to accumulate and lower the pH. Smaller amounts of carbon dioxide are either dissolved in the plasma or bound to hemoglobin.

to prevent electrical imbalance from the shift of the negatively charged bicarbonate ion into the plasma, a chloride ion diffuses from the plasma into the red blood cell. This is called the chloride shift.

Additionally, the formation of hydrogen ions through this reaction enhances oxygen unloading at the level of the tissue. Through this mechanism, hemoglobin acts as a buffer, binding and neutralizing the H^+ and thus preventing any significant acidification of the blood. Acid–base balance is discussed in more detail in chapter 7.

When the blood enters the lungs, where the PCO_2 is lower, the H^+ and bicarbonate ions rejoin to form carbonic acid, which then dissociates into carbon dioxide and water:

$$H^+ + HCO_3^- \rightarrow H_2CO_3 \rightarrow CO_2 + H_2O.$$

The carbon dioxide that is thus re-formed can enter the alveoli and be exhaled.

Dissolved Carbon Dioxide

Part of the carbon dioxide released from the tissues is dissolved in plasma; but only a small amount, typically just 7% to 10%, is transported this way. This dissolved carbon dioxide comes out of solution where the PCO_2 is low, as in the lungs. There it diffuses from the pulmonary capillaries into the alveoli to be exhaled.

Carbaminohemoglobin

Carbon dioxide transport also can occur when the gas binds with hemoglobin, forming carbaminohemoglobin. The compound is so named because carbon dioxide binds with amino acids in the globin part of the hemoglobin molecule, rather than with the heme group as oxygen does. Because carbon dioxide binding occurs on a different part of the hemoglobin molecule than does oxygen binding, the two processes do not compete. However, carbon dioxide binding varies with the oxygenation of the hemoglobin (deoxyhemoglobin binds carbon dioxide more easily than oxyhemoglobin) and the partial pressure of CO_2. Carbon dioxide is released from hemoglobin when PCO_2 is low as it is in the lungs. Thus, carbon dioxide is readily released from the hemoglobin in the lungs, allowing it to enter the alveoli to be exhaled.

GAS EXCHANGE AT THE MUSCLES

We have considered how the respiratory and cardiovascular systems bring air into our lungs, exchange oxygen and carbon dioxide in the alveoli, and transport oxygen to the muscles and carbon dioxide to the lungs. We now consider the delivery of oxygen from the capillary blood to the muscle tissue.

Arterial–Venous Oxygen Difference

At rest, the oxygen content of arterial blood is about 20 ml of oxygen per 100 ml of blood. As shown in figure 6.11*a*, this value decreases to 15 to 16 ml of oxygen per 100 ml after the blood has passed through the capillaries into the venous system. This difference in oxygen content between arterial and venous blood is referred to as the **arterial–mixed venous oxygen difference,** or **$(a\text{-}\bar{v})O_2$ difference.** The term *mixed venous* ($\bar{v}$) refers to the oxygen content of blood in the right atrium, which comes from all parts of the body, both active and inactive. The difference between arterial and mixed venous oxygen content reflects the 4 to 5 ml of oxygen per 100 ml of blood taken up by the tissues. The amount of oxygen taken up is proportional to its use for oxidative energy production. Thus, as the rate of oxygen use increases, the $(a\text{-}\bar{v})O_2$ difference also increases. It can increase to 15 to 16 ml per 100 ml of blood during maximal levels of endurance exercise (figure 6.11*b*). However, at the level of the contracting muscle, the **$(a\text{-}v)O_2$ difference** during intense exercise

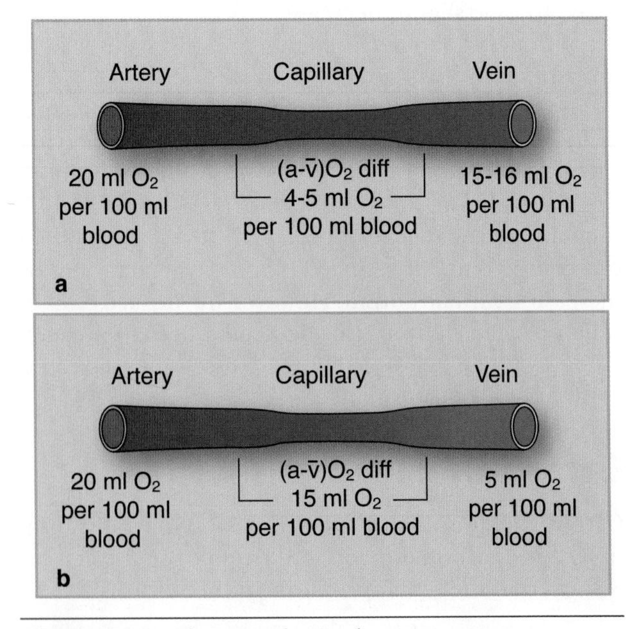

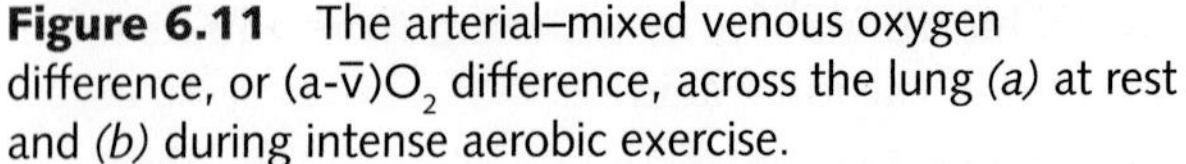

Figure 6.11 The arterial–mixed venous oxygen difference, or $(a\text{-}\bar{v})O_2$ difference, across the lung *(a)* at rest and *(b)* during intense aerobic exercise.

Reprinted, by permission, from S.K. Powers and E.T. Howley, 2004, *Exercise physiology: Theory and application to fitness and performance*, 5th ed. (New York: McGraw-Hill Companies), 206. With permission from The McGraw-Hill Companies.

can increase to 17 to 18 ml per 100 ml of blood. Note that there is not a bar over the *v* in this instance because we are now looking at local muscle venous blood, not mixed venous blood in the right atrium. During such an effort, more oxygen is unloaded to the active muscles because the PO_2 in the muscles is substantially lower than in arterial blood.

Oxygen Transport in the Muscle

Oxygen is transported in the muscle to the mitochondria by a molecule called **myoglobin** where it is used in oxidative metabolism. Myoglobin is similar in structure to hemoglobin, but myoglobin has a much greater affinity for oxygen than hemoglobin. This concept is illustrated in figure 6.12. At PO_2 values less than 20, the myoglobin dissociation curve is much steeper than the dissociation curve for hemoglobin. Myoglobin releases its oxygen content only under conditions in which the PO_2 is very low. Note from figure 6.12 that at a PO_2 at which venous blood is unloading oxygen, myoglobin is loading oxygen. It is estimated that the PO_2 in the mitochondria of an exercising muscle may be as low as 1 to 2 mmHg; thus myoglobin readily delivers oxygen to the mitochondria.

In focus

The $(a\text{-}\bar{v})O_2$ difference increases from a resting value of about 4 to 5 ml per 100 ml of blood up to values of 15 to 16 ml per 100 ml of blood during intense exercise. This increase reflects an increased extraction of oxygen from arterial blood by active muscle, thus decreasing the oxygen content of the venous blood. It is important to remember that the blood returning to the right atrium is coming from all parts of the body, active and inactive. Therefore, the mixed venous oxygen content will not decrease to values much lower than 4 to 5 ml of oxygen per 100 ml of venous blood.

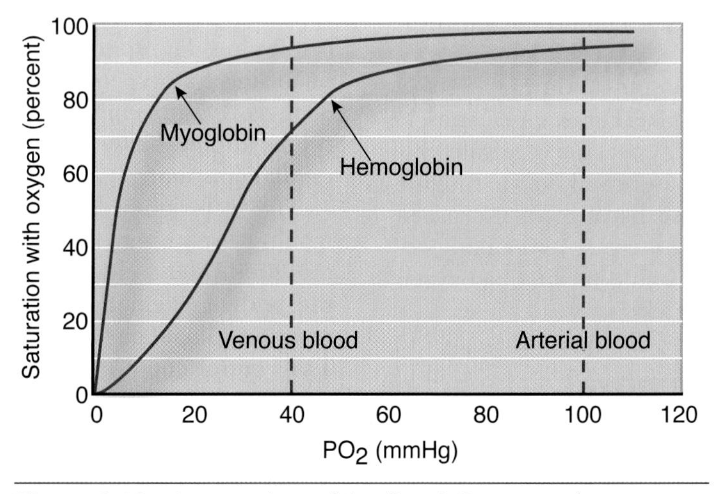

Figure 6.12 A comparison of the dissociation curves for myoglobin and hemoglobin.

Reprinted, by permission, from S.K. Powers and E.T. Howley, 2004, *Exercise physiology: Theory and application to fitness and performance,* 5th ed. (New York: McGraw-Hill Companies), 207. With permission from The McGraw-Hill Companies.

In review

- The $(a\text{-}\bar{v})O_2$ difference is the difference in the oxygen content of arterial and mixed venous blood throughout the body. This measure reflects the amount of oxygen taken up by the tissues.
- Oxygen delivery to the tissues depends on the oxygen content of the blood, blood flow to the tissues, and local conditions (e.g., tissue temperature and PO_2).
- Carbon dioxide exchange at the tissues is similar to oxygen exchange, except that carbon dioxide leaves the muscles, where it is formed, and enters the blood to be transported to the lungs for clearance.

Factors Influencing Oxygen Delivery and Uptake

The rates of oxygen delivery and uptake depend on three major variables:

- Oxygen content of blood
- Blood flow
- Local conditions (e.g., pH, temperature)

With exercise, each of these variables is adjusted to ensure increased oxygen delivery to active muscle. Under normal circumstances, hemoglobin is about 98% saturated with oxygen. Any reduction in the blood's normal oxygen-carrying capacity would hinder oxygen delivery and reduce cellular uptake of oxygen. Likewise, a reduction in the PO_2 of the arterial blood would lower the partial pressure gradient, limiting the unloading of oxygen at the tissue level. Exercise increases blood flow through the muscles. As more blood carries oxygen through the muscles, less oxygen must be removed from each 100 ml of blood (assuming the demand is unchanged). Thus, increased blood flow improves oxygen delivery.

Many local changes in the muscle during exercise affect oxygen delivery and uptake. For example, muscle activity increases muscle acidity because of lactate production. Also, muscle temperature and carbon dioxide concentration both increase because of increased metabolism. All these changes increase oxygen unloading from the hemoglobin molecule, facilitating oxygen delivery and uptake by the muscles.

Carbon Dioxide Removal

Carbon dioxide exits the cells by simple diffusion in response to the partial pressure gradient between the tissue and the capillary blood. For example, muscles generate carbon dioxide through oxidative metabolism, so the PCO_2 in muscles is relatively high compared with that in the capillary blood. Consequently, CO_2 diffuses out of the muscles and into the blood to be transported to the lungs.

REGULATION OF PULMONARY VENTILATION

Maintaining homeostatic balance in blood PO_2, PCO_2, and pH requires a high degree of coordination between the respiratory and circulatory systems. Much of this coordination is accomplished by involuntary regulation of pulmonary ventilation. This control is not yet fully understood, although many of the intricate neural controls have been identified.

Mechanisms of Regulation

The respiratory muscles are under the direct control of motor neurons, which are in turn regulated by **respiratory centers** (inspiratory and expiratory) located within the brain stem (in the medulla oblongata and pons). These centers establish the rate and depth of breathing by sending out periodic impulses to the respiratory muscles. The cortex can override these centers if voluntary control of respiration is desired. Additionally, input from other parts of the brain occurs under certain conditions.

The inspiratory area of the brain (dorsal respiratory group) contains cells that intrinsically fire and control the basic rhythm of ventilation. The expiratory area is quiet during normal quiet breathing (recall that expiration is a passive process at rest). However, during forceful breathing such as during exercise, the expiratory area actively sends signals to the muscles of expiration. Two other brain centers aid in the control of respiration. The apneustic area has an excitatory effect on the inspiratory center and results in prolonged firing of the inspiratory neurons. Finally, the pneumotaxic center inhibits or "switches off" inspiration, helping to regulate inspiratory volume.

The respiratory centers do not act alone in controlling breathing. Breathing also is regulated and modified by the changing chemical environment in the body. For example, sensitive areas in the brain respond to changes in carbon dioxide and H^+ levels. The central chemoreceptors in the brain are stimulated by an increase in H^+ ions in the cerebrospinal fluid. The blood–brain barrier is relatively impermeable to H^+ ions or bicarbonate. However, CO_2 readily diffuses across the blood–brain barrier and then reacts to increase H^+ ions. This, in turn, stimulates the inspiratory center, which then activates the neural circuitry to increase the rate and depth of respiration. This increase in respiration, in turn, increases the removal of carbon dioxide and H^+.

Chemoreceptors in the aortic arch (the aortic bodies) and in the bifurcation of the common carotid artery (the carotid bodies) are sensitive primarily to blood changes in PO_2 but also respond to changes in H^+ concentration and PCO_2. The carotid chemoreceptors are more sensitive to changes in H^+ concentrations and PCO_2. Overall, PCO_2 appears to be the strongest stimulus for the regulation of breathing. When carbon dioxide levels become too high, carbonic acid forms, then quickly dissociates, giving off H^+. If H^+ accumulates, the blood becomes too acidic (pH decreases). Thus, an increased PCO_2 stimulates the inspiratory center to increase respiration—not to bring in more oxygen but to rid the body of excess carbon dioxide and limit further pH changes.

In addition to the chemoreceptors, other neural mechanisms influence breathing. The pleurae, bronchioles, and alveoli in the lungs contain stretch receptors. When these areas are excessively stretched, that information is relayed to the expiratory center. The expiratory center responds by shortening the duration of an inspiration, which decreases the risk of overinflating the respiratory structures. This response is known as the Hering-Breuer reflex.

Many control mechanisms are involved in the regulation of breathing, as shown in figure 6.13. Such simple

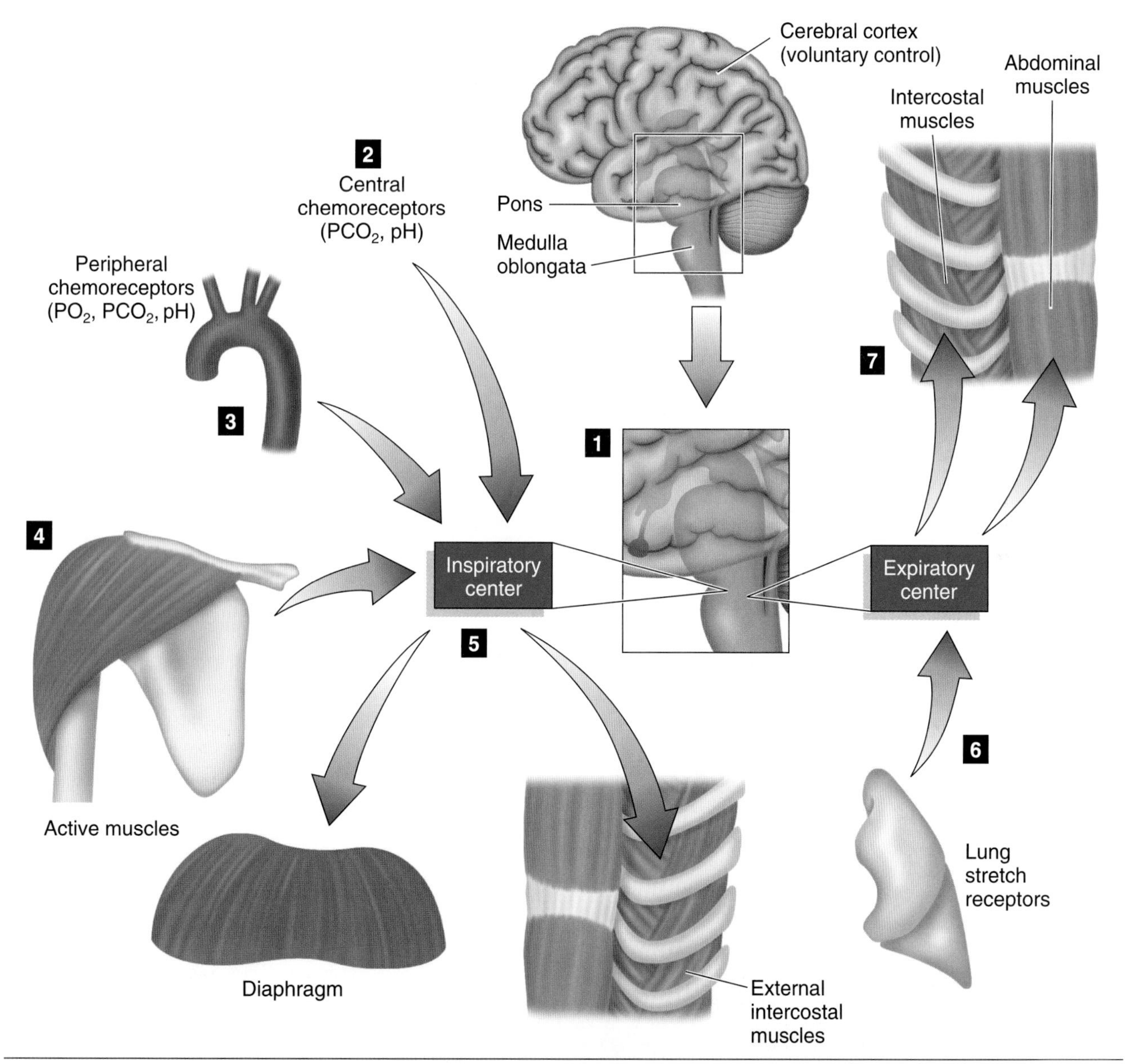

Figure 6.13 An overview of the processes involved in respiratory regulation. (1) The medulla oblongata contains the inspiratory and expiratory centers. When (2) central chemoreceptors, (3) peripheral chemoreceptors, and (4) active muscles stimulate the inspiratory center, the inspiratory center stimulates (5) the external intercostal and diaphragm muscles to contract to increase the volume of the thorax, thereby drawing air into the lungs. (6) This stretching of the lungs triggers the expiratory center to contract (7) the intercostal and abdominal muscles, causing the thoracic volume to decrease and force air out of the lungs.

stimuli as emotional distress or an abrupt change in the temperature of the surroundings can affect breathing. But all these control mechanisms are essential. The goal of respiration is to maintain appropriate levels of the blood and tissue gases and to maintain proper pH for normal cellular function. Small changes in any of these, if not carefully controlled, could impair physical activity and jeopardize health.

In Closing

In chapter 5, we discussed the role of the cardiovascular system during exercise. In this chapter we looked at the role played by the respiratory system. In the next chapter, we examine how the cardiovascular and respiratory systems respond to an acute bout of exercise.

KEY TERMS

alveoli
arterial–mixed venous oxygen difference, or $(a-\bar{v})O_2$ difference
arterial–venous oxygen difference, or $(a-v)O_2$ difference
Boyle's gas law
expiration
external respiration
Fick's law
hemoglobin
Henry's law
inspiration
internal respiration
myoglobin
oxygen diffusion capacity
partial pressure
pulmonary diffusion
pulmonary ventilation
residual volume (RV)
respiratory centers
respiratory membrane
respiratory pump
spirometry
tidal volume
total lung capacity (TLC)
vital capacity (VC)

STUDY QUESTIONS

1. Describe and differentiate between external and internal respiration.
2. Describe the mechanisms involved in inspiration and expiration.
3. What is a spirometer? Describe and define the lung volumes measured using spirometry.
4. Explain the concept of partial pressures of respiratory gases—oxygen, carbon dioxide, and nitrogen. What is the role of gas partial pressures in pulmonary diffusion?
5. Where in the lung does the exchange of gases occur with the blood? Describe the role of the respiratory membrane.
6. How are oxygen and carbon dioxide transported in the blood?
7. How is oxygen unloaded from the arterial blood to the muscle and carbon dioxide removed from the muscle into the venous blood?
8. What is meant by the arterial–mixed venous oxygen difference? How and why does this change from resting to exercise conditions?
9. Describe how pulmonary ventilation is regulated. What are the chemical stimuli that control the depth and rate of breathing? How do they control respiration during exercise?

STUDY GUIDE ACTIVITIES

In addition to the activities listed in the chapter opening outline on page 143, two other activities are available in the online study guide, located at

www.HumanKinetics.com/PhysiologyOfSportAndExercise.

The chapter **SUMMARY** reviews the key concepts covered in the chapter, and the end-of-chapter **QUIZ** tests your understanding of the material covered in the chapter.

CHAPTER 7

Cardiorespiratory Responses to Acute Exercise

In this chapter and in the online study guide

Completing a full 26.2 mi (42 km) marathon is a major accomplishment, even for those who are young and extremely fit. On May 5, 2002, Greg Osterman completed the Cincinnati Flying Pig Marathon, his sixth full marathon, finishing in a time of 5 h and 16 min. This is certainly not a world-record time, or even an average time for fit runners. However, in 1990 at the age of 35, Greg had contracted a viral infection that went right to his heart and progressed to heart failure. In 1992, he received a heart transplant. In 1993, his body started rejecting his new heart and he also contracted leukemia, not an uncommon response to the antirejection drugs given transplant patients. He miraculously recovered and started his quest to get physically fit. He ran his first race (15K) in 1994, followed by five marathons in Bermuda, San Diego, New York, and Cincinnati in 1999 and 2001. Greg is an excellent example of both human resolve and physiological adaptability.

After reviewing the basic anatomy and physiology of the cardiovascular and respiratory systems, this chapter looks specifically at how these systems respond to the increased demands placed on the body as it goes from rest to exercise. With exercise, oxygen demand by the active muscles increases significantly, and more nutrients are used. Metabolic processes speed up, so more waste products are created. During prolonged exercise or exercise in a hot environment, body temperature increases. In intense exercise, H^+ concentration increases in the muscles and blood, lowering their pH.

We then see how these changes are integrated to provide for the exercising body's needs.

CARDIOVASCULAR RESPONSES TO ACUTE EXERCISE

Numerous cardiovascular changes occur during dynamic exercise. All share a common goal: to meet the increased demands necessary to perform exercise. To better understand the changes that occur, we must look more closely at specific cardiovascular functions. In this section we examine changes in all components of the cardiovascular system, looking specifically at the following:

- Heart rate
- Stroke volume
- Cardiac output
- Blood pressure
- Blood flow
- Blood

Heart Rate

Heart rate (HR) is one of the simplest and yet most informative of the cardiovascular parameters. Measuring HR involves simply taking the subject's pulse, usually at the radial or carotid artery. Heart rate is a good indicator of the intensity of exercise.

Resting Heart Rate

Resting heart rate (RHR) averages 60 to 80 beats/min in most individuals. In highly conditioned, endurance-trained athletes, resting rates as low as 28 to 40 beats/min have been reported. This is mainly due to an increase in vagal tone that accompanies endurance exercise training. Resting heart rate can also be affected by environmental factors; for example, it increases with extremes in temperature and altitude.

Just before the start of exercise, preexercise HR usually increases above normal resting values. This is called an anticipatory response. This response is mediated through release of the neurotransmitter norepinephrine from the sympathetic nervous system and the hormone epinephrine from the adrenal medulla. Vagal tone probably also decreases. Because preexercise HR is elevated,

Preexercise HR is not a reliable estimate of RHR because of the anticipatory HR response.

The Fick Principle and the Fick Equation

In the 1870s, a cardiovascular physiologist by the name of Adolph Fick developed a principle critical to our understanding of the basic relationship between metabolism and cardiovascular function: that oxygen consumption of a tissue is dependent on blood flow to that tissue and the amount of oxygen extracted from the blood by the tissue. The Fick principle can be applied to the whole body or to regional circulations. Oxygen consumption is the product of blood flow and the difference in concentration of oxygen in the blood between the arterial blood supplying the tissue and the venous blood draining out of the tissue—the $(a\text{-}\bar{v})O_2$ difference. Whole-body oxygen consumption ($\dot{V}O_2$) is calculated as the product of the cardiac output ($\dot{Q}$) and $(a\text{-}\bar{v})O_2$ difference.

$$\text{Fick equation: } \dot{V}O_2 = \dot{Q} \times (a\text{-}\bar{v})O_2 \text{ diff},$$

which can be rewritten as

$$\dot{V}O_2 = HR \times SV \times (a\text{-}\bar{v})O_2 \text{ diff}.$$

This basic relationship comes up frequently throughout the remainder of this book.

reliable estimates of the true RHR should be made only under conditions of total relaxation, such as early in the morning before the subject rises from a restful night's sleep.

Heart Rate During Exercise

When exercise begins, HR increases directly in proportion to the increase in exercise intensity (figure 7.1), until near-maximal exercise is achieved. As maximal exercise intensity is approached, HR begins to plateau even as the exercise workload continues to increase. This indicates that HR is approaching a maximum value. The **maximum heart rate (HR_{max})** is the highest HR value achieved in an all-out effort to the point of exhaustion. This is a highly reliable value that remains constant from day to day. However, this value changes slightly from year to year due to the normal age-related decline in HR_{max}.

HR_{max} is often estimated based on age because HR_{max} shows a slight but steady decrease of about one beat per year beginning at 10 to 15 years of age. Subtracting one's age from 220 provides an approximation of one's average HR_{max}. However, this is only an estimate—individual values vary considerably from this average value. To illustrate, for a 40-year-old person, HR_{max} would be estimated to be 180 beats/min ($HR_{max} = 220 - 40$). However, 68% of all 40-year-olds have actual HR_{max} values between 168 and 192 beats/min (mean ± 1 standard deviation), and 95% fall between 156 and 204 beats/min (mean ± 2 standard deviations). This demonstrates the potential for error in estimating a person's HR_{max}. A more accurate equation has been developed to estimate HR_{max} from age. In this equation, $HR_{max} = 208 - (0.7 \times \text{age})$.[5]

> ### In focus
> To estimate HR_{max}:
>
> $HR_{max} = 220 - \text{age in years}$
>
> or
>
> $HR_{max} = 208 - (0.7 \times \text{age in years})$.

When the rate of work is held constant at a submaximal intensity, HR increases fairly rapidly until it reaches a plateau. This plateau is the **steady-state heart rate,** and it is the optimal HR for meeting the circulatory demands at that specific rate of work. For each subsequent increase in intensity, HR will reach a new steady-state value within 2 to 3 min. However, the more intense the exercise, the longer it takes to achieve this steady-state value.

The concept of steady-state heart rate forms the basis for several tests that have been developed to estimate physical fitness. In one such test, individuals are placed on an exercise device, such as a cycle ergometer, and then perform exercise at two or three standardized exercise intensities. Those in better physical condition (i.e., those with better cardiorespiratory endurance capacity) will have a lower steady-state HR at each exercise intensity than those who are less fit. Thus, steady-state HR is a valid predictor of cardiorespiratory fitness: A lower steady-state HR reflects greater cardiorespiratory fitness.

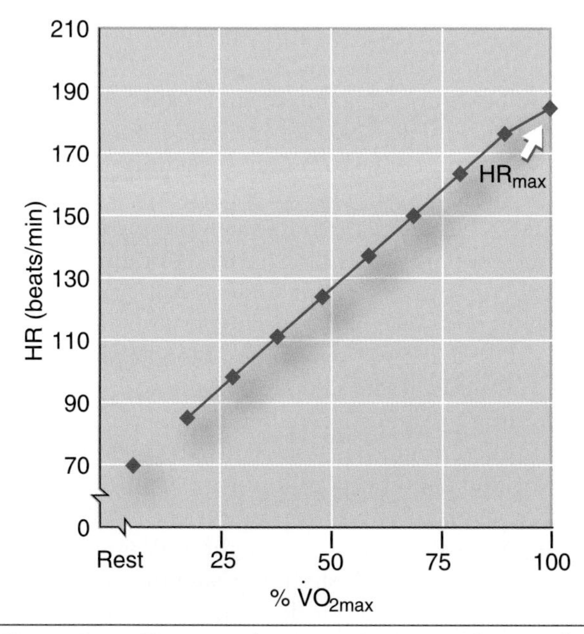

Figure 7.1 Changes in heart rate (HR) as a subject walks, jogs, then runs on a treadmill. Heart rate is plotted against exercise intensity shown as percent of the subject's $\dot{V}O_{2max}$, where HR begins to plateau. The HR at this plateau is the subject's maximal HR or HR_{max}.

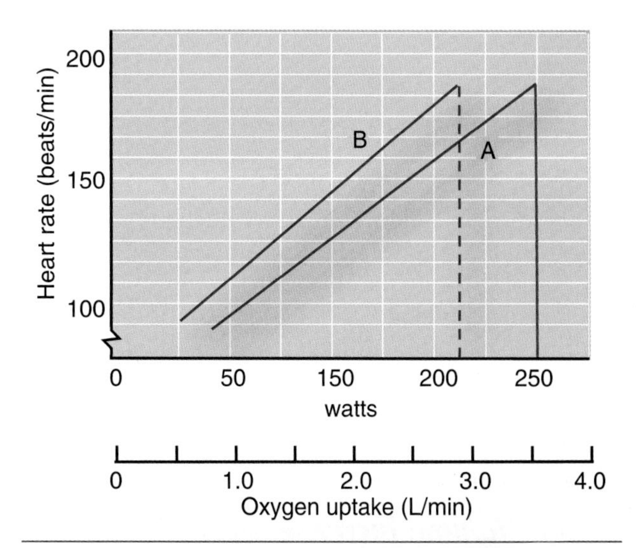

Figure 7.2 The increase in heart rate with increasing power output and oxygen uptake is linear within a wide range. The predicted maximal oxygen uptake can be extrapolated through use of the subject's estimated maximum heart rate as demonstrated here for two subjects with similar estimated maximum heart rate values, but quite different maximal capacities.

Reprinted, by permission, from P.O. Åstrand et al., 2003, *Textbook of work physiology*, 4th ed. (Champaign, IL: Human Kinetics), 285.

Figure 7.2 illustrates results from a submaximal exercise test performed by two different individuals of the same age. Steady-state HR is measured at three to four distinct workloads, and a line of best fit is drawn through the data points. Because there is a consistent relationship between intensity and energy demand, steady-state HR can be plotted versus the corresponding energy ($\dot{V}O_2$) required to do work on the cycle ergometer. The resultant line can be extrapolated to the age-predicted HR_{max} to estimate an individual's maximal exercise capacity. In this figure, subject A has a higher fitness level than subject B because (1) at any given submaximal intensity, his HR is lower; and (2) extrapolation to age-predicted HR_{max} yields a higher estimated maximal exercise capacity ($\dot{V}O_{2max}$).

Stroke Volume

Stroke volume (SV) also changes during acute exercise to allow the heart to meet the demands of exercise. At near-maximal and maximal exercise intensities, SV is a major determinant of cardiorespiratory endurance capacity.

Stroke volume is determined by four factors:

1. The volume of venous blood returned to the heart (the heart can only pump what returns: preload)
2. Ventricular distensibility (the capacity to enlarge the ventricle, for maximal filling)
3. Ventricular contractility (the inherent capacity of the ventricle to contract)
4. Aortic or pulmonary artery pressure (the pressure against which the ventricles must contract: afterload)

The first two factors influence the filling capacity of the ventricle, determining how much blood is available for filling the ventricle and the ease with which the ventricle is filled at the available pressure. This is referred to as **preload.** The last two factors influence the ventricle's ability to empty, determining the force with which blood is ejected and the pressure, or **afterload,** against which it must be expelled into the arteries. These four factors directly control the alterations in SV in response to increasing exercise intensity.

Stroke Volume Increase With Exercise

Stroke volume increases above resting values during exercise. Most researchers agree that SV increases with increasing rates of work, but only up to exercise intensities somewhere between 40% and 60% of maximal capacity. At that point, SV typically plateaus, remaining essentially unchanged up to and including the point of exhaustion as shown in figure 7.3. However, other

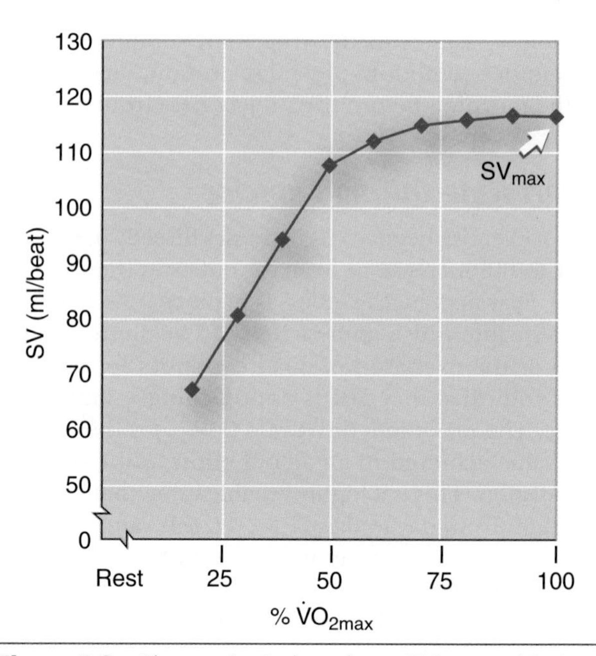

Figure 7.3 Changes in stroke volume (SV) as a subject exercises on a treadmill at increasing intensities. Stroke volume is plotted as a function of percent $\dot{V}O_{2max}$. The SV increases with increasing intensity up to approximately 40% to 60% of $\dot{V}O_{2max}$, then reaches a maximum (SV_{max}).

researchers have reported that SV continues to increase up through maximal exercise intensities. This is discussed in more detail in the sidebar on page 166.

When the body is in an upright position, SV can almost double from resting to maximal values. For example, in active but untrained individuals, SV increases from about 60 to 70 ml/beat at rest to 110 to 130 ml/beat during maximal exercise. In highly trained endurance athletes, SV can increase from 80 to 110 ml/beat at rest to 160 to 200 ml/beat during maximal exercise. During supine exercise, such as recumbent cycling, SV also increases but usually by only about 20% to 40%—not nearly as much as in an upright position. Why does body position make such a difference?

When the body is in the supine position, blood does not pool in the lower extremities. Blood returns more easily to the heart in a supine posture, which means that resting SV values are higher in the supine position than in the upright position. Thus, the increase in SV with maximal exercise is not as great in the supine position as in the upright position because SV starts out higher. Interestingly, the highest SV attainable in upright exercise is only slightly greater than the resting value in the reclining position. The majority of the SV increase during low to moderate intensities of exercise in the upright position appears to be compensating for the force of gravity that causes blood to pool in the extremities.

Explanations for the Stroke Volume Increase

One explanation for the increase in SV with exercise is that the primary factor determining SV is increased preload or the extent to which the ventricle fills with blood and stretches. When the ventricle stretches more during filling, it subsequently contracts more forcefully. For example, when a larger volume of blood enters and fills the ventricle during diastole, the ventricular walls stretch. To eject that greater volume of blood, the ventricle responds by contracting more forcefully. This is referred to as the **Frank-Starling mechanism.** Additionally, SV can increase during exercise if the ventricle's contractility is enhanced by an increase in neural stimulation, an increased release of circulating catecholamines (epinephrine, norepinephrine), or both, even without an increased end-diastolic volume. These mechanisms combine to increase SV during dynamic exercise.

Some clinically used cardiovascular diagnostic techniques have made it possible to determine exactly how SV changes with exercise. Echocardiography (using sound waves) and radionuclide techniques ("tagging" of red blood cells with radioactive tracers) have elucidated how the heart chambers respond to increasing oxygen demands during exercise. With either technique, continuous pictures can be taken of the heart at rest and up to near-maximal intensities of exercise.

Figure 7.4 illustrates the results of one study of normally active but untrained subjects.[3] In this study, participants were tested during both supine and upright cycle ergometry under four conditions, which are depicted on the *x*-axis:

- Rest
- Low-intensity exercise
- Intermediate-intensity exercise
- Peak-intensity exercise

Going from resting conditions to exercise of increasing intensities, there is an increase in left ventricular end-diastolic volume (a greater filling or preload), indicating that the Frank-Starling mechanism is operating. There is also a decrease in the left ventricular end-systolic volume (greater emptying), indicating an increased degree of contractility.

This figure shows that both the Frank-Starling mechanism and increased contractility are important in increasing SV during exercise. The Frank-Starling mechanism appears to have its greatest influence at lower exercise intensities, and contractility has its greatest effects at higher exercise intensities.

Recall that HR also increases with exercise intensity. The plateau or small decrease in left ventricular end-diastolic volume at higher exercise intensities could be caused by reduced ventricular filling time. One study showed that ventricular filling time decreased from

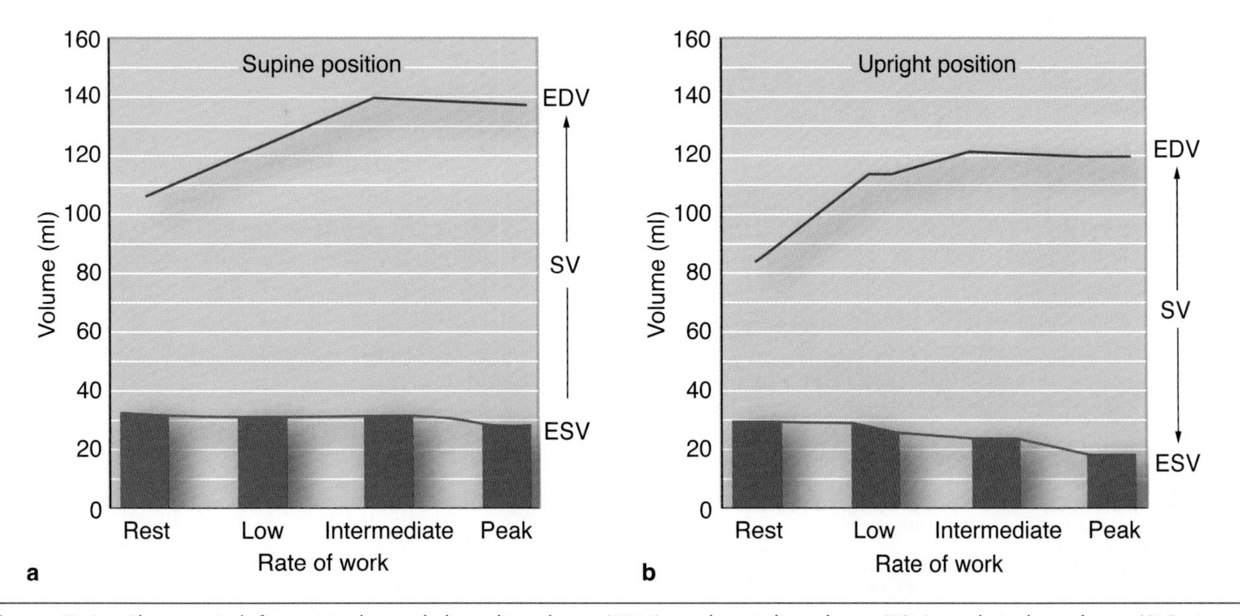

Figure 7.4 Changes in left ventricular end-diastolic volume (EDV), end-systolic volume (ESV), and stroke volume (SV) at rest and during low-, intermediate-, and peak-intensity exercise when the subject is *(a)* in the supine position and *(b)* in the upright position. Note that SV = EDV – ESV.

Reprinted, by permission, from L. R. Poliner et al., 1980, "Left ventricular performance in normal subjects: A comparison of the responses to exercise in the upright and supine position," *Circulation* 62: 528-534.

Conflicting Research on Stroke Volume During Exercise

Although researchers agree that SV increases as work rates increase up to around 40% to 60% of maximum, reports about what happens after that point differ. A review of studies conducted between the 1960s and the early 1990s reveals no clear pattern of SV increase beyond the 40% to 60% work rate range. Several studies have shown a plateau in SV at approximately 50% of $\dot{V}O_{2max}$, with little or no change occurring with further increases, while other studies have shown that SV continues to increase beyond that rate.

This apparent disagreement might result from differences among studies in the mode of exercise testing or the participants' training status. Studies that show plateaus in the 40% to 60% $\dot{V}O_{2max}$ range typically have used cycle ergometers, and studies have demonstrated that blood is trapped in the legs during cycle ergometer exercise. Thus, the plateau in SV might be unique to exercise on cycle ergometers, resulting from decreased venous return of blood from the legs.

In those studies in which SV continued to increase up to maximal rates of exercise, subjects were generally highly trained athletes. Many highly trained athletes, including highly trained cyclists tested on a cycle ergometer, can continue to increase their SV after exceeding the 40% to 60% $\dot{V}O_{2max}$ level, perhaps because of adaptations to training. The increases in cardiac output and SV with increasing rates of work, as represented by increasing HR, in elite athletes, trained university distance runners, and untrained university students, are illustrated in the figure in this sidebar.

Finally, SV is difficult to assess, particularly at higher exercise intensities, so differences between studies could result from differences in the techniques used to measure cardiac output or SV as well as the accuracy of these techniques at higher exercise intensities.

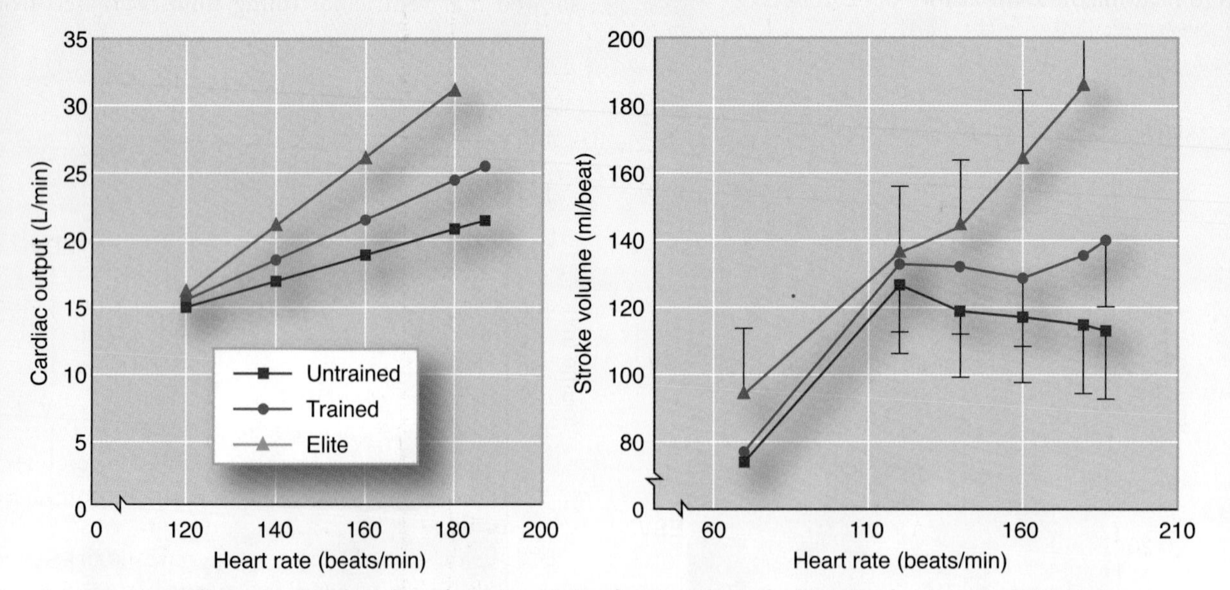

Increases in cardiac output and stroke volume in untrained university students, trained university distance runners, and elite athletes.

Adapted, by permission, from B. Zhou et al., 2001, "Stroke volume does not plateau during graded exercise in elite male distance runners," *Medicine and Science in Sports and Exercise* 33: 1849-1854.

about 500 to 700 ms at rest to about 150 ms at higher HRs (about 150-200 beats/min).[6] Therefore, with increasing work rates approaching HR_{max}, the diastolic filling time could be shortened enough to limit filling. As a result, end-diastolic volume might plateau or start to decrease.

For the Frank-Starling mechanism to work, left ventricular end-diastolic volume must increase by an increase in venous blood return to the heart. This happens through redistribution of blood by sympathetic activation of arteries and arterioles in inactive areas of the body such as in the renal and splanchnic circulations. Also, the muscle and respiratory pumps aid in increasing venous return.

Thus, two factors that can contribute to an increase in SV with increasing intensity of exercise are increased venous return (preload) and increased ventricular contractility. A third factor also contributes to the increase in SV during exercise—a decrease in afterload placed on the heart through a decrease in total peripheral resistance. Total peripheral resistance decreases because of vasodilation of the blood vessels going to exercising skeletal muscle. This decrease in afterload allows the left ventricle to expel blood against less resistance, facilitating emptying of the blood from this chamber.

Cardiac Output

If we recall that cardiac output is the product of heart rate and stroke volume ($\dot{Q} = HR \times SV$), cardiac output predictably increases with increasing exercise intensity (figure 7.5). Resting cardiac output is approximately 5.0 L/min, but varies in proportion to the size of the person. Maximal cardiac output varies between 20 (sedentary person) and 40 (elite endurance athlete) L/min and is a function of both body size and endurance training. The linear relationship between cardiac output and intensity of exercise is predictable because the major purpose of the increase in cardiac output is to meet the muscles' increased demand for oxygen. Like $\dot{V}O_{2max}$, when cardiac output approaches maximal exercise intensities it may reach a plateau (figure 7.5). In fact, it is likely that $\dot{V}O_{2max}$ is limited by this leveling off of cardiac output.

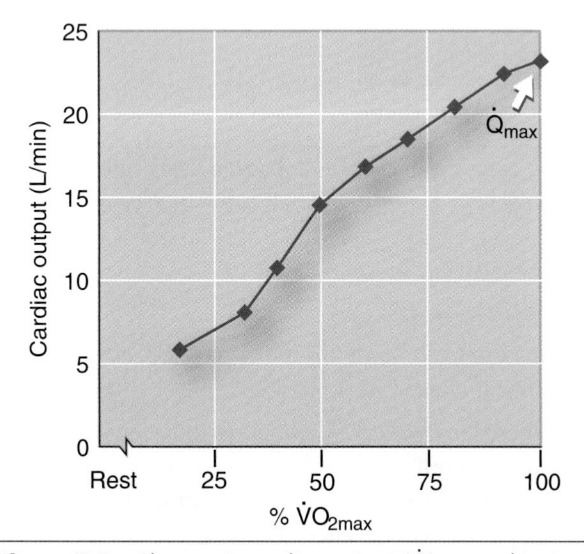

Figure 7.5 Changes in cardiac output ($\dot{Q}$) as a subject exercises on a treadmill at increasing intensities. Cardiac output is plotted as a function of percent $\dot{V}O_{2max}$. The cardiac output increases in direct proportion to increasing intensity, eventually reaching a maximum ($\dot{Q}_{max}$).

The Integrated Cardiac Response to Exercise

To see how HR, SV, and $\dot{Q}$ vary under various conditions of rest and exercise, consider the following example. An individual first transitions from a reclining position to a sitting position and then to standing. Next the person begins walking, then jogging, and finally breaks into a fast-paced run. How does the heart respond?

In a reclined position, HR is ~50 beats/min; it increases to about 55 beats/min during sitting and to about 60 beats/min during standing. When the body shifts from a reclining to a sitting position and then to a standing position, gravity causes blood to pool in the legs, which reduces the volume of blood returning to the heart and thus decreases SV. To compensate for the reduction in SV, HR increases in order to maintain cardiac output; that is, $\dot{Q} = HR \times SV$.

During exercise, cardiac output increases to match the need for increased blood flow to deliver oxygen to exercising muscles.

During the transition from rest to walking, HR increases from about 60 to about 90 beats/min. Heart rate increases to 140 beats/min with moderate-paced jogging and can reach 180 beats/min or more with a fast-paced run. The initial increase in HR up to about 100 beats/min is mediated by a withdrawal of vagal tone. Further increases in HR are mediated by the sympathetic nervous system. Stroke volume also increases with exercise, further increasing cardiac output. These relationships are illustrated in figure 7.6.

During the initial stages of exercise in untrained individuals, increased cardiac output is caused by an increase in both HR and SV. When the level of exercise exceeds 40% to 60% of the individual's maximal exercise capacity, SV either plateaus or continues to increase, but at a much slower rate. Thus, further increases in cardiac output are largely the result of increases in HR. Stroke volume increases are likely to contribute more to the rise in cardiac output during the higher intensities of exercise in those people who are highly trained.

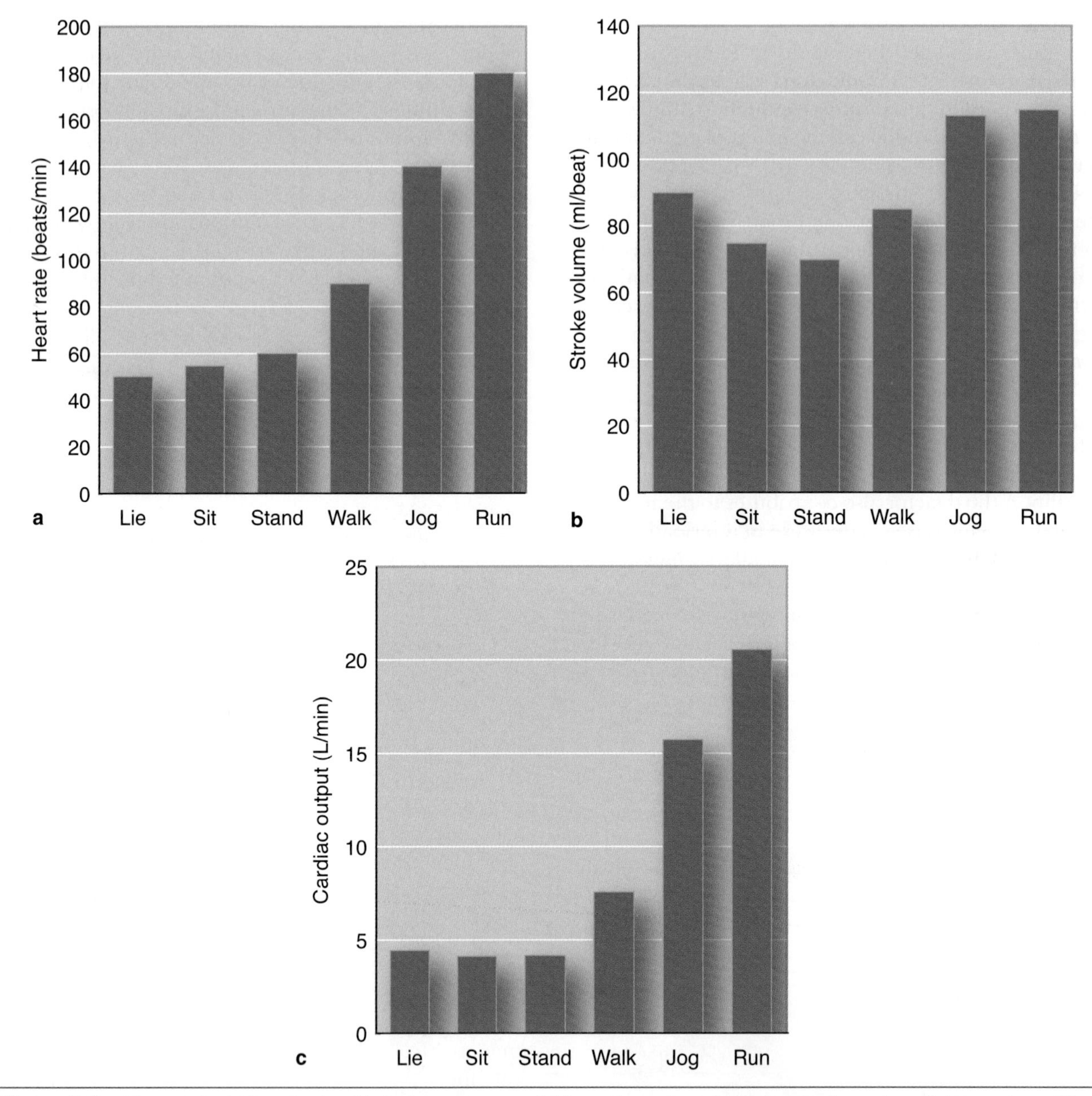

Figure 7.6 Changes in *(a)* heart rate, *(b)* stroke volume, and *(c)* cardiac output with changes in posture (lying supine, sitting, and standing upright) and with exercise (walking at 5 km/h [3.1 mph], jogging at 11 km/h [6.8 mph], and running at 16 km/h [9.9 mph]).

In review

- As exercise intensity increases, HR increases proportionately, up to maximal exercise intensities (HR_{max}).
- Stroke volume (the amount of blood ejected with each contraction) also increases proportionately with increasing exercise intensity but usually achieves its maximal value at about 40% to 60% of $\dot{V}O_{2max}$ in untrained individuals. Highly trained individuals can continue to increase SV, even up to maximal exercise intensities.
- Increases in HR and SV combine to increase cardiac output. Thus, more blood is pumped during exercise, ensuring that adequate supplies of oxygen and nutrients reach the exercising muscles and that the waste products of muscle metabolism are quickly cleared away.

Blood Pressure

During dynamic exercise, mean arterial blood pressure increases substantially. However, systolic and diastolic blood pressure do not increase to a similar degree. With whole-body endurance exercise, systolic blood pressure increases in direct proportion to the increase in exercise intensity. However, diastolic pressure does not change significantly, and may even decrease. A systolic pressure that starts out at 120 mmHg in a normal healthy person at rest can exceed 200 mmHg at maximal exercise. Systolic pressures of 240 to 250 mmHg have been reported in normal, healthy, highly trained athletes at maximal intensities of aerobic exercise.

Increased systolic blood pressure results from the increased cardiac output ($\dot{Q}$) that accompanies increasing rates of work. This increase in pressure helps facilitate the increase in blood flow through the vasculature. Also, blood pressure (that is, hydrostatic pressure) determines how much plasma leaves the capillaries, entering the tissues and carrying needed supplies. Thus increased systolic pressure aids substrate delivery to working muscles.

Blood pressure reaches a steady state during submaximal steady-state endurance exercise. As work intensity increases, so does systolic blood pressure. If steady-state exercise is prolonged, the systolic pressure might start to decrease gradually, but diastolic pressure remains constant. The slight decrease in systolic blood pressure, if it occurs, is a normal response and simply reflects increased arteriole dilation in the active muscles, which decreases the **total peripheral resistance** or **TPR** (since blood pressure = cardiac output × total peripheral resistance).

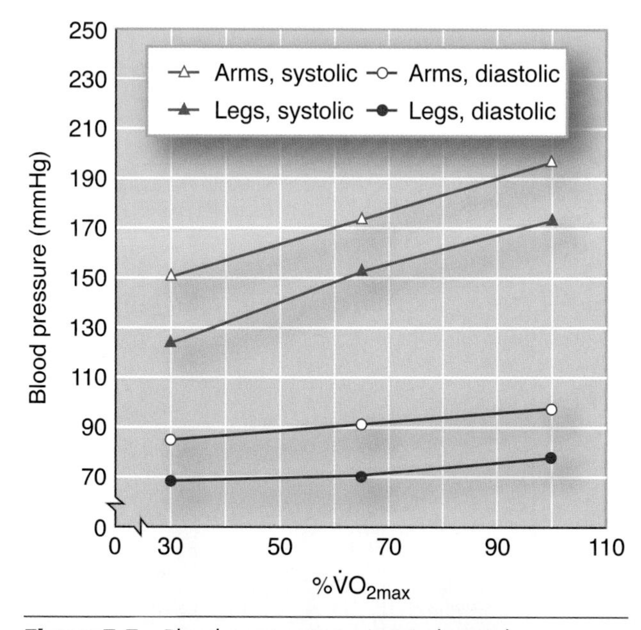

Figure 7.7 Blood pressure response to leg and arm cycling at the same relative rates of oxygen consumption (% $\dot{V}O_{2max}$).

Adapted, by permission, from P.O. Åstrand et al., 1965, "Intraarterial blood pressure during exercise with different muscle groups," *Journal of Applied Physiology* 20: 253-256.

Diastolic blood pressure changes little during submaximal dynamic exercise; however, at maximal exercise intensities, diastolic blood pressure increases slightly. Remember that diastolic pressure reflects the pressure in the arteries when the heart is at rest (diastole). With dynamic exercise there is an overall increase in sympathetic neural tone to the vasculature, causing overall vasoconstriction. However, this vasoconstriction is blunted in the exercising muscles by the release of local vasodilators. Thus, there is a balance between vasoconstriction to inactive regions and vasodilation in the active skeletal muscle; therefore diastolic pressure does not change substantially. However, in some cases of cardiovascular disease, increases in diastolic pressure of 15 mmHg or more occur in response to exercise and are one of several indications for immediately stopping a diagnostic exercise test. Figure 7.7 illustrates a typical blood pressure response to leg and arm cycling exercise with increasing exercise intensities.

As seen in figure 7.7, upper body exercise causes a greater blood pressure response than leg exercise at the same absolute rate of energy expenditure. This is most likely attributable to the smaller muscle mass and vasculature of the upper body compared with the lower body, plus an increased energy demand to stabilize the body during arm exercise. This difference in the systolic blood pressure response to upper and lower body exercise has important implications for the heart. Myocardial oxygen uptake and myocardial blood flow are directly related to the product of HR and systolic blood pressure. This value is referred to as the rate–pressure product or double product (DP = HR × SBP). With static or dynamic resistance exercise or upper body work, the rate–pressure product is elevated, indicating increased myocardial oxygen demand. This relationship between rate–pressure product and myocardial oxygen demand is important in clinical exercise testing.

Blood pressure responses to resistance exercise, such as weightlifting, are exaggerated. With high-intensity resistance training, blood pressure can reach 480/350 mmHg. In such exercise, use of the Valsalva maneuver is quite common. This maneuver occurs when a person tries to exhale while the mouth, nose, and glottis are closed. This action causes an enormous increase in intrathoracic pressure. Much of the subsequent blood pressure increase results from the body's effort to overcome the high internal pressures created during the Valsalva maneuver.

Blood Flow

Acute changes in cardiac output and blood pressure during exercise allow for increased total blood flow to the body. These responses facilitate getting blood to areas where it is needed, primarily the exercising muscles. Additionally, sympathetic control of the cardiovascular system can redistribute blood so that areas with the greatest metabolic need receive more blood than areas with low demands.

Redistribution of Blood During Exercise

Blood flow patterns change markedly in the transition from rest to exercise. Through the action of the sympathetic nervous system, blood is redirected away from areas where elevated flow is not essential to those areas that are active during exercise (refer back to figure 5.11). Only 15% to 20% of the resting cardiac output goes to muscle, but during high-intensity exercise, the muscles may receive 80% to 85% of the cardiac output. This shift in blood flow to the muscles is accomplished primarily by reducing blood flow to the kidneys and the so-called splanchnic circulation that includes the liver, stomach, and intestines. Figure 7.8 illustrates a typical distribution of cardiac output throughout the body at rest and during heavy exercise. Because cardiac output increases greatly with increasing intensity of exercise, the values are shown both as relative percentages of the total blood available and as absolute flows.

Although several physiological mechanisms are responsible for the redistribution of blood flow through the body during exercise, they work together. To illustrate this, consider what happens to blood flow during exercise, focusing on the primary driver of the response, namely the blood flow requirements of the skeletal muscles.

As exercise begins, the active skeletal muscles rapidly sense the need for increased oxygen delivery. This need is met through sympathetic stimulation of vessels in those areas to which blood flow is to be reduced (e.g., the splanchnic and renal circulations). The resulting vasoconstriction in those areas allows for more of the increasing cardiac output to be redistributed to the exercising skeletal muscles. In the skeletal muscles, sympathetic stimulation to the constrictor fibers in the arteriolar walls also increases; however local vasodilating substances are released from the exercising muscle and overcome sympathetic vasoconstriction, producing an overall vasodilation in the muscle.

Many local vasodilating substances are released in exercising skeletal muscle. As the metabolic rate of the muscle tissue increases during exercise, metabolic waste products begin to accumulate. Increased metabolism causes an increase in acidity (increased hydrogen ions and lower pH), carbon dioxide, and temperature in the muscle tissue. These are some of the local changes that trigger vasodilation of, and increasing blood flow through, the arterioles feeding local capillaries. Local vasodilation is also triggered by the low partial pressure of oxygen in the tissue or a reduction in oxygen bound to hemoglobin (increased oxygen demand), the act of muscle contraction, and possibly other vasoactive

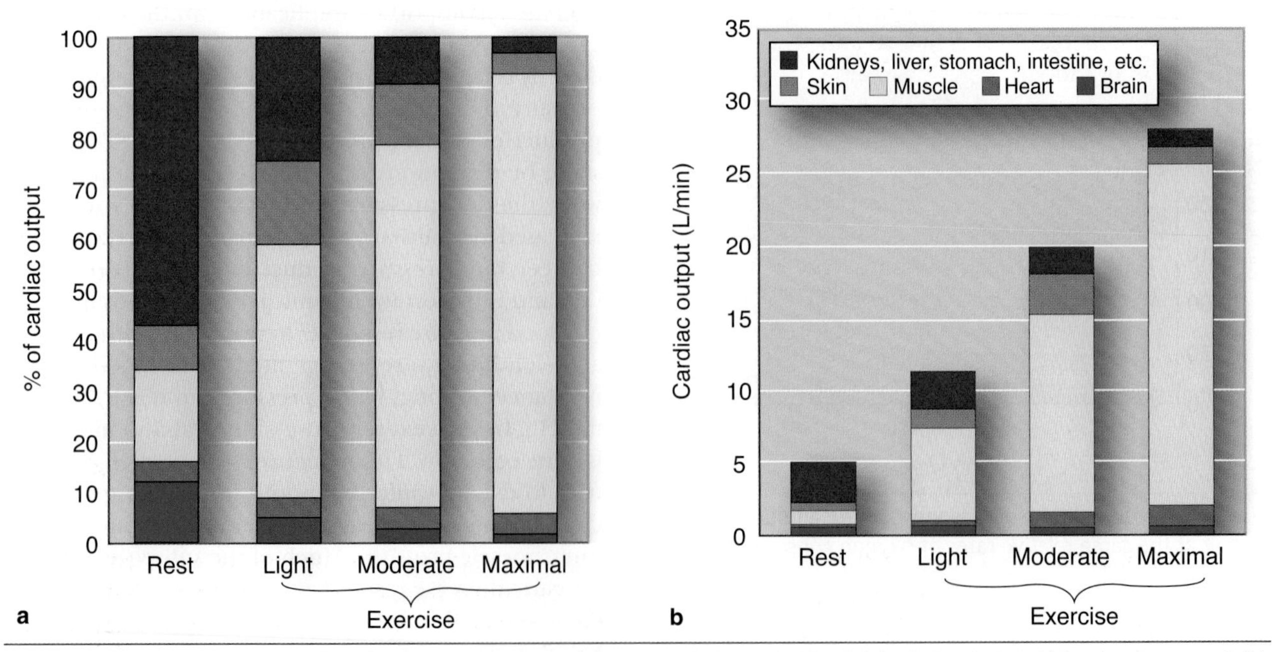

Figure 7.8 The distribution of cardiac output at rest and during exercise, expressed *(a)* relative to total blood volume and *(b)* in absolute values.

Data from A.J. Vander, J.H. Sherman, & D.S. Luciano, 1985, *Human physiology: The mechanisms of body function*, 4th ed. (New York: McGraw-Hill).

substances (including adenosine) released as a result of skeletal muscle contraction.

When exercise is performed in a hot environment, there is an increase in blood flow to the skin to help dissipate the body heat. The sympathetic control of skin blood flow is unique in that there are both typical sympathetic vasoconstrictor fibers (similar to skeletal muscle) and sympathetic active vasodilator fibers interacting. During dynamic exercise, skin blood flow is initially and transiently decreased by an increase in sympathetic vasoconstrictor activity. As body core temperature rises, there is a reduction in this sympathetic vasoconstriction, causing a passive vasodilation. Finally, at a specific body core temperature threshold, skin blood flow begins to dramatically increase through the action of the sympathetic active vasodilator system. The increase in skin blood flow during exercise promotes heat loss, because metabolic heat from deep in the body can be released when blood moves close to the skin. This allows maintenance of body temperature, although body temperature does increase with exercise, as discussed in more detail in chapter 11.

Cardiovascular Drift

With prolonged aerobic exercise or aerobic exercise in a hot environment, at a constant exercise intensity, SV gradually decreases and HR increases. Cardiac output is well maintained, but arterial blood pressure also declines. These alterations, illustrated in figure 7.9, have been referred to collectively as **cardiovascular drift,** and they are generally associated with increasing body temperature. Cardiovascular drift is associated with a progressive increase in the fraction of cardiac output directed to the vasodilated skin to facilitate heat loss and attenuate the increase in body core temperature. With more blood in the skin for the purpose of cooling the body, less blood is available to return to the heart, thus decreasing preload. There is also a small decrease in blood volume resulting from sweating and from a generalized shift of plasma across the capillary membrane into the surrounding tissues. These factors combine to decrease ventricular filling pressure, which decreases venous return to the heart and reduces the end-diastolic volume. With the reduction in end-diastolic volume, SV is reduced (SV = EDV – ESV). In order to maintain cardiac output ($\dot{Q}$ = HR × SV), HR increases to compensate for the decrease in SV.

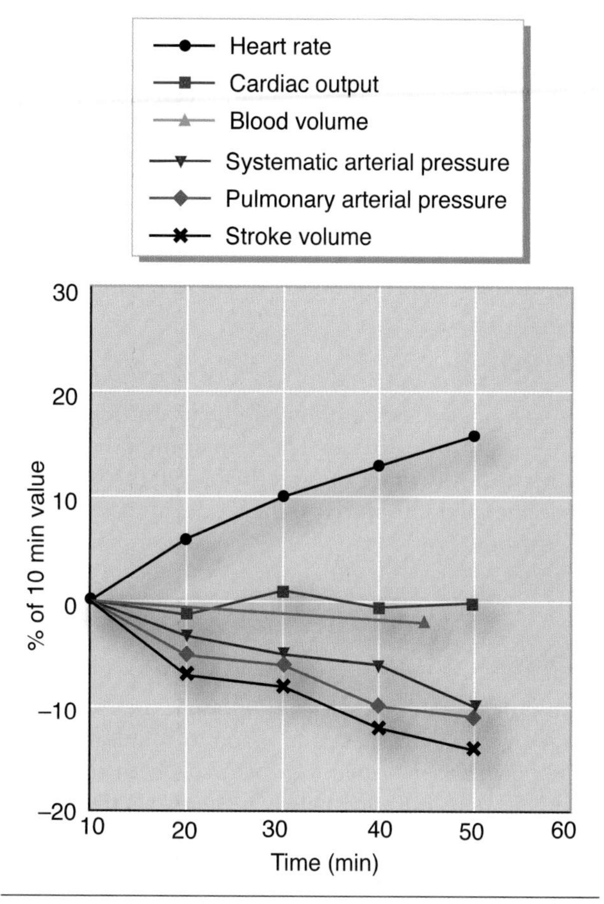

Figure 7.9 Circulatory responses to prolonged, moderately intense exercise in the upright position in a neutral 20 °C environment, illustrating cardiovascular drift. Values are expressed as the percentage of change from the values measured at the 10 min point of the exercise.

Competition for Blood Supply

When the demands of exercise are added to blood flow demands for all other systems of the body, competition for a limited available cardiac output can occur. This competition for available blood flow can develop among several vascular beds, depending on the specific conditions. For example, there is a competition for blood flow between active skeletal muscle and the gastrointestinal system following a meal. McKirnan and coworkers[2] studied the effects of feeding versus fasting on the distribution of blood flow during exercise in miniature pigs. The pigs were divided into two groups. One group fasted for 14 to 17 h before exercise. The other group ate their morning ration in two feedings: Half the ration was fed 90 to 120 min before exercise and the other half 30 to 45 min before exercise. Both groups of pigs then ran at approximately 65% of their $\dot{V}O_{2max}$.

Blood flow to the hindlimb muscles during exercise was 18% lower, and gastrointestinal blood flow was 23% higher, in the fed group than in the fasted group. Similar results in humans suggest that the redistribution of gastrointestinal blood flow to the working muscles is attenuated after a meal. As a practical application, these findings suggest that athletes should be cautious in timing their meals before competition to maximize blood flow to the active muscles during exercise.

In focus

Blood is redistributed in the body primarily to meet the demands of active tissues.

Another example of the competition for blood flow is seen in exercise in a hot environment. In this scenario, the competition for available cardiac output is set up between the skin circulation for thermoregulatory purposes and the exercising muscles. This will be discussed in more detail in chapter 11.

Blood

We have now examined how the heart and blood vessels respond to exercise. The remaining component of the cardiovascular system is the blood: the fluid that carries needed substances to the tissues and clears away waste products of metabolism. As metabolism increases during exercise, the functions of the blood become more critical for optimal performance.

Oxygen Content

At rest, the blood's oxygen content varies from 20 ml of oxygen per 100 ml of arterial blood to 14 ml of oxygen per 100 ml of venous blood returning to the right atrium. The difference between these two values (20 ml – 14 ml = 6 ml) is referred to as the **arterial–mixed venous oxygen difference,** or **$(a\text{-}\bar{v})O_2$ difference.** This value represents the extent to which oxygen is extracted, or removed, from the blood as it passes through the body.

With increasing exercise intensity, the $(a\text{-}\bar{v})O_2$ difference increases progressively and can increase approximately threefold from rest to maximal exercise intensities (see figure 7.10). This increased difference really reflects a decreasing venous oxygen content, because arterial oxygen content changes little from rest up to maximal exertion. With exercise, more oxygen is required by the active muscles; therefore more oxygen is extracted from the blood. The venous oxygen content decreases, approaching zero in the active muscles. However, mixed venous blood in the right atrium of the heart rarely decreases below 4 ml of oxygen per 100 ml of blood because the blood returning from the active tissues is mixed with blood from inactive tissues as it returns to the heart. Oxygen use in the inactive tissues is far lower than in the active muscles.

Plasma Volume

With the onset of exercise, there is an almost immediate loss of plasma from the blood to the interstitial fluid space. The movement of fluid out of the capillaries is dictated by the pressures inside the capillaries, which include the hydrostatic or blood pressure and the osmotic pressure exerted by the proteins in the blood, mostly albumin. The pressures that influence fluid movement outside the capillaries are the pressure provided by the surrounding tissue as well as the osmotic pressures from proteins in the interstitial fluid (figure 7.11). As blood pressure increases with exercise, the hydrostatic pressure within the capillaries also increases. Thus, the increase in blood pressure forces water from the vascular compartment to the interstitial compartment. Also, as metabolic waste products build up in the active muscle, intramuscular osmotic pressure increases, which draws fluid out of the capillaries to the muscle.

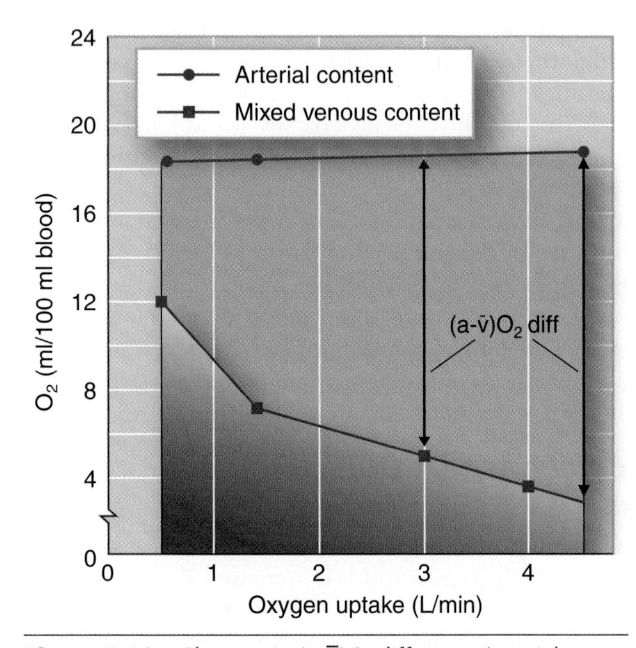

Figure 7.10 Changes in $(a\text{-}\bar{v})O_2$ difference (arterial–mixed venous oxygen difference) from low levels to maximal levels of exercise.

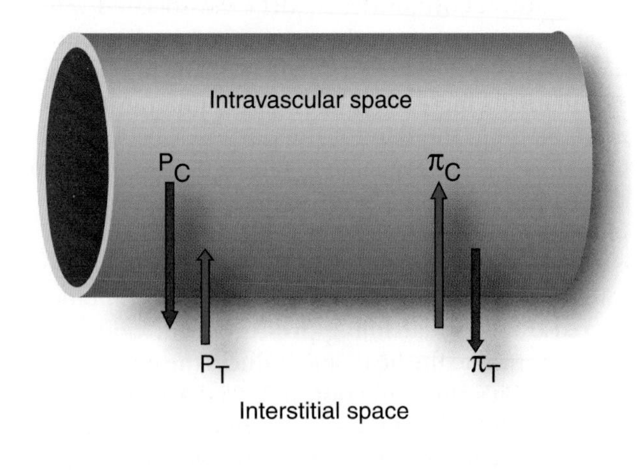

Figure 7.11 Filtration of plasma from the microvasculature. Both the blood pressure (P_C) inside the blood vessel and the osmotic pressure (π_T) in the tissue cause plasma to flow from the intravascular space to the interstitial space. The pressure that the tissue (P_T) exerts on the blood vessel and the osmotic pressure of the blood (π_C) inside the blood vessel cause plasma to be reabsorbed. Net filtration of plasma can be determined by summing the outward forces ($P_C + \pi_T$) and subtracting the inward forces ($P_T - \pi_C$); net capillary filtration = $(P_C + \pi_T) - (P_T - \pi_C)$.

Approximately a 10% to 15% reduction in plasma volume can occur with prolonged exercise. Similarly, 15% to 20% decreases in plasma volume have been observed in 1 min bouts of exhaustive exercise. With resistance training, the plasma volume loss is proportional to the intensity of the effort, with losses of from 10% to 15%.

If exercise intensity or environmental conditions cause sweating, additional plasma volume losses may occur. Although the major source of fluid for sweating is the interstitial fluid, this fluid space will be diminished as sweating continues. This increases the osmotic pressure in the interstitial space (since proteins do not move with the fluid), causing even more plasma to move out of the vascular compartment into the interstitial space. Intracellular fluid volume is impossible to measure directly and accurately, but research suggests that fluid is also lost from the intracellular compartment and even from the red blood cells, which may shrink in size.

A reduction of plasma volume will impair performance under many circumstances. For long-duration activities in which dehydration occurs and heat loss is a problem, the total flow of blood to active tissues must be reduced to allow increasingly more blood to be diverted to the skin in an attempt to lose body heat. Note that a decrease in muscle blood flow occurs only in conditions of dehydration and only at high intensities. Severely reduced plasma volume also increases blood viscosity, which can impede blood flow and thus limit oxygen transport, especially if the hematocrit exceeds 60%.

In activities that last a few minutes or less, body fluid shifts are of little practical importance. As exercise

Central Regulation of the Cardiorespiratory System During Dynamic Exercise

The cardiovascular and respiratory adjustments to dynamic exercise are profound and rapid. Within 1 s of the initiation of muscle contraction, HR dramatically increases by vagal withdrawal, and respiration increases. Increases in cardiac output and blood pressure increase blood flow to the active skeletal muscle to meet its metabolic demands. What causes these extremely rapid early changes in the cardiovascular system?

Over the years there has been considerable debate over what causes the cardiovascular system to be "turned on" at the onset of exercise. One explanation entails **central command** theory, which involves parallel activation of both the motor and the cardiovascular control centers of the brain. Activation of central command rapidly increases HR and blood pressure. In addition to central command, the cardiovascular responses to exercise are modified by mechanoreceptors, chemoreceptors, and baroreceptors. As discussed in chapter 5, baroreceptors are sensitive to stretch and send information back to the cardiovascular control centers about blood pressure. Signals from the periphery are sent back to the cardiovascular control centers through the stimulation of mechanoreceptors that are sensitive to the stretch of the skeletal muscle, and through the chemoreceptors that are sensitive to an increase in metabolites in the muscle. Feedback about blood pressure and the local muscle environment helps to fine-tune and adjust the cardiovascular response. These relationships are illustrated in the figure in this sidebar.

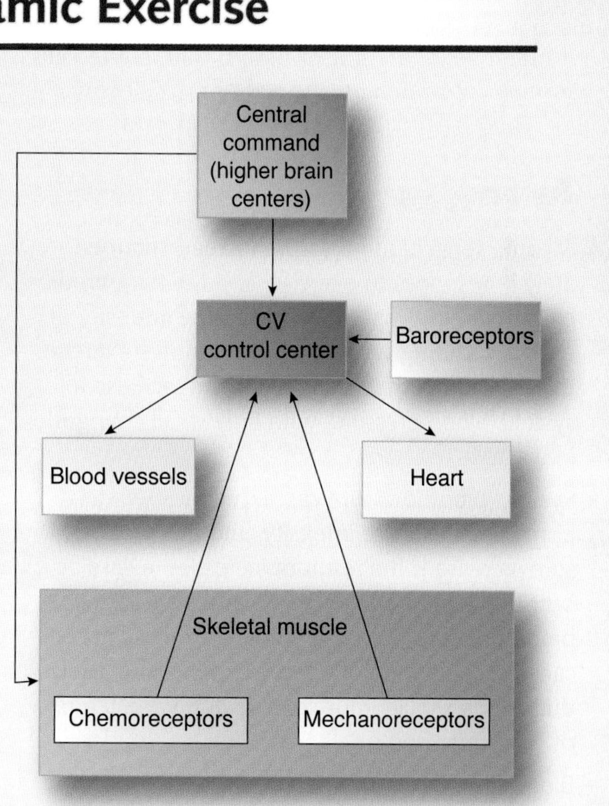

A summary of cardiovascular (CV) control during exercise.

Adapted, by permission, from S.K. Powers & E.T. Howley, 2004. *Exercise Physiology: Theory and Application to Fitness and Performance.* New York, McGraw-Hill, p.188.

duration increases, however, body fluid changes and temperature regulation become important for performance. For the football player, the Tour de France cyclist, or the marathon runner, these processes are crucial, not only for competition but also for survival. Deaths have occurred from dehydration and hyperthermia during or as a result of various sport activities. These issues are discussed in detail in chapter 11.

Hemoconcentration

When plasma volume is reduced, **hemoconcentration** occurs: The fluid portion of the blood is reduced, and the cellular and protein portions represent a larger fraction of the total blood volume. That is, they become more concentrated in the blood. This hemoconcentration increases red blood cell concentration substantially—by up to 20% or 25%. Hematocrit can increase from 40% to 50%. However, the total number and volume of red blood cells do not change substantially.

The net effect, even without an increase in the total number of red blood cells, is to increase the number of red blood cells per unit of blood; that is, the cells are more concentrated. As the red blood cell concentration increases, so does the blood's per unit hemoglobin content. This substantially increases the blood's oxygen-carrying capacity, which is advantageous during exercise and provides a distinct advantage at altitude at rest and during submaximal exercise, as we will see in chapter 12.

Integration of the Exercise Response

As is evident from all of the changes in cardiovascular function that take place at the onset of exercise, the cardiovascular system is extremely complex but responds exquisitely to deliver oxygen to meet the demands of exercising muscle. Figure 7.12 is a simplified flow diagram that illustrates how the body integrates all these cardiovascular responses to provide for its needs during exercise. Key areas and responses are labeled and summarized to help illustrate how these complex control mechanisms are coordinated. It is important to note that although the body attempts to meet the blood flow needs of the muscle, it can do so only if blood pressure is not compromised. Maintenance of arterial blood pressure appears to be the highest priority of

In review

- Mean arterial blood pressure increases immediately in response to exercise, and the magnitude of the increase is proportional to the intensity of exercise. During whole-body endurance exercise, this is accomplished primarily by an increase in systolic blood pressure, with only small changes in diastolic pressure.
- Systolic blood pressure can exceed 200 mmHg at maximal exercise intensity, and this pressure increase is the result of increases in cardiac output. Upper body exercise causes a greater blood pressure response than leg exercise at the same absolute rate of energy expenditure, likely due to the smaller muscle mass and vasculature of the upper body.
- Blood flow is redistributed during exercise from inactive or low-activity areas of the body to exercising muscles to meet their increased metabolic needs.
- With prolonged aerobic exercise, or aerobic exercise in the heat, SV gradually decreases and HR increases, but cardiac output is maintained. This is referred to as cardiovascular drift and is associated with a progressive increase in blood flow to the vasodilated skin to facilitate heat loss and attenuate the increase in core temperature.
- The changes that occur in the blood during exercise include the following:
  1. The $(a\text{-}\bar{v})O_2$ difference increases. This happens because the venous oxygen concentration decreases during exercise, reflecting increased extraction of oxygen from the blood for use by the active tissues.
  2. Plasma volume decreases. Plasma is pushed out of the capillaries by increased hydrostatic pressure as blood pressure increases, and fluid is drawn into the muscles by the increased osmotic pressure that results from waste accumulation. However, with prolonged exercise or exercise in hot environments, increasingly more plasma volume is lost through sweating, often causing dehydration.
  3. Hemoconcentration occurs as plasma volume (water) decreases. Although the actual number of red blood cells stays relatively constant, the relative number of red blood cells per unit of blood increases, which increases oxygen-carrying capacity.

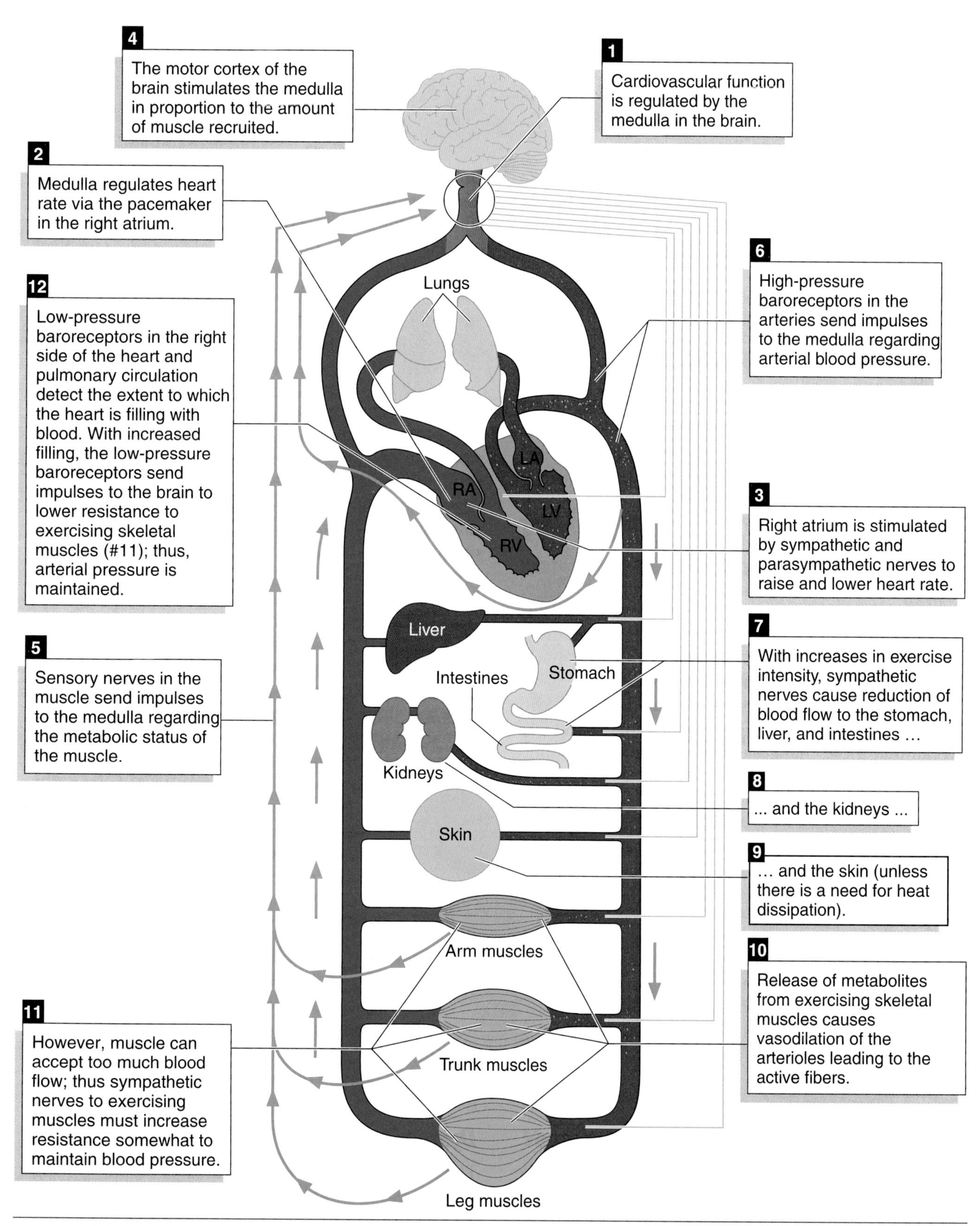

Figure 7.12 Integration of the cardiovascular system's response to exercise.

Adapted, by permission, from E.F. Coyle, 1991, "Cardiovascular function during exercise: Neural control factors," *Sports Science Exchange* 4(34): 1-6. Copyright 1991 by Gatorade Sports Science Institute.

CHAPTER 8

Principles of Exercise Training

In this chapter and in the online study guide

Roman Šebrle, an outstanding athlete from the Czech Republic, won the gold medal in track and field's decathlon during the 2004 Olympic Games in Athens, Greece. In 2001, he had set the world record for the decathlon, scoring over 9,000 points. Decathletes are considered by many to be the "ultimate" athletes, since they have to compete in events that test their speed, strength, power, agility, and endurance. The decathlon is a two-day event, comprising the 100 m sprint, long jump, shot put, high jump, and 400 m run on the first day, and the 110 m hurdles, discus, pole vault, javelin, and 1,500 m run on the second day. As we will see in this chapter and the following two chapters, training is very specific to the sport or event. Intense muscular power training to increase the distance one can heave a 16 lb shot put will do little if anything to improve one's 1,500 m run time. Decathletes must spend countless hours training specifically for each of their 10 events, fine-tuning their training techniques to maximize performance in each event.

When examining the acute response to exercise, we are concerned with the body's immediate response to a single exercise bout. We now investigate how the body responds over time to the stress of repeated exercise bouts. When one performs regular exercise over a period of weeks, the body adapts. The physiological adaptations that occur with chronic exposure to exercise or training improve both exercise capacity and efficiency. With resistance training, the muscles become stronger. With aerobic training, the heart and lungs become more efficient and endurance capacity increases. With high-intensity anaerobic training, the neuromuscular, metabolic, and cardiovascular systems adapt, allowing the person to generate more adenosine triphosphate (ATP) per unit of time, thus increasing muscular endurance and speed of movement over short periods of time. These adaptations are highly specific to the type of training done. Before investigating the specific adaptations the body makes to training, we must first look at the basic terminology and principles used in exercise training.

TERMINOLOGY

Before discussing the principles of exercise training, we first define key terms that will be used throughout the rest of this book.

Muscular Strength

The maximal force that a muscle or muscle group can generate is termed **strength.** Someone with a maximal capacity to bench press 100 kg (220 lb) has twice the strength of someone who can bench press 50 kg (110 lb). In this example, maximal capacity, or strength, is defined as the maximal weight the individual can lift just once. This is referred to as the **1-repetition maximum,** or the **1RM.** To determine their 1RM, people select a weight that they know they can lift at least one time. After a proper warm-up, they try to execute several repetitions. If they can perform more than one repetition, they add weight and try again to execute several repetitions. This continues until the person is unable to lift the weight more than a single repetition. This last weight that can be lifted only once is the 1RM for that particular exercise.

Muscle strength can be accurately measured in the research laboratory through use of specialized equipment that allows quantification of static strength and dynamic strength at various speeds and at various angles in the joint's range of motion (see figure 8.1). Gains in muscular strength involve changes in the structure and neural control of muscle. These will be discussed in the following chapter (chapter 9).

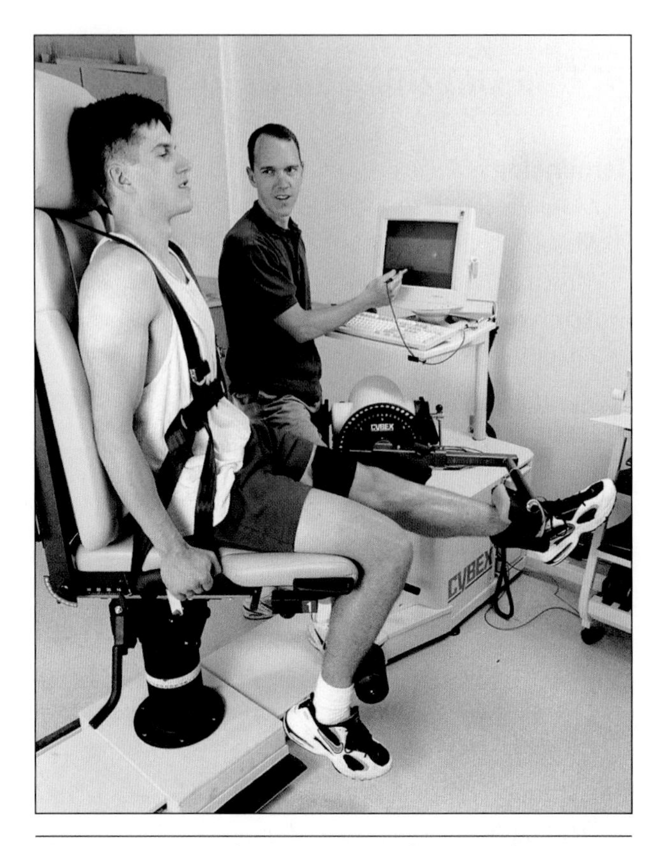

Figure 8.1 An isokinetic testing and training device.

Muscular Power

Power is defined as the rate of performing work, thus the product of force and velocity. Maximal muscular power, generally referred to as power, is the explosive aspect of strength, the product of strength and speed of movement.

$$\text{power} = \text{force} \times \text{distance/time}$$

where force = strength and distance/time = speed.

Consider an example. Two individuals can each bench press 200 kg (441 lb), moving the weight the same distance, from where the bar touches the chest to the position of full extension. But the one who can do it in half the time has twice the power of the slower individual. This is illustrated in table 8.1.

In focus

Maximal muscular power is the functional application of both strength and speed of movement. It is the key component for most athletic performances.

Although absolute strength is an important component of performance, power is even more important for most activities. In football, for example, an offensive lineman with a bench press 1RM of 200 kg (441 lb) may be unable to control a defensive lineman with a

TABLE 8.1 Strength, Power, and Muscular Endurance of Three Athletes Performing the Bench Press

Component	Athlete A	Athlete B	Athlete C
Strength[a]	100 kg	200 kg	200 kg
Power[b]	100 kg lifted 0.6 m in 0.5 s = 120 kg · m/s = 1,177 J/s or 1,177 W	200 kg lifted 0.6 m in 2.0 s = 60 kg · m/s = 588 J/s or 588 W	200 kg lifted 0.6 m in 1.0 s = 120 kg · m/s = 1,177 J/s or 1,177 W
Muscular endurance[c]	10 repetitions with 75 kg	10 repetitions with 150 kg	5 repetitions with 150 kg

[a]Strength was determined by the maximum amount of weight the athlete could bench press just once (i.e., the 1RM).

[b]Power was determined as the athlete performed the 1RM test as explosively as possible. Power was calculated as the product of force (weight lifted) times the distance lifted from the chest to full arm extension (0.6 m or about 2 ft), divided by the time it took to complete the lift.

[c]Muscular endurance was determined by the greatest number of repetitions that could be completed using 75% of the 1RM.

bench press 1RM of only 150 kg (330 lb) if the defensive lineman can move his 1RM at a much faster speed. The offensive lineman is 50 kg (110 lb) stronger, but the defensive lineman's faster speed coupled with adequate strength could give him the performance edge. Although field tests are available to estimate power, these tests are generally not very specific to power, in that their results are affected by other factors besides power. Power can be measured, however, through use of an electronic device as pictured in figure 8.1.

Throughout this book, the primary concern is with issues of muscular strength, with only brief mention of muscular power. Recall that power has two components: strength and speed. Speed is a more innate quality that changes little with training. Thus, power is increased almost exclusively through gains in strength.

Muscular Endurance

Many sporting activities depend on the muscles' ability to repeatedly develop and sustain near-maximal or maximal forces. This capacity to sustain repeated muscle contractions, as when one is performing sit-ups or push-ups, or to sustain fixed or static muscle contractions for an extended period of time, as when one is attempting to pin an opponent in wrestling, is termed **muscular endurance.** Although several excellent laboratory techniques are available to directly measure muscular endurance, a simple way to estimate it is to assess the maximum number of repetitions one can perform at a given percentage of the 1RM. For example, a man who can bench press 100 kg (220 lb) could evaluate his muscular endurance independently of his muscular strength by noting how many repetitions he could perform at, say, 75% of that load (75 kg, or 165 lb). Muscular endurance is increased through gains in muscular strength and through changes in local metabolic and circulatory function. Metabolic and circulatory adaptations that occur with training are discussed in chapter 10.

Table 8.1 illustrates the functional differences between strength, power, and muscular endurance in three athletes. The actual values have been greatly exaggerated for the purpose of illustration. From this table we can see that although Athlete A has half the strength of Athletes B and C, he has twice the power of Athlete B and is equal in power to Athlete C. Therefore, because of his fast speed of movement, his lack of strength does not seriously limit his power output. Also, for purposes of designing training programs, the analysis of these three athletes indicates that Athlete A should focus training on developing strength, without losing speed; Athlete B should focus training on developing speed of movement, although this is unlikely to change much; and Athlete C should focus training on developing muscular endurance. These recommendations are made assuming that each athlete needs to optimize performance in each of these three areas.

In focus

- Muscular strength is the maximum force a muscle or muscle group can generate.
- Muscular power is the product of strength and the speed of a movement. Although two individuals may have the same strength, if one requires less time than the other to move an identical load the same distance, she or he has more power.
- Muscular endurance is the ability of the muscles to sustain repeated muscle contractions or a single static contraction.

form or mode of training and then manipulating the following primary variables to fit the sport and athlete:

- Rate of the exercise interval
- Distance of the exercise interval
- Number of repetitions and sets during each training session
- Duration of the rest or active recovery interval
- Type of activity during the active recovery interval
- Frequency of training per week

Rate of the Exercise Interval

One can determine the rate of the exercise interval, or its intensity, either by establishing a specific duration for a set distance, as illustrated in our previous example for set 1 (i.e., 75 s for 400 m), or by using a fixed percentage of the athlete's maximal heart rate (HR_{max}). Setting a specific duration is more practical, particularly for short sprints. One typically determines this by using the athlete's best time for the set distance and then adjusting the duration according to the relative intensity that the athlete wants to achieve, with 100% equal to the athlete's best time. As an example, to develop the ATP-PCr system, the intensity should be near maximal (e.g., 90-98%); to develop the anaerobic glycolytic system, it should be high (e.g., 80-95%); and to develop the aerobic system, it should be moderate to high (e.g., 75-85%). These estimated percentages are only approximations and are dependent on the athlete's genetic potential and fitness level, duration of the interval (e.g., 10 s vs. 10 min), number of repetitions and sets, and duration of the active recovery interval.

Using a fixed percentage of the athlete's HR_{max} might provide a better index of the physiological stress experienced by the athlete. Heart rate monitors are now readily available and relatively inexpensive (see figure 8.5). HR_{max} can be determined during a maximal exercise test in the laboratory as described in chapter 7, or during an all-out run on the track using the heart rate monitor. Training the ATP-PCr system will require training at very high percentages of the athlete's HR_{max} (e.g., 90-100%), as will training to develop the anaerobic glycolytic system (e.g., 85-100% HR_{max}). To develop the aerobic system, the intensity should be moderate to high (e.g., 70-90% HR_{max}).

Figure 8.6 illustrates changes in blood lactate concentration in a single runner using interval training at three different intensities corresponding to those intensities needed to train the ATP-PCr system, the glycolytic system, and the oxidative system. The runner performed five repetitions in a single set at each intensity

Figure 8.5 A runner outfitted with a heart rate monitor. The receiving unit, attached to the chest strap, picks up and transmits electrical impulses from the heart to the digital monitor and memory device worn on the wrist. After the workout, the contents of the memory device can be downloaded to a computer.

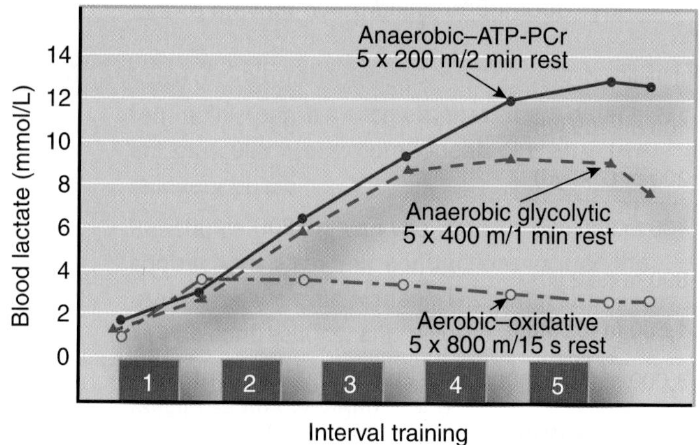

Figure 8.6 Blood lactate concentrations in a single runner following a single set of five repetitions of interval training at three different paces, each on different days, corresponding to the appropriate pace for training each energy system.

on different days, and the lactate values were obtained from a blood sample taken after the last repetition of each intensity.

Distance of the Exercise Interval

The distance of the exercise interval is determined by the requirements of the event, sport, or activity. Athletes who run or sprint short distances, such as track sprinters, basketball players, and football players, will utilize short intervals of 30 m to 200 m (33-219 yd), although a 200 m sprinter will frequently run overdistances of 300 to 400 m (328-437 yd). A 1,500 m runner may run intervals as short as 200 m to increase speed; but most of his or her training would be at distances of 400 to 1,500 m (437-1,640 yd), or even longer distances, to increase endurance and decrease fatigue or exhaustion in the race.

Number of Repetitions and Sets During Each Training Session

The number of repetitions and sets will also be largely determined by the needs of the sport, event, or activity. Generally, the shorter and more intense the interval, the greater should be the number of repetitions and sets. As the training interval is lengthened in both distance and duration, the number of repetitions and sets is correspondingly reduced.

Duration of the Rest or Active Recovery Interval

The duration of the rest or active recovery interval will depend on how rapidly the athlete recovers from the exercise interval. The extent of recovery is best determined by the reduction of the athlete's heart rate to a predetermined level during the rest or active recovery period. For younger athletes (30 years of age or younger), heart rate is generally allowed to drop to between 130 and 150 beats/min before the next exercise interval begins. For those over 30 years, since HR_{max} decreases ~1 beat/min per year, we subtract the difference between the athlete's age and 30 years from both 130 and 150. So, for a 45-year-old, we would subtract 15 beats/min to obtain the athlete's recovery range of 115 to 135 beats/min. The recovery interval between sets can be established in a similar manner, but generally the heart rate should be below 120 beats/min.

Type of Activity During the Active Recovery Interval

The type of activity performed during the active recovery interval for land-based training can vary from slow walking to rapid walking or jogging. In the pool, slow swimming using alternative strokes or the primary stroke is appropriate. In some cases, usually in the pool, total rest can be used. Generally, the more intense the exercise interval, the lighter or less intense the activity performed in the recovery interval. As the athlete becomes better conditioned, he or she will be able to increase the intensity or decrease the duration of the rest interval, or both.

Frequency of Training per Week

The frequency of training will depend largely on the purpose of the interval training. The world-class sprinter or middle-distance runner will need to work out five to seven days a week, although not every workout will include interval training. The swimmer will use interval training almost exclusively. The athlete who plays team sports can benefit from two to four days of interval training per week when interval training is used only as a supplement to a general conditioning program.

The coach or athlete who is interested in the specific details of how to organize and administer an interval training program should refer to the excellent text by Fox and Mathews (1974).[4] These authors have provided many excellent examples of how interval training can be utilized for various types of sports.

Continuous Training

Continuous training involves continuous activity without rest intervals. This can vary from long, slow distance (LSD) training to high-intensity endurance training. Continuous training is structured primarily to affect the oxidative and glycolytic energy systems. High-intensity continuous activity is usually performed at intensities representing 85% to 95% of the athlete's HR_{max}. For swimmers and track and cross country athletes, this could be above, or at or near, race pace. This pace would likely match or exceed the pace associated with the athlete's lactate threshold. Scientific evidence has clearly demonstrated that marathon runners typically race at, or very close to, their lactate threshold.

LSD training became extremely popular in the 1960s. With this form of training, introduced by Dr. Ernst Van Auken, a German physician and coach, in the 1920s, the athlete typically trains at relatively low intensities, between 60% and 80% of HR_{max}, which is approximately the equivalent of 50% to 75% of $\dot{V}O_{2max}$. Distance, rather than speed, is the main objective. Distance runners may train 15 to 30 mi (24-48 km) per day using LSD techniques, with weekly distances of 100 to 200 mi (161-322 km). The pace of the run is substantially slower than the runner's maximal pace. While less stressful to the cardiovascular and respiratory systems, extreme distances can result in overuse injuries and general breakdown of muscles and joints. Further, the serious runner needs to train at or near race pace on a regular basis to develop leg speed and strength. Thus, most runners will vary their workout from one day to the next, from week to week, and from month to month.

Long, slow distance training is probably the most popular and safest form of aerobic endurance conditioning for the nonathlete who just wants to get into shape and stay in shape for health-related purposes. More vigorous or burst types of activity generally are not encouraged in older, sedentary people. Long, slow distance is also a good training program for athletes in team sports to maintain aerobic endurance during the season as well as during the off-season.

Fartlek training, or speed play, is another form of continuous exercise that has a flavor of interval training. This form of training was developed in Sweden in the 1930s and is used primarily by distance runners. The athlete varies the pace from high speed to jogging speed at his or her discretion. This is a free form of training in which fun is the primary goal, and distance and time are not even considered. Fartlek training is normally performed in the countryside where there are a variety of hills. Many coaches have used Fartlek training to supplement either high-intensity, continuous training, or interval training, since it provides variety to the normal training routine.

Interval-Circuit Training

Introduced in the Scandinavian countries in the 1960s and 1970s, **interval-circuit training** combines interval and circuit training into one workout. The circuit may be 3,000 to 10,000 m in length, with stations every 400 to 1,600 m (437-1,750 yd). The athlete jogs, runs, or sprints the distance between stations; stops at each station to perform a strength, flexibility, or muscular endurance exercise in a manner similar to that in actual circuit training; and continues on, jogging, running, or sprinting to the next station. These courses are typically located in parks or in the county where there are many trees and hills.

In review

- Anaerobic and aerobic power training programs are designed to train the three metabolic energy systems: ATP-PCr system, anaerobic glycolytic system, and oxidative system.
- Interval training consists of repeated bouts of high- to moderate-intensity exercise interspersed with periods of rest or reduced-intensity exercise. For short intervals, the rate or pace of activity and the number of repetitions are usually high, and the recovery period is usually short. Just the opposite is the case for long intervals.
- Both the exercise rate and the recovery rate can be closely monitored with use of a heart rate monitor.
- Interval training is appropriate for all sports.
- Continuous training has no rest intervals and can vary from LSD training to high-intensity training. Long, slow distance training is very popular for general fitness training.
- Fartlek training, or speed play, is an excellent activity for recovering from several days or more of intense training.
- Interval-circuit training combines interval training and circuit training into one workout.

In Closing

In this chapter, we reviewed general principles of training and the terminology used to describe these principles. We then learned the essential ingredients of successful resistance training and anaerobic and aerobic power training programs. With this background, we can now focus on how the body adapts to these different types of training programs. In the next chapter, we will see how the body responds to resistance training.

KEY TERMS

1-repetition maximum (1RM)
aerobic power
anaerobic power
continuous training
eccentric training
electrical stimulation training
Fartlek training
free weights
interval-circuit training
interval training
isokinetic training
isometric training
LSD training
muscular endurance
needs analysis
plyometrics
power
principle of hard/easy
principle of individuality
principle of periodization
principle of progressive overload
principle of reversibility
principle of specificity
static-contraction resistance training
strength
stretch-shortening cycle exercise
variable-resistance training

STUDY QUESTIONS

1. Define and differentiate the terms *strength, power,* and *muscular endurance.* How does each component relate to athletic performance?
2. Define aerobic and anaerobic power. How does each relate to athletic performance?
3. Describe and provide examples for the principles of individuality, specificity, reversibility, progressive overload, hard/easy, and periodization.
4. What factors need to be considered when one is conducting a needs analysis for designing a resistance training program?
5. What would be the appropriate range for resistance and repetitions when one is designing a resistance training program targeted to develop strength? Muscular endurance? Muscular power? Hypertrophy?
6. What is the optimal number of sets for increasing strength, and how might this vary depending on the individual's state of training?
7. Describe the various types of resistance training and explain the advantages and disadvantages of each.
8. What type of training program would likely provide the greatest improvement for sprinters? Marathon runners? Football players?
9. Describe the various forms of interval and continuous training programs and discuss the advantages and disadvantages of each. Indicate the sport or event most likely to benefit from each one.

STUDY GUIDE ACTIVITIES

In addition to the activities listed in the chapter opening outline on page 187, two other activities are available in the online study guide, located at

www.HumanKinetics.com/PhysiologyOfSportAndExercise.

The chapter **SUMMARY** reviews the key concepts covered in the chapter, and the end-of-chapter **QUIZ** tests your understanding of the material covered in the chapter.

CHAPTER 9

Adaptations to Resistance Training

In this chapter and in the online study guide

We know that athletes who use resistance training become much stronger. For a number of years, Dr. William Gonyea and his colleagues at the University of Texas Health Sciences Center Dallas have been trying to determine how the athlete's muscle gets stronger with resistance training. But Dr. Gonyea and his colleagues have been working with a different type of athlete—cats! The cats receive food rewards for their daily workouts that entice them to work very hard "pumping iron." Like their human counterparts, these cats experience substantial increases in strength and in muscle size. Dogs, beware—these are not cats that you want to mess around with! This chapter discusses the results of Dr. Gonyea's studies as he and his colleagues have challenged traditional thinking on how a muscle increases its size.

With chronic exercise, many adaptations occur in the neuromuscular system. The extent of the adaptations depends on the type of training program followed: Aerobic training, such as jogging or swimming, results in little or no gain in muscular strength and power, but major neuromuscular adaptations occur with **resistance training.**

Resistance training was once considered inappropriate for athletes except those in competitive weightlifting, weight events in track and field, and—on a limited basis—football, wrestling, and boxing. Women were essentially banned from the weight room! But in the late 1960s and early 1970s, coaches and researchers discovered that strength and power training are beneficial for almost all sports and activities, and for women as well as men. Finally, in the late 1980s and early 1990s, health professionals began to recognize the importance of resistance training to overall health and fitness.

Most athletes now include strength and power training as important components of their overall training programs, including female athletes. Much of this attitude change is attributable to research that has proven the performance benefits of resistance training and to innovations in training techniques and equipment. Resistance training is now an important part of the exercise prescription for those who seek the health-related benefits of exercise (figure 9.1).

Figure 9.1 Resistance training is recognized as important for athletes and nonathletes alike.

RESISTANCE TRAINING AND GAINS IN MUSCULAR FITNESS

Throughout this book we see how important muscular fitness is to athletic performance as well as general health. How do we get stronger, and how do we increase muscle power and muscle endurance? Maintaining an active lifestyle is important in maintaining muscular fitness, but resistance training programs are necessary to increase strength, power, and endurance. In this section, we briefly review the changes that result from resistance training. We focus on strength, with only a brief mention of power and muscular endurance—topics that are discussed in more detail later in this book.

The neuromuscular system is one of the most responsive systems in the body to training. Resistance training programs can produce substantial strength gains. Within three to six months, one can see from 25% to 100% improvement, sometimes even more. These estimates of percentage gains in strength are, however, somewhat misleading. Most subjects in strength training research studies have never lifted weights or participated in any other form of resistance training. Much of their early gains in strength are the result of learning how to more effectively produce force and produce a true maximal movement, such as moving a barbell from the chest to a fully extended position in the bench press. This learning effect can account for as much as 50% of the overall gain in strength.[18] Related to this learning effect, the largest gains are seen during the first few weeks of training, with the rate of increase in strength declining with continued training. As an example, in their position stand as outlined in "Progression Models in Resistance Training for Healthy Adults," the American College of Sports Medicine concluded that muscular strength will increase about 40% in previously untrained people, 20% in moderately trained people, 16% in trained people, 10% in those who are advanced (more than two years of training), and 2% in those who are elite (high-level athletes) over periods of from four weeks to two years.[2]

Gains in strength appear to be similar when we compare women to men, children to adults, and elderly people to young and middle-aged adults, when these gains are expressed as a percentage of their initial strength. However, the increase in the absolute weight lifted will generally be greater in men compared to women, in adults compared to children, and in young adults compared to older adults. For example, after 20 weeks of resistance training, assume that a 12-year-old boy and a 25-year-old man each improves his bench press strength by 50%. If the man's initial bench press strength (1-repetition maximum; 1RM) was 50 kg (110 lb), he would have improved by 25 kg (55 lb) to a new 1RM of 75 kg (165 lb). If the boy's initial 1RM was 25 kg, he would have improved by 12.5 kg (28 lb) to a new 1RM of 37.5 kg (83 lb). This will be discussed in more detail in chapters 16 through 18.

Muscle is very plastic, increasing in size and strength with exercise training and decreasing in size and strength when immobilized. The remainder of this chapter details how these changes occur. How do people become stronger? What physiological adaptations occur that allow them to exert greater strength? What causes the acute pain in the specific muscles trained during the first week or two of training?

MECHANISMS OF GAINS IN MUSCLE STRENGTH

For many years, strength gains were assumed to result directly from increases in muscle size (**hypertrophy**). This assumption was logical because most who strength trained regularly were men, and they often developed large, bulky muscles. Also, muscles associated with a limb immobilized in a cast for weeks or months start to decrease in size (**atrophy**) and lose strength almost immediately. Gains in muscle size are generally paralleled by gains in strength, and losses in muscle size correlate highly with losses in strength. Thus, it is tempting to conclude that a direct cause-and-effect relationship exists between muscle size and muscle strength. While there is a relationship between size and strength, muscle strength involves far more than mere muscle size.

Numerous media reports indicate that people can perform superhuman feats of strength during great psychological stress. Straitjackets were designed specifically to control patients in mental hospitals who suddenly go berserk and are impossible to restrain. Even the world of sport boasts isolated examples of superhuman athletic performances, such as Bob Beamon's long jump of 29 ft 2 1/2 in. (8.9 m) at the 1968 Olympic Games—a jump that exceeded the previous world record by nearly 2 ft (0.6 m)! World records are usually broken by inches or centimeters or, more often, mere fractions of inches or centimeters. Beamon's record stood unbroken until 1991.

Women experience similar, or even greater, percentage increases in strength compared with men who participate in the same training program, but women generally do not experience as much hypertrophy (see chapter 18). Similar findings have been reported in children.

This does not mean that muscle size is unimportant in the ultimate strength potential of the muscle. Size is extremely important, as revealed by the existing men's and women's world records for competitive weightlifting, shown in figure 9.2. As weight classification

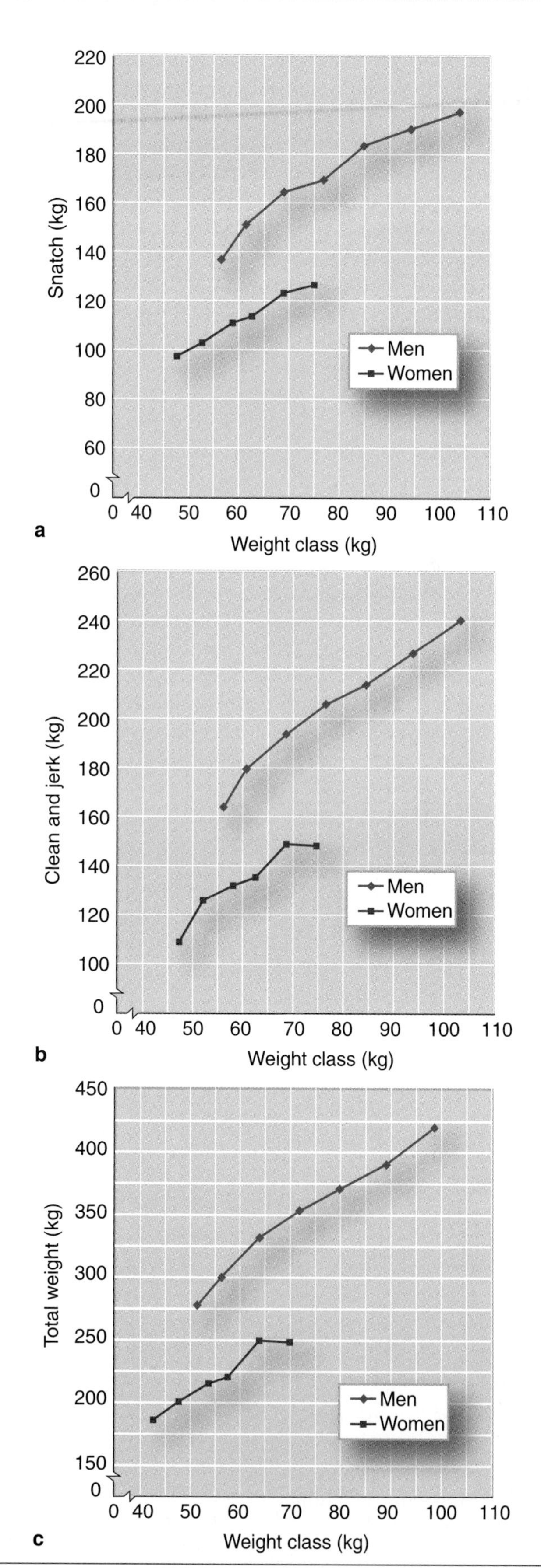

Figure 9.2 World records for *(a)* the snatch, *(b)* the clean and jerk, and *(c)* total weight for men and women through May 2006.

increases (implying increased muscle mass), so does the record for the total weight lifted. However, examples of superhuman strength and studies on women and children indicate that the mechanisms associated with strength gains are very complex and are not completely understood at this time. How, then, can we explain strength gains with training? Obviously, increased muscle size is important, but there is increasing evidence that the neural control of the trained muscle is also altered, allowing a greater force production from the muscle.

Neural Control of Strength Gains

An important neural component explains at least some of the strength gains that result from resistance training. Enoka has made a convincing argument that strength gains can be achieved without structural changes in muscle but not without neural adaptations.[9] Thus, strength is not solely a property of muscle. Rather, it is a property of the motor system. Motor unit recruitment, stimulation frequency, and other neural factors are also quite important to strength gains. They may well explain most, if not all, strength gains that occur in the absence of hypertrophy, as well as episodic superhuman feats of strength.

Synchronization and Recruitment of Additional Motor Units

Motor units are generally recruited asynchronously; they are not all called on at the same instant. They are controlled by a number of different neurons that can transmit either excitatory or inhibitory impulses (see chapter 3). Whether the muscle fibers contract or stay relaxed depends on the summation of the many impulses received by the given motor unit at any one time. The motor unit is activated and its muscle fibers contract only when the incoming excitatory impulses exceed the inhibitory impulses and the threshold is met.

Strength gains may result from changes in the connections between motor neurons located in the spinal cord, allowing motor units to act more synchronously, facilitating contraction, and increasing the muscle's ability to generate force. There is good evidence to support increased motor unit synchronization with resistance training, but there is still controversy as to whether synchronization of motor unit activation produces a more forceful contraction. It is clear, however, that synchronization does improve the rate of force development and the capability to exert steady forces.[7]

An alternate possibility is simply that more motor units are recruited to perform the given task, independent of whether these motor units act in unison. Such improvement in recruitment patterns could result from an increase in neural drive to the α-motor neurons during maximal contraction. This increase in neural drive could also increase the frequency of discharge (rate coding) of the motor units. It is also possible that the inhibitory impulses are reduced, allowing more motor units to be activated, or to be activated at a higher frequency.

Increased Rate Coding of Motor Units

The increase in neural drive of α-motor neurons could also increase the frequency of discharge, or rate coding, of their motor units. Recall from chapter 1 that as the frequency of stimulation of a given motor unit is increased, the muscle eventually reaches a state of tetanus, producing the absolute peak force or tension of the muscle fiber or motor unit (see figure 1.12). There is limited evidence that rate coding is increased with resistance training. Rapid movement or ballistic-type training appears to be particularly effective in stimulating increases in rate coding.

Autogenic Inhibition

Inhibitory mechanisms in the neuromuscular system, such as the Golgi tendon organs, might be necessary to prevent the muscles from exerting more force than the bones and connective tissues can tolerate. This control is referred to as **autogenic inhibition.** During superhuman feats of strength, major damage often occurs to these structures, suggesting that the protective inhibitory mechanisms are overridden.

The function of Golgi tendon organs was discussed in chapter 3. When the tension on a muscle's tendons and internal connective tissue structures exceeds the threshold of the embedded Golgi tendon organs, motor neurons to that muscle are inhibited; that is, autogenic inhibition occurs. Both the reticular formation in the brain stem and the cerebral cortex function to initiate and propagate inhibitory impulses.

Training can gradually reduce or counteract these inhibitory impulses, allowing the muscle to reach greater levels of strength.[1] Thus, strength gains may be achieved by reduced neurological inhibition. This theory is attractive because it can at least partially explain superhuman feats of strength and strength gains in the absence of hypertrophy.

Autogenic inhibition may be attenuated with resistance training, allowing a greater force production from trained muscles independent of increases in muscle mass.

Other Neural Factors

In addition to increasing motor unit recruitment or decreasing neurological inhibition, other neural factors can contribute to strength gains with resistance training. One of these is referred to as coactivation of agonist and antagonist muscles (the agonist muscles are the primary movers, and the antagonist muscles act to impede the agonists). If we use forearm flexor concentric contraction as an example, the biceps is the primary agonist and the triceps is the antagonist. If both were contracting with equal force development, no movement would occur. Thus, to maximize the force generated by an agonist, it is necessary to minimize the amount of coactivation.[10] Reduction in coactivation could explain a portion of strength gains attributed to neural factors, but its contribution likely would be small.

Changes also have been noted in the morphology of the neuromuscular junction, with both increased and decreased activity levels that might be directly related to the muscle's force-producing capacity.

Muscle Hypertrophy

How does a muscle's size increase? Two types of hypertrophy can occur: transient and chronic. **Transient hypertrophy** is the increased muscle size that develops during and immediately following a single exercise bout. This results mainly from fluid accumulation (edema) in the interstitial and intracellular spaces of the muscle that comes from the blood plasma. Transient hypertrophy, as its name implies, lasts only for a short time. The fluid returns to the blood within hours after exercise.

Chronic hypertrophy refers to the increase in muscle size that occurs with long-term resistance training. This reflects actual structural changes in the muscle that can result from an increase in the size of existing individual muscle fibers **(fiber hypertrophy)**, in the number of muscle fibers **(fiber hyperplasia)**, or both. Controversy surrounds the theories that attempt to explain the underlying cause of this phenomenon. Of importance, however, is the finding that the eccentric component of training is important in maximizing increases in muscle fiber cross-sectional area. A number of studies have shown greater hypertrophy and strength resulting solely from eccentric contraction training as compared to concentric contraction or combined eccentric and concentric contraction training. Further, higher-velocity eccentric training appears to result in greater hypertrophy and strength gains than slower-velocity training.[23] These greater increases appear to be related to disruptions in the sarcomere Z-lines. This disruption had originally been labeled as muscle damage, but is now thought to represent fiber protein remodeling.[23] Thus, training with only concentric actions could limit muscle hypertrophy and increases in muscle strength. Let's now look at the two postulated mechanisms for increasing muscle size with resistance training: fiber hypertrophy and fiber hyperplasia.

Fiber Hypertrophy

Early research suggested that the number of muscle fibers in each of a person's muscles is established by birth or shortly thereafter and that this number remains fixed throughout life. If this were true, then whole-muscle hypertrophy could result only from individual muscle fiber hypertrophy. This could be explained by

- more myofibrils,
- more actin and myosin filaments,
- more sarcoplasm,
- more connective tissue, or
- any combination of these.

As seen in the micrographs in figure 9.3, intense resistance training can significantly increase the cross-sectional area of muscle fibers. In this example, fiber hypertrophy is probably caused by increased numbers of myofibrils and actin and myosin filaments, which would provide more cross-bridges for force production during maximal contraction. Such dramatic enlargement of muscle fibers does not occur, however, in all cases of muscle hypertrophy.

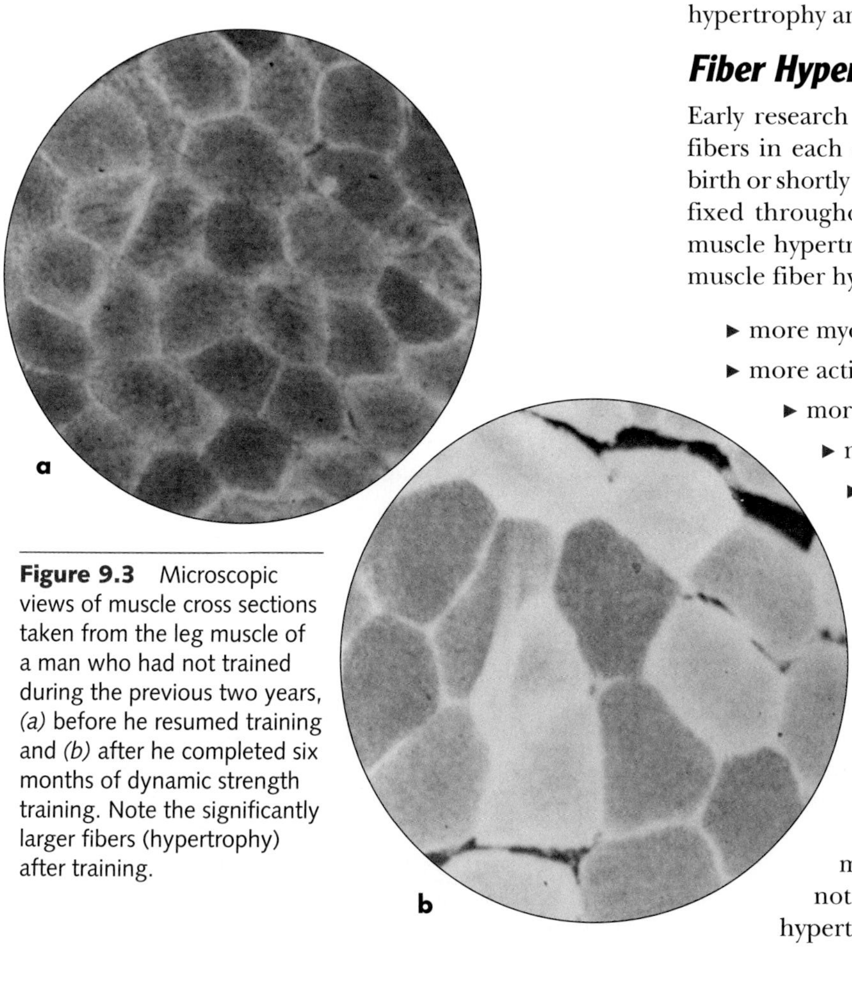

a

b

Figure 9.3 Microscopic views of muscle cross sections taken from the leg muscle of a man who had not trained during the previous two years, *(a)* before he resumed training and *(b)* after he completed six months of dynamic strength training. Note the significantly larger fibers (hypertrophy) after training.

Individual muscle fiber hypertrophy from resistance training appears to result from a net increase in muscle protein synthesis. The muscle's protein content is in a continual state of flux. Protein is always being synthesized and degraded. But the rates of these processes vary with the demands placed on the body. During exercise, protein synthesis decreases, while protein degradation apparently increases. This pattern reverses during the postexercise recovery period, even to the point of a net synthesis of protein. The provision of a carbohydrate and protein supplement immediately after a training bout can create a more positive nitrogen balance, facilitating protein synthesis.

The hormone testosterone is thought to be at least partly responsible for these changes, because one of its primary functions is the promotion of muscle growth. For example, males experience a significantly greater increase in muscle growth starting at puberty, which is largely due to a 10-fold increase in testosterone production. Testosterone is a steroidal hormone with major anabolic functions. It has been well established that massive doses of anabolic steroids coupled with resistance training markedly increase muscle mass and strength (see chapter 15).

Fiber Hyperplasia

Research on animals suggests that hyperplasia may also be a factor in the hypertrophy of whole muscles. Studies on cats provide fairly clear evidence that fiber splitting occurs with extremely heavy weight training.[11] Cats were trained to move a heavy weight with a forepaw to get their food (figure 9.4). They learned to generate considerable force. With this intense strength training, selected muscle fibers appeared to actually split in half, and each half then increased to the size of the parent fiber. This is seen in the cross-sectional cuts through the muscle fibers shown in figure 9.5.

Subsequent studies, however, demonstrated that hypertrophy of selected muscles in chickens, rats, and mice that resulted from chronic exercise overload was attributable solely to hypertrophy of existing fibers, not hyperplasia. In these studies, each fiber in the whole muscle was actually counted. These direct fiber counts revealed no change in fiber number.

This finding led the scientists who conducted the initial cat experiments to conduct an additional resistance training study with cats. This time they used actual

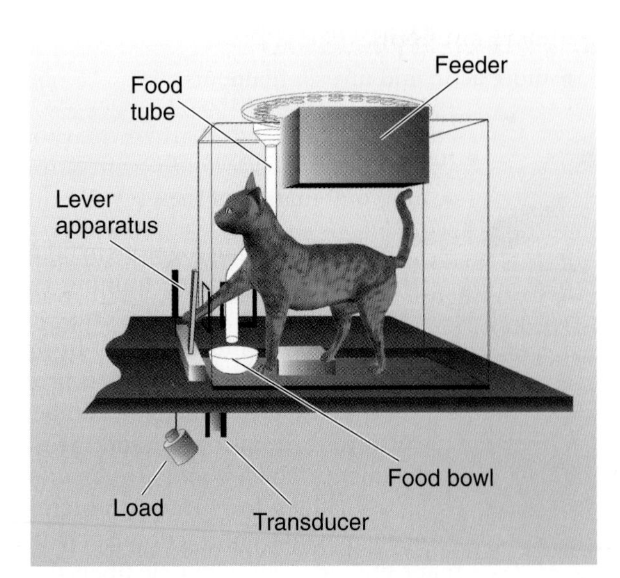

Figure 9.4 Heavy resistance training in cats.

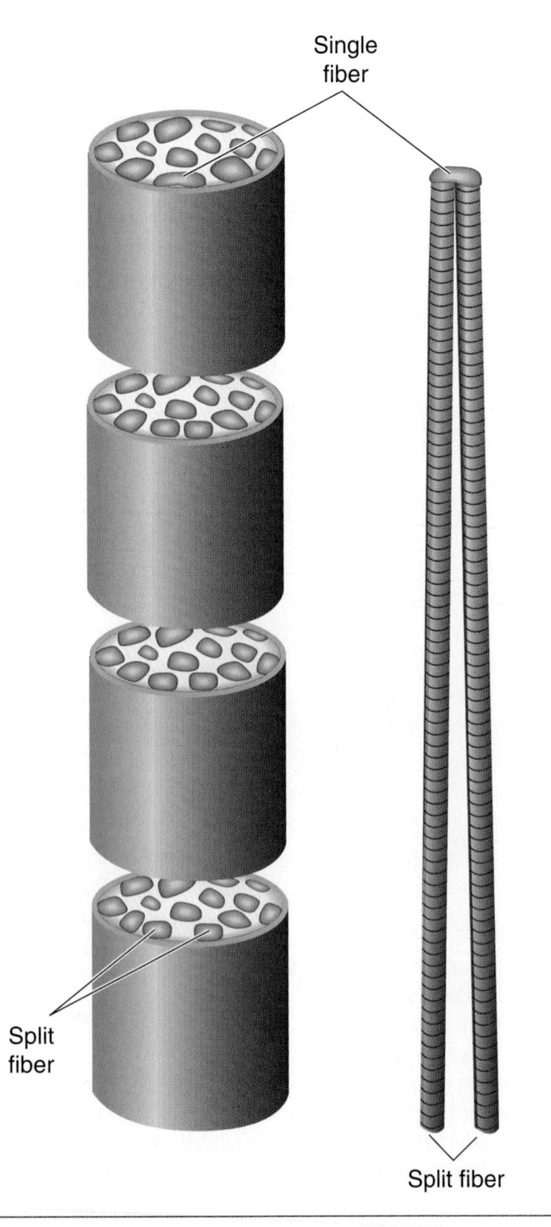

Figure 9.5 Muscle fiber splitting. The drawn models were extrapolated from a series of microscopic slides.

fiber counts to determine if total muscle hypertrophy resulted from hyperplasia or fiber hypertrophy.[12] Following a resistance training program of 101 weeks, the cats were able to perform one-leg lifts of an average of 57% of their body weight, resulting in an 11% increase in muscle weight. Most important, the researchers found a 9% increase in the total number of muscle fibers, confirming that muscle fiber hyperplasia did occur.

The difference in results between the cat studies and those with rats and mice most likely is attributable to differences in the manner in which the animals were trained. The cats were trained with a pure form of resistance training: high resistance and low repetitions. The other animals were trained with more endurance-type activity: low resistance and high repetitions.

One additional animal model has been used to stimulate muscle hypertrophy associated with hyperplasia. Scientists have placed the anterior latissimus dorsi muscle of chickens in a state of chronic stretch by attaching weights to it, with the other wing serving as the normal control condition. In many of the studies that have used this model, the chronic stretch has resulted in substantial hypertrophy and hyperplasia.

Researchers are still uncertain about the roles played by hyperplasia and individual fiber hypertrophy in increasing human muscle size with resistance training. Most evidence indicates that individual fiber hypertrophy accounts for most whole-muscle hypertrophy. However, results from selected studies indicate that hyperplasia is possible in humans.

In several studies of bodybuilders, swimmers, and kayakers, substantial muscle hypertrophy has been observed in trained muscles, but in the absence of individual fiber hypertrophy when compared to values in untrained control subjects. This suggests that there is a greater number of muscle fibers in the trained muscles than in the corresponding muscles of untrained control subjects. However, other studies have shown individual fiber hypertrophy in highly trained athletes compared to untrained controls.

In a study of seven previously healthy young men who had suffered sudden accidental death, the investigators compared cross sections of autopsied right and left tibialis anterior muscles (lower leg). Right-hand dominance is known to lead to greater hypertrophy of the left leg. In fact, the average cross-sectional area of the left muscle was 7.5% larger. This was associated with a 10% greater number of fibers in the left muscle. There was no difference in mean fiber size.[24]

In focus

Muscle fiber hyperplasia has been clearly shown to occur in animal models with the use of resistance training to induce muscle hypertrophy. Only a few studies, on the other hand, suggest evidence of hyperplasia in humans.

The differences among these studies might be explained by the nature of the training load or stimulus. Training at high intensities or high resistances is thought to cause greater fiber hypertrophy, particularly of the type II (fast-twitch) fibers, than training at lower intensities or resistances.

Only one longitudinal study demonstrated the possibility of hyperplasia in men who had previous recreational resistance training experience.[19] Following 12 weeks of intensified resistance training, the muscle fiber number in the biceps brachii of several of the 12 subjects appeared to increase significantly. It appears from this study that hyperplasia can occur in humans, but possibly only in certain subjects or under certain training conditions.

From the preceding information, it appears that fiber hyperplasia can occur in animals and possibly in humans. How are these new cells formed? As shown in figure 9.5, it is postulated that individual muscle fibers have the capacity to divide and split into two daughter cells, each of which can then develop into a functional muscle fiber. It has more recently been established that satellite cells, which are the myogenic stem cells involved in skeletal muscle regeneration, are likely involved in the generation of new muscle fibers. These cells are typically activated by muscle stretching and injury, and, as we see later in this chapter, muscle injury results from intense training, particularly eccentric-action training. Muscle injury can lead to a cascade of responses, in which satellite cells become activated and proliferate, migrate to the damaged region, and fuse to existing myofibers or combine and fuse to produce new myofibers.[17] This is illustrated in figure 9.6 on page 210.

Integration of Neural Activation and Fiber Hypertrophy

Research on resistance training adaptations indicates that early increases in voluntary strength, or maximal force production, are associated primarily with neural adaptations resulting in increased voluntary activation of muscle. This was clearly demonstrated in a study of both men and women who participated in an eight-week, high-intensity resistance training program, training twice per week.[25] Muscle biopsies were obtained at the beginning of the study and every two weeks during the training period. Strength, measured according to the 1RM, increased substantially over the eight weeks of training, with the greatest gains coming after the second week. Muscle biopsies, however, revealed only small, statistically insignificant increases in muscle fiber cross-sectional area by the end of the eight weeks of training. Thus, the strength gains were largely the result of increased neural activation.

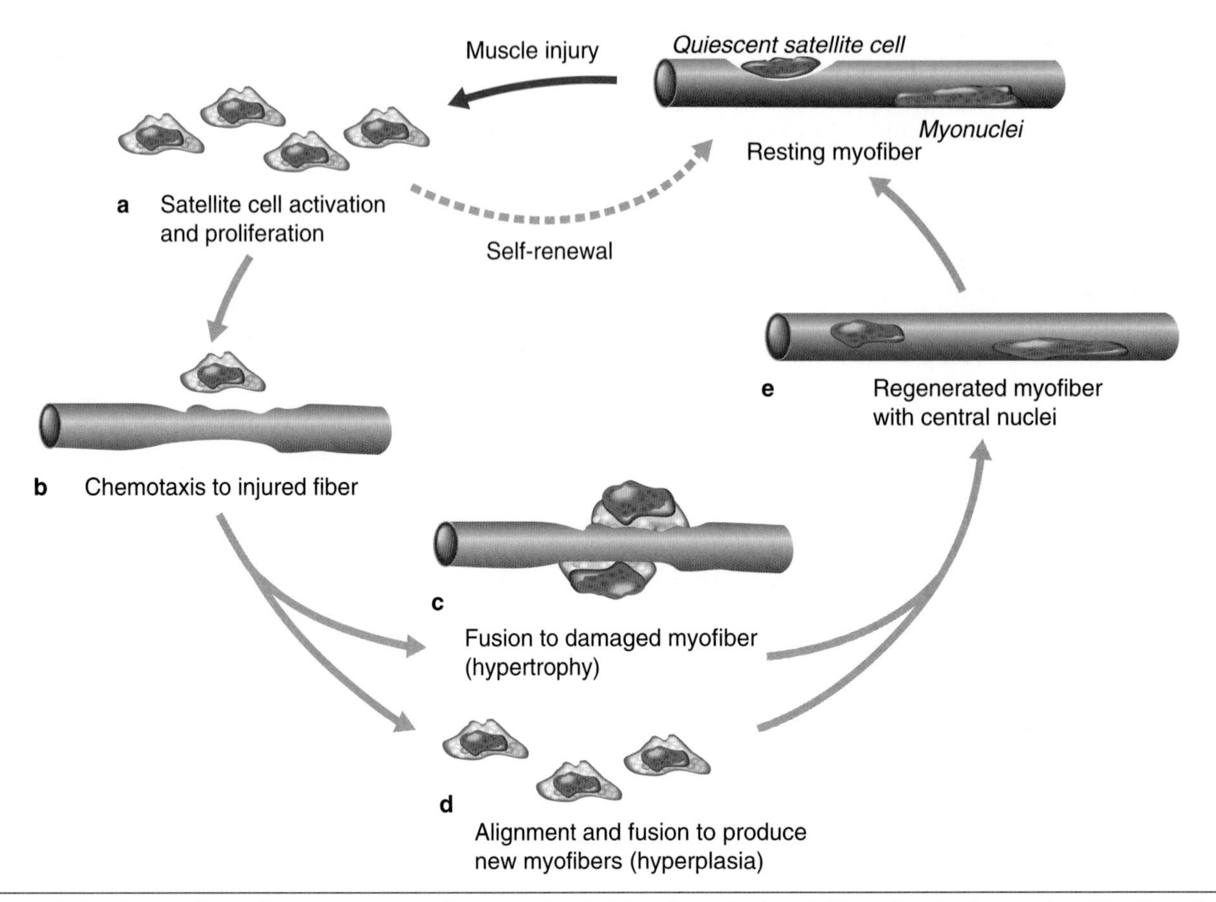

Figure 9.6 The satellite cell response to muscle injury. Muscle injury leads to *(a)* satellite cell activation and proliferation. *(b)* The satellite cells migrate to the damaged region and *(c)* fuse to the damaged myofiber or *(d)* align and fuse to produce a new myofiber, either of which leads to *(e)* a regenerated or new myofiber.

Reprinted, by permission, from T.J. Hawke and D.J. Garry, 2001, "Myogenic satellite cells: Physiology to molecular biology," *Journal of Applied Physiology* 91: 534-551.

Long-term increases in strength generally are associated with hypertrophy of the trained muscle. It takes time to build protein through a decrease in protein degradation, an increase in protein synthesis, or both. Notable exceptions to this generalization have been found. A six-month study of strength-trained athletes showed that neural activation explained most of the strength gains during the most intensive training months and that hypertrophy was not a major factor.[16] It appears that neural factors make their greatest contribution during the first 8 to 10 weeks of training. Hypertrophy contributes little during the initial weeks of training but progressively increases its contribution, becoming the major contributor after 10 weeks of training.

In focus

Early gains in strength appear to be more influenced by neural factors, but later long-term gains are largely the result of hypertrophy.

Muscle Atrophy and Decreased Strength With Inactivity

When a normally active or highly trained person reduces his or her level of activity or ceases training altogether, major changes occur in both muscle structure and function. This is illustrated by the results of two types of studies: studies in which entire limbs have been immobilized and studies in which highly trained people stop training.

Immobilization

When a trained muscle suddenly becomes inactive through immobilization, major changes are initiated within that muscle in a matter of hours (chapter 13). During the first 6 h of immobilization, the rate of protein synthesis starts to decrease. This decrease likely initiates the start of muscular atrophy, which is the wasting away or decrease in the size of muscle tissue. Atrophy results from lack of muscle use and the consequent loss of muscle protein that accompanies the inactivity. Strength decreases are most dramatic during the first week of immobilization, averaging 3% to 4% per day.[4] This is associated with the atrophy but also with decreased neuromuscular activity of the immobilized muscle.

Immobilization appears to affect both type I and type II fibers. From various studies, researchers have observed disintegrated myofibrils, streaming Z-disks (discontinuity of Z-disks and fusion of the myofibrils), and mitochondrial damage. When muscle atrophies, the cross-sectional fiber area decreases. Several studies have shown the effect to be greater in type I fibers, including a decrease in the percentage of type I fibers, thereby increasing the percentage of type II fibers.

Muscles can and often do recover from immobilization when activity is resumed. The recovery period is substantially longer than the period of immobilization but shorter than the original training period.

Cessation of Training

Similarly, significant muscle alterations can occur when people stop training. In one study, women resistance trained for 20 weeks and then stopped training for 30 to 32 weeks. Finally they retrained for six weeks.[26] The training program focused on the lower extremity, using a full squat, leg press, and leg extension. Strength increases were dramatic, as seen in figure 9.7. Compare the women's strength after their initial training period (post-20) with their strength after detraining (pre-6). This represents the strength loss they experienced with cessation of training.

During the two training periods, increases in strength were accompanied by increases in the cross-sectional area of all fiber types and a decrease in the percentage of type IIx fibers. Detraining had relatively little effect on fiber cross-sectional area, although the type II fiber areas tended to decrease (figure 9.8 on page 212).

To prevent losses in the strength gained through resistance training, basic maintenance programs must be established once the desired goals for strength development have been achieved. Maintenance programs are designed to provide sufficient stress to the muscles to maintain existing levels of strength while allowing a reduction in intensity, duration, or frequency of training.

In one study, men and women resistance trained with knee extensions for either 10 or 18 weeks and then spent an additional 12 weeks with either no training or reduced training.[13] Knee extension strength increased 21.4% following the training period. Subjects who then stopped training lost 68% of their strength gains during the weeks they didn't train. But subjects who reduced their training (from three days per week to two, or from two to one) did not lose strength. Thus, it appears that

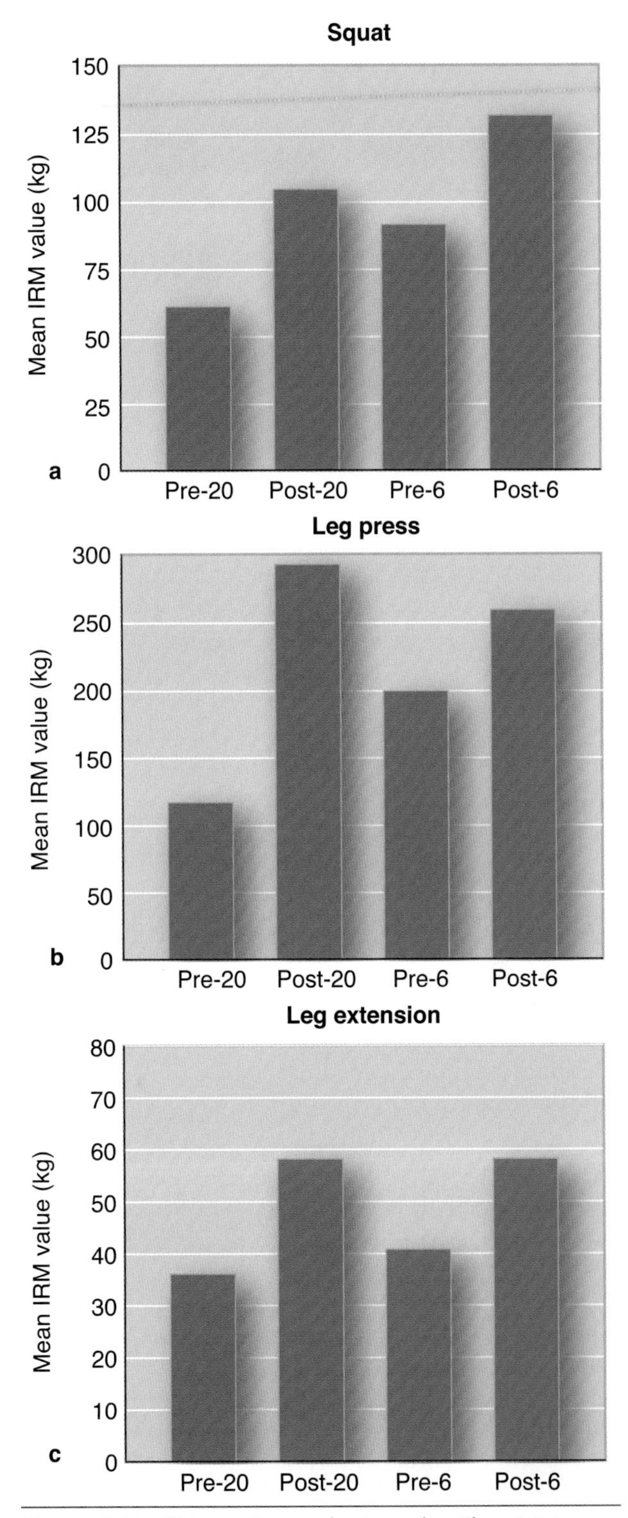

Figure 9.7 Changes in muscle strength with resistance training in women. Pre-20 values indicate strength before starting training; post-20 values indicate the changes following 20 weeks of training; pre-6 values indicate the changes following 30 to 32 weeks of detraining; and post-6 values indicate the changes following six weeks of retraining.

Adapted, by permission, from R.S. Staron et al., 1991, "Strength and skeletal muscle adaptations in heavy-resistance-trained women after detraining and retraining," *Journal of Applied Physiology* 70: 631-640.

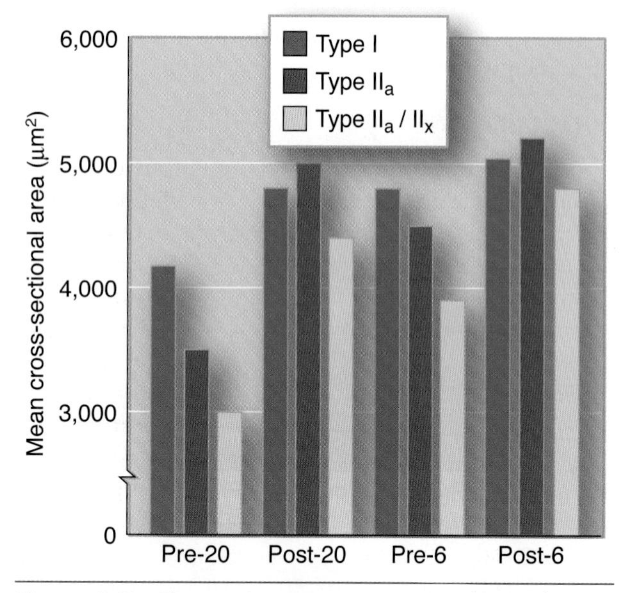

Figure 9.8 Changes in mean cross-sectional areas for the major fiber types with resistance training in women over periods of training (post-20), detraining (pre-6), and retraining (post-6). Type IIa/IIx is an intermediate fiber type. See figure 9.7 caption for more details.

strength can be maintained for at least up to 12 weeks with reduced training frequency.

Fiber Type Alterations

Can muscle fibers change from one type to another through resistance training? The earliest research concluded that neither speed (anaerobic) nor endurance (aerobic) training could alter the basic fiber type, specifically from type I to type II, or from type II to type I. These early studies did show, however, that fibers begin to take on certain characteristics of the opposite fiber type if the training is of the opposite kind (e.g., type II fibers might become more oxidative with aerobic training).

Research with animals has shown that fiber type conversion is indeed possible under conditions of cross-innervation, in which a type II motor unit is artificially innervated by a type I motor neuron or vice versa. Also, chronic, low-frequency nerve stimulation transforms type II motor units into type I motor units within a matter of weeks. Muscle fiber types in rats have changed in response to 15 weeks of high-intensity treadmill training, resulting in an increase in type I and type IIa fibers and a decrease in type IIx fibers.[14] The transition of fibers from type IIx to type IIa and from type IIa to type I was confirmed by several different histochemical techniques.

Staron and coworkers found evidence of fiber type transformation in women as a result of heavy resistance training.[27] Substantial increases in static strength and in the cross-sectional area of all fiber types were noted following a 20-week heavy resistance training program for the lower extremity. The mean percentage of type IIx fibers decreased significantly, but the mean percentage of type IIa fibers increased. The transition of type IIx fibers to type IIa fibers with resistance training has been consistently reported in a number of subsequent studies. More recent studies have shown that a combination of high-intensity resistance training and short-interval speed work can lead to a conversion of type I to type IIa fibers.[3]

In review

- Neural adaptations always accompany the strength gains that result from resistance training, but hypertrophy might or might not be present.
- Neural mechanisms leading to strength gains can include an increase in frequency of stimulation, or rate coding; recruitment of more motor units; more synchronous recruitment of motor units; and decreases in autogenic inhibition from the Golgi tendon organs.
- Transient muscle hypertrophy is the pumped-up feeling people get immediately after an exercise bout. It results from edema and is short-lived.
- Chronic muscle hypertrophy occurs from repeated resistance training and reflects actual structural changes in the muscle.
- Although most muscle hypertrophy probably results from an increase in the size of individual muscle fibers (fiber hypertrophy), some evidence suggests that an increase in the number of muscle fibers (hyperplasia) also might be involved.
- Muscles atrophy (decrease in size and strength) when they become inactive, as with injury, immobilization, or cessation of training.
- Atrophy begins very quickly if training is stopped; but training can be reduced, as in a maintenance program, without resulting in atrophy or loss of strength.
- With resistance training there is a transition of type IIx to type IIa fibers. Evidence indicates that one fiber type can actually be converted to the other type (e.g., type I to type II, or vice versa) as a result of cross-innervation or chronic stimulation, and possibly with training.

MUSCLE SORENESS

Muscle soreness generally results from exhaustive or very high intensity exercise. This is particularly true when people perform a specific exercise for the first time. While muscle soreness can be felt at any time, there is generally a period of mild muscle soreness that can be felt during and immediately after exercise, and then a more intense soreness felt a day or two later.

Acute Muscle Soreness

Pain felt during and immediately after exercise can result from accumulation of the end products of exercise, such as H^+, and from tissue edema, mentioned earlier, which is caused by fluid shifting from the blood plasma into the tissues. Edema is the cause of the pumped-up sensation that people feel after heavy endurance or strength training. The pain and soreness usually disappear within a few minutes to several hours after the exercise. Thus, this soreness is often referred to as **acute muscle soreness.**

Delayed-Onset Muscle Soreness and Injury

Muscle soreness felt a day or two after a heavy bout of exercise is not totally understood, yet researchers are continuing to give us greater insight into this phenomenon. Because this pain does not occur immediately, it is referred to as **delayed-onset muscle soreness (DOMS).** In the following sections, we discuss some theories that attempt to explain this form of muscle soreness.

Almost all current theories acknowledge that eccentric action is the primary initiator of DOMS. This was clearly demonstrated in a study of the relationship of muscle soreness to eccentric, concentric, and static actions.[28] This study showed that a group who trained solely with eccentric actions experienced extreme muscle soreness, whereas the static- and concentric-action groups experienced little soreness. This idea was further explored in studies in which subjects ran on a treadmill for 45 min on two separate days, one day on a level grade and the other day on a 10% downhill grade.[20, 21] No muscle soreness was associated with the level running. But the downhill running, which required extensive eccentric action, resulted in considerable soreness within 24 to 48 h, even though blood lactate levels, previously thought to cause muscle soreness, were much higher with level running.

> **In focus**
>
> Delayed-onset muscle soreness results primarily from eccentric action and is associated with actual muscle disruption or damage.

Let's examine some of the proposed explanations for exercise-induced DOMS.

Structural Damage

The presence of increased concentrations of several specific muscle enzymes in blood after intense exercise suggests that some structural damage may occur in the muscle membranes. These enzyme levels increase from 2 to 10 times their normal levels following bouts of heavy training. Recent studies support the idea that these changes might indicate various degrees of muscle tissue breakdown. Examination of tissue from the leg muscles of marathon runners has revealed remarkable damage to the muscle fibers after both training and marathon competition. The onset and timing of these muscle changes parallel the degree of muscle soreness experienced by the runners.

The electron micrograph in figure 9.9 shows muscle fiber damage as a result of marathon running.[15] In this case, the cell membrane appears to have been totally ruptured, allowing the cell's contents to float freely between the other normal fibers. Fortunately, not all damage to muscle cells is as severe.

Figure 9.10 on page 214 shows changes in the contractile filaments and Z-disks before and after a marathon race. Recall that Z-disks are the points of contact

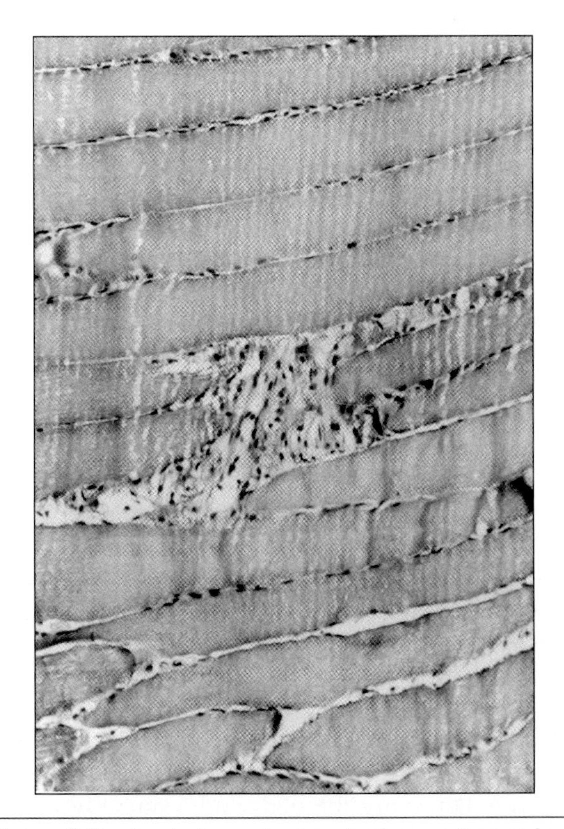

Figure 9.9 An electron micrograph of a muscle sample taken immediately after a marathon, showing the disruption of the cell membrane in one muscle fiber.

Figure 9.10 *(a)* An electron micrograph showing the normal arrangement of the actin and myosin filaments and Z-disk configuration in the muscle of a runner before a marathon race. *(b)* A muscle sample taken immediately after a marathon race shows moderate Z-disk streaming and major disruption of the thick and thin filaments in a parallel group of sarcomeres caused by the eccentric actions of running.

for the contractile proteins. They provide structural support for the transmission of force when the muscle fibers are activated to shorten. Figure 9.10*b,* after the marathon, shows moderate Z-disk streaming and major disruption of the thick and thin filaments in a parallel group of sarcomeres as a result of the force of eccentric actions or stretching of the tightened muscle fibers.

Although the effects of muscle damage on performance are not fully understood, experts generally agree that this damage is responsible in part for the localized muscle pain, tenderness, and swelling associated with DOMS. However, blood enzyme levels might increase and muscle fibers might be damaged frequently during daily exercise that produces no muscle soreness. Also, remember that muscle damage appears to be a precipitating factor for muscle hypertrophy.

Inflammatory Reaction

White blood cells serve as a defense against foreign materials that enter the body and against conditions that threaten the normal function of tissues. The white blood cell count tends to increase following activities that induce muscle soreness. This observation led some investigators to suggest that soreness results from inflammatory reactions in the muscle. But the link between these reactions and muscle soreness has been difficult to establish.

Researchers have tried to use drugs to block the inflammatory reaction, but these efforts have been unsuccessful in reducing either the amount of muscle soreness or the degree of inflammation. Because both effects remain, conclusions about the role of inflammation in muscle soreness cannot be drawn from this research. More recent studies, however, are beginning to establish a link between muscle soreness and inflammation. For example, it is now recognized that substances released from injured muscle can act as attractants, initiating the inflammatory process. Mononucleated cells in muscle are activated by the injury, providing the chemical signal to circulating inflammatory cells. Neutrophils (a type of white blood cell) invade the injury site and release cytokines (immunoregulatory substances), which then attract and activate additional inflammatory cells. Neutrophils possibly also release oxygen free radicals that can damage cell membranes. Macrophages (another type of cell of the immune system) then invade the damaged muscle fibers, removing debris through a process known as phagocytosis. Last, a second phase of macrophage invasion occurs, which is associated with muscle regeneration.[29]

Sequence of Events in DOMS

In 1984, Armstrong[5] reviewed possible mechanisms for exercise-induced DOMS. He concluded that DOMS is associated with

- elevations in plasma enzymes,
- myoglobinemia (presence of myoglobin in the blood), and
- abnormal muscle histology and ultrastructure.

He developed a model of DOMS that proposed the following sequence of events:

1. High tension in the contractile-elastic system of muscle results in structural damage to the muscle and its cell membrane.
2. The cell membrane damage disturbs calcium homeostasis in the injured fiber, resulting in necrosis (cell death) that peaks about 48 h after exercise.
3. The products of macrophage activity and intracellular contents (such as histamine, kinins, and K^+) accumulate outside the cells. These substances then stimulate the free nerve endings in the muscle. This process appears to be accentuated in eccentric exercise, in which large forces are distributed over relatively small cross-sectional areas of the muscle.

Recent comprehensive reviews have provided much greater insight into the cause of muscle soreness. We now are confident that muscle soreness results from injury or damage to the muscle itself, generally the muscle fiber and possibly the plasmalemma.[6] This damage sets up a chain of events that includes the release of intracellular proteins and an increase in muscle protein turnover. The damage and repair process involves calcium ions, lysosomes, connective tissue, free radicals, energy sources, inflammatory reactions, and intracellular and myofibrillar proteins. But the precise cause of skeletal muscle damage and the mechanisms of repair are not well understood. As we have discussed previously, some evidence suggests that this process is an important step in muscle hypertrophy.

Exercise-Induced Muscle Cramps

Few things frustrate athletes more than muscle cramps. Skeletal muscle cramps can come during the height of competition, immediately after competition, or at night in the deep of sleep. Muscle cramps are equally frustrating to scientists, because they have not been able to totally determine the cause of muscle cramping or how to treat or prevent cramps. Although muscle cramps can be the result of rare medical conditions, most exercise-induced or exercise-associated muscle cramps are unrelated to disease or medical disorder. Exercise-associated muscle cramps (EAMCs) have been defined as painful, spasmodic, involuntary contractions of skeletal muscles that occur during or immediately after exercise.[22] Nocturnal muscle cramps may or may not be associated with exercise.

Early research suggested that muscle cramps were caused by disturbances in fluid and electrolyte (particularly sodium) balance, associated with high rates of sweating, as is the case with heat cramps. Although this might be the case for some EAMCs, more recent research proposes that EAMCs result from sustained α-motor neuron activity, which is caused by aberrant control at the spinal level.[22] Muscle fatigue appears to cause this lack of control through an effect on both Golgi tendon organs and muscle spindles. Spindle activity increases and tendon organ activity decreases.

Treatment includes rest, passive stretching of the affected muscle or muscle groups, and holding the muscle in a stretched position until muscle activation is relieved. Fluids also should be taken in if dehydration and electrolyte loss are suspected. To prevent EAMCs, the athlete should

- be well conditioned, to reduce the likelihood of muscle fatigue;
- regularly stretch the muscle groups prone to EAMCs;
- maintain fluid and electrolyte balance and carbohydrate stores; and
- reduce exercise intensity and duration if necessary.[22]

Up to this point, our discussion of DOMS has focused on muscle injury. Edema, or the accumulation of fluids in the muscular compartment, also can lead to DOMS. This edema is likely the result of muscle injury but could occur independently of muscle injury. An accumulation of interstitial or intracellular fluid increases the tissue fluid pressure within the muscle compartment, which in turn activates pain receptors within the muscle.

Delayed-Onset Muscle Soreness and Performance

With DOMS comes a reduction in the force-generating capacity of the affected muscles. Whether the DOMS is the result of injury to the muscle or edema independent of muscle injury, the affected muscles are not able to exert as much force when the person is asked to apply maximal force, such as in the performance of a 1RM strength test. Maximal force-generating capacity gradually returns over days or weeks. It has been proposed that the loss in strength is the result of three factors:[30]

1. The physical disruption of the muscle as illustrated in figures 9.9 and 9.10
2. Failure within the excitation–contraction coupling process
3. Loss of contractile protein

Failure in excitation–contraction coupling appears to be the most important, particularly during the first five days. This is illustrated in figure 9.11.

Muscle glycogen resynthesis also is impaired when a muscle is damaged. Resynthesis is generally normal for the first 6 to 12 h after exercise, but it slows or stops completely as the muscle undergoes repair, thus limiting the fuel-storage capacity of the injured muscle. Figure 9.12 illustrates the time sequence of the various factors associated with intense eccentric exercise, including pain, edema, plasma creatine kinase (a plasma enzyme marker of muscle fiber damage), glycogen depletion, ultrastructural damage in the muscle, and muscular weakness.

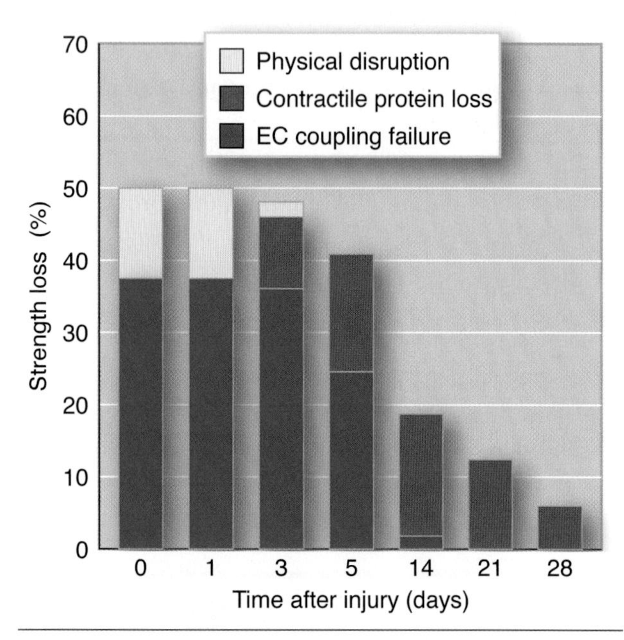

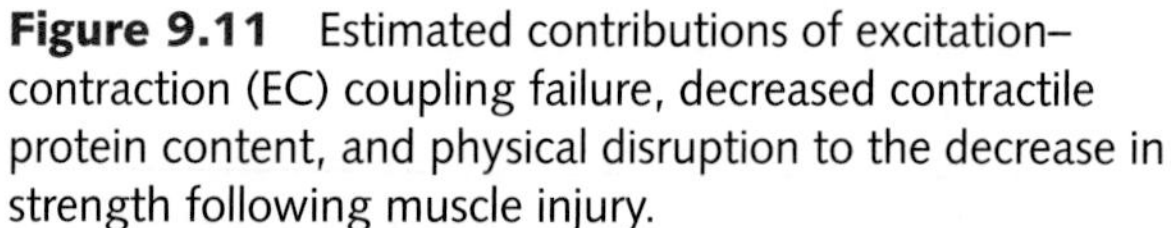
Figure 9.11 Estimated contributions of excitation–contraction (EC) coupling failure, decreased contractile protein content, and physical disruption to the decrease in strength following muscle injury.

Reprinted, by permission, from G.L. Warren et al., 2001, "Excitation-contraction uncoupling: Major role in contraction-induced muscle injury," *Exercise and Sport Sciences Reviews* 29: 82-87.

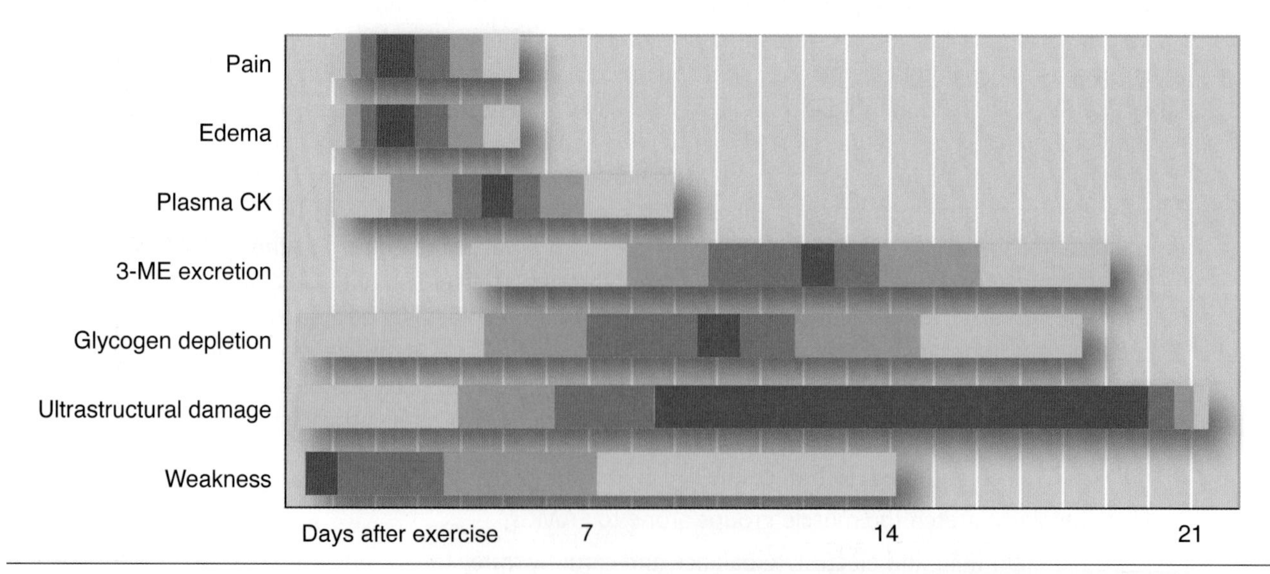

Figure 9.12 The delayed response to eccentric exercise of various physiological markers. The density of the shaded bar corresponds to the intensity of the response at the indicated time. CK = creatine kinase.

Adapted, by permission, from W.J. Evans and J.G. Cannon, 1991, "The metabolic effects of exercise induced muscle damage," *Exercise and Sport Sciences Reviews* 19: 99-125.

In review

- Acute muscle soreness occurs late in an exercise bout and during the immediate recovery period.
- Delayed-onset muscle soreness usually peaks and lasts a day or two after the exercise bout. Eccentric action seems to be the primary instigator of this type of soreness.
- Proposed causes of DOMS include structural damage to muscle cells and inflammatory reactions within the muscles.
- Armstrong's proposed model of the sequence of events that cause DOMS includes
    1. structural damage,
    2. impaired calcium homeostasis leading to necrosis,
    3. accumulation of irritants, and
    4. increased macrophage activity.
- Muscle soreness can be prevented or minimized by
    1. reducing the eccentric component of muscle action during early training;
    2. starting training at a low intensity and gradually increasing it; or
    3. beginning with a high-intensity, exhaustive bout of eccentric-action exercise, which will cause much soreness initially but will decrease future pain.
- Muscle strength is reduced in muscles injured by eccentric contractions and is likely the result of three factors: physical disruption of the muscle, failure of the excitation–contraction coupling process, and loss of contractile protein.
- Muscle soreness may be an important part of maximizing the resistance training response.
- Exercise-associated muscle cramps are attributable to either or both fluid and electrolyte imbalances and muscle fatigue-associated sustained α-motor neuron activity, with increased muscle spindle activity and decreased Golgi tendon organ activity. Rest, passive stretching, and holding the muscle in the stretched position can be effective in treating EAMCs; and proper conditioning, stretching, and nutrition are possible prevention strategies.

Reducing the Negative Effects of DOMS

Reducing the negative effects of DOMS is important for maximizing training gains. The eccentric component of muscle action could be minimized during early training, but this is not possible for athletes in most sports. An alternative approach is to start training at a very low intensity and progress slowly through the first few weeks. Yet another approach is to initiate the training program with a high-intensity, exhaustive training bout. Muscle soreness would be great for the first few days, but evidence suggests that subsequent training bouts would cause considerably less muscle soreness.[8] Because the factors associated with DOMS are also potentially important in stimulating muscle hypertrophy, DOMS is most likely necessary to maximize the training response.

RESISTANCE TRAINING FOR SPECIAL POPULATIONS

Until the 1970s, resistance training was widely regarded as appropriate only for young, healthy, male athletes. This narrow concept led many people to overlook the benefits of resistance training when planning their own activities. In this section, we first consider sex and age, and then we summarize the importance of this form of training for all athletes, regardless of their sex, age, or sport.

Sex and Age Differences

In recent years, considerable interest has focused on training for women, children, and people who are elderly. As mentioned earlier in this chapter, the widespread use of resistance training by women, either for sport or for health-related benefits, is rather recent. Substantial knowledge has developed since the early 1970s revealing that women and men have the same ability to develop strength but that, on average, women may not be able to achieve peak values as high as those attained by men. This difference in strength is attributable primarily to muscle size differences related to sex differences in anabolic hormones. Resistance training techniques developed for and applied to men's training seem equally appropriate for women's training. Issues of

In 1984, the University of Arizona was the first NCAA Division I school to hire a woman as head strength coach for both the men's and women's athletic programs. The position went to Meg Ritchie Stone, former Scottish discus thrower and shot-putter for Great Britain's Olympic team.

strength and resistance training for women are covered in more detail in chapter 18.

The wisdom of resistance training for children and adolescents has long been debated. The potential for injury, particularly growth plate injuries from the use of free weights, has caused much concern. Many people once believed that children would not benefit from resistance training, based on the assumption that the hormonal changes associated with puberty are necessary for gaining muscle strength and mass. We now know that children and adolescents can train safely with minimal risk of injury if appropriate safeguards are followed. Furthermore, they can indeed gain both muscular strength and muscle mass (chapter 16).

Interest in resistance training procedures for elderly people has also increased. A substantial loss of fat-free body mass accompanies aging. This loss reflects mainly the loss of muscle mass, largely because most people become less active as they age. When a muscle isn't used regularly, it loses function, with predictable atrophy and loss of strength.

Can resistance training in elderly people reverse this process? People who are elderly can indeed gain strength and muscle mass in response to resistance training. This fact has important implications for both their health and the quality of their lives (chapter 17). With maintained or improved strength, they are less likely to fall. This is a significant benefit because falls are a major source of injury and debilitation for elderly people and often lead to death.

Resistance Training for Sport

Gaining strength, power, or muscular endurance simply for the sake of being stronger, being more powerful, or having greater muscular endurance is of relatively little importance to athletes unless it also improves their athletic performance. Resistance training by field-event athletes and competitive weightlifters makes intuitive sense. The need for resistance training by the gymnast, distance runner, baseball player, high jumper, or ballet dancer is less obvious.

We do not have extensive research to document the specific benefits of resistance training for every sport or for every event within a sport. But clearly each has basic strength, power, and muscular endurance requirements that must be met to achieve optimal performance. Training beyond these requirements may be unnecessary.

Training is costly in terms of time, and athletes can't afford to waste time on activities that won't result in better athletic performances. Thus, some performance measurement is imperative to evaluate any resistance training program's efficacy. To resistance train solely to become stronger, with no associated improvement in performance, is of questionable value. However, it should also be recognized that resistance training can reduce the risk of injury for most sports, because fatigued individuals are at an increased risk of injury.

In review

- Resistance training can benefit almost everyone, regardless of his or her sex, age, or athletic involvement.
- Most athletes in most sports can benefit from resistance training if an appropriate program is designed for them. But to ensure that the program is working, performance should be assessed periodically and the training regime adjusted as needed.

In Closing

In this chapter we have carefully considered the role of resistance training in increasing muscular strength and improving performance. We have examined how muscle strength is gained through both muscular and neural adaptations, what factors can lead to muscle soreness, and how resistance training is of importance for both health and sport, irrespective of age or sex. In the next chapter, we turn our attention away from resistance training and begin exploring how the body adapts to aerobic and anaerobic training.

KEY TERMS

acute muscle soreness
atrophy
autogenic inhibition
chronic hypertrophy
delayed-onset muscle soreness (DOMS)
fiber hyperplasia
fiber hypertrophy
hypertrophy
resistance training
transient hypertrophy

STUDY QUESTIONS

1. What is a reasonable expectation for percentage strength gains following a six-month resistance training program? How do these percentage gains differ by age, sex, and previous resistance training experience?
2. Discuss possible mechanisms that might account for superhuman feats of strength.
3. Discuss the different theories that have attempted to explain how muscles gain strength with training.
4. What is autogenic inhibition? How might it be important to resistance training?
5. Differentiate transient and chronic muscle hypertrophy.
6. What is fiber hyperplasia? How might it occur? How might it be related to gains in size and muscle strength with resistance training?
7. What is the physiological basis for hypertrophy?
8. What is the physiological basis for atrophy?
9. What is the physiological basis for delayed-onset muscle soreness?

STUDY GUIDE ACTIVITIES

In addition to the activities listed in the chapter opening outline on page 203, two other activities are available in the online study guide, located at

www.HumanKinetics.com/PhysiologyOfSportAndExercise.

The chapter **SUMMARY** reviews the key concepts covered in the chapter, and the end-of-chapter **QUIZ** tests your understanding of the material covered in the chapter.

Adaptations to Aerobic and Anaerobic Training

In this chapter and in the online study guide

Lance Armstrong, 33-year-old cyclist from the United States, won his seventh consecutive Tour de France in 2005. "The Tour" is considered not only the world's most prestigious bicycle race but also one of the most demanding, grueling endurance sporting events. Armstrong completed the 21-day, 3,608 km (2,241 mi) ordeal in a total time of 86 h 15 min at an average speed of 41.8 km/h (~26 mph). The cyclists who compete have to endure multiple long, steady mountain climbs and risky high-speed descents, as well as sprint stages cycling at breakneck speeds, as they compete in individual and team time trials. How are these athletes able to compete in this race? While there is little doubt that they are genetically gifted with a high $\dot{V}O_{2max}$, rigorous training is also required specifically to develop their cardiorespiratory endurance capacities.

During a single bout of exercise, the human body exquisitely adjusts its cardiovascular and respiratory functioning to meet the heightened demands of active muscles. When these systems are challenged repeatedly, as with regular exercise training, they adapt in ways that allow the body to improve $\dot{V}O_{2max}$ and overall endurance performance. The physiological and metabolic processes that bring oxygen into the body, distribute it, and allow it to be used by active tissues improve. **Aerobic training,** or cardiorespiratory endurance training, improves central and peripheral blood flow and enhances the capacity of the muscle fibers to generate greater amounts of adenosine triphosphate (ATP). In this chapter, we examine adaptations in cardiovascular and respiratory function in response to endurance training and how such adaptations affect an athlete's endurance capacity and performance. Additionally, we examine adaptations to anaerobic training. **Anaerobic training** increases anaerobic metabolic function; short-term, high-intensity endurance capacity; tolerance for acid–base imbalances; and possibly muscle strength. Both aerobic and anaerobic training induce a variety of adaptations that benefit exercise and sport performance.

The need to train one's cardiovascular and respiratory endurance, or what has been termed cardiorespiratory or aerobic endurance, is well known to endurance athletes like runners, cyclists, and swimmers but is often ignored by other types of athletes. Training programs for many nonendurance athletes often ignore the aerobic endurance factor. This is understandable, because for maximum improvement in performance, training should be highly specific to the particular sport or activity in which the athlete participates, and endurance is frequently not recognized as important to nonendurance activities. The reasoning is, Why waste valuable training time if the result is not improved performance?

The problem with this reasoning is that most nonendurance sports do indeed have an endurance, or aerobic, component. For example, in football, players and coaches might fail to recognize the importance of cardiorespiratory endurance as part of the total training program. From all outward appearances, football is an anaerobic, or burst-type, activity consisting of repeated bouts of high-intensity work of short duration. Seldom does a run exceed 40 to 60 yd (37-55 m), and even this is usually followed by a substantial rest interval. The need for endurance may not be readily apparent. What athletes and coaches might fail to consider is that this burst-type activity must be repeated many times during the game. With a higher aerobic endurance level, an athlete could maintain the quality of each burst activity throughout the game and would still be relatively "fresh" (less drop-off in performance, fewer feelings of fatigue) during the fourth quarter.

Similar questions have been asked concerning the importance of including resistance training as a part of the total training program for sports that do not demand high levels of strength, or high-intensity sprint training for sports that do not require speed or high anaerobic capacities. Yet athletes in almost all endurance sports are doing some resistance training to increase, or at least maintain, basic strength levels, as well as some sprint training to facilitate their ability to sustain speed when needed (e.g., sprinting to the finish line at the end of a marathon).

Chapters 8 and 13 cover the principles of training for sport performance—the "how," "when," and "how much" questions about training. The focus here is on those physiological changes that occur within the body systems when aerobic or anaerobic exercise is repeated regularly to induce a training response.

ADAPTATIONS TO AEROBIC TRAINING

Improvements in endurance that accompany regular (daily, every other day, etc.) aerobic training, such as running, cycling, or swimming, result from multiple adaptations to the training stimulus. Some adaptations occur within the muscles themselves, promoting more efficient transport and utilization of oxygen and fuel substrates. Still other important changes occur in the cardiovascular system, improving circulation to and within the muscles. Pulmonary adaptations, as will be noted later, occur to a lesser extent.

Endurance: Muscular Versus Cardiorespiratory

Endurance is a term that refers to two separate but related concepts: muscular endurance and cardiorespiratory endurance. Each makes a unique contribution to athletic performance, and each differs in its importance to different athletes.

For sprinters, endurance is the quality that allows them to sustain a high speed over the full distance of, for example, a 100 or 200 m race. This component of fitness is termed muscular endurance, the ability of a single muscle or muscle group to sustain high-intensity, repetitive, or static exercise. This type of endurance is also exemplified by the weightlifter doing multiple repetitions, the boxer, and the wrestler. The exercise or activity can be rhythmic and repetitive in nature, such as multiple repetitions of the bench press for the weightlifter and jabbing for the boxer. Or the activity can be more static, such as a sustained muscle action when a wrestler attempts to pin an opponent to the mat. In either case, the resulting fatigue is confined to a specific muscle group, and the activity's duration is usually no more than 1 or 2 min. Muscular endurance is highly related to muscular strength and to anaerobic power development.

While muscular endurance is specific to individual muscles or muscle groups, **cardiorespiratory endurance** relates to the entire body's ability to sustain prolonged, dynamic exercise using large muscle groups. This type of endurance is used by the cyclist, distance runner, or endurance swimmer who completes long distances at a fairly fast pace. Cardiorespiratory endurance is related to the development of the cardiovascular and respiratory systems' ability to maintain oxygen delivery to working muscles during prolonged bouts of exercise, as well as the muscles' ability to utilize energy aerobically (discussed in chapters 2 and 4). This is why the terms cardiorespiratory endurance and aerobic endurance are sometimes used synonymously.

In focus

Cardiorespiratory endurance, or aerobic endurance, is the ability of the whole body to sustain prolonged exercise involving relatively large muscle groups.

Evaluating Cardiorespiratory Endurance Capacity

To study training effects on endurance, there needs to be an objective, repeatable means of measuring an individual's cardiorespiratory endurance capacity. In that way, the exercise scientist, coach, or athlete can monitor improvements as physiological adaptations occur during the training program.

Maximal Endurance Capacity: $\dot{V}O_{2max}$ or Aerobic Power

Most exercise scientists regard $\dot{V}O_{2max}$, sometimes called maximal aerobic power or maximal aerobic capacity, as the best objective laboratory measure of maximal cardiorespiratory endurance. Recall from chapter 4 that $\dot{V}O_{2max}$ is defined as the highest rate of oxygen consumption attainable during maximal or exhaustive exercise. $\dot{V}O_{2max}$ as defined by the Fick equation is dictated by maximal cardiac output (delivery of oxygen and blood flow to working muscles) and the maximal $(a\text{-}\bar{v})O_2$ difference (the ability of the active muscles to extract and use the oxygen). As exercise intensity increases, oxygen consumption eventually either plateaus or decreases slightly, even with further increases in workload, indicating that a truly maximal $\dot{V}O_2$ has been achieved.

With endurance training, more oxygen can be delivered to, and consumed by, active muscles than in an untrained state. Previously untrained subjects demonstrate average increases in $\dot{V}O_{2max}$ of 15% to 20% after a 20-week training program. These improvements allow individuals to perform endurance activities at a higher intensity, improving their performance potential. Figure 10.1 illustrates the increase in $\dot{V}O_{2max}$ after 12 months of aerobic training in a previously untrained individual. In this example, $\dot{V}O_{2max}$ increased by about 30%. Note that the $\dot{V}O_2$ "cost" of running at a certain submaximal intensity did not change but that higher running speeds could be attained after training.

Submaximal Endurance Capacity

In addition to increasing maximal endurance capacity, endurance training increases **submaximal endurance capacity**, which is much more difficult to evaluate. Steady-state submaximal heart rate at the same exercise intensity measured before and after training is one physiological variable that can be used to objectively quantify the effect of training. Additionally, exercise scientists have used performance measures to quantify submaximal endurance capacity. For example, one test used to determine submaximal endurance capacity is the average peak absolute power output a person can maintain over a fixed period of time on a cycle ergometer. For running, the average peak speed or velocity a person can maintain during a fixed period of time would be a similar type of test. Generally, these tests will last 30 min to an hour.

Submaximal endurance capacity is more closely related to actual competitive endurance performance than $\dot{V}O_{2max}$, and is likely determined by both the

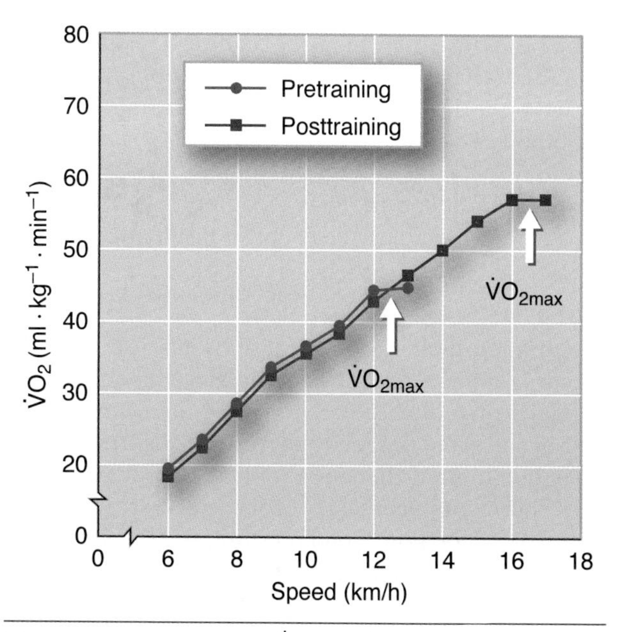

Figure 10.1 Changes in $\dot{V}O_{2max}$ with 12 months of endurance training. $\dot{V}O_{2max}$ increased from 44 to 57 ml · kg^{-1} · min^{-1}, a 30% increase. Peak speed during the treadmill test increased from 13 km/h (8 mph) to 16 km/h (~10 mph).

person's $\dot{V}O_{2max}$ and the threshold for his or her onset of blood lactic acid accumulation (OBLA)—that point at which lactate begins to appear at a disproportionate rate in the blood (see chapter 4). With endurance training, submaximal endurance capacity increases.

Cardiovascular Adaptations to Training

Numerous cardiovascular adaptations occur in response to exercise training, including changes in the following cardiovascular variables:

- Heart size
- Stroke volume
- Heart rate
- Cardiac output
- Blood flow
- Blood pressure
- Blood volume

To fully understand adaptations in these variables, it is important to review how these components relate to oxygen transport.

Oxygen Transport System

Cardiorespiratory endurance is related to the cardiovascular and respiratory systems' ability to deliver sufficient oxygen to meet the needs of metabolically active tissues.

Recall from chapter 7 that the ability of the cardiovascular and respiratory systems to deliver oxygen to active tissues is defined by the Fick equation. The **Fick equation** states that systemic oxygen consumption is determined by both the delivery of oxygen (cardiac output) via blood flow and the amount of oxygen extracted by the tissues, the $(a\text{-}\bar{v})O_2$ difference. The product of cardiac output and the $(a\text{-}\bar{v})O_2$ difference determines the rate at which oxygen is being consumed:

$$\dot{V}O_2 = \text{stroke volume} \times \text{heart rate} \times (a\text{-}\bar{v})O_2 \text{ diff.}$$

The oxygen demand of exercising muscles increases with increasing exercise intensity. Aerobic endurance depends on the cardiorespiratory system's ability to deliver sufficient oxygen to these active tissues to meet their heightened demands for oxygen for oxidative metabolism. As maximal levels of exercise are achieved, heart size, blood flow, blood pressure, and blood volume can all potentially limit the maximal ability to transport oxygen. Endurance training elicits numerous changes in these components of the **oxygen transport system** that enable it to function more effectively.

Heart Size

As an adaptation to the increased work demand, heart mass and volume increase with training. Cardiac muscle, like skeletal muscle, undergoes morphological adaptations as a result of chronic endurance training. At one time, **cardiac hypertrophy** induced by exercise—**"athlete's heart,"** as it was called—caused some concern because experts incorrectly believed that enlargement of the heart always reflected a pathological state, as sometimes occurs with severe hypertension. Training-induced cardiac hypertrophy is now recognized as a normal adaptation to chronic endurance training.

The left ventricle, as discussed in chapter 5, does the most work and thus undergoes the greatest adaptation in response to endurance training. The extent and location of heart size adaptations depend on the type of exercise training performed. For example, during resistance training, the left ventricle must contract against increased afterload from the systemic circulation. It was postulated that to overcome this high afterload, the heart muscle compensated by increasing left ventricular wall thickness, thereby increasing its contractility. From chapter 7 we learned that blood pressure during resistance exercise can exceed 480/350 mmHg. This presents a considerable resistance that must be overcome by the left ventricle. Thus, the increase in its muscle mass is in direct response to repeated exposure to the increased afterload with resistance training.

With endurance training, left ventricular chamber size increases. This allows for increased left ventricular filling and consequently an increase in stroke volume. The increase in left ventricular dimensions is largely attributable to a training-induced increase in plasma volume (discussed later in this chapter) that increases left ventricular end-diastolic volume (increased preload). In concert with this, a decrease in heart rate at rest caused by increased parasympathetic tone, and during exercise at the same rate of work, allows a longer diastolic filling period. The increases in plasma volume and diastolic filling time increase left ventricular chamber size at the end of diastole.

It was originally hypothesized that this increase in left ventricular dimensions was the only change in the left ventricle caused by endurance training. Additional research has revealed that myocardial wall thickness also increases with endurance training, rather than just with resistance training.[11] Using magnetic resonance imaging, Milliken and colleagues[22] found that highly trained endurance athletes (competitive cross-country skiers, endurance cyclists, and long-distance runners) had greater left ventricular masses than did non-endurance-trained control subjects. Left ventricular mass was highly correlated with $\dot{V}O_{2max}$, or aerobic power.

Fagard[11] conducted the most extensive review of the existing research literature in 1996, focusing on long-distance runners (135 athletes and 173 controls), cyclists (69 athletes and 65 controls), and strength athletes (178 athletes, including weight- and powerlifters, bodybuilders, wrestlers, throwers, and bobsledders, and 105 controls). For each group, the athletes were matched by age and body size with a group of sedentary control subjects. For each group of runners, cyclists, and strength athletes, the internal diameter of the left ventricle (LVID, an index of chamber size) and the total left ventricular mass (LVM) were greater in the athletes compared with their age- and sized-matched controls (figure 10.2). Thus, data from this large cross-sectional study support the hypothesis that both left ventricular chamber size and wall thickness increase with endurance training.

Most studies of heart size changes with training have been cross-sectional, comparing trained individuals with sedentary, untrained individuals. We can learn much from cross-sectional studies, but they do not provide us with the same information that we could get from studying a group of untrained people who train for months or years, focusing on their changes from the untrained to the trained state. Certainly a portion of the differences that we see in figure 10.2 can be attributed to genetics, not training. However, a number of studies have followed individuals from an untrained state to a trained state, and others have followed individuals from a trained state to an untrained state. These studies have reported increases in heart size with training and decreases with detraining. So, training does bring about changes, but they are likely not as large as the differences we see in figure 10.2.

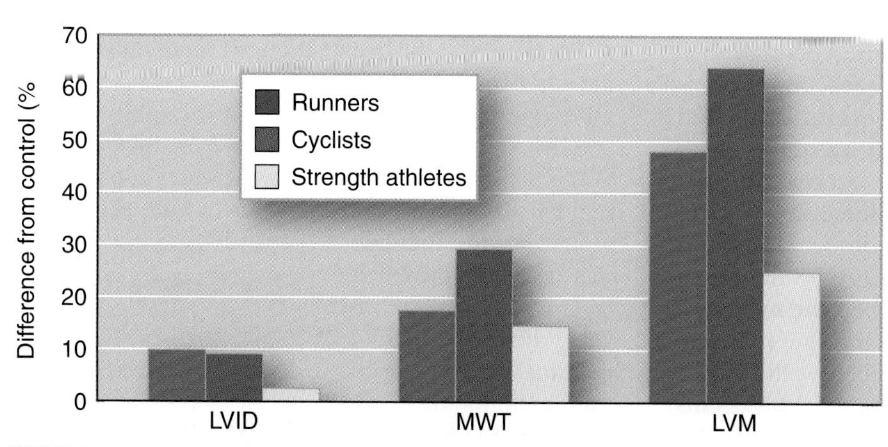

Figure 10.2 Percentage differences in heart size between three groups of athletes (runners, cyclists, and strength athletes) and their age- and size-matched sedentary controls. Percentage differences are presented for left ventricular internal diameter (LVID), mean wall thickness (MWT), and left ventricular mass (LVM). Data are from Fagard (1996).

In review

- Cardiorespiratory endurance refers to the ability to sustain prolonged, dynamic exercise. It is highly related to aerobic power.
- Exercise scientists regard $\dot{V}O_{2max}$—the highest rate of oxygen consumption obtainable during maximal or exhaustive exercise—as the best indicator of cardiorespiratory endurance.
- Cardiac output represents how much blood leaves the heart each minute, whereas $(a-\bar{v})O_2$ difference indicates how much oxygen is extracted from the blood by the tissues. According to the Fick equation, the product of these values is the rate of oxygen consumption: $\dot{V}O_2$ = stroke volume × heart rate × $(a-\bar{v})O_2$ difference.
- Of the chambers of the heart, the left ventricle adapts the most in response to endurance training.
- The internal dimensions of the left ventricle increase, mostly in response to an increase in ventricular filling secondary to an increase in plasma volume and diastolic filling time.
- Left ventricular wall thickness and mass also increase, allowing for greater contractility.

Stroke Volume

As a result of endurance training, stroke volume increases. Stroke volume at rest is substantially higher after an endurance training program than it is before training. This endurance training-induced increase is also seen at a given submaximal exercise intensity and at maximal exercise. This increase is illustrated in figure 10.3, which shows the changes in stroke volume of a subject who exercised at increasing intensities up to maximal intensity before and after a six-month aerobic endurance training program. Typical values for stroke volume at rest and during maximal exercise in untrained, trained, and highly trained athletes are listed in table 10.1. The wide variation

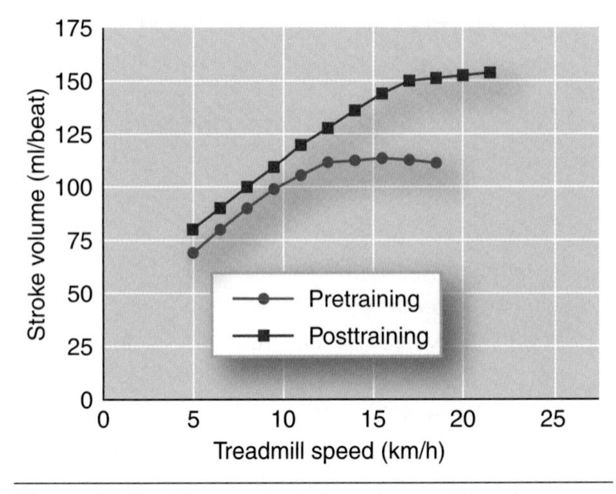

Figure 10.3 Changes in stroke volume with endurance training during walking, jogging, and running on a treadmill at increasing velocities.

TABLE 10.1 Stroke Volumes (SV) for Different States of Training

Subjects	SV_{rest} (ml/beat)	SV_{max} (ml/beat)
Untrained	50-70	80-110
Trained	70-90	110-150
Highly trained	90-110	150- >220

in stroke volume values for any given cell within this table is largely attributable to differences in body size. Absolute stroke volume at rest and during exercise is not merely a function of a person's state of training but also reflects differences in body size. Larger people typically have larger hearts and a greater blood volume, and thus higher stroke volumes—an important point when one is comparing stroke volumes of different people.

After aerobic training, the left ventricle fills more completely during diastole than it does in an untrained state. Plasma volume expands with training, which allows for more blood to enter the ventricle during diastole, increasing end-diastolic volume (EDV). The heart rate of a trained heart is also lower at rest and at the same absolute exercise intensity than that of an untrained heart, allowing more time for the increased diastolic filling. More blood entering the ventricle increases the stretch on the ventricular walls; by the Frank-Starling mechanism (see chapter 7), this results in an increase in force of contraction.

The thickness of the posterior and septal walls of the left ventricle also increases slightly with endurance training. Increased ventricular muscle mass results in increased contractile force, in turn causing end-systolic volume to decrease. More blood is forced out of the heart, leaving less blood in the left ventricle after systole. The decrease in end-systolic volume is augmented by the decrease in peripheral resistance that occurs with training. Increased contractility resulting from an increase in left ventricular thickness and greater diastolic filling (Frank-Starling mechanism), coupled with the reduction in systemic peripheral resistance, increases the ejection fraction [equal to (EDV – ESV)/EDV] in the trained heart. More blood enters the left ventricle, and a greater percentage of what enters is forced out with each contraction, resulting in an increase in stroke volume.

Adaptations in stroke volume during endurance training are illustrated by a study in which older men endurance trained for one year.[9] Their cardiovascular function was evaluated before and after training. The subjects performed running, treadmill, and cycle ergometer exercise for 1 h each day, four days per week. They exercised at intensities of 60% to 80% of $\dot{V}O_{2max}$, with brief bouts of exercise exceeding 90% of $\dot{V}O_{2max}$. End-diastolic volume increased at rest and throughout submaximal exercise. The ejection fraction increased, which was associated with a decreased end-systolic volume, suggesting increased contractility of the left ventricle. $\dot{V}O_{2max}$ increased by 23%, indicating a substantial improvement in endurance.

It is clear that central stroke volume adaptations occur with endurance training, but there are also peripheral adaptations that contribute to the increase in $\dot{V}O_{2max}$, at least in middle-aged exercisers. This was demonstrated in a unique longitudinal study involving both exercise training and a bed rest deconditioning model.[21] Five 20-year-old men were tested (baseline values), placed on bed rest for 20 days (deconditioning), and then trained for 60 days, starting immediately at the conclusion of bed rest. These same five men were restudied 30 years later at the age of 50; they were tested at baseline in a relatively sedentary state and after six months of endurance training. The average percentage increases in $\dot{V}O_{2max}$ were similar for the subjects at age 20 (18%) and at age 50 (14%). However, the increase in $\dot{V}O_{2max}$ at age 20 was explained by increases in both maximal cardiac output and maximal $(a\text{-}\bar{v})O_2$ difference; at age 50, the increase was explained primarily by an increase in $(a\text{-}\bar{v})O_2$ difference, while maximal cardiac output was unchanged. Maximal stroke volume was increased after training at both age 20 and age 50 but to a lesser degree at age 50 (+16 ml/beat at age 20 vs. +8 ml/beat at age 50).

Increased left ventricular dimensions, reduced systemic peripheral resistance, and a greater blood volume account for the increases in resting, submaximal, and maximal stroke volumes after an endurance training program.

Measuring Heart Size

The measurement of heart size has been of interest to cardiologists for years because a hypertrophied, or enlarged, heart is typically a pathological condition indicating the presence of cardiovascular disease. More recently, exercise scientists have been interested in heart size as it relates to the training state and performance of the athlete or exercising individual. Initial studies used X rays to estimate the size of the heart; but the techniques were not very accurate, and the information obtained was limited to only a few crude measurements. Since the 1970s, studies of athletes and people who participate in endurance training have used echocardiography to more accurately measure the size of the heart and its chambers. Echocardiography involves the technique of ultrasound, which utilizes high-frequency sound waves directed through the chest wall to the heart. These sound waves are emitted from a transducer placed on the chest; and once they contact the various structures of the heart, they rebound back to a sensor, which is able to capture the deflected sound waves and provide a moving picture of the heart. A trained physician or technician can visualize the size of the heart's chambers, thicknesses of its walls, and heart valve action. There are several forms of echocardiography: M-mode echocardiography, which provides a one-dimensional view of the heart; two-dimensional echocardiography; and Doppler echocardiography, which is used more often to measure blood flow. The figures in this sidebar illustrate two-dimensional echocardiography being conducted and the resulting echocardiogram.

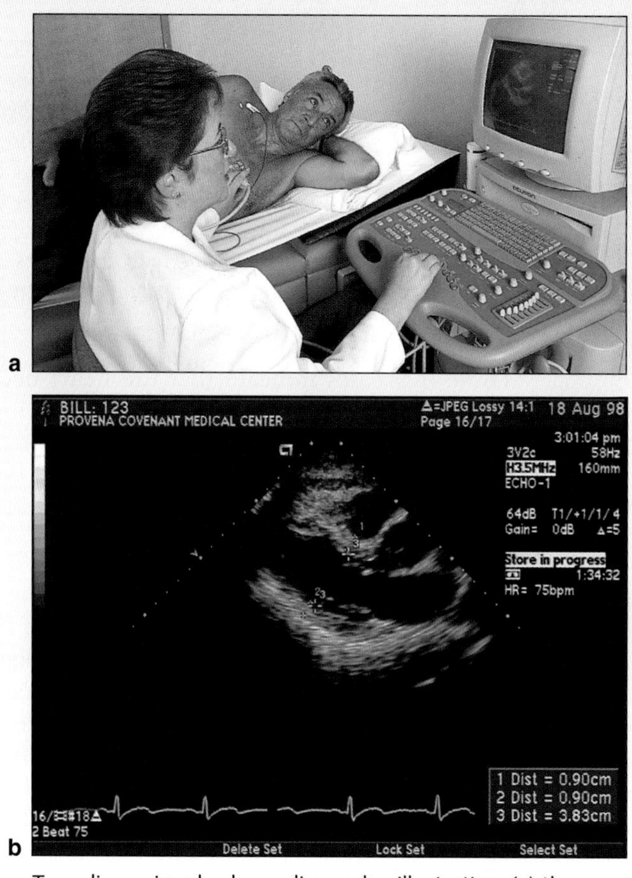

Two-dimensional echocardiography, illustrating *(a)* the procedure and *(b)* the resulting echocardiogram.

In review

- Following endurance training, stroke volume is increased at rest and during submaximal and maximal exercise when compared to pretraining values.
- A major factor leading to the stroke volume increase is an increased EDV caused by an increase in plasma volume and greater diastolic filling time (lower heart rate).
- Another major factor is increased left ventricular force of contraction. This is caused by hypertrophy of the cardiac muscle and increased ventricular stretch resulting from an increase in diastolic filling (increased preload) leading to greater elastic recoil (Frank-Starling mechanism).
- Reduced systemic vascular resistance (decreased afterload) also contributes to the increased volume of blood pumped from the left ventricle with each beat.

Heart Rate

Aerobic training has a major impact on heart rate at rest, during submaximal exercise, and during the postexercise recovery period. The effect of aerobic training on maximal heart rate is not as clear.

Resting Heart Rate

The heart rate at rest can decrease markedly as a result of endurance training. A sedentary individual with an initial resting heart rate of 80 beats/min can decrease resting heart rate by approximately 1 beat/min with each week of aerobic training, at least for the first few weeks. After 10 weeks of moderate endurance training, resting heart rate can decrease from 80 to 70 beats/min or lower. The actual mechanisms responsible for this decrease are not entirely understood, but training appears to increase parasympathetic activity in the heart while decreasing sympathetic activity. It is important to recognize that several well-controlled studies with large numbers of subjects have shown much smaller decreases in resting heart rate, that is, fewer than 5 beats/min following up to 20 weeks of aerobic training.

> **In focus**
>
> Highly conditioned endurance athletes often have resting heart rates lower than 40 beats/min., and some have values lower than 30 beats/min. Lance Armstrong, whose exploits were detailed at the beginning of this chapter, has a resting heart rate of 32 to 34 beats/min.

Recall from chapter 5 that bradycardia is a clinical term indicating a heart rate of fewer than 60 beats/min. In untrained individuals, bradycardia is usually the result of abnormal cardiac function or a diseased heart. Therefore, it is necessary to differentiate between training-induced bradycardia, which is a natural response to endurance training, and pathological bradycardia, which can be a serious cause for concern.

Submaximal Heart Rate

During submaximal exercise, aerobic conditioning results in proportionally lower heart rates at a given absolute exercise intensity. This is illustrated in figure 10.4, which shows the heart rate of an individual exercising on a treadmill both before and after training. At each specified intensity, indicated here by the speed at which the subject is walking or running, the posttraining heart rate is lower than the heart rate before training. After a six-month endurance training program of moderate intensity, decreases in heart rate of 10 to 30 beats/min are common at the same absolute submaximal workload, the training-induced decrease being greater at higher intensities.

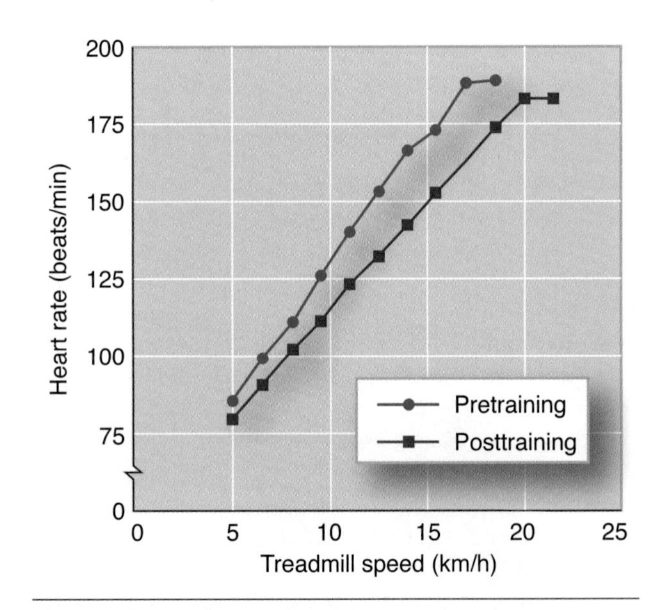

Figure 10.4 Changes in heart rate with endurance training during walking, jogging, and running on a treadmill at increasing velocities.

These decreases indicate that the heart becomes more economical through training. In carrying out its necessary functions, a trained heart performs less work (lower heart rate, higher stroke volume) than an unconditioned heart at the same absolute workload.

Maximum Heart Rate

A person's maximal heart rate (HR_{max}) tends to be stable and typically remains relatively unchanged after endurance training. However, several studies have suggested that for people whose untrained HR_{max} values exceed 180 beats/min, HR_{max} might be slightly lower after training. Also, highly conditioned endurance athletes tend to have lower HR_{max} values than untrained individuals of the same age, although this is not always the case. Athletes over 60 years old sometimes have higher HR_{max} values than untrained people of the same age.

Interactions Between Heart Rate and Stroke Volume

During exercise, the product of heart rate and stroke volume provides a cardiac output appropriate to the intensity of the activity being performed. At maximal or near-maximal intensities, heart rate may change to provide the optimal combination of heart rate and stroke volume to maximize cardiac output. If heart rate is too fast, diastolic filling time is reduced, and stroke volume might be compromised. For example, if HR_{max} is 180 beats/min, the heart beats three times per second. Each cardiac cycle thus lasts for only 0.33 s. Diastole is as short as 0.15 s or less. This fast heart rate allows very little time for the ventricles to fill. As a consequence, stroke volume may decrease at high heart rates at which filling time is compromised.

However, if the heart rate slows, the ventricles would have longer to fill. This has been proposed as the reason highly trained endurance athletes tend to have lower HR_{max} values: Their hearts have adapted to training by drastically increasing their stroke volumes, so lower HR_{max} values can provide optimal cardiac output.

Which comes first? Does increased stroke volume result in a decreased heart rate, or does a lower heart rate result in an increased stroke volume? This question remains unanswered. In either case, the combination of increased stroke volume and decreased heart rate is a more efficient way for the heart to meet the metabolic demands of the body, especially during exercise. The heart expends less energy by contracting less often but more forcefully than it would if contraction frequency were increased. Reciprocal changes in heart rate and stroke volume in response to training share a common goal: to allow the heart to pump the maximal amount of oxygenated blood at the lowest energy cost.

Heart Rate Recovery

During exercise, as discussed in chapter 5, heart rate must increase to increase cardiac output to meet the blood flow demands of active muscles. When the exercise bout is finished, heart rate does not instantly return to its resting level. Instead, it remains elevated for a while, slowly returning to its resting rate. The time it takes for heart rate to return to its resting rate is called the heart rate recovery period.

Following a period of training, as shown in figure 10.5, heart rate returns to its resting level much more quickly after exercise than it does before training. This is true after standardized submaximal exercise as well as after maximal exercise.

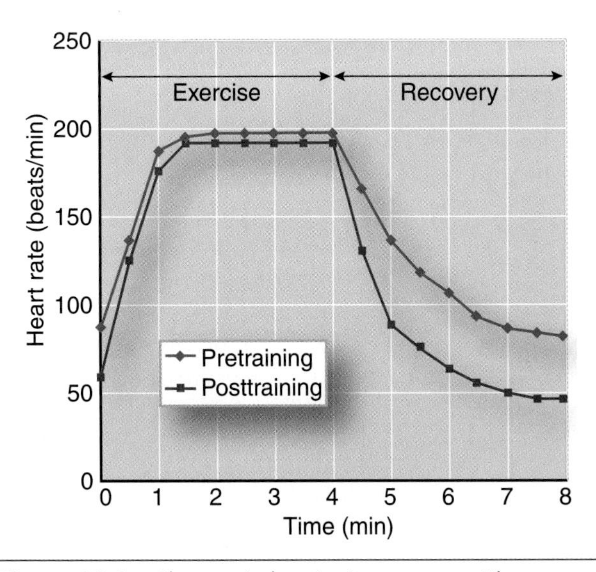

Figure 10.5 Changes in heart rate recovery with endurance training following completion of an all-out maximal work bout of 4 min duration.

Because the heart rate recovery period is shortened after endurance training, this measurement has been proposed as an indirect index of cardiorespiratory fitness. In general, a more fit person recovers faster after a standardized rate of work than a less fit person, so this measure may have some utility in field settings when more direct measures of endurance capacity are not possible or feasible. However, factors other than training can also affect heart rate recovery time. For example, exercise in hot environments or at high altitudes can prolong heart rate elevation. Some people undergo a stronger sympathetic nervous system response during exercise than others, and this also could prolong heart rate elevation.

The heart rate recovery curve is a useful tool for tracking a person's progress during a training program. But because of the potential influence of other factors, it should not be used to compare between individuals.

> ***In focus***
>
> Resting heart rate is typically lower (by more than 10 beats/min) following aerobic endurance training. After endurance training, submaximal heart rates are likewise lower during exercise at the same absolute workload, generally by 10 to 20 beats/min or more. Maximal heart rates generally do not change with endurance training.

Cardiac Output

We have looked at the effects of training on the two components of cardiac output: stroke volume and heart rate. While stroke volume increases, heart rate generally decreases at rest and during exercise at a given absolute intensity.

Because the magnitude of these reciprocal changes is similar, cardiac output at rest and during submaximal exercise at a given exercise intensity does not change much following endurance training. In fact, cardiac output can decrease slightly. This is likely the result of an increase in the $(a\text{-}\bar{v})O_2$ difference (reflecting greater oxygen extraction by the tissues) or a decrease in the rate of oxygen consumption (reflecting an increased mechanical efficiency). Generally, cardiac output matches the oxygen consumption required for any given intensity of effort.

Maximal cardiac output, however, increases considerably at maximal exercise intensity in response to aerobic training, as seen in figure 10.6, and is largely responsible for the increase in $\dot{V}O_{2max}$. This increase in cardiac output must result from an increase in maximal stroke volume, because HR_{max} changes little, if any. Maximal cardiac output ranges from 14 to 20 L/min in untrained individuals and from 25 to 35 L/min in trained individuals, and can be 40 L/min or more in highly conditioned endurance athletes. These absolute values, however, are greatly influenced by body size.

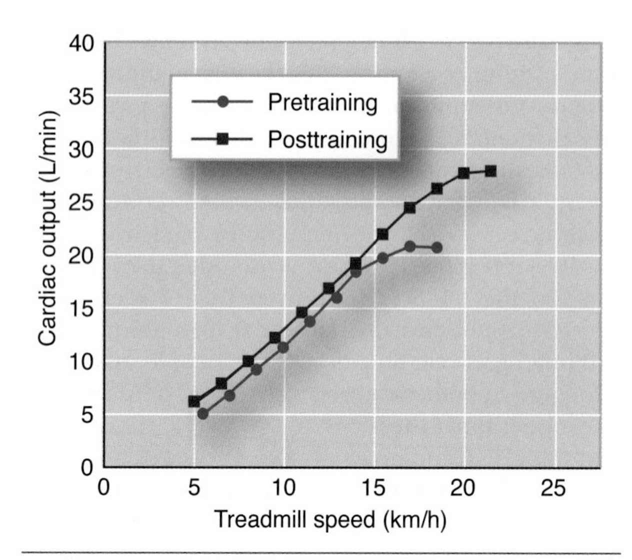

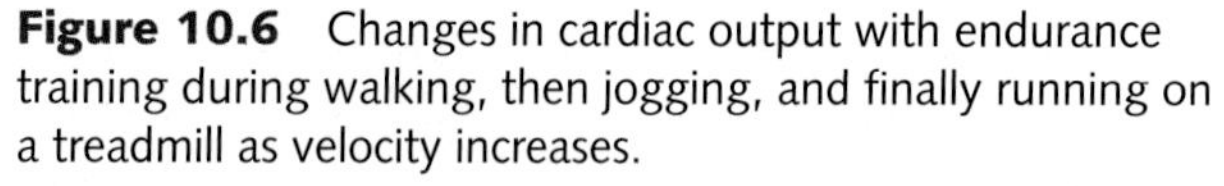

Figure 10.6 Changes in cardiac output with endurance training during walking, then jogging, and finally running on a treadmill as velocity increases.

In review

- Resting heart rate decreases as a result of endurance training. In a sedentary person, the decrease is typically about 1 beat/min per week during the initial weeks of training, but smaller decreases have been reported. Highly trained endurance athletes may have resting heart rates of 40 beats/min or lower.
- Heart rate during submaximal exercise is also lower, and the magnitude of decrease is greater at higher exercise intensities.
- Maximal heart rate either remains unchanged or decreases slightly with training.
- The heart rate during the recovery period decreases more rapidly after training, making it an indirect but convenient way of tracking the adaptations that occur with training. However, this value is not useful for comparing fitness levels of different people.
- Cardiac output at rest and at submaximal levels of exercise remains unchanged or decreases slightly after endurance training.
- Cardiac output at maximal levels of exercise increases considerably and is largely responsible for the increase in $\dot{V}O_{2max}$. The increased maximal cardiac output is the result of the substantial increase in maximal stroke volume, made possible by training-induced changes in cardiac structure and function.

Blood Flow

Active muscles need considerably more oxygen and nutrients than inactive ones. To meet these increased needs, more blood must be delivered to these muscles during exercise. With endurance training, the cardiovascular system adapts to increase blood flow to exercising muscles to meet their higher demand for oxygen and metabolic substrates. Four factors account for this enhanced blood flow to muscle following training:

- Increased capillarization of trained muscles
- Greater recruitment of existing capillaries in trained muscles
- More effective blood flow redistribution from inactive regions
- Increased blood volume

To permit increased blood flow, new capillaries develop in trained muscles. This allows the blood flowing into skeletal muscle from arterioles to more fully perfuse the active tissues. This increase in capillaries usually is expressed as an increase in the number of capillaries per muscle fiber, or the **capillary-to-fiber ratio.** Table 10.2 illustrates the differences in capillary-to-fiber ratios between well-trained and untrained men, both before and after exercise.[13]

Not all capillaries are open at any given time in tissues, including muscle. In addition to new capillarization, existing capillaries in trained muscles can

TABLE 10.2 Muscle Fiber Capillarization in Well-Trained and Untrained Men

	Capillaries per mm^2	Muscle fibers per mm^2	Capillary-to-fiber ratio	Diffusion distance[a]
		Well trained		
Preexercise	640	440	1.5	20.1
Postexercise	611	414	1.6	20.3
		Untrained		
Preexercise	600	557	1.1	20.3
Postexercise	599	576	1.0	20.5

Note. This table illustrates the larger size of the muscle fibers in the well-trained men in that they had fewer fibers for a given area (fibers per mm^2). They also had an approximately 50% higher capillary-to-fiber ratio than the untrained men.

[a]Diffusion distance is expressed as the average half-distance between capillaries on the cross-sectional view expressed in micrometers.

Adapted, by permission, from L. Hermansen and M. Wachtlova, 1971, "Capillary density of skeletal muscle in welltrained and untrained men," *Journal of Applied Physiology* 30: 860-863.

be recruited and open to flow, which increases blood flow to muscle fibers. The increase in new capillaries with endurance training and increased capillary recruitment combine to increase the cross-sectional area for exchange between the vascular system and the metabolically active muscle fibers. Because endurance training also increases blood volume, shifting more blood into the capillaries will not severely compromise venous return.

A more effective redistribution of cardiac output also can increase blood flow to the active muscles. Blood flow is directed to the active musculature and shunted away from areas that do not need high flow. Blood flow can increase to the more active fibers even within a specific muscle group. Armstrong and Laughlin[1] demonstrated that endurance-trained rats could redistribute blood flow to their most active tissues during exercise better than untrained rats could. The researchers used radiolabeled microspheres, radioactive particles that are injected into the bloodstream. The distribution of these microspheres was then determined by using a Geiger counter, which measures the amount of radioactive material throughout the area of interest (in this case exercising muscle). The total blood flow to the hindlimbs did not differ between the trained and untrained rats during exercise. However, the trained rats distributed more of their blood to the most oxidative muscle fibers, effectively redistributing the blood flow away from the glycolytic muscle fibers. These findings are difficult to replicate in humans because of the technical difficulty (microspheres cannot be used in humans) of measuring blood flow to specific muscle fiber types, as well as the fact that human skeletal muscle is a mosaic with mixed fiber types among individual muscles.

In focus

The increase in blood flow to muscle is one of the most important factors supporting increased aerobic endurance capacity and performance. This increase is attributable to improved capillary supply (both new capillaries and greater capillary recruitment), diversion of a larger portion of the cardiac output to the active muscles, and increased blood volume.

Finally, the body's total blood volume increases with endurance training, providing more blood to meet the body's many blood flow needs during endurance activity. The mechanisms responsible for this are discussed later in this chapter.

Blood Pressure

Following endurance training, arterial blood pressure is reduced at a given submaximal exercise intensity; but at maximal exercise capacity, systolic blood pressure is increased and diastolic pressure is decreased. Resting blood pressure in response to endurance training does not change significantly in healthy subjects but is generally lowered in borderline or moderately hypertensive individuals. This reduction occurs in both systolic and diastolic blood pressure. Drops in blood pressure average approximately 6 to 7 mmHg for both systolic and diastolic pressure in hypertensive subjects and slightly less in borderline hypertensives. The mechanisms underlying this reduction are unknown.

Although resistance-type exercise can cause large increases in both systolic and diastolic blood pressure during lifting of heavy weights, chronic exposure to these high pressures does not elevate resting blood pressure. Hypertension is not common in competitive weightlifters or in strength and power athletes. In fact, a few studies have even shown that resistance training may lower resting systolic blood pressure. Hagberg and coworkers[12] followed a group of borderline-hypertensive adolescents through five months of weight training. The subjects' resting systolic blood pressures decreased significantly. These reductions were somewhat greater than those resulting from endurance training.

Blood Volume

Endurance training increases blood volume, and this effect is larger as training intensity increases. Furthermore, the effect occurs rapidly. This increased blood volume results primarily from an increase in plasma volume, but there is also an increase in red blood cells. The time course for the increase of each of these is quite different.[28]

Plasma Volume

The increase in plasma volume is thought to result from two mechanisms. The first mechanism, which has two phases, results in increases in the amount of plasma proteins, particularly albumin. Recall from chapter 7 that plasma proteins are the major source of osmotic pressure in the vasculature. As plasma protein concentration increases, so does osmotic pressure, and fluid is reabsorbed from the interstitial fluid into the vasculature. During an intense bout of exercise, proteins leave the vascular space and move into the interstitial space. They are then returned in greater amounts through the lymph system. It is likely that the first phase of rapid plasma volume increase is the result of the increased plasma albumin, which is noted within the first hour of recovery from the first training bout. In the second phase, protein synthesis is turned on (upregulated) by repeated exercise, and new proteins are formed. With the second mechanism, exercise increases the release of antidiuretic hormone and aldosterone, hormones that cause increased reabsorption of water and sodium in the kidneys, which increases blood plasma. That increased fluid is kept in the vascular space by the oncotic pressure exerted by the proteins. Nearly all of the increase in blood volume during the first two weeks of training can be explained by the increase in plasma volume.

Red Blood Cells

An increase in red blood cell volume with endurance training also contributes to the overall increase in blood volume, but this increase is an inconsistent finding. Although the actual number of red blood cells may increase, the hematocrit—the ratio of the red blood cell volume to the total blood volume—may actually decrease. Figure 10.7 illustrates this apparent paradox. Notice that the hematocrit is reduced even though there has been a slight increase in red blood cells. A trained athlete's hematocrit can decrease to a level where the athlete appears to be anemic on the basis of a relatively low concentration of red cells and hemoglobin ("pseudoanemia").

The increased ratio of plasma to cells resulting from a greater increase in the fluid portion reduces the blood's viscosity, or thickness. Reduced viscosity may facilitate blood movement through the blood vessels, particularly through the smallest vessels such as the capillaries. One of the physiological benefits of decreasing blood viscosity is that it enhances oxygen delivery to the active muscle mass.

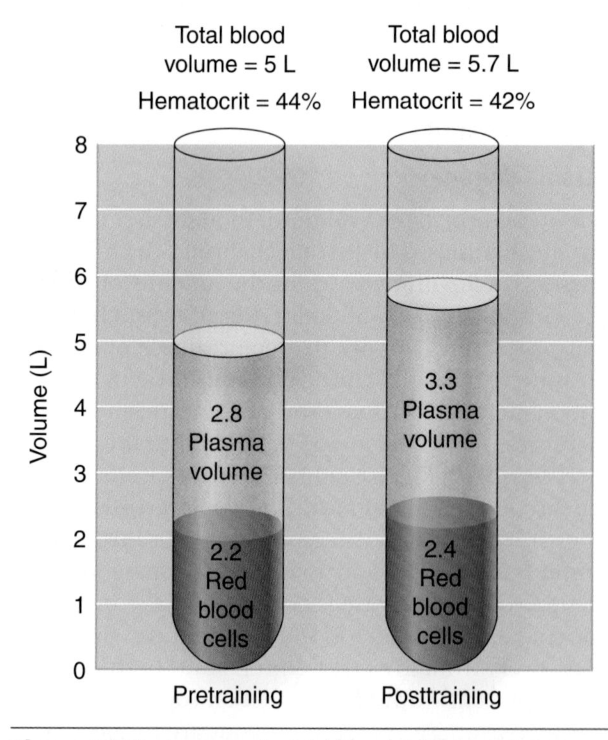

Figure 10.7 Increases in total blood volume and plasma volume with endurance training. Note that although the hematocrit (percentage of red blood cells) decreases from 44% to 42%, the total volume of red blood cells increases by 10%.

Both the total amount (absolute values) of hemoglobin and the total number of red blood cells are typically elevated in highly trained athletes, although these values relative to total blood volume are below normal. This ensures that the blood has more than ample oxygen-carrying capacity. The turnover rate of red blood cells also may be higher with intense training.

The increase in blood volume following aerobic endurance training is attributable to increases in both plasma volume and red blood cell volume; both changes facilitate the delivery of oxygen to active muscles.

In review

- Blood flow to muscles is increased by endurance training.
- Increased blood flow results from four factors:
  1. Increased capillarization
  2. Greater opening of existing capillaries (capillary recruitment)
  3. More effective blood flow distribution
  4. Increased blood volume
- Resting blood pressure generally is reduced by endurance training in those with borderline or moderate hypertension but not in healthy, normotensive exercisers.
- Endurance training results in a reduction in blood pressure during submaximal exercise at the same exercise intensity, but at maximal exercise intensity the systolic blood pressure is increased and diastolic blood pressure is decreased compared to pretraining values.
- Blood volume increases as a result of endurance training.
- The increase in blood volume initially is caused by an increase in plasma volume.
- Plasma volume is expanded through increased protein content (returned from lymph and upregulated protein synthesis) and supported by fluid conservation hormones.
- Red blood cell volume also increases, but the increase in plasma volume is typically higher.
- Increased plasma volume decreases blood viscosity, which can improve circulation and oxygen availability.

Respiratory Adaptations to Training

No matter how adept the cardiovascular system is at supplying adequate amounts of blood to tissues, endurance would be hindered if the respiratory system were not able to bring in enough oxygen to meet oxygen demands. Respiratory system function does not usually limit performance because ventilation can be increased to a much greater extent than cardiovascular function. But, as with the cardiovascular system, the respiratory system undergoes specific adaptations to endurance training to maximize its efficiency.

Pulmonary Ventilation

After training, pulmonary ventilation is essentially unchanged at rest. Although endurance training does not change the structure or basic physiology of the lung, it does decrease ventilation during submaximal exercise by as much as 20% to 30% at a given submaximal intensity. Maximal pulmonary ventilation is substantially increased from a rate of about 100 to 120 L/min in untrained sedentary individuals to about 130 to 150 L/min or more following endurance training. Pulmonary ventilation rates typically increase to about 180 L/min in highly trained athletes and can exceed 200 L/min in very large, highly trained endurance athletes. Two factors can account for the increase in maximal pulmonary ventilation following training: increased tidal volume and increased respiratory frequency at maximal exercise.

Ventilation is usually not considered a limiting factor for endurance exercise performance. However, some evidence suggests that at some point in a highly trained person's adaptation, the pulmonary system's capacity for oxygen transport may not be able to meet the demands of the limbs and the cardiovascular system.[8] This results in what has been termed exercise-induced arterial hypoxemia, in which arterial oxygen saturation decreases below 96%. As discussed in chapter 6, this desaturation in highly trained elite athletes likely results from the large right heart cardiac output directed to the lung during exercise and consequently a decrease in the time the blood spends in the lung.

Pulmonary Diffusion

Pulmonary diffusion, or gas exchange occurring in the alveoli, is unaltered at rest and at standardized submaximal exercise intensities following training. However, it increases during maximal exercise. Pulmonary blood flow (blood coming from the heart to the lungs) appears to increase following training, particularly the flow to the upper regions of the lungs when a person is sitting or standing. This increases lung perfusion. More blood is brought into the lungs for gas exchange, and at the same time ventilation increases so that more air is brought into the lungs. This means that more alveoli will be involved in pulmonary diffusion. The net result is that pulmonary diffusion increases.

Arterial–Venous Oxygen Difference

The oxygen content of arterial blood changes very little with endurance training. Even though total hemoglobin is increased, the amount of hemoglobin per unit of blood is the same or even slightly reduced. The $(a-\bar{v})O_2$ difference, however, does increase with training, particularly at maximal exercise intensity. This increase results from a lower mixed venous oxygen content, which means that the blood returning to the heart (which is a mixture of venous blood from all body parts, not just the active tissues) contains less oxygen than it would in an untrained person. This reflects both greater oxygen extraction at the tissue level and a more effective distribution of blood flow to active tissue. The increased extraction results in part from an increase in oxidative capacity of active muscle fibers as described later in this chapter.

In summary, the respiratory system is quite adept at bringing adequate oxygen into the body. For this reason, the respiratory system seldom limits endurance performance. Not surprisingly, the major training adaptations noted in the respiratory system are apparent mainly during maximal exercise, when all systems are being maximally stressed.

In focus

Although the largest part of the increase in $\dot{V}O_{2max}$ results from the increases in cardiac output and muscle blood flow, an increase in $(a-\bar{v})O_2$ difference also plays a key role. This increase in $(a-\bar{v})O_2$ difference is attributable to a more effective distribution of arterial blood away from inactive tissue to the active tissue and an increased ability of active muscle to extract oxygen.

Adaptations in Muscle

Repeated use of muscle fibers with endurance training stimulates changes in their structure and function. Our main interest here is in aerobic training and the changes it produces in muscle fiber type, mitochondrial function, and oxidative enzymes.

Muscle Fiber Type

As noted in chapter 1, aerobic activities such as jogging and low- to moderate-intensity cycling rely extensively on the slow-twitch (type I) fibers. In response to aerobic training, type I fibers become larger. That is, they develop a larger cross-sectional area, although the

In review

- Unlike what occurs in the cardiovascular system, endurance training has little effect on lung structure and function.
- To support increases in $\dot{V}O_{2max}$, there is an increase in pulmonary ventilation during maximal effort following training as both tidal volume and respiratory rate increase.
- Pulmonary diffusion at maximal intensity increases, probably because of increased ventilation and increased lung perfusion, especially to upper regions of the lung that are not normally perfused.
- The $(a\text{-}\bar{v})O_2$ difference increases with training, reflecting increased oxygen extraction by the tissues and more effective blood distribution to the active tissues.

magnitude of change depends on the intensity and duration of each training bout and the length of the training program. Increases of up to 25% have been reported. Fast-twitch (type II) fibers, because they are not being recruited to the same extent, generally do not increase cross-sectional area.

Most early studies showed no change in the percentage of type I and type II fibers following aerobic training, but subtle changes were noted among type II fiber subtypes. Type IIx fibers are used less often than IIa fibers, and for that reason they have a lower aerobic capacity. Long-duration exercise may eventually recruit these fibers into action, demanding them to perform in a manner normally expected of the IIa fibers. This can cause some IIx fibers to take on the characteristics of the more oxidative IIa fibers. Recent evidence suggests that not only is there a transition of type IIx to IIa fibers, but there can also be a transition of type II to type I fibers. The magnitude of change is generally small, not more than a few percentage points. As an example, in the HERITAGE Family Study,[25] a 20-week program of aerobic training increased type I fibers from 43.2% pretraining to 46.7% posttraining and decreased type IIx fibers from 20.0% to 15.1%, with type IIa remaining essentially unchanged. These more recent studies have included larger numbers of subjects and have taken advantage of improved measurement technology; both might explain why changes are now being recognized.

Capillary Supply

One of the most important adaptations to aerobic training is an increase in the number of capillaries surrounding each muscle fiber. Table 10.2 illustrates that endurance-trained men can have considerably more capillaries in their leg muscles than sedentary individuals.[13] With long periods of aerobic training, the number of capillaries has been shown to increase by more than 15%.[25] Having more capillaries allows greater exchange of gases, heat, wastes, and nutrients between the blood and the working muscle fibers. In fact, the increase in capillary density (i.e., increase in capillaries per muscle fiber) is potentially one of the most important alterations in response to training that allows the increase in $\dot{V}O_{2max}$. It is now clear that the diffusion of oxygen from the capillary to the mitochondria is a major factor limiting the maximal rate of oxygen consumption. Increasing capillary density facilitates this diffusion, thus maintaining an environment well suited to energy production and repeated muscle contractions.

Aerobic training increases both the number of capillaries per muscle fiber and the number of capillaries for a given cross-sectional area of muscle. Both of these changes improve blood perfusion through the muscles, thereby enhancing the exchange of gases, wastes, and nutrients between the blood and muscle fibers.

Myoglobin Content

When oxygen enters the muscle fiber, it binds to **myoglobin,** a compound similar to hemoglobin. This iron-containing compound shuttles the oxygen molecules from the cell membrane to the mitochondria. The type I fibers contain large quantities of myoglobin, which gives these fibers their red appearance (myoglobin is a pigment that turns red when bound to oxygen). The type II fibers, on the other hand, are highly glycolytic, so they require (and have) little myoglobin—hence their whiter appearance. More important, their limited myoglobin supply limits their oxygen capacity, resulting in poor aerobic endurance for these fibers.

Myoglobin stores oxygen and releases it to the mitochondria when oxygen becomes limited during muscle action. This oxygen reserve is used during the transition from rest to exercise, providing oxygen to the mitochondria during the lag between the beginning of exercise and the increased cardiovascular delivery of oxygen.

Myoglobin's precise contributions to oxygen delivery are not yet fully understood. But aerobic training has been shown to increase muscle myoglobin content by 75% to 80%. This adaptation would be expected only if myoglobin enhances a muscle's capacity for oxidative metabolism.

Mitochondrial Function

As noted in chapter 2, aerobic (oxidative) energy production takes place in the mitochondria. Not surprisingly, then, aerobic training also induces changes in mitochondrial function that improve the muscle fibers' capacity to produce ATP. The ability to use oxygen and produce ATP via oxidation depends on the number and size of the muscle mitochondria. Both increase with aerobic training.

During one study that involved endurance training in rats, the actual number of mitochondria increased approximately 15% during 27 weeks of exercise.[14] Average mitochondrial size also increased by about 35% over that training period. As the volume of aerobic training increases, so do the number and size of the mitochondria.

In focus

Skeletal muscle mitochondria increase in both number and size with aerobic training, providing the muscle with an increased capacity for oxidative metabolism.

Oxidative Enzymes

Regular endurance exercise has been shown to induce major adaptations in skeletal muscle, including an increase in the number and size of the muscle fiber mitochondria, as just discussed. These changes are further enhanced by an increase in mitochondrial capacity. The oxidative breakdown of fuels and the ultimate production of ATP depend on the action of **mitochondrial oxidative enzymes,** the special proteins that catalyze (i.e., speed up) the breakdown of nutrients to form ATP. Aerobic training increases the activity of these important enzymes.

Figure 10.8 illustrates the changes in the activity of succinate dehydrogenase (SDH), a key muscle oxidative enzyme, over seven months of gradually increased swim training. While the increases in $\dot{V}O_{2max}$ leveled off after the first two months of training, activity of this key oxidative enzyme continued to increase throughout the entire training period. This suggests that $\dot{V}O_{2max}$ might be more influenced by the circulatory system's limitations with respect to transporting oxygen than by the muscles' oxidative potential.

The activities of muscle enzymes such as SDH and citrate synthase are dramatically influenced by aerobic training. This is seen in figure 10.9 on page 236, which compares the activities of these enzymes in untrained people, moderately trained joggers, and highly trained runners.[7] Even moderate amounts of daily exercise increase these enzyme activities and thus the muscles' aerobic capacity. For example, jogging or cycling for as little as 20 min per day has been shown to increase SDH activity in leg muscles by more than 25%. Training more vigorously, for example for 60 to 90 min per day, produces a two- to threefold increase in this activity.

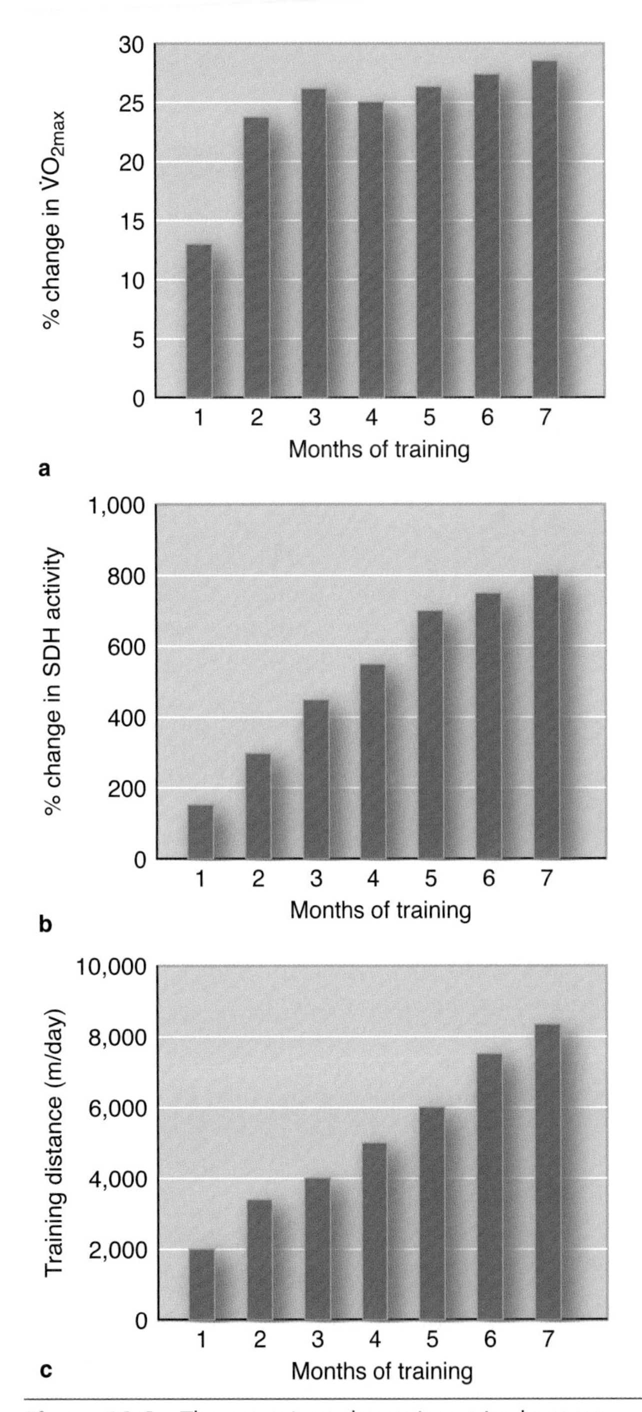

Figure 10.8 The percentage change in maximal oxygen uptake ($\dot{V}O_{2max}$) and the activity of succinate dehydrogenase (SDH), one of the muscles' key oxidative enzymes, during seven months of swimming training. Interestingly, although this enzyme activity continues to increase with increasing levels of training, the swimmers' maximal oxygen uptake appears to level off after the first 8 to 10 weeks of training. It appears that the mitochondrial enzyme activities are not a direct indication of one's aerobic capacity and might not be the best indicators of whole-body endurance capacity.[8]

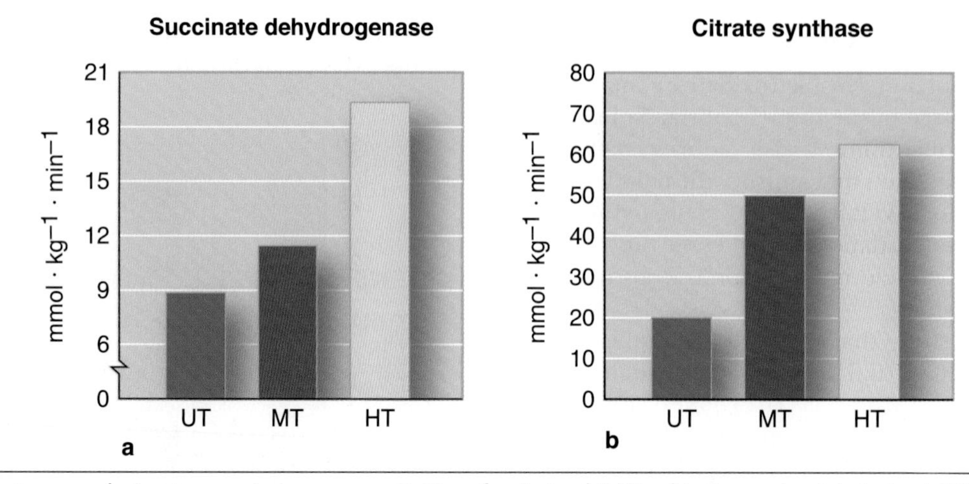

Figure 10.9 Leg muscle (gastrocnemius) enzyme activities of untrained (UT) subjects, moderately trained (MT) joggers, and highly trained (HT) marathon runners. Enzyme levels shown are for *(a)* succinate dehydrogenase and *(b)* citrate synthase, two of many enzymes that participate in the oxidative production of adenosine triphosphate.

Adapted, by permission, from D.L. Costill et al., 1979, "Lipid metabolism in skeletal muscle of endurance-trained males and females." *Journal of Applied Physiology* 28: 251-255 and from D.L. Costill et al., 1979, "Adaptations in skeletal muscle following strength training," *Journal of Applied Physiology* 46: 96-99.

One metabolic consequence of mitochondrial changes induced by aerobic training is **glycogen sparing**, a slower rate of utilization of muscle glycogen and enhanced reliance on fat as a fuel source at a given exercise intensity. This increase in the oxidative enzymes with aerobic training most likely improves the ability to sustain a higher exercise intensity, such as maintaining a faster race pace in a 10 km run.

In review

- Aerobic training recruits type I muscle fibers substantially more than type II fibers. Consequently, the type I fibers tend to enlarge with training.
- There appears to be a small increase in the percentage of type I fibers in addition to a transition of type IIx to type IIa fibers.
- The number of capillaries supplying each muscle fiber increases with training.
- Aerobic training increases muscle myoglobin content by about 75% to 80%. Myoglobin stores oxygen.
- Aerobic training increases both the number and the size of muscle fiber mitochondria.
- Activities of many oxidative enzymes are increased with aerobic training.
- These changes occurring in the muscles, combined with adaptations in the oxygen transport system, enhance the capacity of oxidative metabolism and improve endurance performance.

Metabolic Adaptations to Training

Now that we have discussed training changes in both the cardiovascular and respiratory systems, as well as skeletal muscle adaptations, we are ready to examine how these integrated adaptations are reflected by changes in three important physiological variables related to metabolism:

- Lactate threshold
- Respiratory exchange ratio
- Oxygen consumption

Lactate Threshold

Lactate threshold is a physiological marker that is closely associated with aerobic endurance performance—the higher the lactate threshold, the better the aerobic performance. Figure 10.10*a* illustrates the difference in lactate threshold between an endurance-trained individual and an untrained individual. This figure also accurately represents the changes in lactate threshold that would occur following a 6- to 12-month program of endurance training. In either case, in the trained state, one can exercise at a higher percentage of one's $\dot{V}O_{2max}$ before lactate begins to accumulate in the blood. Because race pace in aerobic endurance events is closely associated with lactate threshold, this translates into a much faster race pace (see figure 10.10*b*). The reduction in lactate values at a given rate of work is likely attributable to a combination of reduced lactate production and increased lactate clearance.

The concentration of lactate in the blood following a fixed-pace swim or run provides an excellent means of monitoring the physiological changes that

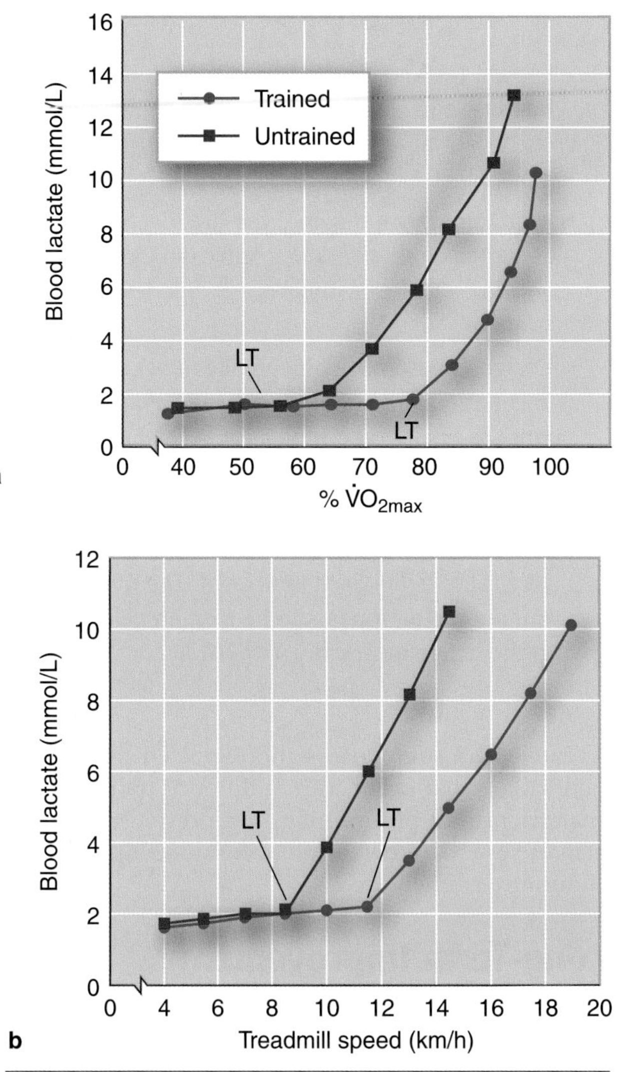

Figure 10.10 Changes in lactate threshold (LT) with training expressed as *(a)* a percentage of maximal oxygen uptake (% $\dot{V}O_{2max}$) and *(b)* an increase in speed on the treadmill. Lactate threshold (LT) occurs at a speed of 8.4 km/h (5.2 mph) in the untrained state and at 11.6 km/h (7.2 mph) in the trained state.

In focus

An increase in the lactate threshold is a major factor in the improved performance of aerobically trained endurance athletes.

occur with training. As athletes become better trained, their blood lactate concentrations are lower for the same rate of work. This suggests that they are developing greater aerobic power, a reduced reliance on the glycolytic system for energy, or perhaps both.

Respiratory Exchange Ratio

Recall from chapter 4 that the **respiratory exchange ratio (RER)** is the ratio of carbon dioxide released to oxygen consumed during metabolism. The RER reflects the type of substrates being used as an energy source.

After training, the RER decreases at both absolute and relative submaximal exercise intensities. These changes are attributable to a greater utilization of free fatty acids instead of carbohydrate at these work rates following training.

Resting and Submaximal Oxygen Consumption

Oxygen consumption ($\dot{V}O_2$) at rest is unchanged following endurance training. While a few cross-sectional comparisons have suggested that training elevates resting $\dot{V}O_2$, the HERITAGE Family Study—with a large number of subjects and with duplicate measures of resting metabolic rate both before and after 20 weeks of training—showed no evidence of an increased resting metabolic rate after training.[31]

During submaximal exercise at a given exercise intensity, $\dot{V}O_2$ is either unchanged or slightly reduced following training. In the HERITAGE Family Study, with more than 700 participants, training reduced submaximal $\dot{V}O_2$ by 3.5% at a work rate of 50 W. There was a corresponding reduction in cardiac output at 50 W, reinforcing the strong interrelationship between $\dot{V}O_2$ and cardiac output.[30] A decrease in $\dot{V}O_2$ during submaximal exercise could result from an increase in exercise economy (performing the same exercise intensity with less extraneous movement).

Maximal Oxygen Consumption

$\dot{V}O_{2max}$ is the best indicator of cardiorespiratory endurance capacity and increases substantially in response to endurance training. While small and very large increases have been reported, an increase of 15% to 20% is typical for a previously sedentary person who trains at 50% to 85% of his or her $\dot{V}O_{2max}$ three to five times per week, 20 to 60 min per day, for six months. For example, the $\dot{V}O_{2max}$ of a sedentary individual could reasonably increase from 35 ml · kg^{-1} · min^{-1} to 42 ml · kg^{-1} · min^{-1} as a result of such a program. This is far below the values we see in world-class endurance athletes, whose values generally range from 70 to 94 ml · kg^{-1} · min^{-1}. The more untrained an individual is when starting an exercise program, the larger the increase in $\dot{V}O_{2max}$.

What Limits Aerobic Power and Endurance Performance?

A number of years ago, exercise scientists were divided on what major physiological factor or factors actually limit $\dot{V}O_{2max}$. Two contrasting theories had been proposed.

One theory held that endurance performance was limited by the lack of sufficient concentrations of oxidative enzymes in the mitochondria. Endurance training programs substantially increase these oxidative enzymes, allowing active tissue to use more of the available oxygen, resulting in a higher $\dot{V}O_{2max}$. In addition, endurance training increases both the size and number of muscle mitochondria. Thus, this theory argued, the main limitation of maximal oxygen consumption is the inability of the existing mitochondria to use the available oxygen beyond a certain rate. This theory was referred to as the utilization theory.

The second theory proposed that central and peripheral circulatory factors limit endurance capacity. These circulatory factors would preclude delivery of sufficient amounts of oxygen to the active tissues. According to this theory, improvement in $\dot{V}O_{2max}$ following endurance training results from increased blood volume, increased cardiac output (via stroke volume), and a better perfusion of active muscle with blood.

Evidence strongly supports the latter theory. In one study, subjects breathed a mixture of carbon monoxide (which irreversibly binds to hemoglobin, limiting hemoglobin's oxygen-carrying capacity) and air during exercise to exhaustion.[23] $\dot{V}O_{2max}$ decreased in direct proportion to the percentage of carbon monoxide breathed. The carbon monoxide molecules bonded to approximately 15% of the total hemoglobin; this percentage agreed with the percentage reduction in $\dot{V}O_{2max}$. In another study, approximately 15% to 20% of each subject's total blood volume was removed.[10] $\dot{V}O_{2max}$ decreased by approximately the same relative amount. Reinfusion of the subjects' packed red blood cells approximately four weeks later increased $\dot{V}O_{2max}$ well above baseline or control conditions. In both studies, the reduction in the oxygen-carrying capacity of the blood—via either blocking hemoglobin or removing whole blood—resulted in the delivery of less oxygen to the active tissues and a corresponding reduction in $\dot{V}O_{2max}$. Similarly, studies have shown that breathing oxygen-enriched mixtures, in which the partial pressure of oxygen in the inspired air is substantially increased, increases endurance capacity.

These and subsequent studies indicate that the available oxygen supply is the major limiter of endurance performance. Saltin and Rowell[27] reviewed this topic and concluded that oxygen transport to the working muscles, not the available mitochondria and oxidative enzymes, limits $\dot{V}O_{2max}$. They argued that increases in $\dot{V}O_{2max}$ with training are largely attributable to increased maximal blood flow and increased muscle capillary density in the active tissues. The major skeletal muscle adaptations (including increased mitochondrial content and respiratory capacity of the muscle fibers) appear more closely related to the ability to perform prolonged, high-intensity, submaximal exercise.

In review

- Lactate threshold increases with endurance training, allowing performance of higher exercise intensities without significantly increasing blood lactate concentration.
- With endurance training, the RER decreases at submaximal work rates, indicating greater utilization of free fatty acids as an energy substrate.
- Oxygen consumption generally remains unchanged at rest and decreases slightly or remains unaltered during submaximal exercise following endurance training.
- $\dot{V}O_{2max}$ increases substantially following endurance training, but the amount of increase possible is limited in each individual. The major limiting factor appears to be oxygen delivery to the active muscles.

Table 10.3 summarizes the expected physiological changes that occur with endurance training. The changes pre- to posttraining in a previously inactive man are compared with values for a world-class male endurance runner.

Long-Term Improvement in Aerobic Power and Cardiorespiratory Endurance

Although an individual's highest attainable $\dot{V}O_{2max}$ is usually achieved within 12 to 18 months of intense endurance conditioning, endurance *performance* continues to improve with continued training for many additional years. Improvement in endurance performance without improvement in $\dot{V}O_{2max}$ is likely attributable to improvements in the ability to perform at increasingly higher percentages of $\dot{V}O_{2max}$ for extended periods.

Consider, for example, a young male runner who starts training with an initial $\dot{V}O_{2max}$ of 52.0 ml · kg^{-1} · min^{-1}. He reaches his genetically determined peak $\dot{V}O_{2max}$ of 71.0 ml · kg^{-1}· min^{-1} after two years of intense training, after which no further increases occur, even with more frequent or more intense workouts. At this point, as shown in figure 10.11, the young runner is able to run at 75% of his $\dot{V}O_{2max}$ (0.75 × 71.0 = 53.3 ml · kg^{-1}· min^{-1}) in a 10 km (6.2 mi) race. After an additional two years of intensive training, his $\dot{V}O_{2max}$ is unchanged, but he is now able to compete at 88% of his $\dot{V}O_{2max}$ (0.88 × 71.0 = 62.5 ml · kg^{-1} · min^{-1}). Obviously, by being able to sustain an oxygen uptake of 62.5 ml · kg^{-1} · min^{-1}, he is able to run at a much faster race pace.

TABLE 10.3 Expected Effects of Endurance Training in a Previously Inactive Man Along With Values for a Male World-Class Endurance Athlete

	Sedentary male subject		
Variables	Pretraining	Posttraining	World-class endurance runner
Cardiovascular			
HR_{rest} (beats/min)	75	65	45
HR_{max} (beats/min)	185	183	174
SV_{rest} (ml/beat)	60	70	100
SV_{max} (ml/beat)	120	140	200
$\dot{Q}$ at rest (L/min)	4.5	4.5	4.5
$\dot{Q}_{max}$ (L/min)	22.2	25.6	34.8
Heart volume (ml)	750	820	1,200
Blood volume (L)	4.7	5.1	6.0
Systolic BP at rest (mmHg)	135	130	120
Systolic BP_{max} (mmHg)	200	210	220
Diastolic BP at rest (mmHg)	78	76	65
Diastolic BP_{max} (mmHg)	82	80	65
Respiratory			
$\dot{V}_E$ at rest (L/min)	7	6	6
$\dot{V}_{Emax}$ (L/min)	110	135	195
TV at rest (L)	0.5	0.5	0.5
TV_{max} (L)	2.75	3.0	3.9
VC (L)	5.8	6.0	6.2
RV (L)	1.4	1.2	1.2
Metabolic			
$(a-\bar{v})O_2$ diff at rest (ml/100 ml)	6.0	6.0	6.0
$(a-\bar{v})O_2$ diff max (ml/100 ml)	14.5	15.0	16.0
$\dot{V}O_2$ at rest ($ml \cdot kg^{-1} \cdot min^{-1}$)	3.5	3.5	3.5
$\dot{V}O_{2max}$ ($ml \cdot kg^{-1} \cdot min^{-1}$)	40.7	49.9	81.9
Blood lactate at rest (mmol/L)	1.0	1.0	1.0
Blood lactate max (mmol/L)	7.5	8.5	9.0
Body composition			
Weight (kg)	79	77	68
Fat weight (kg)	12.6	9.6	5.1
Fat-free weight (kg)	66.4	67.4	62.9
Fat (%)	16.0	12.5	7.5

Note. HR = heart rate; SV = stroke volume; $\dot{Q}$ = cardiac output; BP = blood pressure; $\dot{V}_E$ = ventilation; TV = tidal volume; VC = vital capacity; RV = residual volume; $(a-\bar{v})O_2$ diff = arterial–mixed venous oxygen difference; $\dot{V}O_2$ = oxygen consumption.

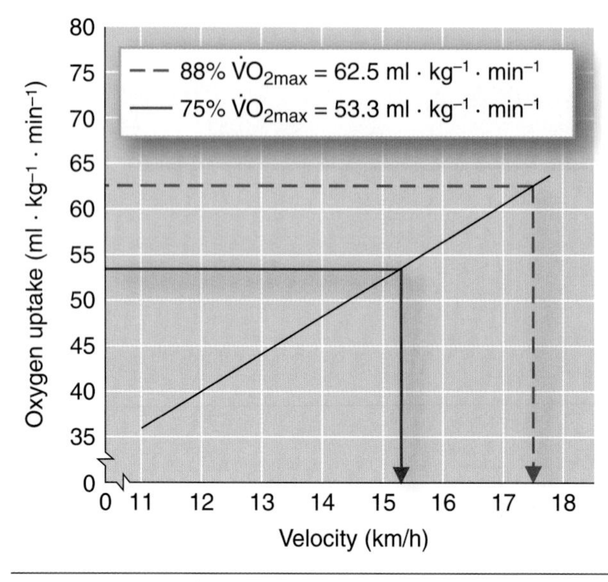

Figure 10.11 Change in race pace with continued training after maximal oxygen uptake stops increasing beyond 71.0 ml · kg^{-1} · min^{-1}.

This ability to sustain exercise at a higher percentage of $\dot{V}O_{2max}$ is the result of an increase in the ability to buffer lactate, because race pace is directly related to the $\dot{V}O_2$ value at which lactate begins to accumulate.

Factors Affecting an Individual's Response to Aerobic Training

We have discussed general trends in adaptations that occur in response to endurance training. However, we must always remember that we are talking about adaptations in individuals and that everyone does not respond in the same manner. Several factors that can affect individual response to aerobic training must be considered.

Level of Conditioning and $\dot{V}O_{2max}$

The higher the initial state of conditioning, the smaller the relative improvement for the same program of training. For example, if two people, one sedentary and the other partially trained, undergo the same endurance training program, the sedentary person will show the greatest relative improvement.

In fully mature athletes, the highest attainable $\dot{V}O_{2max}$ is reached within 8 to 18 months of heavy endurance training, indicating that each athlete has a finite maximal attainable level of oxygen consumption. This finite range may potentially be influenced by training in early childhood during the development of the cardiovascular system.

Heredity

The ability to increase maximal oxygen consumption levels is genetically limited. This does not mean that each individual has a preprogrammed $\dot{V}O_{2max}$ that cannot be exceeded. Rather, a range of $\dot{V}O_{2max}$ values seems to be predetermined by an individual's genetic makeup, and that individual's highest attainable $\dot{V}O_{2max}$ should fall in that range. Each individual is born into a predetermined genetic window, and that individual can shift up or down within that window with exercise training or detraining, respectively.

Research into the genetic basis of $\dot{V}O_{2max}$ began in the late 1960s and early 1970s.[17] Recent research has shown that identical (monozygous) twins have similar $\dot{V}O_{2max}$ values, whereas the variability for dizygous (fraternal) twins is much greater.[5] Figure 10.12 illustrates this. Each symbol represents a pair of brothers. Brother A's $\dot{V}O_{2max}$ value is indicated by the symbol's position on the *x*-axis, and brother B's $\dot{V}O_{2max}$ value is on the *y*-axis. Similarity in the siblings' $\dot{V}O_{2max}$ values is noted by comparing the *x* and *y* coordinates of the symbol (i.e., how close it falls to the diagonal line $x = y$ on the graph). Similar results were found for endurance capacity, determined by the maximal amount of work performed in an all-out, 90 min ride on a cycle ergometer.

Bouchard and colleagues[4] concluded that heredity accounts for between 25% and 50% of the variance in

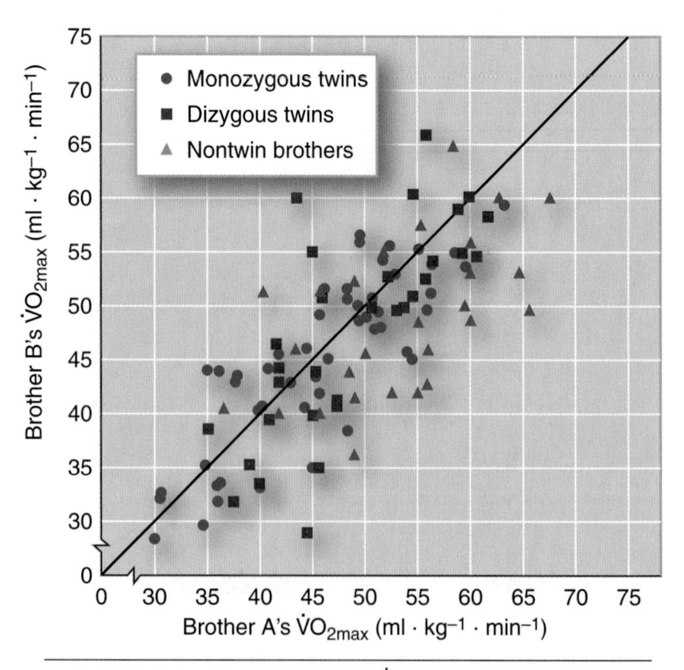

Figure 10.12 Comparisons of $\dot{V}O_{2max}$ in twin (monozygous and dizygous) and nontwin brothers.

Adapted, by permission, from C. Bouchard et al., 1986, "Aerobic performance in brothers, dizygotic and monozygotic twins," *Medicine and Science in Sports and Exercise* 18: 639-646.

$\dot{V}O_{2max}$ values. This means that of all factors influencing $\dot{V}O_{2max}$, heredity alone is responsible for one-quarter to one-half of the total influence. World-class athletes who have stopped endurance training continue for many years to have high $\dot{V}O_{2max}$ values in their sedentary, deconditioned state. Their $\dot{V}O_{2max}$ values may decrease from 85 to 65 ml · kg^{-1} · min^{-1}, but this deconditioned value is still very high.

Heredity also potentially explains the fact that some people have relatively high $\dot{V}O_{2max}$ values yet have no history of endurance training. In a study that compared untrained men who had $\dot{V}O_{2max}$ values below 49 ml · kg^{-1} · min^{-1} with untrained men who had $\dot{V}O_{2max}$ values above 62.5 ml · kg^{-1} · min^{-1}, those with high values were distinguished by having higher blood volume values, leading to higher stroke volume and cardiac output values at maximal rates of exercise. The higher blood volumes in the high $\dot{V}O_{2max}$ group were possibly genetically determined.[19]

In focus

Heredity is a major determinant of aerobic power, accounting for as much as 25% to 50% of the variation in $\dot{V}O_{2max}$.

Thus, both genetic and environmental factors influence $\dot{V}O_{2max}$ values. The genetic factors probably establish the boundaries for the athlete, but endurance training can push $\dot{V}O_{2max}$ to the upper limit of these boundaries. Dr. Per-Olof Åstrand, one of the most highly recognized exercise physiologists during the second half of the 20th century, stated on numerous occasions that the best way to become a champion Olympic athlete is to be selective when choosing one's parents!

Sex

Healthy untrained girls and women have significantly lower $\dot{V}O_{2max}$ values (20-25% lower) than healthy untrained boys and men. Highly conditioned female endurance athletes have values much closer to those of highly trained male endurance athletes (i.e., only about 10% lower). This is discussed in greater detail in chapter 18. Representative ranges of $\dot{V}O_{2max}$ values for athletes and nonathletes are presented in table 10.4 by age, sex, and sport.

High Responders and Low Responders

For years, researchers have found wide variations in improvement of $\dot{V}O_{2max}$ with aerobic training. Studies have demonstrated individual improvements in $\dot{V}O_{2max}$ ranging from 0% to 50% or more, even in similarly fit subjects completing exactly the same training program.

TABLE 10.4 Maximal Oxygen Uptake Values (ml · kg^{-1} · min^{-1}) for Nonathletes and Athletes

Group or sport	Age	Males	Females
Nonathletes	10-19	47-56	38-46
	20-29	43-52	33-42
	30-39	39-48	30-38
	40-49	36-44	26-35
	50-59	34-41	24-33
	60-69	31-38	22-30
	70-79	28-35	20-27
Baseball/Softball	18-32	48-56	52-57
Basketball	18-30	40-60	43-60
Bicycling	18-26	62-74	47-57
Canoeing	22-28	55-67	48-52
Football	20-36	42-60	—
Gymnastics	18-22	52-58	36-50
Ice hockey	10-30	50-63	—
Jockey	20-40	50-60	—
Orienteering	20-60	47-53	46-60
Racquetball	20-35	55-62	50-60
Rowing	20-35	60-72	58-65
Skiing, alpine	18-30	57-68	50-55
Skiing, Nordic	20-28	65-94	60-75
Ski jumping	18-24	58-63	—
Soccer	22-28	54-64	50-60
Speed skating	18-24	56-73	44-55
Swimming	10-25	50-70	40-60
Track and field, discus	22-30	42-55	—
Track and field, running	18-39	60-85	50-75
	40-75	40-60	35-60
Track and field, shot put	22-30	40-46	—
Volleyball	18-22	—	40-56
Weightlifting	20-30	38-52	—
Wrestling	20-30	52-65	—

In the past, exercise physiologists have assumed that these variations result from differing degrees of compliance with the training program. Good compliers should have the highest percentage of improvement, and poor compliers should show little or no improvement—and that certainly is the case. However, given the same training stimulus and full compliance with the program, substantial variations occur in the percentage improvements in $\dot{V}O_{2max}$ values of different people.

It is now evident that the response to a training program is also genetically determined.[2] This is illustrated in figure 10.13. Ten pairs of identical twins completed a 20-week endurance training program; the improvements in $\dot{V}O_{2max}$, expressed as percentages, are plotted for each twin pair—Twin A on the *x*-axis and Twin B on the *y*-axis.[24] Notice the similarity in response for each twin pair. Yet across twin pairs, improvement in $\dot{V}O_{2max}$ varied from 0% to 40%. These results, and those from other studies, indicate that there will be **high responders** (large improvement) and **low responders** (little or no improvement) among groups of people who participate in identical training programs.

Results from the HERITAGE Family Study also support a strong genetic component affecting the magnitude of increase in $\dot{V}O_{2max}$ with endurance training. Families, including the natural mother and father and three or more of their children, trained three days a week for 20 weeks, initially exercising at a heart rate equal to 55% of their $\dot{V}O_{2max}$ for 35 min per day and progressing to a heart rate equal to 75% of their $\dot{V}O_{2max}$ for 50 min per day by the end of the 14th week, which they maintained for the last six weeks.[3] The average increase in $\dot{V}O_{2max}$ was about 17% but varied from 0% to more than 50%. Figure 10.14 illustrates the improvement in $\dot{V}O_{2max}$ for each subject in each family. Maximal heritability was estimated at 47%. Note that subjects who are

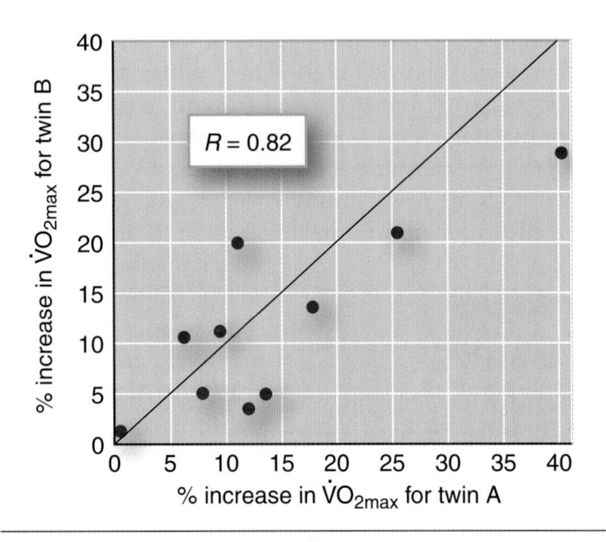

Figure 10.13 Variations in the percentage increase in $\dot{V}O_{2max}$ for identical twins undergoing the same 20-week training program.

From D. Prud'homme et al., 1984, "Sensitivity of maximal aerobic power to training is genotype-dependent," *Medicine and Science in Sports and Exercise* 16(5): 489-493. Copyright 1984 by American College of Sports Medicine. Adapted by permission.

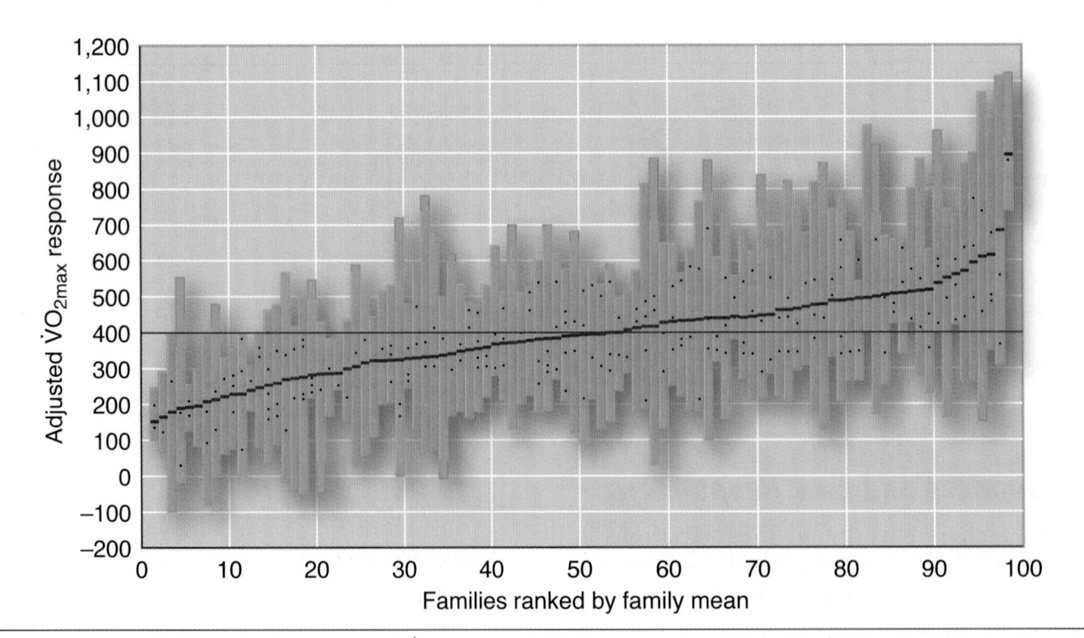

Figure 10.14 Variations in the improvement in $\dot{V}O_{2max}$ following 20 weeks of endurance training by family (the HERITAGE Family Study). Values represent the changes in $\dot{V}O_{2max}$ in ml/min, with an average increase of 393 ml/min. Data for each family are enclosed within a bar, and each family member's value is represented as a dot within the bar.

Adapted, by permission, from C. Bouchard et al., 1999, "Familial aggregation of $\dot{V}O_2$max response to exercise training. Results from HERITAGE Family Study," *Journal of Applied Physiology* 87: 1003-1008.

In focus

Individual differences cause substantial variation in subjects' responses to a given training program. Genetics accounts for much of this variation in response.

high responders tend to be clustered in the same families as those who are low responders.

It is clear that this is a genetic phenomenon, not a result of compliance or noncompliance. One must consider this important point when conducting training studies and designing training programs. Individual differences must be accounted for.

Cardiorespiratory Endurance and Performance

Many people regard cardiorespiratory endurance as the most important component of physical fitness. It is an athlete's major defense against fatigue. Low endurance capacity leads to fatigue, even in the more sedentary sports or activities. For any athlete, regardless of the sport or activity, fatigue represents a major deterrent to optimal performance. Even minor fatigue can hinder the athlete's total performance:

Individual Differences in Response to Training: The HERITAGE Family Study

It has been clearly established that individuals differ considerably in their responses to a specific intervention, such as a drug or diet. The same is true for the response of $\dot{V}O_{2max}$ to aerobic training, with studies showing improvements ranging from 0% to 50% or more in response to exactly the same training program.

The HERITAGE Family Study was funded by the National Institutes of Health in 1992 to study the possible genetic control of these variations in $\dot{V}O_{2max}$ in response to aerobic training and the associated variations in the major risk factors for cardiovascular disease and type 2 diabetes. Funding continued through the year 2004. Families were recruited for this study, including the natural mother and natural father and three or more of their children. Some families were included that did not meet these criteria.

Dr. Claude Bouchard.

Led by Dr. Claude Bouchard, executive director of the Pennington Biomedical Research Center, Louisiana State University, and a pioneer in genetics research in the exercise sciences, a team of researchers from several universities (Arizona State University/Indiana University, Laval University in Canada, the University of Minnesota, and the University of Texas at Austin) recruited a total of 742 sedentary subjects from families of both white and black descent who completed the study. Another university, Washington University, was responsible for data quality control and analyses. The subjects completed a comprehensive battery of tests both before and after a 20-week program of aerobic training, with each training session supervised by an exercise physiologist. The test battery included physiological measures associated with aerobic fitness and clinical medicine markers associated with risk for cardiovascular disease and diabetes. The average increase in $\dot{V}O_{2max}$ expressed per kilogram of body weight was 18%, but the range of increases varied from 0% to 53%. The increase in $\dot{V}O_{2max}$ was influenced by genetics (maximal heritability = 47%; see figure 10.14) but was influenced very little by age, sex, and race.[3] It is important to recognize from these data that each individual responds differently to the exact same exercise stimulus. We cannot expect the same improvement in all people! The mere fact that a person has a low response to training does not mean the person did not follow the training program. For more information on the HERITAGE Family Study, go to its Web site (www.pbrc.edu/heritage/home.htm).

- Muscular strength is decreased.
- Reaction and movement times are prolonged.
- Agility and neuromuscular coordination are reduced.
- Whole-body movement speed is slowed.
- Concentration and alertness are reduced.

The decline in concentration and alertness associated with fatigue is particularly important. The athlete can become careless and more prone to serious injury, especially in contact sports. Even though these decrements in performance might be small, they can be just enough to cause an athlete to miss the critical free throw in basketball, the strike zone in baseball, or the 20 ft (6 m) putt in golf.

Summary of Cardiovascular Adaptation to Chronic Endurance Exercise

Physiologists commonly establish models to help explain how various physiological factors or variables work together to affect a specific outcome or component of performance. Dr. Donna H. Korzik, an exercise physiologist at The Pennsylvania State University, has created a unifying figure to model the factors that contribute to the cardiovascular adaptation to chronic endurance training (see the figure in this sidebar).

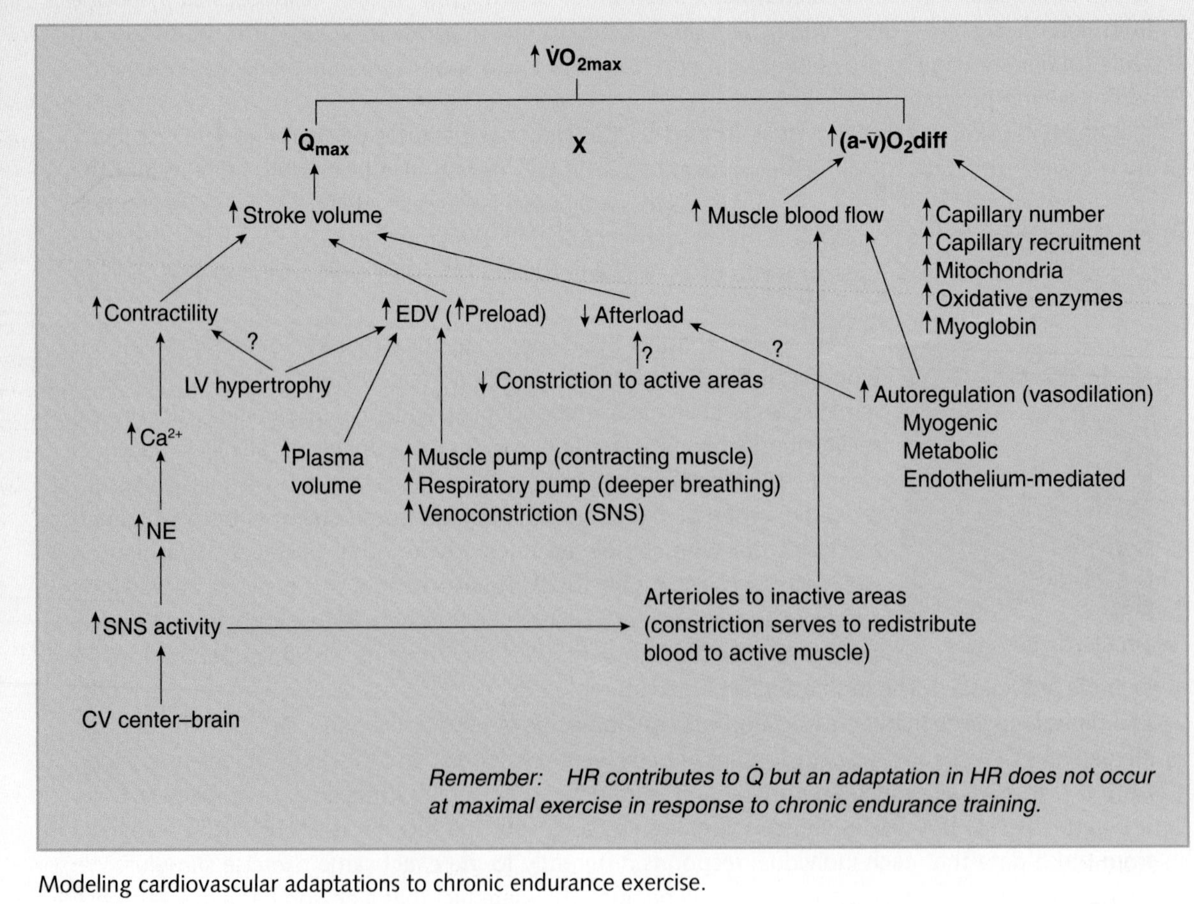

Modeling cardiovascular adaptations to chronic endurance exercise.

Adapted, by permission, from Donna H. Korzick, Pennsylvania State University, 2006.

All athletes can benefit from maximizing their endurance. Even golfers, whose sport is relatively sedentary, can improve. Improved endurance can allow golfers to complete a round of golf with less fatigue and to better withstand long periods of walking and standing.

For the sedentary, middle-aged adult, numerous health factors indicate that cardiovascular endurance should be the primary emphasis of training. Training for health and fitness is discussed at length in part VII of this book.

The extent of endurance training needed varies considerably from one athlete to the next. It depends on the athlete's current endurance capacity and the endurance demands of the chosen activity. The marathon runner uses endurance training almost exclusively, with limited attention to strength, flexibility, and speed. The baseball player, however, places very limited demands on endurance capacity, so endurance conditioning is not as highly emphasized. Nevertheless, baseball players could gain substantially from endurance running, even if only at a moderate pace for 5 km (3.1 mi) per day, three days a week. As a benefit of training, baseball players would have little or no leg trouble (a frequent complaint), and they would be able to complete a doubleheader with little or no fatigue.

Adequate cardiovascular conditioning must be the foundation of any athlete's general conditioning program. Many athletes in nonendurance activities never incorporate even moderate endurance training into their training programs. Those who have done so are well aware of their improved physical condition and its impact on their athletic performance.

In review

- Although $\dot{V}O_{2max}$ has an upper limit, endurance performance can continue to improve for years with continued training.
- An individual's genetic makeup predetermines a range for his or her $\dot{V}O_{2max}$ and accounts for 25% to 50% of the variance in $\dot{V}O_{2max}$ values. Heredity also largely explains individual variations in response to identical training programs.
- Highly conditioned female endurance athletes have $\dot{V}O_{2max}$ values only about 10% lower than those of highly conditioned male endurance athletes.
- All athletes can benefit from maximizing their cardiorespiratory endurance.

ADAPTATIONS TO ANAEROBIC TRAINING

In muscular activities that require near-maximal force production, such as sprinting, cycling, and swimming, much of the energy needs are met by the ATP-phosphocreatine (PCr) system and the anaerobic breakdown of muscle glycogen (glycolysis). The following sections focus on the trainability of these two systems.

Changes in Anaerobic Power and Anaerobic Capacity

Exercise scientists have had difficulty agreeing on an appropriate laboratory or field test of anaerobic power. Unlike the situation with aerobic power, for which $\dot{V}O_{2max}$ is generally agreed to be the gold standard measurement, no single test adequately measures anaerobic power. Most research has been conducted through use of three different tests of either or both anaerobic power and anaerobic capacity: the Wingate anaerobic test, the critical power test, and the maximal accumulated oxygen deficit test (see chapter 4). Of these three, the Wingate test has been the most widely used.

With the Wingate anaerobic test, the subject pedals a cycle ergometer at maximal speed for 30 s against a high braking force. The braking force is determined by the person's weight, sex, age, and level of training. Power output can be determined instantaneously throughout the 30 s test but is generally averaged over 3 to 5 s intervals. The peak power output is the highest mechanical power achieved at any stage in the test; it is generally achieved during the first 5 to 10 s and is considered an index of anaerobic power. The mean power output is computed as the average power output over the total 30 s period, and one obtains total work simply by multiplying the mean power output by 30 s. Mean power output and total work have both been used as indexes of anaerobic capacity.

With anaerobic training, such as sprint training on the track or on a cycle ergometer, there are increases in both peak anaerobic power and anaerobic capacity. However, results have ranged widely across studies, from those that showed only minimal increases to those showing increases of up to 25%.

Adaptations in Muscle With Anaerobic Training

With anaerobic training, which includes sprint training and resistance training, there are changes in skeletal muscle that specifically reflect muscle fiber recruitment for these types of activities. As discussed in chapter 1, at higher intensities, type II muscle fibers are recruited to a greater extent, but not exclusively, because type I fibers continue to be recruited. Overall, sprint and resistance activities use the type II muscle fibers significantly more than do aerobic activities. Consequently, both type IIa and type IIx muscle fibers undergo an increase in their cross-sectional areas. The cross-sectional area of type I fibers also is increased but usually to a lesser extent. Furthermore, with sprint training there appears to be a reduction in the percentage of type I fibers and an increase in the percentage of type II fibers, with the greatest change in type IIa fibers. In two of these studies, in which subjects performed 15 s or 15 s and 30 s all-out sprints, the type I percentage decreased from 57% to 48% and type IIa increased from 32% to 38%.[15,16] This shift of type I to type II fibers usually is not seen with resistance training.

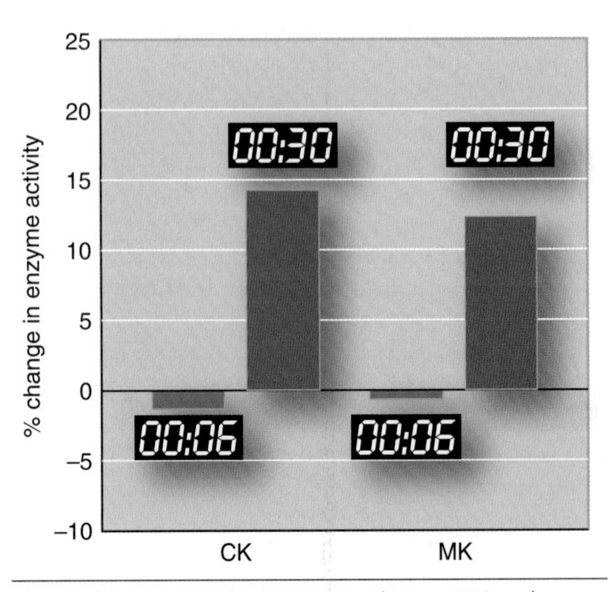

Figure 10.15 Changes in creatine kinase (CK) and muscle myokinase (MK) activities as a result of 6 s and 30 s bouts of maximal anaerobic training.

Adaptations in the Energy Systems

Just as aerobic training produces changes in the aerobic energy system, anaerobic training alters the ATP-PCr and anaerobic glycolytic energy systems. These changes are not as obvious or predictable as those that result from endurance training, but they do improve performance in anaerobic activities.

Adaptations in the ATP-PCr System

Activities that emphasize maximal muscle force production, such as sprinting and weightlifting events, rely most heavily on the ATP-PCr system for energy. Maximal effort lasting less than about 6 s places the greatest demands on the breakdown and resynthesis of ATP and PCr. Costill and coworkers reported their findings from a study of resistance training and its effects on the ATP-PCr system.[6] Their participants trained by performing maximal knee extensions. One leg was trained using 6 s maximal work bouts that were repeated 10 times. This type of training preferentially stressed the ATP-PCr energy system. The other leg was trained with repeated 30 s maximal bouts, which instead preferentially stressed the glycolytic system.

The two forms of training produced the same muscular strength gains (about 14%) and the same resistance to fatigue. As seen in figure 10.15, the activities of the anaerobic muscle enzymes creatine kinase and myokinase increased as a result of the 30 s training bouts but were almost unchanged in the leg trained with repeated 6 s maximal efforts. This finding leads us to conclude that maximal sprint bouts (6 s) might improve muscular strength but contribute little to the mechanisms responsible for ATP and PCr breakdown. Data have been published, however, that show improvements in ATP-PCr enzyme activities with training bouts lasting only 5 s.

Regardless of the conflicting results, these studies suggest that the major value of training bouts that last only a few seconds (sprints) is the development of muscular strength. Such strength gains enable the individual to perform a given task with less effort, which reduces the risk of fatigue. Whether these changes allow the muscle to perform more anaerobic work remains unanswered, although a 60 s sprint-fatigue test suggests that short sprint-type anaerobic training does not enhance anaerobic endurance.[6]

Adaptations in the Glycolytic System

Anaerobic training (30 s bouts) increases the activities of several key glycolytic enzymes. The most frequently studied glycolytic enzymes are phosphorylase, phosphofructokinase (PFK), and lactate dehydrogenase. The activities of these three enzymes increased 10% to 25% with repeated 30 s training bouts but changed little with short (6 s) bouts that stress primarily the ATP-PCr system.[6] In a more recent study, 30 s maximal all-out sprints significantly increased hexokinase (56%) and PFK (49%) but not total phosphorylase activity or lactate dehydrogenase.[18]

Because both PFK and phosphorylase are essential to the anaerobic yield of ATP, such training might enhance glycolytic capacity and allow the muscle to develop greater tension for a longer period of time. However, as seen in figure 10.16, this conclusion is not supported by results of a 60 s sprint performance test, in which the

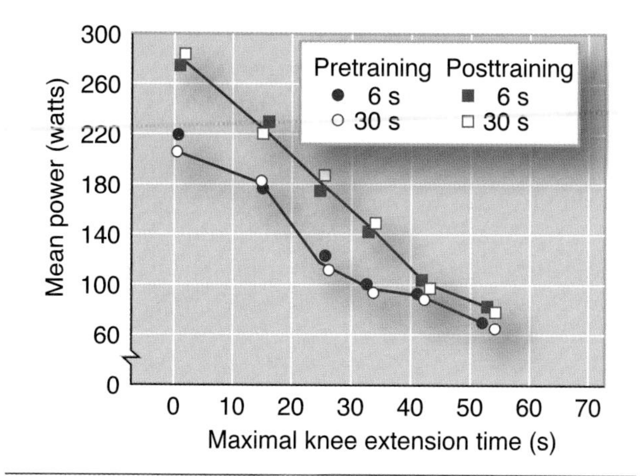

Figure 10.16 Performance in a 60 s sprint bout after training with 6 s and 30 s anaerobic bouts. Subjects are the same as in figure 10.15.

In focus

Anaerobic training increases the ATP-PCr and glycolytic enzymes but has no effect on the oxidative enzymes. Conversely, aerobic training increases the oxidative enzymes but has little effect on the ATP-PCr or glycolytic enzymes. This fact reinforces a recurring theme: Physiological alterations that result from training are highly specific to the type of training.

subjects performed maximal knee extension and flexion. Power output and the rate of fatigue (shown by a decrease in power production) were affected to the same degree after sprint training with either 6 or 30 s training bouts. Thus, we must conclude that performance gains with these forms of training result from improvements in strength rather than improvements in the anaerobic yield of ATP.

In review

- Anaerobic training bouts improve both anaerobic power and anaerobic capacity.
- The performance improvement noted with sprint-type anaerobic training appears to result more from strength gains than from improvements in the functioning of the anaerobic energy systems.
- Anaerobic training increases the ATP-PCr and glycolytic enzymes but has no effect on the oxidative enzymes.

SPECIFICITY OF TRAINING AND CROSS-TRAINING

Physiological adaptations in response to physical training are highly specific to the nature of the training activity. Furthermore, the more specific the training program is to a given sport or activity, the greater the improvement in performance in that sport or activity. The concept of **specificity of training** is very important for all physiological adaptations.

This concept is also important in testing of athletes. As an example, to accurately measure endurance improvements, athletes should be tested while engaged in an activity similar to the sport or activity in which they usually participate. Consider one study of highly trained rowers, cyclists, and cross-country skiers. Their $\dot{V}O_{2max}$ values were tested while they performed two types of work: uphill running on a treadmill and maximal performance of their specific sport activity.[29] The important finding, shown in figure 10.17, was that the $\dot{V}O_{2max}$ values attained by all the athletes during their sport-specific activity were as high as or higher than the values obtained on the treadmill. For many of these athletes, $\dot{V}O_{2max}$ values were substantially higher during their sport-specific activity.

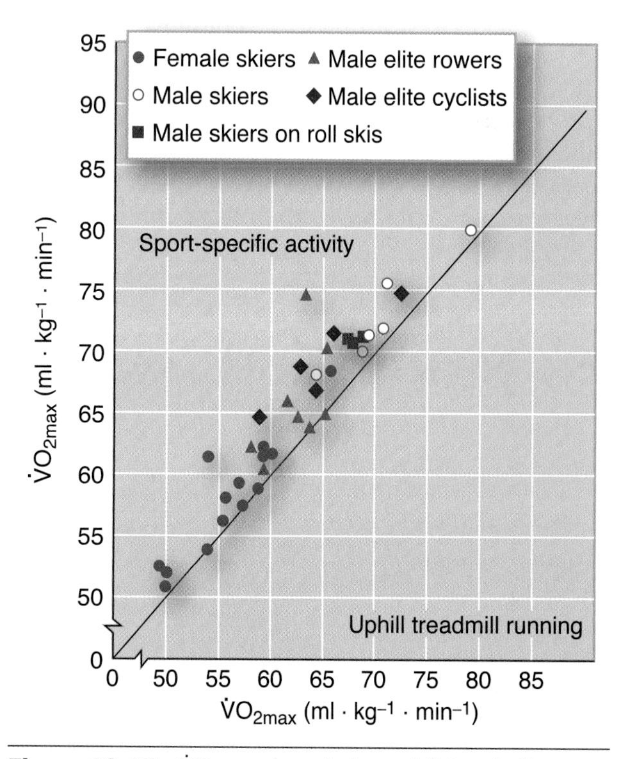

Figure 10.17 $\dot{V}O_{2max}$ values during uphill treadmill running versus sport-specific activities in selected groups of athletes.

Adapted, by permission, from S.B. Strømme, F. Ingjer, and H.D. Meen, 1977, "Assessment of maximal aerobic power in specifically trained athletes," *Journal of Applied Physiology* 42: 833-837.

A highly creative design for studying the concept of specificity of training involves one-legged exercise training, with the untrained opposite leg used as the control. In one study, subjects were placed in three groups: a group that sprint trained one leg and endurance trained the other, a group that sprint trained one leg and did not train the other, and a group that endurance trained one leg and did not train the other.[26] Improvement in $\dot{V}O_{2max}$ and lowered heart rate and blood lactate response at submaximal work rates were found only when exercise was performed with the endurance-trained leg.

Much of the training response occurs in the specific muscles that have been trained, possibly even in individual motor units in a specific muscle. This observation applies to metabolic as well as cardiorespiratory responses to training. Table 10.5 shows the activities of selected muscle enzymes from the three energy systems for untrained, anaerobically trained, and aerobically trained men. The table shows that aerobically trained muscles have significantly lower glycolytic enzyme activities. Thus, they might have less capacity for anaerobic metabolism or might rely less on energy from glycolysis. More research is needed to explain the implications of the muscular changes accompanying both anaerobic and aerobic training, but this table clearly illustrates the high degree of specificity to a given training stimulus.

In focus

Close attention must be given to selecting an optimal training program. The program must be carefully matched with the athlete's individual needs to maximize the physiological adaptations to training, thereby optimizing performance.

Cross-training refers to training for more than one sport at the same time or training for several different fitness components (such as endurance, strength, and flexibility) at one time. The athlete who trains by swimming, running, and cycling in preparation for competing in a triathlon is an example of the former, and the athlete involved in heavy resistance training and high-intensity cardiorespiratory training at the same time is an example of the latter.

For the athlete training for cardiorespiratory endurance and strength at the same time, the studies conducted to date indicate that gains in strength, power, and endurance can result. However, the gains in muscular strength and power are less when strength training is combined with endurance training than when strength training alone is done. The opposite does not appear to be true: Improvement of aerobic power with endurance training does not appear to be attenu-

TABLE 10.5 Selected Muscle Enzyme Activities (mmol · g^{-1} · min^{-1}) for Untrained, Anaerobically Trained, and Aerobically Trained Men

	Untrained	Anaerobically trained	Aerobically trained
	Aerobic enzymes		
Oxidative system			
Succinate dehydrogenase	8.1	8.0	20.8[a]
Malate dehydrogenase	45.5	46.0	65.5[a]
Carnitine palmityl transferase	1.5	1.5	2.3[a]
	Anaerobic enzymes		
ATP-PCr system			
Creatine kinase	609.0	702.0[a]	589.0
Myokinase	309.0	350.0[a]	297.0
Glycolytic system			
Phosphorylase	5.3	5.8	3.7[a]
Phosphofructokinase	19.9	29.2[a]	18.9
Lactate dehydrogenase	766.0	811.0	621.0

[a]Significant difference from the untrained value.

ated by inclusion of a resistance training program. In fact, short-term endurance can be increased with resistance training. Although earlier studies supported the conclusion that concurrent strength and endurance training limits gains in strength and power, one well-controlled study did not show this. McCarthy and colleagues[20] reported similar gains in strength, muscle hypertrophy, and neural activation in a group of previously untrained subjects who underwent concurrent high-intensity strength training and cycle endurance training compared with a group who performed only high-intensity strength training.

In review

- For athletes to maximize cardiorespiratory gains from training, the training should be specific to the type of activity that an athlete usually performs.
- Resistance training in combination with endurance training does not appear to restrict improvement in aerobic power and may increase short-term endurance, but it can limit improvement in strength and power when compared with gains from resistance training alone.

In Closing

In this chapter, we examined how the cardiovascular, respiratory, and metabolic systems adapt to aerobic and anaerobic training. We concentrated on how these adaptations can improve both aerobic and anaerobic performance. This chapter concludes our review of how body systems respond to both acute and chronic exercise. Now that we have completed our examination of how the body responds to internal changes, we can turn our attention to the external world. In the next part of the book, we focus on the body's adaptations to varying environmental conditions, beginning in the next chapter by considering how external temperature can affect performance.

KEY TERMS

aerobic training
anaerobic training
athlete's heart
capillary-to-fiber ratio
cardiac hypertrophy
cardiorespiratory endurance
cross-training
Fick equation
glycogen sparing
high responders
low responders
mitochondrial oxidative enzymes
muscular endurance
myoglobin
oxygen transport system
respiratory exchange ratio (RER)
specificity of training
submaximal endurance capacity
$\dot{V}O_{2max}$

STUDY QUESTIONS

1. Differentiate between muscular endurance and cardiovascular endurance.
2. What is maximal oxygen uptake ($\dot{V}O_{2max}$)? How is it defined physiologically, and what determines its limits?
3. Of what importance is $\dot{V}O_{2max}$ to endurance performance? Why does the competitor with the highest $\dot{V}O_{2max}$ not always win?
4. Describe the changes in the oxygen transport system that occur with endurance training.
5. What is possibly the most important adaptation that the body makes in response to endurance training, which allows for an increase in both $\dot{V}O_{2max}$ and performance? Through what mechanisms do these changes occur?
6. What metabolic adaptations occur in response to endurance training?
7. Explain the two theories that have been proposed to account for improvements in $\dot{V}O_{2max}$ with endurance training. Which of these has the greatest validity today? Why?
8. How important is genetic potential in a developing young athlete?
9. What adaptations have been shown to occur in muscle fibers with anaerobic training?

10. Discuss specificity of anaerobic training with respect to enzyme changes in muscle.

11. Why is cross-training beneficial to endurance athletes? How does it benefit sprint and power athletes?

STUDY GUIDE ACTIVITIES

In addition to the activities listed in the chapter opening outline on page 221, two other activities are available in the online study guide, located at

www.HumanKinetics.com/PhysiologyOfSportAndExercise.

The chapter **SUMMARY** reviews the key concepts covered in the chapter, and the end-of-chapter **QUIZ** tests your understanding of the material covered in the chapter.

Environmental Influences on Performance

In previous sections, we discussed how the various body systems coordinate their activities to allow us to perform physical activity. We also saw how these systems adapt when exposed to the stress of various types of training. In part IV, we turn our attention to how the body responds and adapts when challenged to exercise under unusual environmental conditions. In chapter 11, "Exercise in Hot and Cold Environments: Thermoregulation," we examine the mechanism by which the body can regulate its internal temperature both at rest and during exercise. Then we consider how the body responds and adapts to exercise in the heat and cold, along with health risks that are associated with physical activity in each environment. In chapter 12, "Exercise at Altitude," we discuss the unique challenges that the body faces when performing physical activity under conditions of reduced atmospheric pressure (altitude) and how the body adapts to extended periods at altitude. We then discuss the best way to prepare for competing at altitude and how the use of altitude training might help people perform better at sea level.

CHAPTER 11

Exercise in Hot and Cold Environments

Thermoregulation

In this chapter and in the online study guide

Korey Stringer, a 335 lb (152 kg) right tackle for the Minnesota Vikings professional football team, was one of the National Football League's top offensive linemen. On the first day of training camp in 2001, Korey failed to complete the full team workout, leaving the practice field with symptoms of severe heat strain, including nausea, dizziness, and vomiting. On the second day of camp, July 31, 2001, Stringer collapsed on the practice field during an intense practice at the Vikings' training facility in Mankato, Minnesota. He had been training on a hot, cloudless day wearing a full football uniform and helmet. Having lost consciousness, he was taken to an air-conditioned trailer and eventually was taken by ambulance to a local hospital, where a core body temperature of 108.8 °F was recorded. He died approximately 13 h later—a victim of heatstroke. Athletes are susceptible to heatstroke, particularly athletes who start training for their sport during the hottest months of the year. In addition to a lack of acclimation to the heat, the padding and uniforms worn by football players limit their ability to lose heat, and this strain on the body may be enhanced if they do not drink sufficient fluids.

Source: Sports Illustrated; July 29, 2002; 97 (4): 56-60.

CHAPTER 12

Exercise at Altitude

In this chapter and in the online study guide

Athletic competitions held at high-altitude venues have traditionally yielded poorer performances. As a result, there were many complaints when it was announced that the 1968 Olympic Games would be held in Mexico City, at an altitude of 2,240 m (7,350 ft) above sea level. Ethiopian Mamo Wolde won the marathon, but his time of 2:20:26 was slower than those of previous Olympic winners. Australian distance runner Ron Clarke was the favorite and world-record holder in the 10,000 m. With two laps to go in Mexico City, Clarke had positioned himself for a final burst to the finish; however with 500 m to go, he began to stagger, dropped to sixth place, then collapsed unconscious after finishing. But at least two athletes who participated in those games were glad they had performed in the rarified air of Mexico City. Bob Beamon soared almost 0.6 m (2 ft) farther than the previous world record in the long jump, and Lee Evans beat the world record in the 400 m run by nearly 0.24 s. These records stood for nearly 20 years, leading some sport scientists to suggest that the low air density that accompanies the high-altitude conditions of Mexico City likely contributed to the stellar performances in these relatively short-duration, explosive events.

In Closing

Activities are seldom conducted under ideal environmental conditions. Heat, cold, humidity, and altitude—alone or in combination—present unique problems that are superimposed on the physiological demands of exercise. This chapter and the preceding chapter summarized the nature of these common environmental stresses and how we can cope with them.

Much of our discussion thus far has dealt with how physiological variables and environmental stress can hinder our performance. In the next part of the book, we examine various ways to optimize performance. We begin by looking at the importance of the amount of training, considering what happens when we train either too much or too little.

KEY TERMS

acute altitude (mountain) sickness
barometric pressure
Cheyne-Stokes breathing
erythropoietin (EPO)
high-altitude cerebral edema (HACE)
high-altitude pulmonary edema (HAPE)
hypobaric
hypoxemia
hypoxia
partial pressure of oxygen (PO_2)
polycythemia
respiratory alkalosis

STUDY QUESTIONS

1. Describe the conditions at altitude that could limit physical activity.
2. What types of activities are detrimentally influenced by exposure to high altitude and why?
3. When someone ascends to an altitude of over 1,500 m, describe the physiological adjustments that occur within the first 24 h.
4. Describe the physiological adjustments that accompany acclimatization to altitude over a period of three weeks.
5. Would an endurance athlete who trained at altitude be able to perform better during subsequent sea-level performance? Why or why not?
6. Describe the theoretical advantage of living high and training low.
7. What are the best strategies for preparing athletes for high-altitude competition?
8. What are the health risks associated with acute exposure to high altitude?

STUDY GUIDE ACTIVITIES

In addition to the activities listed in the chapter opening outline on page 279, two other activities are available in the online study guide, located at

www.HumanKinetics.com/PhysiologyOfSportAndExercise.

The chapter **SUMMARY** reviews the key concepts covered in the chapter, and the end-of-chapter **QUIZ** tests your understanding of the material covered in the chapter.

Optimizing Performance in Sport

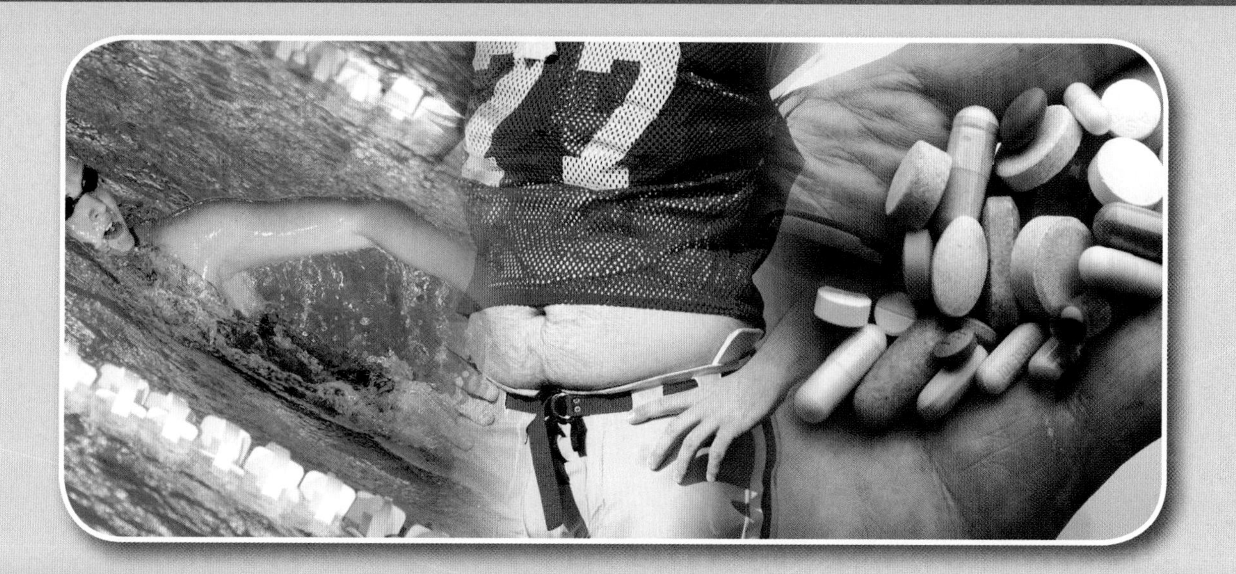

Previous chapters have explained how the body responds to an acute bout of exercise, how it adapts to chronic training, and how it adjusts to environmental extremes. We can now apply that knowledge to optimizing athletic performance. In part V, we focus on how athletes can best prepare for competition. In chapter 13, "Training for Sport," we discuss how to optimize the athlete's training stress and explore how too much or too little training can impair performance. In chapter 14, "Body Composition and Nutrition for Sport," we address the issues of assessing body composition, relating body composition to sport performance, and the use of weight standards. We then evaluate the athlete's dietary needs and consider how nutritional supplementation and dietary manipulation might improve performance. In chapter 15, "Ergogenic Aids and Sport," we discuss various pharmacological, hormonal, and physiological agents that have been proposed to improve performance. We examine the potential benefits, proven effects, and health risks associated with their use.

Training for Sport

In this chapter and in the online study guide

Throughout his college career, Eric had trained at swimming 4 h each day, covering as much as 13.7 km (8.5 mi) per day. Despite this effort, his performance time for the 200 yd (183 m) butterfly event had not improved since his freshman year. With a best performance of 2 min 15 s for the event, he was seldom given a chance to compete because several teammates could perform the event in less than 2 min 5 s. During Eric's senior year, his coach made a major change in the team's training plan. The swimmers trained only 2 h per day and swam an average of 4.5 to 4.8 km (2.8-3.0 mi) per day. Further, they swam each interval at a faster pace and had a longer rest period between intervals. Suddenly Eric's performance began to improve. After three months, his time dropped to 2 min 10 s, still not good enough to make him a major contender. But as a reward for Eric's improvement, the coach chose him to swim the 200 yd butterfly event at the conference championship meet, which was preceded by three weeks of reduced training (tapering) of only 1.6 km (1 mi) per day. Subsequently, with less volume of training than in previous years and well rested after the taper, Eric was able to make the finals of the event at the championship meet. His preliminary time was 2 min 1 s. In the finals, he improved even further, posting a third-place finish with a time of 1 min 57.7 s, an impressive performance for a swimmer who performed better with lower-volume but higher-quality training.

Repeated days and weeks of training can be considered positive stress because training improves one's capacity for energy production, tolerance of physical stress, and exercise performance. The major physical changes associated with training occur in the first 6 to 10 weeks. The magnitude of these adaptations depends on the volume and intensity of exercise performed during training, which has led many coaches and athletes to believe erroneously that the athlete who undertakes the greatest volume (quantity) and intensity (quality) of training will be the best performer. Unfortunately, we often mistakenly consider quantity and quality of training to be synonymous. Too often, training sessions are judged by the total volume (e.g., distance run, cycled, or swum) performed in each training session, leading coaches to design training programs that are not specific for the sport and often impose unrealistic demands on the athlete.

> **In focus**
>
> A person's rate of adaptation to training is genetically limited and cannot be forced beyond his or her body's capacity for development. Each individual responds differently to the same training stress, so what might be excessive training for one person could be well below the capacity of another. Therefore, it is important that trainers recognize and account for individual differences when designing training programs.

The rate at which an individual adapts to training is limited and cannot be forced beyond the body's genetic capacity for development. Too much training can reduce the athlete's optimal potential for improvement and in some cases can cause a breakdown in the adaptation process, eventually reducing performance. When training is taken to extremes, serious illness or injury can occur.

Although the volume of work performed in training is an important stimulus for physical conditioning, a proper balance should be established between volume and intensity. Training can be overdone, leading to chronic fatigue, illness, overuse injury, overtraining syndrome, and performance decrements. In contrast, proper rest and reaching the appropriate balance between training volume and intensity can enhance performance. Much effort has been directed toward determining the appropriate volume and intensity required to achieve optimal adaptation. Exercise physiologists have tested numerous training regimens to determine both the minimal and maximal stimuli needed for cardiovascular and muscular improvements. The next section examines those factors that can affect the response to a given training program, developing a model for optimizing the training stimulus.

OPTIMIZING TRAINING: A MODEL

All well-designed training programs incorporate the principle of progressive overload. As discussed in chapter 8, this principle holds that to continue to accrue the benefits of training, the training stimulus must be progressively increased as the body adapts to the current stimulus. The only way to continue to improve with training is to progressively increase the training stimulus. When this concept is carried too far, the training may become excessive, pushing the body beyond its ability to adapt, producing no additional improvement in conditioning or performance, and can lead to performance decrements. Conversely, if the volume or the intensity of training is too low, optimal performance will not be achieved. Thus, the coach and athlete face the challenge of determining the optimal training stimulus for each particular athlete, recognizing that what works for one athlete might not work for another.

Figure 13.1 provides a model demonstrating the continuum of training stages that a competitive ath-

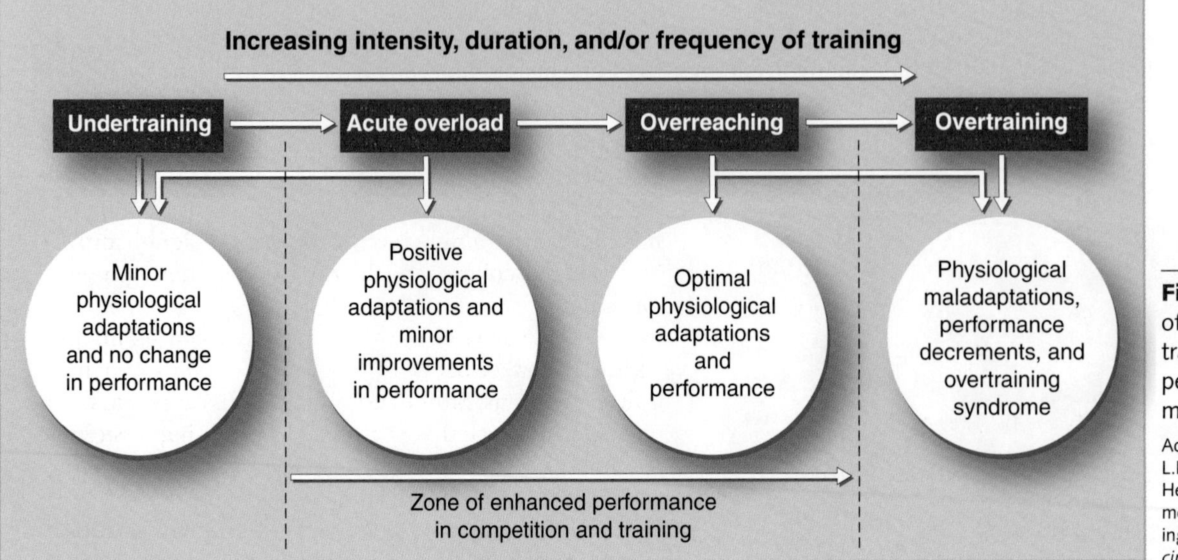

Figure 13.1 Model of the continuum of training stages in a periodized training mesocycle.

Adapted, by permission, from L.E. Armstrong and J.L. VanHeest, 2002, "The unknown mechanism of the overtraining syndrome," *Sports Medicine* 32(1): 185-209.

lete might go through during a full year. This model is based on the principle of periodization, which was described in chapter 8 and which is illustrated and described in figure 13.2. In this model, **undertraining** represents the type of training an athlete would undertake between competitive seasons or during active rest. Generally, physiological adaptations will be minor, and there will be no improvement in performance. **Acute overload** represents what might be considered an "average" training load, whereby the athlete is stressing the body to the extent necessary to improve both physiological function and performance. **Overreaching** is a relatively new term that refers to a brief period of heavy overload without adequate recovery, thus exceeding the athlete's adaptive capacity. There will be a brief performance decrement, from several days to several weeks, but eventually performance will improve. Finally, **overtraining** refers to that point at which an athlete experiences physiological maladaptations and chronic performance decrements. This generally leads to the **overtraining syndrome.**[1] **Excessive training,** not shown in the model, refers to training that is well above what is needed for peak performance but does not strictly meet the criteria for either overreaching or overtraining.

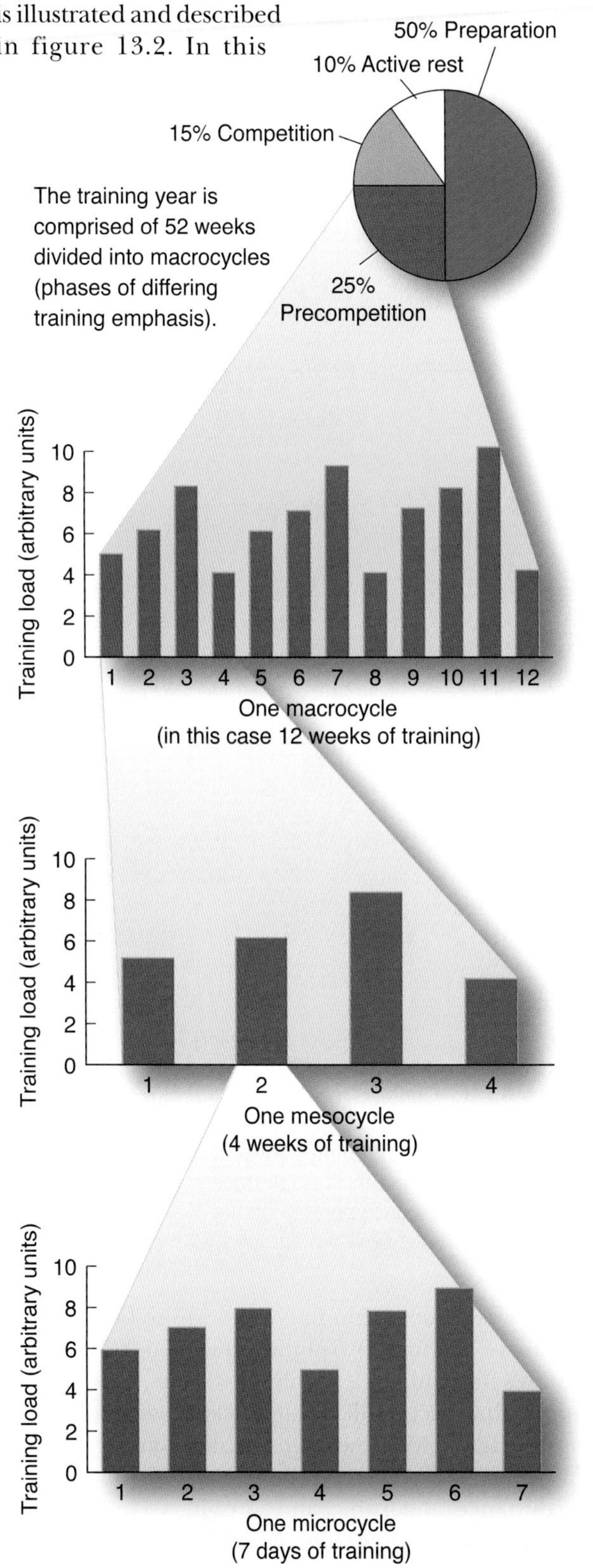

Figure 13.2 The structure of a periodized training program.

Adapted, by permission, from R.W. Fry, A.R. Morton, and D. Keast, 1991, "Overtraining in athletes: An update," *Sports Medicine* 12: 32-65.

EXCESSIVE TRAINING

With excessive training, either or both volume of training and intensity of training are increased to extreme levels. The "more is better" philosophy drives the training schedule. For many years, athletes were undertrained. As coaches and athletes became bolder and started to push the envelope by increasing both training volume and intensity, they found that athletes responded well, and world records began to tumble. However, one can only take this philosophy so far. At a certain point, performance begins to either plateau or decline. Let's take a look at some examples of this.

Most of the research on excessive training has been conducted on swimmers. For that reason, the material in this section concerns swimmers, but it also applies to most other forms of training.

One can increase training volume by increasing either the duration or the frequency of training bouts. But does increased volume translate into increased performance? Research shows that swim training 3 to 4 h per day, five or six days each week, provides no greater benefits than training only 1 to 1.5 h per day.[6] In fact, such excessive training has been shown to significantly decrease muscular strength and sprint swimming performance.

Few studies have compared the physical conditioning and performance benefits of single versus multiple daily training sessions. Studies conducted thus far reveal no scientific evidence that multiple daily training sessions enhance fitness and performance more than a single daily session. This is illustrated by the data in figure 13.3 on page 300, which show the responses of two groups of swimmers who trained once per day (group 1) or twice per day (group 2) for a period of six weeks during a 25-week training program.[5] All swimmers began the program following the same training regimen: one time per day. But from the beginning of the 5th week through the end of the 10th week, group 2 increased its training to twice per day. After six weeks on the different regimens, both groups returned to the once-daily program. All the swimmers' heart rates and blood lactate values decreased dramatically when training began, and no significant differences were seen in the two groups' results in response to the change in training volume. The swimmers who trained twice per day showed no additional improvements over those who trained only once per day. In fact, their blood lactate concentrations (figure 13.3*a*) and heart rates (figure 13.3*b*) appeared to be slightly higher for the same fixed-pace swim.

To determine the influence of long-term, excessive training, performance improvements of swimmers who trained twice daily for a total distance of more than 10,000 m (10,936 yd) per day (the LS, or long-swim, group) were compared with improvements of those who swam approximately half that distance in a single session each day (the SS, or short-swim, group).[6] Changes in performance time for the 100 yd (91 m) front crawl were examined over a four-year period for both groups. The LS swimmers and SS swimmers experienced an identical average improvement of 0.8% per year. Similar findings also were observed for competitors in other events, such as the 200, 500, and 1,650 yd (183, 457, and 1,509 m) front crawl.

The concept of training specificity (see chapter 8) implies that several hours of daily training will not provide the adaptations needed for athletes who participate in events of short duration. Most competitive swimming events last less than 2 min. How can training for 3 to 4 h per day at speeds that are markedly slower than competitive pace prepare the swimmer for the maximal efforts of competition? Such a large training volume prepares the athlete to tolerate a high volume of training but likely does little to benefit actual performance.

The need for long daily workouts (high volume) is now being seriously questioned by researchers. For certain sports, it appears that training volume could be reduced significantly, possibly by as much as one-half in some sports, without reducing the benefits, and with less risk of overtraining athletes to the point of decreased performance. The principle of training specificity suggests that low-intensity, high-volume training does not improve sprint-type performance.

Training intensity is also an important factor and refers to both the relative force of muscle action (i.e., resistance training) and the relative stress placed on the metabolic and cardiovascular systems (i.e., anaerobic and aerobic training). There is a strong interaction between training intensity and training volume: As intensity is reduced, training volume must be increased to achieve adaptation. Training at very high intensities requires substantially less training volume, but the adaptations that occur will be significantly different from those achieved with low-intensity, high-volume training. This concept applies to all three types of training, that is, resistance, anaerobic, and aerobic.

High-intensity, low-volume training can be tolerated only for brief periods. While this type of training does increase muscular strength in resistance training and total body speed and anaerobic capacity in high-inten-

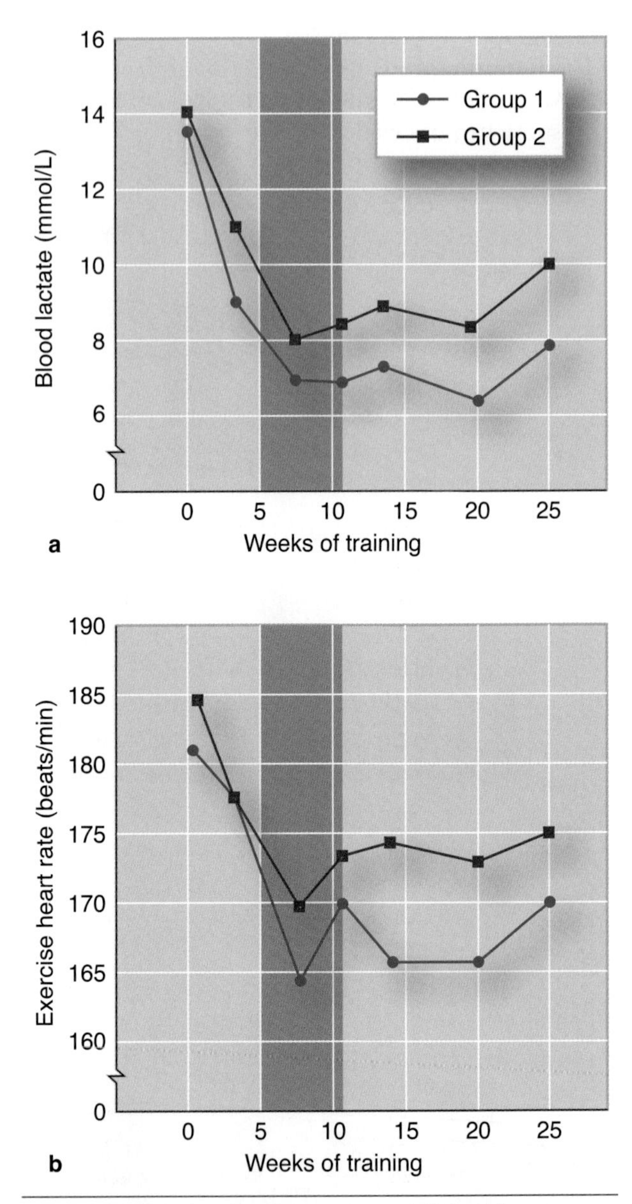

Figure 13.3 Changes in swimmers' *(a)* blood lactate concentrations and *(b)* heart rates during a standardized 366 m (400 yd) swim during 25 weeks of training. From the beginning of the 5th through the end of the 10th week, one group trained once per day (group 1) whereas the other group trained twice each day (group 2).

In review

- Optimal training involves following a model that incorporates the principles of periodization, because the body needs to systematically go through stages of undertraining, acute overload, and overreaching to maximize performance.
- Excessive training refers to training that is done with an unnecessarily high volume, intensity, or both. It provides little or no additional improvements in conditioning or performance and can lead to decreased performance and health problems.
- Training volume can be increased through increase in the duration or frequency of training bouts or both. Many studies have shown no significant differences in improvement between athletes who train with typical training volumes and those who train with twice the volume (training conducted for twice the duration or twice a day instead of once a day).
- Training intensity determines the specific adaptations that occur in response to the training stimulus. As training intensity increases, training volume must be reduced, and vice versa.

sity interval training, it provides little or no improvement in aerobic capacity. Conversely, low-intensity, high-volume training stresses the oxygen transport and oxidative metabolic systems, causing greater gains in aerobic capacity, but has little or no effect on muscular strength, anaerobic capacity, or total body speed.

Attempts to perform large amounts of high-intensity training can have negative effects on adaptation. The energy needs of high-intensity exercise place greater demands on the glycolytic system, rapidly depleting muscle glycogen. If such training is attempted too often, for example daily, the muscles can become chronically depleted of their energy reserves, and the person might demonstrate signs of chronic fatigue or overtraining, as we discuss later in this chapter.

OVERREACHING

In contrast to excessive training, overreaching is a systematic attempt to intentionally overstress the body, allowing the body to adapt to the training stimulus beyond the level of adaptation attained during a period of acute overload. As with overtraining, there is a brief decrement in performance lasting several days to several weeks, followed by both increased physiological function and increased performance. Obviously, this is the critical phase of training—on the edge, leading either to improved physiological function and performance or to overtraining, if one goes too far. With overreaching, the period of full recovery from training takes several days to several weeks. If one overtrains the body, recovery can take many months or, in some cases, years. The key to overreaching is to push the athlete hard enough to accomplish the desired positive physiological and performance improvements but to avoid going into the stage of overtraining. This is not an easy task to accomplish!

OVERTRAINING

During periods of intense overload or overreaching training, athletes may experience an unexplained decline in performance and physiological function that extends over weeks, months, or years. This condition is termed overtraining and has been attributed to both psychological and physiological causes. The precise cause or causes for this breakdown in performance and physiological function are not fully understood. Further, overtraining occurs with each of three major forms of training—resistance, anaerobic, and aerobic training—so it is likely that the cause or causes and symptoms will vary by the type of training.

Athletes experience various levels of fatigue during repeated days and weeks of training, so not all fatigue-producing situations can be classified as overtraining (as we noted previously with overreaching). Fatigue that follows one or more exhaustive training sessions usually is relieved by a few days of reduced training or rest and a carbohydrate-rich diet. Overtraining, on the other hand, is characterized by a sudden decline in performance and physiological function that cannot be remedied by a few days of reduced training, rest, or dietary manipulation.

Effects of Overtraining: The Overtraining Syndrome

Most of the symptoms that result from overtraining, collectively referred to as the overtraining syndrome, are subjective and identifiable only after the individual's performance and physiological function have suffered. Unfortunately, these symptoms can be highly individualized, making it very difficult for athletes, trainers, and coaches to recognize that performance decrements are brought on by overtraining. A decline in physical performance with continued training is usually the first indication of the overtraining syndrome (see figure 13.4). The athlete senses a loss of muscular strength, coordination, and working capacity and generally feels fatigued. Other primary signs and symptoms of the overtraining syndrome include[1]

- change in appetite;
- body weight loss;
- sleep disturbances;
- irritability, restlessness, excitability, anxiousness;
- loss of motivation and vigor;
- lack of mental concentration;
- feelings of depression; and
- lack of appreciation for things that normally are enjoyable.

In focus

Few athletes are undertrained but, unfortunately, many are overtrained, often erroneously in the belief that more training always produces improvement. As their performance declines with overtraining, they train even harder in an effort to compensate. The importance of designing training programs to include both rest and variation of training intensity and volume to avoid overtraining and chronic fatigue cannot be overstated.

Physiological changes also indicate the presence of the overtraining syndrome. We address these later in this section.

The underlying causes of overtraining syndrome are often a complex combination of emotional and physiological factors. Hans Selye[22] noted that a person's stress tolerance can break down as often from a sudden increase in anxiety as from an increase in physical distress. The emotional demands of competition, the desire to win, the fear of failure, unrealistically high goals, and others' expectations can be sources of intolerable emotional stress. Because of this, overtraining is typically accompanied by a loss of competitive desire and a loss of enthusiasm for training. Furthermore, Armstrong and VanHeest[1] make the important observation that the overtraining syndrome and clinical depression involve remarkably similar signs and symptoms, brain structures, neurotransmitters, endocrine pathways, and immune responses, suggesting that they have similar etiologies.

In focus

The symptoms of overtraining syndrome are highly individualized and subjective, so they cannot be universally applied. The presence of one or more of these symptoms is sufficient to alert the coach or trainer that an athlete might be overtrained.

The physiological factors responsible for the detrimental effects of overtraining are not fully understood. However, abnormal responses have been reported that suggest that overtraining is associated with alterations in the nervous, endocrine, and immune systems. Although a cause-and-effect relationship between these changes and the symptoms of overtraining has not been clearly established, these symptoms can help determine whether an individual is overtrained. In the following discussion, we focus on some of the observed changes associated with overtraining and on potential causes of the overtraining syndrome.

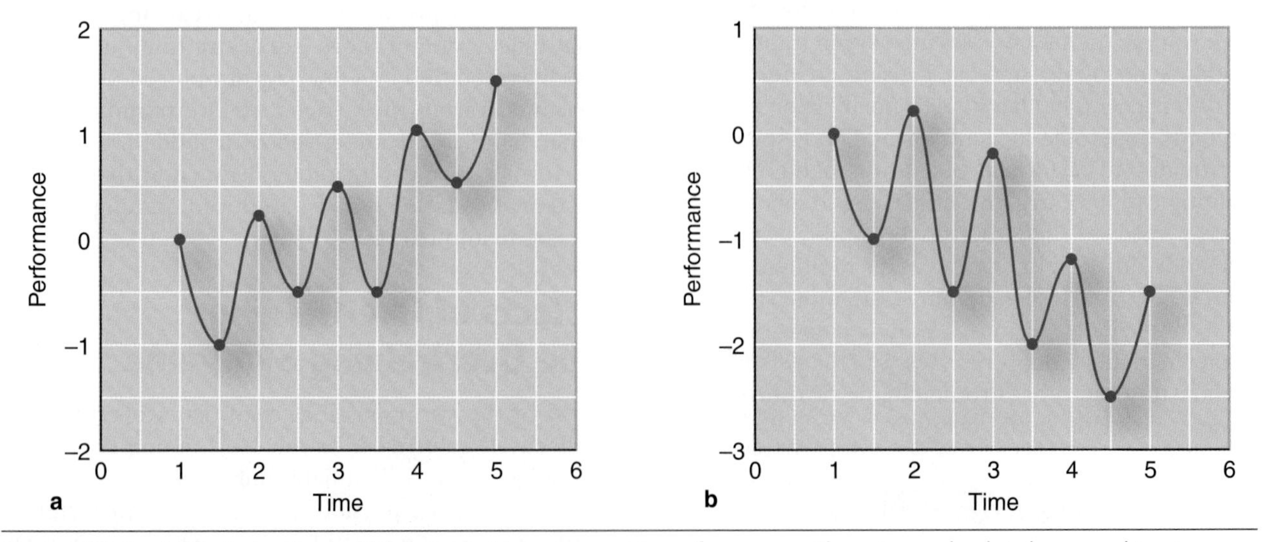

Figure 13.4 *(a)* Typical pattern of the expected improvement in performance with acute overload and overreaching in contrast to *(b)* the pattern seen with overtraining.

Reprinted, by permission, from M.L. O'Toole, 1998, Overreaching and overtraining in endurance athletes. In *Overtraining in sport,* edited by R.B. Krieder, A.C. Fry, and M.L. O'Toole (Champaign, IL: Human Kinetics), 10, 13.

Autonomic Nervous System Responses to Overtraining

Some studies suggest that overtraining is associated with abnormal responses in the autonomic nervous system. Physiological symptoms accompanying the decline in performance often reflect changes in those organs or systems that are controlled by either the sympathetic or parasympathetic branches of the autonomic nervous system (see chapter 3). Sympathetic overtraining can lead to

- increased resting heart rate,
- increased blood pressure,
- loss of appetite,
- decreased body mass,
- sleep disturbances,
- emotional instability, and
- elevated basal metabolic rate.

This form of overtraining occurs predominantly among athletes who emphasize highly intense or resistance training methods.

Other studies suggest that the parasympathetic nervous system might be dominant in some cases of overtraining, usually in endurance athletes. In these cases, the performance decrements markedly differ from those associated with sympathetic overtraining. Signs of parasympathetic overtraining, assumed to be the result of volume overload, include

- early onset of fatigue,
- decreased resting heart rate,
- rapid heart rate recovery after exercise, and
- decreased resting blood pressure.

Thus, it appears that athletes in different sports or events will likely exhibit unique signs and symptoms of the overtraining syndrome that are related to their training regimens. In fact, some authorities have named these forms of overtraining "intensity related" and "volume related," recognizing that specific training stressors result in unique signs and symptoms when applied excessively.[14]

Some of the symptoms associated with autonomic nervous system overtraining are also seen in people who are not overtrained. For this reason, we cannot always assume that the presence of these symptoms confirms overtraining. Of the two conditions, symptoms of sympathetic overtraining are the most frequently observed. Although there is not strong scientific evidence to support the autonomic nervous system overtraining theory, the autonomic nervous system definitely is affected by overtraining.

Hormonal Responses to Overtraining

Measurements of various blood hormone concentrations during periods of overreaching suggest that marked disturbances in endocrine function accompany excessive stress. As shown in figure 13.5, when swimmers increase their training 1.5- to 2-fold, blood concentrations of thyroxine and testosterone usually decrease and blood concentrations of cortisol increase. The ratio of testosterone to cortisol is thought to regulate anabolic processes in recovery, so a change in this ratio is considered an important indicator, and perhaps a cause, of the overtraining syndrome. Decreased testosterone coupled with increased cortisol might lead to more protein catabolism than anabolism in the cells. Other research, however, suggests that although cortisol concentrations increase with overreaching and the early stages of overtraining, both resting and exercise cortisol concentrations eventually decrease in the overtraining syndrome. Further, most overtraining studies have been conducted on aerobically trained endurance athletes. Fewer studies exist on anaerobically trained and resistance-trained athletes. Using the terminology introduced in the last section, intensity-related overtraining (anaerobic and resistance training) does not appear to alter resting hormonal concentrations.[14]

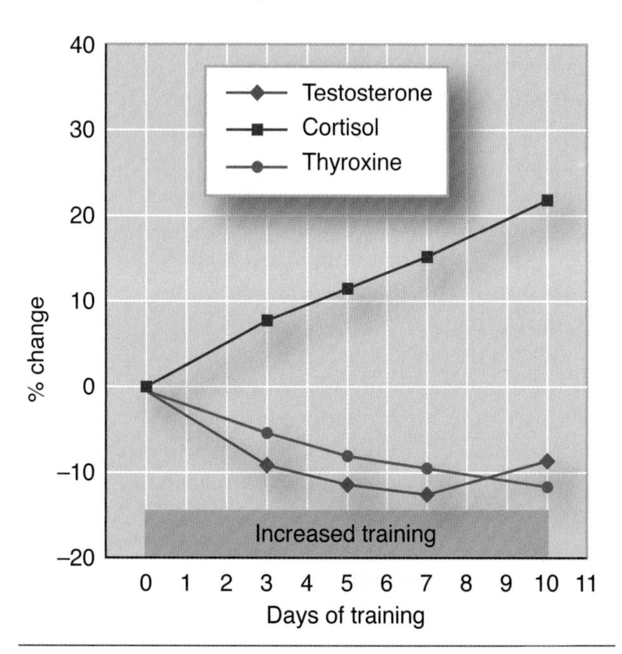

Figure 13.5 Changes in resting blood concentrations of testosterone, cortisol, and thyroxine over a period of intensified training. During the 10-day period shown here, swimmers increased their training from about 4,000 m to 8,000 m (4,374-8,749 yd) per day. These data show that resting cortisol concentrations increased in response to the added stress, whereas testosterone and thyroxine showed a remarkable decline during this period.

Overtrained athletes often have higher blood concentrations of urea, and because urea is produced by the breakdown of protein, this indicates increased protein catabolism. This mechanism is thought to be responsible for the loss in body mass seen in overtrained athletes.

Resting blood concentrations of epinephrine and norepinephrine have also been reported to be elevated during periods of intensified aerobic or volume training. These two hormones elevate heart rate and blood pressure. This has led some exercise physiologists to suggest that the blood concentrations of these catecholamines should be measured to confirm overtraining. However, other studies have found no change in these catecholamines during intensified training, and some have even found decreased resting values.

Acute overload training and overreaching often produce most of the same hormonal changes reported in overtrained athletes. For this reason, measuring these and other hormones might not provide valid confirmation of overtraining. Athletes whose hormone concentrations appear abnormal may simply be experiencing the normal effects of hard training. Further, the time interval between the last training bout and the resting blood sample is very important. Some potential markers remain elevated for more than 24 h and might not reflect a true resting state. These hormonal changes simply might reflect the stress of training rather than a breakdown in the adaptive process. Consequently, many experts have now concluded that no blood marker conclusively defines the overtraining syndrome.

Armstrong and VanHeest[1] proposed that the various stressors associated with the overtraining syndrome act primarily through the hypothalamus. They postulated that these stressors activate the following two predominant hormonal axes involved in the body's response to stressors:

- The sympathetic–adrenal medullary axis (SAM), involving the sympathetic branch of the autonomic nervous system; and
- The hypothalamic-pituitary-adrenocortical axis (HPA)

This is illustrated in figure 13.6*a*. Figure 13.6*b* illustrates the brain and immune system interactions with these two axes. These two figures are quite complex and go well beyond the scope of an introductory-level

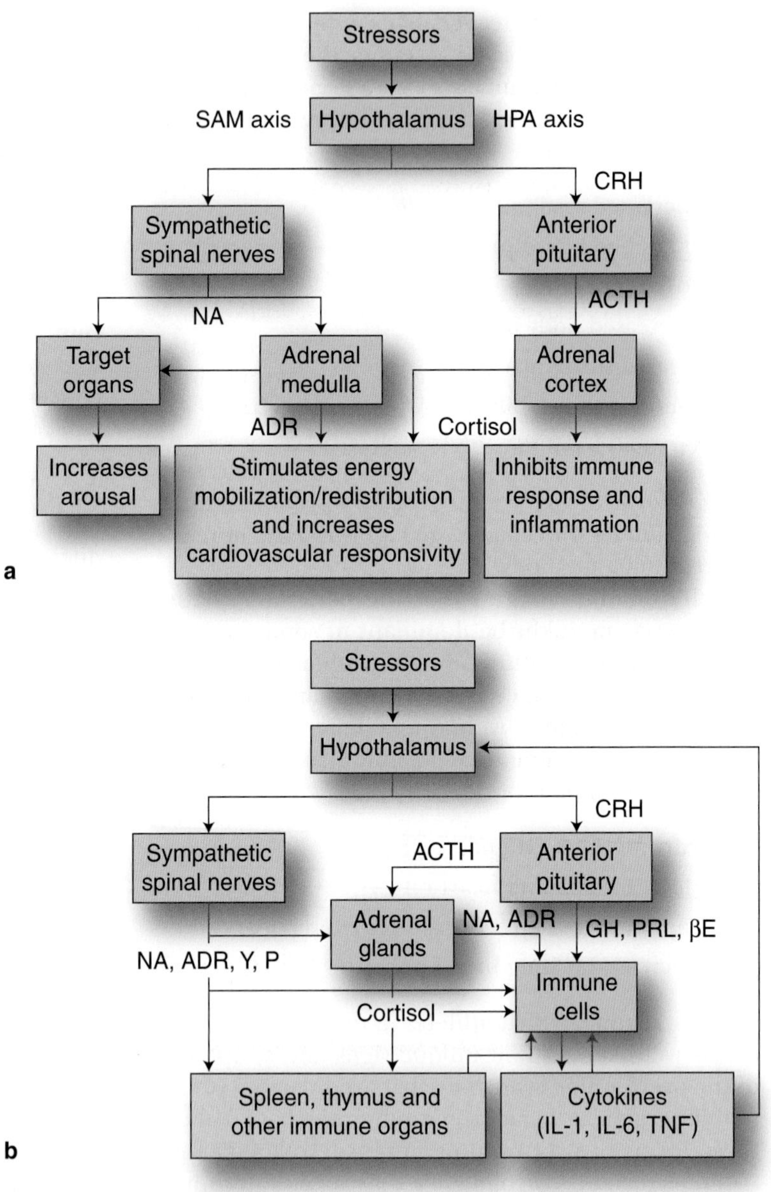

Figure 13.6 *(a)* The role of the hypothalamus and the sympathetic–adrenal medullary (SAM) and hypothalamic-pituitary-adrenocortical (HPA) axes as possible mediators of the overtraining syndrome. *(b)* The brain–immune system interactions with this model, with the cytokines playing a potentially major role in mediating overtraining. Symbols: ACTH = adrenocorticotropin; ADR = adrenaline (epinephrine); CRH = corticotropin-releasing hormone; GH = growth hormone; IL-1 = interleukin-1; IL-6 = interleukin-6; NA = noradrenaline (norepinephrine); P = substance P; PRL = prolactin; TNF = tumor necrosis factor; Y = neuropeptide Y; βE = β-endorphin.

Adapted, by permission, from L.E. Armstrong and J.L. VanHeest, 2002, "The unknown mechanism of the overtraining syndrome," *Sports Medicine* 32: 185-209.

exercise physiology text. However, a cursory study of the interactions depicted in these figures will give one an appreciation of the complexity of this syndrome. Importantly, note that the stressors have their initial effect on the brain (hypothalamus). Thus, it is highly likely that brain neurotransmitters play an important role in the overtraining syndrome. Serotonin is a major neurotransmitter that is suspected to play a significant role in the overtraining syndrome. Unfortunately, plasma concentrations of this important neurotransmitter do not accurately reflect those concentrations in the brain. Advances in technology should provide the necessary tools to help us better understand what is going on inside the brain.

A major role for cytokines in the overtraining syndrome recently has been proposed,[24] providing support for the Armstrong and VanHeest model in figure 13.6*b*. Elevated circulating cytokines result from infection as well as from skeletal muscle, bone, and joint trauma associated with overtraining. They appear to be a normal part of the body's inflammatory response to infection and injury. It is theorized that excessive musculoskeletal stress, coupled with insufficient rest and recovery, sets up a cascade of events whereby a local acute inflammatory response evolves into chronic inflammation and eventually into systemic inflammation. Systemic inflammation activates circulating monocytes, which can then synthesize large quantities of cytokines. Cytokines then act on most of the brain and body functions in a manner consistent with symptoms expressed in the overtraining syndrome.[24]

Immunity and Overtraining

The immune system provides a line of defense against invading bacteria, parasites, viruses, and tumor cells. This system depends on the actions of specialized cells (such as lymphocytes, granulocytes, and macrophages) and antibodies. These primarily eliminate or neutralize foreign invaders that might cause illness (pathogens). Unfortunately, one of the most serious consequences of overtraining is the negative effect it has on the body's immune system. In fact, from the model proposed in figure 13.6, compromised **immune function** is potentially a major factor in the initiation of the overtraining syndrome.

Many studies have shown that excessive training suppresses normal immune function, increasing the overtrained athlete's susceptibility to infections. This is illustrated in figure 13.7 on page 306. Studies also show that short bouts of intense exercise can temporarily impair the immune response, and successive days of heavy training can amplify this suppression. Several investigators have reported an increased incidence of illness following a single, exhaustive exercise bout, such as running a full 42 km (26.2 mi) marathon. This immune suppression is characterized by abnormally low concentrations of both lymphocytes and antibodies. Invading organisms or substances are more likely to cause illness when these concentrations are low. Also, intense exercise during illness might decrease one's ability to fight off the infection and increase the risk of even greater complications.[18]

Overtraining syndrome appears to be associated with systemic inflammation and increased synthesis of large quantities of cytokines. These changes are associated with depressed immune function. This may place the athlete at an increased risk for infection and disease.

The Overtraining, Chronic Fatigue, and Fibromyalgia Syndromes

Chronic fatigue syndrome is very similar to the overtraining syndrome.[23] There is likely considerable overlap between the two. Furthermore, there is considerable overlap between chronic fatigue syndrome and the **fibromyalgia syndrome.** Chronic fatigue and fibromyalgia, however, do occur in nonathletes and in individuals who are not physically active. Other than that, there are many similarities in symptoms across the three syndromes. These similarities can include chronic fatigue at rest and during exercise, psychological distress, immune system dysfunction, hormonal dysfunction, HPA axis dysfunction, and neurotransmitter dysfunction. Furthermore, all three syndromes are difficult to diagnose, and generally the specific cause or causes remain unknown.

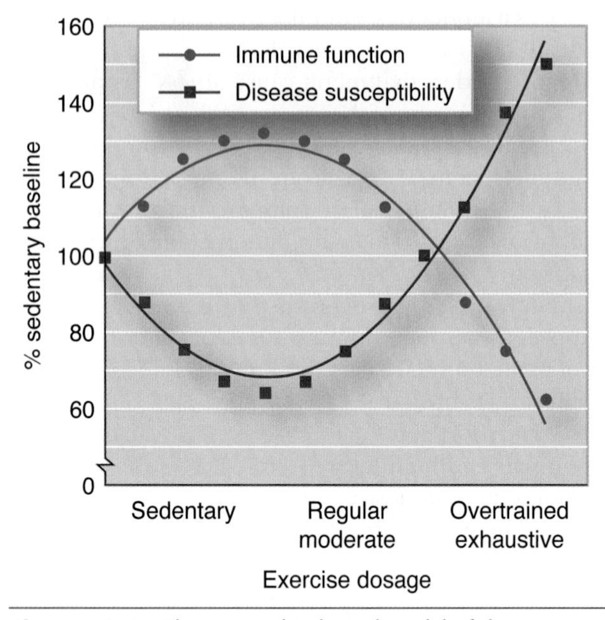

Figure 13.7 The inverted J-shaped model of the relationship between amount of exercise and immune function. This model suggests that moderate exercise may lower the risk of infection or disease, whereas overtraining may increase the risk.

Data from D.C. Nieman, 1997.

Predicting the Overtraining Syndrome

We must remember that the underlying cause or causes of the overtraining syndrome are not fully known, although it is likely that physical or emotional overload, or a combination of the two, might trigger this condition. Trying not to exceed an athlete's stress tolerance by regulating the amount of physiological and psychological stress experienced during training is difficult. Most coaches and athletes use intuition to determine training volume and intensity, but few can accurately assess the true impact of a workout on the athlete. No preliminary symptoms warn athletes that they are on the verge of becoming overtrained. By the time coaches realize that they have pushed an athlete too hard, it is often too late. The damage done by repeated days of excessive training or stress can be repaired only by days, and in some cases weeks or months, of reduced training or complete rest.

Numerous investigators have tried to identify markers of the overtraining syndrome in its early stages by using assorted physiological and psychological measurements. A list of potential markers is provided in table 13.1. Unfortunately, none has proven totally effective. It is often difficult to determine whether the measurements obtained are related to overtraining or whether they simply reflect normal responses to overload or overreaching training.

Possibly the best method to identify the overtraining syndrome is to monitor the athlete's heart rate during a standardized workout, such as a fixed-paced run or swim, using a digital heart rate monitor (figure 8.5 on p. 198). The data presented in figure 13.8 illustrate a runner's heart rate response during a 1 mi (1.6 km) run performed at a fixed pace of 6 min/mi (3.7 min/km), or 10 mph (16 km/h). This response was monitored when the runner was untrained (UT), after the runner

TABLE 13.1 Potential Markers of OR, OT, and OTS

Marker	Response	Potential marker for		
Physiological and psychological		OR	OT	OTS
HR_{rest} and HR_{max}	Decreased		X	X
HR_{submax} and $\dot{V}O_{2submax}$	Increased	X		X
$\dot{V}O_{2max}$	Decreased			X
Anaerobic metabolism	Impaired		X	
Basal metabolic rate	Increased			X
$RER_{submax,\ max}$	Decreased		X	X
Nitrogen balance	Negative			X
Nerve excitability	Increased			X
Sympathetic nervous response	Increased			X
Psychological mood states	Altered	X		
Risk of infection	Increased	X		

Marker	Response	Potential marker for		
Blood		OR	OT	OTS
Hematocrit and hemoglobin	Decreased		X	
Leukocytes and immunophenotypes	Decreased		X	
Serum iron and ferritin	Decreased		X	
Serum electrolyte levels	Decreased			X
Serum glucose and free fatty acids	Decreased		X	
Plasma lactate concentration, submax, max	Decreased		X	X
Ammonia	Increased		X	X
Serum testosterone and cortisol	Decreased	X		
ACTH, growth hormone, prolactin	Decreased			X
Catecholamines, rest, night	Decreased			X
Creatine kinase	Increased			X

OR = overreaching; OT = overtraining; OTS = overtraining syndrome; HR = heart rate; RER = respiratory exchange ratio; ACTH = adrenocorticotropic hormone.

Adapted from Armstrong and VanHeest, *Sports Medicine,* 2002.

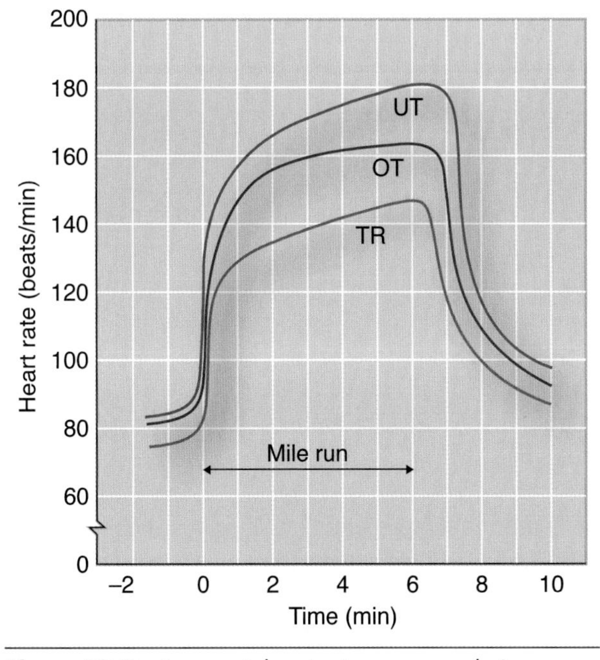

Figure 13.8 A runner's heart rate responses during a fixed-pace treadmill run at 10 mph (16 km/h) performed before training (UT), after training (TR), and when the runner showed symptoms of overtraining (OT).

had trained (TR), and during a period when the runner demonstrated symptoms of overtraining syndrome (OT). This figure shows that heart rate was higher when the runner was in the overtrained state than when the runner was responding well to training. Similar findings have been reported for swimmers.[5] Such a test provides a simple and objective way to monitor training and can possibly provide a warning sign of the onset of the overtraining syndrome.

Possibly the best predictor of the overtraining syndrome is the heart rate response to a standardized bout of work. Performance decrements are also good indicators.

Reducing the Risk and Treating the Overtraining Syndrome

Recovery from the overtraining syndrome is possible with a marked reduction in training intensity or complete rest. Although most coaches recommend a few days of easy training, overtrained athletes require considerably more time for full recovery. This might necessitate the total cessation of training for a period of weeks or months. In some cases, counseling might be needed to help the athletes cope with other stress in their lives that might contribute to this condition.

The best way to minimize the risk of overtraining is to follow periodization training procedures, alternating easy, moderate, and hard periods of training as discussed in chapter 8. Although individual tolerance varies tremendously, even the strongest athletes have periods when they are susceptible to the overtraining syndrome. As a rule, one or two days of intense training should be followed by an equal number of easy training days. Likewise, a week or two of hard training should be followed by a week of reduced effort with little or no emphasis on anaerobic exercise.

Endurance athletes (such as swimmers, cyclists, and runners) must pay particular attention to their carbohydrate intake. Repeated days of hard training gradually reduce muscle glycogen. Unless these athletes consume extra carbohydrate during these periods, their muscle and liver glycogen reserves can be depleted. As a consequence, the most heavily recruited muscle fibers are not able to generate the energy needed for exercise.

In review

- Overtraining stresses the body beyond its capacity to adapt, decreasing performance and physiological capacity.
- The symptoms of overtraining syndrome are subjective and vary from individual to individual; and many also accompany regular training, which makes prevention or diagnosis of the overtraining syndrome difficult.
- Possible explanations for the overtraining syndrome include changes in the functioning of the divisions of the autonomic nervous system, altered endocrine responses, suppressed immune function, and altered brain neurotransmitters.
- Many potential signs and symptoms of overtraining have been proposed to diagnose overtraining in its earliest stages. However, at this time, the heart rate response to a fixed-pace exercise bout appears to be the easiest and most accurate technique.
- Overtraining syndrome is treated by a marked reduction in training intensity or complete rest, for weeks or months. Prevention can best be accomplished through use of periodization training procedures that vary training intensity and volume.
- For endurance athletes, it is important to ensure adequate carbohydrate intake to meet energy needs.

Exertional Rhabdomyolysis

Rhabdomyolysis is an acute disease, which can be fatal, that is distinguished by the breakdown of skeletal muscle fibers. Exertional rhabdomyolysis refers to muscle fiber breakdown that occurs in response to unaccustomed, strenuous exercise. Normally, it is not dangerous and the symptoms are tolerable, similar to delayed-onset muscle soreness (see chapter 9). However, exertional rhabdomyolysis becomes clinically relevant when the muscle damage is severe. This usually occurs when proteins leak out of the damaged muscle and precipitate in the kidneys, causing acute renal failure and even death. Clinically relevant exertional rhabdomyolysis (CRER) represents a relatively small percentage of all rhabdomyolysis cases. Rhabdomyolysis can also be caused by the use of alcohol, heroin, and cocaine.

Clinically relevant exertional rhabdomyolysis is usually reported in case studies, since it is such a rare disease. The case studies that have been reported to date describe incidents that have occurred during military training and during normal fitness training. Two cases reported in 2003 involved healthy adults who were encouraged to perform excessive strenuous exercise in a health club setting. Both developed symptoms and required hospitalization to prevent renal failure.[25] Symptoms include severe muscle pain and weakness, muscle swelling, and dark brown urine indicating the presence of myoglobin. In severe cases, patients may develop fever, leukocytosis, renal failure, and electrolyte abnormalities. Clinically relevant exertional rhabdomyolysis can be precipitated by high-intensity exercise, particularly excessive eccentric exercise, and is exacerbated by exercising in the heat or at altitude.[15] Particular caution must be taken when training unfit people not to push them beyond their comfort level. It is best to start at low intensities and volumes and gradually increase the training stress over weeks or months.

TAPERING FOR PEAK PERFORMANCE

Peak performance requires maximal physical and psychological tolerance for the stress of the activity. But periods of intense training reduce muscular strength, decreasing athletes' performance capacity. For this reason, to compete at their peak, many athletes reduce their training intensity and volume before a major competition to give their bodies and minds a break from the rigors of intense training. This practice is referred to as **tapering.** The **taper period,** during which intensity and volume are reduced, should provide adequate time for healing of tissue damage caused by intense training and for the body's energy reserves to be fully replenished. Research suggests that, for example, taper periods range from 4 to 28 days or longer,[17] depending on the sport, the event, and the athlete's needs. Tapering is not appropriate for all sports, particularly those where competition occurs once a week or more frequently. Still, rested athletes generally perform better.

The most notable change during the taper period is a marked increase in muscular strength, which explains at least part of the performance improvement that occurs. It is difficult to determine whether strength improvements result from changes in the muscles' contractile mechanisms or improved muscle fiber recruitment. However, examination of individual muscle fibers taken from swimmers' arms before and after 10 days of intensified training showed that the type II (fast-twitch) fibers exhibited a significant reduction in their maximal shortening velocity.[8] This change has been attributed to changes in the fibers' myosin molecules. In these cases, the myosin in the type II fibers became more like that in the type I fibers. We assume from this finding that such changes in the muscle fibers cause the power loss that swimmers and runners experience during prolonged periods of intense training. We can also assume that the recovery of strength and power that occurs with tapering is linked to modifications of the muscles' contractile mechanisms. Tapering also allows time for the muscle to repair any damage incurred during intense training and for the energy reserves (i.e., muscle and liver glycogen) to be restored.

In focus

Tapering for competition is crucial to a person's best performance. Intense training tears down the body, so reduced training volume and intensity, coupled with quality rest, are needed to allow the body to repair itself and to restore its energy reserves for competition.

Although tapering is widely practiced in a variety of sports, many coaches fear that reduced training for such a long period before a major competition will decrease conditioning and impair performance. But numerous studies clearly show that this fear is unwarranted. Developing optimal $\dot{V}O_{2max}$ initially requires a considerable amount of training, but once it has been developed, much less training is needed to maintain it at its highest level. In fact, the training level of $\dot{V}O_{2max}$ can be maintained even when training frequency is reduced by two-thirds.[11]

Runners and swimmers who reduce their training by about 60% for 15 to 21 days show no losses in $\dot{V}O_{2max}$ or endurance performance.[4, 12] One study showed that swimmers' blood lactate concentrations after a standard swim were lower after a taper period than before. More important, the swimmers experienced a 3.1% improvement in performance as a result of the reduced training and demonstrated a 17.7% to 24.6% increase in arm strength and power.[4]

In a study of distance runners, those runners who went through a seven-day taper decreased their running time in a 5 km time trial by 3% compared to no improvement in those who did not taper. Submaximal oxygen uptake during running at 80% $\dot{V}O_{2max}$ was decreased by 6% in those who tapered, indicating a greater economy of effort. Blood lactate concentrations at 80% of $\dot{V}O_{2max}$ were unchanged, as were $\dot{V}O_{2max}$ and leg extension peak force.[13]

Unfortunately, little information is available to demonstrate the influence of tapering on performance in team sports and in long-duration endurance events such as cycling and marathon running. Before guidelines can be offered for athletes in these sports, research is needed to demonstrate that similar benefits can be generated by such periods of reduced training.

In review

- Many athletes decrease their training intensity and volume before a competition to increase strength, power, and performance capacity. This practice is called tapering.
- The optimal duration of the taper is between 4 and 28 days, or longer, and is dependent on the sport or event and the athlete's needs.
- Muscular strength increases significantly during the tapering period.
- Tapering allows time for the muscle to repair any damage incurred during intense training and for the energy reserves (i.e., muscle and liver glycogen) to be restored.
- Less training is needed to maintain previous gains than was originally needed to attain them, so tapering does not decrease conditioning.
- Performance improves by an average of about 3% with proper tapering.

TABLE 13.2 Blood Lactate, pH, and Bicarbonate (HCO_3^-) Values in Collegiate Swimmers Undergoing Detraining

	Weeks of detraining			
Measurement	0[a]	1[b]	2	4
Lactate (mmol/L)	4.2	6.3	6.8	9.7[c]
pH	7.26	7.24	7.24	7.18[c]
HCO_3^- (mmol/L)	21.1	19.5[c]	16.1[c]	16.3[c]
Swim time (s)	130.6	130.1	130.5	130.0

Note. Measurements were taken immediately after a fixed-pace swim.
[a]The values at week 0 represent the measurements taken at the end of five months of training.
[b]The values for weeks 1, 2, and 4 are the results obtained after one, two, and four weeks of detraining, respectively.
[c]Significant difference from the value at the end of training.

flexibility, and cardiorespiratory endurance. Consequently, losses of speed and agility that occur with inactivity are relatively small. Also, peak levels of both can be maintained with a limited amount of training. But this does not imply that the track sprinter can get by with training only a few days a week. Success in actual competition relies on factors other than basic speed and agility, such as correct form, skill, and the ability to generate a strong finishing sprint. Many hours of practice are required to tune performance to its optimal level, but most of this time is spent developing performance qualities other than speed and agility.

Flexibility, on the other hand, is lost rather quickly during inactivity. Stretching exercises should be incorporated into both in-season and off-season training programs. Reduced flexibility has been proposed to increase athletes' susceptibility to serious injury.

Cardiorespiratory Endurance

The heart, like other muscles in the body, is strengthened by endurance training. Inactivity, on the other hand, can substantially decondition the heart and the cardiovascular system. The most dramatic example of this is seen in a study conducted on subjects undergoing long periods of total bed rest; they weren't allowed to leave their beds, and physical activity was kept to an absolute minimum.[21] Cardiovascular and metabolic function were assessed at a constant submaximal rate of work and at maximal rates of work both before and after the 20-day period of bed rest. The cardiovascular effects that accompanied bed rest included

- a considerable increase in submaximal heart rate,
- a 25% decrease in submaximal stroke volume,
- a 25% reduction in maximal cardiac output, and
- a 27% decrease in maximal oxygen consumption.

The reductions in cardiac output and $\dot{V}O_{2max}$ appear to result from reduced stroke volume; this appears to be largely attributable to a decreased plasma volume, with reductions in heart volume and ventricular contractility playing a smaller role.

It is interesting that the two most highly conditioned subjects in this study (the two who had the highest $\dot{V}O_{2max}$ values) experienced greater decrements in $\dot{V}O_{2max}$ than the three less fit people, as shown in figure 13.11. Furthermore, the untrained subjects regained their initial conditioning levels (before bed rest) in the first 10 days of reconditioning, but the well-trained

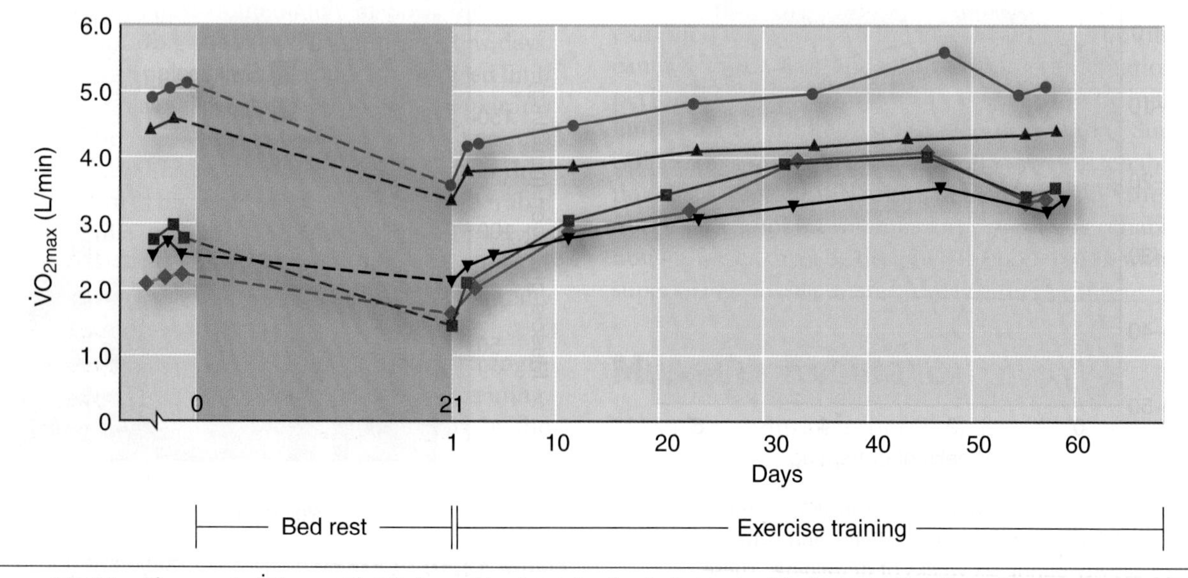

Figure 13.11 Changes in $\dot{V}O_{2max}$ with 20 days of bed rest for five individual subjects. Note that the subjects who were least fit (lowest $\dot{V}O_{2max}$ values) at the start of bed rest showed smaller decrements with inactivity and greater gains when they trained after bed rest. Highly fit individuals, on the other hand, were far more affected by the period of inactivity.

Adapted, by permission, from B. Saltin et al., 1968, "Response to submaximal and maximal exercise after bed rest and training," *Circulation* 38(7): 75.

subjects needed about 40 days for full recovery. This suggests that highly trained individuals cannot afford long periods with little or no endurance training. The athlete who totally abstains from endurance training at the completion of the season will experience great difficulty getting back into top aerobic condition when the new season begins.

Inactivity can significantly reduce $\dot{V}O_{2max}$. How much activity is needed to prevent such considerable losses of

In focus

The body rapidly loses many of the benefits of training if training is discontinued. Some minimal level of training is necessary to prevent these losses. Research indicates that at least three training sessions per week at an intensity of at least 70% $\dot{V}O_{2max}$ are needed to maintain aerobic conditioning.

Detraining in Space

As astronauts orbit the earth, they are in an environment where the gravitational forces are considerably less than those on earth, that is, microgravity. While astronauts experience a sense of weightlessness in orbit, the gravitational forces of earth (i.e., 1 *g*) do not reach 0 *g*. During an extended stay in microgravity, astronauts undergo physiological changes that are nearly identical to those of detraining. But what could be perceived as maladaptations might in fact be necessary adaptations to microgravity. Let's briefly review the changes that take place when astronauts leave the 1 *g* environment on earth to spend weeks or months in space.

Muscle mass and strength decline in microgravity, particularly in the postural muscles, that is, those muscles that maintain the body upright countering the force of gravity. The cross-sectional area of both type I and type II muscle fibers is decreased as well. The extent of decline depends on the muscle group, the duration of the flight, and the type and extent of the in-flight exercise program. Microgravity also affects bone, with average bone mineral losses of about 4% in the weight-bearing bones; but the magnitude of this loss depends on the length of exposure to microgravity.

The cardiovascular system also undergoes major adaptations to microgravity. When the body is in microgravity, there is a reduction in hydrostatic pressure so that blood no longer pools in the lower extremities as it does at 1 *g*. As a consequence, more blood returns to the heart, which leads to transient increases in stroke volume. Reductions in plasma volume occur over time but are likely due to reduced fluid intake rather than increased production of urine (diuresis) by the kidneys. Transcapillary fluid shifts between the microcirculation and surrounding tissues can also account for some of the reduction in plasma volume, most likely upper body capillary filtration; for example, blood relocates into the facial tissues, creating a puffy facial appearance. Red cell mass also decreases, so total blood volume is reduced as well.[27] The reduced blood volume serves astronauts well while they remain in microgravity. But it presents a serious problem on their return to a 1 *g* environment, where the body is again subjected to the hydrostatic pressure effect. Astronauts have experienced postural (orthostatic) hypotension and fainting during their first few hours back in a normal 1 *g* environment because their blood volume was insufficient to meet all their circulatory needs.

Maximal aerobic power ($\dot{V}O_{2max}$) is generally reduced immediately postflight, likely due to the reduction in plasma volume and leg strength during flight. However, data are limited on directly measured $\dot{V}O_{2max}$ in astronauts preflight, during flight, and postflight. Head-down tilt (–6°) bed rest, used as an earth-based model of spaceflight, shows consistent reductions in $\dot{V}O_{2max}$ that are associated with reductions in total blood volume, plasma volume, and, consequently, maximal stroke volume. The head-down tilt model has been shown to provide $\dot{V}O_{2max}$ ($\dot{V}O_{2peak}$) data comparable to actual pre- to postflight changes.[26]

Importantly, having an understanding of the general decline in physiologic function that occurs during spaceflight has led the scientific and medical community to the realization that in-flight exercise programs are essential to preserve the long-term health of astronauts. Research is now under way to design the most appropriate programs and exercise equipment to meet this objective.

physical conditioning? Although a decrease in training frequency and duration reduces aerobic capacity, the losses are significant only when frequency and duration are reduced by two-thirds of the regular training load. However, training intensity apparently plays a more crucial role in maintaining aerobic power during periods of reduced training. Training at 70% $\dot{V}O_{2max}$ appears to be necessary to maintain maximal aerobic capacity.[11]

In review

- Detraining is defined as the partial or complete loss of training-induced adaptations in response to either the cessation of training or a substantial decrement in the training load. The effects of stopping training are quite minor compared with those from immobilization. In general, the greater the gains during training, the greater the losses during detraining simply because the well-trained person has more to lose than the untrained person.
- Detraining causes muscle atrophy, which is accompanied by losses in muscular strength and power. However, muscles require only minimal stimulation to retain these qualities during periods of reduced activity.
- Muscular endurance decreases after only two weeks of inactivity. Possible explanations for this are
  1. decreased oxidative enzyme activity,
  2. decreased muscle glycogen storage, or
  3. disturbance of the acid–base balance.
- Detraining losses in speed and agility are small, but flexibility seems to be lost quickly.
- With detraining, losses of cardiorespiratory endurance are much greater than losses of muscular strength, power, and endurance over the same time period.
- To maintain cardiorespiratory endurance, training must be conducted at least three times per week, and training intensity should be at least 70% of $\dot{V}O_{2max}$.

In Closing

In this chapter we have examined how the quantity of training can affect performance. We saw that too much training, in the form of either excessive training or overtraining, can actually impair performance. Then we looked at the effects of too little training—detraining—as a result of either inactivity or immobilization after an injury. Finally, we saw that with detraining, many of the gains achieved during regular training are quickly lost, especially cardiovascular endurance.

Now that we have dispelled the myth that more training always means better performance, in what other ways can athletes try to optimize their performance? In the next chapter, we turn our attention to optimal body composition and nutrition for the serious athlete.

KEY TERMS

acute overload
chronic fatigue syndrome
detraining
excessive training
fibromyalgia syndrome
immune function
overreaching
overtraining
overtraining syndrome
rhabdomyolysis
tapering
taper period
undertraining

STUDY QUESTIONS

1. Describe the model used to optimize training. Define the terms undertraining, acute overload, overreaching, and overtraining.
2. What is excessive training? How does it relate to the model for optimizing training?
3. Define and describe the overtraining syndrome. What are the general symptoms of the overtraining syndrome? How do these differ between sympathetic and parasympathetic overtraining?
4. How might the hypothalamus be involved in the overtraining syndrome? What role might cytokines play?
5. Describe the relationship between physical activity and immune function and disease susceptibility.
6. What appears to be the best predictor of the overtraining syndrome?
7. How do we treat the overtraining syndrome?
8. What physiological changes occur during the taper period that can be credited with improvements in performance?
9. What alterations occur in strength, power, and muscular endurance with physical detraining?
10. What alterations occur in speed, agility, and flexibility with physical detraining?
11. What changes occur in cardiovascular function as one becomes deconditioned?
12. What similarities do we see between spaceflight and detraining? Why does the body make these adaptations during spaceflight?

STUDY GUIDE ACTIVITIES

In addition to the activities listed in the chapter opening outline on page 297, two other activities are available in the online study guide, located at

www.HumanKinetics.com/PhysiologyOfSportAndExercise.

The chapter **SUMMARY** reviews the key concepts covered in the chapter, and the end-of-chapter **QUIZ** tests your understanding of the material covered in the chapter.

CHAPTER 14

Body Composition and Nutrition for Sport

In this chapter and in the online study guide

A former Major League Baseball player made minimum salary during his first few years in the majors. Early preseason polls projected his team to finish the season at the bottom of its division, but the team ended up in the World Series. This player became one of the best at his position in the National League during that season, and once the World Series was over, he asked management for a substantial salary increase ($75,000 in the mid-1970s). Management agreed to his salary demands contingent on his loss of 25 lb (11 kg)! The player refused to lose the weight, so the parties were deadlocked.

The team physician suggested sending the player to a major university for an accurate body composition assessment, and both parties agreed. A hydrostatic weighing was performed, and the results showed that the player had less than 6% body fat, with a total of only 11 lb (5 kg) of fat! Because 3% to 4% body fat is necessary for survival, this player had only 4 to 5 lb (about 2 kg) of fat to lose, and that loss wasn't advised because he was already at the lower range of acceptable body fat levels recommended for athletes. Management was satisfied, this player received his salary increase, and he did not have to lose weight.

This athlete's weight was well above the weight range recommended for his height, and he had a peculiar gait commonly referred to as a waddle. The combination of being overweight by the standard height–weight charts and having a waddle led management to demand the 25 lb (11 kg) weight loss. Had the athlete agreed to management's demand, he would have likely destroyed his career as a professional athlete. How many athletes have been faced with a similar situation? How many gave in?

Coaches and athletes today are acutely aware of the importance of achieving and maintaining optimal body weight for peak performance in sport. Appropriate size, build, and body composition are critical to success in almost all athletic endeavors. Compare the specific performance requirements of the 152 cm, 45 kg (5 ft, 100 lb) Olympic gymnast and those of the 206 cm, 147 kg (6 ft 9 in., 325 lb) defensive lineman in professional football. Although size and body build can be altered only slightly, body composition can change substantially with dieting and exercise. Resistance training can substantially increase muscle mass, and sound dieting combined with vigorous exercise can significantly decrease body fat. Such changes can be of major importance in achieving optimal athletic performance.

Peak performance also requires a careful dietary balance of the essential nutrients. The U.S. government has established standards for optimal nutrient intake that are termed Dietary Reference Intakes (DRIs). The DRIs provide estimates of the range of intakes of various food substances needed to maintain good health.

The nutritional needs of very active athletes can exceed the DRIs considerably. Individual caloric needs are quite variable, depending on the athlete's size, sex, and sport choice. Cyclists competing in the Tour de France and Norwegian cross-country skiers during training have been reported to expend up to 9000 kcal per day. One ultra-endurance runner expended an average of 10,750 kcal per day over a 5.2-day 600 mi (966 km) race![22] Also, some competitive sports require adherence to rigid weight standards. Athletes who participate in these sports must closely monitor their weight and thus their caloric intake. Too often, this leads to nutritional abuses, drug use, dehydration, and serious health risks. In addition, the dietary tactics used by some athletes to achieve excessive weight loss are of increasing concern because of the potential association with eating disorders, such as anorexia nervosa and bulimia nervosa.

In focus

The fat-free mass is composed of all of the body's nonfat tissue, including bone, muscle, organs, and connective tissue.

BODY COMPOSITION IN SPORT

Body composition refers to the body's chemical composition. Figure 14.1 illustrates three models of body composition. The first two divide the body into its various chemical or anatomical components; the last one simplifies body composition into two components, the fat mass and the fat-free mass, which is the model used in this book. **Fat mass** is often discussed in terms of **relative body fat,** which is the percentage of the total body mass that is composed of fat. **Fat-free mass** simply refers to all body tissue that is not fat.

Assessing Body Composition

Assessment of body composition provides additional information beyond the basic measures of height and weight to both the coach and the athlete. As an example, if the center fielder of a Major League Baseball team is 188 cm (6 ft 2 in.) tall and weighs 91 kg (200 lb), is he at his ideal playing weight? Knowing that 4.5 kg (10 lb) of a total weight of 91 kg (200 lb) is fat weight and that the remaining 86.5 kg (190 lb) is fat-free weight provides considerably more insight than knowing weight alone. In this example, only 5% of his body weight is fat, which is about as low as any athlete should go, as discussed in our chapter opening. Armed with this knowledge, both athlete and coach realize that this athlete's body composition is ideal. They should not be concerned with weight loss, even though standard height–weight charts indicate that the athlete is overweight. However, another baseball player of the same height and weight who has 23 kg (50 lb) of fat would be 25% fat.

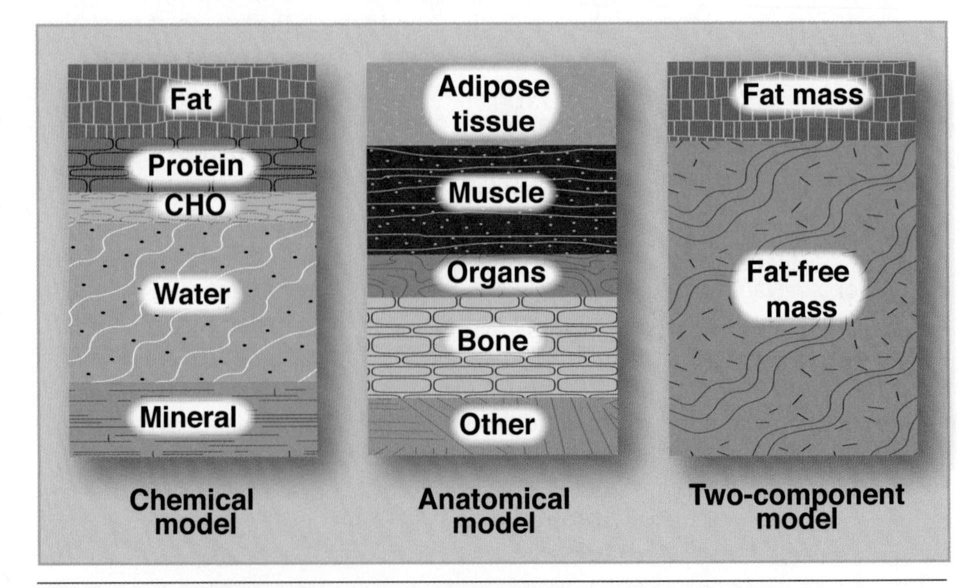

Figure 14.1 Three models of body composition.

Adapted, by permission, from J.H. Wilmore, 1992, Body weight and body composition. In Brownell, Rodin, and Wilmore (Eds.) *Eating, body weight, and performance in athletes: Disorders of modern society* (Philadelphia, PA: Lippincott, Williams, and Wilkins), 77-93.

In focus

Although total body size and weight are important for most athletes, an athlete's body composition is generally of greater concern. Standard height–weight tables do not provide accurate estimates of what an athlete should weigh because they do not take into account the composition of the weight. An athlete can be overweight according to these tables yet have very little body fat.

This would constitute a serious weight problem for an elite athlete: He would be overfat. In most sports, the higher the percentage of body fat, the poorer the performance. An accurate assessment of the athlete's body composition provides valuable insight into the weight that allows optimal performance. But how do we determine an athlete's body composition?

Densitometry

Densitometry involves measuring the density of the athlete's body. Density (D) is defined as mass (M) divided by volume (V):

$$D_{body} = M_{body} \div V_{body}.$$

The mass of the body is the athlete's scale weight. Body volume can be obtained by several different techniques, but the most common is **hydrostatic weighing,** also called underwater weighing, in which the athlete is weighed while totally immersed in water. The difference between the athlete's scale weight and underwater weight, when corrected for the density of water, equals the body's volume. This volume must be further corrected to account for air trapped in the body. The amount of air trapped in the gastrointestinal tract is small, difficult to measure, and usually ignored. The gas trapped in the lungs, however, must be measured because its volume is generally large.

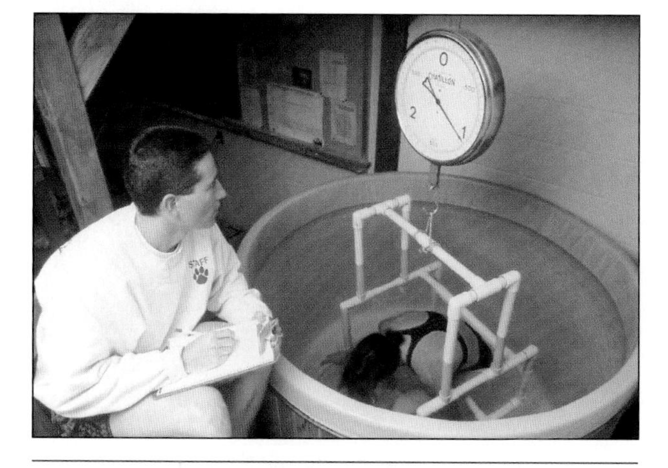

Figure 14.2 The underwater weighing technique to determine the density of the body.

Figure 14.2 shows the hydrostatic weighing technique. The density of the fat-free mass is higher than the density of water, while that of fat is lower than the density of water. Observe a swimming pool full of people of varying body types. Those who have an abundance of body fat have a low total body density and can easily float, while those who are very lean have a higher total body density and tend to sink. This is an oversimplification, but hopefully helpful for understanding the concept.

Densitometry has long been the technique of choice for assessing body composition. New techniques typically are compared against densitometry to determine their accuracy. However, densitometry has its limitations. If body weight, underwater weight, and lung volume during underwater weighing are measured correctly, the resulting **body density** value is accurate. Densitometry's major weakness is in the conversion of body density to an estimate of relative body fat. Accurate estimates of the individual densities of fat mass and fat-free mass are required when the two-component model of body composition is used. The equation most often used to convert body density to an estimate of relative or percentage body fat is the standard equation of Siri:

$$\% \text{ body fat} = (495 \div D_{body}) - 450.$$

This equation assumes that the densities of the fat mass and the fat-free mass are relatively constant in all people. Indeed, the density of fat at different sites is very consistent in the same individual and relatively consistent between people. The value generally used is 0.9007 g/cm^3. But determining the density of the fat-free mass (D_{FFM}), which the equation of Siri assumes is 1.100, is more problematic. To determine this density, we must make two assumptions:

1. The density of each tissue constituting the fat-free mass is known and remains constant.
2. Each tissue type represents a constant proportion of the fat-free mass (e.g., we assume that bone always represents 17% of the fat-free mass).

Exceptions to either of these assumptions cause error when we convert body density to relative body fat. Unfortunately, the density of the fat-free mass does vary between people.

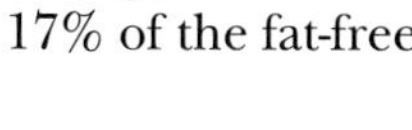

Inaccuracies in densitometry largely reflect the variation in the density of the fat-free mass from one individual to another.

Other Laboratory Techniques

Many other laboratory techniques are available for assessing body composition. These include radiography, computed tomography (CT), magnetic resonance imaging (MRI), hydrometry (for measuring total body water), total body electrical conductivity, and neutron activation. Most of these techniques are complex and require expensive equipment. None of them are likely to be used for the general assessment of athletic populations, so we will not discuss them further in this chapter; however, we do discuss CT in chapter 21. Two other techniques hold considerable promise—dual-energy X-ray absorptiometry and air displacement.

Dual-energy X-ray absorptiometry (DXA) evolved from the earlier single- and dual-photon absorptiometry techniques used between 1963 and 1984. The earlier techniques were used to estimate regional bone mineral content and bone mineral density, primarily in the spine, pelvis, and femur. The newer DXA technique (see figure 14.3) allows the quantification of not only bone but also soft tissue composition. Furthermore, it is not limited to regional estimates but can provide total body estimates. Research to date suggests that DXA provides precise and reliable estimates of body composition. The advantages of DXA over the underwater weighing technique include the ability to estimate bone density and bone mineral content in addition to fat and fat-free mass. Furthermore, it is a passive technique whereby the athlete simply lies on a table during the scan, as opposed to having to be submerged underwater multiple times. The disadvantage is the cost of the equipment and technical support.

Air plethysmography is a densitometric technique. Volume is determined by air displacement rather than by water immersion. This technique, developed in the early 1900s, was used largely in research laboratories up until the 1990s, when a commercial model became available that is now widely used (see figure 14.4). The principle of operation is rather simple. It involves a closed chamber of room air at atmospheric pressure, which has a known volume. The individual to be tested opens the chamber door, enters the chamber, sits in a fixed position, and then closes the chamber door, forming an airtight seal. The new volume of the air in the chamber is determined, which is then subtracted from the total volume of the chamber to provide an estimate of the person's volume.

Although this is a relatively simple technique for the subject, it requires considerable accuracy in controlling for changes in temperature, gas composition, and the subject's breathing while in the chamber. Studies have confirmed the accuracy of this technique under most conditions. It appears to provide a relatively precise measurement of body volume. Just as with the underwater weighing technique, one can obtain relatively accurate measurements of total body volume and thus obtain accurate estimates of total body density. However, one still must use the subject's body density in an equation to estimate relative body fat, recognizing the uncertainties of the D_{FFM} for that subject.

Field Techniques

Several field techniques are also available for assessing body composition. These techniques are more accessible than laboratory techniques because the equipment is less costly and cumbersome; they are techniques that therefore can be used more easily by the coach, the trainer, or even the athlete, outside the laboratory.

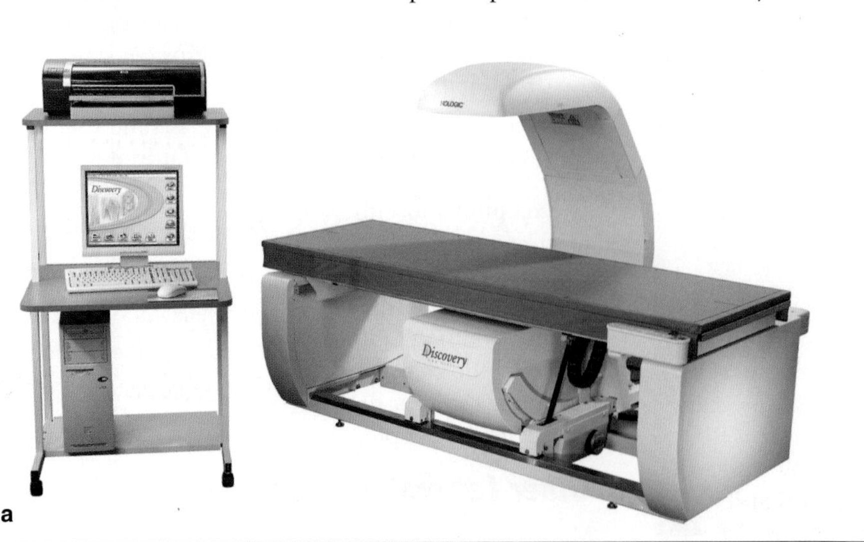

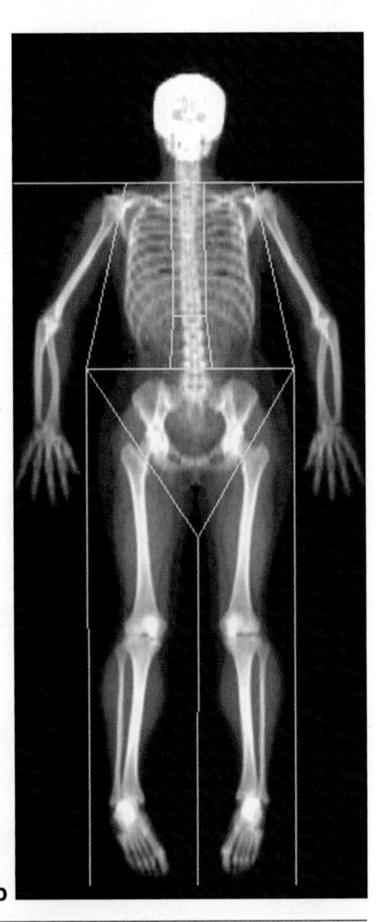

Figure 14.3 The dual-energy X-ray absorptiometry (DXA) machine used to estimate bone density and bone mineral content as well as total body composition (fat mass and fat-free mass): *(a)* the machine, *(b)* a regional scan of the body.

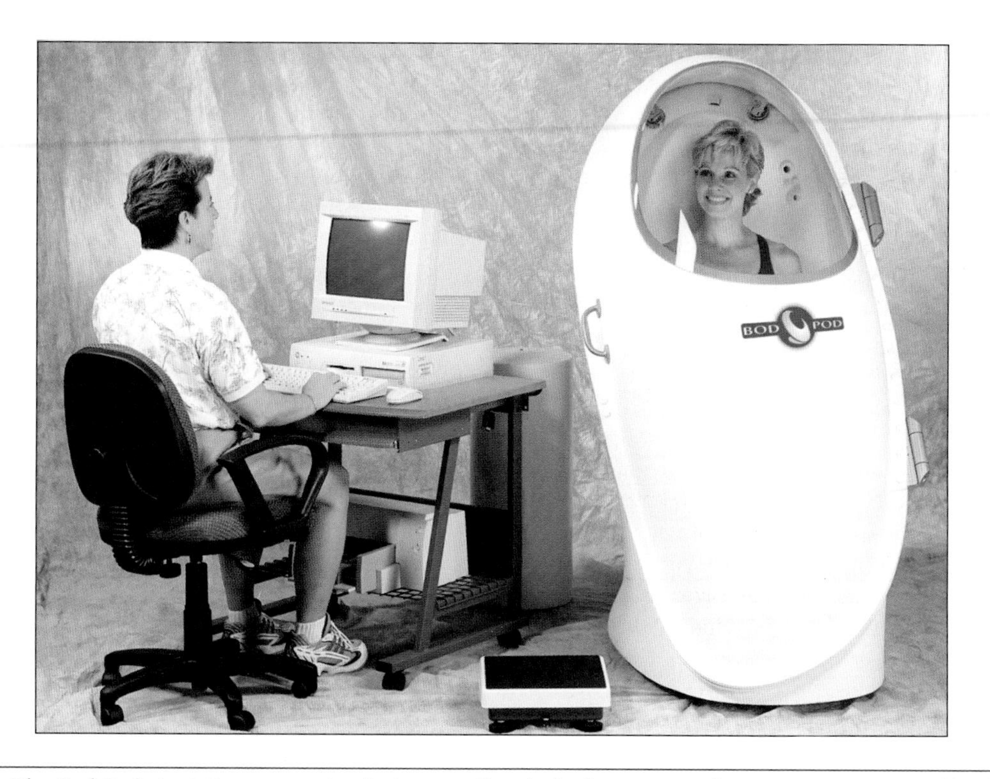

Figure 14.4 The Bod Pod air plethysmography device uses the air displacement technique to estimate total body volume.

Skinfold Fat Thickness

The most widely applied field technique involves measuring the **skinfold fat thickness** (see figure 14.5) at one or more sites and using the values obtained to estimate body composition. It generally is recommended that the sum of the measurements from three or more skinfold sites be used in a quadratic, curvilinear equation to estimate body density. A curvilinear equation more accurately describes the relationship between the sum of skinfold measurements and body density. Linear equations underestimate the density of lean people, which causes overestimation of body fat. Just the opposite happens for obese people: Body density is overestimated, and body fat is underestimated. Skinfold fat thickness measurements that use quadratic equations provide reasonably accurate estimates of total body fat or relative fat.

Bioelectric Impedance

Bioelectric impedance is a simple procedure introduced in the 1980s that takes just a few minutes to perform. Four electrodes are attached to the body, at the ankle, the foot, the wrist, and the back of the hand, as shown in figure 14.6 on page 322. An undetectable current is passed through the distal electrodes (hand and foot). The proximal electrodes (wrist and ankle) receive the current flow. Electrical conduction through the tissues between the electrodes depends on the water

Figure 14.5 Measuring skinfold fat thickness at the triceps skinfold site.

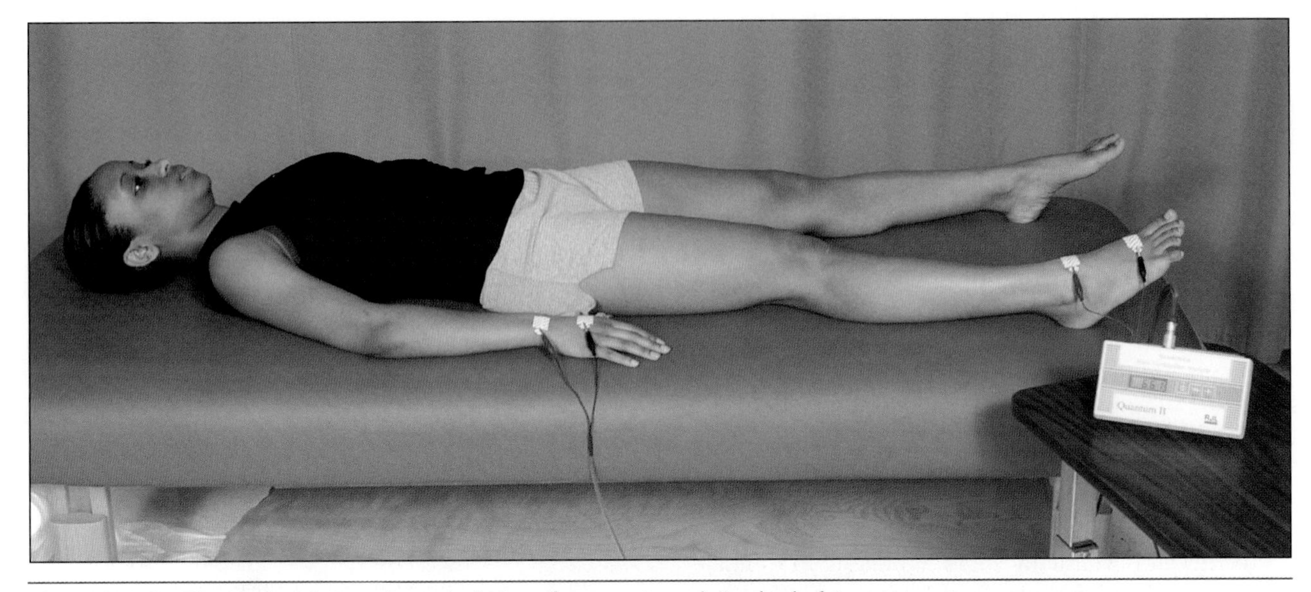

Figure 14.6 The bioelectric impedance technique for assessing relative body fat.

and electrolyte distribution in that tissue. Fat-free mass contains almost all the body water and the conducting electrolytes, so conductivity is much greater in the fat-free mass than in the fat mass. The fat mass has a much greater impedance, meaning that it is much more difficult for the current to flow through the fat mass. Thus, the amount of current flow through the tissues reflects the relative amount of fat contained in those tissues.

With the bioelectric impedance technique, measurements of the impedance, the conductivity, or both are transformed into estimates of relative body fat. Estimates of relative body fat based on bioelectric impedance highly correlate with body fat measurements obtained through hydrostatic weighing. However, the relative body fat in lean athletic populations tends to be overestimated with bioelectric impedance because of the nature of the equations used. More sport-specific equations are being developed; and a newer technique, multifrequency bioelectric impedance spectroscopy, could possibly improve the accuracy of measurement in these lean athletic populations.

Body Composition and Sport Performance

Many athletes in certain sports (e.g., football and basketball) believe that they must be big to be good in their sport because size traditionally has been associated with performance quality: The bigger the athlete, the better the performance. But big does not always mean better. In certain other sports, smaller and lighter are considered better for performance (e.g., gymnastics, figure skating, and diving). Yet this can be taken to extremes, compromising the athlete's health and performance. In the following sections, we consider how performance can be affected by body composition.

In review

- Knowing a person's body composition is more valuable for predicting performance potential than merely knowing height and weight.
- Densitometry is one of the best methods for assessing body composition and has long been considered the most accurate, although it does carry certain risks of error. It involves calculating the density of the athlete's body by dividing body mass by body volume, which is typically determined via hydrostatic weighing or air displacement. Body composition can be calculated, although there is some margin of error.
- Dual-energy X-ray absorptiometry, originally developed for estimating bone density and bone mineral content, is now capable of providing accurate estimates of total body composition—fat mass and fat-free mass.
- Field techniques for assessing body composition include measuring skinfold fat thickness and bioelectric impedance. These techniques are less costly and more accessible to the athlete and the coach than are laboratory techniques.

Fat-Free Mass

Rather than be concerned with total body size or weight, most athletes should be concerned specifically with fat-free mass. Maximizing fat-free mass is desirable for athletes involved in activities that require strength, power, and muscular endurance. But increased fat-free mass is likely to be undesirable for the endurance athlete, such as a distance runner, who must move his or her total body mass horizontally for extended periods. A higher fat-free mass is an additional load that must be carried and might impair the athlete's performance. This might also be true for the high jumper, long jumper, triple jumper, and pole-vaulter, who must maximize their vertical or horizontal distances or both. Additional weight, even though it is active fat-free mass, could decrease rather than facilitate performance in these events.

Techniques eventually will be available to estimate not only athletes' fat mass and fat-free mass but also the potential for increasing their fat-free mass. Such techniques would allow athletes to design training programs that would develop their fat-free mass to this projected maximum while maintaining their fat mass at relatively low levels. Combining resistance training with the ingestion of carbohydrate, or carbohydrate and protein, during recovery from resistance training appears to be effective for increasing the fat-free mass.[18, 30] This routine appears to stimulate the release of the anabolic hormones.

Relative Body Fat

Relative body fat is a major concern of athletes. Adding more fat to the body just to increase the athlete's weight and overall size is generally detrimental to performance. Many studies have shown that the higher the percentage of body fat, above optimal values, the poorer the person's performance. This is true of all activities in which the body weight must be moved through space, such as running and jumping. It is less important for more stationary activities, such as archery and shooting. In general, leaner athletes perform better.

Endurance athletes try to minimize their fat stores because excess weight has been proven to impair their performance. Both absolute fat and relative body fat can profoundly influence running performance in highly trained distance runners. Less fat generally leads to better performance.

Heavyweight weightlifters might be exceptions to the general rule that less fat is better. These athletes add large amounts of fat weight just before competition under the premise that the additional weight will lower their center of gravity and give them a greater mechanical advantage in lifting. The sumo wrestler is another notable exception to the theory that overall size is not the major determinant of athletic success. In this sport, the larger individual has a decided advantage; but even so, the wrestler with the higher fat-free mass should have the best overall success. Performance in swimming also seems to be an exception to this general rule. Body fat might provide some advantage to the swimmer because it improves buoyancy, which can reduce body drag in the water and reduce the metabolic cost of staying on the surface.

In review

- The ideal body composition varies with different sports, but in general, the less fat mass, the greater the performance.
- Maximizing fat-free mass is desirable for athletes in sports that require strength, power, and muscular endurance but could be a hindrance to endurance athletes, who must be able to move their total body mass for extended periods, and jumpers, who must move their body mass vertically or horizontally for distance.
- The degree of fatness has more influence on performance than does total body weight. In general, the greater the relative body fat, the poorer the performance. Possible exceptions include heavyweight weightlifters, sumo wrestlers, and swimmers.

Weight Standards

Weight standards have been used in several sports for many years. More recently, their use has become more widespread, and most sports have now adopted weight standards intended to ensure that athletes have the optimal body size and composition for maximal performance. Unfortunately, this is not always the result.

The elite athlete has long been esteemed as representing the most desirable physical and physiological characteristics for performance in a sport or activity. Theoretically, the elite athlete's genetic foundation and years of intense training have combined to provide the ultimate athletic profile for the given sport. These elite athletes set the standards toward which others aspire. However, this can be misleading as we see in figure 14.7.[28] The figure displays the percentage body fat values of elite female track and field athletes. If we look just at the distance runners, many of the best were below 12% body fat. The two top distance runners had only about 6% fat. One of these had won six consecutive international cross country championships,

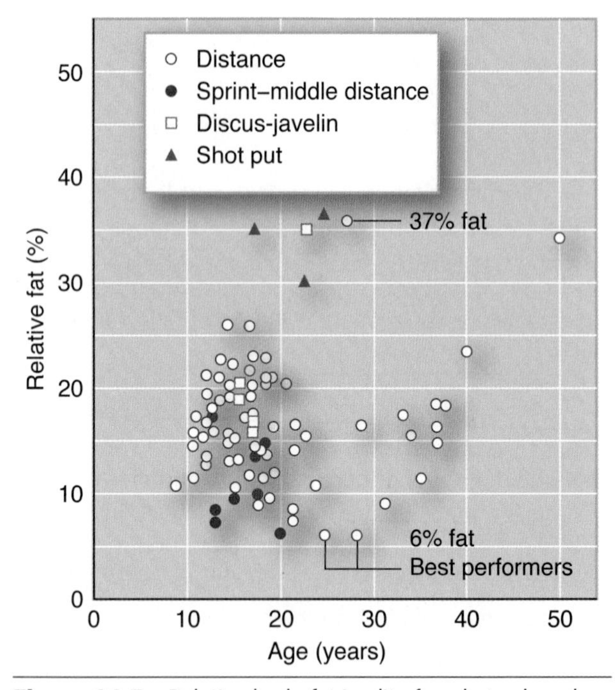

Figure 14.7 Relative body fat in elite female track and field athletes.

Data from Wilmore et al. 1977.

and the other held what was then the best time in the world for the marathon. From these results, we could be tempted to suggest that any female distance runner should have between 6% and 12% relative body fat if she has world-class aspirations. However, one of the best distance runners in the United States at that time, who was within two years of taking over the top spot, had a relative body fat of 17%. Furthermore, one of the women in this study had a relative body fat of 37%, and she set the best time in the world for the 50 mi (80 km) run within six months of her evaluation! More than likely, neither of these women would have gained an advantage if she had been forced to decrease her weight to achieve 12% body fat or lower.

Inappropriate Use of Weight Standards

Weight standards have been seriously abused. Coaches have seen that athletes' performances generally improve as body weight decreases. This has led some coaches to adopt the philosophy that if small weight losses improve performance a little, then major weight losses should improve it even more. Not only coaches are guilty of making this assumption: Athletes and their parents also are drawn into this way of thinking. As an example, a university athlete, considered one of the best in the United States in her sport, had dieted and exercised down to such a low body weight that her relative body fat was less than 5%. If anyone who joined the team appeared to be leaner, she would work even harder to reduce her weight and fat content. This woman's athletic performance began to deteriorate, and she started to develop injuries that never seemed to heal. She was eventually diagnosed with anorexia nervosa (chapter 18) and underwent professional treatment. But her career as an elite athlete was over.

Making Weight: Risks With Severe Weight Loss

Many schools, districts, or state- and national-level organizations organize sports (e.g., wrestling) on the basis of size, with weight as the predominant factor. Athletes in these sports often attempt to achieve the lowest possible weight to gain an advantage over opponents. In so doing, many have jeopardized their health. In the following sections, we examine a few of the consequences of severe weight loss in athletes, both male and female.

Dehydration

Fasting or very low calorie diets lead to large amounts of weight loss, primarily through dehydration. As we discuss later in this chapter, for every gram of carbohydrate stored, there is an obligatory gain of 2.6 g of water. When carbohydrates are used for energy, that water is lost. Thus, with fasting and very low calorie diets, carbohydrate stores are substantially depleted during the first few days. This results in a significant loss of weight attributable to the loss of body water.

Furthermore, athletes trying to make weight might exercise in rubberized sweat suits, sit in steam and sauna baths, chew on towels to lose saliva, and minimize their fluid intake. Such severe water losses compromise kidney and cardiovascular function and are potentially dangerous. Losses of 2% to 4% of the athlete's weight through dehydration can impair performance.

Chronic Fatigue

Pushing body weight too low can have major repercussions. When weight drops below a certain optimal level, the athlete is likely to experience performance decrements and increased incidence of illness and injury. The performance decrements can be attributable to many factors, including chronic fatigue that often accompanies major weight losses. The causes of this fatigue have not been established, but there are several likely possibilities.

The symptoms of an athlete who is chronically underweight (below optimal competitive weight) mimic those seen with overtraining (chapter 13). Both neural and hormonal components are involved in the phenomenon of overtraining. In some cases, the sympathetic nervous system appears to be inhibited, and the parasympathetic system dominates. In addition, the hypothalamus does not function normally, and immune function is likely impaired. These alterations lead to a cascade of symptoms that include chronic fatigue.

This chronic fatigue also could be attributed to substrate depletion. Energy for almost all athletic activities is derived predominantly from carbohydrate. Carbohydrate also represents the smallest source of stored energy. The combined carbohydrate storage in muscle, liver, and extracellular fluid accounts for approximately 2,500 kcal of stored energy. When athletes are training hard and are not eating an adequate diet (when they are deficient either in total calories or in total carbohydrate calories), their carbohydrate energy stores become depleted. Most important to the athlete, liver and muscle glycogen levels decrease, which in turn reduces blood glucose levels. The combined effect of these decreases can be chronic fatigue and considerable declines in performance. In addition, under these conditions the body also uses its protein stores as an energy substrate for exercise. This can, over time, gradually deplete muscle protein.

During the 1990s, we started to see athletes diagnosed with **chronic fatigue syndrome.** This syndrome might or might not be related to what we have termed chronic fatigue. At this time, very little is known about chronic fatigue syndrome, although it does appear to involve immune system dysfunction (see chapter 13). Patients have incapacitating fatigue, and the symptoms may vary in severity over time but generally last for months or years. The symptoms include prolonged, debilitating fatigue; sore throat; muscle tenderness or pain (myalgia); and cognitive dysfunction.

Eating Disorders

The constant attention given to achieving and maintaining a prescribed weight goal, particularly if the weight goal is inappropriate, can lead to disordered eating. A high proportion of athletes, predominantly females, have disordered eating. This term can simply refer to restricting food intake to levels that are well below energy expenditure, but disordered eating can also involve pathological behaviors to control body weight, such as self-induced vomiting and laxative abuse. Disordered eating can lead to clinical eating disorders, such as anorexia nervosa or bulimia nervosa. These disorders have become prevalent among female athletes. Each has strict criteria for diagnosis that set it apart from disordered eating in general.

More than 90% of people with eating disorders are women. Among athletes, those in appearance sports (such as gymnastics, figure skating, diving, and dancing) and in endurance sports (such as running and swimming) seem at greatest risk. On certain teams, particularly in these appearance and endurance sports, the prevalence of eating disorders might approach or even exceed 50% at the elite or world-class level. Athletes and coaches must realize the potential link between weight standards and eating disorders. The issue of eating disorders in athletes is a major focus of chapter 18.

> **In focus**
>
> Eating disorders appear to be prevalent among elite female athletes. It is important to establish weight standards that maximize performance but minimize the risk of initiating an eating disorder.

A female athlete who is prone to disordered eating is open to a triad of disorders that are likely interrelated: anorexia nervosa or bulimia nervosa, menstrual dysfunction, and bone mineral disorders. This group of disorders is now referred to as the **female athlete triad.** This is discussed in much more detail in chapter 18.

Menstrual Dysfunction

Menstrual dysfunction, or abnormal menstruation, is widely recognized in female athletes. High prevalences of oligomenorrhea (infrequent or scant menstrual flow), amenorrhea (cessation of menstrual flow), and delayed menarche (first period) have been associated with sports that emphasize low body weight or low body fat. The combination of caloric restriction and a vegetarian diet is common among female endurance athletes. Substantial weight loss induced by either or both of these is associated with a shortened luteal phase and menstrual dysfunction. Most likely, menstrual dysfunction is the body's natural adaptation to a prolonged energy deficit, in which energy intake remains below energy expenditure over long periods of time. This is discussed more fully in chapter 18.

A strong link exists between anorexia nervosa and menstrual dysfunction. In fact, amenorrhea is one of the strict criteria necessary for the diagnosis of anorexia nervosa in females. A similar relationship has not yet been established with bulimia (see chapter 18), but an increasing number of athletes are found to be both bulimic and amenorrheic.

Bone Mineral Disorders

Bone mineral disorders are recognized as a potentially serious consequence of menstrual dysfunction. The link between the two was first reported in 1984. Now a number of scientists are researching the relationship between athletic-induced amenorrhea and low bone mineral content or density. Past studies suggested that bone density increases with the resumption of normal menses (menstruation), but more recent observations suggest that the amount of bone that is regained might be limited and that bone density might remain well below normal even with reestablishment of normal menstrual function. The long-term consequences of chronically low bone densities in athletic populations have not yet been established.

Establishing Appropriate Weight Standards

The potential for abuse of weight standards is clearly established. If standards are not set appropriately, athletes could be pushed well below optimal body weight. Thus, it is critically important to properly set weight standards.

Body weight standards should be based on an athlete's body composition. Thus, establishing weight standards should translate into establishing standards of relative body fat for each sport and, where appropriate, for each event within a sport. With this in mind, what is the recommended relative body fat for an elite athlete in any given sport? For each sport, an optimal range of values for relative body fat should be established, outside of which the athlete's performance is likely impaired. And because fat distribution shows definite sex differences, the weight standards should be sex specific. Representative ranges for men and women in various sports are presented in table 14.1. In most cases, these values represent the elite athletes in those sports.

The recommended values might not be appropriate for all athletes who engage in a specific activity. The existing techniques for measuring body composition include inherent errors, as we discussed earlier. Even with the better laboratory techniques, measurement of body density can introduce a 1% to 3% error, and an even greater error is associated with converting that density to relative body fat. In addition, we must understand the concept of individual variability. Not every female distance runner will have her best performance at 12% body fat or lower. While some will improve performance with these low values, others won't be able to get down to such low relative fat values, or they will find that their performance starts to decline before they reach the suggested values. For these reasons, a range of values should be set for males and females in specific activities, recognizing individual variability, methodological error, and sex differences.

Achieving Optimal Weight

Many athletes discover that they are considerably above their assigned playing weight with only a few weeks remaining before they report to training camp. Consider a 25-year-old professional football player who realizes that his weight is 9 kg (20 lb) above his playing weight in the previous season. He must lose this excess weight by the start of the preseason training camp, a mere four weeks away. Failure will cost him a fine of $1000 per day for each pound over his assigned weight. Exercise alone is of little value because he would need 9 to 12 months to lose this much weight through that means. Will this athlete accomplish his goal?

Avoiding Fasting and Crash Dieting

Our football player must lose 2.3 kg (5 lb) per week for the next four weeks, so he decides to embark on a crash diet, selecting whatever diet is in vogue, knowing that a person can lose 2.7 to 3.6 kg (6-8 lb) per week with such diets. He is not unique. Many athletes find themselves overweight and out of shape because of overeating and reduced activity during the off-season, and they typically wait until the last few weeks before their reporting dates to attack the problem. In our example, the football player might be able to shed 9 kg (20 lb) in four weeks with his crash diet. But much of this weight loss would be from body water and very little from stored fat. Several studies have shown that substantial weight losses occur with very low calorie diets (500 kcal per day or less) over

In review

- Many sports enforce weight standards with the goal of ensuring that the athletes are of optimal body size for participation. Unfortunately, athletes often turn to questionable, ineffective, or even dangerous methods of weight loss to reach their weight goal.
- Severe weight loss in athletes can cause potential health problems, such as dehydration, chronic fatigue, disordered eating, menstrual dysfunction, and bone mineral disorders.
- The chronic fatigue symptoms that often accompany severe weight loss mimic those of overtraining. This fatigue also can be caused by substrate depletion.
- Body weight standards should be based on body composition. Thus, these standards should emphasize relative body fat rather than total body mass.
- For each sport, a range of values should be established, recognizing the importance of individual variation, methodological error, and sex differences.

TABLE 14.1 Ranges of Relative Body Fat Values for Male and Female Athletes in Various Sports

	% fat			% fat	
Group or sport	Men	Women	Group or sport	Men	Women
Baseball or softball	8-14	12-18	Rugby	6-16	–
Basketball	6-12	10-16	Skating	5-12	8-16
Bodybuilding	5-8	6-12	Skiing (alpine and Nordic)	7-15	10-18
Canoeing or kayaking	6-12	10-16	Ski jumping	7-15	10-18
Cycling	5-11	8-15	Soccer	6-14	10-18
Fencing	8-12	10-16	Swimming	6-12	10-18
Football	6-18	–	Synchronized swimming	–	10-18
Golf	10-16	12-20	Tennis	6-14	10-20
Gymnastics	5-12	8-16	Track and field, field events	8-18	12-20
Horse racing (jockey)	6-12	10-16	Track and field, running events	5-12	8-15
Ice or field hockey	8-16	12-18	Triathlon	5-12	8-15
Orienteering	5-12	8-16	Volleyball	7-15	10-18
Pentathlon	–	8-15	Weightlifting	5-12	10-18
Racquetball	6-14	10-18	Wrestling	5-16	–
Rowing	6-14	8-16			

the first several weeks, but that of the weight lost, more than 60% comes from the body's fat-free tissue and less than 40% from fat depots.

While much of our football player's weight is lost from water stores, a substantial amount of protein is lost as well. Also, most crash diets are based on a major reduction in carbohydrate intake. This reduced intake is insufficient to supply the body's needs for carbohydrate, and as a result the body's carbohydrate stores become depleted. Because water storage accompanies carbohydrate storage, the water stores are also reduced as the carbohydrate stores diminish, as discussed earlier in this chapter.

In addition, the body relies more heavily on free fatty acids for energy because its carbohydrate stores are depleted. As a result, ketone bodies, a by-product of fatty acid metabolism, accumulate in the blood, causing a condition known as ketosis. This condition further increases water loss. Much of this water loss occurs during the first week of the diet. Athletes that have attempted this ill-advised shortcut to rapid weight loss will lose substantial weight, but, because much of the weight loss is from the fat-free mass, their performance has been severely compromised.

Optimal Weight Loss: Decreasing Fat Mass and Increasing Fat-Free Mass

The sensible approach to reducing body fat stores is to combine moderate dietary restriction with increased exercise.

When athletes exceed the upper end of the weight range for their sport, they should work toward achieving the upper-end goal weight slowly, losing no more than 0.5 to 1 kg (less than 2.2 lb) per week. Losing more weight than that per week leads to losses in fat-free mass, which is usually not the desired outcome. When the upper limit of the range is reached, further weight loss should be undertaken only with close supervision of the coach, athletic trainer, or team physician. This weight loss should be achieved at an even slower rate—less than 0.5 kg (1.1 lb) per week—to ensure that performance is not negatively affected. The rate of this loss should be reduced still more, or the weight loss program terminated, if performance is affected or if medical symptoms are noted.

Decreasing caloric intake by 200 to 500 kcal per day will allow weight losses of about 0.5 kg (1.1 lb) per week, particularly if combined with a sound exercise program. This is a realistic goal, and such losses add up to a substantial weight loss over time. When trying to reduce weight, athletes should consume their total daily calories over at least three meals per day. Many athletes make the mistake of eating only one or two meals per day, skipping breakfast, lunch, or both, and then consuming a large dinner. Research in animals has shown that, given the same number of total calories, the animals that eat their daily food ration in one or two meals gain more weight than those that nibble their ration throughout the day. Human research is less clear.

The purpose of weight loss programs is to lose body fat, not fat-free mass. Therefore the combination of diet and exercise is the preferred approach. Combining increased activity with caloric reduction prevents any significant loss of fat-free mass. In fact, body composition can be significantly altered with physical training. Chronic exercise can increase fat-free mass and decrease fat mass. The magnitude of these changes varies with

In focus

Athletes who are above their weight standard should lose weight gradually, not more than 0.5 to 1 kg (about 1-2 lb) per week, to preserve their fat-free mass. They should accomplish this by integrating a good diet containing 200 to 500 kcal less than their daily energy expenditure with a reasonable increase in resistance and endurance activities.

the type of exercise used for training. Resistance training promotes gains in fat-free mass, and both resistance training and endurance training promote loss of fat mass. To lose weight, athletes should combine a moderate resistance and endurance training program with modest caloric restriction.

As a final point, a balanced diet is, of course, essential to ensure that the athlete receives all necessary vitamins and minerals. Vitamin supplementation might or might not be necessary: Results of research thus far are in conflict. But if the nutritional adequacy of the diet is at all questionable, a simple multivitamin that meets the person's needs is suggested.

In review

- When severe (very low calorie) diets are followed, much of the weight loss that occurs is from water, not fat.
- Most severe diets limit carbohydrate intake, depleting carbohydrate stores. Water is lost along with the carbohydrates, exacerbating the problem of dehydration. Also, the increased reliance on free fatty acids can lead to ketosis, which further increases water loss.
- The combination of diet and exercise is the preferred approach to optimal weight loss.
- Athletes should lose no more than about 0.5 to 1.0 kg (1.1-2.2 lb) per week until reaching the upper end of the desired weight range. After that, weight loss should be less than 0.5 kg (1 lb) per week until goal weight is reached. More rapid weight losses result in loss of fat-free mass. People can accomplish weight loss at the recommended rate by reducing dietary intake 200 to 500 kcal per day, especially in combination with a sound exercise program.
- For fat loss, moderate resistance and endurance training is most effective. Resistance training also promotes gains in fat-free mass.

NUTRITION AND SPORT

Having established weight and body composition standards, we now turn to the nutritional aspects of preparing the athlete for optimal performance. As we will see in this section, it is important to maintain a diet that will provide general health benefits, maintain an appropriate weight and body composition, and maximize athletic performance.

A person's diet should contain a relative balance of carbohydrate, fat, and protein. Of the total calories consumed, the recommended balance for most people is

- carbohydrate—55% to 60%;
- fat—no more than 35% (less than 10% saturated); and
- protein—10% to 15%.

Interestingly, this recommended percentage distribution of total calories consumed appears to be optimal for both athletic performance and health. A similar distribution of caloric intake is recommended for the prevention of cardiovascular disease, diabetes, obesity, and cancer as discussed in greater detail later in this chapter. Although all foods ultimately can be broken down to carbohydrate, fat, or protein, these nutrients are not all that the body needs, as we see in the next section.

Classification of Nutrients

Energy from ingested foods is essential to our ability to sustain physical activity, but we rely on foods for much more than energy. Food can be categorized into six classes of nutrients, each with specific functions in the body:

- Carbohydrate
- Fat (lipid)
- Protein
- Vitamins
- Minerals
- Water

The following discussion examines the physiological importance to the athlete of each class of nutrients.

Carbohydrate

A **carbohydrate** (CHO) is classified as either a monosaccharide, disaccharide, or polysaccharide. Monosaccharides are the simple one-unit sugars, such as glucose, fructose, and galactose, that cannot be reduced to a simpler form. Disaccharides (such as sucrose, maltose, and lactose) are composed of two monosaccharides. For example, sucrose (table sugar) consists of glucose and fructose. Oligosaccharides are short chains of 3 to 10 monosaccharides linked together. Polysaccharides are composed of large chains of linked monosaccharides. Glycogen is the polysaccharide found in animals, including man, and is stored in muscle and liver. Starch and fiber are the two plant polysaccharides and are commonly referred to as complex carbohydrates. Simple carbohydrates are carbohydrates derived from processed foods or foods high in sugar. All carbohydrates must be broken down to monosaccharides before the body can use them.

Carbohydrate serves many functions in the body:

- It is a major energy source, particularly during high-intensity exercise.
- Its presence regulates fat and protein metabolism.
- The nervous system relies exclusively on carbohydrate for energy.
- Muscle and liver glycogen are synthesized from carbohydrate.

Major sources of carbohydrate include grains, fruits, vegetables, milk, and concentrated sweets. Refined sugar, syrup, and cornstarch are nearly pure carbohydrates. Many concentrated sweets such as candy, honey, jellies, molasses, and soft drinks contain few if any other nutrients.

Carbohydrate Consumption and Glycogen Storage

The body stores excess carbohydrate, primarily in the muscles and liver, as glycogen. Because of this, carbohydrate consumption directly influences muscle glycogen storage and the ability to train and compete in endurance events. As shown in figure 14.8 on page 330, athletes who trained intensely over three consecutive days and ate a low-carbohydrate diet (40% of total calories) experienced a day-to-day decrease in muscle glycogen.[8] When these same athletes consumed a high-carbohydrate diet (70% of total calories), their muscle glycogen levels recovered almost completely within the 22 h between training bouts. In addition, athletes perceive training as easier when

A Major Transition: From the RDAs to the DRIs

In the early 1940s, the Food and Nutrition Board of the National Academy of Sciences established the United States Recommended Daily Allowance (RDA) guidelines for all food nutrients. The latest edition of the RDAs in their original form was released in 1989. The RDAs provide estimates of safe and adequate daily dietary intakes and estimated minimum requirements for selected vitamins and minerals. A major revision of the RDAs was initiated in the early 1990s. The RDAs have been replaced by new recommendations called Dietary Reference Intakes. The DRIs reflect a joint effort between the United States and Canada to provide dietary intake recommendations grouped by nutrient function and classification.

The new DRIs were released in a series of reports starting in 1997 and continuing through 2005. They include four different reference values:

- Estimated Average Requirement (EAR)—the intake value estimated to meet the requirement for 50% of healthy individuals in an age- and sex-specific group.
- Recommended Dietary Allowance (RDA)—the intake value sufficient to meet the nutrient requirements of nearly all (97-98%) individuals in a specific group.
- Tolerable Upper Intake Level (UL)—the highest level of daily nutrient intake that is unlikely to pose no risk of adverse health effects to almost all individuals in a specific group.
- Adequate Intake (AI)—a recommended intake value based on observed or experimentally determined approximations or estimates of nutrient intake by healthy individuals in a specific group that are assumed to be adequate. This is used when an RDA cannot be determined.

For further information on DRIs and specific recommendations for each of the nutrient classifications by sex and age, refer to the Food and Nutrition Information Center, U.S. Department of Agriculture, at www.nal.usda.gov/fnic.

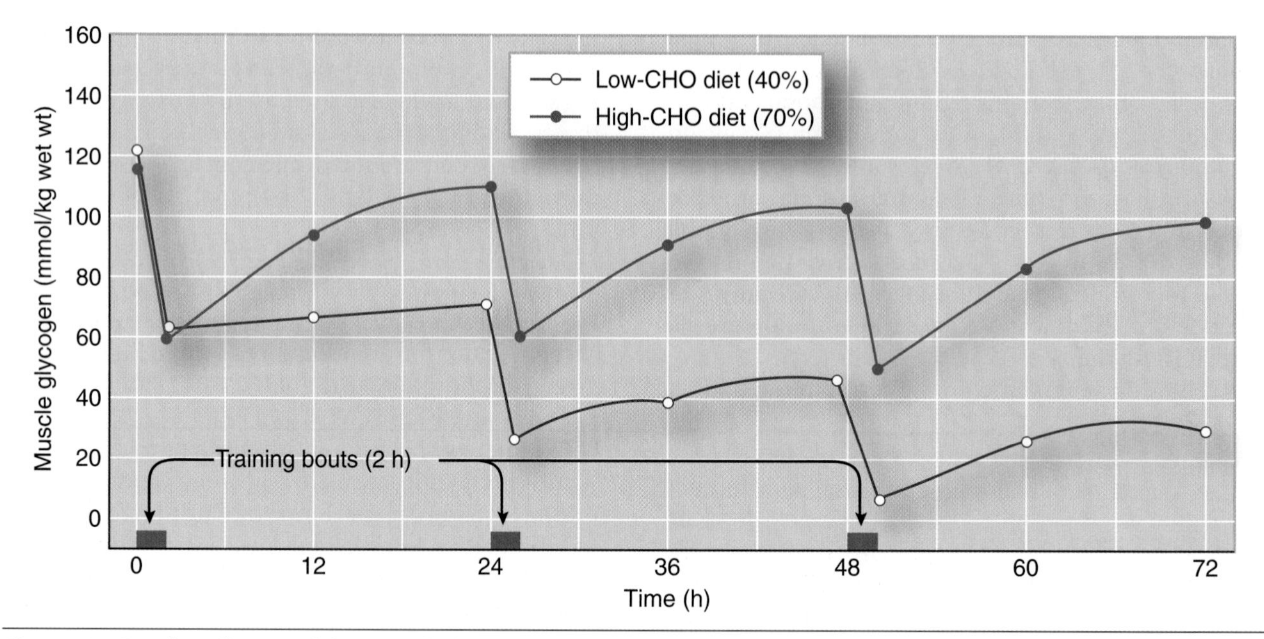

Figure 14.8 The influence of dietary carbohydrates (CHO) on muscle glycogen stores during repeated days of training. Note that when a low-CHO diet was consumed, muscle glycogen gradually declined over the three days of study, whereas the CHO-rich diet was able to return the glycogen to near normal each day.

D.L. Costill and J.M. Miller, "Nutrition for endurance sport: Carbohydrate and fluid balance," 1980, *International Journal of Sports Medicine*, 1: 2-14. Reprinted by permission.

their muscle glycogen is maintained throughout a workout.

Early studies demonstrated that when men eat a diet containing a normal amount of carbohydrate (about 55% of total calories ingested), their muscles store about 100 mmol of glycogen per kilogram of muscle. One study showed that diets containing less than 15% carbohydrate led to storage of only 53 mmol/kg, but carbohydrate-rich diets (60-70% CHO) led to storage of 205 mmol/kg. When subjects exercised to exhaustion at 75% of their maximal oxygen uptake, their exercise times were proportional to the amount of muscle glycogen stored before the test, as shown in figure 14.9.

Most studies have shown that glycogen storage replacement is not determined simply by carbohydrate intake. Exercise with an eccentric (muscle lengthening) component, such as running and weightlifting, can induce some muscle damage and impair glycogen resynthesis. In these situations, muscle glycogen levels can appear quite normal during the first 6 to 12 h after exercise, but glycogen resynthesis slows or stops completely as muscle repair begins.

The precise cause for this response is unknown, but conditions in the muscle could inhibit muscle glucose uptake and glycogen storage. For example, within 12 to 24 h after intense eccentric exercise, damaged muscle fibers are infiltrated with inflammatory cells (leukocytes, macrophages) that remove cellular debris

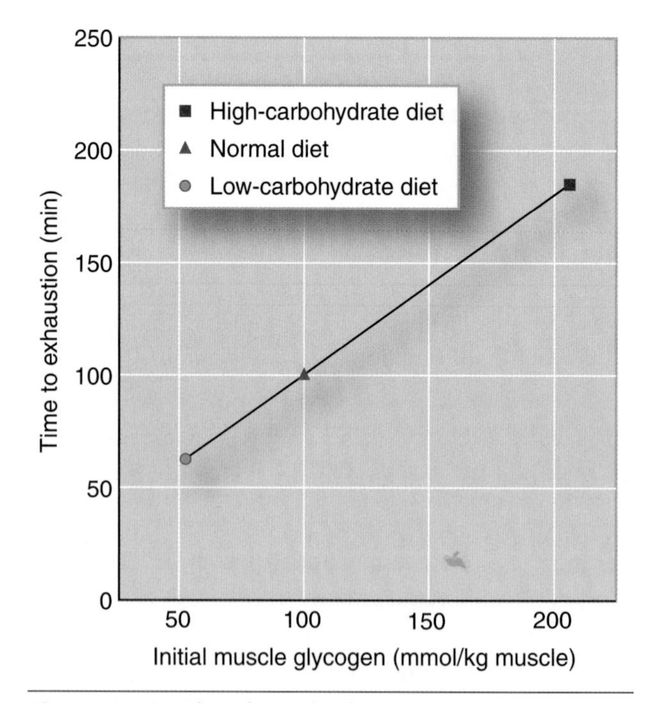

Figure 14.9 The relationship between preexercise muscle glycogen content and exercise time to exhaustion. The exercise time to exhaustion and muscle glycogen were nearly four times greater when the subjects ate a carbohydrate-rich diet than when the diet was composed mostly of fat and protein.

Adapted from Åstrand, P.-O. (1967). Diet and athletic performance. *Federation Proceedings*, 26, 1772-1777.

In focus

Carbohydrate is the primary fuel source for most athletes and should constitute at least 50% of their total caloric intake. For endurance athletes, carbohydrate intake as a percentage of total caloric intake might need to be higher: 55% to 65%. However, most important is the total number of grams of carbohydrate ingested. It appears that athletes need from 5 to 13 g/kg of body weight per day in order to maintain glycogen stores. This wide range is necessary to account for the training intensity and total daily energy expenditure, sex, and environmental conditions. For example, during periods of moderate-intensity training, 5 to 7 g/kg per day should be adequate. However, with long-duration, high- and extremely high-intensity training, intake should be increased to 7 to 10 g/kg per day and 10 to 13 g/kg per day, respectively.[22]

resulting from damage to the cells' membranes (see chapters 9 and 13). This repair process can require a significant amount of the blood glucose, reducing the amount of glucose available for resynthesizing muscle glycogen. In addition, some evidence suggests that eccentrically exercised muscle is less sensitive to insulin, which would limit muscle fiber uptake of glucose. Perhaps future studies will more fully explain why eccentric-type activities delay glycogen storage. But for now, we can only observe that glycogen recovery from various forms of exercise can differ and that this should be considered for optimal diet, training, and competition.

When athletes eat only as much food as hunger dictates, they often fail to consume enough carbohydrate to compensate for the amount used during training or competition. This imbalance between glycogen use and carbohydrate intake might explain in part why some athletes become chronically fatigued and need 48 h or more to restore normal muscle glycogen levels. Athletes who train exhaustively on successive days require a diet rich in carbohydrate to reduce the heavy, tired feeling associated with muscle glycogen depletion.

The Glycemic Index

It has long been known that the rapid increase in blood sugar levels (hyperglycemia) with the intake of carbohydrate usually is associated with simple carbohydrates, such as glucose, sucrose, fructose, and high-fructose corn syrup. However, this is not always the case. Scientists have discovered that the glycemic response (i.e., increase in blood sugar) to carbohydrate intake varies considerably for both simple and complex carbohydrates. This has led to the use of what has been termed the glycemic index of foods (GI). The ingestion of glucose or white bread leads to a rapid increase in blood sugar. Their response is used as a standard and has been arbitrarily assigned a GI of 100. The glycemic response for all other foods is referenced against the response for glucose or white bread, using 50 g of both the test food and glucose or white bread as the standard. The GI is calculated as follows: GI = 100 × (blood glucose response over 2 h to 50 g of test food/blood glucose response over 2 h to 50 g of glucose or white bread). Three categories of GI have been established:[22]

- High glycemic index foods (GI >70) such as sport drinks, jelly beans, baked potato, french fries, popcorn, cornflakes, Corn Chex, and pretzels
- Moderate glycemic index foods (GI 56-70) such as pastry, pita bread, boiled white rice, bananas, Coca-Cola, and regular ice cream
- Low glycemic index foods (GI ≤55) such as white boiled spaghetti, kidney and baked beans, milk, grapefruit, apples, pears, peanuts, M&M's, and yogurt

Note. Food items were classified according to the 2002 International Table of Glycemic Index and Glycemic Load Values.[12]

While the GI is a useful tool for rating foods, it is not without controversy. First, the GI for a given food can vary considerably between individuals as well as between mean values for research studies with large numbers of subjects. Second, some complex carbohydrates have high GIs. Third, adding small amounts of fat to a high-GI carbohydrate can greatly reduce the GI of that food. Finally, GI values differ substantially depending on whether glucose or white bread is used as the reference food, with white bread producing substantially higher values.[12, 22] An additional index has been proposed that might be of importance during exercise. The glycemic load (GL) considers both the GI and the amount of carbohydrate (CHO) in a single serving, and is calculated as follows: GL = (GI × CHO, g)/100.

With this as background, we can now consider the implications of the GI for sport nutrition. Before exercise, low-GI foods would be preferred to reduce the likelihood of hyperinsulinemia. However, high-GI foods should be an advantage during exercise by helping maintain blood glucose levels. This should also be the case during recovery from intense and prolonged exercise, as the higher blood sugar level should increase muscle and liver glycogen storage.

Carbohydrate Intake and Performance

As noted earlier, muscle glycogen provides a major source of energy during exercise. Muscle glycogen depletion has been shown to be a major cause of fatigue and ultimate exhaustion in high-intensity exercise of short duration or in moderate-intensity exercise lasting more than an hour. This is clearly illustrated in figure 14.10, which shows the marked depletion of muscle glycogen at very high intensities (150% and 120% of $\dot{V}O_{2max}$) for durations of less than 30 min and at lower intensities (83%, 64%, and 31% of $\dot{V}O_{2max}$) for durations of an hour up to 3 h. The original data for this figure were from Gollnick, Piehl, and Saltin.[15] Scientists speculated that loading the muscle with extra glycogen before starting exercise should enhance performance.

Early studies demonstrated that men who ate a carbohydrate-rich diet for three days stored nearly twice their normal amounts of muscle glycogen.[3] When they were asked to exercise to exhaustion at 75% of $\dot{V}O_{2max}$, their exercise times significantly increased (see figure 14.9). This practice, called **glycogen loading** or **carbohydrate loading,** is widely used by distance runners, cyclists, and other athletes who must perform exercise for several hours. We discuss this practice in greater detail later in this chapter.

Blood glucose levels become low (hypoglycemia) during exhaustive high-intensity, long-duration exercise, and this might contribute to fatigue. Numerous studies have shown that subjects' performances improve when they are given carbohydrate feedings during exercise lasting 1 to 4 h. Comparisons of subjects receiving carbohydrate feedings and those receiving placebos generally reveal no performance differences during the early stages of exercise; but during the final stages, performance is greatly improved with carbohydrate feedings.

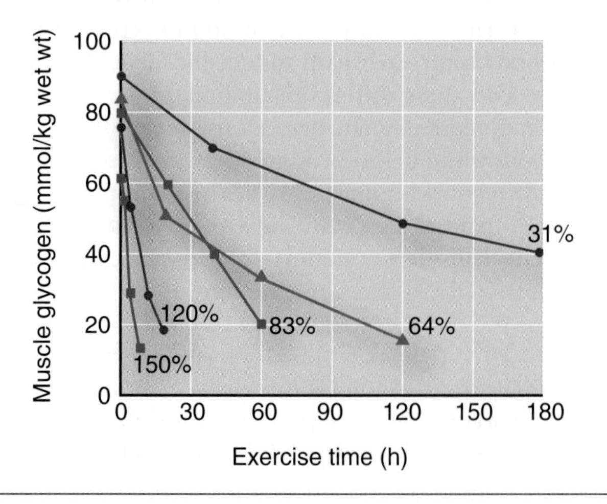

Figure 14.10 The influence of exercise intensity (31%, 64%, 83%, 120%, and 150% of $\dot{V}O_{2max}$) on muscle glycogen stores. At relatively high intensities, the rate of muscle glycogen use is extremely high compared to that at the moderate and lower intensities.

Adapted, by permission, from A. Jeukendrup and M. Gleeson, 2004, *Sport nutrition: An introduction to energy production and performance* (Champaign, IL: Human Kinetics). Original data from Gollnick, Piehl, and Saltin.

Although we don't fully understand how carbohydrate feedings improve performance, most scientists believe that maintaining blood glucose near normal levels allows the muscles to obtain more energy from blood glucose. Carbohydrate feedings during exercise generally do not spare muscle glycogen use, although this is not always the case. Instead, they may preserve liver glycogen or even promote glycogen synthesis during exercise, enabling the exercising muscles to rely more on blood glucose for energy late in the exercise. Carbohydrate feedings might also enhance central nervous system function, reducing the perception of effort. Endurance performance (more than 1 h) can be enhanced when carbohydrate is consumed within 5 min before the exercise begins, more than 2 h before exercise (such as during the precompetition meal), and at frequent intervals during the activity.

An athlete should use caution when ingesting carbohydrate foods during the period from 15 to 45 min before exercise because this could cause hypoglycemia shortly after the exercise begins, which could lead to early exhaustion by depriving the muscle of its primary energy sources. Carbohydrate ingested during that period stimulates insulin secretion, elevating insulin when the activity begins.[9] In response to the elevated insulin level, glucose uptake by the muscles reaches an abnormally high rate, leading to hypoglycemia and early fatigue (figure 14.11). Not everyone experiences this reaction, but sufficient evidence indicates that high-GI carbohydrates (those that cause a large increase in blood insulin) should be avoided or moderated in the period from 15 to 45 min before exercise.

Why don't carbohydrate feedings during exercise produce the same hypoglycemic effects observed with preexercise feedings? Sugar feedings during exercise result in smaller increases in both blood glucose and insulin, lessening the threat of an overreaction that leads to a sudden decrease in blood glucose. This finer control of blood glucose during exercise might be caused by increased muscle fiber permeability that decreases the need for insulin, or insulin-binding sites may be altered during muscular activity. Regardless of the cause, carbohydrate intake during exercise appears to supplement the carbohydrate supply needed for muscular activity.

Finally, it is important to consume carbohydrates immediately after high-intensity and long-duration exercise during which carbohydrate stores have been reduced or depleted. Rates of glycogen resynthesis are very high during the first 2 h of recovery, and progressively decrease thereafter. In a study by Ivy and colleagues,[20] cyclists exercised continuously for 70 min on a cycle ergometer on two separate occasions, a

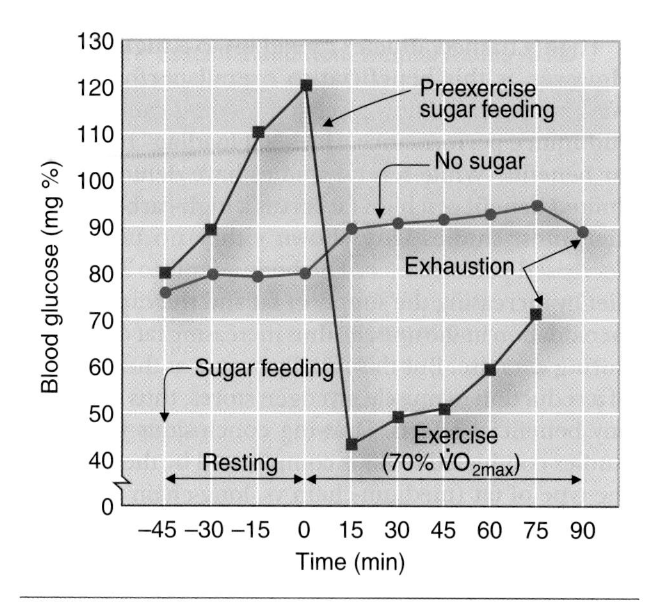

Figure 14.11 The effects of preexercise carbohydrate (sugar) feeding on blood glucose levels during exercise. Note the decrease in blood glucose to hypoglycemic levels with sugar feeding 45 min before exercise. Also, subjects during the sugar-feeding trial were unable to complete the full 90 min at 70% of $\dot{V}O_{2max}$, achieving only 75 min.

Adapted, by permission, from D.L. Costill et al., 1977, "Effects of elevated plasma FFA and insulin on muscle glycogen usage during exercise," *Journal of Applied Physiology,* 43(4): 695-699.

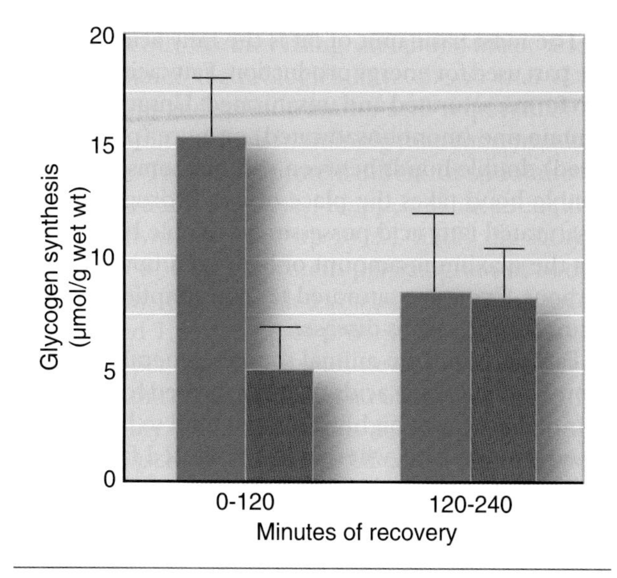

Figure 14.12 Replenishment of muscle glycogen stores following 70 min of muscle glycogen-depleting exercise using two different regimens of carbohydrate replacement. In the trial in which the carbohydrate solution was provided immediately following exercise (left), muscle glycogen storage was three times higher during the first 2 h of recovery compared to the other trial in which the carbohydrate solution was not given until after 2 h of recovery (right). There was no difference in muscle glycogen storage during the next 2 h.

Adapted, by permission, from J.L. Ivy et al., 1988, "Muscle glycogen synthesis after exercise: Effect of time of carbohydrate ingestion," *Journal of Applied Physiology,* 64: 1480-1485.

week apart, at moderate to high work rates to deplete the active muscles' glycogen stores. During one trial, a 25% carbohydrate solution was ingested immediately after exercise, while in the other trial the solution was ingested after 2 h of recovery. Glycogen storage rates were three times higher during the first 2 h in the trial in which the solution was provided immediately after exercise compared to the trial in which the solution was not provided until after 2 h of recovery. The storage rates were the same for the two trials during the second 2 h (see figure 14.12). More recently, it has been shown that adding protein to the carbohydrate supplement enhances the replenishment of muscle glycogen stores during the recovery period. Adding protein to the carbohydrate supplement maximizes glycogen synthesis with less frequent supplementation and less carbohydrate.[18, 19] Further, it also appears to stimulate muscle tissue repair.

The importance of maximizing carbohydrate storage in the liver and muscles prior to exercise, and of providing carbohydrate during and immediately following exercise, has led food and nutrition companies to develop products to meet these needs, as discussed at the end of this chapter.

Fat

Fat, also termed lipid, is a class of organic compounds with limited water solubility. It exists in the body in many forms, such as triglycerides, free fatty acids (FFAs), phospholipids, and sterols. The body stores most fat as triglycerides, composed of three molecules of fatty acids and one molecule of glycerol. Triglycerides are our most concentrated source of energy.

Dietary fat, especially cholesterol and triglycerides, plays a major role in cardiovascular disease (chapter 20); and excessive fat intake also has been linked to other diseases such as cancer, diabetes, and obesity. But despite the negative publicity, fat serves many vital functions in the body:

- It is an essential component of cell membranes and nerve fibers.
- It is a primary energy source, providing up to 70% of our total energy in the resting state.
- It supports and cushions vital organs.
- All steroid hormones in the body are produced from cholesterol.
- Fat-soluble vitamins gain entry into, are stored in, and are transported through the body via fat.
- Body heat is preserved by the insulating subcutaneous fat layer.

In review

- Carbohydrates are sugars and starches. They exist in the body as monosaccharides, disaccharides, oligosaccharides, and polysaccharides. All carbohydrates must be broken down into monosaccharides before the body can use them as a fuel.
- Insufficient intake of carbohydrate during periods of intense training can lead to depletion of glycogen stores. Conversely, muscle glycogen loading by consumption of a diet rich in carbohydrate offers major benefits to performance.
- Endurance performance can be enhanced when carbohydrates are consumed up to an hour before exercise, within 5 min of starting exercise, and during exercise. People can replenish carbohydrate stores rapidly by ingesting carbohydrate during the first 2 h of recovery. This can be facilitated by the addition of protein to the carbohydrate supplement.
- Fats, or lipids, exist in the body as triglycerides, FFAs, phospholipids, and sterols. They are stored primarily as triglycerides, which are the body's most concentrated energy source. A triglyceride molecule can be broken down into one glycerol and three fatty acid molecules. Only the FFAs are used by the body for energy production.
- Although fat is a major energy source, the use of high-fat diets to enhance endurance performance by sparing glycogen has generally been unsuccessful.
- The smallest unit of protein is an amino acid. All proteins must be broken down to amino acids before the body can use them. Only the nonessential amino acids can be synthesized in our bodies. The essential amino acids must be attained through our diets.
- Protein is not a primary energy source in our bodies, but it can be used for energy production during endurance exercise.
- The current RDA for protein (0.8 g/kg per day) may be too low for athletes involved in intense resistance training (1.6-1.7 g/kg per day) or for endurance athletes (1.2-1.6 g/kg per day). During the initial days of training or during periods of very intense training the requirement might be higher. However, extremely high protein diets offer no additional benefits and could offer a health risk to normal kidney function.
- Protein supplementation during recovery from resistance training can stimulate muscle protein synthesis.

can be classified into one of two major categories: fat soluble or water soluble. The fat-soluble vitamins, A, D, E, and K, are absorbed from the digestive tract bound to lipids (fats). These vitamins are stored in the body, so excessive intake can cause toxic accumulations. The B-complex vitamins, biotin, pantothenic acid, folate, and vitamin C are water soluble. They are absorbed from the digestive tract along with water. Any excess of these vitamins is excreted, mostly in the urine, but vitamin toxicity has been reported with some of these. Table 14.3 lists the various vitamins and their RDA values, or AI values when the RDA values are not available.

Most vitamins have some function important to the athlete:

- Vitamin A is crucial for normal growth and development because it plays an integral role in bone development.
- Vitamin D is essential for intestinal absorption of calcium and phosphorus and thus for bone development and strength. By regulating calcium absorption, this vitamin also has a key role in neuromuscular function.
- Vitamin K is an intermediate in the electron transport chain, making it important for oxidative phosphorylation.

Of all the vitamins, though, only the B-complex vitamins and vitamins C and E have been extensively investigated for their potential to facilitate athletic performance. In the following sections, we briefly consider these vitamins.

B-Complex Vitamins

The B-complex vitamins were once thought to be a single vitamin. Now more than a dozen B-complex vitamins have been identified. These vitamins' essential roles in cellular metabolism cannot be overemphasized. Among their diverse functions, they serve as cofactors in various enzyme systems involved in the oxidation of food and the production of energy. Consider just a few examples. Vitamin B_1 (thiamin) is needed for the conversion of pyruvic acid to acetyl coenzyme A. Vitamin B_2 (riboflavin) becomes flavin adenine dinucleotide (FAD), which acts as a hydrogen acceptor during oxidation. Vitamin B_3 (niacin) is a

TABLE 14.3 RDAs or AIs for Vitamins and Minerals

Vitamins	Dose	Age: 9-13 years		Age: 14-18 years		Age: 19-50 years		Age: 51-70 years	
		Male	Female	Male	Female	Male	Female	Male	Female
A (retinol)	μg/day	600	600	900	700	900	700	900	700
B_1 (thiamine)	mg/day	0.9	0.9	1.2	1.0	1.2	1.1	1.2	1.1
B_2 (riboflavin)	mg/day	0.9	0.9	1.3	1.0	1.3	1.1	1.3	1.1
B_3 (niacin)	mg/day	12	12	16	14	16	14	16	14
B_6	mg/day	1.0	1.0	1.3	1.2	1.3	1.3	1.7	1.5
B_{12}	μg/day	1.8	1.8	2.4	2.4	2.4	2.4	2.4	2.4
C	mg/day	45	45	75	65	90	75	90	75
D	μg/day	5[a]	5[a]	5[a]	5[a]	5[a]	5[a]	10[a]	10[a]
E	mg/day	11	11	15	15	15	15	15	15
Biotin (H)	μg/day	20[a]	20[a]	25[a]	25[a]	30[a]	30[a]	30[a]	30[a]
K	μg/day	60[a]	60[a]	75[a]	75[a]	120[a]	90[a]	120[a]	90[a]
Folate	μg/day	300	300	400	400	400	400	400	400
Pantothenic acid	mg/day	4[a]	4[a]	5[a]	5[a]	5[a]	5[a]	5[a]	5[a]
Minerals									
Calcium	mg/day	1,300[a]	1,300[a]	1,300[a]	1,300[a]	1,000[a]	1,000[a]	1,200[a]	1,200[a]
Chloride	g/day	2.3[a]	2.3[a]	2.3[a]	2.3[a]	2.3[a]	2.3[a]	2.0[a]	2.0[a]
Chromium	μg/day	25[a]	21[a]	35[a]	24[a]	35[a]	25[a]	30[a]	20[a]
Copper	μg/day	700	700	890	890	900	900	900	900
Fluoride	mg/day	2[a]	2[a]	3[a]	3[a]	4[a]	3[a]	4[a]	3[a]
Iodine	μg/day	120	120	150	150	150	150	50	150
Iron	mg/day	8	8	11	15	8	18	8	8
Magnesium	mg/day	240	240	410	360	410[b]	315[b]	420	320
Manganese	mg/day	1.9[a]	1.6[a]	2.2[a]	1.6[a]	2.3[a]	1.8[a]	2.3[a]	1.8[a]
Molybdenum	μg/day	34	34	43	43	45	45	45	45
Phosphorus	mg/day	1,250	1,250	1,250	1,250	700	700	700	700
Potassium	g/day	4.5[a]	4.5[a]	4.7[a]	4.7[a]	4.7[a]	4.7[a]	4.7[a]	4.7[a]
Selenium	μg/day	40	40	55	55	55	55	55	55
Sodium	g/day	1.5[a]	1.5[a]	1.5[a]	1.5[a]	1.5[a]	1.5[a]	1.3[a]	1.3[a]
Zinc	mg/day	8	8	11	9	11	8	11	8

[a]AI (RDA is not available).

[b]Men: Age 19-30 years = 400 and age 31-50 years = 420; women: Age 19-30 years = 310 and age 31-50 years = 320.

Full reports can be obtained at the following U.S. government Web site: www.nal.usda.gov/fnic/.

Note: Values are also available for infants and small children, and for pregnancy and lactation.

Food and Nutrition Board of the National Academy of Sciences, and Health Canada: 1997-2005.

component of nicotinamide adenine dinucleotide phosphate (NADP), a coenzyme in glycolysis. Vitamin B_{12} has a role in amino acid metabolism and is also needed for the production of red blood cells, which transport oxygen to the cells for oxidation. The B-complex vitamins have such a close interrelationship that a deficiency in one can impair utilization of the others. Symptoms of deficiencies vary with the vitamins involved.

Several studies have shown that supplementation of one or more of the B-complex vitamins facilitates performance. However, most researchers agree that this is true only if the individual being studied suffers a preexisting B-complex deficiency. Creating a deficiency in one or more of the B-complex vitamins usually impairs performance, but this is reversed when the deficiency is corrected with supplementation. No compelling evidence supports supplementation when there is no deficiency.

Vitamin C

Vitamin C (ascorbic acid) is common in our foods, but deficiencies can occur in people who smoke, use oral contraceptives, have surgery, or run a fever. This vitamin is important for the formation and maintenance of collagen, a crucial protein found in connective tissue, so it is essential for healthy bones, ligaments, and blood vessels. Vitamin C also functions in

- the metabolism of amino acids;
- synthesis of some hormones, such as the catecholamines (epinephrine and norepinephrine) and the anti-inflammatory corticoids; and
- promotion of iron absorption from the intestines.

Many people also believe that vitamin C assists healing, combats fever and infection, and prevents or cures the common cold. Although evidence to date is inconclusive, the role of vitamin C in the fight against disease is an area of major interest.

Vitamin C supplementation to enhance performance has produced equivocal findings in the research conducted to date. However, those who have reviewed this area generally agree that, even with the increased requirements of training, vitamin C supplementation does not improve performance when no deficiency exists. As noted in the sidebar, it has been suggested that vitamins, including vitamin C, also may function as antioxidants to combat the cellular damage created by the metabolically generated free radicals.

Vitamin E

Vitamin E is stored in muscle and fat. This vitamin's functions are not well established, although it is known to enhance the activity of vitamins A and C by preventing their oxidation. Indeed, the most important role of vitamin E is its action as an antioxidant. It disarms free radicals (highly reactive molecules) that could otherwise severely damage cells and disrupt metabolic processes. Exercise has been shown to cause DNA damage to the cell, whereas supplementing the intake of vitamin E reduces DNA damage. In addition, with the supplementation of vitamin E for 14 days before an exhaustive run, exercise-induced damage was reduced.[16] However, investigators found no benefit of 30 days of vitamin E supplementation on the muscle damage that resulted from 240 maximal isokinetic eccentric knee flexion/extension actions (24 sets of 10 repetitions each) when compared with a placebo control condition.[5]

Vitamin E has received considerable media attention over the years as a potential miracle vitamin that might prevent or alleviate a number of medical conditions, such as rheumatic fever, muscular dystrophy, coronary artery disease, sterility, menstrual disorders, and spontaneous abortions. It also has been suggested that vitamin E supplements may prevent lung damage from many of the pollutants that we inhale. Such claims generally lack supporting scientific evidence.

Many athletes have consumed supplementary doses of vitamin E since it has been postulated to benefit performance through its relationship with oxygen use and energy supply. However, research reviews generally conclude that vitamin E supplementation does not improve athletic performance.

Minerals

A number of inorganic substances known as minerals are essential for normal cellular functions. Minerals account for approximately 4% of body weight. Some are present in high concentrations in the skeleton and teeth, but minerals are also found throughout the body, in and around every cell, dissolved in the body's fluids. They can be present either as ions or combined with various organic compounds. Mineral compounds that can dissociate into ions in the body are called **electrolytes.**

By definition, **macrominerals** are those of which the body needs more than 100 mg per day. **Microminerals,** or **trace elements,** are those needed in smaller amounts. Table 14.3 lists the essential minerals and their RDAs or AIs.

Mineral intake is less likely to be supplemented by athletes than vitamin intake. Far less concern has been shown by athletes for their mineral status, possibly because far fewer claims have been made about the performance-enhancing qualities of specific minerals. Of the minerals, calcium and iron have been most frequently investigated.

Free Radicals and Antioxidants

Most of the oxygen consumed during aerobic exercise is used in the mitochondria for oxidative phosphorylation and is reduced to water. However, a small number of univalently produced oxygen intermediates, termed **free radicals,** may leak out of the electron transport chain. Laboratory studies have shown that free radical generation increases after acute exercise, and this has been theorized to coincide with oxidative tissue damage. Because these free radicals are highly reactive, they are theorized to modulate muscle function and accelerate the fatigue process. Fortunately, under normal conditions, muscle fibers are equipped with antioxidant enzymes that serve as an efficient defense system to prevent the damaging accumulation of free radicals. In addition, the dietary intake of antioxidants, such as vitamin E and β-carotene, also directly traps free radicals, preventing them from interfering with cellular function. Some researchers suggest that these dietary supplements may help block the negative effects of exercise-induced free radical release. Consequently, the importance of antioxidant vitamins has become a popular topic of discussion and study in the fields of nutrition and cellular biology.

Calcium

Calcium is the most abundant mineral in the body, constituting approximately 40% of the total mineral content. Calcium is well known for its importance in building and maintaining healthy bones, and that is where most of it is stored. But it is also essential for nerve impulse transmission. Calcium plays major roles in enzyme activation and regulation of cell membrane permeability, both important for metabolism. And this mineral is also essential for normal muscle function: Recall from chapter 1 that calcium is stored in the sarcoplasmic reticulum of muscles and released when the muscle fibers are stimulated. It is required for formation of the actin-myosin cross-bridges that cause the fibers to contract.

Sufficient calcium intake is critical to our health. If we do not consume enough calcium, it will be removed from its storage sites in the body, especially the bones. This condition is called osteopenia. It weakens the bones and can lead to osteoporosis, a common problem in postmenopausal women and aging men and women. Unfortunately, few studies have been conducted on calcium supplementation, and their results suggest that supplementation is of no value in the presence of an adequate (RDA) dietary intake of calcium.

Phosphorus

Phosphorus is closely linked to calcium. It constitutes approximately 22% of the body's total mineral content. About 80% of this phosphorus is combined with calcium (calcium phosphate), providing strength and rigidity to the bones. Phosphorus is an essential part of metabolism, cell membrane structure, and the buffering system to maintain constant blood pH. Phosphorus plays a major role in bioenergetics: It is an essential component of adenosine triphosphate. There is no evidence to suggest the need for supplementation in athletes.

Iron

Iron—a micromineral—is present in the body in relatively small amounts (35-50 mg/kg of body weight). It plays a crucial role in oxygen transportation: Iron is required for the formation of both hemoglobin and myoglobin. Hemoglobin, located in the red blood cells, binds with oxygen in the lungs and then transports it to the body tissues via the blood. Myoglobin, found in muscle, combines with oxygen and stores it until needed.

Iron deficiency is prevalent throughout the world. By some estimates, as much as 25% of the world's population is iron deficient. In the United States, approximately 20% of women and 3% of men are iron deficient, as are 50% of pregnant women. The major problem associated with this condition is iron-deficiency anemia, in which hemoglobin levels are reduced, decreasing the blood's oxygen-carrying capacity. This causes fatigue, headaches, and other symptoms. Iron deficiency is a more common problem in women than in men because both menstruation and pregnancy cause iron losses that must be replenished. This problem is compounded by the fact that women generally consume less food, and thus less iron, than men.

Iron has received much attention in the research literature. Women are considered anemic only when their hemoglobin concentration is below 10 g per 100 ml of blood. For men, the value is 12 g per 100 ml of blood.

Studies generally suggest that 22% to 25% of female athletes and 10% of male athletes are iron deficient. But these numbers may be conservative. These studies also indicate that hemoglobin is not the only marker of anemia or necessarily the best. Plasma ferritin levels provide a good marker of the body's iron stores. Values below 20 to 30 μg/L indicate low body iron stores.

When iron supplements are given to those who are iron deficient (i.e., with low plasma ferritin levels), performance measures, particularly aerobic capacity, typically improve.[13] However, supplementation of iron in those who are not deficient appears to have little or no benefit. In fact, iron supplements can be a health risk, because excess iron is toxic for the liver, and ferritin levels higher than 200 μg/L are associated with an increased risk for coronary artery disease.

Sodium, Potassium, and Chloride

Sodium, potassium, and chloride are distributed throughout all body fluids and tissues. Sodium and chloride are found primarily in the fluid outside of the cells and in the blood plasma, but potassium is located mainly inside the cells. This selective distribution of these three minerals establishes the separation of electrical charge across neuron and muscle cell membranes. Thus, these minerals enable neural impulses to control muscle activity (see chapter 3). In addition, they are responsible for maintaining the body's water balance and distribution, normal osmotic equilibrium, acid–base balance (pH), and normal cardiac rhythm.

Western diets are replete with sodium, so dietary deficiency is unlikely. However, minerals are lost with sweating, so any condition that causes excessive sweating, such as extreme exertion or exercise in a hot environment, can deplete these minerals. When discussing mineral imbalances, we often focus on deficiencies. However, many of these minerals also have negative effects when taken in excess. In fact, excess potassium can cause heart failure! Individual needs vary, but megadoses are never advisable.

To conclude this section on vitamins and minerals, we can say that while physical activity increases vitamin and mineral requirements, this is generally countered by an increase in food intake. For athletes who eat balanced meals in response to their bodies' increased caloric needs, it is highly likely that all vitamin and mineral needs will be met and supplementation will have no performance benefits. However, for those who are intentionally consuming a low-energy or unbalanced diet, supplementation may be necessary to maintain performance. If there is any question about the adequacy of an athlete's diet, a low-dose multivitamin/mineral supplement may be appropriate. Also, the new DRIs have a category for upper limits for most micronutrients that can be used as guidelines for excesses.

In review

- Vitamins perform numerous functions in our bodies and are essential for normal growth and development. Many are involved in metabolic processes, such as those leading to energy production.
- Vitamins A, D, E, and K are fat soluble. These can accumulate to toxic levels in the body. B-complex vitamins, biotin, pantothenic acid, folate, and vitamin C are water soluble. Excesses of these are excreted, so toxicity is rarely a problem. Several of the B-complex vitamins are involved in the processes of energy production.
- Macrominerals are minerals of which we require more than 100 mg per day. Microminerals (trace elements) are those we require smaller amounts of.
- Minerals are required for numerous physiological processes, such as muscle contraction, oxygen transport, fluid balance, and bioenergetics. Minerals can dissociate into ions, which can participate in numerous chemical reactions. Minerals that can dissociate into ions are called electrolytes.
- Vitamins and minerals do not appear to have any special performance-enhancing value. Taking them in amounts greater than the RDA will not improve performance and could be counterproductive.

Water

Seldom is water thought of as a nutrient because it has no caloric value. Yet its importance in maintaining life is second only to oxygen's. Water constitutes about 60% of a typical young man's and 50% of a typical young woman's total body weight; but this varies with body composition, since the fat-free mass has a much higher water content (~73% water) than the fat mass (~10% water). It has been estimated that we can survive losses of up to 40% of our body weight in fat, carbohydrate, and protein. But a water loss of only 9% to 12% of body weight can be fatal.

Athletes commonly lose between 1% to 6% of their body water during intense, prolonged exercise. However, a water loss of 9% to 12% of a person's total body weight can lead to death.

Approximately two-thirds of the water in our bodies is contained in our cells and is referred to as **intracellular fluid.** The remainder is outside the cells, referred to as the **extracellular fluid.** Extracellular fluid includes the interstitial fluid surrounding the cells, the blood plasma, lymph, and other body fluids.

Water plays several critical roles in exercise. Among its most important functions, water provides transportation between and delivery to the body's various tissues, regulates body temperature, and maintains blood pressure for proper cardiovascular function. In the next sections, we more closely examine the role of water in exercise and performance.

Water and Electrolyte Balance

For optimal performance, the body's water and electrolyte contents should remain relatively constant. Unfortunately, this doesn't always happen during exercise. In the next sections, we examine water content and electrolyte balance at rest, how exercise affects these, and the impact on performance when water or electrolyte balance is disturbed.

Water Balance at Rest

Under normal resting conditions, the body's water content is relatively constant: Water intake equals water output. About 60% of our daily water intake is obtained from the fluids we drink and about 30% is from the foods we consume. The remaining 10% is produced in our cells during metabolism (recall from chapter 2 that water is a by-product of oxidative phosphorylation). Metabolic water production varies from 150 to 250 ml per day, depending on the rate of energy expenditure: Higher metabolic rates produce more water. The total daily water intake from all sources averages about 33 ml per kilogram of body weight per day. For a 70 kg (154 lb) person, average intake is 2.3 L per day. Water output, or water loss, occurs from four sources:

- Evaporation from the skin
- Evaporation from the respiratory tract
- Excretion from the kidneys
- Excretion from the large intestine

Human skin is permeable to water. Water diffuses to the skin's surface, where it evaporates into the environment. In addition, the gases we breathe are constantly being humidified by water as they pass through the respiratory tract. These two types of water loss (from the skin and respiration) occur without our sensing them. Thus, they are termed insensible water losses. Under cool, resting conditions, these losses account for about 30% of daily water loss.

The majority of our daily water loss—60% at rest—occurs from our kidneys, which excrete water and waste products as urine. Under resting conditions, the kidneys excrete about 50 to 60 ml of water per hour. Another 5% of the water is lost by sweating (although this is often considered along with insensible water loss), and the remaining 5% is excreted from the large intestine in the feces. The sources of water gain and water loss at rest are depicted in figure 14.13 on page 342.

Water Balance During Exercise

Water loss accelerates during exercise, as seen in table 14.4 (p. 342). The ability to lose the heat generated during exercise depends primarily on the formation and evaporation of sweat. As body temperature increases, sweating increases in an effort to prevent overheating (see chapter 11). But at the same time, more water is produced during exercise because of increased oxidative metabolism. Unfortunately, the amount produced even during the most intense effort has only a small impact on the **dehydration,** or water loss, that results from heavy sweating.

In general, the amount of sweat produced during exercise is determined by

- environmental temperature, radiant heat load, humidity, and air velocity;
- body size; and
- metabolic rate.

These factors influence the body's heat storage and temperature. Heat is transferred from warmer areas to cooler ones, so heat loss from the body is impaired by high environmental temperatures, radiation, high humidity, and still air. Body size is important because large individuals generally expend more energy to do a given task, so they typically have higher metabolic rates and produce more heat. But they also have more surface area (skin), which allows more sweat formation and evaporation. As exercise intensity increases, so does the metabolic rate. This increases body heat production, which in turn increases sweating. To conserve water during exercise, blood flow to the kidneys decreases in an attempt to prevent dehydration; but like the increase in metabolic water production, this too may be insufficient. During high-intensity exercise under environmental heat stress, sweating and respiratory evaporation can cause losses of as much as 2 to 3 L of water per hour. (Chapter 11 contains additional information about body water losses during exercise in warm environments.)

During an event such as the marathon, sweating and water loss from respiration may reduce body water content by 6% or more.

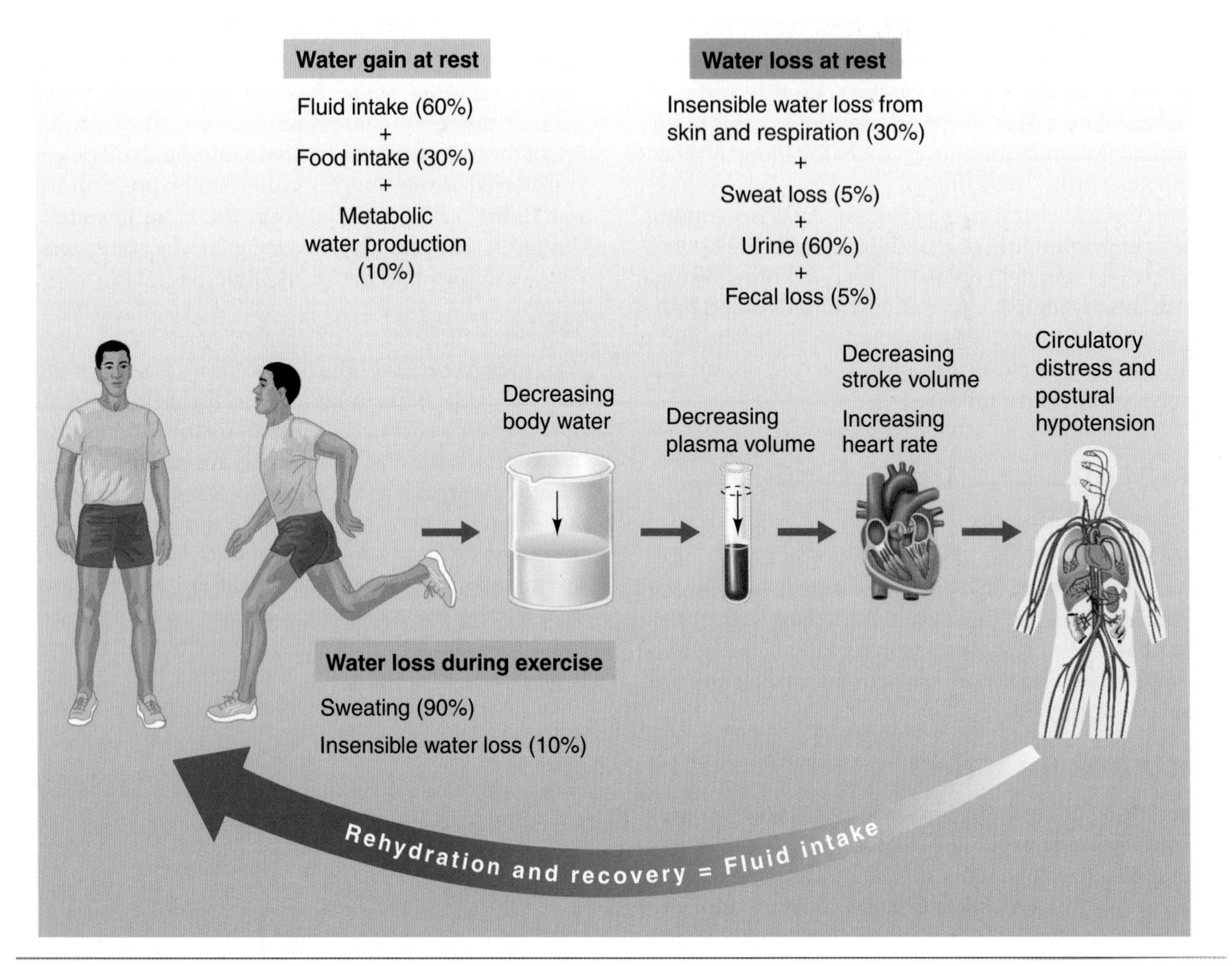

Figure 14.13 The body's fluid compartments and sources of body water gains and losses at rest and during exercise.

TABLE 14.4 Typical Values of Water Loss From the Body at Rest in a Cool Environment and During Prolonged Exhaustive Exercise

	Resting		Prolonged exercise	
Source of loss	ml/h	% total	ml/h	% total
Insensible loss				
Skin	15	15	15	1
Respiration	15	15	100	7
Sweating	4	5	1200	91
Urine	58	60	10	1
Feces	4	5	–	0
Total	96	100	1325	100

Dehydration and Exercise Performance

Even minimal changes in the body's water content can impair endurance performance. Without adequate fluid replacement, an athlete's exercise tolerance shows a pronounced decrease during long-term activity because of water loss through sweating. The impact of dehydration on the cardiovascular and thermoregulatory systems is quite predictable. Fluid loss decreases plasma volume. This decreases blood pressure, which in turn reduces blood flow to the muscles and skin. In an effort to overcome this, heart rate increases. Because less blood reaches the skin, heat dissipation is hindered, and the body retains more heat. Thus, when a person is dehydrated by 2% of body weight or more, both heart rate and body temperature are elevated during exercise above values observed when normally hydrated.

As one might expect, these physiological changes will decrease exercise performance. Figure 14.14 illustrates the effects of an approximate 2% decrease in body weight attributable to dehydration from the use of a diuretic on distance runners' performance in 1500 m, 5000 m, and 10,000 m time trials on an outdoor track.[2] The dehydration condition resulted in plasma volume decreases between 10% and 12%. Although the average $\dot{V}O_{2max}$ did not differ between the normally hydrated and dehydrated trials, mean running velocity decreased by 3% in the 1500 m run and by more than 6% in the 5000 and 10,000 m runs. The greater the duration of the performance, the greater is the expected decline in performance for the same degree of dehydration. These trials were conducted in relatively cool weather. The higher the temperature, humidity, and radiation, the greater the expected decrement in performance for the same degree of dehydration. The decrement in performance would be progressively greater with greater levels of dehydration.

The effect of dehydration on performance in muscular strength, muscular endurance, and anaerobic types of activities is not as clear. Decrements have been seen in some studies, whereas other studies have shown no change in performance. In one of the best-controlled studies, researchers at Penn State University reported that 2% dehydration resulted in significant deterioration of basketball skills in 12- to 15-year-old boys who were skilled basketball players.[10]

Wrestlers and other weight-category athletes commonly dehydrate to get a weight advantage during the weigh-in for a competition. Most rehydrate after the weigh-in before the competition and experience only small decrements in performance. A summary of the effects of dehydration on exercise performance is shown in table 14.5 page 344.

Electrolyte Balance During Exercise

Normal body function depends on a balance between water and electrolytes. We have discussed the effects of water loss on performance. Now we turn our attention to the effects of the other component of this delicate balance: electrolytes. When large amounts of water are lost from the body, as during exercise, the balance between water and electrolytes can be disrupted quickly. In the next sections, we examine the effects of exercise

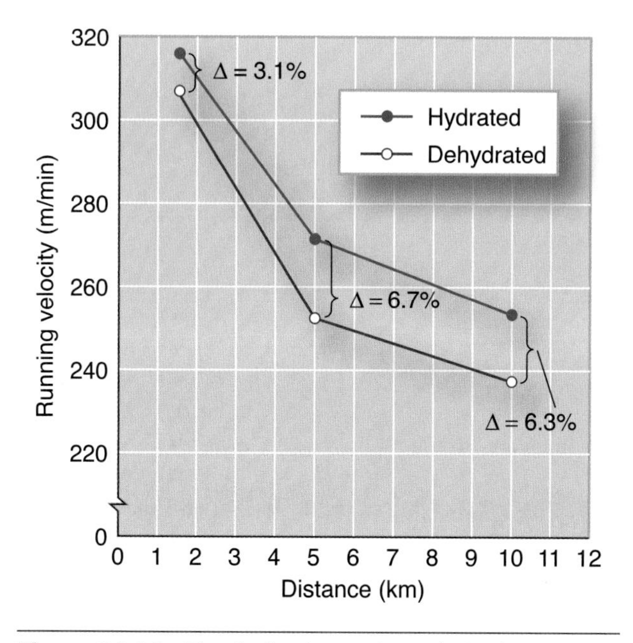

Figure 14.14 The decline in running velocity (meters per minute) with dehydration of about 2% of body weight for 1,500 m, 5,000 m, and 10,000 m time trials compared with velocity in the normally hydrated condition.

Reprinted, by permission, from L.E. Armstrong, D.L. Costill, and W.J. Fink, 1985, "Influence of diuretic-induced dehydration on competitive running performance," *Medicine and Science in Sports and Exercise*, 17: 456-461.

In review

- Water balance depends on electrolyte balance, and vice versa.
- At rest, water intake equals water output. Water intake includes water ingested from foods and fluids and produced as a metabolic by-product. The majority of water output at rest is generated by the kidneys, but water also is lost from the skin, from the respiratory tract, and in the feces.
- During exercise, metabolic water production increases as the metabolic rate increases.
- Water loss during exercise increases because as heat in the body increases, more water is lost with increased sweating. Sweat becomes the primary avenue for water loss during exercise. In fact, the kidneys decrease their excretion in an effort to prevent dehydration.
- When dehydration reaches 2% of body weight, aerobic endurance performance is notably impaired. Heart rate and body temperature increase in response to dehydration.

TABLE 14.5 Alterations in Physiological Function and Performance From Dehydration

Variables	Dehydration
Physiological function	
Cardiovascular	
Blood volume/Plasma volume	Decreased
Cardiac output	Decreased
Stroke volume	Decreased
Heart rate	Increased
Metabolic	
Aerobic capacity ($\dot{V}O_{2max}$)	No change or decreased
Anaerobic power (Wingate test)	No change or decreased
Anaerobic capacity (Wingate test)	No change or decreased
Blood lactate, peak value	Decreased
Buffer capacity of the blood	Decreased
Lactate threshold, velocity	Decreased
Muscle and liver glycogen	Decreased
Blood glucose during exercise	Possibly decreased
Protein degradation with exercise	Possibly increased
Thermoregulation and fluid balance	
Electrolytes, muscle and blood	Decreased
Exercise core temperature	Increased
Sweat rate	Decreased, delayed onset
Skin blood flow	Decreased
Performance	
Muscular strength	No change or decreased
Muscular endurance	No change or decreased
Muscular power	Unknown
Speed of movement	Unknown
Run time to exhaustion	Decreased
Total work performed	Decreased
Wrestling simulation test	Decreased

Note. Data for this table were derived from the following reviews: M. Fogelholm, 1994, "Effects of bodyweight reduction on sports performance," *Sports Medicine* 18: 249-267; C.A. Horswill, 1994, Physiology and nutrition for wrestling, in D.R. Lamb, H.G. Knutten, & R. Murray (Eds.), *Physiology and nutrition for competitive sport* (Vol 7, pp. 131-174); H.L. Keller, S.E. Tolly, & P.S. Freedson, 1994, "Weight loss in adolescent wrestlers," *Pediatric Exercise Science* 6: 211-224; and R. Opplinger, H. Case, C. Horswill, G. Landry, & A. Shelter, 1996, "Weight loss in wrestlers: An American College of Sports Medicine position stand,"*Medicine and Science in Sports and Exercise* 28: ix-xii.

on electrolytc balance. Our focus is on the two major routes for electrolyte loss: sweating and urine production.

Electrolyte Loss in Sweat

Human sweat is a filtrate of blood plasma, so it contains many substances found there, including sodium (Na^+), chloride (Cl^-), potassium (K^+), magnesium (Mg^{2+}), and calcium (Ca^{2+}). Although sweat tastes salty, it contains far fewer minerals than the plasma and other body fluids. In fact, sweat is 99% water.

Sodium and chloride are the predominant ions in sweat and blood. As indicated in table 14.6, the concentrations of sodium and chloride in sweat are about one-third those found in plasma and five times those found in muscle. Each of these three fluids' **osmolarity,** which is the ratio of solutes (such as electrolytes) to fluid, is also shown. Sweat's electrolyte concentration can vary considerably between individuals. It is strongly influenced by genetics, the rate of sweating, the state of training, and the state of heat acclimatization.

At the elevated rates of sweating reported during endurance events, sweat contains large amounts of sodium and chloride but little potassium, calcium, and magnesium. Based on estimates of the athlete's total body electrolyte content, such losses would lower the body's sodium and chloride content by only about 5% to 7%. Total body levels of potassium and magnesium, two ions principally confined to the insides of cells, would decrease by about 1%. These losses probably have no measurable effect on an athlete's performance.

As electrolytes are lost in sweat, the remaining ions are redistributed among the body tissues. Consider potassium. It diffuses from active muscle fibers as they contract, entering the extracellular fluid. The increase this causes in extracellular potassium levels does not equal the amount of potassium that is released from active muscles, because potassium is taken up by inactive muscles and other tissues while the active muscles are losing it. During recovery, intracellular potassium levels normalize quickly. Some researchers suggest that these muscle potassium disturbances during exercise might contribute to fatigue by altering the membrane potentials of neurons and muscle fibers, making it more difficult to transmit impulses.

TABLE 14.6 Electrolyte Concentrations and Osmolarity in Sweat, Plasma, and Muscle of Men Following 2 h of Exercise in the Heat

	Electrolytes (mEq/L)				
Site	Na^+	Cl^-	K^+	Mg^{2+}	Osmolarity (mOsm/L)
Sweat	40-60	30-50	4-6	1.5-5	80-185
Plasma	140	101	4	1.5	295
Muscle	9	6	162	31	295

Note. mEq/L = milliequivalents per liter (thousandths of 1 g of solute per 1 L of solvent).

Electrolyte Loss in Urine

In addition to clearing wastes from the blood and regulating water levels, the kidneys also regulate the body's electrolyte content. Urine production is the other major source of electrolyte loss. At rest, electrolytes are excreted in the urine as necessary to maintain homeostatic levels, and this is the primary route for electrolyte loss. But as the body's water loss increases during exercise, urine production rate decreases considerably in an effort to conserve water. Consequently, with very little urine being produced, electrolyte loss by this avenue is minimized.

The kidneys play another role in electrolyte management. If, for example, a person eats 250 mEq of salt (NaCl), the kidneys will normally excrete 250 mEq of these electrolytes to keep the body NaCl content constant. Heavy sweating and dehydration, however, trigger the release of the hormone aldosterone from the adrenal gland. This hormone stimulates renal reabsorption of sodium. Consequently, the body retains more sodium than usual during the hours and days after a prolonged exercise bout. This elevates the body's sodium content and increases the osmolarity of the extracellular fluids.

This increased sodium content triggers thirst, compelling the person to consume more water, which is

In review

- The loss of large amounts of water can disrupt electrolyte balance, although electrolytes are rather dilute in sweat.
- Electrolyte loss during exercise occurs primarily with water loss from sweating. Sodium and chloride are the most abundant electrolytes in sweat.
- At rest, excessive electrolytes are excreted in the urine by the kidneys. But urine production declines tremendously during exercise, so little electrolyte loss occurs by this route.
- Dehydration causes the hormone aldosterone to promote renal retention of Na^+ and Cl^-, increasing their concentrations in the blood. This triggers thirst in an effort to make us consume more fluid to replace what has been lost.

then retained in the extracellular compartment. The increased water consumption reestablishes normal osmolarity in the extracellular fluids but leaves these fluids expanded, which dilutes the other substances present there. This expansion of the extracellular fluids has no negative effects and is temporary. In fact, this is one of the major mechanisms for the increase in plasma volume that occurs with training and with acclimatization to exercise in the heat. Fluid levels return to normal within 48 to 72 h after exercise, providing there are no subsequent exercise bouts.

Replacement of Body Fluid Losses

The body loses more water than electrolytes during heavy sweating. This raises the osmotic pressure in the body fluids because the electrolytes become more concentrated. The need to replace body water is greater than the need for electrolytes because only by replenishing water content can the electrolytes return to normal concentrations. But how does the body know when this is necessary?

Thirst

When people feel thirsty, they drink. The thirst sensation is regulated by the hypothalamus. It triggers thirst when the plasma's osmotic pressure is increased. Unfortunately, the body's **thirst mechanism** doesn't precisely gauge its state of dehydration. It does not sense thirst until well after dehydration begins. Even when dehydrated, people might desire fluids only at intermittent intervals.

The control of thirst is not fully understood. When permitted to drink water as their thirst dictates, people can require 24 to 48 h to completely replace water lost through heavy sweating. In contrast, dogs and burros can drink up to 10% of their total body weight within the first few minutes after exercise or heat exposure, replacing all lost water. Because of our sluggish drive to replace body water and to prevent chronic dehydration, we are advised to drink more fluid than our thirst dictates. Because of the increased water loss during exercise, it is imperative that athletes' water intake be sufficient to meet their bodies' needs, and it is essential that they rehydrate during and after an exercise bout.

Benefits of Fluids During Exercise

Drinking fluids during prolonged exercise, especially in hot weather, has obvious benefits. Water intake will minimize dehydration, increases in body temperature, cardiovascular stress, and declines in performance. As seen in figure 14.15, when subjects became dehydrated during several hours of treadmill running in the heat (40 °C, or 104 °F) without fluid replacement, their heart rates increased steadily throughout the exercise.[4] When they were deprived of fluids, the subjects became exhausted and couldn't complete the 6 h exercise. Ingesting either water or a saline solution in amounts equal to weight loss prevented dehydration and kept subjects' heart rates lower. Even warm fluids (near body temperature) provide some protection against overheating, but cold fluids enhance body cooling because some of the deep body heat is used to warm cold drinks to body temperature.

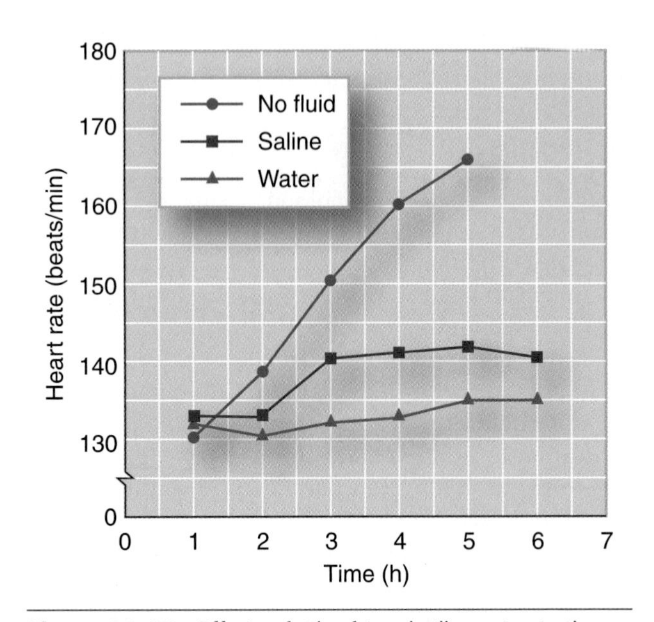

Figure 14.15 Effects of 6 h of treadmill running in the heat on heart rate. Subjects received no fluid, a saline solution, or water. Subjects deprived of fluids became exhausted and could not complete the 6 h exercise.

Data from S.I. Barr et al., 1991.

Hyponatremia

Fluid replacement is beneficial, but too much of a good thing could potentially be bad. In the 1980s, the first cases of hyponatremia were reported in endurance athletes. **Hyponatremia** is clinically defined as a serum sodium concentration below the normal range of 135 to 145 mmol/L. Symptoms of hyponatremia generally appear when serum sodium levels drop below 130 mmol/L. Early signs and symptoms include bloating, puffiness, nausea, vomiting, and headache. As the severity increases, due to increasing cerebral edema (swelling of the brain), the symptoms include confusion, disorientation, agitation, seizures, pulmonary edema, coma, and death.[17] How likely is hyponatremia to occur?

The processes that regulate fluid volumes and electrolyte concentrations are highly effective, so consuming enough water to dilute plasma electrolytes is difficult under normal circumstances. Marathoners who lose 3 to 5 L of sweat and drink 2 to 3 L of water maintain normal plasma concentrations of sodium, chloride, and potassium. And distance runners who run 25 to 40 km

Recommendations for Fluid Replacement Before, During, and After Exercise

Dehydration is a potential problem for athletes who train and compete for long periods of time, as well as in hot and humid environments. To achieve adequate hydration levels, the American College of Sports Medicine, the American Dietetic Association, and the Dietitians of Canada published guidelines to achieve adequate fluid intake before, during, and following exercise. Key recommendations include the following:

- Athletes should drink enough fluid to balance their fluid losses.
- Two hours before exercise, the athlete should consume 400 to 600 ml (14 to 22 oz) of fluid to provide adequate hydration and allow time for excretion of excess ingested water.
- During exercise, the athlete should consume 150 to 350 ml (6 to 12 oz) of fluid every 15 to 20 min beginning at the start of exercise.
- After exercise, the athlete should consume adequate fluids to replace sweat losses during exercise. The volume ingested should be at least 450 to 675 ml (16 to 24 oz) of fluid for every pound (0.5 kg) of body weight lost during exercise.
- Sport drinks containing carbohydrate concentrations of 4% to 8%, and sodium in amounts between 0.5 and 0.7 g/L, are recommended for use during intense exercise events lasting longer than 1 h.
- Including sodium in drinks or eating high-sodium foods during the recovery period can help the rehydration process.[1]

(15.5-24.9 mi) per day in warm weather and do not salt their food don't develop electrolyte deficiencies.

Some research has suggested that during ultramarathon running (more than 42 km, or 26.2 mi), athletes can experience hyponatremia. A case study of two runners who collapsed after an ultramarathon race (160 km, or 100 mi) in 1983 revealed that their blood sodium concentrations had decreased from a normal value of 140 mmol/L to values of 123 and 118 mmol/L.[14] One of the runners experienced a grand mal seizure; the other became disoriented and confused. Examining the runners' fluid intakes and estimating their sodium intakes during the run suggested that they had diluted their sodium contents by consuming too much fluid that contained too little sodium.

The ideal resolution to prevent hyponatremia would be to replace water at the exact rate at which it is being lost or to add sodium to the ingested fluid. The problem with the latter approach is that most sport drinks contain no more than 25 mmol/L of sodium and are apparently too weak to prevent sodium dilution alone, but very strong concentrations cannot be tolerated. Exercise hyponatremia appears to be the result of a fluid overload due to overconsumption, underreplacement of sodium losses, or both.[23] Only a small number of cases have been reported. Thus, it is probably inappropriate

In review

- The need to replace lost body fluid is greater than the need to replace lost electrolytes, because sweat is very dilute.
- The thirst mechanism does not exactly match the body's hydration state, so more fluid should be consumed than thirst dictates.
- Water intake during prolonged exercise reduces the risk of dehydration and optimizes cardiovascular and thermoregulatory function.
- In some rare cases, drinking too much fluid with too little sodium has led to hyponatremia (low plasma levels of sodium), which can cause confusion, disorientation, and even seizures, coma, and death.

to form conclusions from this information to design a fluid replacement regimen for people who must exercise for long periods in the heat.

The Athlete's Diet

Athletes place considerable demands on their bodies every day they train and compete. Their bodies must be as finely tuned as possible. This, by necessity, must include optimal nutrition. Too often, athletes spend considerable time and effort perfecting skills and attaining top physical condition while ignoring proper nutrition and sleep. Performance deterioration often can be traced to poor nutrition.

The previous sections of this chapter have provided guidelines for each of the nutrients and, where appropriate, how these would vary according to the athletes' training requirements. Most athletes need guidance in selecting those foods that will help them meet these requirements. As mentioned at the beginning of the chapter, the Food and Nutrition Information Center, U.S. Department of Agriculture Web site (www.nal.usda.gov/fnic) is an excellent source of information for the coach, trainer, and athlete that will assist them in personalizing diets to meet each athlete's specific nutritional needs.

There are, however, special situations in which additional information is needed. We will now look at vegetarian diets, the precompetition meal, and muscle glycogen replacement and loading.

Vegetarian Diet

In an effort to eat a healthy diet and increase their carbohydrate intake, many athletes have adopted vegetarianism. Vegans are strict vegetarians who eat only food from plant sources. Lactovegetarians also consume dairy products. Ovovegetarians add eggs to their vegetable diets, and lacto-ovovegetarians eat plant foods, dairy products, and eggs.

Can athletes perform well on a vegetarian diet? Athletes who are strict vegans must be very careful in selecting the plant foods they eat to provide a good balance of the essential amino acids, sufficient calories, and adequate sources of vitamin A, riboflavin, vitamin B12, vitamin D, calcium, zinc, and iron. Adequate iron intake is of particular concern in female vegetarian athletes because of the lower bioavailability of iron in plant-based diets and because of women's greater risk for anemia and low iron stores. Some professional athletes have noted significant deterioration in athletic performance after switching to strict vegetarian diets. The problem usually is traced to unwise selection of foods. Including milk and eggs in the diet decreases the risk of nutritional deficiencies. Anyone contemplating switching to a vegetarian diet should either read authoritative material on the subject written by qualified nutritionists or consult a registered dietitian or sport nutritionist.

Precompetition Meal

For years, many athletes have eaten the traditional steak dinner several hours before competition. This practice might have originated from the early belief that muscle consumes itself to fuel its own activity and that steak would provide the necessary protein to counteract this loss. But we now know that steak is probably the worst food an athlete could eat before competing. Steak contains a relatively high percentage of fat, which requires several hours for full digestion. During competition, this would cause the digestive system to compete with the muscles for the available blood supply. Also, nervous tension is typically high before a big competition, so even the choicest steak cannot truly be enjoyed at this time. The steak would be more satisfying and less likely to disturb performance if the athlete were to eat it either the night before or after the competition. But if steak is out, what should the athlete eat before competing?

Although the meal ingested a few hours before competition might contribute little to muscle glycogen stores, it can ensure a normal blood glucose level and prevent hunger. This meal should contain only about 200 to 500 kcal and consist mostly of carbohydrate foods that are easily digested. Foods such as cereal, milk, juice, and toast are digested rather quickly and won't leave the athlete feeling full during competition. In general, this meal should be consumed at least 2 h before competition. The rates at which food is digested and nutrients are absorbed into the body are quite individual, so timing the precompetition meal might depend on prior experience. In one study of endurance cyclists, a prolonged cycling exercise trial to exhaustion at 70% of the subject's $\dot{V}O_{2max}$ was performed under two different conditions, with 14 days between trials: 100 g of carbohydrate breakfast fed 3 h before exercise (Fed) and no feeding before exercise (Fasted). Subjects tested under the Fed condition exercised 136 min before reaching exhaustion compared with 109 min in the fasted trial, indicating the importance of the precompetition meal.[24]

A liquid precompetition meal might be less likely to result in nervous indigestion, nausea, vomiting, and abdominal cramps. Such feedings are commercially available and generally have been found useful both before and between events. Finding time for athletes to eat is often difficult when they must perform in multiple preliminary and final events. Under these circumstances, a liquid feeding that is low in fat and high in carbohydrate might be the only solution.

Muscle Glycogen Replacement and Loading

Earlier in this chapter, we established that different diets can markedly influence muscle glycogen stores and that endurance performance depends largely on these stores. The theory is that the greater the amount of glycogen stored, the better the potential endurance performance because fatigue will be delayed. Thus, an athlete's goal is to begin an exercise bout or competition with as much stored glycogen as possible.

On the basis of muscle biopsy studies conducted in the mid-1960s, Åstrand[3] proposed a plan to help runners store the maximum amount of glycogen. This process is known as glycogen or carbohydrate loading. According to Åstrand's regimen, athletes should prepare for an aerobic endurance competition by completing an exhaustive training bout seven days before the event. For the next three days, they should eat fat and protein almost exclusively to deprive the muscles of carbohydrate, which increases the activity of glycogen synthase, an enzyme responsible for glycogen synthesis and storage. Athletes should then eat a carbohydrate-rich diet for the remaining three days before the event. Because glycogen synthase activity is increased, increased carbohydrate intake results in greater muscle glycogen storage. Training intensity and volume during this six-day period should be markedly reduced to prevent additional muscle glycogen depletion, thus maximizing liver and muscle glycogen reserves. Originally, an additional intense training bout was performed four days prior to competition.

This regimen has been shown to elevate muscle glycogen stores to twice the normal level, but it is somewhat impractical for most highly trained competitors. During the three days of low carbohydrate intake, athletes generally find training difficult. They are also often irritable and unable to perform mental tasks, and they typically show signs of low blood sugar, such as muscle weakness and disorientation. In addition, the exhaustive depletion bouts of exercise performed seven days before the competition have little training value and can impair glycogen storage rather than enhance it. This depletion exercise also exposes athletes to possible injury or overtraining.

Considering these limitations, many proposed that the depletion exercise and the low carbohydrate aspects of Åstrand's regimen be eliminated. Instead, according to Sherman and colleagues,[27] the athlete should simply reduce training intensity a week before competition and eat a normal, mixed diet containing 55% of the calories from carbohydrate until three days before the competition. For these days, training should be reduced to a daily warm-up of 10 to 15 min of activity and accompanied by a carbohydrate-rich diet. Following this plan, as seen in figure 14.16, glycogen will be elevated to nearly 200 mmol/kg of muscle, the same level attained with Åstrand's regimen, and the athlete will be better rested for competition.

It is possible to increase carbohydrate stores rapidly after even a very short near-maximal-intensity bout of exercise. In a study of seven endurance athletes, scientists found that cycling for 150 s at 130% of $\dot{V}O_{2peak}$ followed by 30 s of all-out cycling and 24 h of high-carbohydrate intake was sufficient to nearly double muscle glycogen stores in just one day.[11]

A diet rich in carbohydrates is critical to the success of endurance athletes. Furthermore, carbohydrate loading is a very effective technique for increasing both muscle and liver glycogen stores. The additional stored carbohydrate provides the critical energy source for improved endurance performance.

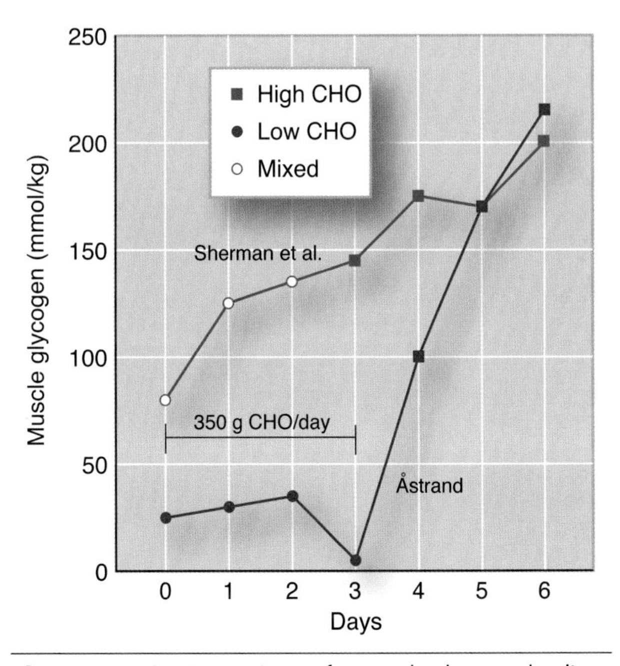

Figure 14.16 Two regimens for muscle glycogen loading. In one regimen, the subjects were depleted of muscle glycogen (day 0) and then ate a low-carbohydrate (CHO) diet for three days. They then switched to a CHO-rich diet, which caused muscle glycogen to increase to about 200 mmol/kg. In the other dietary regimen, the subjects ate a normal, mixed diet and reduced their training volume for the first three days; they then changed to a high-CHO diet and further reduction in training volume for three days, which also resulted in muscle glycogen of about 200 mmol/kg.

Data from P.-O. Åstrand, 1979, Nutrition and physical performance. In *Nutrition and the world food problem,* edited by M. Rechcigl (Basel, Switzerland: S. Karger); and W.M. Sherman, et al., 1981.

Diet is also important in preparing the liver for the demands of endurance exercise. Liver glycogen stores decrease rapidly when a person is deprived of carbohydrates for only 24 h, even when at rest. With only 1 h of strenuous exercise, liver glycogen decreases by 55%. Thus, hard training combined with a low-carbohydrate diet can empty the liver glycogen stores. A single carbohydrate meal, however, quickly restores liver glycogen to normal. Clearly, a carbohydrate-rich diet in the days preceding competition will maximize the liver glycogen reserve and minimize the risk of hypoglycemia during the event.

Water is stored in the body at a rate of about 2.6 g of water with each gram of glycogen. Consequently, an increase or decrease in muscle and liver glycogen generally produces a change in body weight of from 0.5 to 1.4 kg (1-3 lb). Some scientists have proposed monitoring of changes in muscle and liver glycogen stores via recording the athlete's early morning weight immediately after rising—after emptying the bladder but before eating breakfast. A sudden decrease in weight might reflect a failure to replace glycogen, a deficit in body water, or both.

Athletes who must train or compete in exhaustive events on successive days should replace muscle and liver glycogen stores as rapidly as possible. Although liver glycogen can be depleted totally after 2 h of exercise at 70% $\dot{V}O_{2max}$, it is replenished within a few hours when a carbohydrate-rich meal is consumed. Muscle glycogen resynthesis, on the other hand, is a slower process, taking several days to return to normal after an exhaustive exercise bout such as the marathon (see figure 14.17). Studies in the late 1980s revealed that muscle glycogen resynthesis was most rapid when individuals were fed at least 50 g (about 0.7 g/kg body weight) of glucose every 2 h after the exercise.[21] Feeding subjects more than this amount did not appear to accelerate the replacement of muscle glycogen. During the first 2 h after exercise, the rate of muscle glycogen resynthesis is much higher than later in recovery, as discussed earlier in this chapter. Thus, an athlete recovering from an exhaustive endurance event should ingest sufficient carbohydrate as soon after exercise as is practical. Adding protein and amino acids to the carbohydrate ingested during the recovery period enhances muscle glycogen synthesis above that achieved with carbohydrate alone.

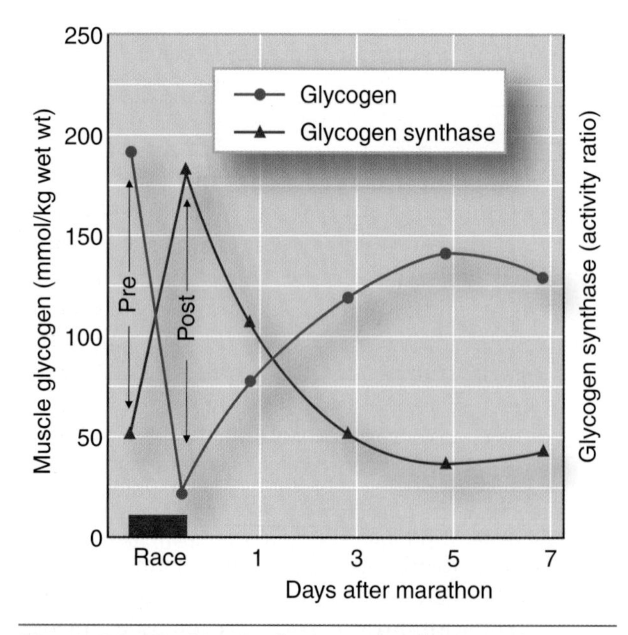

Figure 14.17 Muscle glycogen resynthesis is a slow process, requiring several days to restore normal muscle glycogen storage following exhaustive exercise. Note that when muscle glycogen decreases with hard exercise (race), muscle glycogen synthase is markedly elevated. This triggers the muscle to store glycogen when carbohydrates are eaten, returning glycogen synthase to the baseline level.

In review

- Following the guidelines presented in the earlier sections of this chapter should provide athletes optimal nutrition for their sport.
- Some athletes have adopted vegetarian diets and appear to perform well. However, careful consideration must be given to protein sources and to consuming adequate levels of iron, zinc, calcium, and several vitamins.
- The precompetition meal should be taken no less than 2 h before competition, and it should be low in fat, high in carbohydrate, and easily digestible. A liquid precompetition meal low in fat and high in carbohydrate has advantages.
- Carbohydrate loading greatly increases muscle glycogen content, which, in turn, increases endurance performance.
- After endurance competition or training, it is important to consume substantial carbohydrate to replace the glycogen used during activity. The body seems primed to replace glycogen during the first few hours after training or competing because glycogen synthase levels are at their peak.

The Zone Diet

In the mid-1990s, many athletes were attracted to a new diet proposed to enhance athletic performance, touted in a popular book written by Dr. Barry Sears, *The Zone.*[25] The Zone diet argues against the high-carbohydrate diet typically advocated for the athlete and the general population. The Zone diet centers on the premise that people should take in 1.8 to 2.2 g of protein per kilogram of fat-free mass. The diet approximates a 40% carbohydrate, 30% fat, and 30% protein proportion of total calories consumed. However, for athletes, a much higher percentage of calories from fat is recommended.[26] Supposedly, this low-carbohydrate diet promotes a more favorable insulin-to-glucagon ratio, ultimately improving oxygen delivery to exercising muscle.[6]

Although many anecdotal stories support the performance-enhancing qualities of the Zone diet, its efficacy has yet to be clearly established by well-designed research studies. In fact, a wealth of data in the sport nutrition literature strongly argue against such a diet. The diet promotes an unnecessarily high intake of protein and a relatively low intake of carbohydrate. Furthermore, if the diet is pushed to the extreme, the percentage of total calories from fat increases. So, until controlled research studies support the claims made for this diet, the athlete should follow the dietary recommendations that have been proposed in this chapter, recommendations that have the support of many studies over a number of years.[1]

Sport Drinks

We mentioned earlier that ingesting carbohydrate before, during, and following exercise can benefit performance by ensuring adequate fuel for energy production during exercise and for replenishing glycogen stores following exercise. While selecting a wise diet can provide for most of the athlete's nutritional needs, nutritional supplements can also be of great value. In addition, adequate fluid intake is necessary for preexercise hydration, hydration during exercise, and rehydration following exercise. Sport drinks are uniquely designed to meet both the energy and fluid needs of the athlete. Performance benefits from these drinks have been clearly documented, not only in endurance activities, but in burst activities as well (e.g., soccer and basketball).[7, 10]

Composition of Sport Drinks

Sport drinks differ from one another in a number of ways besides in taste. Of major concern, however, is the rate at which energy and water are delivered. Energy delivery is primarily determined by the concentration of the carbohydrates in the drink, and fluid replacement is influenced by the sodium concentration of the drink.

Energy Delivery—the Carbohydrate Concentration

A major concern is how rapidly the drink leaves the stomach, or the rate of **gastric emptying.** In general, carbohydrate solutions empty more slowly from the stomach than either water or a weak sodium chloride (salt) solution. Research suggests that a solution's caloric content, a reflection of its concentration, might be a major determinant of how quickly it empties from the stomach and is absorbed in the intestine. Since carbohydrate solutions remain in the stomach longer than either water or weak solutions, increasing the glucose concentration of a sport drink significantly reduces the gastric emptying rate. For example, 400 ml (14 oz) of a weak glucose solution (139 mmol/L) is almost completely emptied from the stomach in 20 min, but emptying a similar volume of a strong glucose solution (834 mmol/L) can require nearly 2 h. However, when even a small amount of a strong glucose drink leaves the stomach, it can contain more sugar than a larger amount of a weaker solution simply because of its higher concentration. But, if an athlete is trying to prevent dehydration, this would deliver less water and thus be counterproductive.

Most sport drinks on the market contain about 6 to 8 g of carbohydrate per 100 ml (3.5 oz) of fluid (6% to 8%). The carbohydrate source is generally glucose, glucose polymers, or a combination of glucose and glucose polymers, although fructose or sucrose has also been used.[22] Research studies have confirmed enhanced endurance performance with use of solutions in this range of concentration and with these sources of carbohydrates when compared to water.[1] Carbohydrate solutions above ~6% slow gastric emptying and limit the immediate availability of fluid. However, they can provide a greater amount of carbohydrate in a given period of time to meet the increased energy needs.[1, 22]

Rehydration With Sport Drinks—the Sodium Concentration

Just adding fluid to the body during exercise lessens the risk of serious dehydration. But research indicates that adding glucose and sodium to sport drinks, aside from supplying an energy source, stimulates both water and sodium absorption. Sodium increases both thirst and palatability of the drink. Recall that when sodium is retained, this causes more water to be retained. For rehydration purposes, both during and following exercise, the sodium concentration should range between 20 mmol/L and 60 mmol/L.[22] There is an important loss of sodium from the body with sweating. With high rates of sweating and large volumes of water intake, this can lead to critical reductions in the sodium concentration of the blood and possibly lead to hyponatremia, as discussed earlier in this chapter.

What Works Best?

Athletes will not drink solutions that taste bad. Unfortunately, we all have different taste preferences. To further confound the issue, what tastes good before and after a long, hot bout of exercise will not necessarily taste good during the event. Studies of taste preferences of runners and cyclists during 60 min of exercise showed that most chose a drink with a light flavor and no strong aftertaste. But, will athletes drink more if given a sport drink as compared to water? In one study, runners ran on a treadmill for 90 min and then recovered while seated for an additional 90 min. Both exercise and recovery conditions were controlled in an environmental chamber at a temperature of 32 °C (86 °F), 50% humidity. Three trials were conducted, two with two different sport drinks (6% and 8% carbohydrate) and one with water. Subjects were encouraged to drink throughout each trial. The volume consumed during exercise was similar for all three drinks; but during recovery, the runners drank about 55% more of each of the two sport drinks than water.[29]

In focus

Sport drinks have benefits in addition to those provided by plain water. The addition of carbohydrate to sport drinks provides an important energy source, and the addition of sodium and optimization of taste likely will result in greater fluid consumption, thus delaying dehydration.

In review

- Sport drinks have been shown to reduce the risk of dehydration and provide an important source of energy. They also can improve the performance of the athlete in both endurance and burst activities such as soccer and basketball.
- The carbohydrate concentration of a sport drink generally should not exceed 6% to 8% to maximize both sugar and fluid intake.[1]
- The inclusion of sodium in a sport drink facilitates the intake and storage of water.
- Taste is an important factor when one is considering a sport drink. Most athletes prefer a light flavor without a strong aftertaste. Each athlete should select the drink that tastes best, providing the nutritional ingredients are the same.

In Closing

In this chapter we examined the body composition and nutritional needs of the athlete, considering the importance of optimizing body composition for peak performance and eating wisely to enhance athletic performance. We discovered the importance of each of the six nutrient categories and how they can be adjusted to meet the athlete's training and performance needs. We looked at the precompetition meal, how to effectively replenish and load muscle glycogen stores, and the effectiveness of commercial sport drinks. Now that we have a better understanding of the importance of an appropriate weight and a balanced diet, we turn our attention to another aspect of the athlete's quest for success. In the next chapter, we evaluate those substances that have been proposed to enhance athletic performance—ergogenic aids.

KEY TERMS

air plethysmography
bioelectric impedance
body composition
body density
carbohydrate
carbohydrate loading
chronic fatigue syndrome
dehydration
densitometry
dual-energy X-ray absorptiometry (DXA)
electrolyte
essential amino acids
extracellular fluid
fat
fat-free mass
fat mass
female athlete triad
free radicals
gastric emptying
glycogen loading
hydrostatic weighing
hyponatremia
intracellular fluid
macrominerals
microminerals
nonessential amino acids
osmolarity
protein
relative body fat
skinfold fat thickness
thirst mechanism
trace elements
vitamin

STUDY QUESTIONS

1. Differentiate between body size and body composition.
2. What tissues of the body constitute the fat-free mass?
3. What is densitometry? How is it used to assess the body composition of the athlete? What is the major weakness of densitometry with respect to its accuracy?
4. What are several field techniques for estimating body composition? What are their strengths and weaknesses?
5. What is the relationship of relative leanness and fatness to performance in sport?
6. What guidelines should be used to determine the athlete's goal weight?
7. What are the six categories of nutrients?
8. What role does dietary carbohydrate play in endurance performance? How about fat? Protein?
9. What is an appropriate protein allowance for a normally active adult man? For a woman?
10. Discuss the value of using protein supplements to enhance performance in strength and endurance events.
11. Should the athlete supplement vitamins and minerals?
12. How does dehydration affect exercise performance? What effect does dehydration have on exercise heart rate and body temperature?
13. Describe the recommended precompetition meal.
14. Describe the method used to maximize muscle glycogen storage (glycogen loading).
15. Discuss the value of consuming carbohydrate during and after endurance exercise. What are the potential benefits of sport drinks?

STUDY GUIDE ACTIVITIES

In addition to the activities listed in the chapter opening outline on page 317, two other activities are available in the online study guide, located at

www.HumanKinetics.com/PhysiologyOfSportAndExercise.

The chapter **SUMMARY** reviews the key concepts covered in the chapter, and the end-of-chapter **QUIZ** tests your understanding of the material covered in the chapter.

Ergogenic Aids and Sport

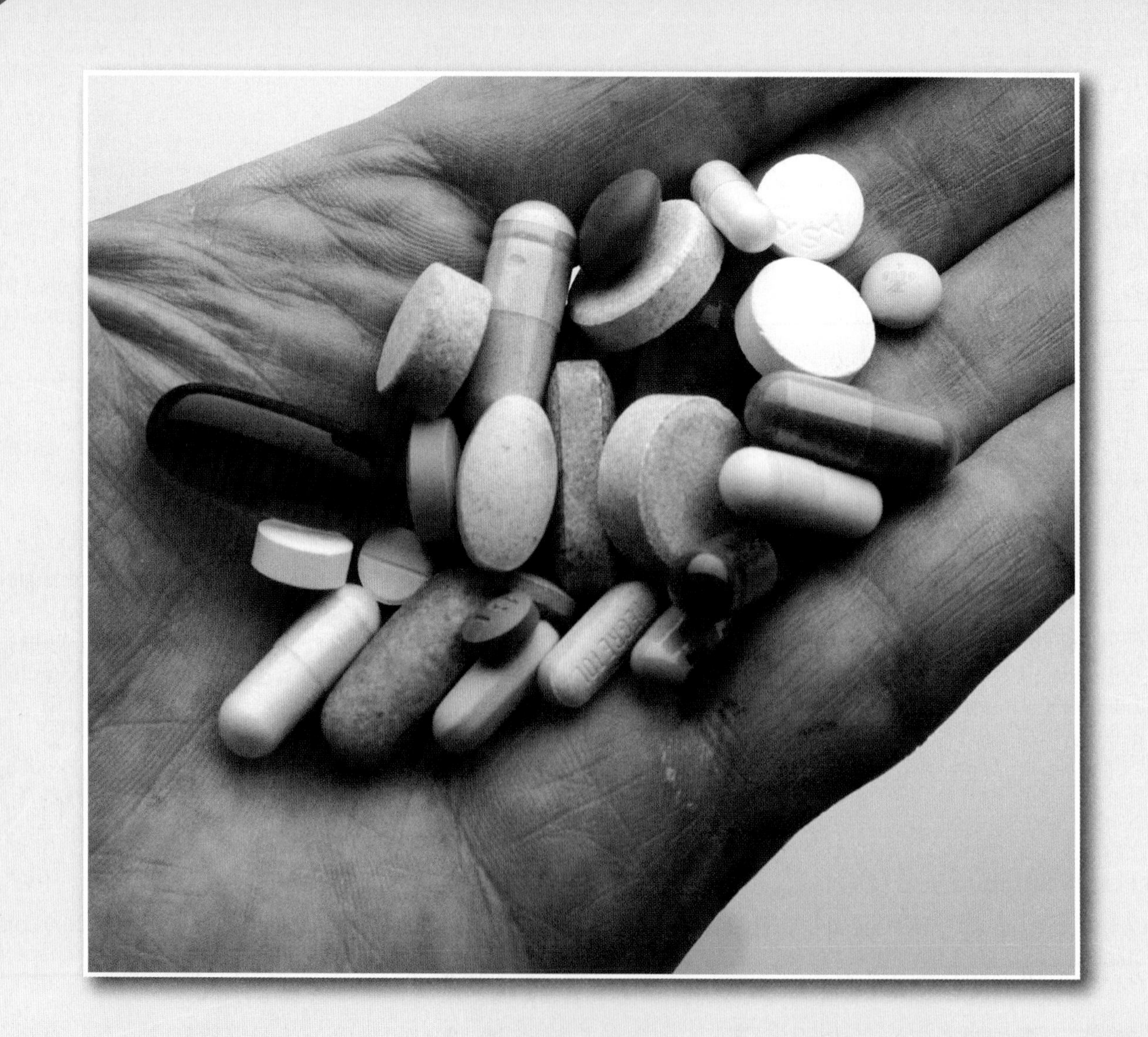

In this chapter and in the online study guide

In May 2006, just weeks before the start of the Tour de France, Spanish police raided the clinic of a Madrid physician. They discovered a number of performance-enhancing substances and drugs, including erythropoietin (EPO), frozen packets of blood, growth hormone, and anabolic steroids. A total of 58 national (Spain) and international elite cyclists were implicated, including 13 slated to compete in the 2006 Tour de France. Two were among the top cyclists in the world, but none of the 13 were allowed to compete. Erythropoietin and blood doping increase the number of red blood cells in the body, increasing the blood's oxygen-carrying capacity. This increase has been shown to increase both $\dot{V}O_{2max}$ and aerobic endurance performance. Growth hormone and anabolic steroids purportedly increase muscle mass and strength, increasing performance in activities requiring muscular strength and endurance. All four have been banned for use in sport by both national and international sport governing bodies. Sadly, the eventual winner of the 2006 Tour de France, Floyd Landis, tested positive for anabolic steroids and could lose his title if the doping charge is upheld. Unfortunately, use of illegal substances to improve performance is not confined to competitive cyclists. Almost all, if not all, sports have to deal with this problem, and abuse occurs not only in adults but in children as well.

The practice of using illegal substances in the hope of improving performance can be found among athletes of many different sports.

In the never-ending quest for glory, athletes often are willing to try anything to improve their performance. Some believe that special nutritional supplements can be the deciding factor. Others might use physiological agents such as oxygen or blood doping. Still others might try certain drugs or hormones.

Substances or phenomena (e.g., hypnosis) that improve an athlete's performance are referred to as ergogenic aids. The variety of potential ergogenic aids is immense, and the effects of many **ergogenic** (work producing) substances are shrouded in myth. Most athletes have received tips about ergogenic aids from a friend or coach and assume that the information is accurate, but this is not always the case. Some athletes experiment with substances, hoping for even a slight performance improvement regardless of possible harmful consequences. In the quest for performance perfection, a concern only with maximizing performance coupled with a lack of knowledge about ergogenic substances can lead an athlete to make unwise decisions as we will see in this chapter.

In focus

An ergogenic aid is any substance or phenomenon that enhances performance. An ergolytic agent is one that has a detrimental effect on performance. Some substances generally thought to be ergogenic are actually ergolytic.

The list of possible ergogenic aids is long, but the number that actually possess ergogenic properties is much shorter. In fact, some allegedly ergogenic substances or phenomena actually can impair performance. These are usually drugs, and Eichner has termed them **ergolytic** (work decreasing) drugs.[17] Ironically and sometimes tragically, several ergolytic agents have been promoted as ergogenic aids.

Many athletes indiscriminately take nutritional supplements and ingest drugs and other substances in the belief that they will improve their performance. In one study of 53 Division I university coaches and trainers, 94% provided their athletes with nutritional supplements.[38] This might seem totally harmless; but as we see later in this chapter, a high percentage of nutritional supplements are contaminated, and some have been found to contain banned substances. Anecdotal stories suggest that anywhere from 20% to 90% of athletes in certain sports are using, or have used, anabolic steroids. Scientific studies, however, suggest a much lower estimate of 6%.[6] Anabolic steroid use has even been reported in the general high school population in the United States, varying from 4% to 11% in boys and up to 3% in girls.[13]

Table 15.1 provides a selected listing of substances and agents proposed to have ergogenic properties that will be discussed in this chapter. This table also lists mechanisms of action by which these ergogenic aids have been proposed to work. These have been studied in sufficient depth to establish their efficacy. Many other substances have been proposed but not adequately researched.

This chapter focuses on pharmacological agents, hormones, physiological agents, and nutritional agents. More general nutritional practices are addressed in chapter 14. Psychological phenomena and mechanical factors are beyond the scope of this book but are reviewed in depth in Williams' book *Ergogenic Aids in Sport*.[45]

RESEARCHING ERGOGENIC AIDS

Assume that a professional athlete consumes a particular substance several hours before game time and then has a successful performance. The athlete likely will attribute the success to this substance, even though there is no proof that ingesting the substance will ensure other athletes similar success.

Anyone can claim that a certain substance is ergogenic—and many substances have been so labeled strictly because of speculation—but before a substance can be legitimately classified as ergogenic, it must be proven to enhance performance. Unfortunately, science, with its carefully controlled investigations, does not have all the answers. Still, scientific studies in this area are essential to differentiate between a true ergogenic response and a pseudoergogenic response, in which performance improves simply because the athlete expects improvement.

Placebo Effect

As we discussed in the introductory chapter, the phenomenon by which one's expectations of a substance determine the body's response to it is known as the **placebo effect.** This effect can seriously complicate the study of ergogenic qualities because researchers must be able to distinguish between the placebo effect and true responses to the substance being tested.

The placebo effect was clearly demonstrated in one of the earliest studies of anabolic steroids.[4] Fifteen male athletes who had been involved in heavy weightlifting for the previous two years volunteered for a weight training experiment using anabolic steroids. They were told that those who made the greatest strength gains over a preliminary four-month weight training period would be selected for the second phase of the study, in which they would receive anabolic steroids.

Following the initial period, 8 of these 15 subjects were randomly selected to enter the treatment phase. Only six of these subjects passed all medical screening

TABLE 15.1 Proposed Ergogenic Aids and Mechanisms Through Which They Might Work

Agent	Influence heart, blood, circulation, and aerobic endurance	Increase oxygen delivery	Supply fuel for muscle and general muscle function	Act on muscle mass and strength	Result in weight loss or weight gain	Counteract or delay onset or sensation of fatigue	Counteract central nervous system inhibition	Aid in relaxation and stress reduction
Pharmacological								
Amphetamines	✓					✓	✓	
β-blockers	✓							✓
Caffeine	✓		✓			✓		
Diuretics	✓				✓			
Hormones								
Anabolic steroids				✓	✓			
Human growth hormone				✓	✓			
Physiological								
Bicarbonate loading						✓		
Blood doping	✓	✓				✓		
Erythropoietin	✓	✓				✓		
Oxygen	✓	✓				✓		
Phosphate loading	✓	✓				✓		
Nutritional								
Amino acids	✓		✓	✓	✓	✓		
Creatine			✓	✓	✓	✓		
L-Carnitine			✓			✓		

tests and were allowed to continue to the treatment phase. This phase consisted of a four-week period in which the subjects were told that they would receive 10 mg per day of Dianabol (an anabolic steroid), when in fact they received a **placebo**—an inactive substance typically provided in a form identical to the genuine drug.

Strength data were collected over the last seven weeks of the four-month pretreatment training period and over all four weeks of the treatment (placebo) period (see figure 15.1 on page 358). Even though the subjects were experienced weightlifters, they continued to gain impressive amounts of strength during the pretreatment training period. However, strength gains while subjects were taking the placebo were substantially greater than during the pretreatment period! The group improved an average of 11 kg (24 lb) during the seven-week pretreatment period but improved 45 kg (~100 lb) during the four-week treatment (placebo) period. This represents an average gain in strength of 1.6 kg (3.5 lb) per week during the pretreatment training period and 11.3 kg (25 lb) per week during the placebo period—a more than seven times greater increase in the rate of strength gain during the placebo (supposed steroid) period over the pretreatment training period. Furthermore, placebos are inexpensive, risk free, and legal for use in sport.

One of the authors of this textbook (JHW) repeatedly witnessed the placebo effect while conducting a large series of studies investigating the effects of β-blocking drugs on the ability to perform single bouts of exercise or to train aerobically. The Human Subjects Committee, a committee mandated by the federal government to oversee all research conducted with human subjects in the United States, requires that all human subjects receive a full disclosure of the risks associated with any experimental intervention so that they can provide informed consent before participating. Before the start of each study, a cardiologist presented a comprehensive background on β-blocking drugs to each subject,

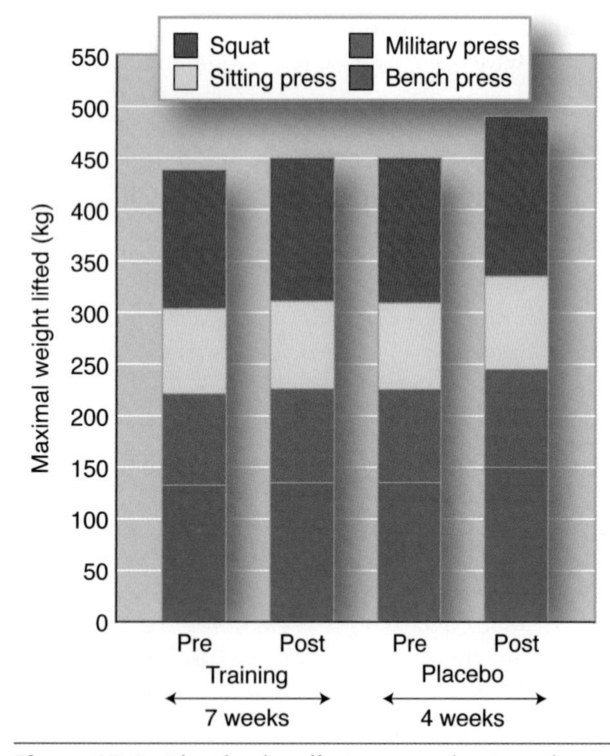

Figure 15.1 The placebo effect on muscular strength gains. The increase in total strength and strength in each of four maximum lifts over the last seven weeks of an intense four-month pretreatment training period is compared with strength increases during a subsequent four-week treatment period, in which the subjects took placebos that they thought were anabolic steroids and continued intense resistance training.

Data from Ariel and Saville 1972.

including the drugs' significance in treating various cardiovascular diseases and potential side effects associated with their use. It was amazing to note that over the course of six years of study, the most serious side effects almost always appeared in the subjects when taking the placebo.

When evaluating a substance for possible ergogenic qualities, researchers must remember that witnessing an ergogenic effect does not necessarily prove that a substance is truly ergogenic. All studies of potential ergogenic substances must include a placebo group so that researchers can compare actual responses resulting from the test substance with those resulting from a placebo. In many studies, a double-blind experimental design is used, in which neither the subject nor the experimenter knows who is getting the proposed ergogenic aid and who is getting the placebo. This is done to eliminate "experimenter bias," whereby the experimenter's beliefs might affect the outcome of the study. With this design, the substances are coded and only an independent person not associated with the project has access to the codes. See the introductory chapter for more information on the proper control of experiments.

In focus

Although the placebo effect has a psychological origin, the body's physical response to a placebo is quite real. This clearly illustrates how effective our mental state can be in altering our physical state.

Limitations of Research

To satisfy the scientific community, scientists often rely on laboratory techniques to evaluate the efficacy of any potential ergogenic aid. Often, however, scientific studies cannot provide absolutely clear answers to the questions under study. With elite athletes, success is defined in fractions of a second or in millimeters. Laboratory tests are often unable to detect such subtle differences in performance.

Scientists can be greatly limited by the accuracy of their equipment or techniques. All research methods have some margin of error. If the results fall within that margin of error, the researcher cannot be certain that the result is an effect of the substance being tested. The results might reflect limitations of the research methodology. Unfortunately, because of measurement error, individual differences, and the day-to-day variability of subjects' responses, a potential ergogenic aid must exert a major effect before scientific tests can prove that it is ergogenic.

The testing situation can also limit accuracy. Performance in a laboratory is considerably different from performance in the usual athletic environment, so laboratory results won't always accurately reflect natural athletic results. Yet an advantage of laboratory testing is that the environment can be carefully controlled. This is not always possible in field studies conducted in the athlete's usual environment, where several uncontrollable variables—such as temperature, humidity, wind, and distractions—can affect the results. Thorough testing of a potential ergogenic aid should include both field and laboratory studies.

Realizing that science is limited in its ability to unequivocally determine the efficacy of a substance, we can now examine some proposed ergogenic aids. We consider substances in four classes:

- Pharmacological agents
- Hormonal agents
- Physiological agents
- Nutritional agents

PHARMACOLOGICAL AGENTS

Numerous **pharmacological agents,** or drugs, have been suggested as having ergogenic properties. The International Olympic Committee (IOC), the United States Olympic Committee (USOC), the International Amateur Athletic Federation (IAAF), and the National Collegiate Athletic Association (NCAA) all publish extensive lists of banned substances, most of which are pharmacological agents. The IOC and the USOC now use the standards established by the World Anti-Doping Agency (WADA). The United States also has an anti-doping agency, USADA. Each athlete, coach, athletic trainer, and team physician must know which drugs are prescribed for and taken by the athlete, and they must check these drugs periodically against the listing of banned substances because the list changes frequently. The USOC has a drug education hotline that provides up-to-date information (1-800-233-0393), and WADA (www.wada-ama.org/en/) and the USADA (www.usantidoping.org/) have Web sites for the same purpose. Athletes have been disqualified and have had to relinquish medals, ribbons, awards, and prizes after testing positive for a banned substance. In many cases, the medication had been used legitimately to treat a known medical condition.

We review here only drugs for which a research base has been established. Many other drugs have been touted as ergogenic, but controlled studies have yet to be conducted to determine if they are effective. The drugs we discuss are

- sympathomimetic amines,
- β-blockers,
- caffeine,
- diuretics, and
- recreationally used drugs.

Sympathomimetic Amines

Amphetamine and its derivatives are central nervous system (CNS) stimulants. They are also considered sympathomimetic amines, which means that their activity mimics that of the sympathetic nervous system. For many years they have been used as appetite suppressants in medically supervised weight loss programs. During World War II, army troops used amphetamines to combat fatigue and to improve endurance. They are now used to treat attention deficit and hyperactivity disorder (ADHD). Also known as "speed," they soon found their way into the athletic arena, where they were considered stimulants with possible ergogenic properties. More recently, two other sympathomimetic amines have been proposed as ergogenic aids—**ephedrine** and **pseudoephedrine.** Ephedrine is derived from ephedra herbs (also known as ma huang) and is used as a decongestant and as a bronchodilator in the treatment of asthma. Pseudoephedrine is used in over-the-counter medications primarily as a decongestant and in the illicit manufacturing of methamphetamine. In the following discussion of sympathomimetic amines, we focus most of our attention on amphetamines, since they have been studied most extensively.

Proposed Ergogenic Benefits

Athletes have found amphetamines readily available even though they are prescription drugs. Amphetamines are used by athletes for many reasons other than weight loss. Psychologically, the drugs are thought to increase concentration and mental alertness. Their stimulating effect decreases mental fatigue. Athletes anticipate more energy and motivation and often feel more competitive when using amphetamines. The drugs also produce a state of euphoria, which is part of their attraction as so-called recreational drugs. Often athletes who use amphetamines report a sense of indestructibility, which they feel spurs them to higher performance levels.

In terms of actual performance, amphetamines are thought to help athletes run faster, throw farther, jump higher, and delay the onset of total fatigue or exhaustion. Athletes who use these drugs expect virtually every aspect of performance to be enhanced. Similar claims and expectations have been associated with the use of ephedrine and pseudoephedrine.

Proven Effects

Generally, for any physiological, psychological, or performance variable that has been investigated, some studies show that amphetamines have no effect, others demonstrate an ergogenic effect, and still others indicate an ergolytic effect. As potent CNS stimulants, amphetamines do increase the state of arousal, which leads to a sense of increased energy, self-confidence, and faster decision making. Athletes who take amphetamines experience a decreased sense of fatigue; increased heart rate, systolic and diastolic blood pressure, and blood flow to skeletal muscles; and elevation of blood glucose and free fatty acids.

Do these effects aid physical performance? Although studies are not in total agreement, the more recent studies, which have used better experimental designs and controls, show that amphetamines can enhance components of athletic performance, specifically

- weight loss;
- reaction time, acceleration, and speed;
- strength, power, and muscular endurance;
- possibly aerobic endurance, but not $\dot{V}O_{2max}$;
- higher maximum heart rate and peak lactate concentrations at exhaustion;
- better focus; and
- fine motor coordination.

In focus

Amphetamines can improve performance in certain sports or activities; but, in addition to being illegal, these drugs carry risks that far outweigh their benefits. Amphetamines can be addictive, and they mask important signals that the body sends out to inform us when we are in potentially dangerous situations. Ephedrine and pseudoephedrine generally do not improve performance but share similar risks with amphetamines.

The results are not as clear for ephedrine and pseudoephedrine. While several studies have shown small improvements in markers of athletic performance with use of these substances, the general conclusion is that performance benefits are inconsistent and probably insignificant for speed, strength, power, and endurance.[1,13]

Risks of Using Sympathomimetic Amines

Deaths have been attributed to excessive amphetamine and ephedrine use. Because heart rate and blood pressure are increased, users place greater stress on their cardiovascular systems. These drugs can trigger cardiac arrhythmias in some susceptible individuals. Also, rather than delaying the onset of fatigue, amphetamines likely delay the sensation of fatigue, enabling the athletes to push dangerously beyond normal limits to the point of circulatory failure. Deaths have occurred when athletes have pushed themselves far beyond the normal point of exhaustion. As just one example, during spring training in 2003, Baltimore Orioles' pitching prospect Steve Bechler collapsed during a workout and died less than 24 h later of complications from heatstroke. He had been taking an over-the-counter supplement containing ephedrine that was linked to his heatstroke by the medical examiner at autopsy.

Amphetamines can be psychologically addictive because of the euphoria and energized feelings they cause. But the drugs also can be physically addictive if taken regularly, and a person's tolerance to them builds with continued use, requiring increasingly larger doses over time to obtain the same effects. Amphetamines also can be toxic. Extreme nervousness, acute anxiety, aggressive behavior, and insomnia are frequently mentioned side effects of regular use. Ephedrine and pseudoephedrine have side effects similar to those of amphetamines and are associated with a high incidence of cardiovascular events.

β-blockers

The sympathetic nervous system influences bodily functions through adrenergic nerves: those that use norepinephrine as their neurotransmitter. Neural impulses traveling through these nerves trigger the release of norepinephrine, which crosses the synapses and binds to adrenergic receptors at the target cells. These adrenergic receptors are classified into two groups: α-adrenergic receptors and β-adrenergic receptors.

β-adrenergic blockers, or **β-blockers,** are a class of drugs that block the β-adrenergic receptors, preventing binding of the neurotransmitter norepinephrine. This greatly reduces the effects of stimulation by the sympathetic nervous system. β-blockers generally are prescribed for the treatment of hypertension, angina pectoris, and certain cardiac arrhythmias. They also are prescribed as a preventive treatment for migraine headaches, to reduce the symptoms of anxiety and stage fright, and for initial recovery from heart attacks.

Proposed Ergogenic Benefits

Because the sympathetic response gears the body up for physical activity (through the fight-or-flight mechanism), it is difficult to understand why athletes might turn to β-blockers as ergogenic aids. β-blocker use in sport has been limited mostly to sports in which anxiety and tremor could impair performance. When a person stands on a force platform (a highly sophisticated device that measures mechanical forces), measurable body movement is detected each time the heart beats. This movement is sufficient to affect a shooter's aim. Accuracy in shooting sports improves if the rifle or pistol can be shot or the arrow released between heartbeats. β-blockers can slow a shooter's heart rate, allowing more time to stabilize the aim before shooting or releasing before the next heartbeat. They have also allegedly been used by golfers to steady their stroke, particularly when putting.

Proven Effects

β-blockers decrease the effects of sympathetic nervous system activity. This is well illustrated by the marked reduction in maximum heart rate with β-blocker administration. It is not unusual for a 20-year-old male athlete with a normal maximum heart rate of 190 beats/min to

β-blockers can enhance performance in shooting types of sports and therefore have been banned.

have a maximum heart rate of only 130 beats/min when taking β-blocking drugs. Resting and submaximal heart rates also are reduced by these drugs. Several studies have confirmed improved scores in shooting sports as a result of this heart rate reduction when subjects used β-blockers. Because of this, the WADA, the IOC, the USOC, and the NCAA have banned the use of β-blockers for these sports.

Risks of β-Blocker Use

Most risks from β-blockers are associated with prolonged use, not isolated incidents of use as in athletics. β-blockers can induce bronchospasm in people with asthma. They can cause cardiac failure in people who have underlying problems with cardiac function. In people with bradycardia, these drugs can lead to heart block. The decreased blood pressure they cause can result in light-headedness. Some people with type 2 diabetes can become hypoglycemic, because β-blockers increase insulin secretion. These drugs, through their various effects, can cause pronounced fatigue, which can inhibit athletic performance and decrease motivation. For athletes who must take β-blockers for a medical condition such as hypertension or arrythmias, β-1 selective blockers are usually preferred because they have fewer negative effects on performance.

Caffeine

Caffeine, one of the most widely consumed drugs in the world, is found in coffee, tea, cocoa, soft drinks, and various foods. This drug is also common in several over-the-counter medications, often even in simple aspirin compounds. Caffeine is a CNS stimulant, and its effects are similar to those noted previously for amphetamines, although weaker.

Proposed Ergogenic Benefits

As with sympathomimetic amines, caffeine generally is touted as improving alertness, concentration, reaction time, and energy level. Athletes taking the drug often feel stronger and more competitive. They believe that they can perform longer before the onset of fatigue and that if they are fatigued beforehand, the fatigue is reduced. Caffeine is known to have metabolic effects on adipose tissue and skeletal muscle as well as on the CNS, and it has been proposed to increase the mobilization and use of free fatty acids, thus sparing muscle glycogen and prolonging endurance activity.

Proven Effects

Because of its effects on the CNS, the general effects of caffeine include

- increased mental alertness,
- increased concentration,
- elevated mood,
- decreased fatigue and delayed onset,
- decreased reaction time (i.e., faster response),
- enhanced catecholamine release,
- increased free fatty acid mobilization, and
- increased use of muscle triglycerides.

In terms of ergogenic properties, caffeine initially was studied for potential effects that could benefit endurance activities. The first studies, conducted by Costill, Ivy, and their colleagues,[14, 28] demonstrated marked improvements in endurance performance when competitive cyclists ingested a caffeinated beverage compared with a placebo beverage. Caffeine increased endurance times in fixed-pace work bouts and decreased times in fixed-distance races.

Although several subsequent studies were unable to replicate these results, the more recent studies have unequivocally demonstrated substantial ergogenic effects of caffeine ingestion for aerobic endurance performance.[24, 32] It was initially postulated that this improvement was the result of an increased mobilization of free fatty acids, sparing muscle glycogen for later use. But the actual mechanisms by which caffeine improves endurance performance appear to be more complex, since glycogen sparing does not always occur. It is now well documented that caffeine lowers the perception of effort at a given rate of work, potentially allowing the athlete to perform at a higher intensity with the same perceived effort. Further, a growing number of studies are suggesting that caffeine has its effect directly on the CNS.[40] Cellular mechanisms within skeletal muscle also are being explored.

Caffeine can enhance performance in endurance sports and may even be of benefit in activities of much shorter duration (e.g., 1 to 6 min). However, some athletes may experience a negative response, in which case caffeine would be considered ergolytic.

Caffeine has also been shown to improve performance in sprint and strength types of activities and in high-intensity team sports. Unfortunately, fewer studies have investigated this area, but caffeine might facilitate calcium exchange at the sarcoplasmic reticulum and increase the activity of the sodium-potassium pump, better maintaining the muscle membrane potential.[24]

Risks of Caffeine Use

In people who are not accustomed to using caffeine, who are sensitive to it, or who consume high doses, caffeine can produce nervousness, restlessness, insomnia, headache, gastrointestinal problems, and tremors. Caffeine also acts as a diuretic, potentially increasing an athlete's risk for dehydration and heat-related illness during performing in hot environments. It can disrupt normal sleep patterns, contributing to fatigue. Caffeine is also physically addictive; abrupt discontinuation of caffeine intake can result in severe headache, fatigue, irritability, and gastrointestinal distress. At one time caffeine, at high doses, was on the list of banned drugs, but it is no longer on this list.

Diuretics

Diuretics affect the kidneys, increasing urine production. Used appropriately, they rid the body of excess fluid and are generally prescribed to control hypertension and reduce edema (water retention) associated with congestive heart failure or other conditions.

Proposed Ergogenic Benefits

Diuretics generally are used as ergogenic aids for weight control. For decades, diuretics have been used by jockeys, wrestlers, and gymnasts to keep their weight down. More recently, they have been used by anorexics and bulimics for weight loss.

Some athletes who are taking banned drugs also have turned to diuretics, but not to enhance their performance. Because diuretics increase fluid loss, these athletes hope that the extra fluid in their urine will dilute the concentration of banned drugs, thus decreasing the likelihood that the banned substances will be detected during drug testing. This practice and other means of altering the urine in an effort to escape drug detection are called masking.

In review

- Amphetamines are CNS stimulants that increase mental alertness, elevate mood, decrease the sense of fatigue, and produce euphoria.
- Studies indicate that amphetamines can increase concentration, reaction time, acceleration, speed, strength, maximum heart rate, peak lactate responses during exhaustive exercise, and time to exhaustion.
- Amphetamines elevate both heart rate and blood pressure and can trigger cardiac arrhythmias. Excessive use of these drugs has been blamed for some deaths, and the drugs can be both psychologically and physically addictive.
- Ephedrine and pseudoephedrine have characteristics similar to amphetamines but are not nearly as effective as ergogenic agents. They also have serious side effects.
- β-blockers block β-adrenergic receptors, preventing neurotransmitter binding.
- β-blockers slow the resting heart rate, which is a distinct advantage for shooters who try to release the arrow or squeeze the trigger between heartbeats to minimize the slight tremor associated with each beat, and could be an advantage for golfers when putting.
- β-blockers cause bradycardia and can even cause heart block, hypotension, bronchospasm, pronounced fatigue, and decreased motivation. Selective β-blockers have fewer side effects than nonselective blockers and usually would be prescribed for an athlete with a medical need for β-blocking drugs.
- Caffeine, one of the most widely consumed drugs in the world, is also a CNS stimulant, and its effects are similar to those of amphetamine but weaker.
- Caffeine increases mental alertness and concentration, elevates mood, decreases fatigue and delays its onset, increases catecholamine release and mobilization of free fatty acids, and is proposed to increase muscle use of free fatty acids to spare glycogen.
- Caffeine can cause nervousness, restlessness, insomnia, tremors, and diuresis. Diuresis increases susceptibility to heat injury.
- Diuretics affect the kidneys, increasing urine production and excretion. They often are used by athletes for temporary weight reduction and also by those trying to mask the use of other drugs during drug testing.
- Weight loss is the only proven ergogenic effect of diuretics, but this weight loss is primarily from the extracellular fluid compartment, including blood plasma. This may lead to dehydration, increased cardiac strain, and electrolyte imbalances.

Proven Effects

Diuretics lead to significant temporary weight loss, but no evidence suggests any other potential ergogenic effects. In fact, several side effects make diuretics ergolytic. The fluid loss results primarily from losses in extracellular fluid, including plasma. For athletes, particularly those who depend on moderate to high levels of aerobic endurance, this reduction in plasma volume reduces maximal cardiac output, which in turn reduces aerobic capacity and impairs performance.

Risks of Diuretic Use

In addition to reducing plasma volume, diuretics may hinder thermoregulation. As internal body heat increases, more blood must be diverted to the skin so that the heat can be lost to the environment. However, when blood plasma volume is diminished, as with diuretic use, more blood must be kept in the central regions of the body to maintain central venous blood pressure and adequate blood supply and blood pressure to the vital organs. Thus, less blood is available to be shunted to the skin, and heat loss may be impaired.

Electrolyte imbalance also can occur. Many diuretics cause fluid loss by ensuring electrolyte loss. A diuretic called furosemide inhibits sodium reabsorption in the kidneys, thus allowing more of it to be excreted in the urine. Because fluid follows the sodium, more fluid also will be excreted. Electrolyte imbalances can occur with losses of either sodium or potassium. These imbalances can cause fatigue and muscle cramping. More serious imbalances can lead to exhaustion, cardiac arrhythmias, and even cardiac arrest. The topic of hyponatremia was discussed in chapter 14. Some athletes' deaths have been attributed to electrolyte imbalances caused by diuretic use.

Recreationally Used Drugs

A class of drugs referred to as "recreational drugs" has been widely used by athletes for both recreation and their potential ergogenic properties. These include alcohol, cocaine, marijuana, and nicotine. None of these have been shown to have ergogenic properties and most are ergolytic, so they will not be further discussed in this chapter.

In focus

Many pharmacological agents do not have ergogenic properties, yet some athletes believe that they do. Several substances are banned not because they are ergogenic but because their use carries high risks. Such bans are intended to keep athletes from trying harmful substances with the erroneous notion that they will enhance performance, when in fact some of these substances can be lethal.

HORMONAL AGENTS

The use of **hormonal agents** as ergogenic aids in competitive athletics began in the late 1940s or early 1950s. Anabolic steroids were the hormones most frequently used by athletes between the 1950s and the 1980s. During the last half of the 1980s, a new potential ergogenic aid emerged with the introduction of synthetic human growth hormone.

Although numerous scientific studies have been conducted on anabolic steroids and sport, much less is known about the effects of human growth hormone on sport performance. Both anabolic steroids and human growth hormone are banned for all sports, and the medical risks associated with their use are high.

Anabolic Steroids

Androgenic-anabolic steroids, commonly referred to simply as **anabolic steroids,** are nearly identical to the male sex hormones. The anabolic (building) properties of these steroid hormones accelerate growth by increasing the rate of bone maturation and the development of muscle mass. For years, anabolic steroids have been given to youngsters with delayed growth patterns to normalize their growth curves. The development of synthetic steroids has allowed alteration of the natural chemical composition of these hormones to reduce their androgenic (masculinizing) properties and increase their anabolic effects on muscle.

Proposed Ergogenic Benefits

Steroid administration is known to increase fat-free mass and strength. Consequently, an athlete who depends on muscle size, body size, or strength might be tempted to take steroids. Early claims that aerobic capacity improves with anabolic steroid use caught the attention of endurance athletes. Anabolic steroids also have been postulated to facilitate recovery from exhaustive training bouts, allowing athletes to train hard on subsequent days. This potential benefit has stirred the interest of athletes from almost all sports.

The potential for anabolic steroid use among athletes is very high, and this continues to be a major problem in sport. Among professional athletes in sports such as football and baseball, it is suspected that the percentage of athletes taking steroids is high. The evidence is anecdotal, but the percentage of anabolic steroid users is at least 20%, and possibly higher.[6] Actual drug testing results indicate a much lower percentage of users, but athletes who take steroids have become very good at "beating the system" when they know that they will be tested. They have used masking agents and even the urine of friends who have not used steroids. New designer steroids introduced in the early 2000s are engineered to help the athlete avoid detection.

Proven Effects

The limitations of scientific research have been apparent in the study of the effects of anabolic steroids. Results of early investigations were almost evenly divided. Many of these studies showed no significant change in body size or physical performance attributable to taking steroids, yet many of the other early studies and all recent studies found steroids to have considerable positive influence on increasing muscle mass and strength.

One basic problem with almost all research conducted in this area to date is the inability to observe in the research laboratory the effects of the drug dosages being used in the athletic world. Athletes are estimated to be taking 5 to 20 times the recommended maximum daily dosage or more.[25] Obviously, it would be unethical to design a study using dosages that exceed the recommended maximum. However, some researchers have been able to observe athletes both when the athletes are taking high doses of steroids and when they are off the drug. Research has demonstrated the following effects of steroids on performance.

Muscle Mass and Strength

In one of the first studies involving athletes who were taking steroids on their own, the effects of relatively high doses were observed in seven male weightlifters.[27] Two treatment periods, each lasting six weeks, were separated by a six-week interval without treatment. Half the subjects received a placebo during the first treatment period and the steroid during the second treatment period. The other half received the medications in reverse order: steroid first, then placebo. When the data from all subjects were analyzed, results showed that while on the steroid, the weightlifters had significant increases in

- body mass and fat-free mass,
- total body potassium and total body nitrogen (markers of fat-free mass),
- muscle size, and
- leg strength.

These increases did not occur during the placebo period. Results of this study are summarized in figure 15.2.

In a second study, Forbes[21] observed body composition changes in a professional bodybuilder and a competitive weightlifter. Both were on self-prescribed high doses of steroids. The bodybuilder had been on the high dose for 140 days and the weightlifter for 125 days. Fat-free body mass increased an average of 19.2 kg (42.3 lb), and fat mass decreased almost 10

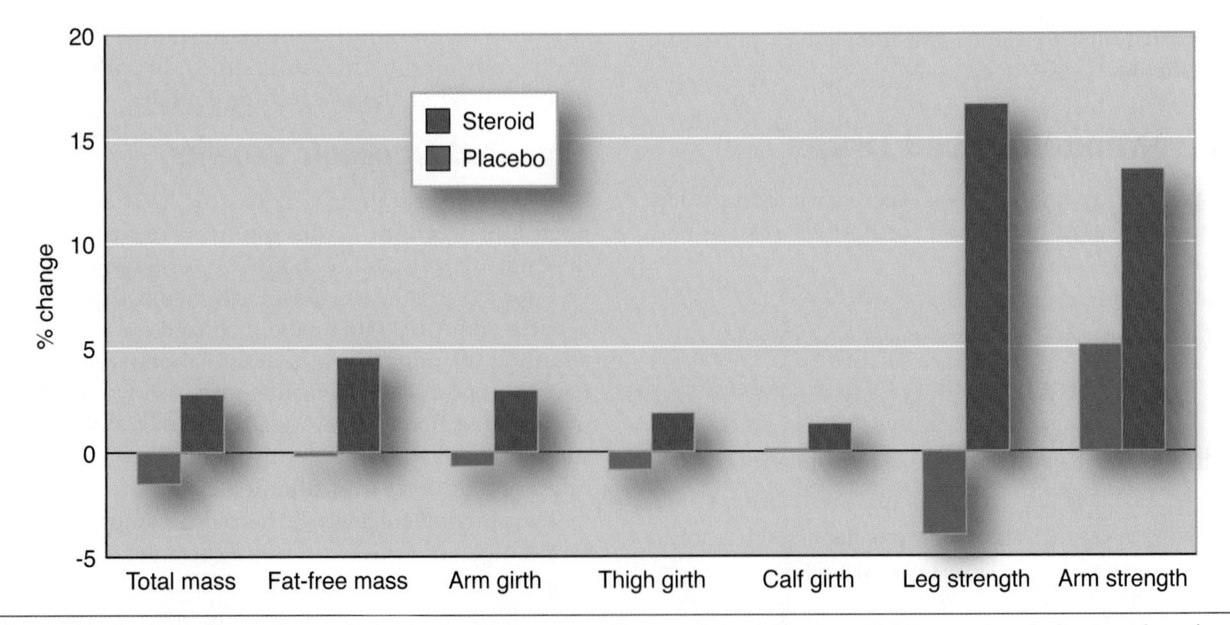

Figure 15.2 Percentage changes in body size, body composition, and strength when athletes used anabolic steroids and a placebo.

Adapted from G.R. Hervey et al., 1981, "Effects of methandienone on the performance and body composition of men undergoing athletic training," *Clinical Science* 60: 457-461.

kg (22 lb). Forbes plotted the results of a number of different studies that used different dosages (figure 15.3). He concluded that only minimal increases of 1 to 2 kg (2.2-4.4 lb) in fat-free body mass occur with low doses of anabolic steroids. But with high doses, fat-free body mass increases markedly. His results show a threshold level for steroid doses, with only very high doses resulting in substantial increases in fat-free body mass.

A third study looked at supraphysiological doses of testosterone on muscle size and strength in men who were nonathletes but were experienced with weightlifting.[7] Forty men completed the study and were assigned to one of the following groups: placebo with no exercise, placebo with exercise, testosterone with no exercise, and testosterone with exercise. The men received either 600 mg of testosterone enanthate or placebo intramuscularly each week for 10 weeks. The exercise groups strength trained three days per week for 10 weeks. Body composition was measured by underwater weighing, triceps and quadriceps muscle size by magnetic resonance imaging, and upper and lower body strength by the 1-repetition maximum technique. The testosterone and exercise group showed the largest increases in fat-free mass, triceps and quadriceps muscle area, and strength, whereas the placebo and no-exercise group remained unchanged (figure 15.4). The placebo and exercise

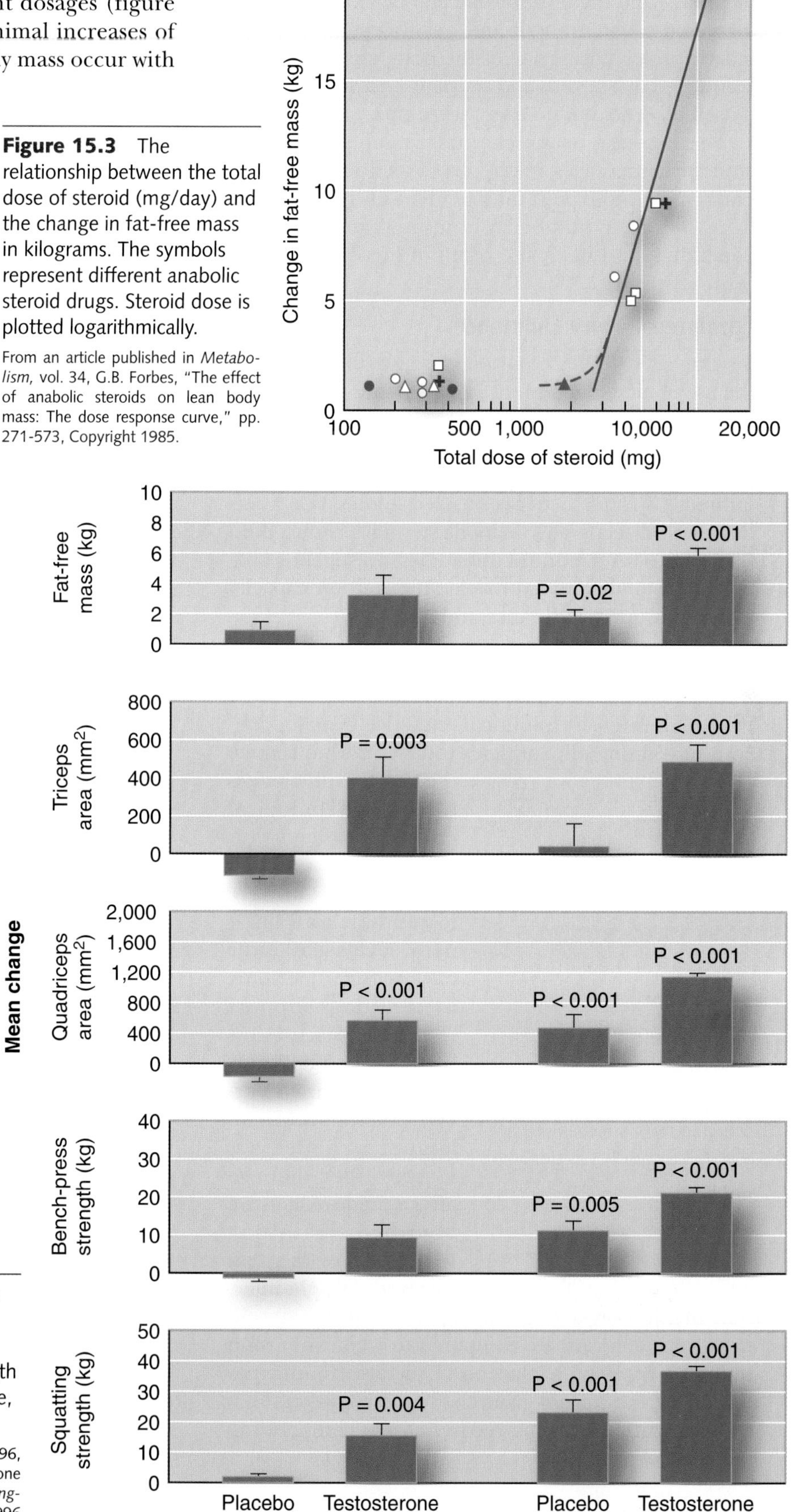

Figure 15.3 The relationship between the total dose of steroid (mg/day) and the change in fat-free mass in kilograms. The symbols represent different anabolic steroid drugs. Steroid dose is plotted logarithmically.

From an article published in *Metabolism*, vol. 34, G.B. Forbes, "The effect of anabolic steroids on lean body mass: The dose response curve," pp. 271-573, Copyright 1985.

Figure 15.4 Changes in fat-free mass and quadriceps and triceps muscle areas from magnetic resonance imaging, and changes in squat and bench press strength over 10 weeks of placebo or testosterone, with or without exercise training.

Reprinted, by permission, from S. Bhasin et al., 1996, "The effects of supraphysiologic doses of testosterone on muscle size and strength in normal men," *New England Journal of Medicine* 335: 1-7. Copyright © 1996 Massachusetts Medical Society. All rights reserved.

group increased strength, quadriceps area, and fat-free mass; and the testosterone and no-exercise group increased squat strength and quadriceps and triceps muscle areas. This is one of the best-designed studies conducted on steroids and resistance exercise because it used placebo and no-exercise groups.

The increase in muscle mass is generally associated with increases in the cross-sectional areas of both type I and type II muscle fibers and an increase in myonuclear number. These increases are dose dependent and likely the result of increased muscle protein synthesis.[20]

Cardiorespiratory Endurance

Several early studies reported increases in $\dot{V}O_{2max}$ with the use of anabolic steroids. These results were consistent with the known effects of steroid administration on increasing red blood cell production and total blood volume. However, in these studies, $\dot{V}O_{2max}$ was estimated indirectly. In later, better-controlled studies, $\dot{V}O_{2max}$ was measured directly, and anabolic steroids produced no benefit. However, none of the studies investigating anabolic steroid use and improvement in aerobic capacity involved trained endurance athletes.

Recovery From Training

The theory that anabolic steroids facilitate recovery from high-intensity training is attractive. A major concern in training elite athletes today is how to reduce the negative physiological and psychological effects associated with high-intensity training, enabling the athlete to continue to train at peak levels day after day. At this time, only limited data support the possibility that steroid use facilitates recovery from exercise. Tamaki and colleagues[41] reported less muscle fiber damage after a single exhaustive bout of weightlifting in a group of rats that received a single injection of the long-acting androgenic anabolic steroid nandrolone decanoate compared to a control group that had received a placebo. They also found that the rats in the steroid group had an increased rate of protein synthesis during recovery when compared to the control group.

Anabolic steroid use does increase muscle mass and strength, which can improve performance in strength-type sports or activities. Aerobic endurance appears to be unaffected by steroid use. Steroid use in sport is illegal and is banned in most sports. Furthermore, the health risks can be considerable.

Risks of Anabolic Steroid Use

Although use of anabolic steroids can be beneficial for certain types of athletic performance, several major issues must be addressed. It is neither moral nor ethical for athletes to use drugs to improve their chances in competition. Most athletes feel that it is wrong for their competitors to artificially improve their performance. Yet many of these same athletes feel compelled to use steroids in an effort to compete with the other athletes in their sport or event who are chronic steroid users. Fair competition is not possible if an individual is the only athlete in a particular competition who has remained steroid free.

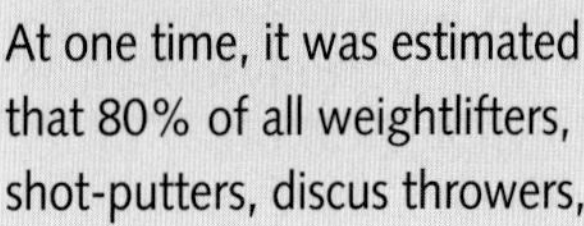

At one time, it was estimated that 80% of all weightlifters, shot-putters, discus throwers, and javelin throwers of national caliber were using anabolic steroids, and this was considered by many to be a conservative number.

Although the pressures on the athlete are great, are the potential gains worth the possible risks associated with steroid use? These drugs are illegal, and athletes who use them risk being banned from their sports. More important, however, are the medical risks associated with steroid use, especially with the massive doses often used by athletes. Steroid use by people who are not physically mature can lead to early closure of the epiphyses of the long bones, so final stature can be reduced. Use of anabolic steroids suppresses the secretion of gonadotropic hormones, which control the development and function of the gonads (testes and ovaries). In males, decreased gonadotropin secretion can cause testicular atrophy, decreased secretion of testosterone, a reduced sperm count, and impotence. Decreased testosterone can lead to enlargement of the male breasts. In females, gonadotropins are required for ovulation and secretion of estrogens, so a decrease in these hormones disrupts those processes and menstruation. In addition, these hormonal disturbances in females can lead to masculinization: breast regression, enlargement of the clitoris, deepening of the voice, and growth of facial hair.

Prostate gland enlargement in males is another possible side effect of steroid use. Liver damage from a form of chemical hepatitis brought on by steroid use has also been identified, and it can lead to liver tumors.

Abnormal cardiac hypertrophy (enlarged heart), cardiomyopathy (diseased heart muscle), myocardial infarction (heart attack), thrombosis, arrhythmia, and hypertension have been reported in chronic steroid users. It is suspected that steroid use was at least partially

"Andro"—Fuel for Power Hitters?

During the 1998 Major League Baseball season, Mark McGwire blasted 70 home runs to eclipse the previous home run record of 61 set by Roger Maris in 1961. During the season, McGwire admitted to using androstenedione ("Andro"), a prohormone, which is a precursor to testosterone. Andro is marketed to increase testosterone levels and, subsequently, skeletal muscle mass. Sales of Andro increased markedly in response to McGwire's revelation. Does Andro work? King and his colleagues at Iowa State University were the first to investigate the combined effects of Andro and resistance training, in twenty 19- to 29-year-old men.[30] Ten subjects were assigned to the Andro and resistance training group and 10 to the placebo and resistance training group. Andro had no effect on serum testosterone levels, and the gains in strength and muscle mass were the same for the Andro and placebo groups. However, there was an unexpected finding: Andro increased serum estradiol and estrone concentrations! Other studies have substantiated these findings in young, middle-aged, and older men.[9, 34] Researchers are now looking at alternate ways of getting Andro into the blood other than through oral ingestion, which can lead to hepatic breakdown of the ingested androgens.

Dehydroepiandrosterone (DHEA) is another steroid hormone, as is its conjugate sulfate (DHEA-S). Bloodborne DHEA can be converted to androstenedione or androstenediol, which can then be converted to testosterone. It has been proposed that DHEA increases muscle mass. Brown and colleagues investigated the potential benefit of DHEA in increasing both serum testosterone and strength in 19 young men participating in an eight-week resistance training program, 9 receiving DHEA and 10 receiving the placebo.[11] As with the Andro study, there were no changes in serum testosterone concentrations as a result of taking DHEA, and the changes in strength and muscle mass were the same for the DHEA and placebo groups. Others have obtained similar results in young and middle-aged men and women.[34] However, in a study of elderly women and men (65-78 years), 10 months of DHEA supplementation did increase strength and thigh muscle volume when combined with strength training during the last four months of the study.[44] At this time, it appears that DHEA supplementation does not affect muscle size or strength in young and middle-aged men and women.

Major League Baseball players must have anticipated the results of these studies on Andro and DHEA. The cover of the June 3, 2002, issue of *Sports Illustrated* announced, "Special Report—Steroids in Baseball: Confessions of an MVP." From this issue, it appeared that a substantial number of professional baseball players had skipped over Andro and DHEA and had gone straight to anabolic steroids. Although the actual percentage of players using steroids is unknown, some players estimate that it was as great as 50% at its peak, prior to mandated drug testing. Curt Schilling, former Arizona Diamondback pitcher and Co-Most Valuable Player of the 2001 World Series, was quoted as saying, "Guys look like Mr. Potato Head, with six or seven body parts that just don't look right."

responsible for the medical condition of one former offensive lineman in the National Football League who was on the waiting list for a heart transplant. Scientists have found markedly depressed high-density lipoprotein cholesterol (HDL-C) levels—reductions of 30% or more—in athletes on even moderate steroid doses. High-density lipoprotein cholesterol has anti-atherogenic properties, meaning that it prevents the development of atherosclerosis. Low levels of HDL-C are associated with a high risk for both coronary artery disease and heart attack (see chapter 20). Furthermore, low-density lipoprotein cholesterol (LDL-C), which has atherogenic properties, appears to be increased with steroid use.

Substantial personality changes have occurred with steroid use. The most notable change is a marked increase in aggressive behavior, or "roid rage." Some teens have become extremely violent and have attributed these drastic mood changes to steroid use. Evidence also suggests that drug dependency can result from steroid use.

It is important to note that not everyone using anabolic steroids is an athlete. In fact, it appears that the majority of those taking steroids are nonathletes who use steroids for "cosmetic" purposes. Further, many self-inject steroids, and of those a significant number share needles.

Neither scientists nor physicians know the potential long-term effects of chronic steroid use. One study of male mice that received four different anabolic steroids of the kinds and doses taken by athletes showed that their life span was markedly decreased.[10] It is critical to develop a large database of former steroid users who can be followed over their life span. We have to remember that most diseases begin many years before symptoms develop. It is possible that the more serious health risks of steroid use won't be apparent for 20 or 30 years.

Recent reviews provide more detail on the potential ergogenic effects and health risks associated with anabolic steroid use.[13, 20, 25, 29, 35, 49] Most governing bodies of sports likely to be affected by anabolic steroid use have developed educational materials for their athletes in hopes of preventing steroid use. Also, national governing bodies for most sports have instituted aggressive year-round testing programs in which athletes are tested randomly for steroid use.

Human Growth Hormone

For years, the medical treatment for hypopituitary dwarfism has been administration of **human growth hormone (hGH),** a hormone secreted by the anterior pituitary gland. Before 1985, this hormone was obtained from cadaver pituitary extracts, and the supply was limited. Since the introduction of genetically engineered hGH in the mid-1980s, availability is no longer an issue, although the cost is still high.

During the 1980s, realizing this hormone's numerous functions, athletes started investigating hGH as a possible substitute for or complement to their use of anabolic steroids. As drug testing for anabolic steroids became more sophisticated, athletes were looking for an alternative for which there was no drug test at the time. Growth hormone appeared to be the current drug of choice for athletes who wanted to increase their strength and muscle mass.

Proposed Ergogenic Benefits

Growth hormone (GH) has six functions of interest to athletes:

- Stimulation of protein and nucleic acid synthesis in skeletal muscle
- Stimulation of bone growth (elongation) if bones are not yet fused (important to young athletes)
- Stimulation of insulin-like growth factor (IGF-1) synthesis
- Increase in lipolysis, leading to an increase in free fatty acids and an overall decrease in body fat
- Increase in blood glucose levels
- Enhancement of healing after musculoskeletal injuries

Athletes have turned to this hormone thinking that it would increase muscle development and thus increase fat-free mass. Often GH is used with anabolic steroids to maximize the anabolic effects.

Proven Effects

Administration of GH to older men (>60 years) has been shown to increase fat-free mass, decrease fat mass, and increase bone density.[37] However, in studies of young men and experienced weightlifters, there appear to be few if any significant benefits.[48]

In one study, young men were randomly assigned to either a GH or a placebo group. After 12 weeks of resistance training, the two groups had similar changes in muscle size, muscle strength, and rate of muscle protein synthesis in the quadriceps muscle. In a second study, experienced weightlifters were placed on GH treatment for 14 days while continuing to weight train. The investigators found that GH did not alter either the rate of muscle protein synthesis or the rate of whole-body protein breakdown, the two factors that would promote an increase in muscle mass.

Some athletes also take other drugs and certain amino acid supplements to stimulate GH release from the pituitary. To date, little evidence suggests that this practice is effective.

In focus

It appears that human growth hormone has no ergogenic properties in young, healthy athletes. However, major health risks are associated with the use of growth hormone.

Risks of Growth Hormone Use

As with steroids, potential medical risks are associated with GH use. Acromegaly can result from taking GH after the bones have fused. This disorder results in bone thickening, which causes broadening of the hands, feet, and face; skin thickening; and soft tissue growth. Internal organs typically enlarge. Ultimately, the victim suffers muscle and joint weakness and often heart disease. Cardiomyopathy is the most common cause of death with GH use. Glucose intolerance, diabetes, and hypertension also can result from GH use.

In review

- Anabolic steroids are more appropriately termed androgenic-anabolic steroids, because in their natural state they include both androgenic (masculinizing) and anabolic (building) properties. Synthetic steroids have been designed to maximize the anabolic effects while minimizing androgenic effects.
- Anabolic steroids have been proposed to increase muscle mass, strength, and endurance capacity and to facilitate recovery from exhaustive training bouts.
- Anabolic steroids can increase muscle mass and strength, but the effect is dose dependent. They do not increase endurance capacity, and their ability to facilitate recovery from exhaustive exercise has limited support.
- Andro and DHEA, precursors of testosterone, have been proposed to have ergogenic properties—increased muscle mass and strength—but most research studies have not supported these claims.
- Potential risks are associated with use of anabolic steroids, including personality changes, "roid rage," testicular atrophy in men, reduced sperm count in men, breast enlargement in men, breast regression in women, prostate gland enlargement in men, masculinization in women, menstrual cycle disruption in women, liver damage, and cardiovascular diseases.
- Growth hormone has not been studied extensively for its potential ergogenic effects. The limited research available supports its ability to increase fat-free mass and decrease fat mass in older men, but GH appears to have little or no effect on increasing muscle mass and strength. Most of the increase in mass and fat-free mass is associated with increased water retention.
- Risks associated with GH use include acromegaly, hypertrophy of internal organs, muscle and joint weakness, diabetes, hypertension, and heart disease.

PHYSIOLOGICAL AGENTS

Many **physiological agents** have been proposed as ergogenic aids. The goal of using these agents is to improve the body's physiological response during exercise. An athlete typically adds to a substance that occurs naturally in the body to try to improve performance. The reasoning is that if natural levels of a substance are beneficial to performance, higher levels should be even better. Several physiological agents have been proven effective but generally only under very specific conditions or for certain events or sports.

As with hormonal agents, many athletes consider use of these substances to be more ethical than use of pharmacological agents because these substances are found naturally in the body. Athletes also often assume that because these substances are normally found in the body, they must be safe at any level. Unfortunately, our bodies can be very unforgiving, and this assumption can be fatal.

We will look at only a few of the major physiological agents now being used as ergogenic aids:

- Blood doping
- Erythropoietin
- Oxygen supplementation
- Bicarbonate loading
- Phosphate loading

Blood Doping

Although altering blood composition in any way can be considered blood doping, the term has taken on a more specific meaning. **Blood doping** refers to any means by which a person's total volume of red blood cells is increased. This often is accomplished by transfusion of red blood cells, previously donated by either the recipient (autologous transfusions) or someone else with the same blood type (homologous transfusions). Blood doping also includes the use of erythropoietin, but this is discussed separately in the next section.

Proposed Ergogenic Benefits

Because oxygen is carried through the body bound to hemoglobin, it seems logical that increasing the number of red blood cells available to ferry the oxygen to the tissues could benefit performance. Increasing the number of oxygen carriers should increase the blood's oxygen-carrying capacity, allowing more oxygen to be delivered to the active tissues. If this happens, aerobic endurance and thus performance could be substantially improved. That is the premise underlying blood doping.

Proven Effects

Ekblom and coworkers[19] created quite a stir in the sport world in the early 1970s. In a landmark study, they withdrew between 800 and 1200 ml of blood from their subjects, refrigerated the blood, then reinfused the red blood cells into those subjects about four weeks later. Results showed a considerable improvement in $\dot{V}O_{2max}$ (9%) and treadmill performance time (23%) after reinfusion. Over the next few years, several studies confirmed these original findings, but several others failed to demonstrate an ergogenic effect.

Thus, the research literature was divided on the effectiveness of blood doping until a major breakthrough occurred in 1980 as a result of a study by Buick and colleagues.[12] Eleven highly trained distance runners were tested at different times during the study: (1) before blood withdrawal, (2) following blood withdrawal after the body was allowed adequate time to reestablish normal red blood cell levels but before reinfusion of the removed blood, (3) following a sham reinfusion of 50 ml of saline (a placebo), (4) following reinfusion of 900 ml of the subject's own blood that originally had been withdrawn and preserved by freezing, and (5) after the elevated red blood cell levels had returned to normal. The sham trial was identical to the blood reinfusion trial, except that a small volume of saline was infused rather than whole blood. This trial served as a placebo to rule out any psychological effects.

As shown in figure 15.5, the researchers found a substantial increase in $\dot{V}O_{2max}$ and treadmill running time to exhaustion after the reinfusion of the red blood cells and no change after the sham reinfusion. This increase in $\dot{V}O_{2max}$ persisted for up to 16 weeks, but the increase in treadmill time decreased within the first seven days.

Maximizing the Benefits

Why was the Buick study a major breakthrough? Gledhill[22] helped explain the controversy arising from the early studies. Many early studies that showed no improvement with blood doping had reinfused only small volumes of red blood cells, and the reinfusion was conducted within three to four weeks of the blood withdrawal. First, it appears 900 ml or more of whole blood must be reinfused to have an effect. Increases in $\dot{V}O_{2max}$ and performance are not as great when smaller volumes are used. In fact, some studies using smaller volumes failed to show any differences.

Second, it appears that it is necessary to wait for at least five to six weeks and possibly as long as 10 weeks before reinfusion. This is based on the time it takes the body to reestablish the blood's original, prewithdrawal hematocrit.

Finally, researchers who conducted early studies refrigerated the withdrawn blood. Maximal storage time for blood under refrigeration is approximately five weeks. Furthermore, when blood is refrigerated, approximately 40% of the red blood cells are destroyed or lost. Later studies used frozen storage. Freezing allows an almost unlimited storage time, and only about 15% of the red blood cells are lost. Gledhill[22] concluded that blood doping significantly improves $\dot{V}O_{2max}$ and endurance performance when done under optimal conditions:

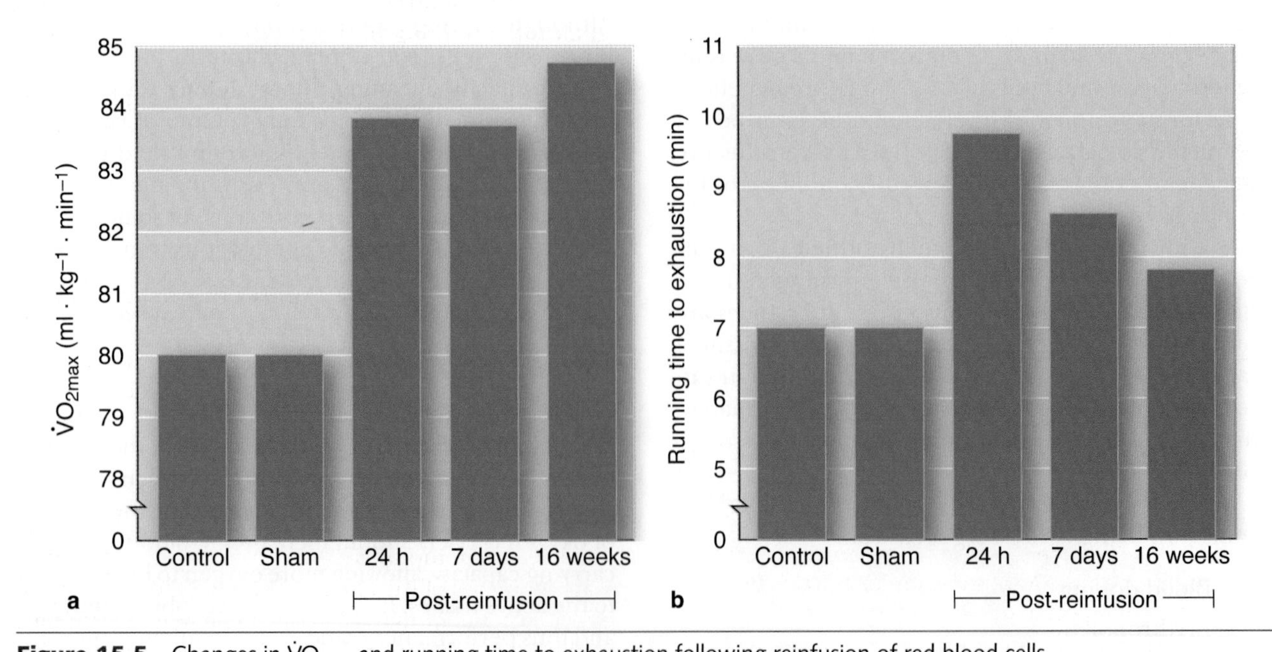

Figure 15.5 Changes in $\dot{V}O_{2max}$ and running time to exhaustion following reinfusion of red blood cells.

Adapted, by permission, from F.J. Buick, N. Gledhill, A.B. Froese, L. Spriet, and E.C. Meyers, 1980, "Effect of induced erythocythemia on aerobic work capacity," *Journal of Applied Physiology* 48: 636-642.

- A minimum of 900 ml of blood reinfusion
- A five- to six-week minimum interval between withdrawal and reinfusion
- Blood storage by freezing

He also showed that these improvements are the direct result of the blood's increased hemoglobin content, not an increased cardiac output caused by an expanded plasma volume.

Blood Doping and Endurance Performance

Does an increase in $\dot{V}O_{2max}$ and treadmill time as a result of blood doping translate into improved endurance performance? Several studies have addressed this issue. Researchers in one study observed 5 mi (8 km) treadmill run times in a group of 12 experienced distance runners.[46] Their times were checked before and after saline (placebo) infusion and before and after blood infusion. The 5 mi (8 km) run times on the treadmill were significantly faster following blood infusion, but this difference became significant only over the last 2.5 mi (4 km). The blood infusion trials were 33 s (3.7%) faster over the last 2.5 mi (4 km) and 51 s (2.7%) faster over the full 5 mi (8 km) than the placebo trials.

A second study looked at 3 mi (4.8 km) run times in a group of six trained distance runners and reported an average decrease of 23.7 s following blood doping when compared to the runners' blind control trials.[23]

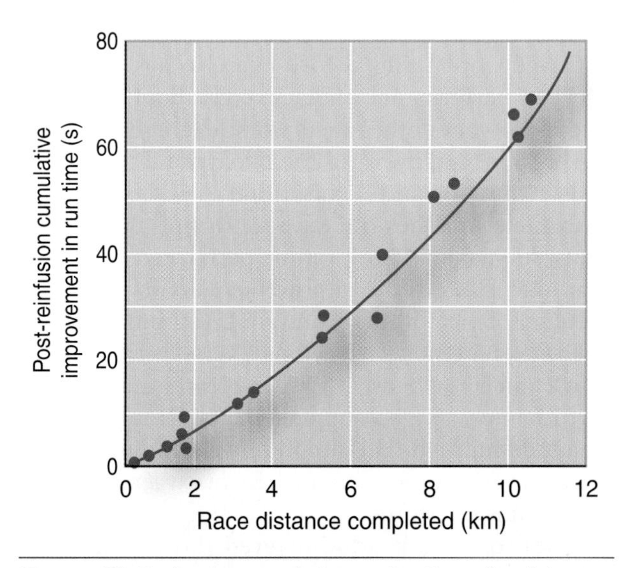

Figure 15.6 Improvements in running times for distances of up to 11 km (6.8 mi) following reinfusion of red blood cells from two units of blood preserved by freezing. Values on the *y*-axis reflect the reduction in time to run a specific distance on the *x*-axis. As an example, for a 10 km (6.2 mi) race, one could expect to run 60 s faster after reinfusion.

Adapted, by permission, from L.L Spriet, 1991, Blood doping and oxygen transport. In *Ergogenics—Enhancement of performance in exercise and sport*, edited by D.R. Lamb and M.H. Williams (Dubuque, IA: Brown & Benchmark), 213-242. Copyright 1991 Cooper Publishing Group, Carmel, IN.

Subsequent studies confirmed improvements in distance running and cross-country skiing performance with blood doping.[39] Figure 15.6 illustrates the improvement in run time with blood doping for distances of up to 11 km (6.8 mi).

Risks of Blood Doping

Although this procedure is relatively safe in the hands of competent physicians, it has inherent dangers. Adding more red blood cells to the cardiovascular system can overload it, causing the blood to become too viscous, which could lead to clotting and possibly heart failure. With autologous blood transfusions, in which the recipient receives his or her own blood, mislabeling of the blood could occur. With homologous blood transfusions, in which blood is received from a matched donor, several other complications can occur. The reinfused blood could be mismatched. An allergic reaction could be triggered. The athlete may experience chills, fever, and nausea. The athlete also risks infection from hepatitis or HIV pathogens.[42] The potential risks of blood doping, even without consideration of the legal, moral, and ethical issues involved, outweigh any potential benefits.

Erythropoietin

As mentioned in the last section, erythropoietin falls into the class of blood doping, but because the mechanism is somewhat different, we examine it separately. **Erythropoietin (EPO)** is a naturally occurring hormone produced by the kidneys. It stimulates red blood cell production. In fact, this hormone is responsible for the elevated red blood cell production seen during training at altitude; training in the presence of a lower partial pressure of oxygen stimulates EPO release.

Human EPO can now be cloned through genetic engineering, so it is widely available. This hormone increases the hematocrit substantially when administered to patients with renal failure.

Proposed Ergogenic Benefits

Theoretically, if administered to athletes, human EPO would have the same effects as the reinfusion of red blood cells. The goal of its use is to increase the red blood cell volume, thus increasing the blood's oxygen-carrying capacity.

Proven Effects

Erythropoietin's ability to increase oxygen-carrying capacity was demonstrated in 1991 when the first study was conducted of the effects of subcutaneous injections of low doses of human EPO on maximal treadmill time and $\dot{V}O_{2max}$.[18] The study involved moderately trained to well-trained subjects. Six weeks after EPO administration, the following effects were observed:

- Both hemoglobin concentration and hematocrit increased 10%.
- $\dot{V}O_{2max}$ increased 6% to 8%.
- Time to exhaustion on the treadmill increased 13% to 17%.

Seven of the 15 subjects had been through a previous study of red blood cell reinfusion conducted four months earlier. Increases in $\dot{V}O_{2max}$ and treadmill time were almost identical in the two studies, and these improvements were attributed directly to the increase in hemoglobin.

In a second study of 20 well-trained male endurance athletes, 10 received EPO injections three times weekly for 30 days or until their hematocrit reached 50%, and 10 received saline injections (placebo).[8] The hematocrit in the EPO group increased from 42.7% to 50.8%, and $\dot{V}O_{2max}$ increased from 63.6 to 68.1 ml · kg^{-1} · min^{-1}. There were no changes in the placebo group. It should be noted that EPO injections stopped once the hematocrit reached 50%. Hematocrit values following EPO injections can greatly exceed this value!

Risks of Erythropoietin Use

Serious consequences can arise from EPO use. Up to 18 deaths among competitive cyclists, reported between 1987 and 1990, were alleged to be linked to EPO use, but this has not been confirmed.[3]

> **In focus**
>
> Blood doping and EPO can enhance aerobic capacity and the performance of aerobic sports or activities. This occurs through an increase in the oxygen-carrying capacity of the blood, primarily attributable to increased red blood cells that result from blood doping and the use of EPO. Both procedures involve extreme risk.

The outcome of EPO use is less predictable than that of red blood cell reinfusion. Once the hormone has been put into the body, no one can predict how much red blood cell production will occur. This places the athlete at great risk of substantial increases in blood viscosity. Known risks include thrombosis (blood clot), myocardial infarction (heart attack), congestive heart failure, hypertension, stroke, and pulmonary embolism.

Oxygen Supplementation

Watch any professional football game on television—the star running back breaks loose for a 35 yd (32 m) touchdown run, struggles back to the bench, grabs a face mask, and starts breathing 100% oxygen to facilitate his recovery. How much does he gain by using this oxygen supplementation instead of just breathing ordinary air?

Proposed Ergogenic Benefits

Obviously, the idea behind taking in more oxygen is to increase the oxygen content of the blood, as with blood doping. The aim of blood doping is to do this by increasing the oxygen-carrying capacity of the blood; with **oxygen supplementation** the idea is to achieve this directly by providing more oxygen to the blood and tissues. By increasing the available oxygen, athletes hope to compete at higher intensities and fend off fatigue for longer periods. This technique also has been suggested as a means to speed recovery between exercise bouts.

Proven Effects

Initial attempts to investigate the ergogenic properties of pure oxygen began in the early 1900s, but it was not until the 1932 Olympic Games that oxygen was considered a potential ergogenic aid for athletic performance. That year, Japanese swimmers won impressive victories, and many attributed their success to breathing pure oxygen before competing. However, it is unclear whether their success was attributable to their use of oxygen or to their athletic abilities.

Of historical note, one of the first studies to observe the effects of breathing oxygen on performance was conducted by Sir Roger Bannister, a physician-scientist who is world renowned for his research in neurological disorders.[5] As an athlete, Dr. Bannister was the first person in the world to break the 4 min mile barrier.

Oxygen can be administered immediately before competition, during competition, during recovery from competition, or at any combination of these times.

Oxygen breathing before exercise has a limited effect on performance of that exercise bout. The total amount of work or the rate of work (exercise intensity) can be increased by breathing of oxygen if the bout is of short duration and occurs within a few seconds after the athlete breathes the oxygen. During these short bouts, submaximal work can be performed at a lower heart rate. However, no improvement occurs unless the exercise follows within seconds of breathing oxygen.

For exercise bouts exceeding 2 min, or when more than 2 min lapses between oxygen breathing and actual performance, oxygen supplementation's influence is greatly diminished. This simply reflects the limits of the body's oxygen-storage potential. Extra oxygen dissipates rapidly; it is not stored.

When oxygen is administered during exercise, definite performance improvements occur. The total amount of work performed and the rate of work performed increase substantially. Likewise, submaximal work is performed more efficiently with lower physiological cost to the individual. Peak blood lactate levels are depressed following exhaustive exercise that the subject performs while breathing oxygen, even though considerably more work can be performed.

Studies have been unable to demonstrate any clear advantage to oxygen breathing during the recovery period. Recovery does not seem to be facilitated, nor does subsequent performance improve. In a study with professional soccer players running on a treadmill, researchers found no improvements in recovery or subsequent performance on the second exhaustive bout as a result of oxygen breathing.[47]

From a practical standpoint, oxygen administration before exercise would have little value because of the relatively short time during which oxygen stores remain elevated. The nature of most sports doesn't allow an athlete to go immediately from oxygen breathing into competition. Regardless of the ergogenic effects of oxygen intake during performance, administration during exercise has limited value for obvious reasons: During which sports or events could someone carry a cylinder of oxygen?

The recovery period seems the only practical time to administer oxygen, but this would be worthwhile only if oxygen administration were known to speed the recovery process, allowing the athlete to reenter the contest more fully recovered. However, such an effect has not been substantiated by research.

In focus

Oxygen supplementation can increase aerobic performance but only if administered during exercise, which is not practical in sport. It has no ergogenic effect during recovery.

Risks of Oxygen Supplementation

At this time, no known risks are associated with oxygen supplementation. More research must be conducted to determine its safety. However, oxygen is flammable, so oxygen equipment should never be near any heat source or flame, nor should it be allowed near anyone who is smoking.

Bicarbonate Loading

Recall from chapter 6 that bicarbonates are an important part of the buffering system necessary to maintain the acid–base balance of body fluids. Scientists naturally began to investigate whether performance in highly anaerobic events, in which large amounts of lactic acid are formed, could be improved by enhancing the body's buffering capacity through elevation of the blood's bicarbonate concentrations, a process called **bicarbonate loading.**

Proposed Ergogenic Benefits

By ingesting agents that increase the bicarbonate concentrations in the blood plasma, such as sodium bicarbonate (baking soda), subjects can increase blood

In review

- Blood doping refers to an artificial increase in a person's total volume of red blood cells. It has been proposed to improve endurance performance by increasing the blood's oxygen-carrying capacity.
- Studies have shown major increases in maximal oxygen uptake, time to exhaustion, and actual performance in cross-country skiing and distance running as a result of blood doping.
- Risks associated with blood doping include major complications such as blood clotting; heart failure; and administration of mislabeled blood and its potential consequences, such as transfusion reactions and transmission of hepatitis and HIV.
- Erythropoietin is the naturally occurring hormone that stimulates red blood cell production. It has been proposed as an ergogenic aid under the premise that increasing the number of red blood cells increases the blood's oxygen-carrying capacity.
- Few well-controlled studies have been conducted on the role of EPO in enhancing aerobic endurance performance, but two have clearly demonstrated increased maximal oxygen consumption and increased time to exhaustion.
- Because we cannot predict the magnitude of the body's response to EPO administration, it can be dangerous. The hormone can lead to death if red blood cells are overproduced, increasing the blood's viscosity. Known risks include thrombosis, myocardial infarction, congestive heart failure, hypertension, stroke, and pulmonary embolism.
- Oxygen administration during exercise improves performance but is too cumbersome to be practical. Administration before or immediately after exercise has not been proven ergogenically effective.
- No serious risks are associated with oxygen breathing.

pH, making the blood more alkaline. It was proposed that increasing plasma bicarbonate levels would provide additional buffering capacity, allowing higher concentrations of lactate in the blood. Theoretically, this could delay the onset of fatigue in short-term, all-out anaerobic work, such as all-out sprinting.

Proven Effects

Oral intake of sodium bicarbonate elevates plasma bicarbonate concentrations. However, this has little effect on intracellular concentrations of bicarbonate in muscle. Therefore, the potential benefits of bicarbonate ingestion were thought to be limited to anaerobic bouts of exercise lasting longer than 2 min, because bouts less than 2 min would be too brief to allow many hydrogen ions (H^+, from the lactic acid) to diffuse out of the muscle fibers into the extracellular fluid where they could be buffered.

In 1990, however, Roth and Brooks[36] described a cell membrane lactate transporter that operates in response to the pH gradient. Increasing the extracellular buffering capacity by ingesting bicarbonate increases the extracellular pH, which in turn increases transport of lactate from the muscle fiber via this membrane transporter to the blood plasma and other extracellular fluids. This should improve anaerobic performances even for events briefer than 2 min.

Although the theory proposing bicarbonate ingestion as an ergogenic aid for anaerobic performance is sound, the research literature is, again, conflicting. However, Linderman and Fahey,[31] in their review of the literature, found several important patterns in the research that might explain these conflicts. They concluded that bicarbonate ingestion had little or no effect on performances of less than 1 min or of more than 7 min but that for performances between 1 and 7 min, the ergogenic effects were evident. Furthermore, they found that the dose was important. Most studies that used a dose of 300 mg/kg of body mass showed a benefit, whereas most studies of lower dosage showed little or no benefit. Thus, it appears that bicarbonate ingestion of 300 mg/kg of body mass can enhance the performance of all-out, maximal anaerobic activities of 1 to 7 min duration.

An example of a study supporting these conclusions is illustrated in figure 15.7. In this study, blood bicarbonate concentrations were artificially elevated by bicarbonate ingestion before and during five sprint-cycling bouts, each lasting 1 min (see figure 15.7*a*).[15] Performance on the final trial improved by 42%! This elevation in blood

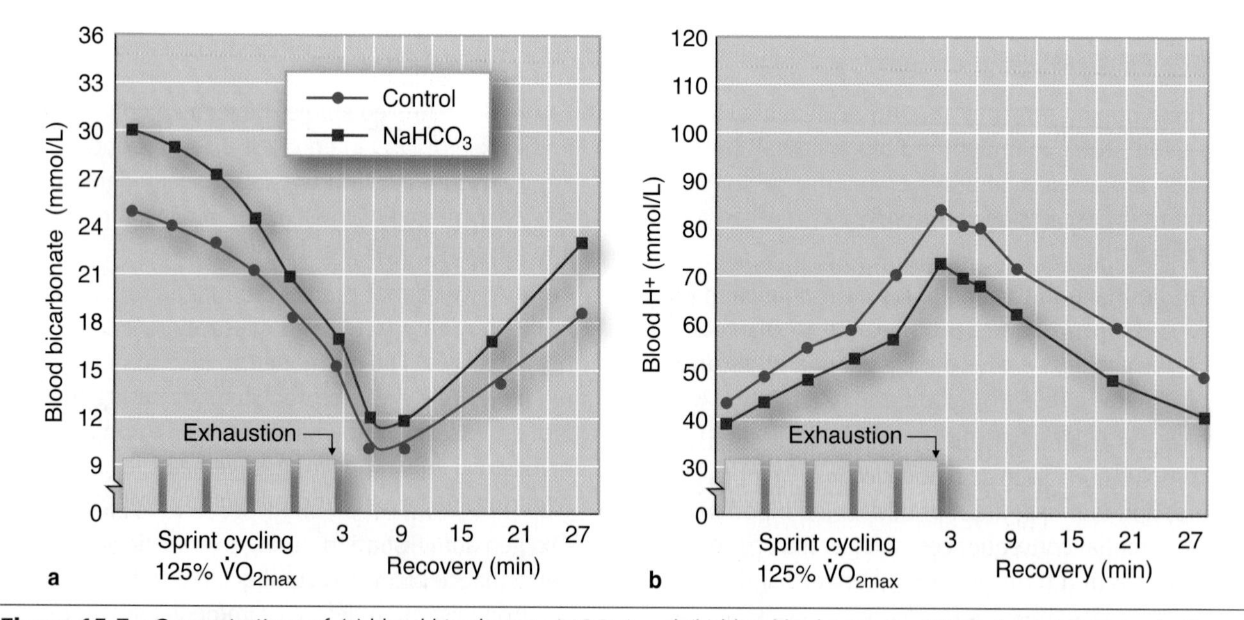

Figure 15.7 Concentrations of *(a)* blood bicarbonate (HCO_3^-) and *(b)* blood hydrogen ion (H^+) before, during, and after five sprint-cycling bouts with and without ingestion of sodium bicarbonate ($NaHCO_3$). The fifth sprint was performed to exhaustion. The elevated blood HCO_3^- concentrations caused attenuation of the elevation in blood H^+, a smaller decline in blood pH, a 42% increase in performance to exhaustion during the fifth sprint, and faster recovery after the sprints.

Adapted, by permission, from D.L. Costill, F. Verstappen, H. Kuipers, E. Janssen, and W. Fink, 1984, "Acid–base balance during repeated bouts of exercise: Influence of HCO_3^-," *International Journal of Sports Medicine* 5: 228-231.

bicarbonate levels reduced the concentration of free H^+ both during and after exercise (see figure 15.7*b*), thereby elevating blood pH. The authors concluded that in addition to improving buffering capacity, the extra bicarbonate appeared to speed the removal of H^+ ions from the muscle fibers, thereby reducing the decrease in intracellular pH. Six years later, in 1990, Roth and Brooks[36] reported the presence of a lactate transporter in the muscle cell membrane, as described earlier, that works precisely as Costill and colleagues[15] had originally postulated.

Risks of Bicarbonate Loading

Although sodium bicarbonate has long been used as a remedy for indigestion, many authors studying bicarbonate loading have reported severe gastrointestinal discomfort in some of their subjects, including diarrhea, cramps, and bloating, from high doses of bicarbonate. These symptoms can be prevented if one ingests as much water as desired and divides the total bicarbonate dosage of at least 300 mg/kg of body mass into five equal parts over a 1 to 2 h period.[31] Also, several studies have shown sodium citrate to have similar effects on buffering capacity and performance without gastrointestinal discomfort.

Phosphate Loading

Since the early 1900s, scientists have been interested in the possibility of increasing the dietary consumption of phosphorus to improve cardiovascular and metabolic function during exercise. Several of the early studies suggested that **phosphate loading,** which involves ingestion of sodium phosphate as a dietary supplement, is an effective ergogenic aid.

Proposed Ergogenic Benefits

Phosphate loading has been proposed to have numerous potential benefits during exercise. These include elevation of extracellular and intracellular phosphate levels, which would increase the availability of phosphate for oxidative phosphorylation and phosphocreatine synthesis, thus improving the body's energy production capacity. Phosphate loading is also thought to enhance 2,3-diphosphoglycerate (2,3-DPG) synthesis in the red blood cells. This substance facilitates the release of oxygen from the red blood cells. This increase in 2,3-DPG would shift the oxygen–hemoglobin dissociation curve to the right, permitting greater oxygen unloading in the active muscles. Thus, phosphate loading has been proposed to improve the cardiovascular response to exercise, improve the body's buffering capacity, and consequently improve endurance capacity and performance.

Proven Effects

Only a few studies have been conducted to determine the ergogenic benefits of phosphate loading. Unfortunately, the results are divided. Several studies showed significant improvements in $\dot{V}O_{2max}$ and time to exhaustion. However, several others showed no effects. There appear to be some potential benefits to phosphate loading, but additional research is needed to confirm this.

Risks of Phosphate Loading

At this time, no known risks are associated with phosphate loading. However, because insufficient research has been conducted to date, more studies are needed to determine its safety.

In review

- Bicarbonate is an important component of the body's buffering system, needed to maintain normal pH by neutralizing excess acid.
- Bicarbonate loading is proposed to increase the blood's alkalinity, thus increasing the buffering capacity so that more lactate can be cleared. This delays the onset of fatigue.
- Bicarbonate ingestion of 300 mg/kg of body weight can delay fatigue and increase performance in all-out bouts of exercise lasting more than 1 min but less than 7 min.
- Bicarbonate loading can cause gastrointestinal distress, including cramping, bloating, and diarrhea.
- Ingestion of sodium phosphate has been postulated to improve general cardiovascular and metabolic functioning. During exercise, phosphate loading has been proposed to elevate phosphate levels throughout the body, which would increase the potential for oxidative phosphorylation and phosphocreatine synthesis, enhance oxygen release to the cells, improve cardiovascular response to exercise, improve the body's buffering capacity, and improve endurance capacity.
- Little research at this time supports the use of phosphate loading as an ergogenic aid. Existing research is conflicting, and the risks of phosphate loading are largely unknown.

Beware—Contamination of Nutritional Supplements

Many, if not most, athletes are taking one or more types of nutritional supplements. Most, if not all, assume that they are ingesting a substance that exactly matches the ingredients listed on the product container. The sport nutrition industry has become so big that there are now specialty stores and Internet Web sites selling sport nutrition products. Unfortunately, the regulations governing the purity of these products are very permissive, and many of the claims for these products have not been substantiated by scientific research studies. This lack of regulation by the U.S. Food and Drug Administration (FDA) has led to a serious problem of supplement contamination. Starting in the year 2000, researchers began to investigate the purity of many of these supplements. Their findings are sobering. In some cases products have not contained the substances listed on the label in measurable amounts, and in other cases there has been up to 150% of the listed dose. Many common supplements have been contaminated with prohibited substances that could lead to positive doping results and banning of the athlete from competition. Contaminants have included anabolic steroids, ephedrine, and caffeine. Numerous studies have now substantiated the extent and critical nature of this problem. For example, in one study conducted at the IOC-accredited laboratory in Cologne, Germany, researchers analyzed 634 nonhormonal nutritional supplements obtained from 13 countries and from 215 different suppliers. Of the 634 samples, 94 samples (14.8%) were found to contain hormones or prohormones that were not declared on the product label, and 23 samples contained compounds related to nandrolone and testosterone. The bottom line: Athletes who use supplements are taking an extremely high risk!

A special thanks to Dr. Ron J. Maughan, at Loughborough University, United Kingdom, for the information contained in this sidebar. For further information and references, refer to Maughan, R.J. (2004). Contamination of dietary supplements and positive drug tests in sport. *Journal of Sports Sciences*, 23: 883-889.

NUTRITIONAL AGENTS

Although basic concepts of nutrition and the specific performance-enhancing properties of carbohydrates, fats, proteins, vitamins, and minerals are discussed in chapter 14, many **nutritional agents** have been proposed to have specific ergogenic properties. It is appropriate to discuss several of these in this chapter because they have received so much publicity and hype from both manufacturers and users. Most of these nutritional agents have not been adequately researched, however, so our discussion of each is brief.

Amino Acids

Specific **amino acids,** or groups of amino acids, have been proposed to have special ergogenic properties. **L-tryptophan,** an essential amino acid, has been proposed to increase aerobic endurance performance through its effects on the CNS, acting as an analgesic and delaying fatigue. L-tryptophan is the first precursor of serotonin, a potent CNS neurotransmitter. Although an initial study of L-tryptophan supplementation indicated dramatic increases in endurance performance, subsequent studies have been unable to confirm these results, showing no improvement in endurance performance.

Branched-chain amino acids (BCAA)—leucine, isoleucine, and valine—have been postulated to work in combination with L-tryptophan to delay fatigue, primarily through CNS mechanisms. There is convincing evidence that exercise-induced increases in the plasma free tryptophan/BCAA ratio are associated with increased brain serotonin and the onset of fatigue during prolonged exercise.[16] Theoretically, increasing the BCAA would reduce the ratio and delay the onset of fatigue. One study dealt with the time to exhaustion on a cycle ergometer at 70% to 75% of $\dot{V}O_{2max}$ under conditions that either increased tryptophan levels, increased BCAA levels, or reduced BCAA levels, all substantially altering the tryptophan/BCAA ratio.[43] Exercise time to exhaustion was not different among treatments (see figure 15.8). This study and others call into question the

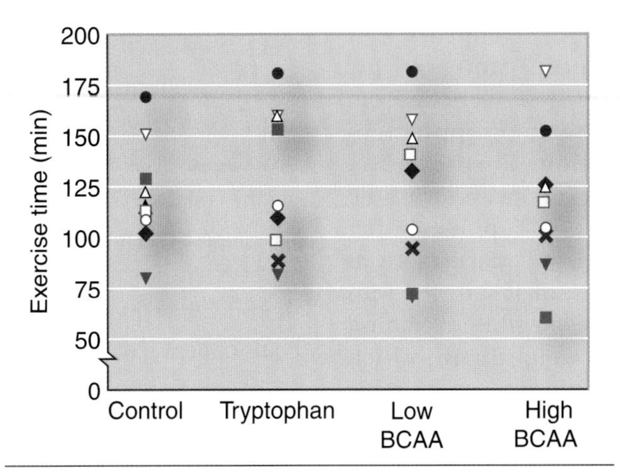

Figure 15.8 Time to exhaustion of 10 subjects (each represented by a different symbol) on a cycle ergometer at 70% to 75% of $\dot{V}O_{2max}$. Each subject was tested under control conditions and after each of three treatments that increased tryptophan, reduced branched-chain amino acid (BCAA), or increased BCAA. Exercise time to exhaustion was not significantly different between control and treatment conditions.

Adapted, by permission, from G. van Hall et al., 1995, "Ingestion of branched-chain amino acids and tryptophan during sustained exercise in man: Failure to affect performance," *Journal of Physiology* 486: 789-794.

efficacy of supplementing either tryptophan or BCAA to improve endurance performance.[16]

Others have postulated that supplementation of specific amino acids increases serum GH release from the anterior pituitary gland. This also has not been clearly substantiated by research. There is, however, strong evidence that supplementation with a metabolite of leucine (β-hydroxy-β-methylbutyrate, or HMB) does increase fat-free mass and strength. It acts by decreasing the breakdown of protein that occurs with resistance training. In a review of studies on six dietary supplements proposed to increase fat-free mass and strength, only HMB and creatine were shown to be effective. There appears to be little risk associated with HMB supplementation; and, in fact, it has been reported to decrease total cholesterol, LDL-C, and systolic blood pressure.[34]

L-Carnitine

Long-chain fatty acids are the major source of energy in the body, and fatty acid oxidation provides energy both at rest and during exercise. **L-carnitine** is important in fatty acid metabolism because it assists in the transfer of fatty acids from the cytosol (the fluid portion of the cytoplasm, exclusive of organelles) across the inner mitochondrial membrane for β-oxidation. This membrane is normally impermeable to long-chain fatty acids, so the availability of L-carnitine may be a limiting factor for the rate of fatty acid oxidation.

It has been theorized that by increasing the availability of L-carnitine, athletes might facilitate the oxidation of lipids. By relying more on fat as an energy source, one could possibly spare glycogen and increase aerobic endurance capacity. The studies of L-carnitine are mixed: Several show indirect evidence of increased fat oxidation with L-carnitine supplementation, but most show no effect on fat oxidation when using either indirect or direct estimates. Studies have shown that L-carnitine supplementation does not increase muscle storage of carnitine, enhance fatty acid oxidation, spare glycogen, or postpone fatigue during exercise; nor has it been shown unequivocally to improve athletes' performance.[26]

Creatine

The use of **creatine** as an ergogenic aid has been widespread among athletes, ranging from the recreational to the professional. The primary use of creatine is based on its role in skeletal muscle, where approximately two-thirds of its total is in the form of phosphocreatine (PCr). Increasing the creatine content of skeletal muscle through creatine supplementation is theorized to increase muscle PCr levels, thus enhancing the adenosine triphosphate (ATP)-PCr energy system by better maintaining muscle ATP levels. This, in turn, theoretically would enhance peak power production during intense exercise and possibly facilitate recovery from high-intensity exercise. Creatine also serves as a buffer, helping to regulate acid–base balance, and is involved in the oxidative metabolic pathways.

Because of the popularity of creatine supplementation in the 1990s and the wide-ranging claims of its ergogenic properties, the American College of Sports Medicine (ACSM) published a consensus statement on "The Physiological and Health Effects of Oral Creatine Supplementation" in 2000.[2] A group of eminent exercise and sport scientists reviewed the existing research literature on creatine and performance at that time to develop a consensus on the actual ergogenic benefits of creatine. From their review, they concluded the following:

- Creatine supplementation can increase muscle phosphocreatine content, but not in all individuals.
- The combination of creatine with a large amount of carbohydrate might increase muscle uptake of creatine.
- Exercise performance in short periods of intense, high power output activity can be enhanced, particularly with repeated bouts, consistent with the role of PCr in this type of activity.

- Maximal isometric strength, the rate of maximal force production, and maximal aerobic capacity are not enhanced by creatine supplementation.
- Creatine supplementation leads to weight gain within the first few days, likely attributable to water accumulation with creatine uptake in the muscle.
- In combination with resistance training, creatine supplementation is associated with greater gains in strength, possibly associated with the increased ability to train at higher intensities.
- The high expectations for performance enhancement exceed the true ergogenic benefits.

Subsequent scientific reviews have been published since the release of this ACSM Consensus Statement and are generally in agreement. As mentioned in our discussion of HMB, creatine is one of two supplements that effectively increase both fat-free mass and strength.[34] With respect to improving athletic performance, studies are mixed. This is likely due to two factors: the physiological demands of the sport or event and the individual variability in response to the supplement. Performance enhancement is more likely to occur in sports involving brief periods of high-intensity exercise. Concerning individual variability, we discussed in chapter 8 the principle of individuality—the fact that there are high responders and low responders to any given intervention. In studies involving only a few subjects (e.g., <10), it is possible that there are more high than low responders represented in the study sample or vice versa. Finally, it is possible that creatine supplementation might enhance muscle growth by stimulating protein synthesis.

So, it appears that there is potential for ergogenic benefits from creatine supplementation. Furthermore, there appears to be little risk in supplementing creatine, particularly at the smaller doses, with the possible exception of weight gain, which would not be desirable for some athletes.

In focus

Creatine supplementation appears to have ergogenic benefits, particularly for increasing the creatine content of skeletal muscle and for improving performance in intense, short-duration maximal exercise bouts of between 30 and 150 s.

In review

- A substantial risk is associated with using nutritional supplements due to the potential problem of contamination of the ingredients.
- Although amino acid supplementation, particularly L-tryptophan and BCAA, has been proposed to have ergogenic properties, little evidence supports this proposal; HMB, however, does appear to have ergogenic benefits.
- Although L-carnitine is important in fatty acid metabolism, most studies show that supplementation does not increase muscle storage of carnitine, enhance fatty acid oxidation, spare glycogen, or delay fatigue during exercise.
- Creatine supplementation has been shown to increase muscle creatine levels and increase performance in sports involving brief periods of high-intensity exercise.

In Closing

In this chapter, we reviewed some common substances and procedures thought to have ergogenic properties. All athletes must recognize the legal, ethical, moral, and medical consequences of using any ergogenic agent. The list of banned substances increases almost daily. Athletes who use banned substances risk disqualification from a particular competition, and they can be banned from competition in their sport for a year or more. In their quest for the perfect performance, athletes can easily get caught up in the hype surrounding various substances and the purported benefits they might bestow. Unfortunately, too many athletes are blinded by ambition and do not consider the consequences of their actions until their careers have been jeopardized or their health seriously impaired.

We have discussed pharmacological, hormonal, physiological, and some specific nutritional ergogenic aids. In the next part of the book, we shift our focus away from athletes in general to the unique characteristics of younger, older, and female athletes within the broader categories of growth and development, aging, and sex differences in exercise performance. We begin in chapter 16 by examining special considerations in the child and adolescent.

KEY TERMS

amino acids
amphetamines
anabolic steroids
β-blockers
bicarbonate loading
blood doping
branched-chain amino acids (BCAA)
caffeine
creatine
diuretics
ephedrine
ergogenic
ergolytic
erythropoietin (EPO)
hormonal agents
human growth hormone (hGH)
L-carnitine
L-tryptophan
nutritional agents
oxygen supplementation
pharmacological agents
phosphate loading
physiological agents
placebo
placebo effect
pseudoephedrine

STUDY QUESTIONS

1. What is the meaning of the term *ergogenic aid*? What is an ergolytic effect?
2. Why is it important to include control groups and placebos in studies of the ergogenic properties of any substance or phenomenon?
3. What is presently known about the use of amphetamines in athletic competition? What are the potential risks of using amphetamines?
4. Under what circumstances might β-blockers be ergogenic aids?
5. How might caffeine improve athletic performance?
6. What is known about alcohol, nicotine, cocaine, and marijuana as ergogenic aids?
7. Are diuretics ergogenic? What are some risks associated with their use?
8. What are the effects of anabolic steroid use on athletic performance? What are some of the medical risks of steroid use?
9. What is known about human growth hormone as a potential ergogenic aid? What are the risks associated with its use?
10. What is blood doping? Does blood doping improve athletic performance?
11. By what mechanism is erythropoietin theorized to benefit performance?
12. How beneficial is breathing oxygen before the start of competition, during competition, and during the recovery from competition?
13. What are the potential ergogenic properties of bicarbonate, phosphate, HMB, and creatine?

STUDY GUIDE ACTIVITIES

In addition to the activities listed in the chapter opening outline on page 355, two other activities are available in the online study guide, located at

www.HumanKinetics.com/PhysiologyOfSportAndExercise.

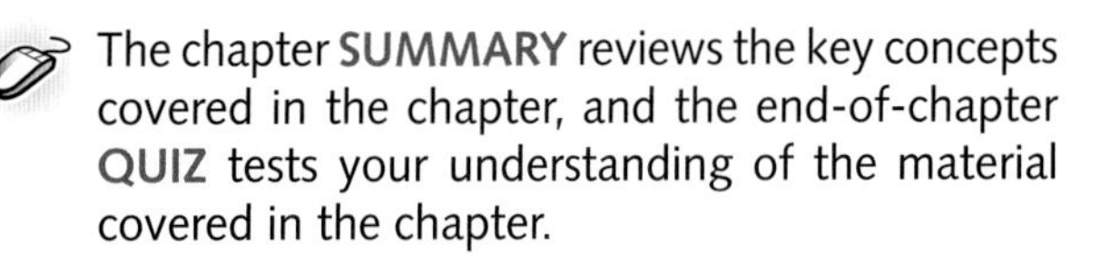

The chapter **SUMMARY** reviews the key concepts covered in the chapter, and the end-of-chapter **QUIZ** tests your understanding of the material covered in the chapter.

Age and Sex Considerations in Sport and Exercise

From the previous parts of this book, we have gained a good understanding of the general principles of exercise and sport physiology. Now we turn our attention to how these principles are specifically applied to children and adolescents, older individuals, and females. In chapter 16, "Children and Adolescents in Sport and Exercise," we examine the processes of human growth and development and how different stages affect a young person's physiological capacity and performance. We also consider how these stages of growth and development might alter our strategies for training young athletes for competition. In chapter 17, "Aging in Sport and Exercise," we discuss how exercise capacity and sport performance change as people age beyond the middle years, asking to what extent these changes are attributable to physiological aging and how much change might be attributable to an increasingly sedentary lifestyle. We discover the important role that training can play in minimizing the loss of performance capacity and physical conditioning that accompanies aging. In chapter 18, "Sex Differences in Sport and Exercise," we examine differences between women's and men's responses to exercise and training and the extent to which these differences are biological. We also focus on physiological and clinical issues specific to female athletes, including menstrual function, pregnancy, osteoporosis, and the high prevalence of eating disorders in female athletes.

CHAPTER 16

Children and Adolescents in Sport and Exercise

In this chapter and in the online study guide

At age 16, Bulgarian weightlifter Naim Suleimanov (formerly Naim Süleymanoğlu) lifted 170 kg (375 lb) from the floor to full arms' length overhead—the clean and jerk. He was able to perform this incredible feat at a body weight of only 56 kg (123 lb)! In other words, he was able to lift more than three times his body weight, something that only one other weightlifter had ever been able to do at that time. At the age of 15, he had set his first adult world record. He was considered at the time to be the best competitive weightlifter in the world. We are discovering that children and adolescents have tremendous physiological capacity in many areas of sport, but we must be cautious in our expectations and training techniques because this is a time of rapid change and possibly increased risk for injury—anatomical, physiological, and emotional.

Naim Suleimanov was only 16 years old when he lifted three times his body weight.

Throughout the previous chapters of this book, we have examined the body's physiological responses to acute bouts of exercise and its adaptations to training and the environment. But the focus has been on the adult. For many years, it was assumed that children and adolescents responded and adapted identically to adults, but few scientists actually had studied children and adolescents. There is now a cadre of researchers—pediatric exercise physiologists—who focus on the exercise responses and adaptations of the child and adolescent. As a result, we now have a much better understanding and appreciation of both the differences and similarities between adults and children and adolescents, which we discuss in this chapter.

GROWTH, DEVELOPMENT, AND MATURATION

Growth, development, and maturation are terms used to describe changes that occur in the body starting at conception and continuing through adulthood. **Growth** refers to an increase in the size of the body or any of its parts. **Development** refers to differentiation of cells along specialized lines of function (e.g., organ systems), so it reflects the functional changes that occur with growth. Finally, **maturation** refers to the process of taking on adult form and becoming fully functional, and it is defined by the system or function being considered. For example, skeletal maturity refers to having a fully developed skeletal system in which all bones have completed normal growth and ossification, whereas sexual maturity refers to having a fully functional reproductive system. The state of a child's or adolescent's maturity can be defined by

- chronological age,
- skeletal age, and
- stage of sexual maturation.

Throughout this chapter we refer to the child and the adolescent. The period of life from birth to the start of adulthood is generally divided into three phases: infancy, childhood, and adolescence. **Infancy** is defined as the first year of life. **Childhood** spans the period of time between the end of infancy (the first birthday) and the beginning of **adolescence** and is usually divided into early childhood (preschool) and middle childhood (elementary school). The period of adolescence is more difficult to define in chronological years, because it varies in both its onset and its termination. Its onset generally is defined as the onset of **puberty,** when secondary sex characteristics develop and sexual reproduction becomes possible, and its termination as the completion of growth and development processes, such as attaining adult height. For most girls, adolescence ranges from 8 to 19 years and for most boys from 10 to 22 years.[16]

With the increasing popularity of youth sport and an emphasis on increasing children's physical fitness, we must understand the physiological aspects of growth and development. Children and adolescents must not be regarded as mere miniature versions of adults. They are unique at each stage in their development. The growth and development of their bones, muscles, nerves, and organs largely dictate their physiological and performance capacities. As children's size increases, so do almost all of their functional capacities. This is true of motor ability, strength, cardiovascular and respiratory function, and aerobic and anaerobic capacity. In the following sections, we examine age-related changes in a child's physical abilities.

BODY COMPOSITION: GROWTH AND DEVELOPMENT OF TISSUES

To understand the physical capabilities of children and the potential impact that sport activity can have on young athletes, we must first consider the physical state of their bodies. In this section, we examine growth and development of selected body tissues.

Height and Weight

Specialists in the field of growth and development have spent considerable time analyzing the changes in height and weight that accompany growth. These two variables are most useful when we examine the rates of change. Change in height is assessed in terms of centimeters per year and change in weight in terms of kilograms per year. Figure 16.1 shows that height increases rapidly during the first two years of life. In fact, the child reaches about 50% of adult height by age 2. After this, height increases at a progressively slower rate throughout childhood; thus, there is a decline in the rate of its change. Just before puberty, the rate of change in height increases markedly, followed by an exponential decrease in rate until full height is attained at a mean or average age of about 16 years in girls and 18 years in boys. Some boys do not reach their full height until their early 20s. The peak rate of growth in height occurs at approximately 12 years in girls and 14 years in boys. The peak rate of growth in body weight occurs at approximately 12.5 years in girls and 14.5 years in boys—slightly later than for height.

Girls mature physiologically about two years earlier than boys do.

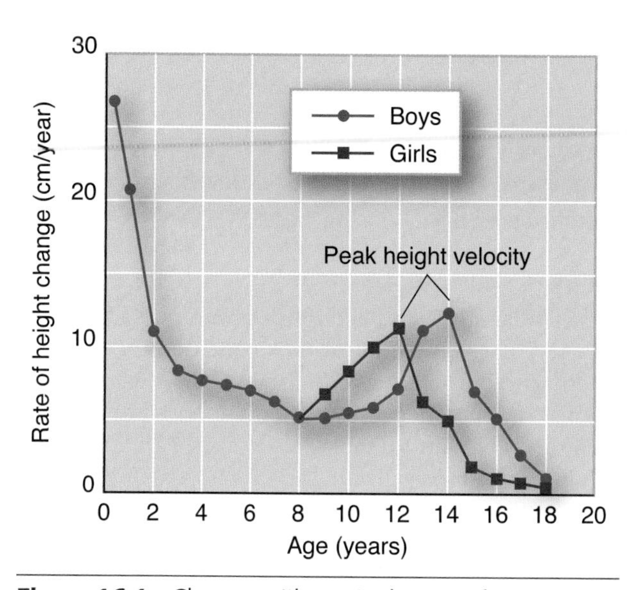

Figure 16.1 Changes with age in the rate of increase in height (cm/year).

Bone

Bones, joints, cartilage, and ligaments form the body's structural support. Bones provide points of attachment for the muscles, protect delicate tissues, and act as reservoirs for calcium and phosphorus; and some are involved in blood cell formation. Early in fetal development, the bones begin to develop from cartilage. Some flat bones, such as those of the skull, develop from fibrous membranes, but the vast majority of bones instead develop from hyaline cartilage. During fetal development, as well as during the initial 14 to 22 years of life, membranes and cartilage are transformed into bone through the process of **ossification,** or bone formation. The average ages at which the different bones in our bodies complete ossification differ widely, but bones typically begin to fuse in the preteens, and all are fused by the early 20s. On average, girls achieve full bone maturity several years before boys.

The structure of mature long bones is complex. Bone is a living tissue that requires essential nutrients, so it receives a rich blood supply. Bone consists of cells distributed throughout a matrix or lattice-type arrangement, and it is dense and hard because of deposits of lime salts, mainly calcium phosphate and calcium carbonate. For this reason, calcium is an essential nutrient, particularly during periods of bone growth and in the later years of life when bone tends to become brittle because of bone mineral loss associated with aging. Bones also store calcium. When our blood calcium level is high, excess calcium can be deposited in our bones for storage; and when calcium levels are too low, bone is resorbed, or broken down, to release calcium into the blood. When injury occurs or when extra stress is placed on a bone, more calcium is deposited. Thus, throughout life, our bones are constantly changing.

Weight-bearing exercise is essential for proper bone growth. Although exercise has little or no influence on bone lengthening, it does increase bone width and bone density by depositing more mineral in the bone matrix, which increases the bone's strength. There is preliminary evidence indicating that the prepubertal years may be the most opportune time to lay down bone in response to an exercise stimulus.[3]

> ***In focus***
>
> Exercise, along with an adequate diet, is essential for proper bone growth. Exercise affects primarily bone width, density, and strength but has little or no effect on length.

> ***In review***
>
> - Growth in height is very rapid during the first two years of life, with a child reaching 50% of adult stature by age 2. After that, the rate is slower throughout childhood until a marked increase occurs near puberty.
> - The peak rate of height growth occurs at age 12 in girls and age 14 in boys. Full height is typically attained at age 16 in girls and age 18 in boys.
> - Growth in weight follows the same trend as height. The peak rate of weight increase occurs at age 12.5 in girls and age 14.5 in boys.

Muscle

From birth through adolescence, the body's muscle mass steadily increases, along with the youngster's weight. In males, the skeletal muscle mass increases from 25% of total body weight at birth to about 40% to 45% or more in young men (20-30 years). Much of this gain occurs when the muscle development rate peaks at puberty. This peak corresponds to a sudden, almost 10-fold increase in testosterone production. Girls don't experience such rapid acceleration of muscle growth at puberty; but their muscle mass does continue to increase, although more slowly than boys', to about 30% to 35% of their total body weight as young adults. This rate difference is largely attributed to hormonal differences at puberty (see chapter 18). These percentage values for both men and women decrease with aging due to loss of muscle mass and gains in fat mass.

Increases in muscle mass with age appear to result primarily from hypertrophy (increase in size) of existing fibers, with little or no hyperplasia (increase in fiber number). Fiber hypertrophy results from increases in the myofilaments and myofibrils. Increases in muscle length as young bones elongate result from increases in the number of sarcomeres (which are added at the junction of the muscle and the tendon) and from increases in the length of existing sarcomeres. Muscle mass peaks in females at age 16 to 20 years and in males at 18 to 25 years, unless it is increased further through exercise, diet, or both.

In focus

The increase in muscle mass with growth and development is accomplished primarily by hypertrophy of individual muscle fibers through increases in their myofilaments and myofibrils. Muscle length increases through the addition of sarcomeres and by increases in the length of existing sarcomeres.

Fat

Fat cells form and fat deposition starts in these cells early in fetal development, and this process continues indefinitely thereafter. Each fat cell can increase in size at any age from birth to death. Early studies investigating fat cell and fat mass development suggested that the number of fat cells becomes fixed early in life. This led many scientists to believe that maintaining a low total body fat content during the early period of development would minimize the total number of fat cells that develop, greatly reducing the likelihood of obesity as an adult. But subsequent studies provided evidence showing that the number of fat cells can continue to increase throughout life. In light of this evidence, it is important to maintain good dietary and exercise habits throughout life.

The amount of fat that accumulates with growth and aging depends on

- diet,
- exercise habits, and
- heredity.

Heredity is unchangeable, but both diet and exercise can be altered to either increase or decrease fat stores.

At birth, 10% to 12% of total body weight is fat. At **physical maturity**, the fat content reaches approximately 15% of total body weight for males and approximately 25% for females. This sex difference, like that seen in muscle growth, is primarily attributable to hormonal differences. When girls reach puberty, their estrogen levels increase, which promotes the deposition of body fat. The trend of body fat increasing with age is shown in figure 16.2, which illustrates the relationship between subcutaneous fat (measured at the triceps and

In focus

Fat storage occurs through increase in the size of existing fat cells and increase in the number of fat cells. It appears that as existing fat cells become full, they signal the need for the development of new fat cells.

subscapular sites) and age for boys and girls from ages 2 to 18 years. The amount of subcutaneous fat is representative of total body fat. Figure 16.3 illustrates the changes in percent body fat, fat mass, and fat-free mass for both males and females from ages 8 to 20 years.[16] It is important to realize that both fat mass and fat-free mass increase during this time period, so an increase in absolute fat does not necessarily mean an increase in relative fat.

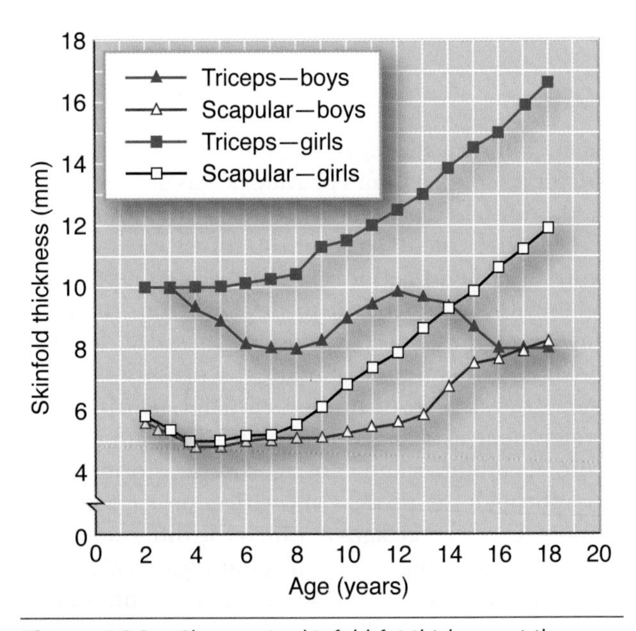

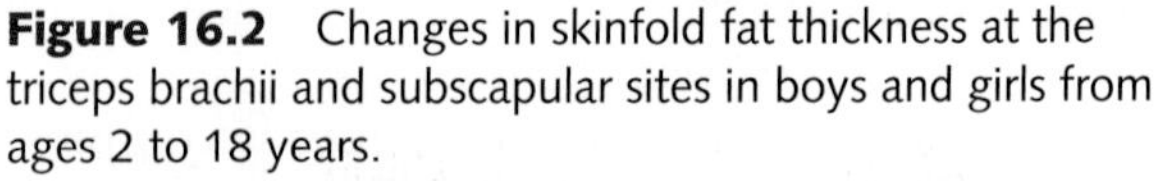

Figure 16.2 Changes in skinfold fat thickness at the triceps brachii and subscapular sites in boys and girls from ages 2 to 18 years.

Data from the NHANES-I, National Center for Health Statistics.

Nervous System

As children grow, they develop better balance, agility, and coordination as their nervous systems develop. Myelination of the nerve fibers must be completed before fast reactions and skilled movement can occur because conduction of an impulse along a nerve fiber is considerably slower if myelination is absent or incomplete (chapter 3). **Myelination** of the cerebral cortex occurs most rapidly during childhood but continues well beyond puberty. Although practicing an activity or skill can improve performance to a certain extent, the full development of that activity or skill depends on full

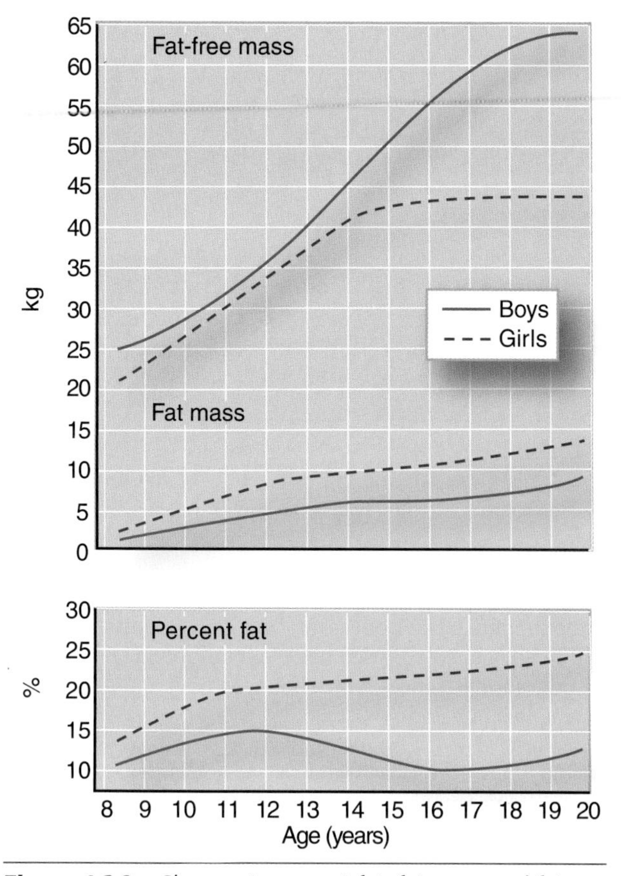

Figure 16.3 Changes in percent fat, fat mass, and fat-free mass for females and males from birth to 20 years of age.

Reprinted, by permission, from R.M. Malina, C. Bouchard, and O. Bar-Or, 2004, *Growth, maturation, and physical activity*, 2nd ed. (Champaign, IL: Human Kinetics), 114.

In review

- Muscle mass increases steadily along with weight gain from birth through adolescence.
- In boys, the rate of muscle mass increase peaks at puberty, when testosterone production increases dramatically. Girls do not experience this sharp increase in muscle mass.
- Muscle mass increases in boys and girls result primarily from fiber hypertrophy with little or no hyperplasia.
- Muscle mass peaks in girls between ages 16 and 20 and in boys between ages 18 and 25, although it can be further increased through diet and exercise.
- Fat cells can increase in size and number throughout life.
- The amount of fat accumulation depends on diet, exercise habits, and heredity.
- At physical maturity, the body's fat content averages 15% in males and 25% in females. The differences are caused primarily by higher testosterone levels in males and higher estrogen levels in females.
- Balance, agility, and coordination improve as children's nervous systems develop.
- Myelination of nerve fibers must be complete before fast reactions and skilled movements are fully developed because myelination speeds the transmission of electrical impulses.

maturation (and myelination) of the nervous system. The development of strength is also likely influenced by myelination.

PHYSIOLOGICAL RESPONSES TO ACUTE EXERCISE

The function of almost all physiological systems improves until full maturity is reached or shortly before. After that, physiological function plateaus for a period of time before starting to decline with advancing age. In this section, we focus on some of the changes in children and adolescents that accompany growth and development, including the following:

- Strength
- Cardiovascular and respiratory function
- Metabolic function, including aerobic capacity, running economy, and anaerobic capacity

Strength

Strength improves as muscle mass increases with age. Peak strength usually is attained by age 20 in women and between ages 20 and 30 in men. The hormonal changes that accompany puberty lead to marked increases in strength in pubescent males because of the increased muscle mass noted before. The extent of development and the performance capacity of muscle also depend on the relative maturation of the nervous system. High levels of strength, power, and skill are impossible if the child has not reached neural maturity. Myelination of many motor nerves is incomplete until sexual maturity, so the neural control of muscle function is limited before that time.

Figure 16.4 on page 388 illustrates changes in leg strength in a group of boys from the Medford Boys' Growth Study.[5] The boys were followed longitudinally from age 7 to 18. The rate of strength gain (slope of

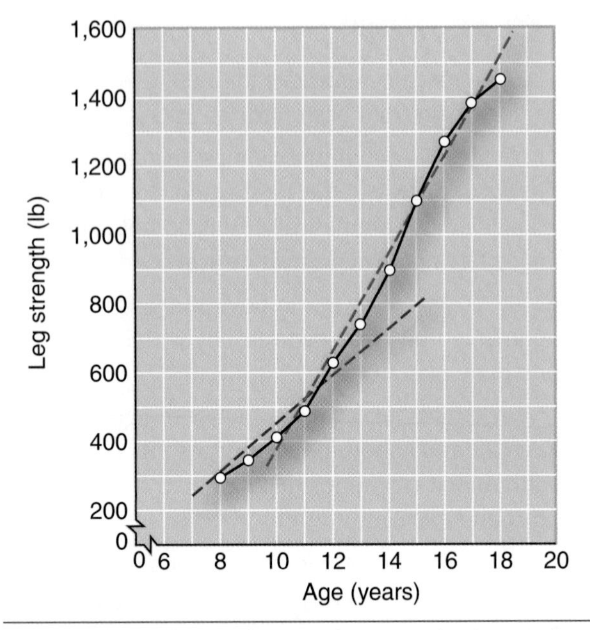

Figure 16.4 Gains with age in leg strength of young boys followed longitudinally over 12 years. Note the increase in the slope of the curve from 12 to 16 years of age.

Data from H.H. Clarke, 1971, *Physical and motor tests in the Medford boys' growth study* (Englewood Cliffs, NJ: Prentice Hall).

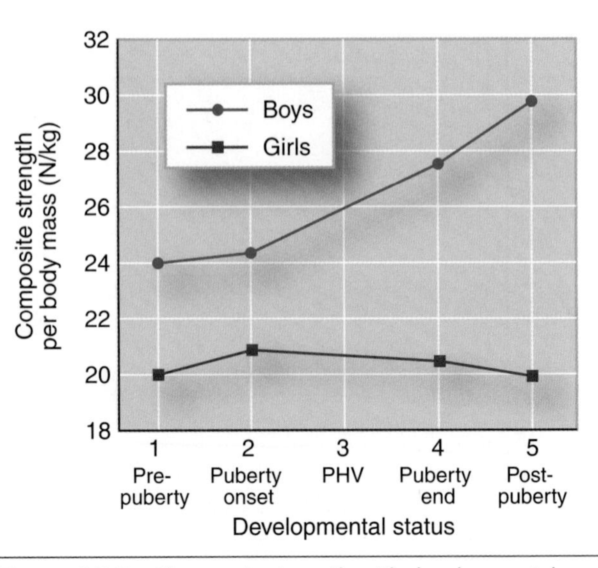

Figure 16.5 Changes in strength with developmental status in boys and girls. Strength is expressed as a composite static strength score from several strength testing sites, and the data are expressed per kilogram of body mass to account for differences in size between boys and girls. PHV = peak height velocity.

Reprinted, by permission, from K. Froberg and O. Lammert, 1996, "Development of muscle strength during childhood." In *The child and adolescent athlete* (London: Blackwell Publishing Company) 28.

line) increased noticeably around age 12, the typical age for onset of puberty. Similar longitudinal data for girls over this same age span are not available. Cross-sectional data, however, indicate that girls experience a more gradual and linear increase in strength and do not exhibit a marked change in their rate of strength gain with puberty,[11] as shown in figure 16.5.

Cardiovascular and Respiratory Function

Cardiovascular function undergoes considerable change as children grow and age. Let's consider some of these changes during submaximal and maximal exercise.

Rest and Submaximal Exercise

Blood pressure at rest and during submaximal levels of exercise is lower in children than in adults but progressively increases to adult values during the late teen years. Blood pressure is also directly related to body size. Larger people generally have higher blood pressures, so size is at least partially responsible for children's lower blood pressure values. In addition, blood flow to active muscles during exercise in children can be greater for a given volume of muscle than in adults because children have less peripheral resistance.

Recall that cardiac output is the product of heart rate and stroke volume. A child's smaller heart size and total blood volume result in a lower stroke volume, both at rest and during exercise, than in an adult. In an attempt to compensate for this, the child's heart rate response to a given rate of submaximal work (such as on a cycle ergometer), where the absolute oxygen requirement is the same, is higher than an adult's. As the child ages, heart size and blood volume increase along with body size. Consequently, stroke volume also increases, as body size increases, for the same absolute rate of work.

However, a child's higher submaximal heart rate cannot completely compensate for the lower stroke volume. Because of this, the child's cardiac output is also somewhat lower than the adult's for a given absolute rate of work or a given oxygen consumption. To maintain adequate oxygen uptake during these submaximal levels of work, the child's arterial–mixed venous oxygen difference, or $(a\text{-}\bar{v})O_2$ difference, increases to further compensate for the lower stroke volume. The increase in $(a\text{-}\bar{v})O_2$ difference is most likely attributable to increased blood flow to the active muscles—a greater percentage of the cardiac output goes to the active muscles.[26] These submaximal relationships are illustrated in figure 16.6, in which the responses of a 12-year-old boy are compared with those of a fully mature man.

Maximal Exercise

Maximum heart rate (HR_{max}) is higher in children than in adults but decreases linearly as children age. Children under age 10 frequently have maximum heart rates

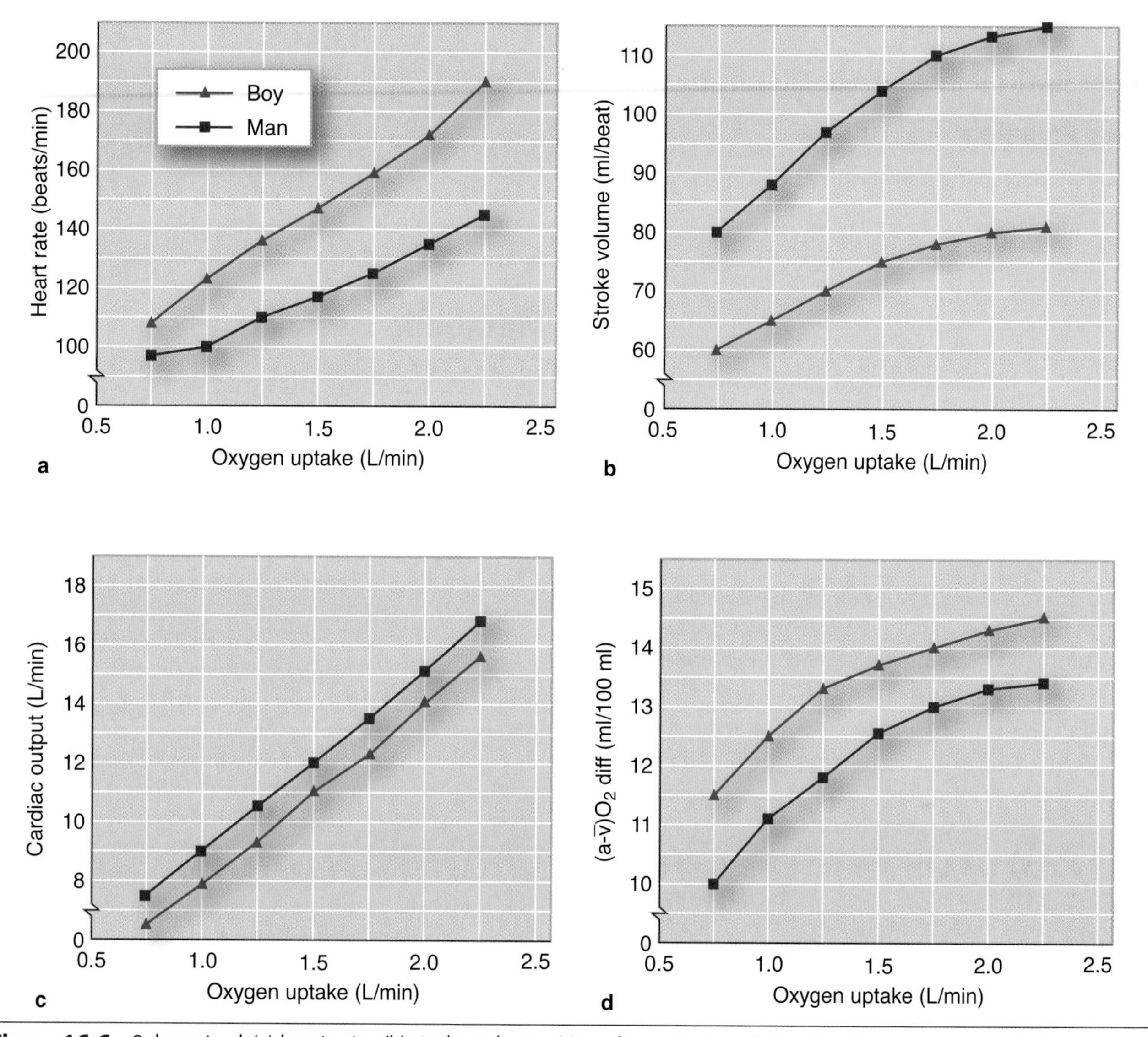

Figure 16.6 Submaximal *(a)* heart rate, *(b)* stroke volume, *(c)* cardiac output, and *(d)* arterial–venous oxygen difference, or $(a\text{-}\bar{v})O_2$ difference, in a 12-year-old boy and a fully mature man at the same rates of oxygen uptake.

exceeding 210 beats/min, whereas the average 20-year-old has a maximum heart rate of approximately 195 beats/min. With further aging (25-30 years and older), however, results of cross-sectional studies suggest that maximum heart rate decreases by slightly less than 1 beat/min per year. Longitudinal studies, on the other hand, suggest that maximum heart rate decreases only 0.5 beats/min per year. Longitudinal studies, in which the same people are followed over time, generally provide more accurate estimates of the true changes.

During maximal exercise, as also seen with submaximal exercise, the child's smaller heart and blood volume limit the maximal stroke volume that he or she can achieve. Again, the high HR_{max} cannot fully compensate for this, leaving the child with a lower maximal cardiac output than the adult. This limits the child's performance at high absolute rates of work (e.g., pedaling at 100 W on a cycle ergometer or trying to achieve the same absolute $\dot{V}O_2$) because the child's capacity for oxygen delivery is less than an adult's. However, for high relative rates of work in which the child is responsible for moving only his or her body mass (e.g., running on a treadmill at the same speed with no grade), this lower maximal cardiac output is not as serious a limitation. In running, for example, a 25 kg (55 lb) child requires (in proportion to body size)

In focus

Heart size is directly related to body size, so children have smaller hearts than adults. As a result of this and a smaller blood volume, children have a smaller stroke volume capacity. A child's higher maximum heart rate can only partially compensate for this lower stroke volume capacity, and thus maximal cardiac output is lower than that of an equally trained adult.

considerably less oxygen than a 90 kg (198 lb) man would require, yet the rate of oxygen consumption when scaled for body size is about the same for both.

Lung Function

Lung function changes markedly with growth. All lung volumes increase until growth is complete. Peak flow rates follow the same pattern. The changes in these volumes and flow rates are matched by the changes in the highest ventilation that can be achieved during exhaustive exercise, which is referred to as maximal expiratory ventilation ($\dot{V}_{Emax}$), or maximal minute ventilation. $\dot{V}_{Emax}$ increases with age until physical maturity and then decreases with aging. For example, cross-sectional data show that $\dot{V}_{Emax}$ averages about 40 L/min for 4- to 6-year-old boys and increases to 110 to 140 L/min at full maturity. Girls follow the same general pattern, but their absolute values are considerably lower post-puberty, primarily because of their smaller body size. These changes are associated with the growth of the pulmonary system, which parallels the general growth patterns of children. As body size increases with growth and development, so do lung size and function.

In review

- Strength improves as muscle mass increases with age.
- Gains in strength with growth also depend on neural maturation because neuromuscular control is limited until myelination is complete, usually around sexual maturity.
- Blood pressure is directly related to body size: It is lower in children than in adults but increases to adult levels in the late teen years, both at rest and during exercise.
- During both submaximal and maximal exercise, a child's smaller heart and blood volume result in a lower stroke volume than in adults. In partial compensation, a child's heart rate is higher than an adult's for the same rate of work or $\dot{V}O_2$.
- Even with increased heart rate, a child's cardiac output remains less than an adult's. In submaximal exercise, an increase in the (a-$\bar{v}$)O_2 difference ensures adequate oxygen delivery to the active muscles. But at maximal work rates, oxygen delivery limits performance in activities other than those in which the child merely needs to move his or her body mass, such as running.
- Lung volumes increase until physical maturity, primarily because of increasing body size.
- Until physical maturity, maximal ventilatory capacity and maximal expiratory ventilation increase in direct proportion to the increase in body size during maximal exercise.

Metabolic Function

Metabolic function also changes as the child and adolescent grow larger, as we would expect from the changes that we have just reviewed in muscle mass and strength and cardiorespiratory function.

Aerobic Capacity

The purpose of the basic cardiovascular and respiratory adaptations that occur in response to varying levels of exercise (rates of work) is to accommodate the exercising muscles' need for oxygen. Thus, increases in cardiovascular and respiratory function that accompany growth suggest that aerobic capacity ($\dot{V}O_{2max}$) similarly increases. In 1938, Robinson[18] demonstrated this phenomenon in a cross-sectional sample of boys and men ranging in age from 6 to 91 years. He found that $\dot{V}O_{2max}$ peaks between ages 17 and 21 and then decreases linearly with age. Other studies subsequently confirmed these observations. Studies of girls and women have shown essentially the same trend, although in females the decrease begins at a much younger age, generally age 12 to 15 (refer to chapter 18), probably attributable to an earlier assumption of a more sedentary lifestyle. The changes in $\dot{V}O_{2max}$ with age, expressed in liters per minute, are illustrated in figure 16.7*a*.

Expressing $\dot{V}O_{2max}$ relative to body weight ($ml \cdot kg^{-1} \cdot min^{-1}$) provides a considerably different picture, as shown in figure 16.7*b*. Values change little in boys from age 6 to young adulthood. For girls, however, little change occurs from age 6 to 13; but after age 13, aerobic capacities show a gradual decrease. Although these observations are of general interest, they might not accurately reflect the development of the cardiorespiratory system as children grow and their physical activity levels change. Several concerns have been raised about the validity of using body weight to account for changes in the size of the cardiorespiratory and metabolic systems, as when one is dividing absolute values by body weight, for example, $\dot{V}O_2$ per kilogram.

Arguments against using body weight to scale $\dot{V}O_{2max}$ for differences in size include the following. First, although $\dot{V}O_{2max}$ values expressed relative to body weight

In focus

Aerobic capacity ($\dot{V}O_{2max}$), when expressed in liters per minute, is lower in children than in adults at similar levels of training. This is attributable primarily to the child's lower maximal cardiac output capacity. When $\dot{V}O_{2max}$ values are expressed to normalize for the differences in body size between children and adults, there is little or no difference in aerobic capacity.

remain relatively stable or decline with age, endurance performance steadily improves. The average 14-year-old boy can run the mile (1.6 km) almost twice as fast as the average 5-year-old boy, yet the two boys' $\dot{V}O_{2max}$ values expressed relative to body weight are similar.[21] Second, although the increases in $\dot{V}O_{2max}$ that accompany endurance training in children are relatively small compared with those in adults, the performance increases in these children are relatively large. Therefore, body weight is likely not the most appropriate way to scale $\dot{V}O_{2max}$ values for differences in body size in children and adolescents. The relationships between $\dot{V}O_{2max}$, body dimensions, and system functions during growth are extraordinarily complex.[6, 22] This is discussed in greater detail later in the chapter.

Running Economy

How do growth-related changes in aerobic capacity affect a child's performance? For any activity that requires a fixed rate of work, such as cycling on an ergometer, the child's lower $\dot{V}O_{2max}$ limits endurance performance. But as noted earlier, for activities in which body weight is the major resistance to movement, such as distance running, children should not be at a disadvantage, because their $\dot{V}O_{2max}$ values expressed relative to body weight are already at or near adult values.

Yet children cannot maintain a running pace as fast as adults can because of basic differences in economy of effort. At a given speed on a treadmill, a child will have a substantially higher submaximal oxygen consumption when expressed relative to body weight than an adult. Even if the child's lactate threshold occurred at the same relative oxygen consumption as the adult's (at the same percentage of their respective $\dot{V}O_{2max}$ values), the child would be running at a much slower pace. Also, as children age, their legs lengthen, their muscles become stronger, and their running skills improve. Running economy increases, and this improves their distance-running pace, even if the children are not training and if their $\dot{V}O_{2max}$ values don't increase.[7, 13] Rowland argues that increased stride frequency as children and adolescents grow is the most important factor in explaining these changes in running economy.[23] It is also possible that scaling oxygen consumption to body weight is inappropriate during growth and development, as discussed in the previous section.[19]

Anaerobic Capacity

Children have a limited ability to perform anaerobic-type activities. This is demonstrated in several ways. Children cannot achieve adolescent or adult concentrations

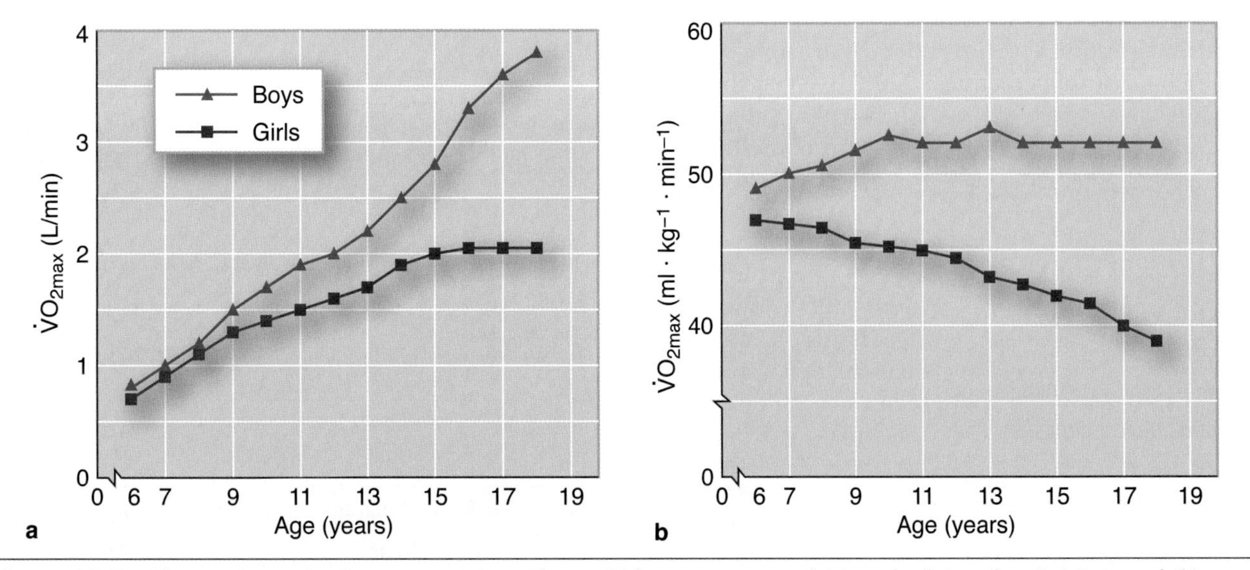

Figure 16.7 Changes in maximal oxygen uptake with age. Values are expressed *(a)* as absolute values in L/min and *(b)* relative to body weight in ml · kg^{-1} · min^{-1}.

In review

- As pulmonary and cardiovascular function improve with continued development, so does aerobic capacity.
- $\dot{V}O_{2max}$, expressed in liters per minute, peaks between ages 17 and 21 years in males and between 12 and 15 years in females, after which it plateaus for several years and then steadily decreases.
- When $\dot{V}O_{2max}$ is expressed relative to body weight, it plateaus in males from ages 6 to 25 years before it begins to decline. In females, the decline in $\dot{V}O_{2max}$ is small from ages 6 to 12 years but becomes more substantial starting at about age 13. However, expressing $\dot{V}O_{2max}$ relative to body weight might not provide an accurate estimate of aerobic capacity. Such $\dot{V}O_{2max}$ values do not reflect the significant gains in endurance performance capacity that are noted with both maturation and training.
- The child's lower $\dot{V}O_{2max}$ value (L/min) limits endurance performance unless body weight is the major resistance to movement, as in distance running.
- When expressed relative to body weight, a child's $\dot{V}O_{2max}$ is similar to an adult's, yet in activities such as distance running, a child's performance is far inferior to adult performance.
- Running economy is lower in children compared with adults when $\dot{V}O_2$ is expressed relative to body weight. One factor has been identified to explain this difference: the difference between children and adults in stride frequency for the same fixed-pace run. Scaling $\dot{V}O_2$ to body weight might also be inappropriate.

of lactate in either muscle or blood for maximal and supramaximal rates of exercise.[4] This suggests that children have a lower glycolytic capacity. The lower lactate levels might reflect a lower concentration of phosphofructokinase, the key rate-limiting enzyme of anaerobic glycolysis. Lactate dehydrogenase activity also seems to be lower in children. However, lactate threshold, when expressed as a percentage of $\dot{V}O_{2max}$, does not appear to be a limiting factor in children because children's lactate thresholds are similar to, if not somewhat higher than, those of similarly trained adults. Also, children's resting levels of adenosine triphosphate (ATP) and phosphocreatine (PCr) are similar to those of adults, so activities of less than 10 to 15 s should not be compromised. Thus, only activities that tax the anaerobic glycolytic system—those from 15 s up to 2 min in duration—will be lower.

In focus

Anaerobic capacity is lower in children than adults, which may reflect children's lower concentration of the key rate-limiting enzyme phosphofructokinase or lactate dehydrogenase.

Children cannot achieve high respiratory exchange ratios during maximal or exhaustive exercise. Maximal respiratory exchange ratios in children are seldom above 1.10 and are sometimes below 1.00, but adult ratios are usually more than 1.10 and often greater than 1.15. This indicates that less carbon dioxide is produced in children for the same oxygen consumption, which in turn indicates less buffering of lactate.

Anaerobic mean and peak power output, as determined by the Wingate anaerobic power test (a 30 s, all-out maximal effort on a cycle ergometer), is also lower in children than in adults. Figure 16.8 illustrates the

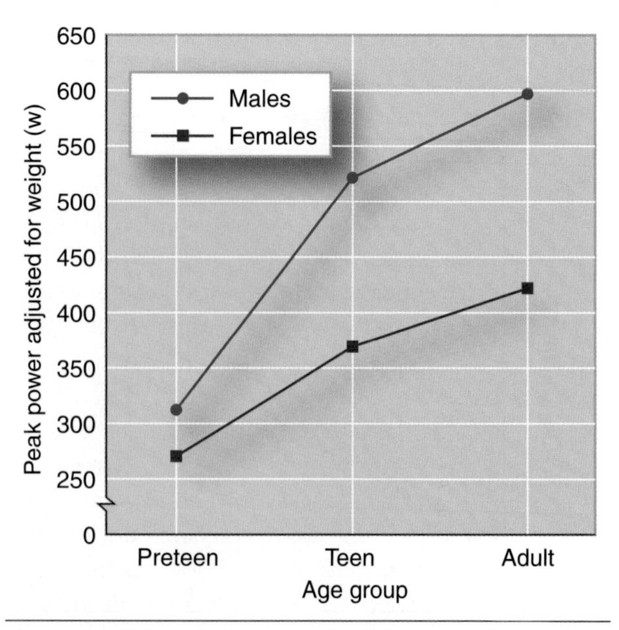

Figure 16.8 Optimal peak power output (anaerobic power) statistically adjusted for body mass in preteenagers (9-10 years old), teenagers (14-15 years old), and adults (mean age of 21 years). These values represent anaerobic power independent of body size.

Data from Santos et al., 2002.

results of a similar cycle ergometer anaerobic power test that is potentially a better discriminator of peak power output capacity.[24] In this figure, peak power is statistically adjusted for body mass to account for differences in body size when we compare values for preteenagers, teenagers, and adults. This figure demonstrates the very low peak power outputs for preteenagers (9-10 years of age) compared with both teenagers (14-15 years of age) and adults (mean age of 21 years). Teenagers were much closer to the values for adults than the preteenagers. Again, these values were adjusted for body size, so they should accurately reflect anaerobic power.

Bar-Or[1] summarized the development of both the aerobic and anaerobic characteristics of boys and girls from ages 9 through 16, using 18 years of age as the criterion for 100% of the adult value. The changes with age are shown in figure 16.9. Aerobic power is represented by the child's $\dot{V}O_{2max}$, whereas anaerobic power is represented by the child's performance on the Margaria step-running test (a field test). Maximal energy expenditure per kilogram represents the maximal energy-generating capacities of the aerobic and anaerobic systems, scaled to body weight to account for body size differences with growth. Notice that aerobic fitness remains constant for the boys but declines for

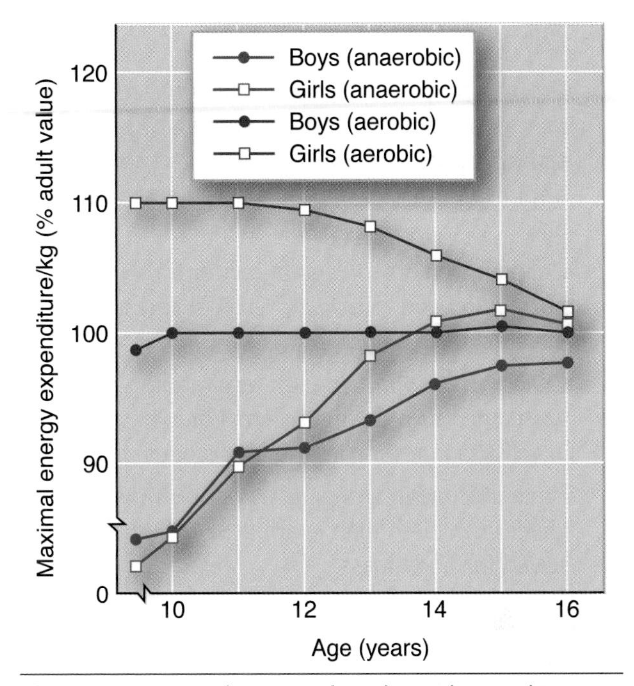

Figure 16.9 Development of aerobic and anaerobic characteristics in boys and girls ages 9 to 16 years. Values are expressed as a percentage of adult values (values at age 18).

Adapted, by permission, from O. Bar-Or, 1983, *Pediatric sports medicine for the practitioner: From physiologic principles to clinical applications* (New York: Springer-Verlag).

Scaling Physiological Data to Account for Size Differences

Throughout this chapter and previous chapters, we have discussed the need to express physiological data relative to the size of the individual. In chapter 4, when we introduced the concept of $\dot{V}O_{2max}$, we mentioned that values normally are expressed relative to body mass, by dividing the absolute $\dot{V}O_{2max}$ (expressed in L/min) by body weight ($ml \cdot kg^{-1} \cdot min^{-1}$). Many scientists believe that dividing by body mass alone does not adequately account for size differences. This becomes a major issue when we compare children's values with those of adults, or men's with women's, as we see in chapter 18.

Strong cases have been made for scaling $\dot{V}O_2$, cardiac output, stroke volume, and other size-related physiological variables relative to body surface area, measured in square meters, or relative to weight, expressed to the 0.67 power or 0.75 power ($wt^{0.67}$ or $wt^{0.75}$). For years, cardiologists have expressed heart volumes relative to body surface area. Recent research suggests that using body surface area ($ml \cdot m^{-2} \cdot min^{-1}$) or $wt^{0.75}$ ($ml \cdot kg^{-0.75} \cdot min^{-1}$) provides the best means by which to express the data to reduce the effect of body size.[19] One study followed young boys longitudinally from 12 to 20 years of age; one group remained untrained but active, and the other group trained.[25] There was little or no increase with run training in $\dot{V}O_{2max}$ expressed in $ml \cdot kg^{-1} \cdot min^{-1}$, whereas submaximal $\dot{V}O_2$ expressed in the same manner decreased with age, suggesting no change in aerobic capacity but an improvement in running economy. When these same data were expressed in $ml \cdot kg^{-0.75} \cdot min^{-1}$, the boys who were training showed increased aerobic capacity with increased training and age but no change in running economy. This latter finding intuitively makes more sense, suggesting the use of $wt^{0.75}$ as the best way of expressing the data.

In review

- Children's ability to perform anaerobic activities is limited. A child has a lower glycolytic capacity, possibly because of a limited amount of phosphofructokinase or lactate dehydrogenase.
- Children have lower lactate concentrations in both blood and muscle at maximal and supramaximal rates of work.
- Children cannot attain high respiratory exchange ratios during maximal or exhaustive exercise, suggesting less lactate production.
- Anaerobic mean and peak power outputs are lower in children than in adults, even when scaled for body mass.

the girls from 12 to 16 years. Nine- to 12-year-old girls have a higher aerobic capacity than the 18-year-old reference adult value; thus, their values are 110% of the adult value. For both boys and girls, anaerobic capacity increases from 9 through 15 years of age.

PHYSIOLOGICAL ADAPTATIONS TO EXERCISE TRAINING

We have seen that, indeed, children are not just miniature adults. The child is physiologically distinct from the adult and must be considered differently. But how should these differences affect individualized training programs for children? Training can improve the strength, aerobic capacity, and anaerobic capacity of children. Generally, youngsters adapt well to the same type of training routine used by adults. But training programs for children and adolescents should be designed specifically for each age-group, keeping in mind the developmental factors associated with that age. In this section, we look at training-induced changes in each of the following:

- Body composition
- Strength
- Aerobic capacity
- Anaerobic capacity

Then, where appropriate, we discuss proper training procedures to optimize performance gains and reduce the risk of injury.

Body Composition

The child and adolescent respond to physical training similarly to adults with respect to changes in body weight and composition. With both resistance and aerobic training, both boys and girls will decrease body weight and fat mass and increase fat-free mass, although the increase in fat-free mass is attenuated in the child compared with the adolescent and adult. There is also evidence of significant bone growth as a result of exercise training, above that seen with normal growth. In fact, Bass[3] suggested that the prepubertal years may be the most opportune time to increase bone mass because of increases in bone density and periosteal expansion of cortical bone.

There is presently an epidemic of obesity in the United States, Canada, much of Europe, and other westernized countries as discussed in chapter 21. This is true not only in adults but in children and adolescents as well. Physical training and an active lifestyle are critical throughout the growing years to maintain a healthy body composition and establish a lifelong habit of exercise and activity.

Strength

For many years, the use of resistance training to increase muscular strength and endurance in prepubescent and adolescent boys and girls was highly controversial. Boys and girls were discouraged from using free weights for fear that they might injure themselves and prematurely stop the growth process. Furthermore, many scientists speculated that resistance training would have little or no effect on the muscles of prepubescent boys because their levels of circulating androgens were still low. Is resistance training in children and adolescents dangerous or risky? Even if it is safe, are there any benefits?

Studies on animals suggest that heavy resistance exercise can lead to stronger, broader, and more compact bones. But these studies have not contributed much to our understanding of the benefits or risks associated with this form of activity for humans because it is nearly impossible to load these animals to the same extent as youngsters can be loaded. Fortunately, several studies have been conducted in which both prepubescent and adolescent children have participated in resistance training. From these studies, Kraemer and Fleck[12] concluded that the risk of injury is very low. In fact, resistance training might offer some protection against injury, for example by strengthening the muscles that cross a joint. Still, a conservative approach is recommended in prescribing resistance exercise for children, particularly preadolescents.

Now that we have established that resistance training is relatively safe, does it increase strength? A number of studies have now been conducted on both children and adolescents and have clearly demonstrated that resistance training is very effective in increasing strength. The increase is largely dependent on the volume and intensity of training. Further, the percentage increases for children and adolescents are similar to those for young adults.[9]

How are these increases in strength accomplished? The mechanisms allowing strength changes in children are similar to those for adults, with one minor exception: Prepubescent strength gains are accomplished largely without any changes in muscle size. A comprehensive study of the mechanisms responsible for strength increases in prepubescent boys concluded that the likely determinants of the strength gains achieved are improved motor skill coordination, increased motor unit activation, and other undetermined neurological adaptations.[17] Strength gains in the adolescent result primarily from neural adaptations and increases in both muscle size and specific tension.

In focus

Prepubescent children can improve their strength with resistance training. These strength gains are attributable largely to neurological factors, with little or no change in the size of the muscle.

For actual training programs, resistance training for children should be prescribed in much the same way as for adults. Specific guidelines have been established by a number of professional organizations, including the American Orthopaedic Society for Sports Medicine, the American Academy of Pediatrics, the American College of Sports Medicine, the National Athletic Trainers' Association, the National Strength and Conditioning Association, the President's Council on Physical Fitness and Sports, the U.S. Olympic Committee, and the Society of Pediatric Orthopaedics. Basic guidelines have been established for the progression of resistance exercise in children, which are presented in table 16.1.[12] Further information on resistance training program designs for children is available.[10, 12]

Any youth resistance training program must be carefully supervised by competent instructors who have been trained specifically to work with children. Furthermore, resistance training should be only one part of a more comprehensive fitness program for this age-group. In the next sections, we consider the value of aerobic and anaerobic training.

TABLE 16.1 Basic Guidelines for Resistance Exercise Progression in Children

Age	Considerations
7 years or younger	Introduce child to basic exercises using little or no weight; develop the concept of a training session; teach exercise technique; progress from body weight calisthenics, partner exercises, and lightly resisted exercises; keep volume low.
8-10 years	Gradually increase the number of exercises; practice exercise technique in all lifts; start gradual progressive loading of exercises; keep exercises simple; gradually increase training volume; carefully monitor tolerance of the exercise stress.
11-13 years	Teach all basic exercise techniques; continue progressive loading of each exercise; emphasize exercise techniques; introduce more advanced exercises with little or no resistance. Progress to more advanced youth programs in resistance exercise; add sport-specific components; emphasize exercise techniques; increase volume.
14-15 years	Progress to more advanced youth programs in resistance exercise; add sport-specific components; emphasize exercise techniques; increase volume.
16 years or older	Move child to entry-level adult programs after all background knowledge has been mastered and a basic level of training experience has been gained.

Note. If a child of any age begins a program with no previous experience, start the child at the level for the previous age category and move him or her to more advanced levels as exercise toleration, skill, amount of training time, and understanding permit.

Reprinted, by permission, from W.J. Kraemer and S.J. Fleck, 2005, *Strength training for young athletes, second edition* (Champaign, IL: Human Kinetics), 5.

Aerobic Capacity

Do prepubescent boys and girls benefit from aerobic training to improve their cardiorespiratory systems? This also has been a highly controversial area because several early studies indicated that training prepubescent children did not change their $\dot{V}O_{2max}$ values.[20] Interestingly, even without significant increases in $\dot{V}O_{2max}$, the running performance of the children studied did improve substantially. They could run a fixed distance faster following the training program. More recent studies have shown small increases in aerobic capacity with training in prepubescent children, but these increases are less than would be expected for adolescents or adults—about 5% to 15% in children compared with about 15% to 25% in adolescents and adults.

More substantial changes in $\dot{V}O_{2max}$ appear to occur once children have reached puberty, although the reason for this is unknown. Because stroke volume appears to be the major limitation to aerobic performance in this age-group, it is quite possible that further increases in aerobic capacity depend on heart growth. Also, as discussed earlier in this chapter, scaling of these variables is an issue. The study of Sjödin and Svedenhag[25] presented in the sidebar on page 393 clearly establishes this as a key factor.

In review

- Body composition changes with training in children and adolescents are similar to those seen in adults—loss of total body weight and fat mass and increase in fat-free mass.
- The risk of injury from resistance training in young athletes is relatively low, and the programs they should follow are much like those for adults.
- Strength gains achieved from resistance training in preadolescents result primarily from improved motor skill coordination, increased motor unit activation, and other neurological adaptations. Unlike adults, preadolescents who resistance train experience little change in muscle size. Mechanisms of strength gains in adolescents are similar to those for adults.
- Aerobic training in preadolescents does not alter $\dot{V}O_{2max}$ as much as would be expected for the training stimulus, possibly because $\dot{V}O_{2max}$ depends on heart size. But endurance performance improves with aerobic training. Adolescents are similar to adults in their improvement.
- A child's anaerobic capacity increases with anaerobic training.
- In general, growth and maturation rates and processes are probably not altered significantly by training.

Anaerobic Capacity

Anaerobic training appears to improve children's anaerobic capacity. Following training, children have

- increased resting levels of PCr, ATP, and glycogen;
- increased phosphofructokinase activity; and
- increased maximal blood lactate levels.[1, 8]

Ventilatory threshold, a noninvasive marker of lactate threshold, also has been reported to increase with endurance training in 10- to 14-year-old boys.[14]

When one is designing aerobic and anaerobic training programs for children and adolescents, it appears that standard training principles for adults can be applied. Children and adolescents have not been well studied, but what we do know suggests that they can be trained in a manner similar to that for adults. Again, because children and adolescents are not adults, it is prudent to be conservative to reduce the risk of injury, overtraining, and loss of interest in sport. The approach outlined earlier for resistance training is a good model to use for aerobic and anaerobic training. This is also an appropriate time in life to focus on learning a variety of motor skills by having children explore a number of activities and sports.

MOTOR ABILITY AND SPORT PERFORMANCE

As shown in figure 16.10, the motor ability of boys and girls generally increases with age for the first 17 years, although girls tend to plateau at about the age of puberty for most items tested. These improvements result primarily from development of the neuromuscular and endocrine systems and secondarily from the increased activity.

The plateau observed in the girls at puberty is likely explained by three factors. First, as mentioned earlier, the increase in estrogen levels at puberty, or in the estrogen/testosterone ratio, leads to increased fat deposition. Performance tends to decrease as fat increases. Second, girls have less muscle mass. Finally, and probably of greater importance, around puberty many girls assume a much more sedentary lifestyle than boys. This is largely a matter of social conditioning, as boys are encouraged to be more active and athletic than girls. As girls become less active, their motor abilities tend to plateau. This trend appears to be changing because of changing social attitudes and more opportunities for sport and activity now available for girls (see chapter 18).

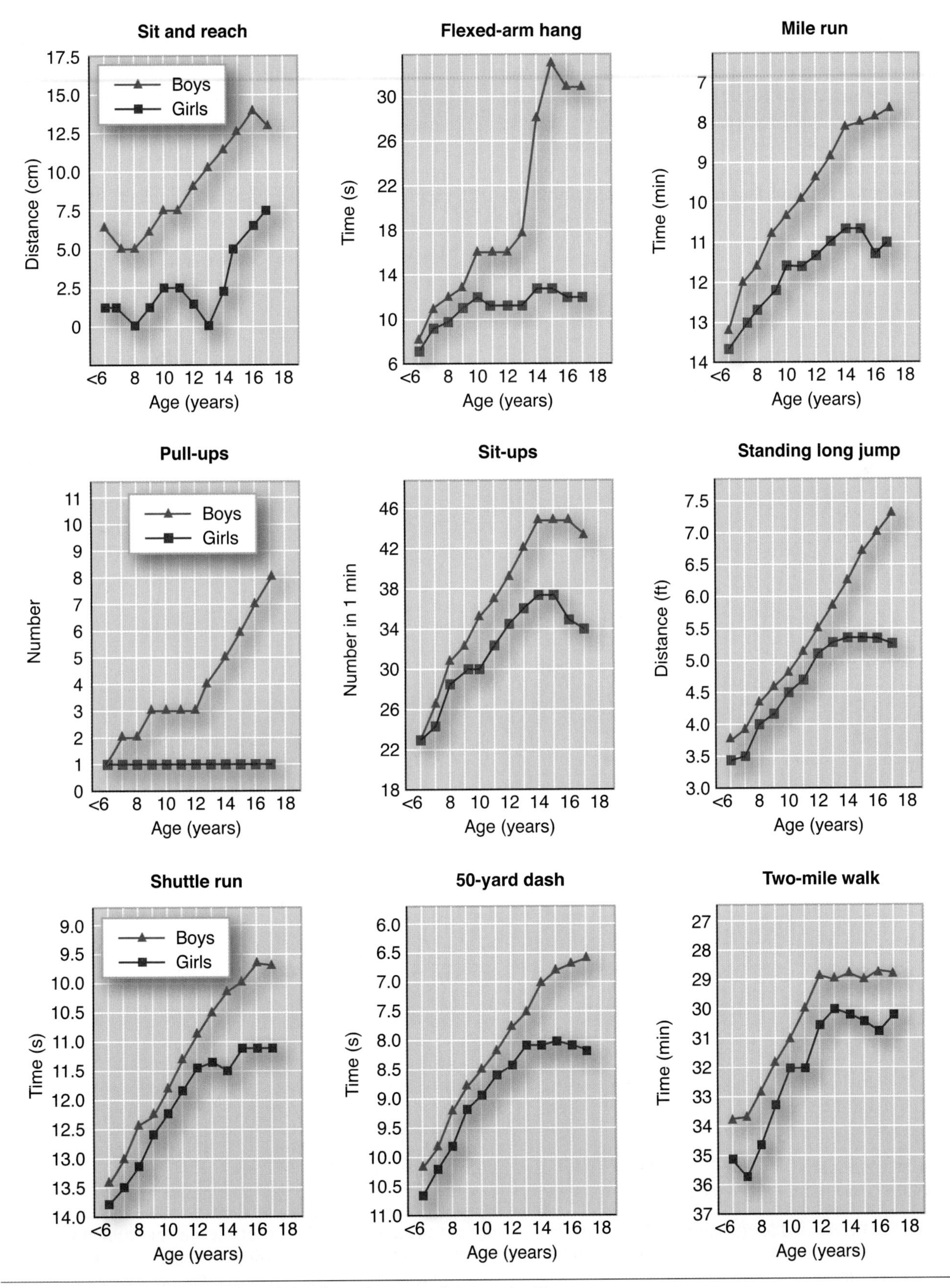

Figure 16.10 Changes in motor ability from the ages of 6 years to 17 years.

Data from the President's Council on Physical Fitness and Sports, 1985.

Sport performance in children and adolescents improves with growth and maturation, as can be seen for age-group records in sports such as swimming and track and field. Figure 16.11 illustrates the improvement in American records for various age-groups.

The figure gives values for the 100 m and 400 m swim and the 100 m and 1,500 m run. These events were selected because they represent a predominantly anaerobic event in swimming and running (100 m swim and run) and a predominantly aerobic activity (400 m swim and 1,500 m run). Both anaerobic and aerobic performance improve progressively with increasing age-groups, with the exception of the 1,500 m run for 17- and 18-year-old girls. Similar age-group records for weightlifting do not appear to be available, because weightlifting competition is organized by weight in

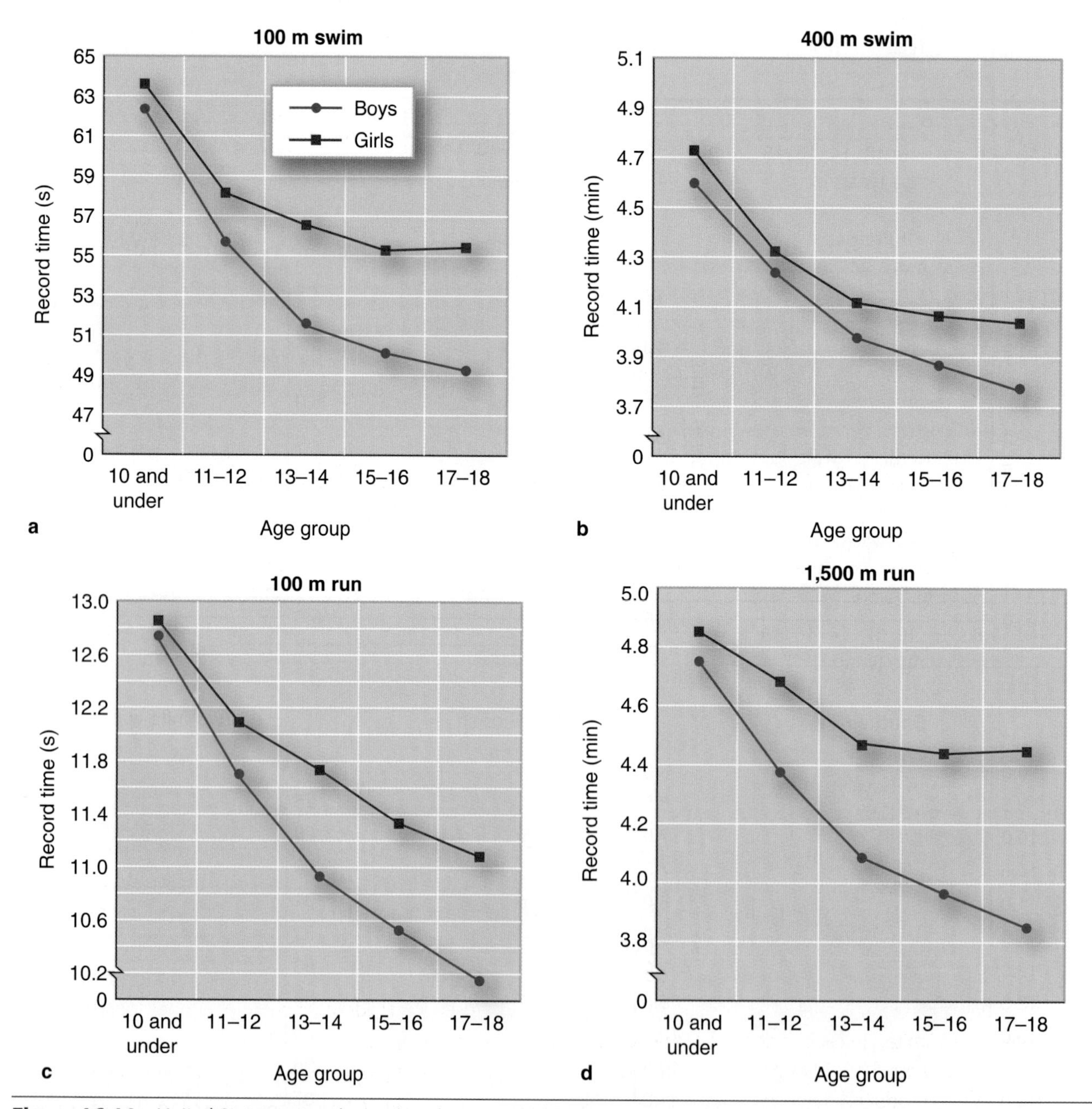

Figure 16.11 United States national record performances for boys and girls from 10 years of age and under up through 17 to 18 years of age in the following swimming and running events: *(a)* 100 m swim, *(b)* 400 m swim, *(c)* 100 m run, and *(d)* 1,500 m run. Records were obtained from USA Track & Field (as of December 2005; www.usatf.org) and United States Swimming (as of March 2006; www.usaswimming.org).

broad classifications such as 16 and under, 17 to 20 years of age, and then adult classifications. On the basis of normal strength gains with growth and development, it is assumed that weightlifting records would increase markedly from late childhood through adolescence, particularly in boys.

SPECIAL ISSUES

During the period of growth and development from childhood through adolescence, these special issues need to be addressed:

- Thermal stress
- Growth and maturation with training

Thermal Stress

Laboratory experiments suggest that children are more susceptible to heat- and cold-induced illness or injury than adults. But the number of reported cases of thermal illness or injury have not supported this theory. A major concern is the child's apparently lower capacity during exercise in the heat to dissipate heat through evaporation. Children appear to rely much more on convection and radiation, which are enhanced through greater peripheral vasodilation.[2] Compared with adults, children have a greater ratio of body surface area to mass, meaning that they have more skin surface area from which to gain or lose heat for each kilogram of body weight. Unless the environment is hot, this is an advantage, because children are better able to lose heat through radiation, convection, and conduction. However, once the environmental temperature exceeds the skin's temperature, children more readily gain heat from the environment, which is a distinct disadvantage. A child's lower capacity for evaporative heat loss is largely the result of a lower sweating rate. Individual sweat glands in children form sweat more slowly and are less sensitive to increases in the body's core temperature than those in adults. Although young boys can acclimatize to exercise in the heat, their rate of acclimatization is slower than that of adults. Acclimatization data are not available for girls.

Only a few studies have focused on children exercising in the cold. From the limited information available, children appear to have greater conductive heat loss than adults because of a larger ratio of surface area to mass. This should be expected to place them at higher risk for hypothermia and to necessitate more clothing layers during exercise in cold temperatures.

Few studies have been conducted on children in relation to either heat or cold stress, and conclusions from existing studies sometimes have been contradictory. More research is needed in this area to determine the risks faced by children who exercise in the heat and cold. In the meantime, a conservative approach is advisable. Children may be at an increased risk of heat- and cold-related injuries compared with adults.[2]

Growth and Maturation With Training

Many people have wondered what effect physical training has on growth and maturation. Does hard physical training slow down or accelerate normal growth and development? In a comprehensive review of this area, Malina made some interesting and relevant

In review

- Laboratory studies suggest that children may be more susceptible to injury or illness from thermal stress because they have a greater ratio of body surface area to mass when compared to adults. However, the number of reported cases does not support this.
- Children are limited in their evaporative heat loss compared to adults because children sweat less (less sweat is produced by each active sweat gland).
- Young boys acclimatize to heat more slowly than adults do. Data are not available for girls.
- Children appear to have greater conductive heat loss than adults, which may place children at greater risk for hypothermia in cold environments.
- Until more is known about children's susceptibility to thermal stress, a conservative approach should be used for children who exercise in temperature extremes.
- Physical training appears to have little or no negative effect on normal growth and development. Its effects on markers of sexual maturation are less clear.

observations.[15] Regular training has no apparent effect on growth in height. It does, however, affect weight and body composition, as discussed earlier in this chapter.

As for maturation, the age at which peak height velocity occurs generally is not affected by regular training, nor is the rate of skeletal maturation. But the data concerning the influence of regular training on indexes of sexual maturation are not as clear. Although some data suggest that menarche (the initial onset of menstruation) is delayed in highly trained girls, these data are confounded by a number of factors that generally have not been properly controlled in each study's analysis. Menarche is discussed in chapter 18.

In Closing

In this chapter, we have discussed children and young athletes. We have seen how children gain more control of movements as their body systems grow and develop. We have seen how their developing systems can sometimes limit performance capacities and how training can improve children's performances.

We have seen that, in general, the ability to perform increases as children approach physical maturity. But as people move beyond the point of physical maturity, their physiological functioning begins to decline. Having considered the developmental process, we are now ready to consider the aging process. How is performance affected as we move beyond our physiological prime? This is our focus in the next chapter as we turn our attention to aging and the older athlete.

KEY TERMS

adolescence	development	infancy	myelination	physical maturity
childhood	growth	maturation	ossification	puberty

STUDY QUESTIONS

1. Explain the concepts of growth, development, and maturation. How do they differ?
2. At what ages do height and weight reach their peak rate of growth in males and in females?
3. What typical changes occur in fat cells with growth and development?
4. How does pulmonary function change with growth?
5. What changes occur in heart rate and stroke volume for a fixed rate of work as a child grows? What factors explain these changes? What changes in these two variables occur with aerobic training?
6. What changes occur in cardiac output for a fixed rate of work as a child grows? What factors explain these changes? What changes occur with aerobic training?
7. What changes occur in maximal heart rate as a child grows?
8. What physiological variables account for the $\dot{V}O_{2max}$ increase from age 6 to 20?
9. How dangerous is resistance training for children? What advice would you give to children if they wanted to improve their strength? Can they improve strength, and if so, how does this occur?
10. What happens to aerobic capacity as a prepubescent child trains aerobically?
11. What happens to anaerobic capacity as a prepubescent child trains anaerobically?
12. How do children differ from adults with respect to thermoregulation?
13. How do physical activity and regular training affect the growth and maturation processes?

STUDY GUIDE ACTIVITIES

In addition to the acivities listed in the chapter opening outline on page 383, two other activities are available in the online study guide, located at

www.HumanKinetics.com/PhysiologyOfSportAndExercise.

The chapter **SUMMARY** reviews the key concepts covered in the chapter, and the end-of-chapter **QUIZ** tests your understanding of the material covered in the chapter.

CHAPTER 17

Aging in Sport and Exercise

In this chapter and in the online study guide

Few athletes continue to compete against younger opponents at a national level into middle and old age. One exception was Clarence DeMar, who won his seventh Boston Marathon at age 42, placed 7th at age 50, and was 78th in a field of 153 runners at age 65. In all, he ran more than 1,000 distance races, including more than 100 marathons between 1909 and 1957, a period when exercising or engaging in competition as an older adult was not popular. His performances at the Boston Marathon alone spanned 48 years, from age 20 to 68. DeMar's last race in 1957, at age 68, was 15 km (9.3 mi), which he ran despite advanced intestinal cancer and a colostomy. His best time for the Boston Marathon was 2:29:42 at age 36. Thereafter, his time gradually slowed to 3:58:37 at age 66.

The number of men and women over age 50 years who exercise regularly or participate in competitive sport has increased dramatically over the past 30 years. Many of these older competitors, often termed masters or senior athletes, engage in competition for fun, general recreation, and fitness, while others train with the same enthusiasm and intensity as Olympians. Opportunities are now available for older athletes to compete in activities ranging from marathon running to powerlifting. The success achieved and the performance records set by many older athletes are phenomenal. However, although these older athletes exhibit strength and endurance capacities that are far greater than those of untrained people their age, even the most highly trained older athlete experiences a decline in performance after the fourth or fifth decade of life.

In modern societies, the level of voluntary physical activity begins to decline soon after people reach physical maturity. Technology has made virtually every aspect of life less physically demanding. Voluntary participation in strenuous physical activity on a regular basis is an unusual pattern of behavior that is not observed in most aging laboratory animals. Studies have shown that humans and other animals tend to decrease their physical activity as they grow older. As shown in figure 17.1, rats that were allowed to eat ad libitum ran an average of more than 4,000 m (4,374 yd) per week in the early months of life but covered less than 1,000 m (1,094 yd) per week during their final months.

Thus, older men and women who choose to participate in competitive sports or to train exhaustively do not follow natural human or animal behavior patterns. Why do some older individuals choose to remain physically active when the natural tendency is to become sedentary? The psychological factors that motivate these older athletes to compete are not clearly defined, but their goals probably do not differ substantially from those of their younger counterparts.

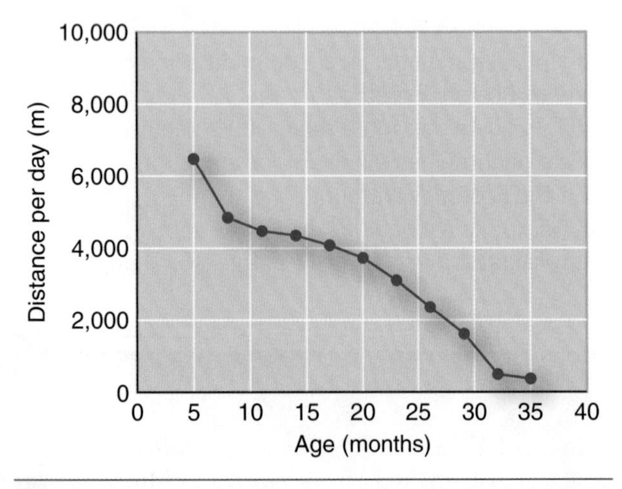

Figure 17.1 Voluntary running activity in rats throughout life.

Adapted, by permission, from J.O. Holloszy, 1997, "Mortality rate and longevity of food-restricted exercising male rats: A reevaluation," *Journal of Applied Physiology* 82: 399-403.

Considering the importance of exercise for maintaining muscular and cardiorespiratory fitness, it is not surprising that inactivity can lead to deterioration of one's capacity for strenuous effort. Because of this, it is difficult to distinguish between the effects of aging and those of the reduced activity that often accompanies aging. Researchers in aging most commonly use cross-sectional designs, but these have some important limitations compared to longitudinal study designs. For example, historical changes in medical care, diet and exercise, and other lifestyle variables may affect age cohorts differently. Selective mortality, that is, the fact that the subject population consists of the survivors of a cohort that has already experienced some degree of mortality, is an issue as well. Finally, when all but apparently healthy older individuals are excluded from exercise studies, it is difficult to apply research findings to the larger aged population with underlying disease, medication use, or both. Therefore, estimating the magnitude of age-related changes in physiological function often depends on the study design, as well as the population being tested.

HEIGHT, WEIGHT, AND BODY COMPOSITION

As we age, we tend to lose height and gain weight, as illustrated in figure 17.2.[33] The reduction in height generally starts at about 35 to 40 years of age and is primarily attributable to compression of the intervertebral disks and poor posture early in aging. At about age 40 to 50 years in women, and 50 to 60 years in men, osteoporosis becomes a factor. **Osteoporosis** refers to a severe loss of bone mass with deterioration of the microarchitecture of bone, leading to increased risk of bone fracture (see chapter 18). Poor diet and exercise habits throughout the life span contribute to the development of osteoporosis in both men and women, while decreased estrogen levels after menopause appear to be responsible for the greater rate of bone loss in women. A gain in weight typically occurs between age 25 and 45 and is attributable to both a decrease in physical activity levels and excess caloric intake. Beyond the age of 45, weight stabilizes for about 10 to 15 years and then decreases as the body loses bone calcium and muscle mass. Many people over 65 to 70 years of age tend to lose their appetite and

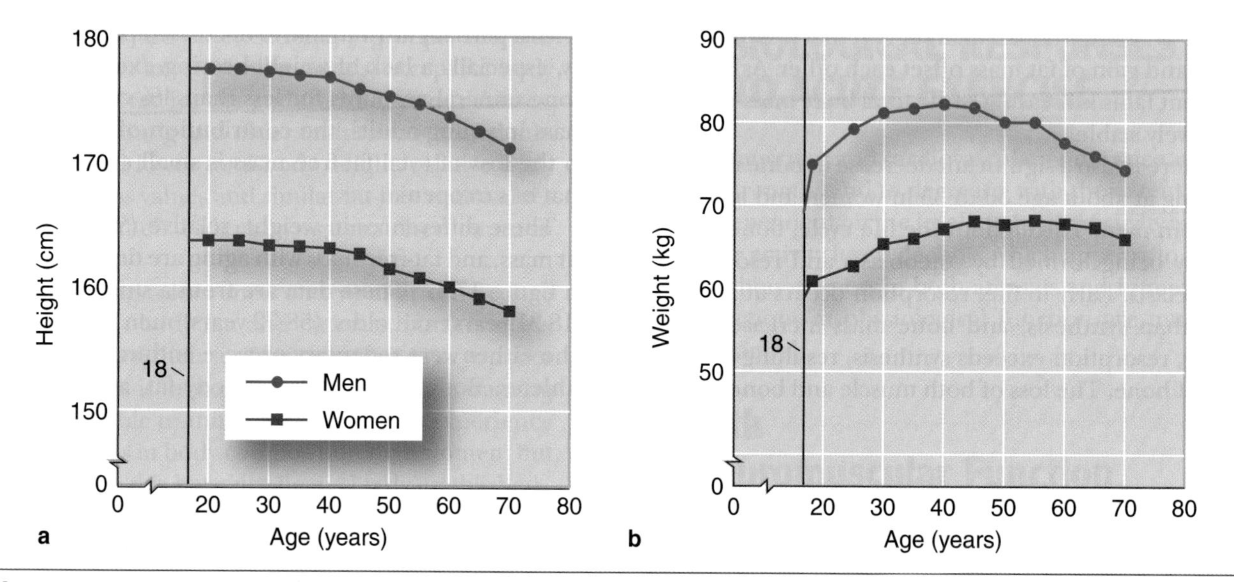

Figure 17.2 Changes in *(a)* body height and *(b)* weight in men and women up to 70 years of age.

Reprinted, by permission, from W.W. Spirduso, 1985, *Physical Dimensions of Aging* (Champaign, IL: Human Kinetics), 59.

thus don't consume sufficient calories to maintain body weight. An active lifestyle, however, tends to help better regulate appetite so that caloric intake more closely approximates caloric expenditure, thereby maintaining weight and preventing frailty in old age.

Beginning at about 20 years of age, as we get older we tend to gain fat. This is largely attributable to three factors: diet, physical inactivity, and reduced ability to mobilize fat stores. As one might anticipate, the body fat content of physically active older people, including older athletes, is significantly lower than that of age-matched sedentary people. However, older athletes have substantially more body fat than younger competitors.

Fat-free mass decreases progressively in both men and women beginning at about the age of 40. This results primarily from decreased muscle and bone mass, with muscle having the greatest effect because it constitutes about 50% of the fat-free mass. **Sarcopenia** is the term used to describe the loss of muscle mass associated with the aging process. **Osteopenia** is a companion term used to describe the less severe loss in bone mass with aging than is seen in osteoporosis. Figure 17.3 illustrates the changes in muscle mass with aging in a cross-sectional study of 468 men and women, aged 18 to 88 years.[18] There is almost no decline in muscle mass until about age 40, at which time the rate of decline increases, with a greater decline in men than women. Obviously, a decline in activity level is a major cause of this decline in muscle mass with aging, but there are other factors. It is now known that the rate of muscle protein synthesis is reduced as we age while muscle protein breakdown rate is unchanged, leading to negative nitrogen balance and net loss of muscle. Muscle protein synthesis rate in 60- to 80-year-olds is about 30% lower than in a 20-year-old. This reduction in muscle protein synthesis rate in older people is likely associated with declines in growth hormone and insulin-like growth

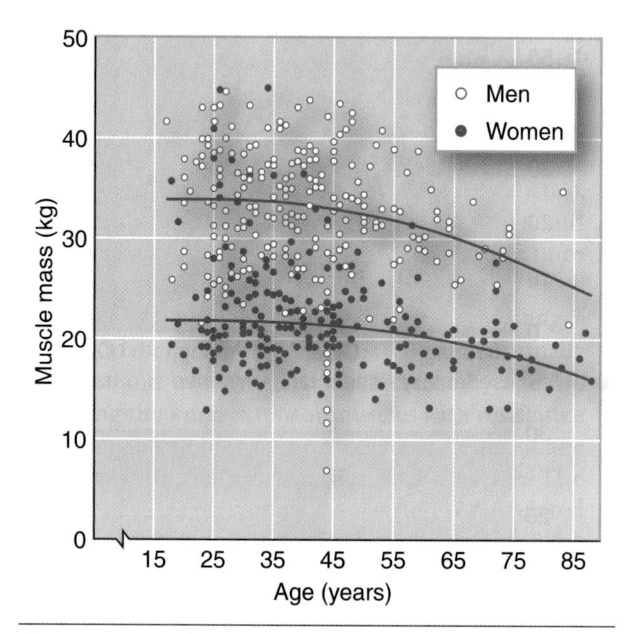

Figure 17.3 Changes in muscle mass with aging in 468 men and women 18 to 88 years of age. The rate of decline is greater in men than in women and is steeper after about 45 years of age.

Adapted, by permission, from I. Janssen et al., 2000, "Skeletal muscle mass and distribution in 468 men and women aged 18-88 yr.," *Journal of Applied Physiology* 89: 81-88.

Figure 17.9 Cross-sectional and longitudinal declines in $\dot{V}O_{2max}$ with age in a group of 86 male and 49 female masters endurance athletes.

Adapted from S.A. Hawkins et al., 2001, "A longitudinal assessment of change in $\dot{V}O_{2max}$ and maximal heart rate in master athletes," *Medicine and Science in Sports and Exercise* 33: 1744-1750.

TABLE 17.2 Changes in Aerobic Capacity and Maximal Heart Rates With Aging in a Group of 10 Highly Trained Masters Distance Runners

		$\dot{V}O_{2max}$		
Age (years)	Weight (kg)	(L/min)	(ml · kg^{-1} · min^{-1})	HR_{max} (beats/min)
21.3 (±1.6)	63.9 (±2.2)	4.41 (±0.09)	69.0 (±1.4)	189 (±6)
46.3 (±1.3)	66.0 (±0.6)	4.25 (±0.05)	64.3 (±0.8)	180 (±6)

Note. Values are mean ± *SE*.

the 25-year period,[38] as shown in table 17.2. Although their maximal oxygen uptake decreased from 69.0 to 64.3 ml · kg^{-1} · min^{-1}, this is a decrease of only 0.19 ml · kg^{-1} · min^{-1} per year or 0.3% per year, and most of that change was attributable to a 2.1 kg (4.6 lb) increase in body weight.

This rate of decrease in these older runners' $\dot{V}O_{2max}$ values is significantly less than that of either sedentary people or those who fitness train at levels and intensities below those of these older runners. One of these runners performed a 4 min 11 s mile and a 2:29 marathon in 1992 at the age of 46! Both of these performances were significantly faster than his best in 1966. Similar findings have been reported for other athletes who continue to train with the same relative intensity and volume as they did in college.

Are these performances exceptions to the natural rules of aging? Can other athletes reduce the effects of aging on their endurance by continuing to train intensely? Much depends on the training adaptability of the individual athlete, a factor that might be determined as much by heredity as by training regimen.

> **In focus**
>
> It is often difficult to differentiate between the results of biological aging and physical inactivity. A natural deterioration in physiological function occurs with aging, but this is compounded by the fact that most people also become more sedentary as they age.

The effects of aging and training on $\dot{V}O_{2max}$ in men are summarized in figure 17.10. Although the number of studies on women is much smaller, a similar trend would be expected. Note that although hard training reduces the normal aging-related decline in $\dot{V}O_{2max}$, aerobic capacity still declines. Thus, it appears that highly intense training has a slowing effect on the rate of loss in aerobic capacity during the early and middle years of adult life (e.g., 30-50 years) but less effect after 50 years of age.

In summary, $\dot{V}O_{2max}$ declines with age, and the rate of decline is approximately 1% per year. Many factors influence this rate of decline, including the following:

ing various sport activities. Here we look briefly environmental stresses and then consider the issu of longevity, injury, and risk of death resulting fro exercise and sport.

Environmental Stress

Because a variety of physiological control process become less effective with aging, we can logically assu that older people are less tolerant of environmen stress than their younger counterparts. As we have se elsewhere in this chapter, it is difficult to determine t separate effects of aging and physical fitness. In the f lowing discussion, we compare the responses of youn and older adults to exercise in the heat and comme on cold stress and altitude exposure in older athlet

Exposure to Heat

Exposure to heat stress presents a problem for old people. A preponderance of deaths during envir mental heat waves occurs in people over the age of The rate of metabolic heat production is related to t absolute exercise intensity, while heat loss mechanis are related to the relative exercise intensity, so mat ing young and older subjects for $\dot{V}O_{2max}$ is importa When subjects are matched for body composition a $\dot{V}O_{2max}$, there is no difference in core temperatu during exercise in the heat. However, when older su jects with a normal $\dot{V}O_{2max}$ for their age are compar to young subjects, they have a higher core temperatu (see figure 17.14).[20]

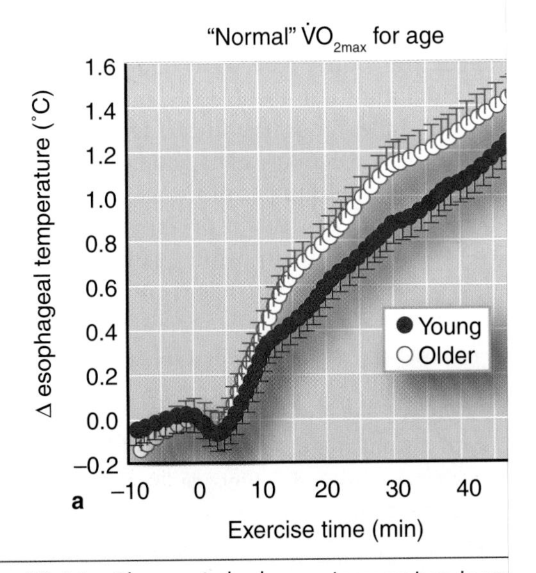

Figure 17.14 Changes in body core temperature in res exercising in a warm environment. When subjects in each temperature rises more sharply in the older individuals. H $\dot{V}O_{2max}$, the difference in core temperature disappears, sug determining this response.

Adapted, by permission, from W.L. Kenney, 1977, "Thermoregulation at 41-77.

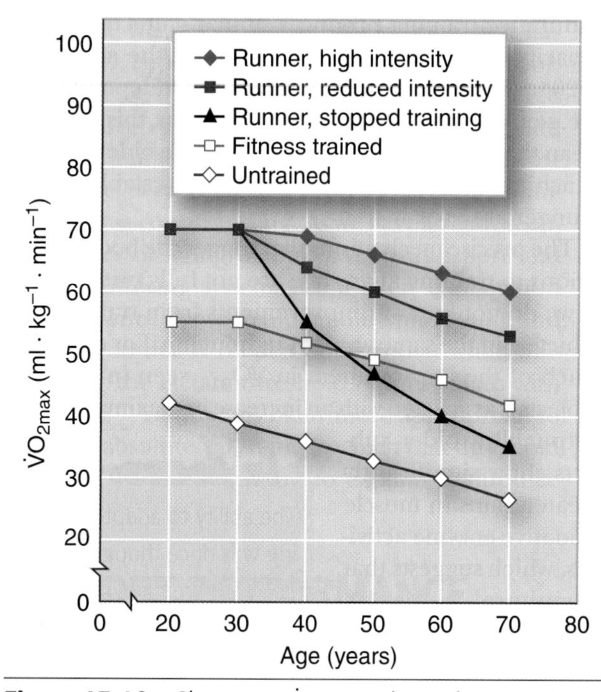

Figure 17.10 Changes in $\dot{V}O_{2max}$ with age for trained and untrained men.

- Genetics
- General activity level
- Intensity of training
- Volume of training
- Increased body weight and body fat mass, and decreased fat-free mass
- Age range, with older individuals experiencing greater declines

There is not universal agreement on which of these factors are most important.

Lactate Threshold

Few studies have addressed the changes in lactate threshold, or anaerobic threshold derived from ventilatory variables, with aging. Lactate threshold was determined in a cross-sectional study of a group of masters endurance runners, 40 to 70+ years of age (111 men and 57 women).[39] Lactate threshold expressed as a percentage of $\dot{V}O_{2max}$ (LT-% $\dot{V}O_{2max}$) provides the best marker relative to endurance running performance in individuals with similar $\dot{V}O_{2max}$ values. Interestingly, LT-% $\dot{V}O_{2max}$ did not differ between men and women, but it did increase with age. Another study reported similar results in 152 untrained men and 146 untrained women.[28] However, in both studies $\dot{V}O_{2max}$ was lower in the older groups, which helps explain the increase in LT-% $\dot{V}O_{2max}$. When the LT at the absolute $\dot{V}O_2$ is compared between age-groups, LT declined with age. This illustrates the concept of differential aging, in which different physiological systems age at different rates. In this case, the rate of aging of the oxygen transport system is different from the rate of aging of the metabolic pathways responsible for lactic acid production and clearance.

In review

- Aerobic capacity generally decreases by about 10% per decade or 1% per year in relatively sedentary men and women.
- Similar results have been obtained for highly trained endurance athletes, although there is a much wider variation in the results of different studies. Also, if athletes begin with a higher $\dot{V}O_{2max}$ and $\dot{V}O_{2max}$ declines at the same rate, it will remain higher than that of sedentary people of the same age.
- Studies of older athletes and less active people of the same age-group indicate that the decrease in $\dot{V}O_{2max}$ is not strictly a function of age. Athletes who continue to train have significantly smaller decreases in $\dot{V}O_{2max}$ as they age, particularly if they train at a high intensity.
- Lactate threshold, expressed as a percentage of $\dot{V}O_{2max}$, increases with aging, but decreases when expressed in relation to the absolute $\dot{V}O_2$ at which it occurs.

PHYSIOLOGICAL ADAPTATIONS TO EXERCISE TRAINING

Despite the decrements in body composition and exercise performance associated with aging, well-trained middle-aged and older athletes are capable of exceptional performances. Furthermore, those who train for general fitness appear to experience gains in muscular strength and endurance similar to those of young adults. In fact, four to six months of aerobic training or two to three months of resistance training in previously sedentary older individuals can restore $\dot{V}O_{2max}$ and maximal strength levels to those seen in individuals 20 years younger.

Strength

As with most other physiological functions, the loss of strength with aging is likely the result of a combination of the natural aging process and reduced physical activity that produces a decline in muscle

Swimming Performance

A retrospective study of freestyle performan U.S. Masters swimming championships betw and 1995 revealed that both men's and wo formances in the 1500 m declined steadily fr to about 70 years, after which swimming tin at a faster rate.[35] However, the rate and ma the declines in both 50 m and 1500 m per with age were found to be greater for wome men.

As shown in figure 17.12, U.S. Master records in the 100 m freestyle decrease by ab year for both men and women from age 25 t 75. Because success in this sport depends on as on strength and endurance, some U.S. Ma mers have achieved their personal best per at 45 to 50 years of age.

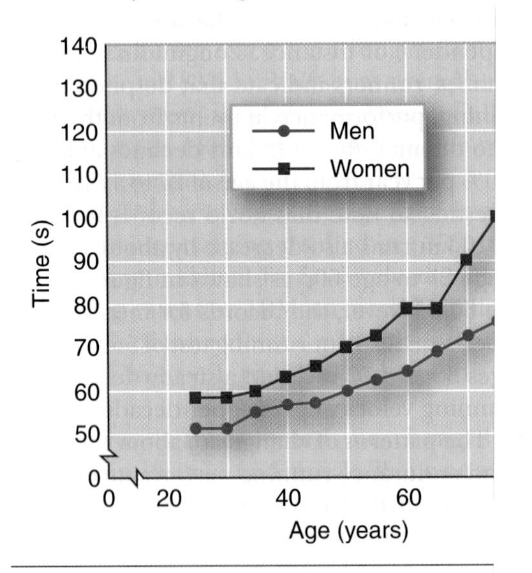

Figure 17.12 Change with age in 100 m free swimming world records among masters-level c

Cycling Performance

As with other strength and endurance spo setting cycling performances are general in the age range of 25 to 35. Male and fem records for 40 km (24.9 mi) races decrease same rate with age, an average of 20 s (app 0.6%) per year. The U.S. national cycling re km (12.4 mi) show a similar pattern for bo women. For this distance, speed decreases s (approximately 0.7%) per year from age age 65.

Weightlifting

In general, maximal muscle strength i between the ages of 25 and 35. Beyond tha as shown in figure 17.13, men's records fo

splanchnic blood flow decreased ~200 ml more than before training, which reduced cardiovascular strain.

Exposure to Cold and to Altitude

In contrast to heat exposure, exercise in the cold typically poses less of a health risk. Due to declining aerobic fitness and loss of muscle mass, older individuals do have a reduced ability to generate metabolic heat. Also, the ability of the cutaneous vasculature to constrict is impaired with aging, which may increase heat loss. As a result of these changes, older individuals may fail to appropriately maintain core temperature during cold stress. However, people can easily offset these decrements by wearing clothing that is appropriate for the environmental conditions and activity level. By performing so-called behavioral thermoregulation, older athletes can offset the decrements in physiological thermoregulation and continue to exercise safely in cold environments.

During exposure to high altitude, there is little reason to expect older athletes to respond differently than their younger counterparts. Unfortunately, data are lacking regarding aging and the rate and magnitude of acclimatization to altitude. Likewise, it is unclear whether aging per se increases the incidence of any of the altitude illnesses. We can expect the performance of an older athlete at altitude to be similar to that of a younger athlete of comparable physical fitness.

In review

- The impaired ability of older individuals to tolerate exercise in the heat is due to reduced aerobic capacity and impaired cardiovascular adaptations rather than a direct effect of aging on thermoregulatory control or sweating.
- Regular exercise training can increase skin blood flow and sweating rate and improve the redistribution of cardiac output in older individuals as well as young men and women.
- Older people generally have an impaired ability to tolerate cold, but they can compensate by wearing appropriate clothing.
- Adaptation to altitude appears to be independent of age.

Longevity and Risk of Injury and Death

Regular physical activity is an important contributor to good health. Does training throughout adulthood affect longevity? Because the aging rate in rats is more rapid than in humans, they have been used as subjects in studies conducted to determine the influence of chronic exercise (training) on **longevity** (the length of one's life). A study by Goodrick[8] demonstrated that rats that exercised freely lived about 15% longer than sedentary rats. But an investigation at Washington University in St. Louis showed no significant increase in the life span of rats that voluntarily ran on an exercise wheel.[15] More of the active rats lived to old age, but on the average, they still died at the same age as their sedentary counterparts. It is interesting that rats that had restricted food intake and maintained a lower body weight lived 10% longer than the freely eating, sedentary rats.

Of course, we cannot directly apply these findings to humans, but these results raise some interesting questions that might be relevant to our health and longevity. Although it is true that an endurance exercise program can reduce a number of the risk factors associated with cardiovascular disease, only limited information supports the contention that people will live longer if they exercise regularly. Data collected from the alumni at Harvard University and the University of Pennsylvania and from participants at the Aerobic Center in Dallas suggest that there is a decrease in mortality rate and a small increase in longevity (about two years) among people who remain physically active throughout life. Therefore, regular physical activity can increase the active life span, that is, the number of years living independently and free of disability. This is sometimes referred to as compression of mortality or rectangularization of the survival curve. Perhaps future longitudinal studies will shed more light on the relationship between lifelong exercise and longevity.

What about the risk of injury and death from exercise as we get older? Studies show that as people get older they are at a greater risk for injuries involving tendons, cartilage, and bone. The most common orthopedic injuries include rotator cuff tears, quadriceps tendon ruptures, Achilles tendon ruptures, degenerative meniscus tears, focal articular cartilage defects and injuries, and stress fractures. Furthermore, when injuries occur, the healing process is usually prolonged and complete recovery can take up to a full year.[25] On the other hand, increasing the strength and endurance of elderly

In review

- An active lifestyle appears to be associated with a small increase in longevity. Just as important, an active lifestyle leads to a higher quality of life!
- There is an increased risk of injury from exercise as people age, and injuries tend to be slower to heal.
- The risk of death during exercise is not increased in those who are regularly active but is increased in those who seldom exercise.

people reduces their risk of falls and related injury, so the benefits of maintaining a regular exercise program with aging outweigh the potential risks.

The risk of death during exercise appears to be no higher in the older athlete compared with the younger and middle-aged athlete. However, the risk in older people who are not regularly active appears to be increased, possibly due to undiagnosed or subclinical disease processes.[32] Importantly, an active lifestyle does, in fact, reduce the risk of death from many chronic diseases, something that we discuss in chapters 19 through 21.

In Closing

In this chapter we examined the effects of aging on physical performance. We evaluated changes in cardiorespiratory endurance and strength with age. We considered the effect of aging on body composition, which we know can affect performance. And yet, in the course of our discussion, it became clear that much of the change that occurs with aging is to a great extent attributable to the inactivity that often accompanies aging. When older people participate in training, most of the changes associated with aging are lessened and the resulting degree of change is similar to that seen in young and middle-aged adults. Thus, we have dispelled many of the myths about the capacity for physical activity of older people.

In the next chapter, we turn our attention to females, who as a group are often considered less capable of physical activity than males. We consider the physiology of girls and women, the impact of this physiology on athletic ability, how performances of female athletes compare with those of male athletes, and special issues associated with being female.

KEY TERMS

cardiovascular deconditioning
forced expiratory volume in 1 s ($FEV_{1.0}$)
longevity
maximal expiratory ventilation ($\dot{V}_{Emax}$)
osteopenia
osteoporosis
peripheral blood flow
residual volume (RV)
sarcopenia
total lung capacity (TLC)
vital capacity (VC)

STUDY QUESTIONS

1. What changes occur in height, weight, and body composition with aging? What accounts for these changes? How do these changes affect maximal oxygen uptake?
2. What changes occur in muscle with aging? How do they affect strength and athletic performance?
3. Describe the changes in HR_{max} with age. How does training alter this relationship?
4. How does aging affect maximal stroke volume and maximal cardiac output? What mechanisms can potentially explain these changes?
5. How does the respiratory system change with aging? What happens to vital capacity, $FEV_{1.0}$, residual volume, total lung capacity, and the ratio RV/TLC?
6. $\dot{V}O_{2max}$ declines with age across the entire population. Describe the physiological mechanisms that account for this decline. How do trained older individuals maintain a relatively high $\dot{V}O_{2max}$?
7. How does age affect anaerobic function?
8. Differentiate between biological aging and physical inactivity.
9. What influence do aging and training have on body composition?
10. Describe the trainability of the older individual for both strength and aerobic endurance.
11. Describe the changes in strength and endurance performance records with aging.
12. What concerns should we have about older people exercising in hot and cold environments or at altitude?
13. Describe the risk of injury and death associated with exercise in the elderly.
14. What is the effect of exercise on longevity?

STUDY GUIDE ACTIVITIES

In addition to the activities listed in the chapter opening outline on page 403, two other activities are available in the online study guide, located at

www.HumanKinetics.com/PhysiologyOfSportAndExercise.

The chapter **SUMMARY** reviews the key concepts covered in the chapter, and the end-of-chapter **QUIZ** tests your understanding of the material covered in the chapter.

CHAPTER 18

Sex Differences in Sport and Exercise

In this chapter and in the online study guide

Girls and women were prohibited from running any race longer than 800 m until the 1960s. They were also barred from official participation in the marathon until 1970. Both of these restrictions resulted from a misconception that women were physiologically unsuited for endurance activity. Yet at the 1984 Los Angeles Olympic Games, American runner Joan Benoit won the gold medal in the first-ever Olympic marathon for women with a time of 2:24:52. Her time would have won 11 of the previous 20 men's Olympic marathons!

Another myth that is falling by the wayside is that pregnant women should not exercise. The following story appeared in the *Austin American Statesman* on June 17, 1995. "When Sue Olsen lines up today for Grandma's Marathon in St. Paul, Minn., along the North Shore of Lake Superior, she will be precisely 16 days short of the due date for her first child. 'I get a lot of advice both ways,' Olsen, 38, said with a laugh. 'There are people who think I'm crazy and shouldn't be out there, and there are a lot of supportive people.' . . . Olsen's husband will drive parallel to the course with a cellular phone in his car, prepared to whisk her to the hospital if the situation arises." Although most would question the wisdom of this, the point is that Olsen was able to complete the marathon in 4 h. The following weekend she completed a 24 h race, delivering a baby boy the next day. In 2005, Olsen's son, John "Miles" Olsen, completed 32.7 mi in a 24 h run, just a few weeks before his 10th birthday. His mother completed 119.6 mi (Doug Grow, *Star Tribune,* Minneapolis, MN, June 7, 2005).

In the recent past, young girls typically were discouraged from participating in vigorous physical activity, while young boys climbed trees, raced against each other, and participated in various sports. The underlying notion was that boys were meant to be active and athletic but girls were weaker and less well suited to physical activity and competition. Physical education classes furthered this idea by having girls exercise differently than boys—by running shorter distances and performing modified push-ups. Less physical activity was expected of girls; and as they progressed through school, most girls could not compete on an equal basis with boys of their own age, even if given the opportunity. In athletics, girls and women were not allowed to run in long-distance races, and basketball was limited to half-court, with each team having only offensive or defensive players.

Now more athletic activities are accessible to girls and women than in the past, and the results have been amazing. Their athletic accomplishments parallel those of boys and men, with performance differences of 15% or less for most sports and events. This is illustrated in table 18.1, where men's and women's world records (2006) are compared for representative events in both track and field and swimming. Do these performance differences represent true biological differences, or are there other factors we need to consider? The focus of this chapter is to determine the extent to which biological differences between females and males affect performance capacity.

TABLE 18.1 Selected Men's and Women's World Records Through 2006

Event	Men	Women	Difference
Track and field			
100 m	9.77 s	10.49 s	7%
1500 m	3:26 min:s	3:50 min:s	12%
10,000 m	26:17 min:s	29:31 min:s	12%
High jump	2.45 m	2.09 m	15%
Long jump	8.95 m	7.52 m	16%
Swimming (freestyle)			
100 m	47.8 s	53.5 s	12%
400 m	3:40 min:s	4:03 min:s	10%
1500 m	14:34 min:s	15:52 min:s	9%

BODY SIZE AND COMPOSITION

Body size and composition are similar in boys and girls during early childhood. During late childhood, as we saw in chapter 16, girls begin to accumulate more fat than boys, and starting in early adolescence boys begin to increase their fat-free mass at a much higher rate than girls (see figure 16.3 on p. 387).

> **In focus**
>
> Major differences in body size and composition between girls and boys do not start to appear until late childhood and early adolescence.

These body composition differences between the sexes occur primarily because of endocrine changes with development. Before puberty, the anterior pituitary gland secretes very small amounts of the gonadotropic hormones: follicle-stimulating hormone (FSH) and luteinizing hormone (LH). These hormones stimulate the gonads (ovaries and testes). During puberty, however, the anterior pituitary begins to secrete significantly greater amounts of both of these hormones. In females, when sufficient quantities of FSH and LH are secreted, the ovaries develop and estrogen secretion begins. In males, these same hormones trigger development of the testes and, in turn, testosterone secretion. Figure 18.1 illustrates these changes in estrogen (estradiol—the most potent form of estrogen) and testosterone from the beginning of puberty (S1) to the end of puberty (S5). **Testosterone** increases bone formation, which leads to larger bones, and increases protein synthesis, which leads to increased muscle mass. As a result, adolescent males are larger and more muscular than females, and these characteristics continue into adulthood. At full maturity, men not only have a greater muscle mass, but the distribution of the muscle mass differs from that of women. Men carry a higher percentage of their muscle mass in the upper body compared with women (42.9% vs. 39.7%).[24] Testosterone also stimulates erythropoietin production by the kidneys, which leads to increased red blood cell production, as discussed later in this chapter.

Estrogen also has a significant influence on body growth by broadening the pelvis, stimulating breast development, and increasing fat deposition, particularly in the thighs and hips. This increase in fat deposition in the thighs and hips is the result of increased lipoprotein lipase activity in these areas. This enzyme is considered the gatekeeper for storing fat in adipose tissue. **Lipoprotein lipase** is produced in the fat cells (adipocytes) but is bound to the walls of the capillaries, where it exerts its influence on the chylomicrons, which are the major transporters of triglycerides in the blood. When lipoprotein lipase activity in any area of the body is high, chylomicrons are trapped and their triglycerides are hydrolyzed and transported into the adipocytes in that area for storage.[5]

Estrogen also increases the growth rate of bone, allowing the final bone length to be reached within two to four years following the onset of puberty. As a result, females grow very rapidly for the first few years following puberty and then cease to grow. Males have a much longer growth phase, allowing them to attain a greater height. Because of these differences, compared with fully mature males, fully mature females are on average nearly

- 13 cm (5 in.) shorter,
- 14 to 18 kg (30-40 lb) lighter in total weight,
- 18 to 22 kg (40-50 lb) lighter in fat-free mass (FFM),
- 3 to 6 kg (7-13 lb) heavier in fat mass, and
- 6% to 10% higher in relative body fat.

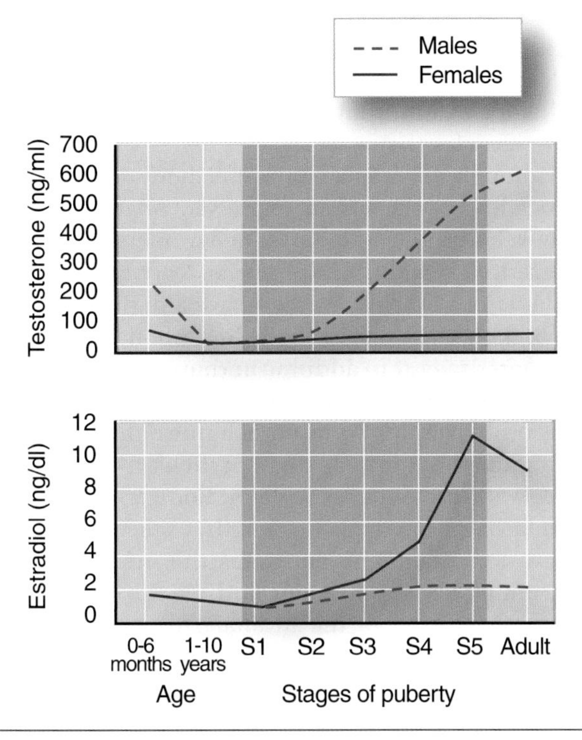

Figure 18.1 Changes in blood concentrations of testosterone and estrogen (estradiol) from birth to adulthood. The symbols S1 through S5 represent stages of puberty based on secondary sex characteristics, with S1 representing the initial stages of puberty and S5 the final stages.

Reprinted, by permission, from R.M. Malina, C. Bouchard, and O. Bar-Or, 2004, *Growth, maturation, and physical activity*, 2nd ed. (Champaign, IL: Human Kinetics), 414.

In review

- Until puberty, girls and boys do not differ significantly in most measurements of body size and composition.
- At puberty, because of the influences of estrogen and testosterone, body composition begins to change markedly.
- Testosterone increases bone formation and protein synthesis, leading to a larger FFM. It also stimulates the production of erythropoietin, which increases red blood cell production.
- Estrogen causes increased fat deposition in females, particularly in the hips and thighs, and an increased rate of bone growth, such that bones in females reach their final length earlier than in males.

Fat Deposition: Why the Hips and Thighs?

Many women are constantly fighting fat deposition on the thighs and hips, but they are usually fighting a losing battle. Lipoprotein lipase activity is very high and lipolytic activity (fat breakdown) is low in the hips and thighs of women compared with women's other fat storage areas and compared with the hips and thighs of men. This results in a rapid storage of fat in women's thighs and hips, and the decreased lipolytic activity makes it difficult for women to lose fat from these areas. During the last trimester of pregnancy and throughout lactation, the activity of lipoprotein lipase decreases and lipolytic activity increases dramatically, which suggests that fat is stored in the hips and thighs for reproductive purposes.[5]

PHYSIOLOGICAL RESPONSES TO ACUTE EXERCISE

When females and males are exposed to an acute bout of exercise, whether an all-out run to exhaustion on the treadmill or a single attempt to lift the maximal weight possible, responses differ between the sexes. Differences between children and adolescent boys and girls were discussed in chapter 16. Here we focus on these differences in adults in the areas of strength and of cardiovascular, respiratory, and metabolic responses to exercise.

Strength

In terms of strength, women have traditionally been regarded as the weaker sex. In fact, women are approximately 40% to 60% weaker than men in upper body strength but only 25% to 30% weaker in lower body strength. However, because of the considerable size difference between the average man and the average woman, it is more appropriate to express strength either relative to body weight (absolute strength ÷ body weight) or relative to FFM, as a reflection of muscle mass (absolute strength ÷ FFM). When lower body strength is expressed relative to body weight, women are still 5% to 15% weaker than men, but when it is expressed relative to FFM, this difference disappears. This suggests that the innate qualities of muscle and its mechanisms of motor control are similar for women and men, a fact that was confirmed by computed tomography (CT) scans of the upper arms and thighs of physical education majors of both sexes and of male bodybuilders.[41] Computed tomography scans allow the actual estimation of muscle mass. Although both groups of men had much greater absolute levels of strength than the women, no differences between the groups were found when strength was expressed per unit of muscle cross-sectional area for both knee extensor muscles (figure 18.2*a*) and elbow flexor muscles (figure 18.2*b*).

Although differences in upper body strength are reduced somewhat when expressed relative to total body weight and FFM in untrained women, substantial differences remain. There are at least two possible explanations for this. Women have a higher percentage of their muscle mass in the lower body when compared with men.[24] In addition and probably related to this muscle mass distribution, women use the muscle mass of their lower bodies much more than they use their upper body muscle mass, particularly when compared with use patterns in men. Some average-sized women have remarkable strength, exceeding that of an average man. This indicates the importance of neuromuscular recruitment and synchronization of motor unit firing in the ultimate determination of strength (chapter 3).

Muscle biopsies have become more common among female athletes, allowing fiber type comparisons with male athletes in the same sport or event. From these

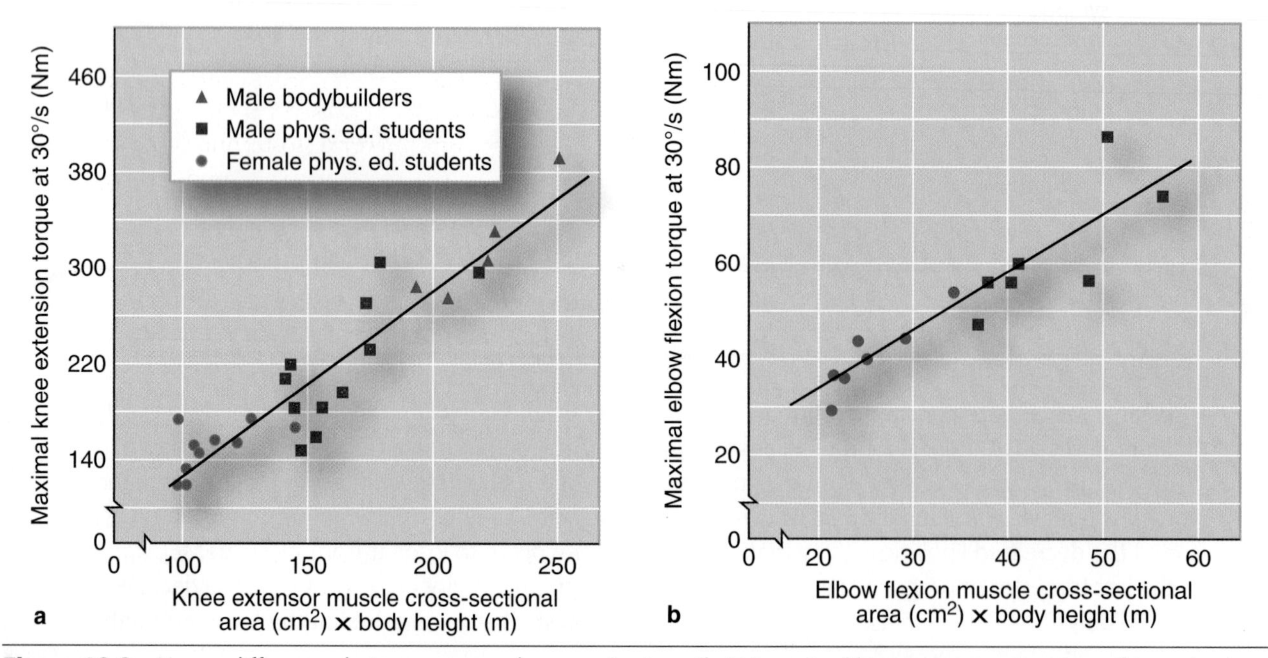

Figure 18.2 No sex differences between men and women in strength [*(a)* maximal knee extension torque or *(b)* maximal elbow flexion torque] are seen when strength is expressed per unit of muscle cross-sectional area.

Reprinted, by permission, from P. Schantz et al., 1983, "Muscle fibre type distribution, muscle cross-sectional area and maximal voluntary strength in humans," *Acta Physiologica Scandinavica* 117: 219-226.

data, men and women have similar fiber type distributions, as shown in figure 18.3, although men in one study reached greater extremes (greater than 90% slow twitch [type I] or greater than 90% fast twitch [type II]). As shown in figure 18.3, when the vastus lateralis muscle was biopsied, male distance and sprint runners varied between approximately 15% and 85% type I fiber type distribution, compared with female distance and sprint runners, whose distributions were between approximately 25% and 75%.[39] However, different results were found in two studies, one of elite female[8] and the other of elite male distance runners.[18] In these elite runners, the extremes for percentages of type I fibers were similar (41-96% for women and 50-98% for men), even though the mean values were different: Women had a mean value of 69% type I fibers compared with 79% for the men. The women had much smaller fiber areas for both type I and type II fibers (mean values of less than 4500 μm^2 in women and greater than 8000 μm^2 in men). Despite smaller fiber size in women, capillarization appears to be similar between men and women.[34]

In focus

For the same amount of muscle, there are no differences in strength between the sexes, although women have smaller muscle fiber cross-sectional areas than men and typically less muscle mass than men.

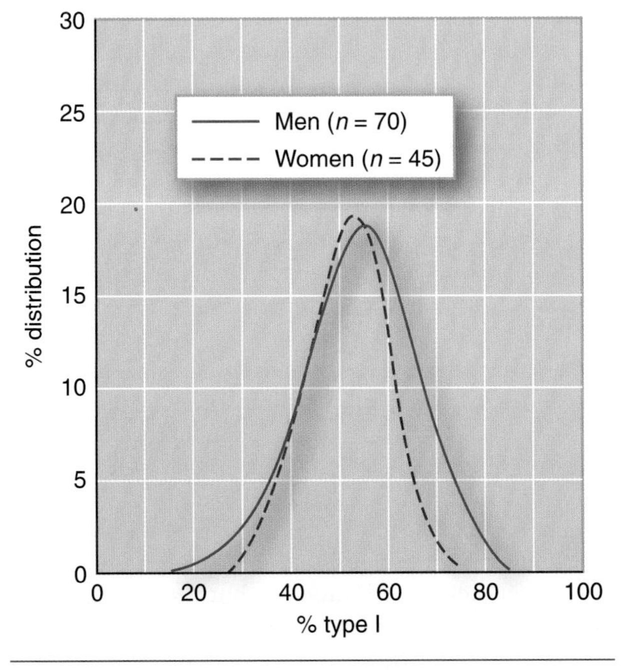

Figure 18.3 Distribution of type I fibers in the vastus lateralis muscle in male and female runners.

Adapted, by permission, from B. Saltin et al., 1977, "Fiber types and metabolic potentials of skeletal muscles in sedentary man and endurance runners," *Annals of the New York Academy of Sciences* 301: 3-29.

Research indicates that women have a greater resistance to fatigue compared with men. Fatigue usually is tested by having subjects maintain a constant force output at a given percentage of their maximal voluntary static action. As an example, women would be able to maintain a constant force output at 50% of their maximal static action for a longer period of time than men at the same 50% of their maximal static action. Men, being stronger, will have to apply a greater absolute amount of force to achieve the same 50% relative force. The reason for this greater resistance to fatigue is not yet known but could be related to the amount of muscle mass recruited and the compression of blood vessels, substrate utilization, muscle fiber type, and neuromuscular activation.[7, 23]

Cardiovascular and Respiratory Function

When placed on a cycle ergometer, where the power output can be precisely controlled independent of body weight (50 W, as an example), women generally have a higher heart rate (HR) response for any absolute level of submaximal exercise. However, maximum heart rate (HR_{max}) is generally the same in both sexes. Stroke volume (SV) is lower in women, but cardiac output ($\dot{Q}$) is usually the same in men and women at any absolute submaximal power output.[35] The higher submaximal HR response in women appears to compensate for the lower SV, allowing a similar $\dot{Q}$ for the same power output, since $\dot{Q} = HR \times SV$. The lower SV results primarily from at least two factors:

- Women have smaller hearts and therefore smaller left ventricles because of their smaller body size and possibly lower testosterone concentrations.
- Women have a smaller blood volume, which also is related to their size (lower FFM).

The average woman may also be less aerobically active and therefore less aerobically conditioned.

When power output is controlled to provide the same relative level of exercise, usually expressed as a fixed percentage of maximal oxygen uptake ($\dot{V}O_{2max}$), women's heart rates are still slightly elevated compared with men's, and their stroke volumes are markedly lower. At 60% $\dot{V}O_{2max}$, for example, a woman's cardiac output, stroke volume, and oxygen consumption are generally less than a man's, and her heart rate is slightly higher. With the exception of HR_{max}, these differences also are seen at maximal levels of exercise.

These relationships between HR, SV, and $\dot{Q}$ for the same absolute power output (50 W) and the same relative power output (60% $\dot{V}O_{2max}$) are illustrated in figure 18.4 on page 428. These data were derived from

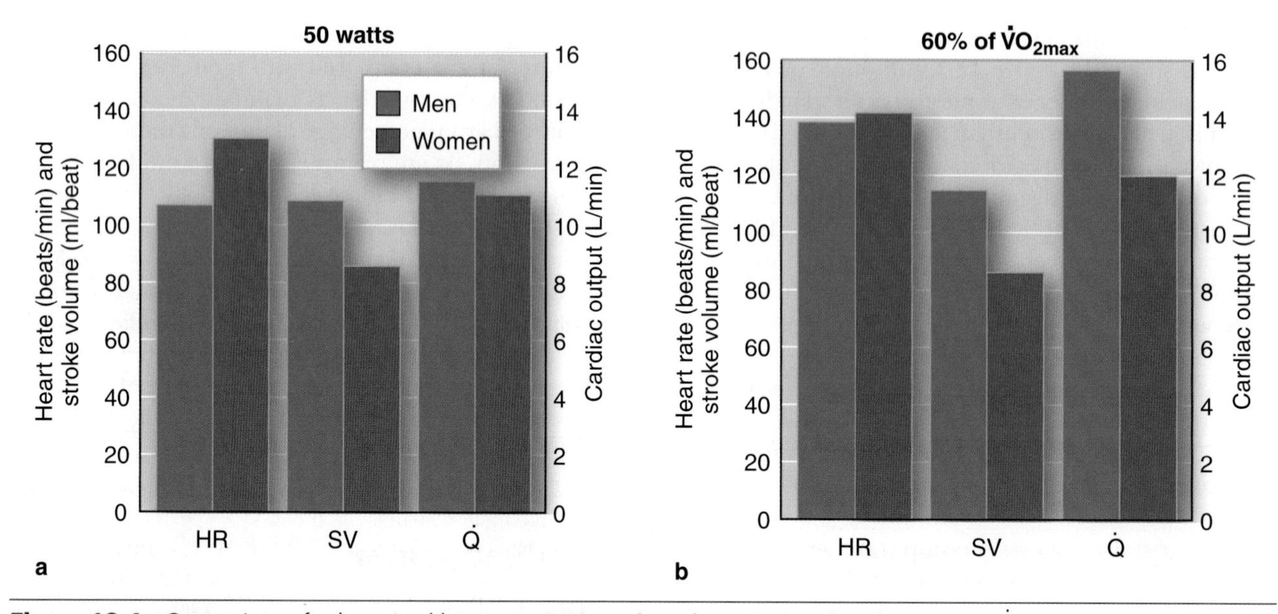

Figure 18.4 Comparison of submaximal heart rate (HR), stroke volume (SV) and cardiac output ($\dot{Q}$) between men and women at *(a)* the same absolute power output (50 W) and *(b)* the same relative power output (60% $\dot{V}O_{2max}$).
Data from Wilmore et al., 2001.

> ***In focus***
>
> At the same submaximal exercise intensity, women generally have cardiac outputs similar to those of men. Thus, their higher heart rates appear to fully compensate for their lower stroke volumes. The lower stroke volume is largely due to a smaller left ventricle and lower blood volume, both the result of women's smaller body size.

the HERITAGE Family Study.[46] Interestingly, when these same relationships are compared in 7- to 9-year-old boys and girls, there are no sex differences.[43]

It should be noted that several early studies, summarized by Åstrand and colleagues,[4] reported $\dot{Q}$ to be higher in women at identical submaximal power outputs, possibly compensating for their lower hemoglobin concentrations. More recent studies, however, have consistently shown no differences between sexes.[20, 35, 46] Apparently, women can compensate for their lower hemoglobin levels with a steeper increase in their arterial–venous oxygen difference, or $(a\text{-}\bar{v})O_2$ difference, for a given power output.[17]

Women also have less potential for increasing their peak $(a\text{-}\bar{v})O_2$ difference. This is likely attributable to their lower hemoglobin content, which results in lower arterial oxygen content and reduced muscle oxidative potential. Lower hemoglobin content is an important contributor to **sex-specific differences** in $\dot{V}O_{2max}$ because less oxygen is delivered to the active muscle for a given volume of blood.

Sex differences in respiratory responses to exercise are largely attributable to body size differences. Breathing frequency during exercise at the same relative power output (e.g., 60% $\dot{V}O_{2max}$) differs little between the sexes. However, at the same absolute power output, women tend to breathe more rapidly than men, probably because when men and women are working at the same absolute power output, women are working at a higher percentage of their $\dot{V}O_{2max}$.

Tidal volume and ventilatory volume are generally smaller in women at the same relative and absolute power outputs, up to and including maximal levels. Most highly trained female athletes have maximal ventilatory volumes below 125 L/min; but highly trained men have maximal values of 150 L/min and higher, some exceeding 250 L/min (figure 18.5). Again, these differences are closely associated with body size.

Metabolic Function

$\dot{V}O_{2max}$ is regarded by most exercise scientists as the single best index of a person's cardiorespiratory endurance capacity. Recall that $\dot{V}O_2$ is the product of cardiac output and $(a\text{-}\bar{v})O_2$ difference. This means that $\dot{V}O_{2max}$ represents that point during exhaustive exercise at which the subject has maximized oxygen delivery and utilization capabilities. The average female tends to reach her peak $\dot{V}O_{2max}$ between ages 12 and 15, but the average male does not reach his peak until ages 17 to 21 (see chapter 16). Beyond puberty, the average woman's $\dot{V}O_{2max}$ is only 70% to 75% of the average man's.

$\dot{V}O_{2max}$ differences between women and men must be interpreted carefully. A classic study published in 1965 showed considerable variability in $\dot{V}O_{2max}$ within each

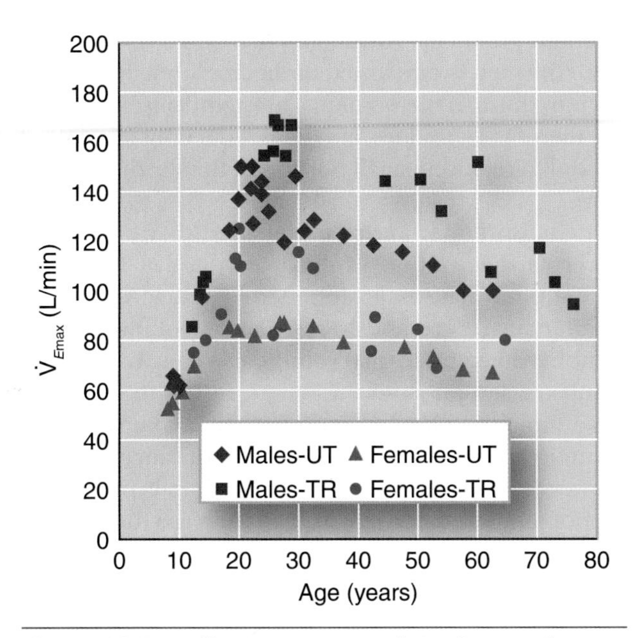

Figure 18.5 Differences in maximal ventilatory volumes with age in untrained (UT) and trained (TR) females and males.

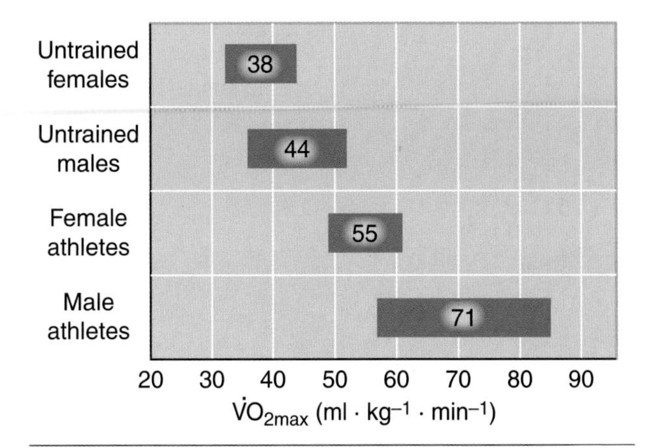

Figure 18.6 Range of $\dot{V}O_{2max}$ values (mean ±2 *SD*) for female nonathletes, male nonathletes, elite female athletes, and elite male athletes. The mean $\dot{V}O_{2max}$ value is presented within each box. This figure illustrates that although differences in the average $\dot{V}O_{2max}$ between groups can be substantial, there can be considerable overlap of one group with another.

Data from Hermansen and Andersen, 1965.

sex and considerable overlap of values between sexes.[22] The study involved a group of women and men 20 to 30 years of age. The investigators divided the group into subgroups:

- Elite female athletes
- Female nonathletes
- Elite male athletes
- Male nonathletes

They compared the subjects' physiological responses to submaximal and maximal exercise. Barbara Drinkwater's[13] further analysis of this 1965 study revealed that 76% of the female nonathletes overlapped 47% of the male nonathletes and that 22% of the female athletes overlapped 7% of the male athletes. The relationships are illustrated in figure 18.6. These data demonstrate the importance of looking beyond mean values to consider both the subjects' levels of physical conditioning and the extent of overlap between the groups being compared.

Although the $\dot{V}O_{2max}$ values of females and males are similar until puberty, comparisons of $\dot{V}O_{2max}$ values of normal nonathletic females and males beyond puberty might not be valid. Such data likely reflect an unfair comparison of relatively sedentary females with relatively active males. Thus, reported differences would reflect the level of conditioning as well as possible sex-specific differences. To overcome this potential problem, investigators began to examine highly trained female and male athletes, with the assumption that the level of training would be similar for both sexes and would allow a more accurate evaluation of true sex-specific differences.

Saltin and Åstrand[38] compared $\dot{V}O_{2max}$ values of female and male athletes from Swedish national teams. In comparable events, the women had 15% to 30% lower $\dot{V}O_{2max}$ values. However, more recent data suggest a smaller difference. The $\dot{V}O_{2max}$ values for a group of elite and good female distance runners are compared in figure 18.7 on page 430 with values for elite male distance runners and values for average, nonathletic women and men.[3, 9, 31, 33, 37, 45] The elite female runners had substantially higher values than untrained men and women. Some women's values were even higher than a few of the elite male runners' values, but when we consider the average for each elite group, the women's values were still 8% to 12% lower than those of the elite male runners.

Several studies have been aimed at scaling $\dot{V}O_{2max}$ values relative to height, weight, FFM, or limb volume in an attempt to more objectively compare women's and men's values. Some of these studies have shown that differences between the sexes disappear when $\dot{V}O_{2max}$ is expressed relative to FFM or active muscle mass, yet others continue to demonstrate differences even when differences in body fat are adjusted for.[12]

In focus

The highest $\dot{V}O_{2max}$ value reported in the literature for a female athlete (a Russian cross-country skier) was 77 ml · kg^{-1} · min^{-1}. The highest value reported for a male athlete (Norwegian cross-country skier) was 94 ml · kg^{-1} · min^{-1}.[4]

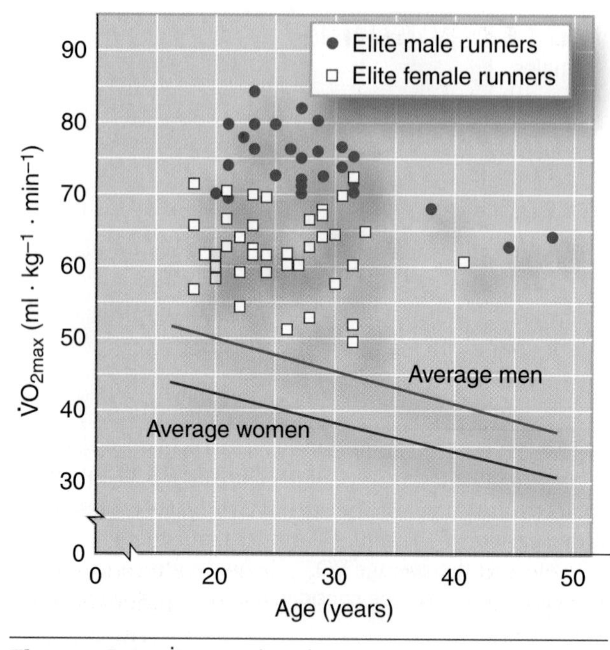

Figure 18.7 $\dot{V}O_{2max}$ values for elite female and male distance runners compared with average values in untrained women and men.

Data from S. Robinson, 1938; I. Åstrand, 1960; D.L. Costill and E. Winrow, 1970; M.L. Pollock, 1977; R.R. Pate, et al., 1987; and J.H. Wilmore and C.H. Brown, 1974.

In one study, researchers used a novel approach to this problem.[11] They examined the responses to submaximal and maximal treadmill runs under various conditions in 10 women and 10 men who regularly engaged in distance running. The women exercised only under normal weight conditions, but the men exercised both at normal weight and with external weight added to their trunks so that the total percentage of excess weight, defined as the men's fat weight plus the added weight, equaled the percentage fat of the matched women. Equalizing the sexes for excess weight reduced the amount of mean sex-specific differences in

- treadmill run time (32%),
- the amount of oxygen required per unit of FFM for running at various submaximal speeds (38%), and
- $\dot{V}O_{2max}$ (65%).

The researchers concluded that women's greater sex-specific essential body fat stores are major determinants of sex-specific differences in metabolic responses to running.

Women have lower hemoglobin levels than men, and this also has been proposed as a factor contributing to their lower $\dot{V}O_{2max}$ values. In one study researchers attempted to equate the hemoglobin concentrations of a group of 10 men and 11 women who were active but not highly trained.[10] An amount of blood was withdrawn from the men to equalize their hemoglobin concentrations to those of the women. This significantly reduced the men's $\dot{V}O_{2max}$ values, but these reductions accounted for only a relatively small portion of the sex differences in $\dot{V}O_{2max}$.

It is also important to understand that a woman's lower cardiac output at maximal rates of work is a limitation to achieving a high $\dot{V}O_{2max}$ value. A woman's smaller heart size and lower plasma volume greatly limit her maximal stroke volume capacity. In fact, the results of several studies have suggested that women have limited ability to increase their maximal stroke volume capacity with high-intensity endurance training. However, more recent studies showed that young, premenopausal women were able to increase their stroke volume with training identically to men. Furthermore, in the untrained state after their plasma volume was artificially increased with a plasma volume expander and after β-blockade (which reduced the heart rate for a given rate of work, allowing more time to fill the left ventricle), women were able to increase their stroke volume to the same extent as untrained men during an acute bout of exercise.[27, 28]

If, instead of looking at $\dot{V}O_{2max}$, we consider submaximal oxygen consumption ($\dot{V}O_2$), little if any difference is found between women and men for the same absolute power output. But remember that at the same absolute submaximal work rate, women usually are working at a higher percentage of their $\dot{V}O_{2max}$. As a result, their blood lactate levels are higher, and lactate threshold occurs at a lower absolute power output. Peak blood lactate values are generally lower in active but untrained women than in active but untrained men. Also, limited data suggest that elite female middle-distance and long-distance runners have peak lactate values that are approximately 45% lower than those of similarly trained elite male runners (8.8 mmol/L vs. 12.9 mmol/L).[31, 33] Such sex differences in peak blood lactate values are unexpected and unexplained.

Lactate threshold appears to be similar between equally trained men and women if expressed in relative (% $\dot{V}O_{2max}$), not absolute, terms. Lactate threshold appears to be closely related to the mode of testing and to the individual's state of training. Thus, sex-specific differences are not expected.

In focus

Compared to men, women generally have lower $\dot{V}O_{2max}$ values expressed in ml · kg^{-1} · min^{-1}. A part of this difference in $\dot{V}O_{2max}$ values between women and men is related to the extra body fat carried by women and, to a lesser extent, to their lower hemoglobin levels, which result in a lower oxygen content in the arterial blood.

In review

- The innate qualities of muscle and the mechanisms of motor control are similar for women and men.
- Women and men do not differ in lower body strength expressed relative to body weight or to FFM. But women have less upper body strength expressed relative to body weight or FFM than men, largely because more of women's muscle mass is below the waist and women use their lower body muscles more than their upper body muscles.
- At submaximal exercise levels, women have higher heart rates than men, but women's submaximal cardiac outputs are similar for the same rate of work. This indicates that women have lower stroke volumes, primarily because they have smaller hearts and less blood volume.
- Women also have a lower capacity to increase $(a\text{-}\bar{v})O_2$ difference, probably because of their lower hemoglobin content, so less oxygen is delivered to their active muscles per unit of blood.
- Sex differences in respiratory responses are primarily attributable to body size differences.
- Beyond puberty, the average woman's $\dot{V}O_{2max}$ is only 70% to 75% that of the average man. However, some of this difference could be attributable to women's traditionally less active lifestyles. Research with highly trained athletes reveals an 8% to 15% difference, and much of this difference is attributable to women's greater fat mass, lower hemoglobin levels, and lower maximal cardiac output.
- Little or no difference in lactate threshold is found between the sexes, but peak lactate concentrations are generally higher in men.

PHYSIOLOGICAL ADAPTATIONS TO EXERCISE TRAINING

Basic physiological function both at rest and during exercise changes substantially with physical training. In this section we investigate how women adapt to chronic exercise, emphasizing areas in which their responses might differ from men's.

Body Composition

With either cardiorespiratory endurance training or resistance training, both women and men experience

- losses in total body mass,
- losses of fat mass,
- losses of relative fat, and
- gains in FFM.

The magnitude of the change in body composition appears to be related more to the total energy expenditure associated with the training activities than to the participant's sex. Significantly more FFM is gained in response to strength training than with endurance training, and the magnitude of these gains is similar between sexes.

Bone and connective tissue undergo alterations with training, but these changes are not well understood. In general, animal studies and limited human studies have shown an increase in the density of the weight-bearing long bones, primarily in growing animals and in children and adolescents. This adaptation appears to be independent of sex. In adults, weight-bearing exercise is critical for maintaining bone mass and density. We discuss some exceptions later in this chapter.

Connective tissue appears to be strengthened with endurance training, and sex-specific differences in this response have not been identified. Higher injury rates in women point to the possibility that women are more susceptible to injury than men while participating in physical activity and sport. This has led to concerns about sex-specific differences in joint integrity and laxity and the strength of ligaments, tendons, and bones. Unfortunately, the research literature contributes little to confirming or denying the validity of such concerns. Where differences in the rate of injury have been observed, the injury could be related more to the level of conditioning than to the participant's sex. Those who are less fit are more prone to injury. This is an extremely difficult area in which to obtain objective data but nevertheless is an important area that needs to be better researched.

Strength

Until the 1970s, prescribing strength training programs for girls and women was not considered appropriate. Women were not believed capable of gaining strength because of their innately low levels of male anabolic hormones. Paradoxically, many people also generally feared that strength training would masculinize women. During the 1960s and 1970s, however, it became evident that many of the United States' better female athletes were not doing well in international competition, largely because they were weaker than their competitors. Slowly, research demonstrated that women can gain considerably from strength training programs even though strength gains are usually not accompanied by large increases in muscle bulk.

In part because of their lower levels of testosterone, women have less total muscle mass than men.[29] If muscle mass is the major determinant of strength, then women have a distinct disadvantage. But if neural factors are as important as or more important than size, women's potential for absolute strength gains is considerable. Also, some women can attain significant muscle hypertrophy. This has been demonstrated in female bodybuilders who have remained free of anabolic steroids. Also, a number of studies have shown similar increases for men and women in FFM and muscle volume, as well as hypertrophy of type I, type IIa, and type IIx muscle fibers following periods of resistance training.

With all of this in mind, it is of interest to look at the men's and women's world records for weightlifting by weight classification. Figure 18.8 illustrates these world records for the total weight lifted (the sum of the snatch and the clean and jerk) as of 2006. As seen from this figure, men are considerably stronger at each weight classification. While the weight classifications differ slightly for men and women, the men generally lifted at least 75 kg (165 lb) more than the women did for an equivalent body weight. Part of the difference can be explained by the fact that, for a given body weight, men most likely have a higher FFM. Furthermore, few women participate in competitive weightlifting. The greater the number of participants, the greater the likelihood of higher world records. Still, the differences are so great that there must be other factors operating.

Women can experience major increases in strength (20-40%) as a result of resistance training, and the magnitude of these changes is similar to that seen in men. These gains are attributable to both muscle hypertrophy and neural factors.

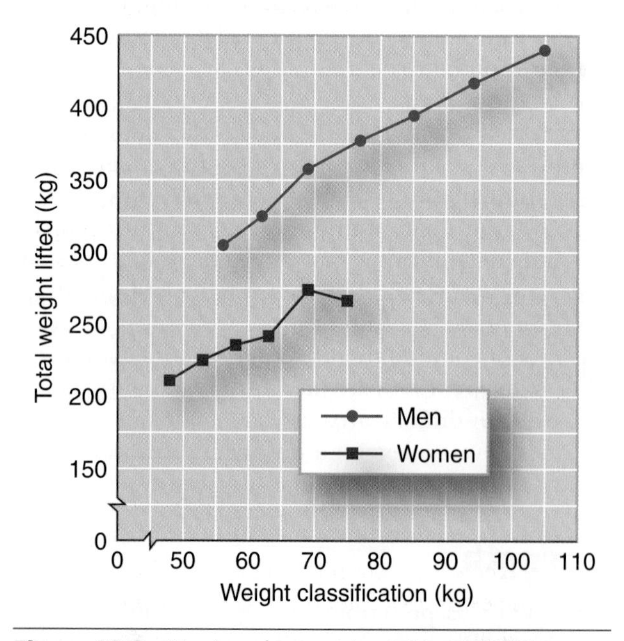

Figure 18.8 Men's and women's world weightlifting records as of 2006 for the total amount of weight lifted (combination of the snatch and the clean and jerk) by weight category. The highest weight category for both men and women is not included because it is not defined by weight. The men's total weight lifted is considerably more than the women's at a similar weight classification.

International Weightlifting Federation, August 2006 (www.iwf.net/events/wr/record_cur.php).

Cardiovascular and Respiratory Function

Major cardiovascular and respiratory adaptations result from cardiorespiratory endurance training, and these adaptations do not appear to be sex specific. Major increases in maximal cardiac output ($\dot{Q}_{max}$) accompany training. Maximum heart rate usually does not change or decreases with training, so this increase in $\dot{Q}_{max}$ is the result of a large increase in stroke volume, which results from two factors. End-diastolic volume (the amount of blood in the ventricles before contraction) increases with training because blood volume increases and venous return is more efficient. In addition, end-systolic volume (the amount of blood remaining in the ventricles after contraction) is reduced with training because the stronger myocardium produces a stronger contraction, ejecting more blood.

At submaximal work rates, cardiac output shows little or no change, although stroke volume is considerably higher for the same absolute rate of work. Consequently, heart rate for any given rate of work is reduced after training. Resting heart rate can be reduced to 50 beats/min or less. Several female distance runners have had resting heart rates below 36 beats/min. This is considered a classic training response and corresponds to an exceptionally high stroke volume.

The increases in $\dot{V}O_{2max}$ that accompany cardiorespiratory endurance training, which are discussed subsequently, result primarily from the large increases in maximal cardiac output, with only small increases in $(a\text{-}\bar{v})O_2$ difference. However, Saltin and Rowell[40] stated that the major limitation to $\dot{V}O_{2max}$ is oxygen transport to the working muscles. Although cardiac output is important to oxygen transport, these researchers believe that the increases in $\dot{V}O_{2max}$ that accompany training are primarily attributable to increased maximal muscle blood flow and muscle capillary density. These changes are firmly established in both women and men. Although women also experience considerable increases in maximal ventilation similar to those of men, reflecting increases in both tidal volume and breathing frequency, these changes are thought to be unrelated to the increase in $\dot{V}O_{2max}$.

Metabolic Function

With cardiorespiratory endurance training, women experience the same average relative increase in $\dot{V}O_{2max}$ that has been observed in men (15% to 20% on average). The magnitude of change noted generally depends on the intensity and duration of the training sessions, the frequency of training, and the length of the study.

In focus

Women experience average increases in $\dot{V}O_{2max}$ similar to men (15-25%) with aerobic training.

After cardiorespiratory endurance training, women's oxygen uptake at the same absolute submaximal work rate does not appear to change, although several studies have reported decreases. Women's blood lactate concentrations are reduced for the same absolute submaximal rates of work; peak lactate concentrations generally are increased; and the lactate threshold increases with training.

From this discussion, it is obvious that women respond to physical training in the same manner as men. Although the magnitudes of their adaptations to training might differ somewhat from those of their male counterparts, the overall trends appear identical. This is important to remember when one is prescribing exercise, something that is discussed in chapter 19.

In review

- With training, women and men experience similar changes in body composition that seem to be related to total energy expenditure during training.
- Women, similar to men, gain considerable strength through strength training, an effect associated with increases in FFM and muscle volume as well as hypertrophy of type I, type IIa, and type IIx muscle fibers.
- Cardiovascular and respiratory changes that accompany cardiorespiratory endurance training do not appear to be sex specific.
- Women experience the same relative increases in $\dot{V}O_{2max}$ as men with cardiorespiratory endurance training.

SPORT PERFORMANCE

Women are outperformed by men in all athletic activities in which performance can be precisely and objectively measured by distance or time. The difference is most pronounced in activities such as the shot put, where high levels of upper body strength are crucial to successful performance. In the 400 m freestyle swim, however, the winning time for women in the 1924 Olympic Games was 19% slower than that for men, but this difference decreased to 15.9% in the 1948 Olympics and to only 7.0% in the 1984 Olympics. The fastest women's 800 m freestyle swimmer in 1979 swam faster than the world record–holding man for the same distance in 1972. From these results, it would appear that the gap between the sexes is narrowing. However, as we see from table 18.1 (p. 424), the difference between the men's and women's world record for the 400 m freestyle was 10.4% in 2006. For the 1,500 m freestyle it was 8.9%. Unfortunately, making valid comparisons through the years has been difficult because the degree to which an activity has been emphasized and its popularity are not constant, and other factors—such as opportunities to participate, coaching, facilities, and training techniques—have differed considerably between the sexes over the years.

As noted at the beginning of this chapter, large numbers of girls and women did not start entering competitive sport until the 1970s. Even then, there was an initial reluctance to train women as hard as men. Once girls and women started training as hard as boys and men, their performance improved dramatically. This is illustrated in figure 18.9 on page 434, which shows world records from 1960 to 2006 for women and men in six running events in track and field. For distances of 100 m through the marathon, women's present world records are consistently 7% to 14% slower than men's. Furthermore, as can be seen in this figure, the improvement in women's records, which was initially quite dramatic, is beginning to level off and to parallel the curves for men's records.

SPECIAL ISSUES

Although the sexes respond to acute exercise and adapt to chronic exercise in much the same manner, several additional areas that are unique to females must be considered. Specifically, we look at

- menstruation and menstrual dysfunction,
- pregnancy,
- osteoporosis,
- eating disorders, and
- environmental factors.

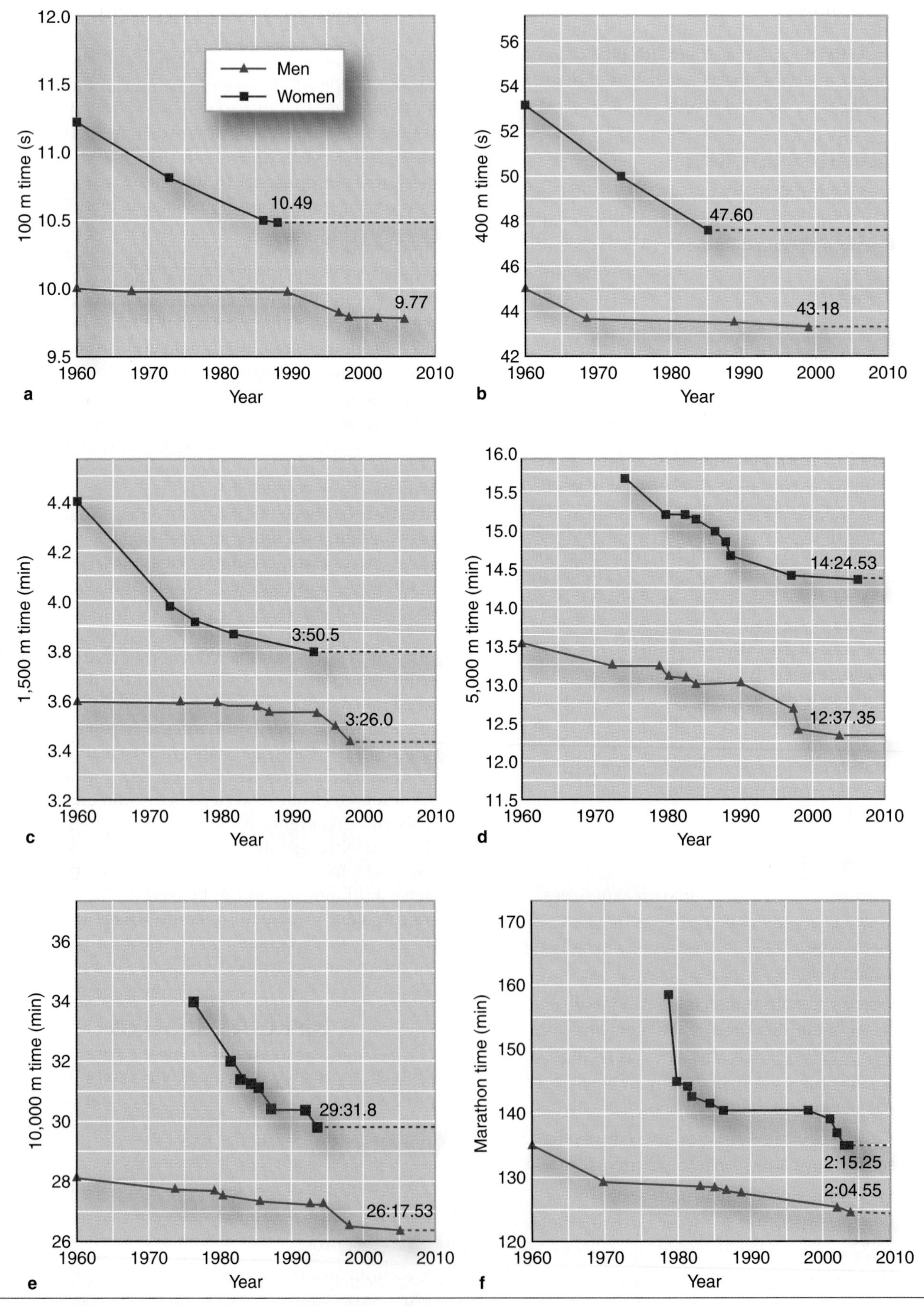

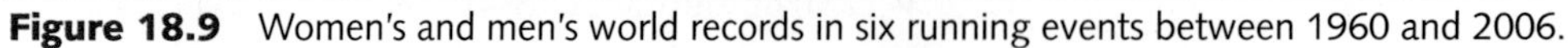

Figure 18.9 Women's and men's world records in six running events between 1960 and 2006.

Menstruation and Menstrual Dysfunction

How does the menstrual cycle or pregnancy influence exercise capacity and performance? How do physical activity and competition influence the menstrual cycle or pregnancy? These are two questions of interest to exercising women, particularly female athletes.

The three major phases of the **menstrual cycle** are illustrated in figure 18.10. The first is the menstrual (flow) phase, which lasts four to five days, during which time the uterine lining (endometrium) is shed and menstrual flow, or bleeding, occurs. The second is the proliferative phase, which prepares the uterus for fertilization and lasts about 10 days. During this phase, the endometrium begins to thicken, and some of the ovarian follicles that house the ova mature. These follicles secrete estrogen. The proliferative phase ends when a mature follicle ruptures, releasing its ovum (ovulation). The menstrual and proliferative phases correspond to the follicular phase of the ovarian cycle.

The third and final phase of the menstrual cycle is the secretory phase, which corresponds to the luteal phase of the ovarian cycle. This phase lasts 10 to 14 days, during which the endometrium continues to thicken, its blood and nutrient supply increases, and the uterus prepares itself for pregnancy. During this time, the empty follicle (now termed a corpus luteum, hence the term *luteal phase*) secretes progesterone, and estrogen secretion also continues. The complete menstrual cycle averages 28 days. However, there is considerable variation in the cycle length among healthy women, from 23 to 38 days.

Menstruation and Performance

Alterations in athletic performance experienced during different phases of the menstrual cycle are subject to considerable individual variability. Some women have no noticeable change in their performance ability at any time during their menstrual cycle, yet others have considerable difficulty in the preflow or early-flow phase, or both.

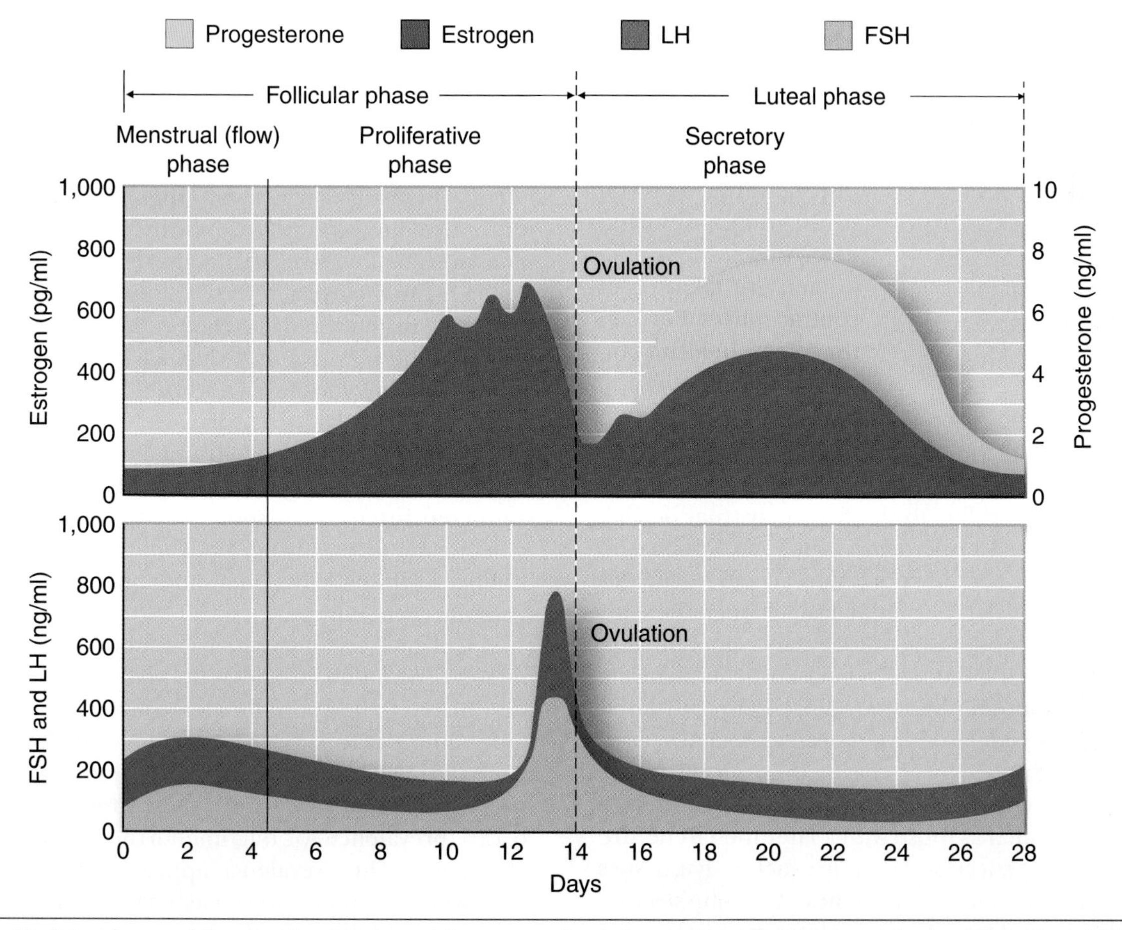

Figure 18.10 Phases of the menstrual cycle and the concomitant changes in progesterone and estrogen (top) and follicle-stimulating hormone (FSH) and luteinizing hormone (LH) (bottom). For most purposes, the cycle is divided into the follicular phase, which starts with the initiating of bleeding during menstrual flow, and the luteal phase, which starts with ovulation.

Little information is available from well-designed and well-controlled research studies. Several studies have suggested that athletic performance is best during the immediate postflow period up to the 15th day of the cycle (the first day of the cycle corresponds to the initiation of the flow, or **menses,** and ovulation occurs on about the 14th day). However, other studies have indicated that performance is best during the flow phase. In fact, some athletes have reportedly set world records during the flow phase.

Several studies have been conducted in research laboratories using physiological measures to gauge performance changes. These generally have shown no performance differences in the various phases of the menstrual cycle.

From currently available information, we can conclude that the performance of some women can be affected by the phase of their menstrual cycle, but that many, if not most, women are not affected. Any woman who has premenstrual syndrome (PMS) or dysmenorrhea (pain or abdominal cramping with menstruation) will likely not perform as well while she is experiencing symptoms. For these women, some degree of control over their menstrual cycle is possible through the use of low-dose oral contraceptives. This might not be wise, however, as several studies have reported decreased $\dot{V}O_{2max}$ while women are taking oral contraceptives. Further, there are health risks associated with the use of oral contraceptives.

There appears to be no general pattern concerning the ability of women to achieve their best performances during any specific phase of their menstrual cycle. However, those who have PMS or dysmenorrhea will likely not perform as well while experiencing symptoms.

Menarche

Menarche, which refers to the first menstrual period, has been reported to be delayed in some young athletes involved in certain sports and activities, such as gymnastics and ballet. The median age of menarche for American girls is about 12.4 to 13.0 years, depending on the population sampled. For gymnasts, the median age appears to be closer to 14.5 years. Frisch[19] hypothesized that menarche is delayed five months for each year of training before menarche, implying that training causes delayed menarche. Malina,[26] however, postulated that late maturers, such as those with a later menarche, are more likely to be successful in sports such as gymnastics because of their small, lean bodies. This implies that those who naturally experience a later menarche have an advantage in, and thus are likely to be involved in, certain sports rather than that their sport involvement delays menarche.

These opposing viewpoints can be summarized by the following questions: Does intense training to achieve the level of an elite performer delay menarche, or does a later menarche provide an advantage that contributes to the success of the elite performer? Stager and colleagues[42] used computer modeling to analyze this issue and concluded that the age of menarche in athletes is later rather than delayed. At this time, evidence is insufficient to support the theory that training delays menarche, so Malina's statement that menarche occurs later in the trained athlete is appropriate.

In focus

Menarche appears to come later in highly trained elite athletes in certain sports such as gymnastics. However, there is no strong evidence to support the contention that the intense training for the sport delays menarche.

Menstrual Dysfunction

Female athletes can experience disruptions of their normal menstrual cycle. These disruptions are collectively referred to as **menstrual dysfunction,** of which there are several types. **Eumenorrhea** is the term for normal menstrual function. **Oligomenorrhea** refers to abnormally infrequent or scant menstruation, or menses (bleeding) occurring at intervals longer than 35 days. **Amenorrhea** refers to the absence of menstruation; **primary amenorrhea** refers to the absence of menarche in girls and women 16 years of age and older—women who never began menstruating. Some athletes with previously normal menstrual function have reported the absence of menstruation for three to six months or longer when they have trained and competed intensely in sports such as figure skating, ballet, gymnastics, diving, bodybuilding, cycling, and distance running. This phenomenon is referred to as **secondary amenorrhea.**

The prevalence of secondary amenorrhea and oligomenorrhea among athletes is not well documented but is estimated to be approximately 5% to 60% or higher, depending on the sport or activity and the level of competition. This is considerably higher than the estimated 2% to 5% prevalence for amenorrhea and 10% to 12% prevalence for oligomenorrhea in the general population. The prevalence appears to be greater in those who train many hours each day and in those who train at very high intensities.

Many women who become amenorrheic are relieved to be free of menstruation each month. Most also assume that they have developed a simple but effective form of birth control. However, athletes have become pregnant while amenorrheic, which indicates that ovulation, and thus fertility, is not always influenced by the absence of menstruation. This point needs to be stressed among female athletes prone to amenorrhea to reduce the possibility of unexpected pregnancy.

Since the 1970s, scientists have been conducting experiments to determine the basic cause of secondary amenorrhea. Some of the factors that have been proposed as potential causes include the following:

- History of menstrual dysfunction
- Acute effects of stress
- High quantity or intensity of training
- Low body weight or body fat
- Hormonal alterations
- Energy deficit through inadequate nutrition and disordered eating

Considerable research has been conducted on each of these proposed factors, and five of the six have been eliminated for consideration as the primary cause. It is tempting to surmise that high-volume or high-intensity training (or a combination of the two) leads to menstrual dysfunction, but this is likely not the case.

Current evidence indicates that inadequate nutrition resulting in an energy deficit is the primary cause of secondary amenorrhea. Studies have shown that inadequate intake of calories, such that the body is not matching caloric intake to caloric expenditure over an extended period of time, is the primary cause of secondary amenorrhea.

Recent research by Dr. Anne Loucks and her colleagues [25, 36] at Ohio University has clearly demonstrated that simply inducing an energy deficit in eumenorrheic women results in significant hormonal alterations that are associated with amenorrhea. Specifically, reducing caloric intake, with or without the added stress of increased energy expenditure from exercise training, reduces LH pulse frequency and the thyroid hormone triiodothyronine (T_3), both of which are associated with disturbed menstrual function. Food deprivation may trigger signals that inhibit LH secretion, signals that may come from different pathways. Insulin and leptin have been proposed as potential signals. Disruption of LH pulsatility and low estrogen levels suggest that there is a disruption of the gonadotropin-releasing hormone (GnRH) pulse generator in the hypothalamus.[36]

Thus, exercise training is likely not directly associated with menstrual dysfunction at all, other than through its contribution to an energy deficit. An energy deficit, in either the absence or presence of exercise training, is associated with these hormonal alterations. Intense or high-volume training most likely is not associated with menstrual dysfunction as long as energy intake matches or exceeds energy expenditure over days, weeks, and months.

The relationship between clinically disordered eating and menstrual dysfunction is a more recent concern; several studies have shown a strong relationship between the two. In one study, 8 of 13 amenorrheic distance runners reported eating disorders, compared with 0 of 19 eumenorrheic distance runners.[21] In another study, 7 of 9 amenorrheic elite middle- and long-distance runners were diagnosed with either anorexia nervosa, bulimia nervosa, or both, compared with 0 of 5 eumenorrheic runners.[47] Eating disorders are discussed in detail later in this chapter, but we can say that eating disorders generally involve an energy deficit.

In review

- The effects of different phases of the menstrual cycle on performance are subject to considerable individual variation. In general, the number of women reporting impaired performance during the flow phase is about the same as the number reporting no difficulty. Any woman who has PMS or dysmenorrhea is likely to not perform as well while experiencing those symptoms.
- Menarche can occur late in some young athletes in certain sports. However, the most likely explanation for this is that late maturers, because of lean body build, are more likely to participate successfully in these activities, not that these activities cause delayed menarche.
- Female athletes can experience menstrual dysfunction, most often secondary amenorrhea or oligomenorrhea. Current evidence implicates inadequate nutrition, or a prolonged energy deficit, as the primary cause of secondary amenorrhea. In addition, hormonal changes from exercise and training might disrupt GnRH secretion, which is needed to direct the normal cycle. This, too, is associated with a prolonged energy deficit.

Pregnancy

What are the effects of exercise during **pregnancy**? Four major physiological concerns are associated with exercise during pregnancy:[49, 50]

1. The acute risk associated with reduced blood flow to the uterus (blood is diverted to the mother's active muscles), leading to fetal hypoxia (insufficient oxygen)
2. Fetal hyperthermia (elevated temperature) associated with the increase in the mother's internal body temperature during prolonged aerobic-type exercise or exercise under conditions of heat stress
3. Reduced carbohydrate availability to the fetus as the mother's body uses more carbohydrate to fuel her exercise
4. The possibility of miscarriage and the final outcome of pregnancy

Each of these is discussed in the following sections.

Reduced Uterine Blood Flow and Hypoxia

There have been reports that uterine blood flow in both animals and humans is reduced by 25% or more during moderate to strenuous exercise. The magnitude of the reduction is directly correlated with exercise intensity and duration.[48] Whether this reduction in uterine blood flow leads to fetal hypoxia is less clear. Apparently, an increase in the uterine $(a\text{-}\bar{v})O_2$ difference at least partially compensates for any reduced blood flow. Increases in fetal heart rate, although not always observed during maternal exercise, have been interpreted as an index of hypoxia in the fetus. Although increased fetal heart rate might reflect hypoxia to a certain degree, it more likely represents the fetal heart's response to increased catecholamine levels in the blood originating from both the fetus and the mother.

Hyperthermia

Fetal hyperthermia is a distinct possibility if the mother's core temperature is elevated substantially during and immediately after exercise. **Teratogenic effects** (abnormal fetal development) have been documented with chronic exposure to thermal stress in animals, and these effects have been documented with maternal fever in humans. Central nervous system defects are the most common result.[50] Although fetal temperature has been shown to increase with exercise in animal studies, it is unclear whether this increase is sufficient to warrant concern.

Carbohydrate Availability

The potential for reduced carbohydrate availability for the fetus during exercise is also not well understood. We know that endurance athletes who train or compete for long durations reduce both liver and muscle glycogen stores and that blood glucose concentrations also can decrease. But whether this is a potential problem in pregnant women is less clear.

Miscarriage and Pregnancy Outcome

Concerns also have been expressed regarding the potential of exercise to induce miscarriage during the first trimester, to induce premature labor, and to alter the normal course of fetal development. Unfortunately, little information is available concerning the risk for miscarriage and premature labor. Regarding pregnancy outcome, data are scarce and conflicting. Although there are some indications of lighter birth weights and shorter gestation periods, most studies have shown either favorable effects of exercise (such as reduced maternal weight gain, shorter postdelivery hospital stays, and fewer cesarean sections) or no differences between the control and exercise groups.

In focus

Although there are several concerns for the health of the fetus during maternal exercise, the risk to the fetus from aerobic exercise during pregnancy appears to be low, particularly if guidelines for exercising during pregnancy are followed.

Recommendations for Exercise During Pregnancy

To summarize, exercise during pregnancy can have associated risks (see table 18.2), but the benefits far

In review

- During exercise, major concerns for the pregnant athlete include the possible risk of fetal hypoxia, fetal hyperthermia, reduced carbohydrate supply to the fetus, miscarriage, premature labor, low birth weight, and abnormal fetal development.
- The benefits of a properly prescribed exercise program during pregnancy outweigh the potential risks. Such an exercise program must be coordinated with the woman's obstetrician.

outweigh the potential risks if caution is taken in designing the exercise program. It is important that the pregnant woman coordinate her exercise program with her obstetrician so that sound medical judgment can be used to determine the most appropriate mode, frequency, duration, and intensity of activity.

The American College of Obstetricians and Gynecologists (ACOG) developed a set of guidelines in 1985, which were revised in 1994. These guidelines were summarized by Pivarnik[32] as follows:

- Pregnant women can derive health benefits from mild to moderate exercise performed at least three days per week.
- Women should avoid supine exercise and motionless standing after the first trimester because this compromises venous return, which in turn compromises cardiac output.
- Women should stop exercising when fatigued, should not exercise to exhaustion, and should modify their routines based on maternal symptoms. Weight-bearing exercises under some circumstances may be continued, but non-weight-bearing activities such as cycling or swimming are encouraged to reduce the risk of injury.
- Care should be taken not to participate in sports or exercises in which falling, a loss of balance, or blunt abdominal trauma may occur.
- Because pregnancy requires an extra 300 kcal (1,255 kJ) of energy per day, an exercising woman should pay particular attention to diet to ensure that she is receiving adequate calories.
- Heat dissipation is of particular concern in the first trimester, so an exercising woman should wear correct clothing, be sure that her fluid intake is sufficient, and select optimal environmental conditions.
- A woman's regular prepregnancy exercise routine should be resumed gradually postpartum, as pregnancy-associated changes may persist four to six weeks.

In 2002, ACOG published a short "Committee Opinion" on exercise during pregnancy and the postpartum period, which essentially supported their previous guidelines.[1] In addition, they supported the current recommendation of the Centers for Disease Control and Prevention and the American College of Sports Medicine for nonpregnant individuals, which states that individuals should accumulate 30 min or more per day of moderate exercise on most, if not all, days of the week (see chapter 19). Furthermore, they state that scuba diving should be avoided throughout pregnancy because the fetus is at increased risk for decompression sickness. Also, there is an increased risk when pregnant women exercise at altitudes in excess of 6,000 ft (1,830 m).

TABLE 18.2 Hypothetical Risks and Postulated Benefits of Exercise During Pregnancy

	Hypothetical risks	Postulated benefits
Maternal	Acute hypoglycemia Chronic fatigue Musculoskeletal injury	Increased energy level (aerobic fitness) Reduced cardiovascular stress Prevention of excessive weight gain Facilitation of labor Faster recovery from labor Promotion of good posture Prevention of lower back pain Prevention of gestational diabetes Improved mood state and body image
Fetal	Acute hypoxia Acute hyperthermia Acute reduction in glucose availability Miscarriage in the first trimester Induction of premature labor Altered fetal development Shortened gestation Reduced birth weight	Fewer complications of a difficult labor

Adapted from L.A. Wolfe et al., 1989, "Prescription of aerobic exercise during pregnancy," *Sports Medicine* 8: 273-301.

Osteoporosis

Maintaining a healthy lifestyle might retard one detrimental aging process that is a major health concern for women: osteoporosis. **Osteoporosis** is characterized by decreased bone mineral content, which causes increased bone porosity (see figure 18.11). **Osteopenia,** as we discovered in chapter 17, refers to a loss of bone mass that occurs with aging. Osteoporosis is a more severe loss of bone mass with deterioration of the microarchitecture of bone, leading to skeletal fragility and increased risk of bone fracture. These changes typically begin in the early 30s. The occurrence rate for fractures associated with osteoporosis increases by two to five times after the onset of menopause. Men also experience osteoporosis but to a lesser degree early in life because of a slower rate of bone mineral loss. Much remains to be learned about the etiology of osteoporosis; however, three major contributing factors common to postmenopausal women are

- estrogen deficiency,
- inadequate calcium intake, and
- inadequate physical activity.

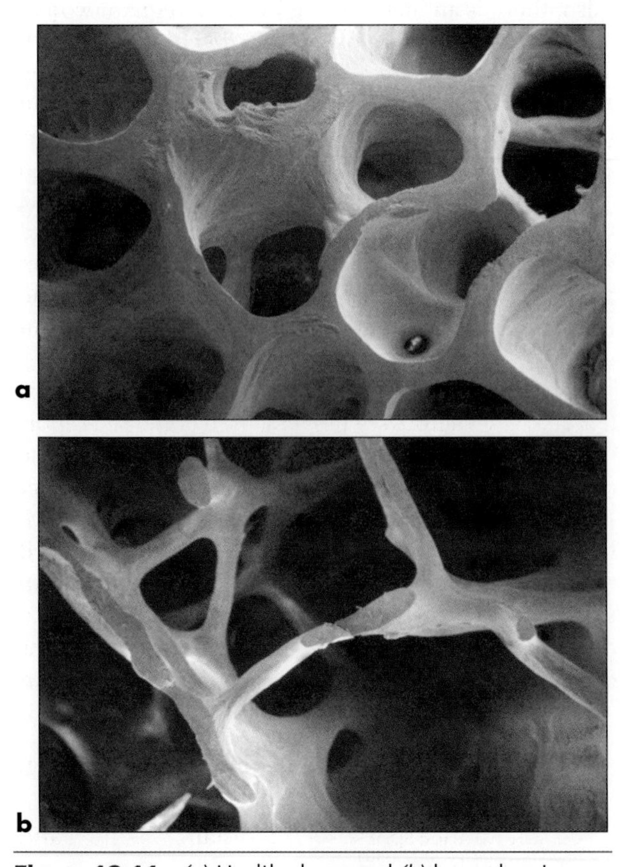

Figure 18.11 *(a)* Healthy bone and *(b)* bone showing increased porosity (decreased density, appearing darker) resulting from osteoporosis.

Although the first of these is a direct result of menopause, the last two reflect dietary and exercise patterns throughout life.

In addition to postmenopausal women, women with amenorrhea and those with anorexia nervosa also suffer from osteoporosis attributable to insufficient calcium intake, low serum estrogen levels, or possibly both. In studies of women with anorexia, investigators found that their bone densities were reduced significantly compared with those of controls. Cann and associates[6] were the first to report a substantially lower than normal bone mineral content in physically active women classified as having hypothalamic amenorrhea.

In another study, the radial and vertebral bone densities of 14 athletic women (mostly runners) with amenorrhea were compared with those of 14 athletic women with normal menstruation (eumenorrhea).[15] Investigators discovered that physical activity did not protect the group with amenorrhea from significant bone density losses. The amenorrheic group's bone density values at a mean age of 24.9 were equivalent to those of normally active women at a mean age of 51.2. In a follow-up study, increases in vertebral bone mineral density were found in the women who previously had been amenorrheic but had resumed menstruation.[16] However, their bone mineral densities remained well below the average for their age-group, even four years after they resumed normal menses.[14]

It generally is assumed that exercise is a positive factor for bone health in that it is associated with an increase in bone mass, or at least with the maintenance of bone mass in young, middle-aged, and older women. Therefore, it is confusing to learn that amenorrheic runners have reduced bone mass. Figure 18.12 is an

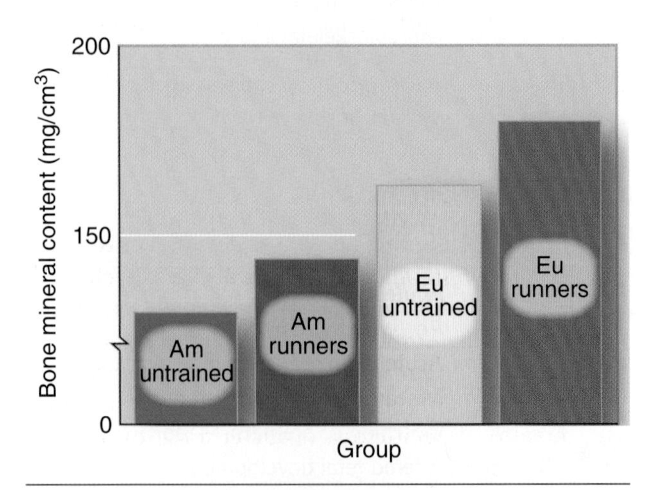

Figure 18.12 Bone mineral content of women runners and of untrained women who are amenorrheic (Am) and eumenorrheic (Eu). Note that when women with the same menstrual status are compared, the runners have higher bone mineral content than the untrained women.

Unpublished data from Dr. Barbara Drinkwater.

In focus

Athletes with secondary amenorrhea are at increased risk for bone mineral loss. This loss does not appear to be totally reversible with the resumption of normal menstrual function.

attempt to clarify this apparent contradiction. From this figure we see that the bone mineral content of normally menstruating runners tends to be higher than that of normally menstruating nonrunning controls. Furthermore, female runners who are amenorrheic have higher bone mineral contents than untrained women who are amenorrheic. Thus, when we compare women of like menstrual status, those who are exercising will have the higher bone mineral content. Caution should be used in interpreting data such as those presented in this section, because the results can be confounded by such factors as body composition, age, height, weight, and diet.

Although the precise mechanism is unknown, estrogen deficiency appears to play a major role in the development of osteoporosis. In the past, estrogen has been prescribed in an effort to reverse the degenerative effects of osteoporosis, but this therapy can have serious side effects, such as an increased risk of endometrial cancer. To reduce this risk, estrogen is given in combination with progestin (hormone replacement therapy, or HRT). However, there is an increased risk of breast cancer, strokes, and heart attacks associated with the use of HRT. Bisphosphonates, considered antiresorptive medications, are also used. Increasing calcium intake to 1,200 to 1,500 mg per day also has been proposed for decreasing the risk of osteoporosis.

Evidence certainly suggests that increased physical activity and adequate calcium intake combined with adequate caloric intake are a sensible approach to preserving the integrity of bone at any age. However, maintaining normal menstrual function is critical for those who have not reached menopause.

Eating Disorders

Eating disorders are a group of disorders that must meet specific criteria established by the American Psychiatric Association.[2] The two most commonly diagnosed eating disorders are anorexia nervosa and bulimia nervosa. **Disordered eating,** on the other hand, refers to patterns of eating that are not considered normal but don't meet the specific diagnostic criteria for a given eating disorder.

Eating disorders in girls and women became the focus of considerable attention beginning in the 1980s. Men constitute about 10% or less of the reported cases. Anorexia nervosa has been considered a clinical syndrome since the late 19th century, but bulimia nervosa was first described in 1976.

Anorexia nervosa is a disorder characterized by

- refusal to maintain more than the minimal normal weight based on age and height,
- distorted body image,
- intense fear of fatness or gaining weight, and
- amenorrhea.

Females from ages 12 to 21 are at greatest risk for this disorder. Its prevalence in this group is likely less than 1%.

Bulimia nervosa, originally termed bulimarexia, is characterized by

- recurrent episodes of binge eating;
- a feeling of lack of control during these binges; and
- purging behavior, which can include self-induced vomiting, laxative use, and diuretic use.

The prevalence of bulimia in the population at greatest risk, again adolescent and young adult females, is generally considered to be about 4% and possibly closer to 1%.

It is important to realize that a person might exhibit disordered eating and yet not meet the strict diagnostic criteria for either anorexia or bulimia. As an example, the diagnosis of bulimia requires that the individual average a minimum of two binge-eating and purging episodes a week for at least three months. What about the person who meets all the criteria, except that bingeing and purging occur only once per week? Although this person cannot technically be diagnosed as having bulimia, her or his eating is certainly disordered and is a potential cause for concern. Thus, the term "disordered eating" has been used to describe those who do not meet the strict criteria for an eating disorder but who do have abnormal eating patterns.

The prevalence of eating disorders in athletes is controversial. Numerous studies have used either self-report or at least one of two inventories developed to diagnose disordered eating: the Eating Disorders Inventory (EDI) and the Eating Attitudes Test (EAT). Results have varied because not all studies used the strict standard diagnostic criteria for either anorexia or bulimia. As in the general population, female athletes are typically at a much higher risk than male athletes, and certain sports carry higher risks than others. The high-risk sports can generally be grouped into three categories:

1. Appearance sports, such as diving, figure skating, gymnastics, bodybuilding, and ballet
2. Endurance sports, such as distance running and swimming
3. Weight-classification sports, such as horse racing (jockeys), boxing, and wrestling

Self-reports or inventories do not always provide accurate results. In a study of 110 elite female athletes representing seven sports, EAT results showed that no athlete fell within the disordered eating range of the inventory. But in the subsequent two-year period, 18 of these athletes received either inpatient or outpatient treatment for eating disorders. In a second study of 14 nationally ranked middle- and long-distance runners who completed the EDI, only three were shown to have possible problems with disordered eating, and none were shown to have eating disorders.[47] In follow-up, seven subjects were subsequently diagnosed as having an eating disorder: four with anorexia nervosa, two with bulimia nervosa, and one with both. People with eating disorders, by their very nature, are secretive. We cannot realistically expect those with eating disorders to identify themselves, even when anonymity is ensured. For the athlete, this need for secrecy might be heightened by fear that a coach or a parent will learn of the eating disorder and not allow the athlete to compete.

Even though research is limited, it seems appropriate to conclude that athletes (figure 18.13) are at higher risk for eating disorders than the general population. Existing evidence likely does not reflect the seriousness of this problem in athletic populations. Although research data are not yet available, the prevalence might be as high as 60% or more in the specific high-risk athletic populations listed earlier.

In focus

Disordered eating has become a major concern in female athletes. Some researchers have estimated the prevalence to be as high as 60% or greater for elite athletes in certain sports.

Eating disorders generally are considered to be addictive disorders and are extremely difficult to treat. The physiological consequences are substantial and can include death. Considering this, along with the emotional distress suffered by the athlete, the extraordinary costs of treatment ($5,000-$25,000 per month for hospital inpatient treatment), and the effect on those closest to the athlete, eating disorders must be considered among the most serious problems facing female athletes today, paralleling the seriousness of anabolic steroid use in male athletes.

In 1990, the National Collegiate Athletic Association developed a list of warning signs for anorexia nervosa and bulimia nervosa (table 18.3). When an eating disorder is suspected, it is important to recognize the seriousness of the disorder and refer the athlete to a person specifically trained in dealing with eating disorders. Most athletic trainers, coaches, and even physicians are not trained to provide professional help to those with

Figure 18.13 Although disordered eating is more prevalent among female athletes, it can also occur in male athletes who participate in sports that use weight classifications.

serious eating disorders. Most of the athletes who experience eating disorders are very intelligent, come from an upper-middle-class or higher socioeconomic level, and are very good at denying that they have a problem. These athletes are unfortunate victims of the unhealthy emphasis on extreme leanness promoted by the media and the challenges of attaining the optimal weight for their sport. Treating eating disorders is extremely difficult, and even the best-trained professionals are not always successful. Some extreme cases end in suicide or premature death from failure of the cardiovascular system. Immediate professional help should be sought for an athlete suspected of having an eating disorder.

TABLE 18.3 Warning Signs for Anorexia Nervosa and Bulimia Nervosa

Anorexia nervosa	Bulimia nervosa
Dramatic loss in weight	A noticeable weight loss or gain
A preoccupation with food, calories, and weight	Excessive concern about weight
Wearing baggy or layered clothing	Bathroom visits after meals
Relentless, excessive exercise	Depressed mood
Mood swings	Strict dieting followed by eating binges
Avoiding food-related social activities	Increased criticism of one's body

Note. The presence of one or two of these signs does not necessarily indicate an eating disorder. Diagnosis should be made by appropriate health professionals.

Adapted from a poster distributed by the National Collegiate Athletic Association, 1990.

Female athletes are at higher risk than nonathletes for disordered eating and eating disorders for several reasons. Perhaps most important, there is tremendous pressure on athletes, particularly female athletes, to get their weight down to very low levels, often below what is appropriate. This weight limit can be imposed by the coach, trainer, or parent or can be self-imposed by the athlete. In addition, the personality of the typical elite female athlete closely matches the profile of the female at high risk for an eating disorder (competitive, perfectionistic, and under the tight control of a parent or other significant figure such as a coach). Furthermore, the nature of the sport or activity largely dictates those at high risk. As previously mentioned, athletes in three categories are at high risk: appearance sports, endurance sports, and weight-classification sports. Added to

Female Athlete Triad

In the early 1990s, it became apparent that there is a reasonably strong association among

- disordered eating,
- secondary amenorrhea, and
- bone mineral disorders.

This group of disorders has been termed the female athlete triad.[44] From the limited research available at this time, it appears that the triad might start with disordered eating. Over a period of time, the length of which has not been well established and might vary considerably from one athlete to another, an athlete who has disordered eating, creating a prolonged energy deficit, starts to experience disordered menstrual function, which eventually leads to secondary amenorrhea. Following an additional period of time, again the length of which has not been defined, secondary amenorrhea leads to bone mineral disorders. A number of researchers have become interested in these intriguing relationships, and considerable research is now under way. We should know much more about this important sequence of events when the results of these research efforts are known. In 1997, the American College of Sports Medicine issued a position stand on the female athlete triad that presents a strong case for the connection of each of the three components of the triad.[30] This report concluded that female athlete triad disorders can decrease physical performance and cause morbidity (illness or disease) and mortality (death). Anyone working with young athletes, particularly female athletes, should be educated about the triad, know how to recognize it, and have a plan of action involving professionals trained in dealing with the female athlete triad to help prevent it, treat it, and reduce its risks.

these risks are the normal pressures imposed by the media and culture on young women, whether they are athletes or not.

Environmental Factors

Exercise in the heat, in the cold, or at altitude provides additional stress or challenge to the body's adaptive abilities (see chapters 11 and 12). Many early studies indicated that women are less tolerant to heat than men are, particularly when exercising. Much of this difference, however, was the result of lower fitness levels of the women included in these studies, because the men and women were tested at the same absolute rate of work. When the rate of work is adjusted relative to individual $\dot{V}O_{2max}$ values, women's responses are almost identical to men's. Women have a delayed onset of sweating and dilation of the skin during the luteal phase of the menstrual cycle (i.e., onset occurs at a higher core temperature). However, this should not affect performance until core temperature approaches 40 °C.[7] Women generally have lower sweat rates for the same exercise and heat stress: Although they possess a larger number of active sweat glands than men do, women produce less sweat per gland. This is a slight disadvantage in hot, dry environments, but a slight advantage in humid conditions in which sweat evaporation is minimal.

When exposed to repeated bouts of heat stress, the body undergoes considerable adaptation (acclimatization) that enables it to withstand future heat stress more efficiently. After acclimatization, the internal temperature at which sweating and vasodilation begin is similarly lowered in women and men. Also, the sensitivity of the sweating response per unit increase in internal temperature increases by a similar amount in the two sexes following both physical training and heat acclimatization. Therefore, most differences noted between women and men in the early studies can be attributed to initial differences in their physical conditioning and acclimation status and not to their sex.

Women generally have slightly lower sweat rates than men for the same heat stress, apparently the result of lower sweat production per sweat gland. However, this reduced sweating capacity does not appear to affect women's ability to tolerate heat.

Women have a slight advantage over men during cold exposure because they have more subcutaneous body fat. But their smaller muscle mass is a disadvantage in extreme cold because shivering is the major adaptation for generating body heat. The greater the active muscle mass, the greater the subsequent heat generation. Muscle also provides an additional insulating layer.

Several studies have reported sex differences in response to altitude hypoxia, both at rest and during submaximal exercise. Maximal oxygen consumption decreases during hypoxic work in both sexes, but these decreases do not seem to adversely affect women's ability to work at high altitude. Studies of maximal exercise at altitude demonstrate no difference in response between the sexes.

In review

- Three major contributing factors to osteoporosis are estrogen deficiency, inadequate calcium intake, and inadequate physical activity.
- Postmenopausal women, amenorrheic women, and those who have anorexia nervosa are at greater risk of osteoporosis. Physical activity and adequate calcium and caloric intake are important to the preservation of bone at any age.
- Eating disorders, such as anorexia nervosa and bulimia nervosa, are much more common in women than in men and are especially common among athletes in appearance sports, endurance sports, and weight-classification sports. Athletes seem to be at a higher risk for eating disorders than the general population.
- When exercise intensity is adjusted relative to an individual's $\dot{V}O_{2max}$, women and men respond almost identically to heat stress. Most differences noted are likely attributable to different initial levels of conditioning.
- Because they have more insulating subcutaneous fat, women have a slight advantage over men during cold exposure, but their smaller muscle mass limits their ability to generate body heat.
- Studies indicate that maximal responses during exercise at altitude do not differ in women and men, but differences might exist at rest and during submaximal exercise.

In Closing

In this chapter, we discussed sex-specific differences in performance. Most true differences between the sexes result from women's smaller body size, lower FFM, and greater relative and absolute body fat. We also considered how women's relatively more sedentary lifestyle, an artifact from a society that traditionally frowned on women's participation in physical activity, has affected research through the years. Making valid comparisons in sport performances has been difficult because an event's popularity and other factors—such as opportunities to participate, coaching, facilities, and training techniques—have differed considerably between the sexes over the years. Finally, we have found that female and male athletes are not as different as many people believe.

With this chapter, we conclude our examination of age and sex considerations in sport and exercise. In the next part of the book, we turn our attention from athletics to a different application of exercise physiology: the use of physical activity for health and fitness. We begin with an examination of exercise prescription.

KEY TERMS

amenorrhea
anorexia nervosa
bulimia nervosa
disordered eating
eating disorders
estrogen
eumenorrhea
lipoprotein lipase
menarche
menses
menstrual cycle
menstrual dysfunction
oligomenorrhea
osteopenia
osteoporosis
pregnancy
primary amenorrhea
secondary amenorrhea
sex-specific differences
teratogenic effects
testosterone

STUDY QUESTIONS

1. How does the body composition of females compare with that of males? How do male and female athletes differ from male and female nonathletes?
2. What are the roles of testosterone and estrogen in the development of strength, fat-free mass, and fat mass?
3. How does women's upper body strength compare with men's? Lower body strength? Fat-free mass? Can women gain strength with resistance training?
4. What differences in $\dot{V}O_{2max}$ exist between average females and males? Between highly trained females and males? What can explain these differences?
5. What cardiovascular differences exist between females and males with respect to submaximal exercise? Maximal exercise?
6. How does the menstrual cycle influence athletic performance?
7. What is the primary reason that some female athletes in intense training stop menstruating for intervals of several months to several years or more?
8. What risks are associated with training during pregnancy? How can these be avoided?
9. What are the effects of amenorrhea on bone mineral? How does exercise training affect bone mineral?
10. What are the two major eating disorders, and what is the level of risk for elite female athletes having these eating disorders? How does this vary by sport?
11. What is the female athlete triad? What factors are involved and how does the triad develop?
12. How do women differ from men in their exercise response when exposed to intense heat and humidity? To cold? To altitude?

STUDY GUIDE ACTIVITIES

In addition to the activities listed in the chapter opening outline on page 423, two other activities are available in the online study guide, located at

www.HumanKinetics.com/PhysiologyOfSportAndExercise.

The chapter **SUMMARY** reviews the key concepts covered in the chapter, and the end-of-chapter **QUIZ** tests your understanding of the material covered in the chapter.

PART VII

Physical Activity for Health and Fitness

In previous parts of this book, we focused on the physiological bases of physical activity, observing the responses to an acute bout of exercise and the adaptations to chronic training, and on how people can improve their performance in sport-related activities. In part VII, we shift our focus away from athletic performance and turn to a special area of exercise and sport physiology: the use of physical activity for health and fitness. In chapter 19, "Prescription of Exercise for Health and Fitness," we discuss how to design an exercise program that can improve health and fitness. We consider the essential components, ways to tailor the program to an individual's specific needs, and the unique role of physical activity for rehabilitation of people who are ill. In chapter 20, "Cardiovascular Disease and Physical Activity," we examine the major types of cardiovascular disease, their physiological bases, and how physical activity can help prevent these diseases. Finally, in chapter 21, "Obesity, Diabetes, and Physical Activity," we examine the causes of obesity and diabetes, the health risks associated with each, and the ways in which physical activity can be used to control both disorders.

CHAPTER 19

Prescription of Exercise for Health and Fitness

In this chapter and in the online study guide

Jason Walker, a 55-year-old executive, went in for his annual physical examination with a vow to start a long overdue exercise program. Because of his obesity, high blood pressure, and one-pack-a-day smoking habit, his physician decided to give Jason a graded exercise test to determine the normality of his electrocardiogram (ECG) during the stress of exercise. As Jason was nearing exhaustion on the treadmill, his doctor noticed changes in the ST segment of his ECG, which are considered indicative of coronary artery disease. The following week, Jason, with fear and trepidation, underwent a coronary arteriogram procedure to check for coronary artery disease. His arteriogram was abnormal, indicating that he had 85% occlusion of the circumflex coronary artery and 90% occlusion of the right coronary artery. He was immediately scheduled for coronary artery bypass surgery and the surgery was successful. Fortunately, Jason had been scared to the point that he stopped smoking, lost weight, and began an exercise program. He is now very fit, competing in 10 km (6.2 mi) races, and has his blood pressure under control.

Patterns of today's living have channeled the average American into an increasingly sedentary existence. Humans, however, were designed and built for movement. Physiologically, we have not adapted well to this inactive lifestyle. In fact, during what appeared to be a fitness boom in the 1970s and 1980s, fewer than 20% of adult Americans were exercising at levels that would increase or maintain their aerobic fitness and strength. Yet research had clearly determined that, for almost everyone, an active lifestyle is important for optimal health.

HEALTH BENEFITS OF EXERCISE: THE GREAT AWAKENING

The 1990s will be remembered as the decade in which the medical profession formally recognized the fact that physical activity is vital to the body's health. It seems rather ironic that it took this long for clinicians and scientists to reach this conclusion, as Hippocrates (460-377 BC), a prominent physician and athlete, had strongly endorsed physical activity and proper nutrition as essential to health more than 2,000 years earlier!

The first acknowledgment from the modern medical profession came in July 1992, when the American Heart Association proclaimed physical inactivity a major risk factor for coronary artery disease, placing it alongside smoking, abnormal blood lipids, and hypertension.[10] In 1994, the Centers for Disease Control and Prevention (CDCP), in collaboration with the American College of Sports Medicine (ACSM), held a press conference to announce to the American public the importance of physical activity as a public health initiative and subsequently published the full text of a consensus statement by a panel of experts in this field in February 1995.[16] The National Institutes of Health (National Heart, Lung, and Blood Institute) released a consensus statement in December 1995, the full text of which was published in 1996, advocating physical activity as important for cardiovascular health.[15] Finally, in July 1996, coinciding with the start of the Olympic Games in Atlanta, the Surgeon General of the United States released a written report on the health benefits of physical activity.[20] This was a landmark report recognizing the importance of physical activity in reducing the risk for chronic degenerative diseases. (See "The Surgeon General's Report on Physical Activity and Health" on p. 451 for the major conclusions of this report.)

Much of the research supporting the benefits of physical activity in reducing the risk of developing chronic degenerative disease has come from the field of epidemiology, in which large populations are studied and associations between activity levels and disease risk determined. In the year 2000, molecular biologists and exercise physiologists started waging war on what they termed the "sedentary death syndrome" by forming an action group advocating governmental support for research into the diseases and disorders associated with a sedentary lifestyle. The group, Researchers Against Inactivity-Related Disorders, or RID, has been very effective in gaining the support of top government leaders for basic research into the role of an active lifestyle in preventing or delaying chronic degenerative diseases. Key scientific articles have been published, several of which are referenced in this chapter,[3, 4] and a Web site has been established (http://hac.missouri.edu/RID/index.htm).

With the health benefits of an active lifestyle so clearly established, what has been the response of the U.S. population in general? We need to go back a few years to get a proper historical perspective. The seeds for a fitness revolution in the United States were planted in the late 1960s with the publication of the book *Aerobics,* written by Dr. Kenneth Cooper (figure 19.1).[7] This book provided a sound medical basis for the importance of exercise, particularly aerobic exercise, to health and fitness. The fitness movement grew throughout the 1970s, possibly peaking in the early 1980s, at which time the media declared that America was in the midst of a fitness boom.

Then, in 1983, came a penetrating article by Kirshenbaum and Sullivan,[12] published in *Sports Illustrated,*

Figure 19.1 Dr. Kenneth H. Cooper, founder of the Cooper Institute and author of numerous books on the health-related benefits of maintaining an active lifestyle.

The Surgeon General's Report on Physical Activity and Health

In July 1996, the U.S. Surgeon General's office released its official report, *Physical Activity and Health*.[20] This detailed report resulted in the following major conclusions:

1. People of all ages, both male and female, benefit from regular physical activity.
2. Significant health benefits can be obtained by including a moderate amount of physical activity (e.g., 30 min of brisk walking or raking leaves, 15 min of running, or 45 min of playing volleyball) on most, if not all, days of the week. Through a modest increase in daily activity, most Americans can improve their health and quality of life.
3. Additional health benefits can be gained through greater amounts of physical activity. People who can maintain a regular regimen of activity that is of longer duration or of more vigorous intensity are likely to derive greater benefit.
4. Physical activity reduces the risk of premature mortality in general and of coronary artery disease, hypertension, colon cancer, and diabetes mellitus in particular. Physical activity also improves mental health and is important for the health of muscles, bones, and joints.
5. More than 60% of American adults are not regularly physically active. In fact, 25% of all adults are not active at all.
6. Nearly half of American youths 12 to 21 years of age are not vigorously active on a regular basis. Moreover, physical activity declines dramatically during adolescence.
7. Daily enrollment in physical education classes declined among high school students from 42% in 1991 to 25% in 1995.
8. Research on understanding and promoting physical activity is at an early stage, but some interventions to promote physical activity through schools, work sites, and health care settings have been evaluated and found to be successful.

that brought everything into proper perspective. The authors questioned the existence of the fitness boom, contending that involvement was basically limited to a small but highly visible segment of the total population. They maintained that the fitness boom included mostly high-income, executive-level, white, college-educated, young to middle-aged adults. The results of several surveys confirmed this analysis.

Things have not gotten any better. According to the U.S. Department of Health and Human Services, as of June 2004,

- nearly 40% of the U.S. population 18 years of age and older reported no leisure-time physical activity of light, moderate, or vigorous intensity for at least 10 min;
- only 22% reported engaging in vigorous physical activity sufficient to promote the development and maintenance of aerobic fitness three or more days per week for 20 or more minutes per occasion; and
- only 20% reported doing physical activities specifically designed to strengthen muscles at least twice a week.[21]

Despite the disappointing statistics, most Americans are aware that exercise is an integral part of preventive medicine. And yet people often equate exercise with jogging 8 km (5 mi) a day or lifting weights until their muscles can do no more. Many believe that high volume and intensity of exercise training are necessary to attain health-related benefits, yet this is not true. This myth was the primary focus of the CDCP/ACSM report published in 1995,[16] which concluded that significant health benefits can be obtained if one includes a moderate amount of physical activity, such as 30 min of brisk walking, 15 min of running, or 45 min of playing volleyball, on most, if not all, days of the week. The major emphasis of this report was that through a modest increase in daily activity, most people can improve their health and quality of life. In fact, in 2006, a study of older adults (70-82 years) showed that just being more active greatly reduced the risk of mortality, independent of a formal exercise program.[14]

The Surgeon General's report,[20] however, emphasized that additional health benefits can be gained through greater amounts of physical activity. Research suggests that people who can maintain a regular regimen of activity that is of longer duration or of more vigorous intensity are likely to derive greater benefit. It is now apparent that the appropriate exercise type and intensity vary, depending on individual characteristics, current fitness level, and specific health concerns. Knowing this, how should people begin exercise programs to improve their general health and fitness? The first step is deciding to take action. The next step is getting medical clearance.

MEDICAL CLEARANCE

Is a medical evaluation really necessary before starting an exercise program? Dr. Per-Olof Åstrand (see figure 19.2), eminent Swedish physician and physiologist who has had a worldwide impact promoting physical activity for health, has suggested in jest that those individuals who elect to remain sedentary should be required to have a medical evaluation to determine if their bodies can withstand the rigors of a sedentary lifestyle. The medical evaluation is perceived as a significant barrier to starting an exercise program for many people, yet it is useful and important for the following reasons:

- Some people either should not exercise at all or are considered at high risk and should be restricted to exercising only under close medical supervision. A comprehensive medical evaluation will help identify these high-risk individuals.
- The information obtained in a medical evaluation can be used to develop the exercise prescription.
- The values obtained for certain clinical measures, such as blood pressure, body fat content, and blood lipid levels, can be used to motivate the person to adhere to the exercise program.
- A comprehensive medical evaluation, particularly of healthy people, can provide a baseline against which any subsequent changes in health status can be compared.
- Children and adults should establish the habit of periodic medical evaluations because many illnesses and diseases, such as cancer and cardiovascular diseases, can be identified in their earliest

Figure 19.2 Dr. Per-Olof Åstrand, eminent Swedish physician and physiologist, bicycling through the woods.

stages when the chances of successful treatment are much higher.

Medical Evaluation

Although a comprehensive medical evaluation is useful and desirable before exercise is prescribed, not all people need one. Many people can't afford the cost of such an evaluation, and the medical system is not prepared to provide this service for the total population, even if money were available. Also, medical evaluation before prescribing exercise for a population presumed healthy has not been proven to reduce the medical risks associated with exercise. For these reasons, guidelines or recommendations have been established that attempt to target moderate- to high-risk individuals.[2, 9] People at moderate risk are those who have two or more risk factors for coronary artery disease (table 19.1); and those who are at high risk are people with one or more signs or symptoms of cardiovascular, pulmonary, or metabolic disease (see sidebar on p. 454).

In focus

Although a general medical evaluation on a regular basis is important and desirable for almost everyone, it simply is not practical to require this for all people who want to start an exercise program.

The ACSM has published specific recommendations for each phase of the medical evaluation.[2] This document should be consulted whenever there is any question as to what should be included. The physical examination should include discussion between the physician and patient of the proposed exercise program in case any medical contraindications are associated with the proposed activity. For example, people with hypertension should be cautioned to avoid activities that use isometric actions. Isometric actions tend to increase blood pressure considerably and usually result in the Valsalva maneuver, in which intra-abdominal and intrathoracic pressures increase to the point of restricting blood flow through the vena cava, limiting venous return to the heart. Both responses can lead to serious medical complications, such as loss of consciousness or stroke. Also, as we have discussed previously (see chapter 7), even dynamic resistance training can cause very high blood pressure during the activity.

TABLE 19.1 Coronary Artery Disease Risk Factors for Targeting At-Risk People

Positive risk factors	Defining criteria
Family history	Myocardial infarction, coronary revascularization or sudden death before 55 years of age in father or other male first-degree relative or before 65 years of age in mother or other female first-degree relative.
Cigarette smoking	Current cigarette smoker or a smoker who quit within the previous six months.
Hypertension	Blood pressure ≥140/90 mmHg, confirmed by measurements on at least two separate occasions, or on antihypertensive medication.
Dyslipidemia	Low-density lipoprotein (LDL) cholesterol >130 mg/dL (3.4 mmol/L) or high-density lipoprotein (HDL) cholesterol <40 mg/dL (1.03 mmol/L), or on lipid-lowering medication. If total serum cholesterol is all that is available, use >200 mg/dL (5.2 mmol/L) rather than LDL >130 mg/dL.
Impaired fasting glucose	Fasting blood glucose ≥100 mg/dL (5.6 mmol/L) confirmed by measurements on at least two separate occasions.
Obesity	Body mass index ≥30 kg/m^2; or waist girth >102 cm for men and >88 cm for women; or waist/hip ratio ≥0.95 for men and ≥0.86 for women.
Sedentary lifestyle	Persons not participating in a regular exercise program or not meeting the minimal physical activity recommendations from the U.S. Surgeon General's report.
Negative risk factors	**Defining criteria**
High serum HDL-C	>60 mg/dL (1.6 mmol/L)

Note. It is common to sum risk factors in making clinical judgments. If high-density lipoprotein cholesterol (HDL-C) is high, subtract one risk factor from the sum of positive risk factors, because high HDL-C decreases coronary artery disease risk.

Adapted, by permission, American College of Sports Medicine, 2006, *ACSM's guidelines for exercise testing and prescription,* 7th ed. (Philadelphia, PA: Lippincott, Williams, and Wilkins), 22.

Major Signs or Symptoms Suggestive of Cardiovascular, Pulmonary, or Metabolic Disease

- Pain or discomfort (or other anginal equivalent) in the chest, neck, jaw, arms, or other areas that may be ischemic in nature
- Shortness of breath at rest or with mild exertion
- Dizziness or syncope
- Orthopnea or paroxysmal nocturnal dyspnea
- Ankle edema
- Palpitations or tachycardia
- Intermittent claudication
- Known heart murmur
- Unusual fatigue or shortness of breath with usual activities

These signs or symptoms must be interpreted within the clinical context in which they appear because they are not all specific for cardiovascular, pulmonary, or metabolic disease.

Adapted, by permission, American College of Sports Medicine, 2006, *ACSM's guidelines for exercise testing and prescription*, 7th ed. (Philadelphia, PA: Lippincott, Williams, and Wilkins), 23-24.

Graded Exercise Testing

Ideally, a comprehensive medical examination will include an exercise test, usually conducted on a motor-driven treadmill. A cycle ergometer can also be used but is not that common for clinical testing in the United States. An **exercise electrocardiogram (ECG)** and blood pressure readings are obtained during exercise, as shown in figure 19.3. The ECG and blood pressure are monitored as the person progresses from low-intensity exercise, such as slow walking, up to maximal-intensity exercise. Maximal intensity might be brisk walking for an older, deconditioned subject or running up a grade for a younger, fit individual. The rate of work is generally increased every 1 to 3 min until the maximal rate of work is achieved. This progression is referred to as a **graded exercise test (GXT)**. The ECG is monitored to detect heart rhythm and electrical conductivity abnormalities. Blood pressure is monitored to determine if there is a normal increase in systolic blood pressure and little or no change in diastolic blood pressure as the rate of work progresses from low intensity to maximal or near-maximal levels. It is also important to interact with the subject, observing signs and symptoms during and immediately following the exercise test, such as chest pain or pressure (angina), unusual shortness of breath, light-headedness or dizziness, and inappropriate heart rate response.

The exercise ECG is an important part of the medical evaluation because a small but significant percentage of the adult population have abnormalities in ECGs taken during or following exercise even though they have normal resting ECGs. These abnormalities include arrhythmias (irregular heart rhythms) and ST-segment changes. Figure 19.4 illustrates a normal ECG and an abnormal ECG in which there is depression of the ST segment. Generally a horizontal (flat) or down-sloping ST segment of 1.0 mm or greater below the isoelectric line, for at least 80 ms, is suggestive of myocardial ischemia (insufficient blood flow to the myocardium) and thus the presence of **coronary artery disease (CAD)**. Results from an exercise test are considered either positive, when there have been abnormal changes in the ECG, or negative, which implies a normal response and that no disease was detected. But some people have normal, or negative, exercise tests, yet have CAD. These are false-negative tests. Others do not have the disease, yet have positive exercise ECGs suggesting disease. These are false-positive tests.

Typically, a person with abnormal changes in his or her exercise ECG will be referred for additional tests. The presence of CAD is normally determined by a coronary arteriogram, in which radiopaque dye (dye that is opaque to X rays) is injected through a catheter into the coronary arteries, allowing visualization of the insides of the arteries. If there is narrowing of one or more of

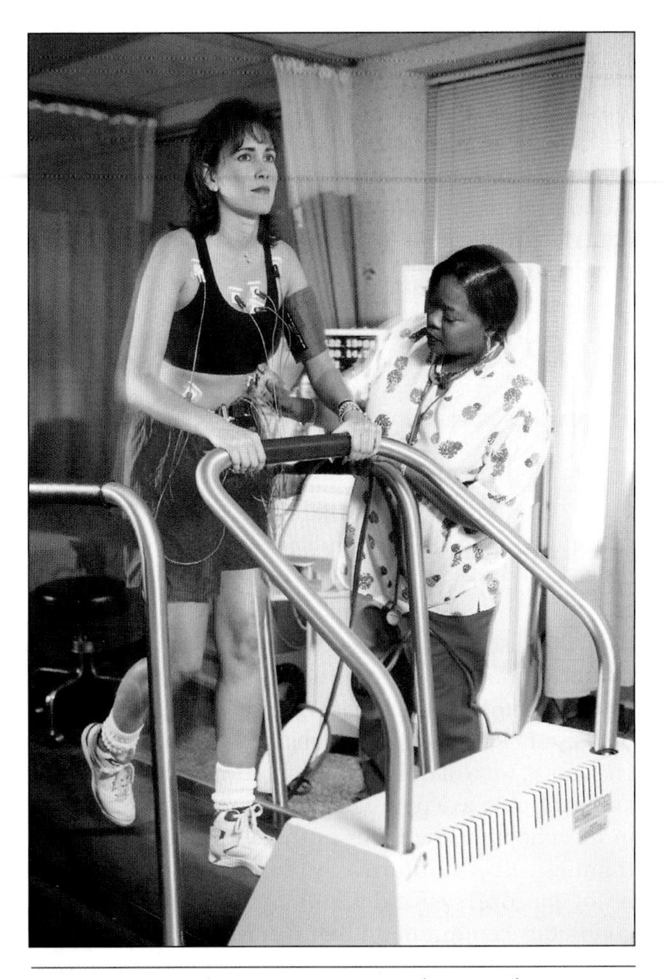

Figure 19.3 Obtaining an exercise electrocardiogram.

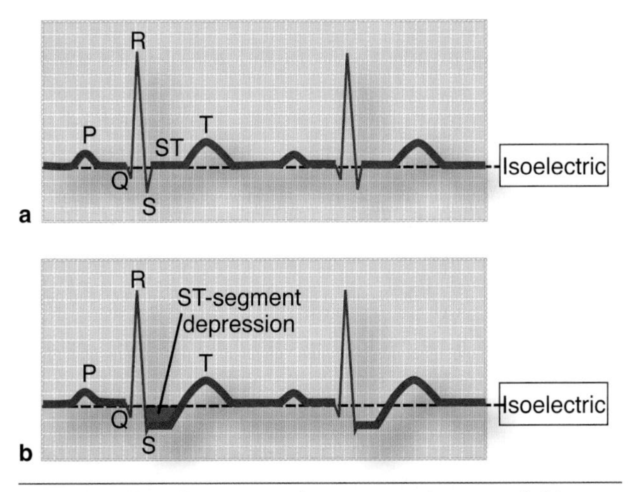

Figure 19.4 Illustration of *(a)* a normal ECG and *(b)* an ECG with ST-segment depression, suggestive of the presence of coronary artery disease.

the coronary arteries, the degree (%) of occlusion or narrowing can be determined.

To determine the accuracy of exercise ECG results, the sensitivity, specificity, and predictive value of an exercise test must be considered. **Sensitivity** refers to the exercise test's ability to correctly identify people who have the disease in question, such as CAD. **Specificity** refers to the test's ability to correctly identify people who do not have the disease. And the **predictive value of an abnormal exercise test** refers to the accuracy with which abnormal test results reflect presence of the disease.

> **In focus**
>
> The sensitivity and predictive value of an abnormal exercise test are generally low in a young, healthy population in which there is a low prevalence of CAD. As a result, the value of using exercise electrocardiography to screen for CAD in this population is questionable.

Unfortunately, both the sensitivity and the predictive value of an abnormal exercise test for detecting CAD are relatively low in healthy populations of people who have no symptoms of this disease. Past studies reveal that sensitivity averages about 66%, indicating that 66% of those with CAD are correctly identified by exercise ECGs as having the disease. Conversely, 34% of those with the disease are incorrectly diagnosed as disease free based on exercise ECGs. Average specificity is about 84%, indicating that 84% of those without disease are correctly identified by the exercise test as being disease free. But this means that 16% are incorrectly identified as having the disease.[9]

The predictive value of an abnormal exercise test varies considerably with the prevalence of CAD in the population. Assuming an average sensitivity of 60% and an average specificity of 90%, in a population with 5% prevalence of CAD, the predictive value of an abnormal test is only 24.0%. This indicates that only 24% of those who have an abnormal exercise ECG actually have CAD. In other words, in this population, more than three of every four people identified with abnormal exercise tests will be labeled as diseased but won't actually have identifiable CAD! But if we consider a population with a prevalence of 50%, the predictive value of an abnormal exercise test is much higher—85.7%.

From this information, we can conclude that exercise testing is of limited value in screening young, apparently healthy individuals before prescribing exercise for them. The accuracy of interpreting the results of the exercise ECG is questionable, particularly in a population with such a low prevalence of disease. Also, the actual risk of death or cardiac arrest during exercise is relatively low. Another important consideration is the expense of conducting clinical exercise tests, generally around $150 to $750 per test. Finally, far too few clinical facilities are equipped to conduct these tests to accommodate testing everyone who should be involved in an exercise program. Fortunately, the ACSM and the

American Heart Association have recommended this exercise test only for the at-risk groups mentioned earlier in this chapter. In fact, in 2005, the American Heart Association stated that there is insufficient evidence at this time to recommend exercise testing as a routine screening modality in asymptomatic adults.[13]

From a medical and legal perspective, however, the question must be raised as to whether a recommendation within a set of national guidelines is tantamount to a standard of practice in the medical community. The most recent guidelines of the ACSM indicate that a medical examination and exercise test are not necessary in people at low or moderate risk if low- to moderate-intensity exercise is undertaken gradually, with no competitive participation.[2] Moderate exercise is defined as exercise that is well within the individual's current capacity and can be sustained comfortably for a prolonged period of time, such as 60 min. In contrast, vigorous exercise is defined as exercise at an intensity greater than 60% of the individual's $\dot{V}O_{2max}$. This seems like a reasonable compromise, because moderate exercise is associated with considerable health benefits and few risks. But, as mentioned earlier, exercise testing offers benefits other than diagnosing CAD: It can provide valuable physiological data, such as a person's blood pressure response to exercise, and much of the data obtained can be used in formulating the exercise prescription.

In review

- Before beginning any exercise program, men over age 45, women over age 55, and anyone who is considered to be at a high risk for CAD should have a comprehensive medical evaluation.
- ACSM guidelines should be followed for each phase of the evaluation, and the physician should be consulted about the proposed exercise activity in case there are any medical contraindications.
- Exercise ECGs should be conducted for anyone in one of the previously mentioned high-risk categories. This test can detect undiagnosed CAD and other cardiac abnormalities.
- Test sensitivity refers to the test's ability to correctly identify people with a given disease. Test specificity refers to its ability to correctly identify people who do not have the disease. The predictive value of an abnormal exercise test refers to the accuracy with which the test reflects presence of the disease in a given population.
- The most recent ACSM guidelines state that a medical examination and exercise test might not be necessary if moderate exercise is undertaken gradually, without competition, in people without symptoms of cardiopulmonary disease.

EXERCISE PRESCRIPTION

The **exercise prescription** involves four basic factors:

1. Mode or type of exercise
2. Frequency of participation
3. Duration of each exercise bout
4. Intensity of the exercise bout

In our discussion, we assume that the goal of the exercise program is to improve aerobic capacity in people who have not been exercising. Since the prescription of resistance training programs is discussed in detail in chapter 8, we will only briefly mention including resistance training as a part of a total exercise program later in this chapter. The focus of this section is on aerobic training. Also, the information contained in this section is not appropriate for designing training programs for competitive endurance athletes or for those who simply wish to gain the health-related benefits of moderate activity but do not wish to improve aerobic capacity. Aerobic training for competitive athletes is covered in chapter 8.

Before examining the components of the exercise prescription, we must consider how much exercise is effective. A minimum **threshold** for frequency, duration, and intensity of exercise must be reached before any aerobic benefits are obtained. But, as we have discussed elsewhere, individual responses to any given training program are highly variable, so the threshold required differs from one person to the next. If we use exercise intensity as an example, a position statement by the ACSM for developing and maintaining aerobic capacity recommends a training intensity of 55% or 60% to 90% of one's maximum heart rate (HR_{max}) or 40% or 50% to 85% of $\dot{V}O_{2max}$.[1] Although this recommendation is appropriate for most healthy adults, some might improve their aerobic capacities at, for example, intensities below 40% of their

A minimal threshold for frequency, duration, and intensity of exercise must be reached to provide aerobic benefits from that exercise. Furthermore, minimal thresholds vary widely, making individualized exercise prescription necessary.

Parallel Careers, Lifelong Impact!

Since the early 1970s, considerable progress has been made in providing a research base for better understanding the relationship between an active lifestyle and reduced risk for chronic debilitating disease and for identifying how much and what types of activity promote health. During this time, two exercise scientists have had a particularly significant impact on helping us better understand the relationship between physical activity and disease prevention through their research, advocacy, and professional leadership. Interestingly, both had their roots in the Los Angeles area of Southern California and both were graduate students at the same time studying for their PhD degrees in exercise physiology at the University of Illinois. Dr. William L. Haskell received his undergraduate training at the University of California, Santa Barbara, while Dr. Michael L. Pollock was playing baseball and completing his undergraduate degree at the University of Arizona. Both served in the U.S. Army before completing their PhD degrees.

Dr. William L. Haskell. Dr. Michael L. Pollock.

During their professional careers, both were deeply committed to their research and to their primary professional organization, the ACSM, each serving as president. These men were instrumental in developing the original and subsequent ACSM Position Stands on the recommended quantity and quality of exercise needed to promote health and prevent chronic disease. These two scientists had parallel careers and a mutual impact on our understanding of the importance of an active lifestyle in promoting health and preventing chronic disease in today's sedentary society.

$\dot{V}O_{2max}$, whereas others would have to exercise at intensities greater than 85% $\dot{V}O_{2max}$ to show improvement. Each individual's threshold for frequency, duration, and intensity must be exceeded to achieve gains in aerobic capacity, and this threshold is likely to increase as aerobic capacity improves.

Exercise Mode

The prescribed exercise program should focus on one or more **modes,** or types, of cardiovascular endurance activities. Traditionally, the activities prescribed most frequently are

- walking,
- jogging,
- running,
- hiking,
- cycling,
- rowing, and
- swimming.

Because these activities do not appeal to everyone, alternative activities (see figure 19.5 on page 458) have been identified that should promote similar cardiovascular endurance gains. Spinning, aerobic dance, box or bench stepping, and most racket sports also have been shown to improve aerobic capacity.

For most competitive sport activities, preconditioning with one of the standard endurance activities, such as jogging or cycling, is advisable before one undertakes serious competition. Some researchers, clinicians, and practitioners believe that for people to successfully compete in certain sports or activities, a basic preconditioning program is essential to bring them up to the level of conditioning needed for the sport or activity and to reduce risk of injury. Rather than using the sport or activity to get in shape, people get in shape—or precondition—before participating in that sport or activity. For example, if the desired activity requires a moderate to high level of cardiovascular endurance, as basketball does, one might engage in jogging, aerobic dance, or a cycling program for several months until the endurance capacity increases to the necessary level. At that

Figure 19.5 Many activities promote gains in cardiovascular endurance, including aerobic dance.

> **In focus**
>
> Sport and recreational activities are appropriate for maintaining desirable fitness levels, but they generally are not appropriate for developing fitness in unfit individuals. Relatively unfit individuals should use conditioning activities to reach the desired level of fitness and then switch to the sport or recreational activity.

time, one switches over to the sport. The sport then acts as a maintenance activity through which one maintains the desired fitness level. In some cases, with intense sports, people can continue to develop their aerobic fitness level.

Individuals should select activities that they enjoy and are willing to continue throughout life. Exercise must be regarded as a lifetime pursuit because, as we saw in chapter 13, the benefits are soon lost if participation stops. Motivation is probably the most important factor in a successful exercise program. Selecting an activity that is fun, provides a challenge, and can produce needed benefits is one of the most crucial tasks in exercise prescription. Having several different activities to pursue is also wise in case of inclement weather, travel, or other barriers. Other considerations include geographic location, climate, and availability of equipment and facilities. Home exercise has become more common as many people are homebound either because of responsibilities, such as child rearing, or because of weather considerations such as heat, humidity, cold, rain, ice, and snow. Exercise videos and home exercise equipment have become popular, but they need to be selected carefully to avoid inappropriate exercise or faulty equipment. Potential purchasers should seek professional advice and, when possible, use the video or equipment during a trial period before purchase.

Exercise Frequency

The frequency of exercise participation, although certainly an important factor to consider, is probably less critical than either exercise duration or intensity. Research studies conducted on exercise frequency show that three to five days per week is an optimal frequency. This does not mean that six or seven days per week won't give additional benefits; but simply for the health-related benefits, the optimal gain is achieved with a time investment of three to five days per week. Exercise initially should be limited to three or four days per week and increased up to five or more days per week only if the activity is enjoyed and physically tolerated. All too often, a person starts out with great intentions, is highly motivated, and exercises every day for the first few weeks, only to stop from utter fatigue, injury, or boredom. Obviously, additional days above the three- to four-day frequency are beneficial for weight loss, but this level should not be encouraged until the exercise habit is firmly established and the injury risk is reduced.

Exercise Duration

Several studies have demonstrated improvement in cardiovascular conditioning with endurance exercise periods as brief as 5 to 10 min per day. More recent research has indicated that 20 to 30 min per day is an optimal amount. Again, "optimal" is used here to reflect the greatest return for time invested, and the specified time refers to the time during which one is at one's appropriate exercise intensity. Exercise duration cannot be discussed appropriately without discussion of exercise intensity also. Similar improvements in aerobic capacity are gained with either a short-duration, high-intensity program or a long-duration, low-intensity program if the minimal threshold is exceeded for both duration and intensity. Similar benefits are also gained whether the daily endurance training session is conducted in multiple shorter bouts (e.g., three 10 min bouts) or a single long one (e.g., a single 30 min bout).

Exercise Intensity

The intensity of the exercise bout appears to be the most important factor. How hard must people push themselves to gain benefits? Former athletes immediately recall the exhaustive workouts they endured to condition themselves for their sport. Unfortunately, this concept also gets carried over into the exercise programs they pursue for health benefits. Evidence now suggests that a modest training effect can be accomplished in some people through training at intensities of 40% or less of their aerobic capacities and possibly could lead to health benefits. For most, however, the appropriate minimum intensity appears to be at least 50% to 60% $\dot{V}O_{2max}$. An upper level for intensity will depend on the purpose for training. Obviously, training for competition will require high-intensity training (see chapter 13). However, training for purposes of attaining and maintaining optimal health would seldom exceed 80% $\dot{V}O_{2max}$.

In review

- The four basic factors in an exercise program are exercise mode, frequency, duration, and intensity. A minimum threshold for the last three must be met to yield any aerobic benefits, and this threshold is quite variable from one individual to another.
- The program should include one or more cardiovascular endurance activities. If the activity involves competition, preconditioning with a standard endurance activity is recommended before sport participation begins to bring the person up to an appropriate level of fitness.
- Activities must be matched with individual needs and likes so that motivation can be maintained.
- Optimal exercise frequency is three to five days of training each week, although greater frequency might provide additional benefits. Exercise should begin with three to four sessions per week and then progress to more if desired.
- Exercise duration of 20 to 30 min working at the appropriate intensity is optimal, but the key is reaching the threshold for both duration and intensity.
- Exercise intensity appears to be the most important factor. For most people, intensity should be at least 50% to 60% $\dot{V}O_{2max}$. However, health benefits might occur at intensities lower than those needed for aerobic conditioning in some people.

MONITORING EXERCISE INTENSITY

Exercise intensity can be quantified on the basis of the training heart rate (THR), the metabolic equivalent (MET), or the rating of perceived exertion (RPE). Let's examine each of these and their strengths and weaknesses in quantifying exercise intensity.

Training Heart Rate

The concept of **training heart rate (THR)** is based on the linear relationship between heart rate and $\dot{V}O_2$ with increasing rates of work, as shown in figure 19.6 on page 460. When people are exercise tested, their heart rate and $\dot{V}O_2$ values are obtained each minute and plotted against each other. The THR is established through use of the heart rate that is equivalent to a set percentage of the $\dot{V}O_{2max}$. For example, if a training level of 75% $\dot{V}O_{2max}$ is desired, 75% of $\dot{V}O_{2max}$ is calculated ($\dot{V}O_{2max} \times 0.75$), and the heart rate corresponding to this $\dot{V}O_2$ then is selected as the THR. An important point is that the exercise intensity necessary to achieve a given percentage of $\dot{V}O_{2max}$ results in a much higher heart rate than that same percentage of HR_{max}. As an example, a THR that is set at 75% of the $\dot{V}O_{2max}$ represents an intensity of 87% of the HR_{max} (see figure 19.6 on page 460).

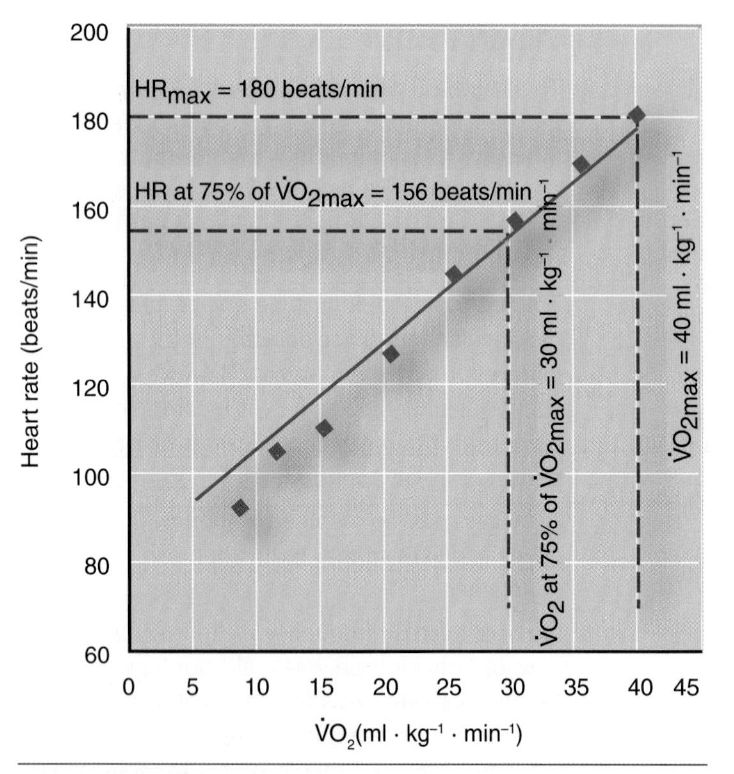

Figure 19.6 The linear relationship between heart rate and oxygen consumption ($\dot{V}O_2$) with increasing rates of work and the heart rate equivalent to a set percentage (75%) of $\dot{V}O_{2max}$.

The Karvonen Method

One can also establish the THR by using what is known as the Karvonen concept of maximal heart rate reserve, or the **Karvonen method.**[11] **Maximal heart rate reserve** is defined as the difference between HR_{max} and the resting heart rate (HR_{rest}):

$$\text{maximal heart rate reserve} = HR_{max} - HR_{rest}.$$

With this method, the THR is calculated by taking a given percentage of the maximal heart rate reserve and adding it to the resting heart rate. Let's consider an example. For 75% of maximal heart rate reserve, the equation would be as follows:

$$THR_{75\%} = HR_{rest} + 0.75\ (HR_{max} - HR_{rest}).$$

The Karvonen method adjusts the THR so that THR as a specific percentage of the maximal heart rate reserve is nearly identical to the heart rate equivalent of that same percentage of $\dot{V}O_{2max}$ at moderate to high intensities.[8] Thus, a THR computed as 75% of maximal heart rate reserve is approximately the same as the heart rate corresponding to 75% of the $\dot{V}O_{2max}$. However, there is a substantial difference between the two at low intensities.[19]

Training Heart Rate Range

More recently, appropriate exercise intensity has been established by setting a THR range, rather than a single THR value. This is a more sensible approach because exercising at a set percentage of $\dot{V}O_{2max}$ can place people above their lactate threshold, making it difficult for them to train for any extended period. With the THR range concept, low and high values are established that will ensure a training response. One starts at the low end of the THR range and progresses through the range as one feels comfortable. To illustrate this, using the Karvonen method for establishing the THR, consider the following example. A 40-year-old man has a resting heart rate of 75 beats/min and a maximum heart rate of 180 beats/min, and he is advised to exercise within a THR range of 50% to 75% of his maximal heart rate reserve. His training heart rate range would be as follows:

$$THR_{50\%} = 75 + 0.50\ (180 - 75)$$
$$= 75 + 53 = 128 \text{ beats/min}.$$

$$THR_{75\%} = 75 + 0.75\ (180 - 75)$$
$$= 75 + 79 = 154 \text{ beats/min}.$$

This same THR range method can be used if one estimates HR_{max} [208 – (0.7 × age)] without losing much accuracy if true HR_{max} has not been determined.

The concept of THR is extremely valuable. Heart rate is highly correlated with the work done by the heart. Heart rate alone is a good index of myocardial oxygen consumption as well as coronary blood flow. With use of the THR method for monitoring exercise intensity, the heart works at the same rate, even though the metabolic cost of the work might vary considerably. As an example, during exercise at high altitudes or in the heat, the heart rate will be elevated significantly if the person attempts to maintain a set rate of work, such as running at a pace of 6 min/km (9 min/mi). With the THR method, one simply trains at a lower rate of work under these extreme environmental conditions to maintain the same heart rate (THR). This is a much safer approach to monitoring exercise intensity, particularly for high-risk patients in whom the work of the heart must be closely regulated. The THR method also allows for improvement in aerobic capacity with training. As people become better conditioned, their heart rate decreases for the same rate of work, which means that they must perform at a higher rate of work to reach their THR.

It is important to come back to a point that was raised in the first paragraph of this section—the point about lac-

In focus

Heart rate is the preferred method for monitoring exercise intensity because it is highly correlated to the work of the heart (or stress on the heart) and allows for a progressive increase in the rate of training with improvements in fitness to maintain the same THR. When one is prescribing exercise intensity, it is appropriate to establish a training heart rate range, with exercise starting at the low end of the range and progressing to the upper end of the range over time.

tate threshold. In chapter 4 we learned that lactate threshold was that point as one increases exercise intensity where the rate of production of lactate exceeds the rate of its clearance, resulting in increased blood lactate levels. When people are at an exercise intensity above their lactate threshold, they limit the length of time they can train comfortably at that intensity. For those just starting a training program, it is important not to exceed the lactate threshold. Having a THR range allows people to set the lower end of the range at an intensity that would be below the expected lactate threshold for an untrained individual. Obviously it would be better to actually measure the lactate threshold so that this range could be more accurately determined. However, this isn't practical because of the difficulty and expense associated with determining lactate threshold.

Metabolic Equivalent

Exercise intensity also has been prescribed on the basis of the **metabolic equivalent (MET)** system. The amount of oxygen the body consumes is directly proportional to the energy expended during physical activity. In this system, it is assumed that the body uses approximately 3.5 ml of oxygen per kilogram (2.2 lb) of body weight per minute ($3.5 \text{ ml} \cdot \text{kg}^{-1} \cdot \text{min}^{-1}$) at rest. However, in the sidebar, we see that this is likely not the case. The MET system, however, is based on this value, and the resting metabolic rate value of $3.5 \text{ ml} \cdot \text{kg}^{-1} \cdot \text{min}^{-1}$ is referred to as 1.0 MET. All activities can be classified by intensity according to their oxygen requirements. An activity that is rated as a 2.0 MET activity would require two times the resting metabolic rate, or $7 \text{ ml} \cdot \text{kg}^{-1} \cdot \text{min}^{-1}$; and an activity that is rated at 4.0 METs would require approximately $14 \text{ ml} \cdot \text{kg}^{-1} \cdot \text{min}^{-1}$. Some activities and their MET values are presented in table 19.2 on page 462. These values are only approximations because of the potential error in using $3.5 \text{ ml} \cdot \text{kg}^{-1} \cdot \text{min}^{-1}$ as a constant resting value. Further, metabolic efficiency varies considerably from one person to the next, and even in the same individual. Although the MET system is useful as a guideline for training, it fails to account for changes in environmental conditions, and it does not allow for changes in physical conditioning as discussed in the previous section.

Prescribing Exercise Intensity Using the $\dot{V}O_2$ Reserve Method

In the ACSM Position Stand on exercise prescription,[1] a slightly different approach to prescribing exercise intensity was proposed. Exercise intensity is prescribed based on what has been termed the $\dot{V}O_2$ reserve method ($\dot{V}O_2R$). Instead of prescribing exercise at a given percentage of $\dot{V}O_{2max}$, one bases the prescription on a given percentage of the $\dot{V}O_2R$, where $\dot{V}O_2R$ is defined as $\dot{V}O_{2max} - \dot{V}O_{2rest}$. This also can be thought of as the $\dot{V}O_{2max}$ reserve. As an example, with a $\dot{V}O_{2max}$ of $40 \text{ ml} \cdot \text{kg}^{-1} \cdot \text{min}^{-1}$ and a $\dot{V}O_{2rest}$ of $3.5 \text{ ml} \cdot \text{kg}^{-1} \cdot \text{min}^{-1}$, $\dot{V}O_2R = 40 - 3.5 \text{ ml} \cdot \text{kg}^{-1} \cdot \text{min}^{-1} = 36.5 \text{ ml} \cdot \text{kg}^{-1} \cdot \text{min}^{-1}$. To prescribe an exercise intensity range of between 60% and 75% of $\dot{V}O_2R$, we simply multiply $\dot{V}O_2R$ by 60% and 75%: $\dot{V}O_2R_{60\%} = 36.5 \text{ ml} \cdot \text{kg}^{-1} \cdot \text{min}^{-1} \times 0.60 = 21.9 \text{ ml} \cdot \text{kg}^{-1} \cdot \text{min}^{-1}$; and $\dot{V}O_2R_{75\%} = 36.5 \text{ ml} \cdot \text{kg}^{-1} \cdot \text{min}^{-1} \times 0.75 = 27.4 \text{ ml} \cdot \text{kg}^{-1} \cdot \text{min}^{-1}$. The major advantage of using the $\dot{V}O_2R$ technique is that one now has an equivalency between the percentage of the maximal heart rate reserve and the percentage of $\dot{V}O_{2max}$ reserve. There is a potential problem with this technique, however, in that using $3.5 \text{ ml} \cdot \text{kg}^{-1} \cdot \text{min}^{-1}$ as a standard value for $\dot{V}O_{2rest}$ assumes that everyone has the same resting value. This, in fact, is not the case. Further, in one study, a large sample of women ($n = 642$) and men ($n = 127$) were found to have average $\dot{V}O_{2rest}$ values of 2.5 and $2.7 \text{ ml} \cdot \text{kg}^{-1} \cdot \text{min}^{-1}$. The range of values varied from 1.6 to $4.1 \text{ ml} \cdot \text{kg}^{-1} \cdot \text{min}^{-1}$.[6]

TABLE 19.2 Selected Activities and Their Respective MET Values

Activity	MET	Activity	MET
Self-care			
Rest, supine	1.0	Propulsion, wheelchair	2.0
Sitting	1.0	Walking, 4 km/h (2.5 mph)	3.0
Standing, relaxed	1.0	Showering	3.5
Eating	1.0	Walking downstairs	4.5
Conversation	1.0	Walking, 5.6 km/h (3.5 mph)	5.5
Dressing and undressing	2.0	Ambulation, braces and crutches	6.5
Washing hands and face	2.0		
Housework			
Hand sewing	1.0	Scrubbing floors	3.0
Machine sewing	1.5	Cleaning windows	3.0
Sweeping floor	1.5	Making beds	3.0
Polishing furniture	2.0	Ironing, standing	3.5
Peeling potatoes	2.5	Mopping	3.5
Scrubbing, standing	2.5	Wringing wash by hand	3.5
Washing small clothes	2.5	Hanging wash	3.5
Kneading dough	2.5	Beating carpets	4.0
Occupational			
Sitting at desk	1.5	Bricklaying and plastering	3.5
Writing	1.5	Heavy assembly work	4.0
Riding in automobile	1.5	Wheeling heavy wheelbarrow	4.0
Watch repair	1.5	Carpentry	5.5
Typing	2.0	Mowing lawn by hand mower	6.5
Welding	2.5	Chopping wood	6.5
Radio assembly	2.5	Shoveling	7.0
Playing musical instrument	2.5	Digging	7.5
Parts assembly	3.0		
Physical conditioning			
Level walking, 3.2 km/h, 1 km in 19 min (2 mph, 1 mi in 30 min)	2.5	Swimming, crawl, 0.3 m/s (1 ft/s)	5.0
Level cycling, 8.9 km/h, 1 km in 6 min 44 s (5.5 mph, 1 mi in 10 min 54 s)	3.0	Level walking, 5.6 km/h, 1 km in 10 min 43 s (3.5 mph, 1 mi in 17 min)	5.5
Level cycling, 9.7 km/h, 1 km in 6 min 12 s (6 mph, 1 mi in 10 min)	3.5	Level walking, 6.4 km/h, 1 km in 9 min 23 s (4.0 mph, 1 mi in 15 min)	6.5
Level walking, 4 km/h, 1 km in 15 min (2.5 mph, 1 mi in 24 min)	3.5	Level jogging, 8 km/h, 1 km in 7 min 30 s (5.0 mph, 1 mi in 12 min)	7.5
Level walking, 4.8 km/h, 1 km in 12 min 30 s (3 mph, 1 mi in 20 min)	4.5	Level running, 12 km/h, 1 km in 5 min (7.5 mph, 1 mi in 8 min)	9.0
Calisthenics	4.5	Level cycling, 20.9 km/h, 1 km in 2 min 52 s (13 mph, 1 mi in 4 min 37 s)	9.0
Level cycling, 15.6 km/h, 1 km in 3 min 50 s (9.7 mph, 1 mi in 6 min 18 s)	5.0	Swimming, crawl 0.6 m/s (2 ft/s)	10.0

Activity	MET	Activity	MET
		Physical conditioning *(continued)*	
Level running, 13.7 km/h, 1 km in 4 min 23 s (8.5 mph, 1 mi in 7 min)	12.0	Level running, 19.3 km/h, 1 km in 3 min 6 s (12 mph, 1 mi in 5 min)	20.0
Level running, 16 km/h, 1 km in 3 min 45 s (10.0 mph, 1 mi in 6 min)	15.0	Level running, 24.1 km/h, 1 km in 2 min 30 s (15 mph, 1 mi in 4 min)	30.0
Swimming, crawl, 0.8 m/s (2.5 ft/s)	15.0	Swimming, crawl, 1.1 m/s (3.5 ft/s)	30.0
Swimming, crawl, 0.9 m/s (3.0 ft/s)	20.0		
		Recreational	
Painting, sitting	1.5	Table tennis	4.5
Playing piano	2.0	Baseball	4.5
Driving car	2.0	Tennis	6.0
Canoeing, 4 km/h (2.5 mph)	2.5	Horseback riding, trot	6.5
Horseback riding, walk	2.5	Folk dancing	6.5
Volleyball, 6-player recreational	3.0	Skiing	8.0
Billiards	3.0	Horseback riding, gallop	8.0
Bowling	3.5	Squash racquets	8.5
Horseshoes	3.5	Fencing	9.0
Golf	4.0	Basketball	9.0
Cricket	4.0	Football	9.0
Archery	4.5	Gymnastics	10.0
Ballroom dancing	4.5	Handball and paddleball	10.0

In focus

One simple way of monitoring exercise intensity is referred to as the talk test, which has been used as an informal guideline for years. Scientists have now confirmed that the highest exercise intensity that just barely allows a person to talk comfortably while exercising is a very consistent method that correlates well with ventilatory threshold (see chapter 7) and is well within the THR range.[17]

6	No exertion at all
7	Extremely light
8	
9	Very light
10	
11	Light
12	
13	Somewhat hard
14	
15	Hard (heavy)
16	
17	Very hard
18	
19	Extremely hard
20	Maximal exertion

Borg RPE scale
© Gunnar Borg, 1970, 1985, 1994, 1998

Figure 19.7 The Borg ratings of perceived exertion scale.

From G. Borg, 1998, *Borg's perceived exertion and pain scales* (Champaign, IL: Human Kinetics), 47.

Ratings of Perceived Exertion

Ratings of perceived exertion (RPE) also have been proposed for use in prescribing exercise intensity. With this method, individuals subjectively rate how hard they feel they are working. A given numerical rating corresponds to the perceived relative intensity of exercise. When the RPE scale is used correctly, this system for monitoring exercise intensity has proven very accurate. Using the **Borg RPE scale,**[5] which is a rating scale ranging from 6 to 20 (figure 19.7), the exercise intensity should be between an RPE of 12 to 13 (somewhat hard) and an RPE of 15 to 16 (hard). Initially, this sounds too simple. However, most people can use the RPE technique very accurately. Studies have shown that when people are asked to select a pace on a treadmill, or a resistance on a cycle ergometer, at a moderate or heavy intensity of exercise (see table 19.3), they are able to select a pace or resistance that gets their heart rates into the appropriate range. This is a more natural way to prescribe exercise and very efficient if the person is able to relate perceptions of intensity accurately.

Table 19.3 compares the various methods for rating exercise intensity. Let's use them to determine a moderate exercise intensity. As the second column shows, one would want to work within a range of 60% to 79% HR_{max}. If, instead, one is monitoring intensity by $\dot{V}O_{2max}$ or using the Karvonen method, this heart rate range is equivalent to 50% to 74% of either $\dot{V}O_{2max}$ or HR_{max} reserve, as shown in the third column. With use of the rating of perceived exertion, shown in the fourth column, this is equivalent to an RPE value of 12 to 13. All these values reflect moderate-intensity exercise.

In focus

Physical activity must be considered a lifetime pursuit! The benefits of a sound exercise program are rapidly lost once that program is discontinued.

TABLE 19.3 Classification of Exercise Intensity Based on 20 to 60 min of Endurance Activity: Comparing Three Methods

	Relative intensity		
Classification of intensity	HR_{max}	$\dot{V}O_{2max}$ or HR_{max} reserve	Rating of perceived exertion
Very light	<35%	<30%	<9
Light	35-59%	30-49%	10-11
Moderate	60-79%	50-74%	12-13
Heavy	80-89%	75-84%	14-16
Very heavy	≥90%	≥85%	>16

Fom an article published in *Exercise in health and disease: Evaluation and prescription for prevention and rehabilitation,* 2nd ed. M.L. Pollock and J.H. Wilmore. Copyright Elsevier 1990.

In review

- Exercise intensity can be monitored on the basis of training heart rate, metabolic equivalent, or rating of perceived exertion.
- Training heart rate can be established through use of the heart rate equivalent to a certain percentage of $\dot{V}O_{2max}$. It can also be determined using the Karvonen method, which takes a given percentage of maximal heart rate reserve and adds it to resting heart rate. With this method, the percentage of maximal heart rate reserve used corresponds to approximately the same percentage of $\dot{V}O_{2max}$ when a person is exercising at moderate to high intensities.
- A sensible approach is to establish a THR range to work within, instead of a single THR, attempting to estimate the low end at an intensity below lactate threshold.
- The amount of oxygen consumed reflects the amount of energy expended during an activity. $\dot{V}O_2$ at rest has been assigned a value of 3.5 $ml \cdot kg^{-1} \cdot min^{-1}$, which equals 1.0 MET. Activity intensities can be classified by their oxygen requirements as multiples of the resting metabolic rate.
- The RPE method requires that a person subjectively rate how difficult the work is, using a numerical scale that is related to exercise intensity. The subject looks at the standard scale to determine the appropriate number.

EXERCISE PROGRAM

Once the exercise prescription has been determined, it is integrated into a total exercise program, which is generally only part of an overall health improvement plan. Individual exercise capacity varies widely even among people of similar ages and physical builds. For this reason, each program must be individualized, based on results of physiological and medical tests and, if possible, individual needs and interests.

The total exercise program consists of the following activities:

- Warm-up and stretching activities
- Endurance training
- Cool-down and stretching activities
- Flexibility training
- Resistance training
- Recreational activities

Generally, the first three activities are performed three to four times each week. Flexibility training can be included as part of the warm-up, cool-down, and stretching exercises, or it can be done at a separate time during the week. Resistance training is usually done on alternate days when endurance training is not done; however, the two can be combined into the same workout.

Warm-Up and Stretching Activities

The exercise session should begin with low-intensity, calisthenic-type and stretching exercises (figure 19.8). Such a warm-up period gradually increases both heart rate and breathing, preparing the exerciser for the efficient and safe functioning of the heart, blood vessels, lungs, and muscles during the more vigorous exercise that follows. A good warm-up can reduce the amount of muscle and joint soreness experienced during the early stages of the exercise program. An acceptable warm-up would begin with 5 to 10 min of stretching, followed by 5 to 10 min of low-intensity activity using the mode of exercise selected for endurance training. For example, someone who trains by running might start with stretching and then do 5 to 10 min of light jogging before starting to run.

Endurance Training

Physical activities that develop cardiovascular endurance are the heart of the exercise program. They are designed to improve both the capacity and efficiency of the cardiovascular, respiratory, and metabolic systems. These activities also help one control or reduce body weight. Activities such as walking, jogging, running, cycling, swimming, rowing, aerobic dancing, box stepping, and hiking are good endurance activities. Sports such as handball, racquetball, tennis, badminton, and basketball also have aerobic potential if they are pursued vigorously. Activities such as golf, bowling, and softball are generally of little value for developing aerobic capacity; but they are fun, have definite recreational value, and may offer health-related benefits. For these reasons, such activities certainly have a place in the overall exercise program.

Cool-Down and Stretching Activities

Every endurance exercise session should conclude with a cool-down period. The best way to accomplish cool-down is to slowly reduce the intensity of the endurance activity during the last several minutes of a workout. After running, for example, a slow, restful walk for several minutes helps prevent blood from pooling in the extremities. Stopping abruptly after an endurance exercise bout causes blood to pool in the legs and can result in dizziness or fainting. Also, catecholamine levels might be elevated during the immediate recovery period, and this can lead to a fatal heart arrhythmia.

Figure 19.8 The warm-up period should include stretching and low-intensity activity, such as walking or light jogging.

After the cool-down period, stretching exercises can be performed to facilitate increased flexibility.

Flexibility Training

Flexibility exercises (see figure 19.9) usually supplement exercises performed during the warm-up or cool-down period and are useful for those who have poor flexibility or muscle and joint problems, such as low back pain. These exercises should be performed slowly. Quick stretching movements are potentially dangerous and can lead to muscle pulls or spasms. At one time it was recommended that these exercises be performed before the endurance conditioning period. However, recently it has been hypothesized that the muscles, tendons, ligaments, and joints are more adaptable and responsive to flexibility exercises when they are done after the endurance conditioning phase. Research has yet to confirm this hypothesis.

Figure 19.9 Flexibility exercises should be performed slowly to prevent injury.

Resistance Training

The importance of resistance training as part of a general health and fitness exercise program has been clearly established. Many health-related benefits can be obtained from resistance training. The ACSM has included resistance training in its recommendations for a general health and fitness program.[1]

Starting a Resistance Training Program

Recall from chapter 8 that the maximum amount of weight one can lift successfully only one time is the 1-repetition maximum, or 1RM. When people begin a resistance training program, they should start with a weight that is exactly one-half of their maximal strength, or 1RM, for each lift. They should attempt to lift that weight 10 consecutive times. If they can lift the weight just 10 times before reaching fatigue, this is the proper starting point. If they can do more repetitions, they should go to the next higher weight for the second set. If, instead, they were able to lift the weight fewer than eight times in the first set, the original weight was too heavy and should be reduced to the next lower weight for the second set.

When a given weight brings the exerciser to fatigue by the 8th to 10th repetition in the first set, this is the appropriate starting weight. People should try to achieve as many repetitions as possible during the second and third sets, but the number of repetitions they can complete in these last sets will probably decrease as their muscles become fatigued. As strength increases, the number of repetitions they can complete per set will increase. When one reaches 15 repetitions on the first set, one is ready to progress to the next higher weight. This training technique is referred to as progressive resistance exercise. See chapter 8 for further details.

One can perform two or three sets of each lift per day, two to three days per week, for weight control purposes. Strength gains, however, appear to be fully achieved with just one set per day in previously untrained people. People should select a variety of exercises that tax most or all of the major muscle groups of the upper body, trunk, and lower body. If time is a factor, it is better to reduce the number of sets to one or two and maintain a full-body workout.

Health Benefits Associated With Resistance Training

Interest has surged in the use of resistance training for promoting improvements in general health. Unfortunately, research findings are not always consistent, and this is likely attributable to differences in the training programs used in different studies. The volume and intensity of the training, the rest interval between sets and exercises, and the selection of exercises all can greatly influence the results of a given study. With this in mind, table 19.4, published by the American Heart Association in 2000, briefly summarizes some of the potential health benefits of resistance training compared with those of aerobic training.[18] This table illustrates that both aerobic and resistance training exercise can provide tremendous health benefits, independently and in combination.

TABLE 19.4 Comparison of the Effects of Aerobic Endurance Training and Strength Training on Health and Fitness Variables

Variable	Aerobic exercise	Resistance exercise
Bone mineral density	↑↑	↑↑
Body composition		
% fat	↓↓	↓
Lean body mass	↔	↑↑
Strength	↔	↑↑↑
Glucose metabolism		
Insulin response to glucose challenge	↓↓	↓↓
Basal insulin levels	↓	↓
Insulin sensitivity	↑↑	↑↑
Serum lipids		
HDL-C	↑↔	↑↔
LDL-C	↓↔	↓↔
Resting heart rate	↓↓	↔
Stroke volume	↑↑	↔
Blood pressure at rest		
Systolic	↓↔	↔
Diastolic	↓↔	↓↔
$\dot{V}O_{2max}$	↑↑↑	↑↔
Endurance performance	↑↑↑	↑↑
Basal metabolism	↑	↑↑

Note. HDL-C = high-density lipoprotein cholesterol; LDL-C = low-density lipoprotein cholesterol; ↑ = increase; ↓ = decrease; ↔ = little or no change. The more arrows, the greater the change.

Reprinted from M.L. Pollock and K.R. Vincent, 1996, "Resistance training for health," *Research Digest: Presidents' Council on Physical Fitness and Sports* 2(8): 1-6.

Recreational Activities

Recreational activities (see figure 19.10) are important to any comprehensive exercise program. Although people engage in these activities primarily for enjoyment and relaxation, many recreational activities can also improve health and fitness. Activities such as hiking, tennis, handball, squash, and certain team sports fall into this category. Guidelines for selecting these activities include the following:

Figure 19.10 The recreational activities in which a person can participate regularly may depend on location, but many communities offer a wide variety of programs.

- Can you learn or perform the activities with at least a moderate degree of success?
- Do the activities include opportunities for social development, if that is desired?
- Are the costs associated with participation reasonable and within your budget?
- Are the activities varied enough to maintain your continued long-term interest?
- Given your current age and health status, is this activity safe for you?

Many excellent opportunities exist for people who have no recreational hobbies or activities but who would like to become involved. Local public recreation centers, park districts, YMCAs, YWCAs, churches and some public schools, community colleges, and universities offer instructional classes in a wide variety of activities at little or no cost. Often the entire family can participate in these classes—an added bonus to a total health improvement program! Also, the number of commercial fitness centers is rapidly growing, and most now employ trained staff who can properly prescribe exercise programs and help individuals get started.

In review

- An exercise session should begin with a warm-up of low-intensity, calisthenic-type and stretching exercises to prepare the cardiovascular, respiratory, and muscle systems to work more efficiently.
- Endurance activities should be performed three to four times each week.
- Each endurance session should be followed by cool-down and stretching to prevent blood pooling in the extremities and muscle soreness.
- Flexibility exercise should be performed slowly, and this phase of the program might be best included immediately after the endurance component.
- Resistance training should begin with a weight of one-half the person's 1RM. This is the proper weight if the person can lift it about 10 times. If he or she can lift it more than that, more weight is needed; if he or she can lift it fewer than eight times, less is needed.
- Recreational activities should be included in an exercise program for enjoyment and relaxation.
- Exercise is a vital part of rehabilitation for most diseases. The type and details of the rehabilitation program depend on the patient, the specific disease involved, and its extent.

EXERCISE AND REHABILITATION OF PEOPLE WITH DISEASES

Exercise has become a major component in **rehabilitation programs** for a number of diseases. Cardiopulmonary rehabilitation programs, which began in the 1950s, have become the most visible. Tremendous advances in cardiopulmonary rehabilitation have led to the formation of a professional association, the American Association of Cardiovascular and Pulmonary Rehabilitation, and a professional research journal, the *Journal of Cardiopulmonary Rehabilitation.*

Exercise is also an important part of the rehabilitation of people with

- cancer;
- obesity;
- diabetes;
- renal disease;
- osteoporosis;
- arthritis, chronic fatigue syndrome, and fibromyalgia; and
- cystic fibrosis.

Most recently, emphasis on the use of exercise in the rehabilitation of transplant patients, including those with heart transplants, liver transplants, and kidney transplants, has increased because exercise helps alleviate some drug side effects and improves general health.

The manner in which exercise is used in the rehabilitation of people with disease is highly specific to the nature and extent of the disease. It is therefore beyond the scope of this chapter to present specific details for any disease; but many resources are now available that provide extensive detail about establishing exercise programs for those with specific diseases and the clinical values of these programs.

In focus

Exercise training has become an extremely important part of rehabilitation programs for a number of diseases. Although the specific physiological mechanisms explaining the benefits of exercise training for each of these diseases have not been clearly defined, exercise training carries many general health benefits that appear to improve the patient's prognosis.

In Closing

In this chapter, we have seen that the medical community now regards a physically active lifestyle as vital to maintaining good health and reducing the risk of disease. We looked at the importance and practicality of both a medical examination and an exercise ECG in screening previously sedentary adults before prescribing exercise. We discussed the components of an exercise prescription and methods of monitoring exercise intensity. Finally, we reviewed the components of an exercise program and the role of exercise in rehabilitating patients with disease.

Now that we have seen the importance of exercise in disease prevention, we will look more closely at physical activity as it relates to specific disease states. In the next chapter, we turn our attention to cardiovascular diseases.

KEY TERMS

Borg RPE scale
coronary artery disease (CAD)
exercise electrocardiogram (ECG)
exercise prescription
graded exercise test (GXT)
Karvonen method
maximal heart rate reserve
metabolic equivalent (MET)
mode
predictive value of an abnormal exercise test
rating of perceived exertion (RPE)
rehabilitation programs
sensitivity
specificity
threshold
training heart rate (THR)

STUDY QUESTIONS

1. How active are adult Americans today? Are we in a fitness revolution?
2. What were the general conclusions of the Surgeon General's report, *Physical Activity and Health*?
3. What role does the graded exercise test to exhaustion play in medical clearance? Is this test essential for all adults?
4. Discuss the concepts of sensitivity and specificity of exercise testing and the predictive value of an abnormal test. Of what value is this information in the establishment of policy mandating who should be exercise tested?
5. How can we get our population to be more active? What levels of exercise do we need to promote to help people gain the health-related benefits associated with exercise?
6. What four factors must be considered in the exercise prescription? Which of these is the most important?
7. Discuss the concept of a minimal threshold for initiating physiological changes with exercise training as it relates to the exercise prescription.
8. Discuss the various ways of monitoring exercise intensity, and give the advantages and disadvantages of each.
9. Describe the components of a good exercise program and their importance in the total program.
10. How does one effectively motivate individuals to maintain regular exercise habits?

STUDY GUIDE ACTIVITIES

In addition to the activities listed in the chapter opening outline on page 449, two other activities are available in the online study guide, located at

www.HumanKinetics.com/PhysiologyOfSportAndExercise.

The chapter **SUMMARY** reviews the key concepts covered in the chapter, and the end-of-chapter **QUIZ** tests your understanding of the material covered in the chapter.

Cardiovascular Disease and Physical Activity

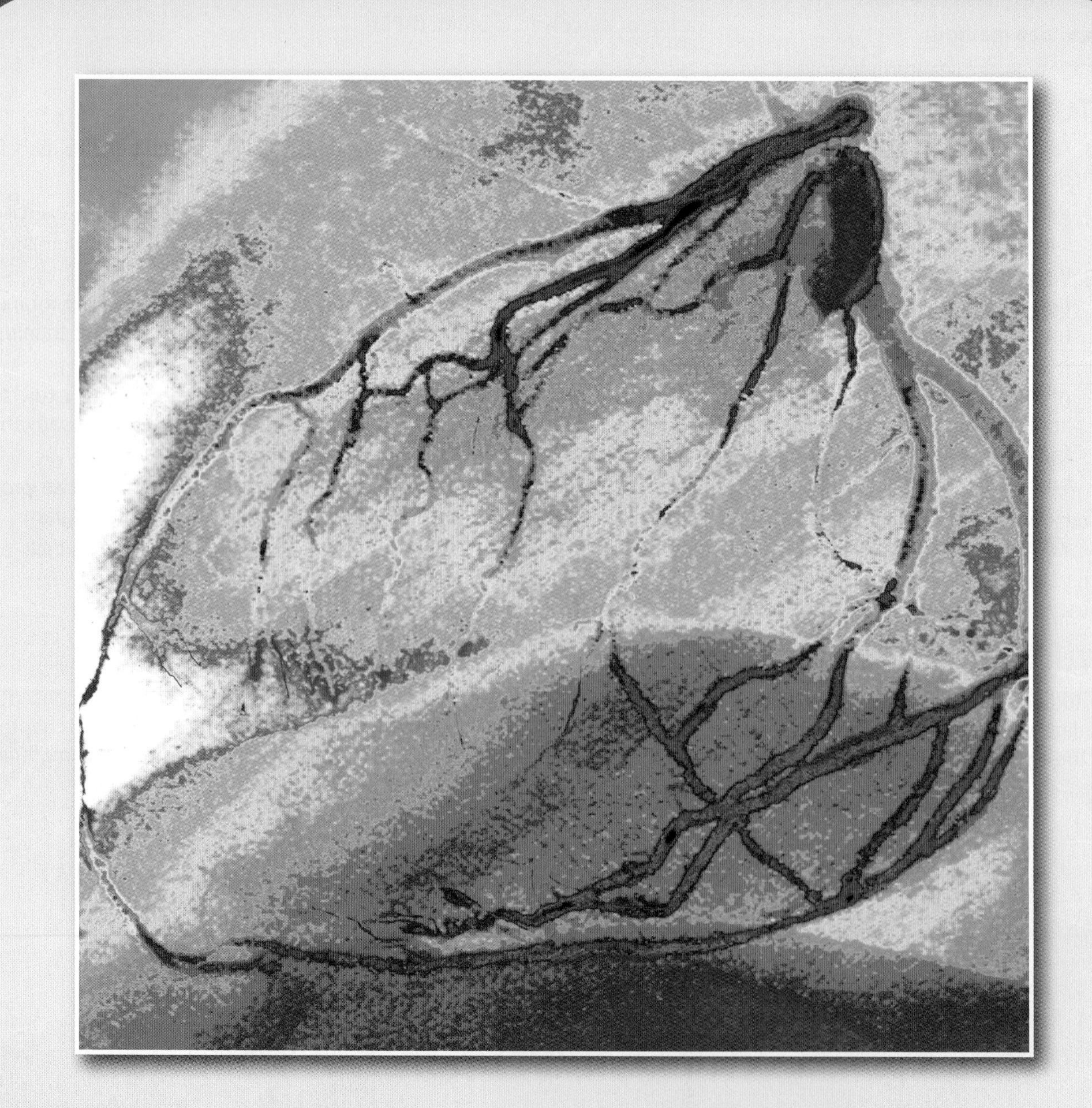

In this chapter and in the online study guide

On Saturday afternoon, June 22, 2002, St. Louis Cardinals pitcher Darryl Kile was found dead in his hotel room in Chicago. The Cardinals were in Chicago to play a three-game series against the Chicago Cubs. He was scheduled to start the last game of the series on Sunday night. He was considered one of the Cardinals' best pitchers and a clubhouse leader. Kile, who was only 33 years old, apparently died of a heart attack caused by coronary atherosclerosis—on autopsy, two of his three major coronary arteries were found to be narrowed by 80% to 90%. Although he had no medical history or symptoms of disease, his father had died of a stroke at the age of 44 years, and Kile had complained of shoulder pain and fatigue during dinner the previous night. This tragedy illustrates the important fact that being an outstanding athlete during youth and young adulthood does not confer lifelong immunity for coronary artery disease (CAD). Although a genetic predisposition to CAD is serious, it doesn't have to result in premature death. Paying close attention to all of the CAD risk factors and knowing how to minimize the risk become extremely important.

Most of us consider ourselves to be healthy until we experience some overt sign of illness. With chronic degenerative diseases, such as heart disease, most people are unaware that the disease process is smoldering and progressing to the point that it could cause major complications, including death. Fortunately, early detection and proper treatment of various chronic diseases can substantially reduce their severity and often avert disability and death. Even more important, decreasing the risk factors for a disease often can either prevent the disease or delay its onset. In this chapter, we look at cardiovascular diseases, focusing primarily on CAD and hypertension.

Chronic and degenerative diseases of the cardiovascular system are the major cause of serious illness and death in the United States (figure 20.1). In 2003, cardiovascular diseases affected 71.3 million Americans, resulting in over 910,000 deaths, at a cost to individuals, government, and private industry of approximately $403 billion.[5]

From the early 1900s to the mid-1960s, the relative number of heart disease deaths, expressed per 100,000 people, increased threefold. The population of the United States more than doubled during that time, so the absolute number of heart disease deaths increased even more dramatically than the relative rate indicates. Cardiovascular disease continues to be a major problem in the United States, where in the year 2003,

- there were more than 1.2 million heart attacks;
- nearly 480,000 died from heart attacks;
- about one of every five deaths was attributable to CAD; and
- one of every 2.7 deaths was attributable to all cardiovascular diseases.

Furthermore, it was estimated that in 2003 there were

- about 467,000 coronary artery bypass surgeries,
- about 1,244,000 coronary angioplasties, and
- over 2,000 heart transplants.

In focus

In the 1970s, cardiovascular diseases accounted for well over 50% of all deaths in the United States! While cardiovascular diseases remain the number-one underlying cause of death in the United States, they accounted for only 37.3% of all deaths in 2003. Furthermore, in 1979, CAD accounted for nearly a third of all deaths in the United States, and in 2003 this had decreased to about 20%.

Fortunately, the death rate from cardiovascular disease and heart attacks has steadily decreased since its peak in the mid-1960s. The reasons for this decline have been heavily debated but likely include a greater focus on disease prevention, for example:

- Improved public awareness of risk factors and symptoms
- Increased use of preventive measures, including lifestyle changes (e.g., nutrition, exercise, and smoking cessation) to reduce individual risk
- Better and earlier diagnosis

Another probable reason is better treatment of patients with disease, for example:

- Improved drugs for specific treatment
- Angioplasty, drug-coated stents, and bypass surgery
- Better emergency care and treatment for heart attack victims

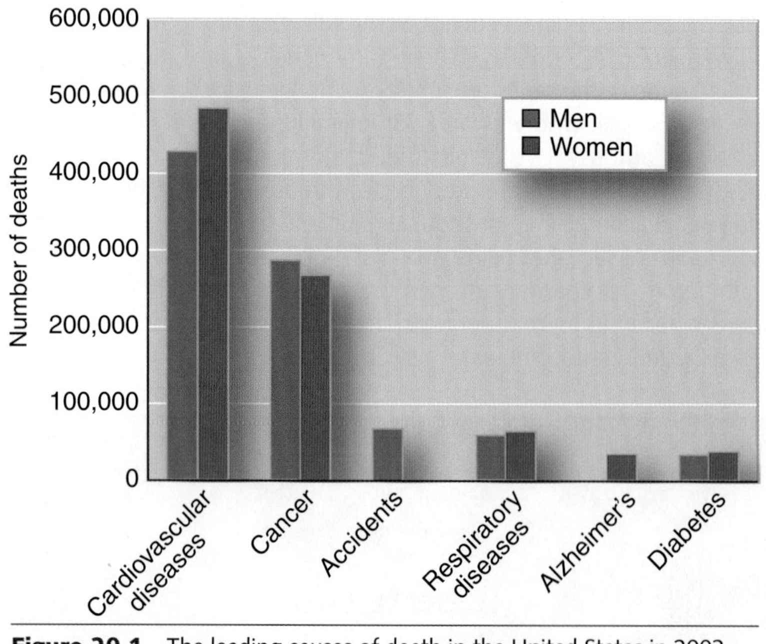

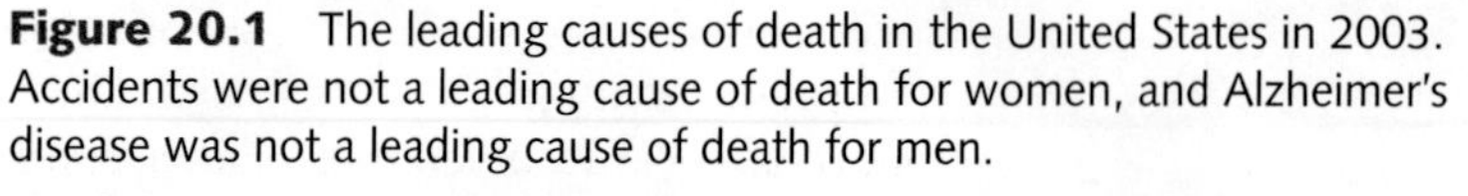

Figure 20.1 The leading causes of death in the United States in 2003. Accidents were not a leading cause of death for women, and Alzheimer's disease was not a leading cause of death for men.

Data from American Heart Association, 2006.

FORMS OF CARDIOVASCULAR DISEASE

There are several different cardiovascular diseases. In this section, we focus primarily on those that are preventable and that affect the largest number of Americans each year; these are illustrated in figure 20.2.

Coronary Artery Disease

As most humans age, their coronary arteries, which supply the myocardium (heart muscle) itself, become progressively narrowed as a result of the formation of fatty **plaque** along the inner wall of the artery, as seen in figure 20.3. This progressive narrowing of the arteries in general is referred to as **atherosclerosis;** and when the coronary arteries are involved, it is termed **coronary artery disease (CAD).** As the disease progresses and the coronary arteries become narrower, the capacity to supply blood to the myocardium is progressively reduced. This is what happened to the baseball pitcher described in the beginning of this chapter.

As the narrowing worsens, the myocardium (heart muscle) eventually can't receive enough blood to meet all of its needs. When this occurs, the portion of the myocardium that is supplied by the narrowed arteries becomes ischemic, meaning that it suffers a deficiency of blood. **Ischemia** of the heart often causes severe chest pain, referred to as *angina pectoris.* This typically is first experienced during periods of physical exertion or stress, when the demands on the heart are greatest.

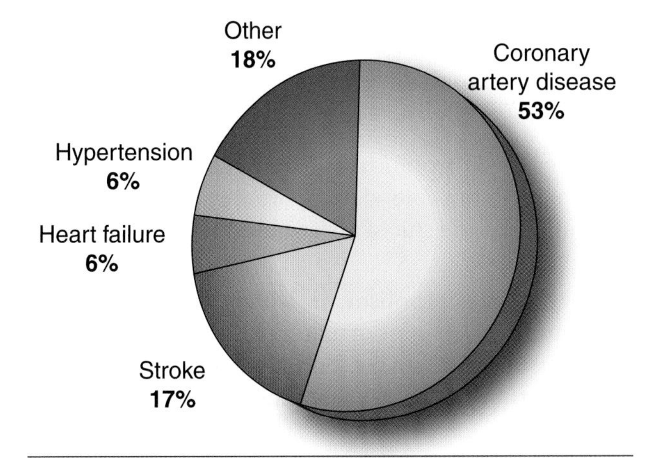

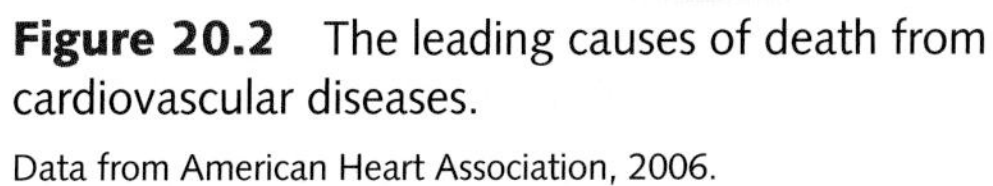

Figure 20.2 The leading causes of death from cardiovascular diseases.

Data from American Heart Association, 2006.

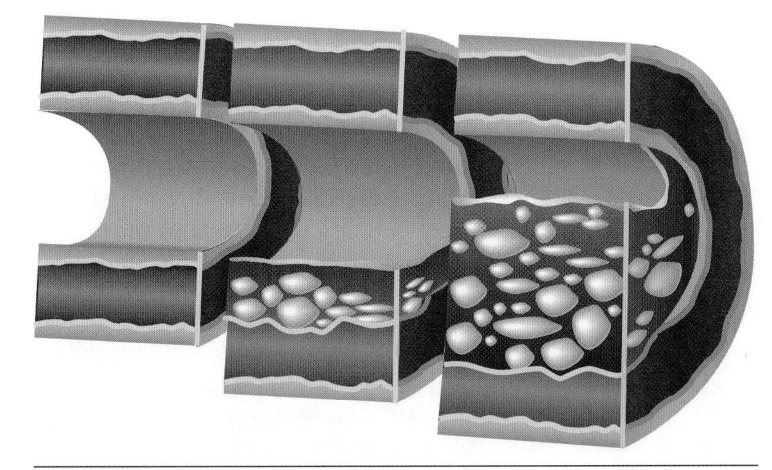

Figure 20.3 The progressive formation of plaque in a coronary artery.

When blood supply to a part of the myocardium is severely or totally restricted, ischemia can lead to a heart attack, or **myocardial infarction,** because cardiac muscle cells that are deprived of blood for several minutes are also deprived of oxygen, which leads to irreversible damage and necrosis (cellular death). This can lead to mild, moderate, or severe disability or even death, depending on the location of the infarction and the extent of the damage. Sometimes a heart attack is so mild that the victim is unaware that it has occurred. In such cases, the heart attack is discovered weeks, months, or even years later when an electrocardiogram is obtained during a routine medical examination.

Atherosclerosis is not a disease of the aged. Rather, it is more appropriately classified as a pediatric disease because the pathological changes that lead to atherosclerosis begin in infancy and progress during childhood.[25] **Fatty streaks,** or lipid deposits, which are thought to be the probable precursors of atherosclerosis, commonly are found in the aortas of children by age 3 to 5. These fatty streaks start to appear in the coronary arteries during the early teens, can develop into fibrous plaques during one's 20s, and can progress to unstable or complicated lesions during one's 40s and 50s.

The rate at which atherosclerosis progresses is determined largely by genetics and lifestyle factors, including smoking history, diet, physical activity, and stress. For some people, the disease progresses rapidly, with a heart attack occurring at a relatively young age—in their 20s or 30s. For others, the disease progresses very slowly, with few or no symptoms throughout their lives. Most people fall somewhere between these two extremes.

> **In focus**
>
> Atherosclerosis begins in childhood and progresses at different rates, depending primarily on heredity and lifestyle choices.

To illustrate this, a study of combat fatalities from the Korean War revealed that 77% of autopsied American soldiers, average age 22.1, already had some gross evidence of coronary atherosclerosis.[19] The extent of disease ranged from fibrous thickening to complete occlusion of one or more of the main branches of the coronary arteries. The autopsied Korean soldiers, however, were free of the disease. Evidence of coronary atherosclerosis also was found in 45% of the American fatalities from the Vietnam War, and 5% exhibited severe manifestations of the disease.[33]

Hypertension

Hypertension is the medical term for high blood pressure, a condition in which blood pressure is chronically elevated above levels considered desirable or healthy for a person's age and size. Blood pressure depends primarily on body size, so children and young adolescents have much lower blood pressures than adults. For this reason, determining what constitutes hypertension in the growing child and adolescent is difficult. Clinically, hypertension in these groups is defined as blood pressure values above the 90th or the 95th percentile for the youth's age. Hypertension is uncommon during childhood but can appear during midadolescence. For adults, the Joint National Committee on Detection, Evaluation, and Treatment of High Blood Pressure has established guidelines, presented in table 20.1, for **systolic blood pressure,** which is the highest pressure in the arteries at any time, and for **diastolic blood pressure,** which is the lowest pressure in the arteries at any time.[24]

Hypertension causes the heart to work harder than normal, because it has to expel blood from the left ventricle against a greater resistance. Furthermore, hypertension places great strain on the systemic arteries and arterioles. Over time, this stress can cause the heart to enlarge and the arteries and arterioles to become scarred, hardened, and less elastic. Eventually, this can lead to atherosclerosis, heart attacks, heart failure, stroke, and kidney failure.

In 2003, at least 65 million American adults, or about 32% of the adult population, were estimated to have high blood pressure (i.e., systolic ≥140 mmHg, diastolic ≥90 mmHg, or both).[5] Prehypertension (i.e., systolic 120-139 mmHg, diastolic 80-89 mmHg, or both) accounted for an additional 28% of the adult population. The age-adjusted death rate from hypertension increased 29.3% from 1993 to 2003, and hypertension was the primary or contributing cause of death in about 277,000 individuals. Compared with white Americans, black Americans develop high blood pressure at an earlier age, and it is more severe at any decade of life. Consequently, black Americans have a 1.3 times greater rate of nonfatal stroke, a 1.8 times greater rate of fatal stroke, a 1.5 times greater rate of heart disease deaths, and a 4.2 times greater rate of end-stage renal disease when compared with white Americans.[5] The estimated age-adjusted prevalence of high blood pressure in American adults age 20 and older was 30.6% for non-Hispanic white males, 31.0% for non-Hispanic white females, 41.8% for non-Hispanic black males, 45.4% for non-Hispanic black females, 27.8% for Mexican American males, and 28.7% for Mexican American females.[5]

TABLE 20.1 Classification of Blood Pressure for Adults, Age 18 Years and Older

Category	Systolic (mmHg)	Diastolic (mmHg)
Normal	<120	<80
Prehypertension	120-139	80-89
Hypertension		
Stage 1	140-159	90-99
Stage 2 (moderate)	≥160	≥100

From the seventh report of the Joint National Committee on Prevention, Detection, Evaluation, and Treatment of High Blood Pressure, 2003 *Journal of the American Medical Association* 289: 2560-2572.

In focus

About one of every three adult Americans has hypertension.

Stroke

Stroke is a form of cardiovascular disease that affects the cerebral arteries, those that supply the brain. Approximately 700,000 strokes occur in the United States each year, and stroke was the underlying or contributing cause of about 273,000 deaths in 2003.[5] As with CAD, the death rate from strokes also has decreased significantly in recent years—an 18.5% reduction between 1993 and 2003.

Strokes generally fall into two categories: **ischemic stroke** and **hemorrhagic stroke.** Ischemic strokes are the most common (~83% of all cases) and result from an obstruction within a cerebral blood vessel that limits the flow of blood to that region of the brain. Obstructions are the result of either

- cerebral thrombosis, the most common, in which a thrombus (blood clot) forms in a cerebral vessel, often at the site of atherosclerotic damage to the vessel; or
- cerebral embolism, in which an embolus (an undissolved mass of material, such as fat globules, bits of tissue, or a blood clot) breaks loose from another site in the body and lodges in a cerebral artery.

In cases of ischemic stroke, blood flow beyond the blockage is restricted, and the part of the brain that relies on that supply becomes ischemic, is oxygen deficient, and can die.

Hemorrhagic strokes are of two major types:

- Cerebral hemorrhage, in which one of the cerebral arteries ruptures in the brain
- Subarachnoid hemorrhage, in which one of the brain's surface vessels ruptures, dumping blood into the space between the brain and the skull

In both cases, blood flow beyond the rupture is diminished because the blood leaves the vessel at the site of injury. Also, as the blood accumulates outside of the vessel, it puts pressure on the fragile brain tissue, which can alter brain function. Brain hemorrhages often result from aneurysms, which arise from weak spots in the vessel wall that balloon outward; and aneurysms often arise because of hypertension or atherosclerotic damage to the vessel wall.

As with a heart attack, a stroke results in death of the affected tissue. The consequences depend largely on the location and extent of the stroke. Brain damage from a stroke can affect the senses, speech, body movement, thought patterns, and memory. Paralysis on one side of the body is common, as is the inability to verbalize thoughts. Most effects of a stroke are indicative of the side of the brain that was damaged.

Heart Failure

Heart failure is a clinical condition in which the heart muscle becomes too weak to maintain an adequate cardiac output to meet the body's oxygen demands. This usually results from either damage to, or overworking of, the heart. Hypertension, atherosclerosis, valvular heart disease, viral infection, and heart attack are among the possible causes of this disorder.

When cardiac output is inadequate, blood begins to back up in the veins. This causes excess fluids to accumulate in the body, particularly in the legs and ankles. This fluid accumulation (edema) also can affect the lungs (pulmonary edema), disrupting breathing and causing shortness of breath. Heart failure can progress to the point of irreversible damage to the heart, and the patient becomes a candidate for a heart transplant.

Other Cardiovascular Diseases

Other cardiovascular diseases include peripheral vascular diseases, valvular heart diseases, rheumatic heart disease, and congenital heart disease.

Peripheral vascular diseases involve the systemic arteries and veins, as opposed to the coronary vessels. **Arteriosclerosis** refers to numerous conditions in which the walls of the arteries become thickened, hard, and less elastic. Atherosclerosis is a form of arteriosclerosis. Arteriosclerosis obliterans, in which an artery becomes completely occluded, is another form. Peripheral venous diseases include varicose veins and phlebitis. Varicose veins result from incompetency of the valves in the veins, allowing blood to back up in the veins and causing them to become enlarged, tortuous, and painful. Phlebitis is inflammation of a vein and is also very painful.

Valvular heart diseases involve one or more of the four valves that control the direction of blood flow into and out of the four heart chambers. **Rheumatic heart disease** is one form of valvular heart disease involving a streptococcal infection that has caused acute rheumatic fever, typically in children between ages 5 and 15. Rheumatic fever is an inflammatory disease of the connective tissue and commonly affects the heart, specifically the heart valves. The damage to the valves usually causes difficulty in their opening, hindering blood flow out of that chamber, or difficulty in their closing, allowing blood to flow back into the previous chamber.

Congenital heart disease includes any heart defects that are present at birth, which are also appropriately termed *congenital heart defects*. These defects occur when the heart or the blood vessels near the heart do not develop normally before birth. These include coarctation of the aorta, in which the aorta is abnormally constricted; valvular stenosis, in which one or more heart valves are narrowed; and septal defects, in which the septum separating the right and left sides of the heart is defective, allowing blood from the systemic side to mix with that in the pulmonary side, and vice versa.

The remainder of this chapter focuses on the two major diseases in this category: CAD and hypertension.

In review

- Atherosclerosis is a process in which arteries become progressively narrowed. Coronary artery disease is atherosclerosis of the coronary arteries.
- When blood flow to the heart is sufficiently blocked, the part of the heart supplied by the diseased artery suffers from lack of blood (ischemia), and the resulting oxygen deprivation can cause myocardial infarction, which results in tissue necrosis.
- Atherosclerotic changes in the arteries actually begin in young children, but the extent and progression of this disease process are quite variable.
- Hypertension is the clinical term for high blood pressure.
- Stroke affects the cerebral arteries so that the part of the brain they supply receives too little blood. Ischemic stroke is the most common form of stroke, usually resulting from cerebral thrombosis or embolism. The other cause of stroke is cerebral hemorrhage (cerebral and subarachnoid).
- Heart failure is a condition in which the cardiac muscle becomes too weak to maintain an adequate cardiac output, causing blood to back up in the veins.
- Peripheral vascular diseases involve systemic, rather than coronary, vessels and include arteriosclerosis, varicose veins, and phlebitis.
- Congenital heart disease includes all heart defects present at birth.

UNDERSTANDING THE DISEASE PROCESS

Pathophysiology refers to the pathology and physiology of a specific disease process or disordered function. Understanding the pathophysiology of a disease gives us insight into how physical activity might affect or alter the disease process. In the following sections, we examine the pathophysiology of CAD and hypertension.

Pathophysiology of Coronary Artery Disease

How does atherosclerosis develop in the coronary arteries? The walls of the coronary arteries are composed of three distinct tunics or layers, as shown in figure 20.4: the tunica intima (inner layer), the tunica media (middle layer), and the tunica adventitia (outer layer). These are referred to more simply as the intima, the media, and the adventitia. The innermost layer of the intima, the endothelium, is formed by a thin lining of endothelial cells that provides a smooth protective coating between the blood flowing through the artery and the intimal layer of the vessel wall. The endothelium provides a protective barrier between toxic substances in the blood and the vascular smooth muscle cells. For vessels larger than 1 mm in diameter, the intima also includes a subendothelial layer, formed from connective tissue. The media consists mainly of the smooth muscle cells, which control the constriction and dilation of the vessel, and elastin. The adventitia is composed of collagen fibers that protect the vessel and anchor it to its surrounding structure.

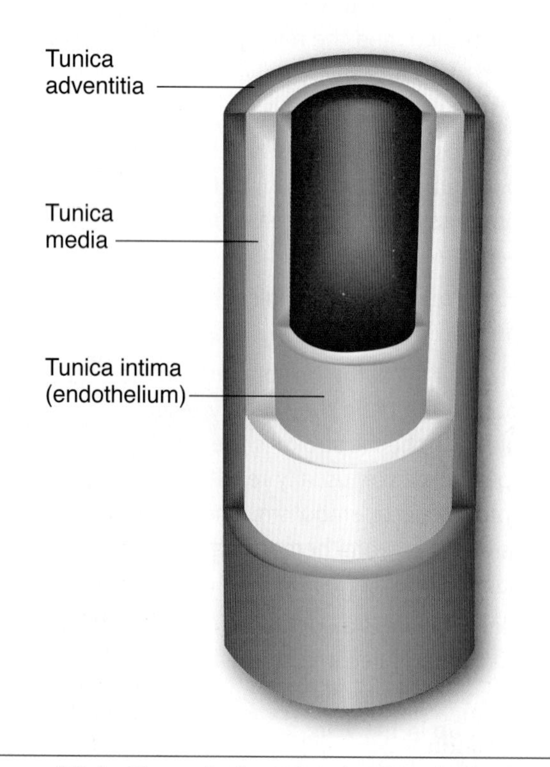

Figure 20.4 The wall of an artery has three layers: tunica intima, tunica media, and tunica adventitia.

According to an early theory of atherosclerosis, which evolved from the work of Dr. Russell Ross and his colleagues at the University of Washington, local injury to, or dysfunction of, endothelial cells appears to be an important factor in the initiation of atherosclerosis (see figure 20.5*a*).[43] Blood platelets and monocytes then are attracted to the site of injury and adhere to the exposed connective tissue (see figure 20.5*b*). These platelets release a substance referred to as **platelet-derived growth factor (PDGF)** that promotes migration of smooth muscle cells from the media into the intima. The intima normally contains few if any smooth muscle cells. A plaque, which is basically composed of smooth muscle cells, connective tissue, and debris, forms at the site of injury (see figure 20.5*c*). Eventually, lipids in the blood, specifically low-density lipoprotein cholesterol (LDL-C), are attracted to and deposited in the plaque (see figure 20.5*d*).

More recently, researchers have theorized that monocytes—white blood cells that are effector cells of the immune system—attach between endothelial cells. These monocytes differentiate into macrophages, which ingest oxidized LDL-C. They slowly become large foam cells and form fatty streaks. Smooth muscle cells then accumulate under these foam cells. The endothelial cells then separate or are sloughed off, exposing the underlying connective tissue and allowing platelets to attach to it.[43] In this modification of the original theory, endothelial injury is not always the precipitating event. Injury or disruption of the endothelium can result from high blood concentrations of the atherogenic form of cholesterol (LDL-C); free radicals caused by cigarette smoking, hypertension, and diabetes; elevated plasma homocysteine; and infectious microorganisms among other factors. In fact, atherosclerosis is now recognized as an inflammatory disease.[44]

The plaque consists of a collection of smooth muscle cells and inflammatory cells (macrophages and T lymphocytes), with both intracellular and extracellular lipid.[10] The plaque also contains a fibrous cap. It is now recognized that the composition of the plaque and its fibrous cap is critical to its stability. Unstable

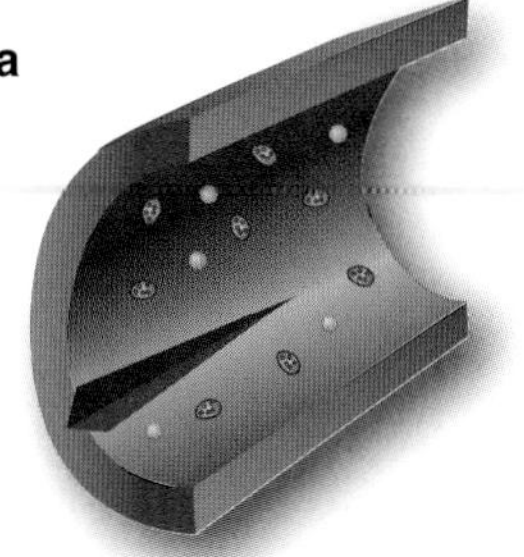

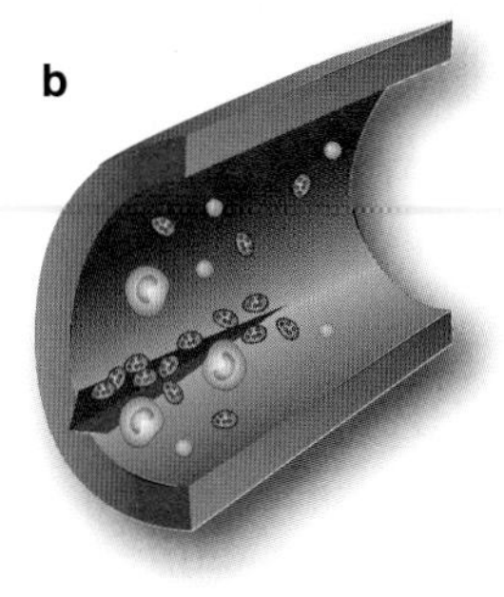

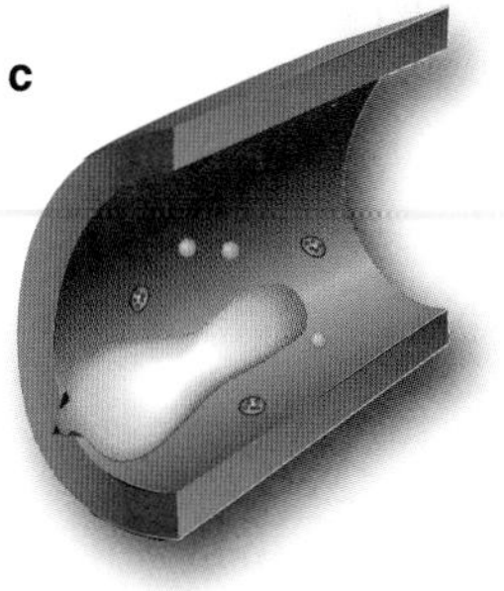

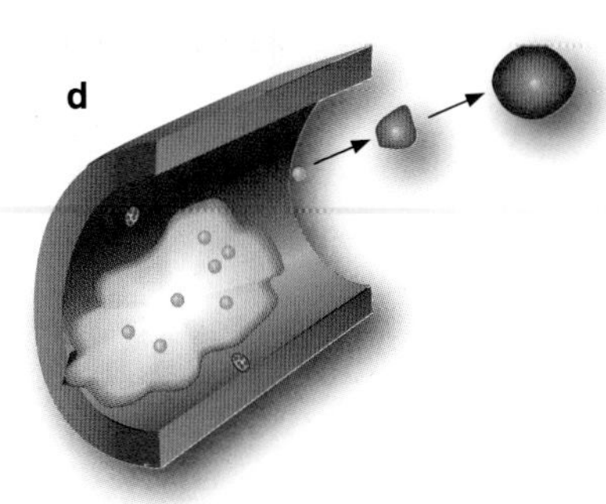

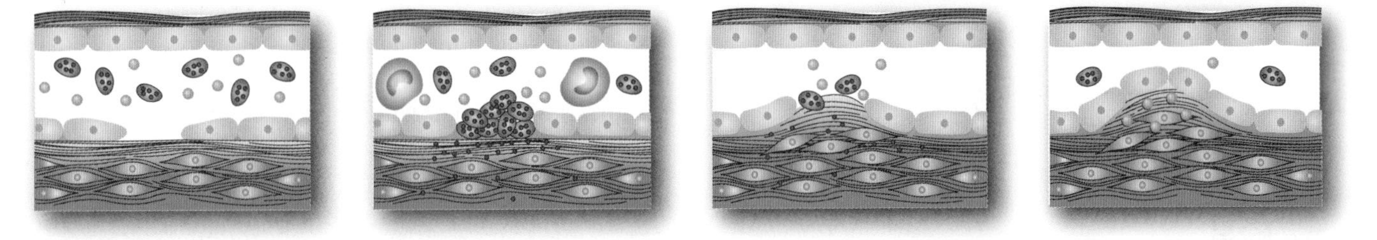

Figure 20.5 Changes in the arterial wall with injury, illustrating the disruption of the endothelium and the subsequent alterations that lead to atherosclerosis.

plaques are those that have thin fibrous caps and are heavily infiltrated by foam cells. These plaques are much more susceptible to rupture; and when rupture occurs, proteolytic enzymes are released, causing a breakdown of the cellular matrix leading to blood clotting (thrombus) as illustrated in figure 20.6 on page 478. The thrombus, depending on its size, can occlude or block the artery, resulting in myocardial infarction (MI, heart attack) and even cardiac arrest. In fact, plaque rupture and thrombosis account for up to 70% of MIs and cardiac arrests. Interestingly, the plaques that do rupture are typically small, causing less than 50% stenosis or narrowing of a coronary artery.[10, 15, 44]

There is now good evidence that plaque is a dynamic structure, undergoing cycles of erosion and repair that are responsible for its growth. Ironically, smooth muscle cells are important to the stability of the plaque, and smooth muscle cell proliferation is potentially beneficial to maintaining the integrity of the plaque. Plaque rupture sites are characterized by a low density of smooth muscle cells.[10]

In focus

The process of atherosclerosis appears to begin with disruption of the endothelial cells lining the intima of coronary arteries. This leads to a chain of events that eventually develops into a full-blown atherosclerotic plaque. The most dangerous plaques are those with a thin fibrous cap that are heavily infiltrated by foam cells.

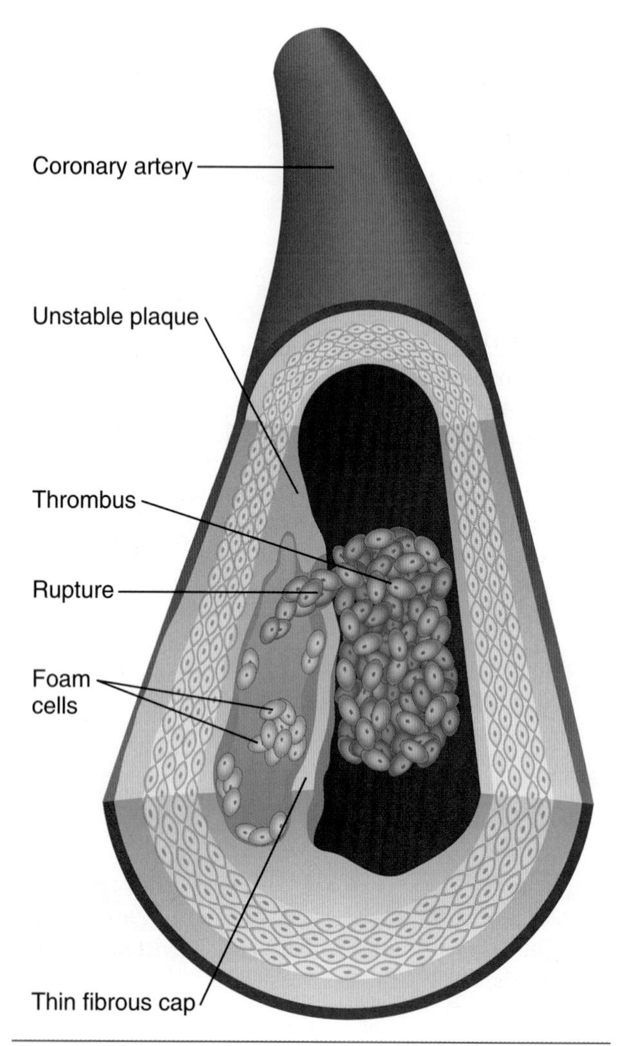

Figure 20.6 Illustration of fissure or rupture of an unstable plaque in a coronary artery, releasing its contents into the bloodstream and stimulating the formation of a thrombus (clot).

Pathophysiology of Hypertension

The pathophysiology of hypertension is not well understood. In fact, it is estimated that 90% or more of those identified with hypertension are classified as having idiopathic hypertension, or hypertension of unknown origin. Idiopathic hypertension is also referred to as essential hypertension. Risk factors for hypertension include

- heredity, including race;
- increasing age and male sex;
- sodium sensitivity;
- excessive alcohol consumption and use of tobacco products;
- obesity and overweight;
- diabetes or insulin resistance or both;
- physical inactivity;
- oral contraceptives;
- pregnancy; and
- stress.[4]

In review

- Pathophysiology refers to the pathology and physiology of a specific disease process or disordered function.
- Early theories held that CAD can be initiated by damage to the smooth endothelial lining of the intimal layer of the arterial wall. This damage attracts platelets to the area, which in turn release PDGF. Platelet-derived growth factor attracts smooth muscle cells; and a plaque, composed of smooth muscle cells, connective tissue, and debris, begins to form. Eventually, lipids are deposited in the plaque.
- More recent research indicates that monocytes, involved with the immune system, can attach between endothelial cells in the intima and begin forming fatty streaks; this then leads to plaque formation. According to this theory, endothelial damage is not necessary for plaque formation.
- It is now clear that the composition of the plaque and its fibrous cap is critical with respect to myocardial infarctions and cardiac arrest. Smaller plaques, where there is typically less than 50% occlusion of the artery, that have thin fibrous caps and are heavily infiltrated with foam cells are the most dangerous.
- The pathophysiology of hypertension is poorly understood.
- More than 90% of people with hypertension have idiopathic, or essential, hypertension, meaning that its cause is unknown. Possible causes include genetic factors, race, increasing age and male sex, sodium sensitivity, excessive alcohol consumption and use of tobacco products, obesity and overweight, diabetes or insulin resistance (or both), oral contraceptives, pregnancy, physical inactivity, and stress.

DETERMINING INDIVIDUAL RISK

Over the years, scientists have attempted to determine the basic etiology, or cause, of both CAD and hypertension. Much of our understanding of these two diseases comes from the field of epidemiology, a science that studies the relationships of various factors to a specific disease or disease process. In several studies, selected members of various communities have been observed for extended periods of time. These observations include periodic medical examinations and clinical tests.

Eventually, some of the participants in such studies become diseased and many die. All those who develop heart disease or hypertension or who die from heart attacks or hypertension are grouped accordingly. Then their previous medical and clinical tests are analyzed to determine shared attributes or factors. Although this approach does not define the causal mechanism of the disease, it does provide researchers with valuable insights into the disease process. As identified in long-term longitudinal population studies, the factors that place individuals at risk for disease are referred to as **risk factors.** Let's examine the risk factors for heart disease and hypertension.

Risk Factors for Coronary Artery Disease

The factors associated with an increased risk for premature development of CAD can be classified into two groups: those over which a person has no control and those that can be altered through basic changes in lifestyle. Those that a person cannot control include heredity (family history of CAD), race, male sex, and advanced age. According to the American Heart Association,[4] the **primary risk factors** that can be controlled or altered include

- tobacco smoke,
- hypertension,
- abnormal blood lipids and lipoproteins,
- physical inactivity,
- obesity and overweight, and
- diabetes and insulin resistance.

When one or more risk factors for a certain disease are present, an individual is at increased risk for morbidity (developing the disease) or mortality (dying from it).

Table 20.2 on page 480 illustrates the ranges of values for some of these risk factors based on the categories of "desirable," "borderline," and "high." These are approximations and vary somewhat by sex and age.

Other CAD risk factors have been proposed, but there is not yet sufficient evidence to support their inclusion as primary risk factors as determined by the American Heart Association. The following are considered good candidates to be added to the list of primary risk factors:

The Framingham Heart Study

In July 1948, the National Institutes of Health (National Heart, Lung, and Blood Institute) started the Framingham Heart Study (FHS). The FHS was designed as a longitudinal investigation aimed at identifying those factors that influence the development of cardiovascular diseases. The original study population of 5209 persons from Framingham, Massachusetts, was examined over a four-year period starting in September 1948 and was reexamined every two years for over 48 years. The FHS pioneered the concept of risk factors associated with the development of CAD. Inclusion of the offspring, or second generation, of the original Framingham group began in 1971, and the third generation study started in 2002. The FHS has been one of the most successful longitudinal studies in the history of medical research and has resulted in over 1200 research-based publications. As just one example, the FHS was the first to indicate the importance of cholesterol as a risk factor for CAD, and subsequently to demonstrate that the true risk was associated with high levels of LDL-C and low levels of high-density lipoprotein cholesterol (HDL-C).

TABLE 20.2 Level of Risk Associated With Selected Risk Factors for Coronary Artery Disease

		Level of risk		
Risk factor		Desirable	Borderline	High
Blood pressure[a]				
Systolic	mmHg	<120	120-139	≥140
Diastolic	mmHg	<80	80-89	≥90
Blood lipids and lipoproteins[a]				
Cholesterol	mg/dL	<200	200-239	≥240
LDL-C	mg/dL	<130	130-159	≥160
HDL-C	mg/dL	≥60	40-59	<40
Triglycerides	mg/dL	<150	150-199	≥200
Overweight/Obesity (BMI)[a]	kg/m^2	18.5-24.9	25.0-29.9	≥30
Fasting plasma glucose[b]	mg/dL	<100	100-125	≥126
Physical inactivity[a, c]	min/day	30-60	15-29	<15

[a]American Heart Association (2006).
[b]American Diabetes Association (2006). www.diabetes.org.
[c]Moderate to vigorous exercise on most days of the week.

- C-reactive protein (CRP): CRP is produced in the liver and smooth muscle cells within coronary arteries in response to injury or infection. C-reactive protein is a marker of inflammation.
- Fibrinogin: Fibrinogin is a blood protein integral to the process of blood clotting. Excessive concentrations can lead to abnormal clumping of blood platelets. Fibrinogin is also an indicator of inflammation.
- Homocysteine: Homocysteine is an amino acid used to make protein and build and maintain body tissues. Excessive levels are associated with increased risk for CAD and other cardiovascular diseases.
- Lipoprotein(a) [Lp(a)]: Lp(a) is similar in structure to LDL-C and might reduce the body's ability to dissolve blood clots. Its specific role in atherosclerosis has not yet been determined, but high levels are associated with increased risk for CAD.

In focus

High concentrations of HDL-C and low concentrations of LDL-C place the individual at the lowest risk for CAD. Low-density lipoprotein cholesterol has been implicated in plaque formation, whereas HDL-C is likely involved in plaque regression.

Lipids and Lipoproteins

The inclusion of elevated **blood lipids** as a primary risk factor needs to be further defined. For many years, cholesterol and **triglycerides** were the only lipids observed in these epidemiological studies. The public was confused by conflicting data and opinions about the role of lipids in the development of atherosclerosis. More recently, scientists have studied the manner in which lipids are transported in the blood. Lipids by themselves are insoluble in blood, so they are packaged with a protein to allow transport through the body. **Lipoproteins** are the proteins that carry the blood lipids. Two classes of lipoproteins of major concern for CAD are low-density lipoprotein (LDL) and high-density lipoprotein (HDL). High levels of **low-density lipoprotein cholesterol (LDL-C)** and low levels of **high-density lipoprotein cholesterol (HDL-C)** place a person at extremely high risk of having a heart attack at a relatively young age—under age 60. Conversely, a high level of HDL-C and a low level of LDL-C place a person at an extremely low risk. Yet a third class of lipoproteins is called very low density lipoproteins (VLDL). **Very low density lipoprotein cholesterol (VLDL-C)** is becoming increasingly implicated as a risk factor for CAD.

In focus

The ratio of Total-C to HDL-C is possibly the most accurate lipid index of risk for CAD, with values of 5.0 or greater indicating increased risk and values of 3.0 or lower representing low risk.

Merely looking at total cholesterol is not sufficient. A person might have a moderately high level of total cholesterol (Total-C) and yet be at a relatively low risk because of a high concentration of HDL-C and low concentration of LDL-C. Conversely, a person might have a moderately low level of Total-C and yet be at a relatively high risk because of a high concentration of LDL-C and a low concentration of HDL-C.

Why are these two cholesterol carriers associated with different risk levels? Low-density lipoprotein cholesterol is theorized to be responsible for depositing cholesterol in the arterial wall. High-density lipoprotein cholesterol, however, is regarded as a scavenger that removes cholesterol from the arterial wall and transports it to the liver to be metabolized. Because of these very different roles, it is essential to know the specific concentrations of both of these lipoproteins when one is determining individual risk. The ratio of Total-C to HDL-C may be the best blood lipid index of risk for CAD. Values of 3.0 or less place a person at low risk, but values of 5.0 or greater place a person at high risk. As an example, a Total-C of 225 mg/dL and an HDL-C of 45 mg/dL would provide a ratio of 5.0 ($225 \div 45 = 5.0$), but with the same Total-C and an HDL-C of 75 mg/dL the ratio would be 3.0 ($225 \div 75 = 3.0$). Others have used the ratio of Total-C to LDL-C or LDL-C to HDL-C to establish the degree of risk. At this time, there is no consensus as to which ratio provides the best estimate of risk, although most use the Total-C to HDL-C ratio.

Early Detection of Risk Factors

Coronary artery disease risk factors can be identified at an early age, and the earlier they are identified, the earlier preventive treatment can begin. In a study of 96 boys ages 8 to 12,

- 19.8% had Total-C values above the suggested high-normal value of 200 mg/dL;
- 5.2% exhibited abnormal resting electrocardiograms;
- 37.5% had more than 20% relative body fat; and
- none had elevated blood pressure.[53]

Similar data were reported in a later study of 13- to 15-year-old boys.[52] Both studies are summarized in table 20.3 on page 482. Individuals with an elevated risk during childhood generally remain at elevated risk as young adults.

In addition, the results of the Bogalusa Heart Study must be considered. This was a longitudinal study of cardiovascular disease risk factor development from birth through age 39. In 204 of the subjects who died prematurely (primarily from accidents, homicides, or suicides), the scientists found a strong relationship between the risk factors and development of fatty streaks; the greater the number of risk factors, the greater the development of aortic and coronary artery fatty streaks.[6]

Risk Factors for Hypertension

The risk factors for hypertension, like those for CAD, can be classified as ones we can control and ones we

Metabolic Syndrome

Metabolic syndrome is a term that has been used to link CAD, hypertension, abnormal blood lipids, type 2 diabetes, and abdominal obesity to insulin resistance and hyperinsulinemia. This syndrome also has been referred to as syndrome X and the insulin resistance syndrome. It is not totally clear where the syndrome starts, but it has been observed that upper body obesity is associated with insulin resistance and that insulin resistance is highly correlated with increased risk for CAD, hypertension, and type 2 diabetes. It appears, however, that obesity or insulin resistance (or a combination of the two) is the trigger that initiates a cascade of events leading to the metabolic syndrome. Systemic inflammation has also been suggested as a causal factor. This became a major topic of research in the 1990s and continues to be today. The results of this research should help us better understand the pathophysiology of these diseases and their interrelationships.

TABLE 20.3 Coronary Artery Disease Risk Factor Prevalence in Boys 8 Through 15 Years of Age

	Percentage of boys with risk factor	
Risk factor	8- to 12-year-olds (*n* = 96)	13- to 15-year-olds (*n* = 308)
Blood lipids		
Total cholesterol (≥200 mg/dL)	20.0	11.0
HDL-C (≤36 mg/dL)	No data	14.6
Triglycerides (≥100 mg/dL)	8.4	25.0
Blood pressure		
Systolic (>90th percentile)	0.0	13.0
Diastolic (>90th percentile)	0.0	4.9
Smoking (≥10 cigarettes a day)	0.0	0.0
Diabetes	0.0	1.3
Obesity (≥25% relative body fat)	12.6	14.9
Physical activity ($\dot{V}O_{2max}$ ≤42 $ml \cdot kg^{-1} \cdot min^{-1}$)	3.2	18.8
Family history (heart attack at ≤60 years of age)	33.7	30.9
Presence of risk factors		
None	36.0	29.9
One	46.0	35.4
Two	14.0	22.1
Three	3.0	10.7
Four	1.0	1.9

Note. Values represent the percentage of the boys with the risk factor. HDL-C = high-density lipoprotein cholesterol.

cannot. Those we cannot control are heredity (family history of hypertension), sex, advanced age, and race (increased risk for people of African or Hispanic ancestry). Risk factors we can control are

- insulin resistance,
- obesity and overweight,
- diet (sodium, alcohol),
- use of tobacco products
- use of oral contraceptives,
- stress, and
- physical inactivity.

Although heredity is a risk factor for hypertension, it probably plays a much smaller role than many of the other proposed factors. We must remember that lifestyle factors are often quite similar within a family.

Recently, scientists have shown great interest in a possible link between hypertension, obesity, type 2 diabetes, and coronary heart disease through the common pathway of insulin resistance or impaired insulin action (see sidebar). But obesity also has been established as an independent risk factor for hypertension. Numerous studies have shown substantial reductions in blood pressure with weight loss in hypertensive patients. Also, although sodium intake traditionally has

In focus

Although the pathways are complex, it is becoming increasingly clear that hypertension, CAD, abnormal blood lipids, obesity, and diabetes might be linked through the common pathway of insulin resistance. It is also possible that obesity is the trigger that starts a cascade of events leading to the metabolic syndrome.

In review

- Risk factors for CAD that we cannot control are heredity (and family history), male sex, and advanced age. Those that we can control are abnormal blood lipids and lipoproteins, hypertension, smoking, physical inactivity, obesity, diabetes, and insulin resistance.
- Low-density lipoprotein cholesterol is thought to be responsible for depositing cholesterol in the arterial walls. Very low density lipoprotein cholesterol is also implicated in the development of atherosclerosis. However, HDL-C acts as a scavenger, removing cholesterol from the vessel walls. Thus, high HDL-C levels provide some degree of protection from CAD.
- The ratio of Total-C to HDL-C might be the best indicator of personal risk for CAD. Values below 3.0 reflect a low risk; values above 5.0 reflect a high risk.
- Risk factors for hypertension that can't be controlled include heredity, advanced age, and race. Those we can control are insulin resistance, obesity, diet (sodium and alcohol), use of tobacco products and oral contraceptives, stress, and physical inactivity.

been linked to hypertension, this relationship is likely limited to those who are salt sensitive.

Physical inactivity is a risk factor for hypertension. Its role has been conclusively established in epidemiological studies. Furthermore, substantial evidence indicates that increasing physical activity tends to reduce elevated blood pressure.[3]

REDUCING RISK THROUGH PHYSICAL ACTIVITY

The role that physical activity might play in preventing or delaying the onset of CAD and hypertension has been of major interest to the medical community for many years. In the following sections, we try to unravel this mystery by examining the following areas:

- Epidemiological evidence
- Physiological adaptations with training that might reduce risk
- Risk factor reduction with exercise training

Reducing the Risk of Coronary Artery Disease

Physical activity has been proven effective in reducing the risk of CAD. In the following sections, we discover what is known about this topic and what physiological mechanisms are involved.

Epidemiological Evidence

Hundreds of research papers have dealt with the epidemiological relationship between physical inactivity and CAD. Generally, studies have shown the risk of heart attack in sedentary male populations to be about two to three times that of men who are physically active in either their jobs or their recreational pursuits. The early studies of Dr. J.N. Morris (see figure 20.7 on p. 484) and his colleagues in England in the 1950s were among the first to demonstrate this relationship.[37] In these studies, sedentary bus drivers were compared with active bus conductors who worked on double-decker buses, and sedentary postal workers were compared with active postal carriers who walked their routes. The death rate from CAD was about twice as high in the sedentary groups as in the active groups. Many studies published over the subsequent 20 years showed essentially the same results: Those who were occupationally sedentary were at approximately twice the risk for death from CAD as those who were active.

Most of these early epidemiologic studies focused exclusively on occupational activity. Not until the 1970s did researchers start looking at leisure-time activity as well. Again, the studies by Dr. Morris and his colleagues[36, 38] were among the first to observe the relationship between leisure-time activity and the risk of CAD: The least active people were at two to three times greater risk. Subsequent studies by epidemiologists such as Drs. Paffenbarger, Leon, and Blair (figure 20.7) have provided similar results.[8, 9, 31, 32, 39] Physical inactivity approximately doubles the risk of having a fatal heart attack.[42] While most of these early studies were conducted on men, subsequent studies have demonstrated similar results in women.[13]

Scientists at the Centers for Disease Control in Atlanta conducted an extensive review of all epidemiologic studies published on physical inactivity and CAD up to the mid-1980s.[42] They used stringent criteria for including studies in their analysis, and the quality of each study also was assessed. They found that the average relative risk of CAD associated with inactivity ranged from 1.5 to 2.4, with a median value of 1.9, meaning that inactive people have about twice the risk of more active people, as we have been discussing. Additionally, these researchers found that the relative risk from physical inactivity is similar to the risk associated with

Figure 20.7 Key exercise epidemiologists whose research activities were instrumental in leading the American Heart Association to include physical inactivity as a major risk factor for coronary artery disease: Drs. Steven Blair, Ralph Paffenbarger, Jerry Morris, and Art Leon.

Exercise Type and Intensity Are Related to CAD Risk

In 2002, a group of scientists from Harvard University reported in the *Journal of the American Medical Association* the results of their epidemiologic study of the relationship of exercise type and intensity to CAD in more than 44,000 men enrolled in the Health Professional's Follow-Up Study.[46] These men were followed every two years from 1986 through 1998 to assess potential CAD risk factors, to identify newly diagnosed cases of CAD, and to assess levels of leisure-time physical activity. Men who ran 6 mph (9.7 km/h) or faster for 1 h or more per week had a 42% risk reduction compared with men who didn't run. Men who trained with weights for 30 min or more per week had a 23% risk reduction when compared with men who didn't weight train. Brisk walking for 30 min or more per day was associated with an 18% risk reduction, as was rowing for 1 or more hours per week. Surprisingly, swimming and cycling were unrelated to risk. This study was the first to show the direct benefits of weight training on CAD risk and to indicate that exercise intensity is also a critical consideration, with higher intensities providing greater risk reduction.

other major risk factors for CAD. Furthermore, the percentage of the total population of the United States who are physically inactive far exceeds the percentage with the other three major risks: those who smoke, are hypertensive, or have elevated cholesterol levels.[12] This is illustrated in figure 20.8. The results of these epidemiological studies played a major role in leading the American Heart Association in 1992 to declare physical inactivity a primary risk factor for CAD.

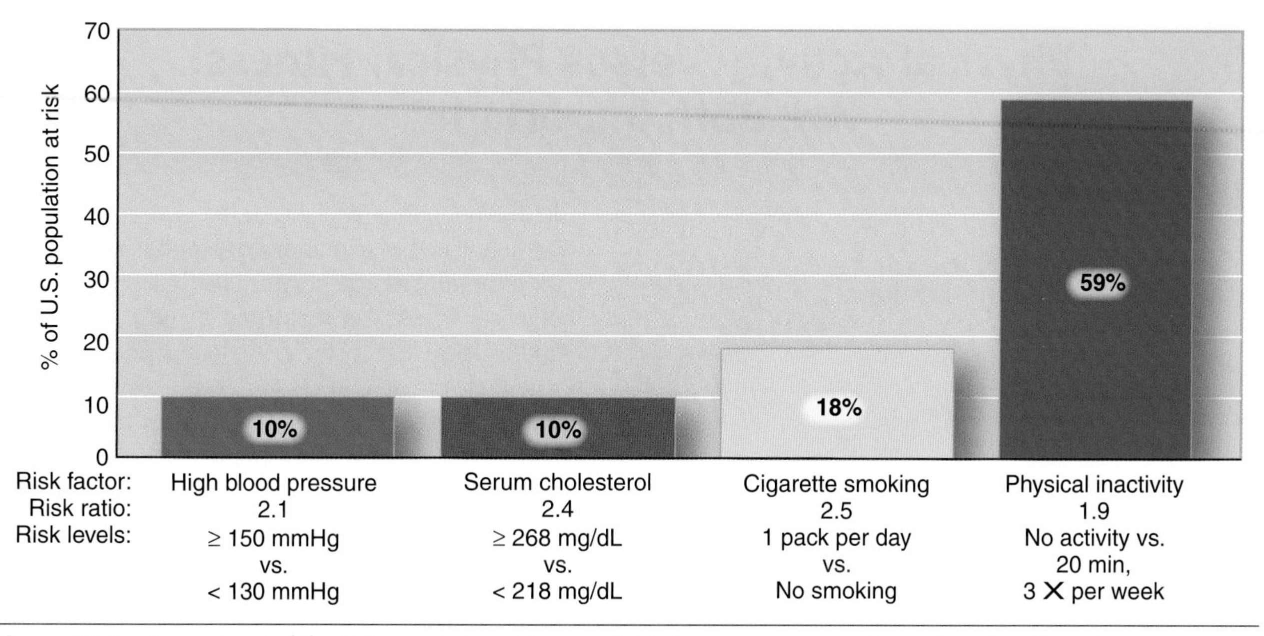

Figure 20.8 Percentages of the U.S. population at increased risk for coronary artery disease based on the primary risk factors.

Reprinted from C.J. Caspersen: Physical activity and coronary heart disease. *Physicians Sportsmedicine* 1987; 15(11): 43-44.

Another important concern was raised in the mid-1980s: What level of physical activity or fitness is necessary to reduce one's risk of CAD? It was not totally clear from the epidemiological studies what level of fitness or activity was effective. In fact, during the mid-1980s, scientists had just started to differentiate between activity level and fitness, defining personal fitness by a person's $\dot{V}O_{2max}$. In retrospect, distinguishing these two terms was crucial because a person can be active yet unfit (low $\dot{V}O_{2max}$) or fit (high $\dot{V}O_{2max}$) yet inactive. Dr. Ronald LaPorte and his colleagues at the University of Pittsburgh were instrumental in redirecting the thinking and subsequent research in this area.[28] Dr. LaPorte pointed out that, based on various epidemiological studies, the levels of activity associated with a lower risk for CAD were generally low and certainly not at the level that would be necessary to increase aerobic capacity. Subsequent studies have supported this.[9, 31, 32] Low levels of activity, such as walking and gardening, can provide considerable benefit by reducing the risk for CAD. More vigorous exercise likely provides even greater benefit.[30,46]

In focus

From epidemiological studies, it has been established that physical inactivity doubles the risk of CAD. However, it is now equally clear that low-intensity activity is sufficient to reduce the risk of this disease. Health benefits do not require high-intensity exercise. But, more vigorous exercise likely provides even greater benefits!

Training Adaptations That Might Reduce Risk

The importance of regular physical activity in reducing the risk of CAD becomes apparent when we consider anatomical and physiological adaptations in response to exercise training. For example, as we learned in chapter 10, exercise training causes the heart to hypertrophy, primarily through an increase in left ventricular chamber size but also through increases in left ventricular wall thickness. This adaptation may be important for improved contractility and increased cardiac work capacity.

The capacity of the coronary circulation appears to increase with training. Studies have shown that the size of major coronary vessels increases, which implies an increased capacity for blood flow to all regions of the heart. In fact, several studies have demonstrated that the peak flow rate in the major coronary arteries increases following an exercise training program. An important study in the early 1980s addressed the effects of moderate exercise training on the development of CAD in monkeys.[27] The monkeys were divided into three groups:

Physical Activity Versus Physical Fitness: Are Both Important?

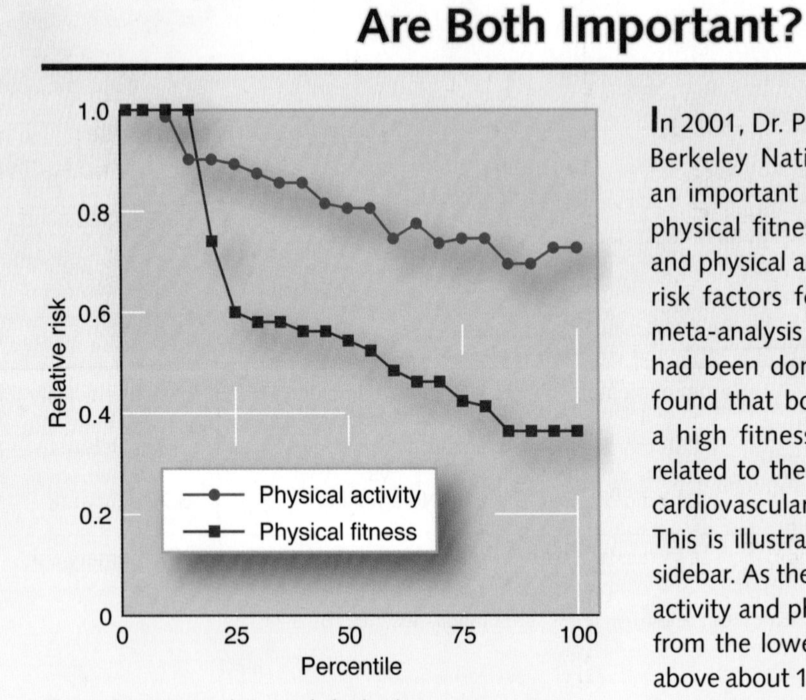

The reduction in relative risk for both coronary artery disease and cardiovascular disease with increasing levels of physical activity and physical fitness. This is a dose–response curve, indicating that the higher the dose, the better the response.

Reprinted, by permission, from P.T. Williams, 2001, "Physical fitness and activity as separate heart disease risk factors: A metaanalysis," *Medicine and Science in Sports and Exercise* 33: 754-761.

In 2001, Dr. Paul Williams of the Lawrence Berkeley National Laboratory published an important article suggesting that both physical fitness, as measured by $\dot{V}O_{2max}$, and physical activity levels are independent risk factors for CAD.[51] He conducted a meta-analysis of a number of studies that had been done on large populations. He found that both being active and having a high fitness level were independently related to the degree of risk for CAD and cardiovascular diseases (CVD) in general. This is illustrated in the figure within this sidebar. As the percentiles for both physical activity and physical fitness increase, going from the lowest percentile to the highest, above about 15% there is a reduction in risk for both CAD and cardiovascular diseases. Being physically fit had a stronger relationship to reduced risk than being physically active. These findings are controversial among epidemiologists, so there will likely be more studies and dialogue on this issue.[8]

1. A control group that ate normal, low-fat monkey chow
2. A nonexercising group that ate an atherogenic (high fat) diet known to induce heart disease
3. An exercising group that also ate the atherogenic diet

The sedentary group that consumed the atherogenic diet developed atherosclerosis. The coronary arteries of the exercising monkeys on this same diet had a greater internal diameter and substantially less atherosclerosis than those in the sedentary monkeys, as shown in figure 20.9. For the exercise group, the cross-sectional area of the lumen (diameter) of all of the major coronary vessels was two to three times larger than in the sedentary monkeys. Similar data have been shown for the coronary arteries of marathon runners versus sedentary men following a nitroglycerin challenge.[22]

Evidence is also starting to accumulate suggesting that exercise training increases endothelial function. Endothelial dysfunction is primarily the result of decreased nitric oxide bioavailability. Exercise training has been shown to increase nitric oxide bioavailability.[49] Further, exercise training has been shown to have an anti-inflammatory effect, and we have just learned that atherosclerosis is an inflammatory disease.[40]

In focus

Aerobic training produces favorable anatomical and physiological changes that decrease the risk of heart attack, including larger coronary arteries, increased heart size, and increased pumping capacity. Aerobic training also has a favorable effect on most of the other risk factors for CAD.

Some evidence also suggests that the heart's collateral circulation

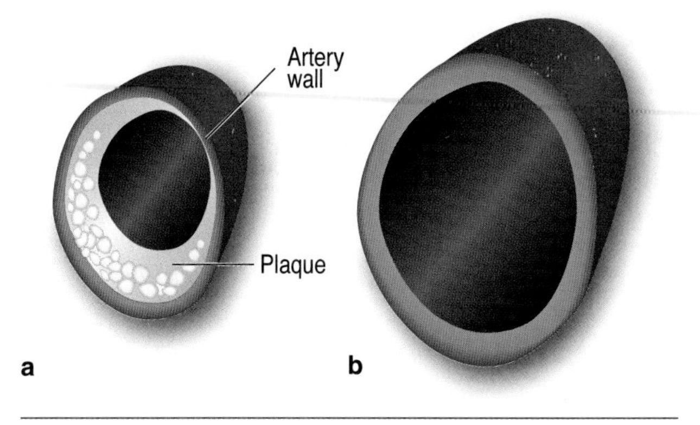

Figure 20.9 Comparison of the left main coronary artery in *(a)* sedentary and *(b)* exercising monkeys on atherogenic diets.

improves with exercise training. The collateral circulation is a system of small vessels that branch off the major coronary vessels and are important in providing blood to all regions of the heart, particularly when there are blockages in the major coronary arteries. It is possible, however, that collateral circulation development is more the result of the blockages and compromised circulation than of exercise training.

Risk Reduction With Exercise Training

Many studies have investigated the role of exercise in altering risk factors associated with heart disease. Let's consider the major risk factors and how exercise might affect them.

Little direct evidence is available to indicate that exercise leads to smoking cessation or reduces the number of cigarettes smoked. Relatively strong data support the effectiveness of exercise in reducing blood pressure in those with mild to moderate hypertension. Endurance training can reduce systolic and diastolic blood pressures by approximately 7 and 6 mmHg, respectively, in individuals who are hypertensive with blood pressures of ≥140 systolic or ≥90 mmHg diastolic.[3] Exercise even can lead to small reductions in both systolic and diastolic blood pressure (by ~2 mmHg each) in those with normal blood pressure.[3] The specific mechanisms responsible for the decreases in blood pressure with endurance training have yet to be determined.

Exercise possibly exerts its most beneficial effect on blood lipid levels.[17] Although the decreases in Total-C and LDL-C with endurance training are relatively small (generally less than 10%), there appear to be relatively major increases in HDL-C and major decreases in triglycerides. Cross-sectional studies of athletes and nonathletes alike show unequivocally that people with greater levels of aerobic activity or higher aerobic capacities have higher HDL-C and lower triglyceride levels. Results of longitudinal training studies, however, are much less clear. Many studies have shown increases in HDL-C and decreases in triglycerides from training, yet others have reported little or no change. Several even have shown decreased HDL-C levels. Almost all studies, however, have shown that the ratios of LDL-C to HDL-C and of Total-C to HDL-C are decreased following endurance training. This implies a reduced risk.

We must consider two confounding factors when we evaluate lipid changes with exercise training because they can have a marked independent effect on such changes. Because plasma lipids are expressed as a concentration (milligrams of lipid per deciliter of blood), any change in plasma volume will affect plasma concentrations independently of the change in total lipid. Recall that training typically increases plasma volume (chapter 10). With this plasma expansion, the absolute amount of HDL-C could increase yet the HDL-C concentration might not change or even could be lowered. In addition, plasma lipid levels are tightly coupled with changes in body weight. When we evaluate the effects of exercise training, the independent effects that a change in body weight could have on plasma lipids must be considered.

With respect to the remaining risk factors, exercise plays an important role in weight reduction and control and in the control of diabetes. These topics are discussed in detail in chapter 21. Exercise also has been reported to be effective for stress reduction and control and for reducing anxiety.[41] Some research supports the use of exercise training in the treatment of depression, although the results are not yet conclusive.[16, 35]

Reducing the Risk of Hypertension

Physical activity's role in reducing the risk of hypertension has not been as well established as its role with respect to CAD. As we saw in the last section, exercise training lowers blood pressure in those with moderate hypertension, but the precise mechanisms allowing this reduction are not yet fully known. Let's consider what is known.

Epidemiological Evidence

Very few epidemiological studies have dealt with the relationship between physical inactivity and hypertension. In the Tecumseh Community Health Study, 1700 males (age 16 and older) completed questionnaires and interviews to provide estimates of their average daily energy expenditures, their peak daily energy expenditures, and the hours they spent in particular activities. The more active men had significantly lower systolic and diastolic blood pressures, irrespective of age.[34] Similar results were obtained when resting blood pressure was analyzed by fitness level in nearly 3,000 adult men and more than 3,900 adult women tested at the Cooper Clinic in Dallas.[14,21] The more fit individuals exhibited lower systolic and diastolic blood pressures. In a follow-up of the participants from the Cooper Clinic study, the investigators reported a relative risk of 1.5 for the development of hypertension in people with low levels of fitness compared with highly fit people.[7] In a sample of 2205 adults (20-49 years) from the National Health and Nutrition Examination Survey (NHANES), newly identified hypertension was associated with low fitness estimated from a submaximal treadmill test, with an odds ratio of 2.12 for women and 1.83 for men.[11] From these limited studies, active people and fit people are at reduced risk for developing hypertension. Epidemiological studies also have shown that higher physical activity levels and aerobic fitness are related to a decreased risk for stroke in both men[29] and women.[23]

Training Adaptations That Might Reduce Risk

A number of physiological adaptations that accompany endurance training could affect blood pressure both at rest and during exercise. One of the most important changes associated with endurance training is the previously mentioned plasma volume increase. We might logically assume that any increase in plasma volume would increase blood pressure, particularly because one of the first lines of drug treatment for hypertension is the prescription of a diuretic to reduce total body water and thus plasma volume. However, recall from chapter 10 that trained muscle has a notable increase in capillaries. Also, the venous system in a trained person has a greater capacity, allowing it to contain more blood. For these reasons, the increased plasma volume following exercise training does not increase blood pressure.

Specific mechanisms responsible for reductions in resting blood pressure with endurance training have not been established. Some studies show that resting cardiac output is reduced and that the body's oxygen demands are met by an increased arterial–mixed venous oxygen difference, or $(a\text{-}\bar{v})O_2$ difference. But other studies have shown cardiac output to remain unchanged. Without a decrease in cardiac output, the observed reductions in resting blood pressure that follow training must result from reductions in peripheral vascular resistance, which may be attributable to an overall reduction of sympathetic nervous system activity. Increased vasodilation and vascular remodeling of existing arteries and new vessel growth are also possible mechanisms of blood pressure reduction. Weight loss also is associated with reductions in blood pressure.[3]

Risk Reduction With Exercise Training

In the previous section on CAD, we determined that exercise training lowers resting blood pressure in normotensives and in those with hypertension, with

In focus

Aerobic training reduces blood pressure in healthy individuals as well as in those who have hypertension. The mechanisms by which exercise reduces blood pressure have not been completely determined.

In review

- Epidemiological studies generally have found that the risk of CAD in sedentary male populations is about two to three times that of men who are physically active, and that physical inactivity approximately doubles a person's risk of a fatal heart attack.
- The levels of activity associated with a reduced risk for CAD can be lower than those needed to increase aerobic capacity.
- Physical training improves the heart's contractility, work capacity, and coronary circulation.
- Exercise may have its major impact on blood lipid levels. Studies show that endurance training decreases the ratios of LDL-C to HDL-C and of Total-C to HDL-C.
- Exercise is anti-inflammatory and appears to improve endothelial function.
- Exercise can help control blood pressure, weight, and blood glucose levels and can help alleviate stress.
- People who are active and those who are fit have reduced risk for developing hypertension.
- Increased plasma volume that accompanies physical training does not increase blood pressure because trained people have more capillaries and greater venous capacity.
- Resting blood pressure is decreased by training in people with hypertension; this is probably attributable to decreased peripheral resistance, but the actual mechanisms are unknown.
- Exercise also reduces body fat, blood glucose levels, and insulin resistance, factors related to an increased risk for hypertension.

greater decreases in hypertensives. These reductions are unrelated to the duration of the training program but might be greater in response to low- to moderate-intensity activity compared with higher-intensity activity.

Not only does exercise reduce blood pressure itself, but it also affects other risk factors. Exercise is important in reducing body fat and can increase muscle mass, which may be important in reducing blood glucose levels and thus assisting in better glycemic (blood sugar) control. This could partially explain the reduction in insulin resistance, another risk factor for hypertension, that has been observed in training studies. Exercise training also has been associated with stress reduction.

RISK OF HEART ATTACK AND DEATH DURING EXERCISE

When a person dies while exercising, the incident usually makes newspaper headlines. Deaths during exercise don't happen often, but they are highly publicized. How safe, or how dangerous, is exercise? It was estimated in a review of the scientific literature prior to 1982 that there would be approximately one death for every 7,620 middle-aged joggers per year.[48] In a study several years later, the estimate was one death for every 18,000 physically active men.[45] In a study published in 2000, the risk with vigorous exercise was one death per 1.42 million hours of exercise.[1] Most of the research results up to this time were predominantly for men. In 2006, a study in women showed a much lower risk of one death for every 36.5 million hours of moderate to vigorous exercise.[50] So, the overall risk of heart attack and death during exercise is very low. Further, although the risk of death increases during a period of vigorous exercise, habitual vigorous exercise is associated with an overall decreased risk of heart attack.[45] This is illustrated in figure 20.10. There is concern, however, with those athletes who pursue ultra-endurance exercise, that is, training bouts or competition in excess of 4 h per session. Theoretically, they are at greater risk for cardiovascular disorders because of the high oxidative stress associated with this type of training or competition.[26] Information is insufficient at this time to allow resolution of this potential issue.

When death during exercise occurs in people age 35 or older, it usually results from a cardiac arrhythmia caused by atherosclerosis of the coronary arteries. On the other hand, those under age 35 are most likely to die from hypertrophic cardiomyopathy (enlarged diseased heart, usually genetically transmitted), congenital coronary artery anomalies, an aortic aneurysm, or myocarditis (inflammation of the myocardium).

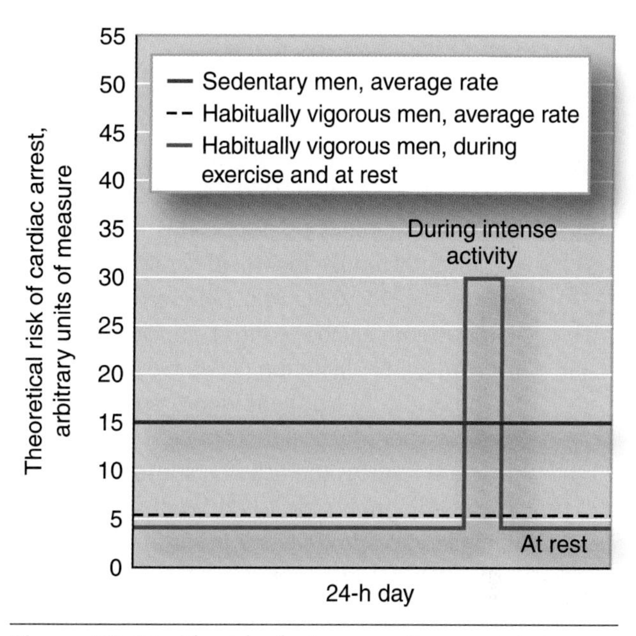

Figure 20.10 The risk of primary cardiac arrest during vigorous exercise and at other times throughout a 24 h period, comparing sedentary men with habitually active men.
Data from Siscovick et al., 1984.

In focus

The risk of heart attack increases during an actual period of exercise. However, over the course of a 24 h period, those who exercise regularly have a much lower risk of a heart attack than those who do not exercise.

In review

- Deaths during exercise are rare, although typically highly publicized.
- Deaths during exercise in people over age 35 usually are caused by a cardiac arrhythmia resulting from atherosclerosis.
- Deaths during exercise in people under age 35 usually are caused by hypertrophic cardiomyopathy, congenital coronary artery anomalies, aortic aneurysm, or myocarditis.

Exercise Training and Rehabilitating Patients With Heart Disease

Can active participation in a cardiac rehabilitation program that has a strong aerobic exercise component help a heart attack survivor to either survive a subsequent attack or avoid one altogether? Endurance training leads to many physiological changes that reduce the work or oxygen demand of the heart. As we have seen, many of these are peripheral, not involving the heart directly. To recap, training increases the capillary/muscle fiber ratio and plasma volume. Because of these changes, blood flow increases to the muscles. In some cases, as mentioned earlier, this allows a reduction in cardiac output, with the body's oxygen demands being met by an increased $(a\text{-}\bar{v})O_2$ difference. Training also possibly can increase or maintain oxygen supply to the heart.

However, significant changes also may occur in the heart itself. Studies of heart disease patients at Washington University in St. Louis have provided dramatic evidence that intense aerobic conditioning not only can substantially change peripheral factors but also can alter the heart itself, possibly increasing blood flow to the heart and increasing left ventricular function.[18]

From our previous discussions in this chapter, it is clear that endurance exercise training can significantly reduce the risk of cardiovascular disease through its independent effect on the individual risk factors for CAD and hypertension. Favorable changes in blood pressure, lipid levels, body composition, glucose control, and stress have been reported in patients undergoing exercise training for cardiac rehabilitation. We have every reason to believe that these changes are just as important to the health of a patient who has had a heart attack as they are for a presumably healthy person.

A number of researchers have tried to determine whether participation in a cardiac rehabilitation program reduces the risk of a subsequent heart attack or of death from a subsequent heart attack. It is nearly impossible to design a study to answer these questions, primarily because it would be necessary to enroll several thousand people into one study to have a large enough sample to prove a statistically significant effect. Consequently, several published reports have combined the results of the most highly controlled of these studies and have used meta-analysis, a special type of statistical analysis, to examine these data. The most recent report (2004), which is a composite covering the patients from previous reports plus >4,000 more patients (8,940 total), concluded that exercise rehabilitation substantially reduces total mortality (20% lower) as well as the risk of death from a subsequent heart attack (26% lower) . However, the effect on reducing the risk for recurrence of a nonfatal heart attack, while substantial (21% lower), was not statistically significant.[47]

The evidence that physical activity is important in the rehabilitation of the cardiac patient is sufficiently clear. The American College of Sports Medicine issued a position stand that concluded, "most patients with CAD should engage in individually designed exercise programs to achieve optimal physical and emotional health" (p. iv).[2] The recommendation was that such programs include a comprehensive preexercise medical evaluation, including a graded exercise test, and an individualized exercise prescription. Such programs should focus on multifactorial risk factor modification by using diet, drugs, and exercise to control blood lipid disorders, diabetes, and hypertension. With an aggressive approach to rehabilitation, it is even possible to see slight regression in the disease.[20]

In Closing

In this chapter, we have seen how important physical activity is in reducing the risk for cardiovascular diseases, especially CAD and hypertension. We discussed the prevalence of these disorders, the risk factors associated with each, and the ways in which physical activity can help reduce our personal risks. In the next chapter, we continue examining the effects of exercise on health as we turn our attention to obesity and diabetes.

KEY TERMS

arteriosclerosis
atherosclerosis
blood lipids
congenital heart disease
coronary artery disease (CAD)
diastolic blood pressure
fatty streaks
heart failure
hemorrhagic stroke
high-density lipoprotein cholesterol (HDL-C)
hypertension
ischemia
ischemic stroke
lipoproteins
low-density lipoprotein cholesterol (LDL-C)
metabolic syndrome
myocardial infarction
pathophysiology
peripheral vascular disease
plaque
platelet-derived growth factor (PDGF)
primary risk factors
rheumatic heart disease
risk factor
stroke
systolic blood pressure
triglycerides
valvular heart disease
very low density lipoprotein cholesterol (VLDL-C)

STUDY QUESTIONS

1. What are currently the major causes of death in the United States?
2. What is atherosclerosis, how does it develop, and at what age does it begin?
3. What is hypertension, how does it develop, and at what age does it begin?
4. What is stroke? How does stroke occur? What are the results of stroke?
5. What are the basic risk factors for coronary artery disease? For hypertension?
6. What is the risk of death from coronary artery disease associated with a sedentary lifestyle as compared with an active lifestyle? How has this been established?
7. What are three basic physiological alterations resulting from exercise training that would reduce the risk of death from coronary artery disease?
8. In what ways does endurance exercise training alter risk factors for heart disease?
9. What is a sedentary individual's risk of developing hypertension compared with an active individual's?
10. What are three basic physiological alterations resulting from exercise training that would reduce the risk for developing hypertension?
11. What changes in blood pressure that result from endurance exercise training occur in hypertensive individuals?
12. Of what value is cardiac rehabilitation in treating a patient who has had a heart attack?
13. What is the risk of death with endurance exercise training?

STUDY GUIDE ACTIVITIES

In addition to the activities listed in the chapter opening outline on page 471, two other activities are available in the online study guide, located at

www.HumanKinetics.com/PhysiologyOfSportAndExercise.

The chapter **SUMMARY** reviews the key concepts covered in the chapter, and the end-of-chapter **QUIZ** tests your understanding of the material covered in the chapter.

Obesity, Diabetes, and Physical Activity

In this chapter and in the online study guide

William "the Refrigerator" Perry, defensive lineman for the Chicago Bears professional football team during the 1980s and early 1990s, reported to the 1988 summer training camp at a weight of 170 kg (375 lb), some 25 kg (55 lb) over his mandated playing weight. Although there was an obvious concern regarding his ability to perform on the football field at this excessive weight, of greater concern are the health risks associated with obesity. Chris Taylor, an Iowa State and U.S. Olympic team wrestler, competed at a weight between 181 and 204 kg (400-450 lb). He died in his sleep at age 29, most likely from obesity-related causes.

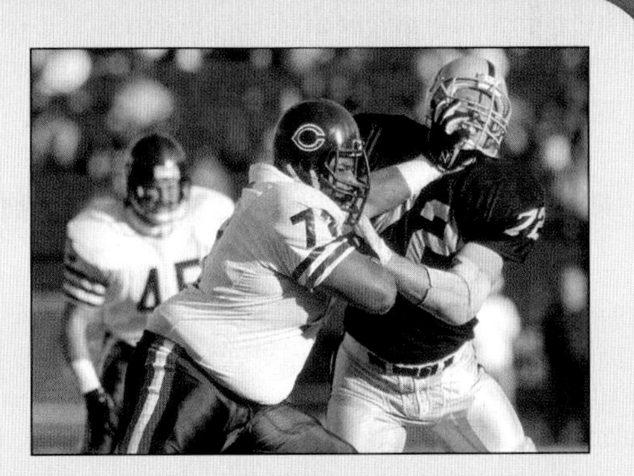

The excessive weight of some professional athletes not only affects their performance but also increases their health risks.

While millions of people are dying of starvation each year in most parts of the world, many Americans are dying as an indirect result of overconsumption of food. Billions of dollars are spent each year overfeeding the American public. This, in turn, leads to the expenditure of billions of dollars more each year on various weight loss methods, and additional billions for the increase in health care costs associated with obesity. Another common disorder in America is diabetes mellitus, which affects about 15 million Americans. This disorder of carbohydrate metabolism centers on insulin. Interestingly, scientists have established a link between insulin resistance and obesity, coronary artery disease, hypertension, and many other diseases or disorders.

A sedentary lifestyle has been associated with an increased risk for both obesity and diabetes, and both are strongly associated with other diseases that have high mortality rates, such as coronary artery disease and cancer. Furthermore, millions of Americans are obese, diabetic, or both. The consequences of these diseases are debilitating, and the costs associated with their treatment are exceedingly high. In this chapter we focus on obesity and diabetes, discussing their prevalence, their etiology, health problems associated with each disease, and general treatment options. Finally, we consider the role that physical activity can play in their prevention and treatment.

OBESITY

The terms *overweight* and *obesity* often are used interchangeably, but technically they have different meanings, as discussed next.

Terminology and Classification

Overweight is defined as a body weight that exceeds the normal or standard weight for a particular person based on height and frame size. These standard weights were established in 1959 but are still widely used. New weight and height tables were introduced in 1983, but their introduction was controversial because many experts believed the weight allowances were too liberal. Many professional health organizations refused to accept the newer tables.

Weight values in the standard tables are based solely on population averages. For this reason, a person can be overweight according to these standards and yet have a lower than normal body fat content. For example, football players frequently are found to be overweight according to standard tables, yet many are typically much leaner than people of the same age, height, and frame size who are of normal weight or even underweight (see chapter 14). Still other people are within the normal range of body weights for their height and frame size by the standard tables and yet are obese.

Obesity refers to the condition of having an excessive amount of body fat. This implies that the actual amount of body fat or its percentage of total weight must be assessed or estimated (see chapter 14 for assessment techniques). Exact standards for allowable fat percentages have not been established. However, men with more than 25% body fat and women with more than 35% should be considered obese. Men with relative fat values of 20% to 25% and women with values of 30% to 35% should be considered borderline obese. Allowances are higher for women due to sex-specific fat depots such as breast tissue and hips, buttocks, and thighs, as will be discussed.

Body mass index (BMI) is now the most widely used clinical standard to estimate obesity. To determine a person's BMI, body weight in kilograms is divided by the square of body height in meters. As an example, a man who weighs 104 kg (230 lb) and is 183 cm (6 ft) tall would have a BMI of 31 kg/m^2 ($104\ kg/(1.83\ m)^2 = 104\ kg/3.35\ m^2 = 31\ kg/m^2$). Generally, the BMI is highly correlated with body fat and usually provides a reasonable estimate of obesity. Table 21.1 provides a simple way of determining BMI from height and weight.

In 1997, the World Health Organization proposed a classification system for underweight, overweight, and obesity based solely on BMI values.[40] This classification system was adopted by the National Institutes of Health in 1998 with several modifications and has been used widely since 2000.[24] In table 21.2 on page 496, BMI values have been divided into five categories: underweight, normal weight, overweight, obesity, and extreme obesity. Within the obesity classification, there are two subclassifications, Class I and Class II. Extreme obesity is Class III. The degree of disease risk also is included and is determined by both BMI and waist circumference. A larger waist circumference increases risk for a given BMI category. Waist circumference reflects the abdominal visceral fat, which plays a major role in increasing the risk for disease. It is now known that racial and ethnic differences affect the relationship between BMI and fatness, necessitating different BMI cut-points for overweight and obesity in specific groups. As an example, a number of studies have indicated that for the same BMI, the health risk is higher in Asian populations.

This classification system has made a major contribution to our understanding of the true prevalence of overweight and obesity. Prior to the adoption of this system, there was a wide range of estimates of the percentage of adults who were overweight or obese or both. This was the result of studies using different cut-points or standards for overweight and obesity, which led to considerable confusion among scientists and the general public about the true prevalence of

TABLE 21.1 Body Mass Index

BMI	19	20	21	22	23	24	25	26	27	28	29	30	31	32	33	34	35
Height	Weight (in pounds)																
4'10" (58")	91	96	100	105	110	115	119	124	129	134	138	143	148	153	158	162	167
4'11" (59")	94	99	104	109	114	119	124	128	133	138	143	148	153	158	163	168	173
5' (60")	97	102	107	112	118	123	128	133	138	143	148	153	158	163	168	174	179
5'1" (61")	100	106	111	116	122	127	132	137	143	148	153	158	164	169	174	180	185
5'2" (62")	104	109	115	120	126	131	136	142	147	153	158	164	169	175	180	186	191
5'3" (63")	107	113	118	124	130	135	141	146	152	158	163	169	175	180	186	191	197
5'4" (64")	110	116	122	128	134	140	145	151	157	163	169	174	180	186	192	197	204
5'5" (65")	114	120	126	132	138	144	150	156	162	168	174	180	186	192	198	204	210
5'6" (66")	118	124	130	136	142	148	155	161	167	173	179	186	192	198	204	210	216
5'7" (67")	121	127	134	140	146	153	159	166	172	178	185	191	198	204	211	217	223
5'8" (68")	125	131	138	144	151	158	164	171	177	184	190	197	203	210	216	223	230
5'9" (69")	128	135	142	149	155	162	169	176	182	189	196	203	209	216	223	230	236
5'10" (70")	132	139	146	153	160	167	174	181	188	195	202	209	216	222	229	236	243
5'11" (71")	136	143	150	157	165	172	179	186	193	200	208	215	222	229	236	243	250
6' (72")	140	147	154	162	169	177	184	191	199	206	213	221	228	235	242	250	258
6'1" (73")	144	151	159	166	174	182	189	197	204	212	219	227	235	242	250	257	265
6'2" (74")	148	155	163	171	179	186	194	202	210	218	225	233	241	249	256	264	272
6'3" (75")	152	160	168	176	184	192	200	208	216	224	232	240	248	256	264	272	279

Evidence Report of Clinical Guidelines on the Identification, Evaluation, and Treatment of Overweight and Obesity in Adults, 1998. NIH/National Heart, Lung, and Blood Institute (NHLBI).

TABLE 21.2 Classification of Overweight and Obesity by BMI, Waist Circumference, and Associated Disease Risk[a]

			Disease risk (relative to normal weight and waist circumference)	
Classification	BMI (kg/m^2)	Obesity class	Men ≤40 in. (102 cm) Women ≤35 in. (88 cm)	Men >40 in. (102 cm) Women >35 in. (88 cm)
Underweight	<18.5		–	–
Normal[b]	18.5-24.9		–	–
Overweight	25.0-29.9		Increased	High
Obesity	30.0-34.9	I	High	Very high
	35.0-39.9	II	Very high	Very high
Extreme obesity	≥40	III	Extremely high	Extremely high

[a]Disease risk for type 2 diabetes, hypertension, and cardiovascular disease.

[b]Increased waist circumference also can be a marker for increased risk even in persons of normal weight.

Adapted, by permission, from World Health Organization, 1998, "Obesity: Preventing and managing the global epidemic." In *Report of a WHO Consultation on Obesity* (Geneva: WHO).

weight disorders. We can now better understand the true prevalence of overweight and obesity and how it has changed over time. Further, having an overweight category has been very useful, providing a buffer zone between normal weight and obesity. People falling in this category can be overweight based on an above-average fat-free body mass, such as the football players mentioned previously, or they can be slightly overfat. Almost all people with a BMI of 30.0 or higher are definitely obese.

Prevalence of Overweight and Obesity in the United States

The prevalence of overweight and obesity in the United States has increased dramatically since the 1970s, as illustrated in figure 21.1. This figure presents data from national surveys of large numbers of men and women representative of the total U.S. population, obtained in 1960-1962, 1971-1974, 1976-1980, 1988-1994, and 1999-2004.[10, 25] The percentage overweight represents the percentage of the total population with a BMI of 25.0 to 29.9 (figure 21.1*a*), and the percentage obese is for those with BMI values of 30.0 or greater (figure 21.1*b*), consistent with the World Health Organization and National Institutes of Health classification system. Figure 21.1*c* presents the data for the combination of those who are overweight and obese (BMI ≥25.0). It is most striking to note that 31% of men and 33% of women in the United States are obese and that nearly 71% of men and 62% of women are overweight or obese.[25] Further, the prevalence of obesity increased by 62% in men and 52% in women between the 1976-1980 and the 1988-1994 data collection periods,[10] and by an additional 34% in men and 31% in women between the 1988-1994 and 1999-2000 periods.[11] Interestingly, the prevalence of overweight remained relatively constant over these same time periods. When we look at these data by race, it is apparent that the problem is much more significant in Mexican American men and women and in black women (figure 21.2 on page 498). These trends are not unique to the United States. Canada, Australia, and most of Europe have seen similar increases but, with few exceptions, not to the extent seen in the United States.[40] The most recent studies are now showing that obesity is spreading to all regions of the world!

Unfortunately, this same trend of increasing prevalence of overweight has been reported in U.S. children and adolescents.[25, 33] Figure 21.3 on page 498 illustrates the trends in the prevalence of overweight from 1963 through 2004 in preadolescent and adolescent boys and girls. Because BMI is much less precise for estimating body fat in children and adolescents, scientists typically use the cut-point for BMI of greater than the 95th percentile, a value that likely indicates that the child is overfat. Similar to the adult trends shown in figure 21.1, the prevalence of overweight remained relatively constant from 1963 through 1980 but has increased dramatically since 1980.

In focus

More than 70% of men and nearly 62% of women in the U.S. adult population are overweight or obese, and the prevalence of overweight in children has increased at an alarming rate since 1980.

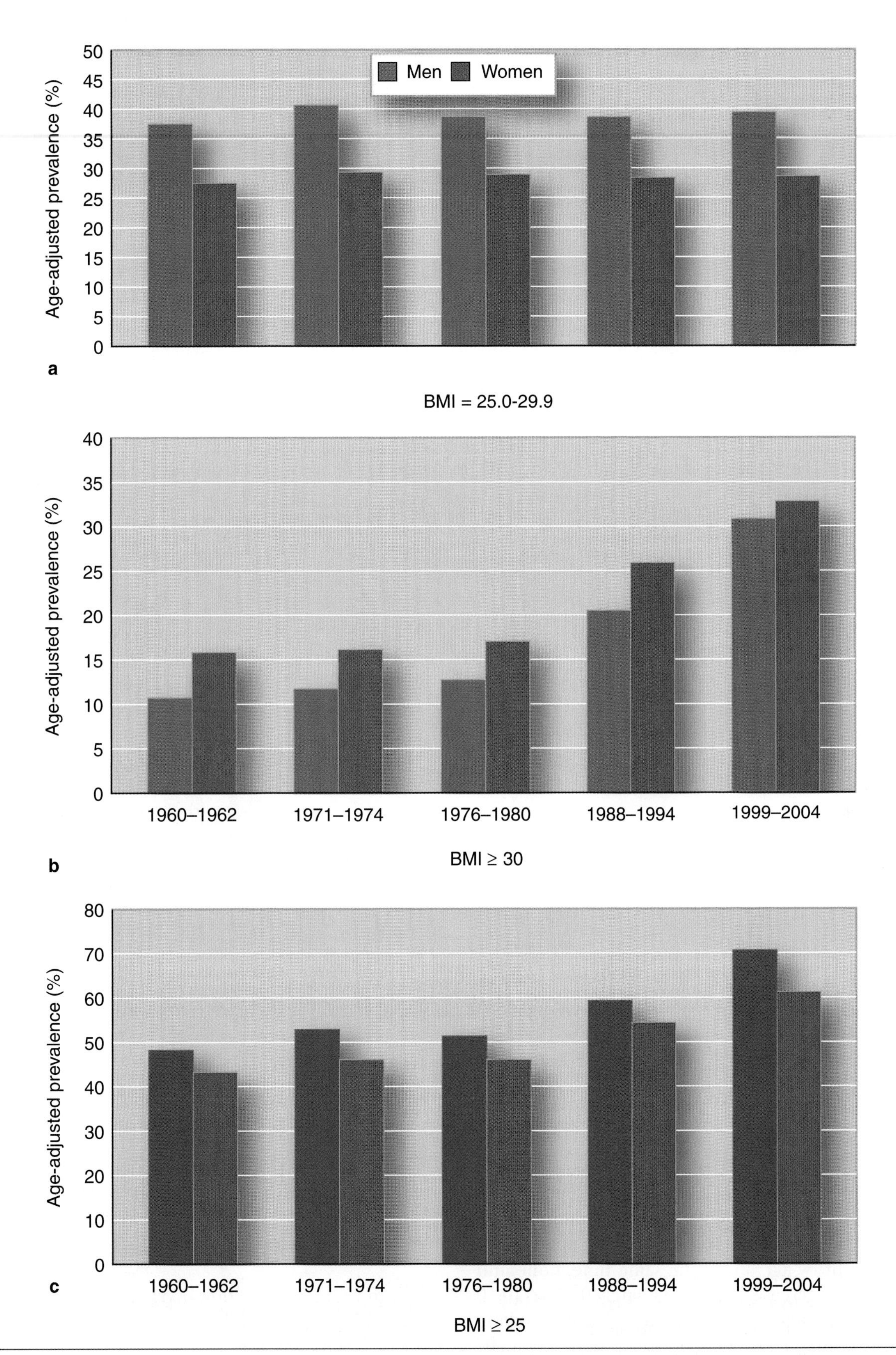

Figure 21.1 The increasing prevalence of *(a)* overweight (body mass index [BMI] = 25.0-29.9), *(b)* obesity (BMI of 30.0 and greater), and *(c)* the combination of overweight and obesity (BMI of 25 and greater) in the United States from 1960 through 2004.

Data from Flegal et al., 1998; Flegal et al., 2002; and Ogden et al., 2006.

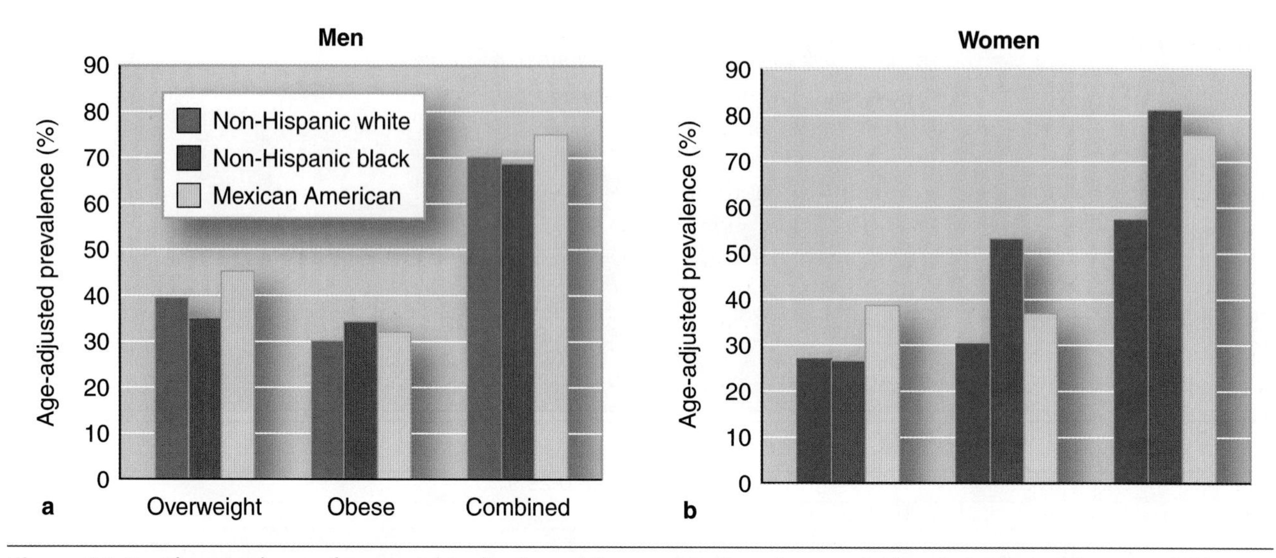

Figure 21.2 The prevalence of overweight, obesity, and the combination of overweight and obesity in *(a)* men and *(b)* women by race (2004).

Data from C.L. Ogden et al., 2002, "Prevalence of overweight and obesity in the United States, 1999-2004," *Journal of the American Medical Association* 295: 1549-1555.

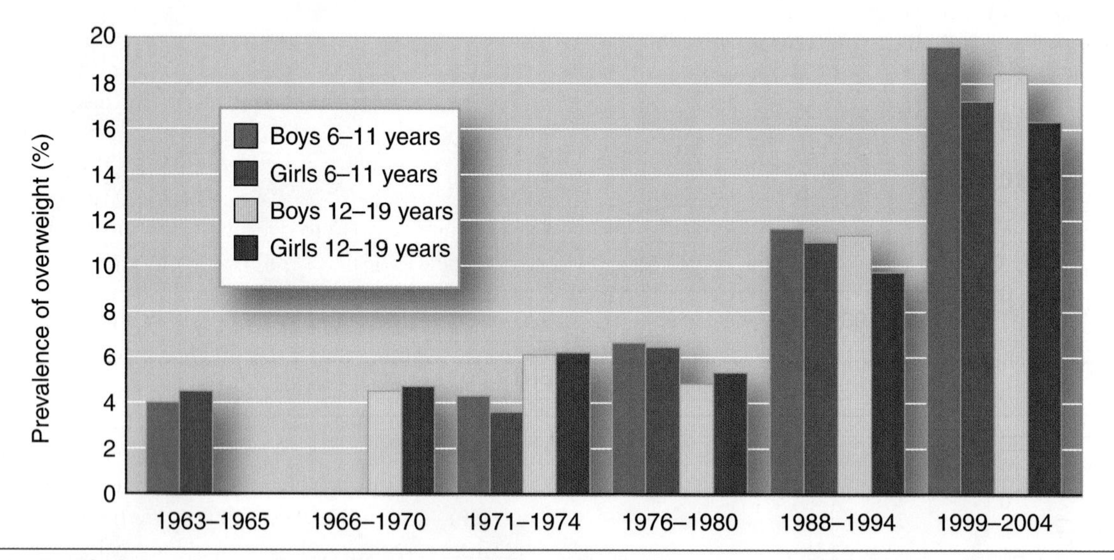

Figure 21.3 The increasing prevalence of overweight (95th percentile) in children and adolescents in the United States from 1962 through 2004.

Data from C.L. Ogden et al., 2002, "Prevalence and trends in overweight among US children and adolescents," *Journal of the American Medical Association* 288: 1728-1732; and from C.L. Ogden et al., 2006, "Prevalence of overweight and obesity in the United States, 1999-2004," *Journal of the American Medical Association* 295: 1549-1555.

The average person in the United States will gain approximately 0.45 to 0.90 kg (1 to 2 lb) of additional weight each year after age 25. Such a seemingly small gain, however, results in 14 to 28 kg (30 to 60 lb) of excess weight by age 55. At the same time, bone and muscle mass decrease by approximately 0.25 kg (0.50 lb) per year because of reduced physical activity. Taking this into account, an average person's fat mass actually increases by 0.7 to 1.4 kg (1.5 to 3.0 lb) each year, which equates to a 20 to 40 kg (45 to 90 lb) fat gain over a 30-year period! It is no wonder that weight loss is an American obsession.

Control of Body Weight

One must understand how body weight is controlled or regulated to better understand how a person becomes obese. Body weight regulation has puzzled scientists for years. The average-sized person takes in about 2,500 kcal per day, or nearly 1 million kcal per year. An

In focus

The body has the ability to balance energy intake and expenditure to within 10 to 15 kcal per day, about the equivalent of one potato chip!

average gain of 0.7 kg (1.5 lb) of fat each year represents an imbalance of only 5,250 kcal per year between energy intake and expenditure (3,500 kcal is the energy equivalent of 0.45 kg, or 1 lb, of adipose tissue). This translates into a surplus of less than 15 kcal per day. Even with a weight gain of 1.5 lb (0.7 kg) of fat per year, the body can balance caloric intake to within one potato chip per day of what is expended, a truly remarkable example of homeostasis.

The body's ability to balance its caloric intake and expenditure to within such a narrow range has led scientists to propose that body weight is regulated around a given set point, similar to the way in which body temperature is regulated. Excellent evidence for this is found in the animal research literature.[18] When animals are force-fed or starved for various periods of time, their weights respectively increase or decrease markedly. But when they go back to their normal eating patterns, they always return to their original weight or to the weight of the control animals (for animals that naturally continue to gain weight throughout their life span).

Similar results have been found in humans, although the number of studies is limited. Subjects placed on semistarvation diets have lost up to 25% of their body weight but regained that weight within months of returning to a normal diet.[19] In a study involving Vermont prisoners, overfeeding resulted in weight gains of 15% to 25%, yet their weights returned to original levels shortly after the experiment ended.[30]

How is the body able to do this? Considering energy expenditure, the total amount of energy expended each day can be expressed as the sum of three components (see figure 21.4):

1. Resting metabolic rate (RMR)
2. The thermic effect of a meal (TEM)
3. The thermic effect of activity (TEA)

Recall that **resting metabolic rate (RMR),** as discussed in chapter 4, is the body's metabolic rate early in the morning following an overnight fast and 8 h of sleep. The term basal metabolic rate (BMR) also is used but generally implies that the person fasted for 12 to 18 h and slept over in the clinical facility where the BMR measurement would be made. Most research today uses the RMR. This value, as we learned in chapter 4, represents the minimal amount of energy expenditure needed to support basic physiological processes. It accounts for about 60% to 75% of the total energy we expend each day.

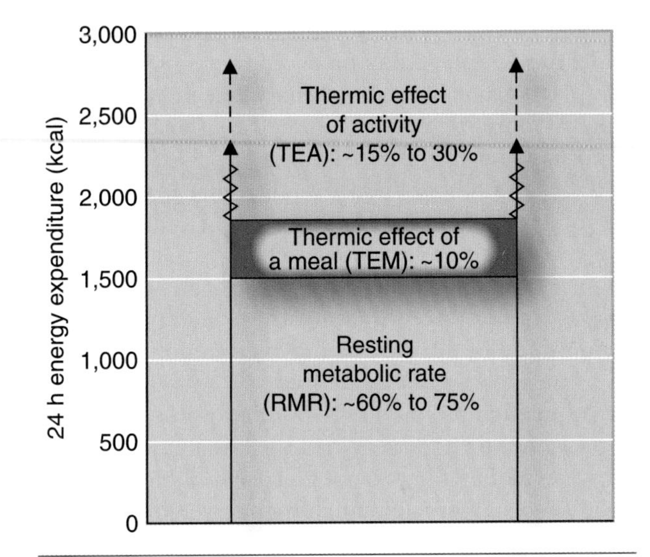

Figure 21.4 The three components of energy expenditure. See the text for a detailed explanation.

Adapted, by permission, from E.T. Poehlman, 1989, "A review: Exercise and its influence on resting energy metabolism in man," *Medicine and Science in Sports and Exercise* 21: 515-525.

The **thermic effect of a meal (TEM)** represents the increase in the metabolic rate that is associated with the digestion, absorption, transport, metabolism, and storage of ingested food. The TEM accounts for approximately 10% of our total energy expenditure each day. This value also includes some energy waste, because the body can increase its metabolic rate above that necessary for food processing and storage. We seldom notice the TEM; however, after a very large holiday meal with family, people will start feeling warm and sleepy, with small beads of sweat forming on the forehead. These changes indicate that the metabolic rate has increased considerably. The TEM component of metabolism might be defective in people with obesity, possibly attributable to a defect in the energy-wastage component, leading to a surplus of calories.

The **thermic effect of activity (TEA)** is simply the energy expended above the RMR to accomplish a given task or activity, whether it is combing one's hair or running a 10 km race. The TEA accounts for the remaining 15% to 30% of our energy expenditure.

The body adapts to major increases or decreases in energy intake by altering the energy expended by each of these three components—RMR, TEM, and TEA. With fasting or very low calorie diets, all three decrease. The body appears to be attempting to conserve its energy stores. This is dramatically illustrated by decreases in RMR of 20% to 30% or more reported within several weeks after patients begin fasting or consuming a very low calorie diet. Conversely, all three components of energy expenditure increase with overeating. In this case, the body appears to be trying to prevent unnecessary storage of the surplus calories. All of these adaptations may be under the control of the sympathetic nervous system and may play a major, if not the primary, role in maintaining weight around a given set point. This remains a most important area for future research.

If the body has a set-point weight, how can we explain the increasing prevalence of overweight and obesity? It appears that the set point can change, at least in the animals that have been most extensively studied. In several studies, the duration of overfeeding and the composition of the diet during overfeeding appear to alter the set-point. For example, the set-point weight of rats maintained on a high-fat diet over a six-month period tends to increase: When the rats are placed back on a low-fat diet, they do not go back to their expected weight but stabilize at a much higher weight. For dieting periods of less than six months, they do tend to go back to their original set point. The composition of the diet is therefore a prime suspect for increasing set-point weight, providing the duration of the intervention is sufficient. The physical activity level is also a potential factor. It is quite possible that an increase in the fat content of the diet and a decrease in physical activity levels over an extended period of time would increase set-point weight. This could at least partially explain the increasing prevalence of overweight and obesity in the United States today. It is also important to note that people, like rats, generally consume more calories per day when they are placed on high-fat diets.

Another related factor evolved during the 1990s—supersizing food portions. Fast food restaurants and many restaurant food chains have started serving much larger portions of food, a trend referred to as *supersizing*. How much difference does it make when one supersizes a meal? At several fast food chains, a single meal including a supersized double cheeseburger (~900 kcal), supersized French fries (~500 kcal), and a supersized 42 oz soft drink (~500 kcal) provides about 1900 kcal of energy. For smaller and less active people, this would be sufficient to supply their total daily caloric needs!

The trends of food availability and food purchasing and preparation in the United States between 1970 and 1998 revealed that the per capita energy availability increased by 15%.[14] Furthermore, Americans are eating more meals outside the home, relying more heavily on convenience foods, and consuming larger food portions. These trends point to a general increase in daily energy intake over the past 30 years.

In focus

The body attempts to defend its weight when overfed or underfed by increasing or decreasing the three components of energy expenditure: RMR, TEM, and TEA.

In review

- Overweight is a body weight that exceeds the standard weight for a certain height and frame size. Obesity refers to excessive body fat, meaning more than 25% body fat for men and more than 35% body fat for women.
- To calculate a person's BMI, we divide body weight in kilograms by the square of height in meters. This value is highly correlated with relative body fat and provides a reasonable estimate of obesity. Body mass index values of 25.0 to 25.9 correspond to overweight, and values of 30.0 or higher correspond to obesity.
- Prevalence of obesity and overweight in the United States has increased dramatically since the 1970s.
- The average person gains 0.45 to 0.90 kg (1 to 2 lb) per year after age 25 but also loses 0.25 kg (0.5 lb) of fat-free mass per year, meaning a net gain of 0.7 to 1.4 kg (1.5 to 3.0 lb) of fat each year.
- Body weight appears to be regulated around a set point.
- Daily energy expenditure is reflected by the sum of the RMR, the TEM, and the TEA. The body adapts to changes in energy intake by adjusting any or all of these components.

Etiology of Obesity

At various times throughout human history, obesity has been thought to be caused by basic hormonal imbalances resulting from failure of one or more of the endocrine glands to properly regulate body weight. At other times, it has been believed that gluttony, rather than glandular malfunction, was the primary cause of obesity. In the first case, a person is perceived as having no control over the situation, and yet in the second, he or she is held directly responsible! Results of recent medical and physiological research show that obesity can be the result of any one or a combination of many factors. Its etiology, or cause, is not as simple as was once believed.

Experimental studies on animals have linked obesity to hereditary (genetic) factors. Studies of humans have also shown a direct genetic influence on height, weight, and BMI. A study from Laval University in Quebec provided possibly the strongest evidence yet of a significant genetic component for obesity.[5] The investigators took 12 pairs of young adult male monozygotic (identical) twins and housed them in a closed section of a dormitory under 24 h observation for 120 consecutive days. The subjects' diets were monitored during the initial 14 days to determine their baseline caloric intake. Over the next 100 days, the subjects were fed 1000 kcal above their baseline consumption for six of every seven days. On the seventh day, the subjects were fed only their baseline diet. Thus, they were overfed by 1,000 kcal per day for 84 of the 100 days. Activity levels were also tightly controlled. At the end of the study period, as shown in figure 21.5, the actual individual weight gained varied widely, from 4.3 to 13.3 kg (9.5-29.3 lb)—a threefold variation in weight gain for overconsumption of the same number of calories. However, the response of both twins in any given twin pair was quite similar; the major variations occurred between different twin pairs. Similar results were found for gains in fat mass, percentage body fat, and subcutaneous fat.

In focus

Recent research confirms that there is a significant genetic component in the etiology of obesity. However, it is possible to be obese, attributable basically to lifestyle choices, in the absence of a family history (genetics) of obesity. It is also possible to be relatively lean, even with a genetic predisposition to obesity, through proper diet and activity levels.

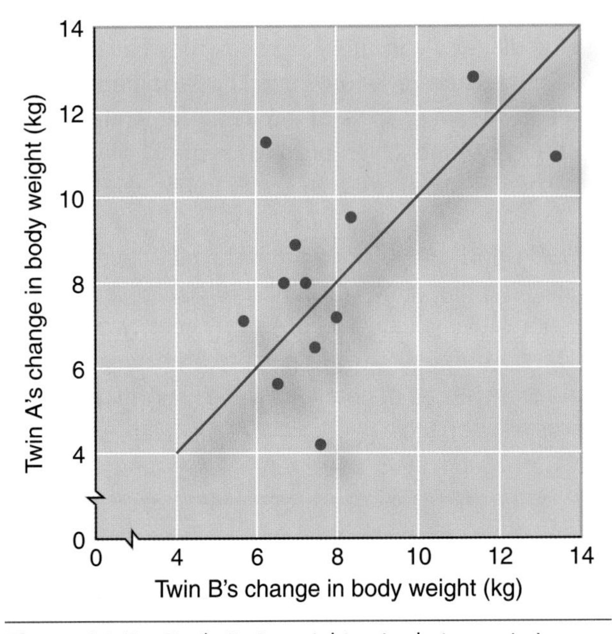

Figure 21.5 Similarity in weight gains between twins in response to a 1,000 kcal increase in dietary intake for 84 days of a 100-day study. A data point represents the weight gain for each twin in a pair; twin A's value is shown on the *y*-axis, twin B's on the *x*-axis. See the text for further explanation of these data.

Adapted, by permission, from C. Bouchard et al., 1990, "The response to long-term overfeeding in identical twins," *New England Journal of Medicine* 322: 1477-1482.

Hormonal imbalances, emotional trauma, and alterations in basic homeostatic mechanisms all have been shown to be either directly or indirectly related to the onset of obesity. Environmental factors, such as cultural habits, inadequate physical activity, and improper diets, are major causes of obesity as discussed in the previous section.

Thus, obesity is of complex origin, and the specific causes undoubtedly differ from one person to the next. Recognizing this is important for treating existing obesity and for preventing its onset. To attribute obesity solely to gluttony is unfair and psychologically damaging to people who are concerned about their problem and are attempting to correct it. In fact, several studies have shown that some obese people actually eat less, although they get far less physical activity, than people of the same sex and similar age with average body fat contents.

Health Problems Associated With Excessive Weight and Obesity

Overweight and obesity are associated with an increased overall rate of death (general excess mortality).[6] This relationship is curvilinear, as shown in figure 21.6 on page 503. A major increase in risk occurs when the BMI exceeds 30 kg/m^2, although BMI values between 25.0 and 30.0 are associated with an increased risk for many diseases. Excess mortality associated with obesity and overweight is linked with the following major diseases:

- Heart disease
- Hypertension
- Type 2 diabetes
- Certain types of cancer
- Gall bladder disease
- Osteoarthritis

With the large increase in the prevalence of obesity in the United States since the 1970s, it is not surprising to also see a very high prevalence of the metabolic syndrome (chapter 20) in U.S. adults. Data from the third National Health and Nutrition Examination Survey (NHANES-3), conducted between 1988 and 1994, demonstrated that 23.7% of all adults in the United States met the criteria for the metabolic syndrome. In adults over 60 years of age, the prevalence was greater than 40%. Consistent with the trends in obesity, the prevalence was highest in Mexican American men and

A Genetic Predisposition to Obesity: The Pima Indians

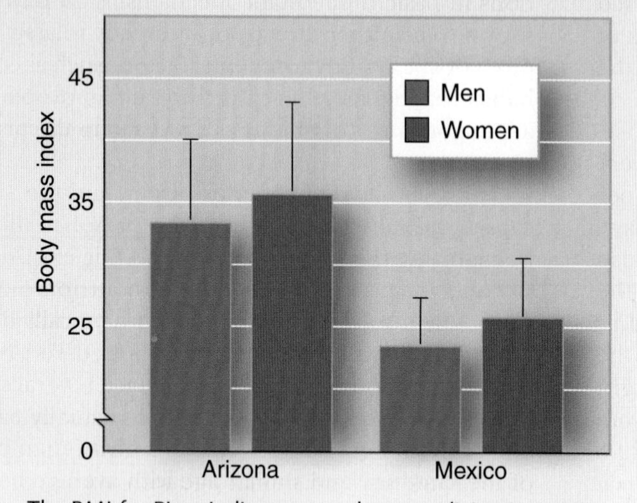

The BMI for Pima Indian men and women living in Arizona and in northern Mexico (2006).

From L.O. Schulz et al., 2006, "Effects of traditional and western environments on prevalence of type 2 diabetes in Pima Indians in Mexico and the U.S.," *Diabetes Care*, 29: 1866-1871.

It has been clearly established that genetics is a major factor in the development of obesity. Dr. Claude Bouchard, director of the Pennington Biomedical Research Center (Louisiana State University System), has conducted many studies on the heritability of obesity and has concluded that the heritability of fat mass or relative body fat (percentage fat) is about 25% of the age- and sex-adjusted variance.[4] Does having a genetic predisposition to obesity mean that someone is destined to be obese? The answer is no! We have learned a great deal from the study of the Pima Indians, who have lived for at least 2,000 years near the Gila River in the Sonora Desert in what is now southern Arizona. Until the early 1900s, the Pima Indians apparently were lean and healthy people who were physically active and ate a healthy diet. As they have moved onto reservations, stopped farming, and started eating a westernized high-fat diet and consuming alcohol, they have become very obese, with a prevalence of obesity of 64% for men and 75% for women.[28] Associated with their obesity is a high prevalence of diabetes—34% of men and 41% of women. The Pima Indians are such an unusual population with respect to their extremely high prevalence of obesity and diabetes that the National Institutes of Health established a special research center in Phoenix, Arizona, just to study them and their health issues.

Interestingly, another group of Pima Indians who live in northern Mexico have stayed active working on farms but use no motorized equipment. Furthermore, they consume a diet high in carbohydrates and low in fat. They have managed to stay relatively lean. The BMIs for these two groups of Indians are seen in the figure in this sidebar. The bottom line is that a person can have a genetic predisposition for obesity but with proper diet and exercise can maintain a relatively normal body weight. The Pima Indians have taught us an important lesson.

women.[12] Furthermore, obesity has been directly related to changes in normal body function, increased risk for certain diseases, detrimental effects on established diseases, and adverse psychological reactions.

Changes in Normal Body Function

The prevalence and extent of changes in body function vary with the individual and with the degree of obesity. Respiratory problems are common among people with obesity, including sleep apnea. These can lead to other common consequences of obesity, such as lethargy (sluggishness), because of increased carbon dioxide levels in the blood, and polycythemia (increased red blood cell production) in response to lower arterial blood oxygenation. These conditions can lead to abnormal blood clotting (thrombosis), enlargement of the heart, and congestive heart failure. Those with obesity typically have a lower exercise tolerance because of these respiratory problems and also because of the increased body mass that must be moved during exercise. Additional weight gains further reduce activity levels, and exercise tolerance decreases even more.

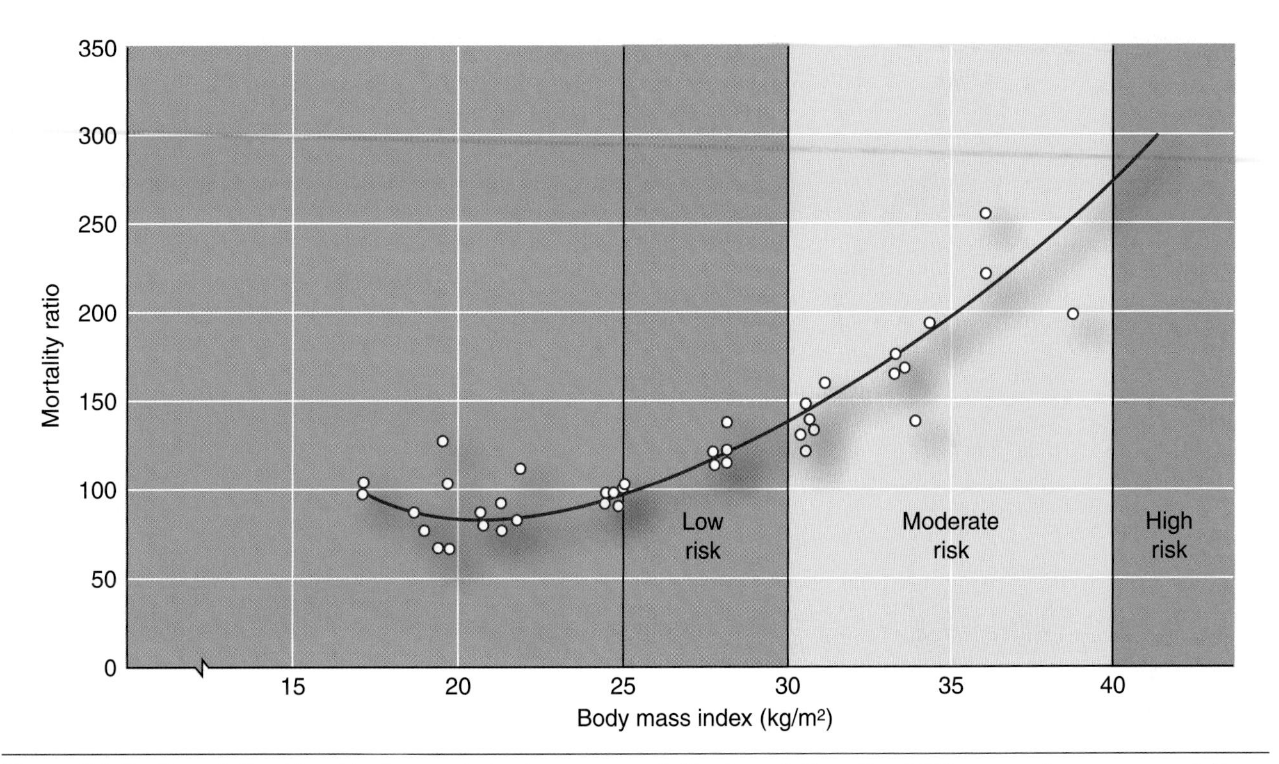

Figure 21.6 The relationship of body mass index to excess mortality. A mortality ratio of 100 represents average mortality. The lowest portion of the curve (body mass indexes under 25) indicates very low risk.

Bray, G.A. "Obesity: Definition, diagnosis and disadvantages." MJA 1985; 142: S2-S8. © Copyright 1985. *The Medical Journal of Australia*—reproduced with permission.

Increased Risk for Certain Diseases

An increased risk of developing certain chronic degenerative diseases also is associated with obesity. Both hypertension and atherosclerosis have been directly linked to obesity (see chapter 20). So have various metabolic and endocrine disorders, such as impaired carbohydrate metabolism and diabetes. Obesity is a problem associated particularly with the onset of type 2 diabetes (non-insulin-dependent diabetes). A major research breakthrough has enabled us to better understand the role of obesity as a risk factor for most of these diseases. Since the 1940s, major sex differences in the way in which fat is stored or patterned on the body have been recognized. As shown in figure 21.7 on page 504, males tend to store fat in the upper body, particularly the abdominal area, whereas females tend to store fat in the lower body, particularly the hips, buttocks, and thighs. Obesity that follows the male pattern is referred to as **upper body (android) obesity,** or apple-shaped obesity, and the female pattern is referred to as **lower body (gynoid) obesity,** or pear-shaped obesity.

Research beginning in the late 1970s and early 1980s established upper body obesity as a risk factor for the following conditions:

- ► Coronary artery disease
- ► Hypertension
- ► Stroke
- ► Elevated blood lipids
- ► Diabetes

Furthermore, upper body obesity appears to be more important than total-body fatness as a risk factor for these diseases. Waist and hip circumference, or girth, measurements can be used to identify people with increased risk. A waist/hip girth ratio greater than 0.90 for men and greater than 0.85 for women indicates increased risk. With upper body obesity, the increased risk may result from the visceral fat depots' close proximity to

In focus

Obesity places an individual at significantly increased risk for hypertension, diabetes, coronary artery disease, and metabolic and digestive diseases. The health risks associated with obesity are most likely associated with the manner in which the fat is distributed on the body, with upper body obesity (representing high levels of visceral fat) posing significantly greater risk.

the portal circulatory system (circulation to the liver). Figure 21.8 shows a young woman being placed into a computed tomography (CT) scanner in order to assess visceral abdominal fat (figure 21.8*a*) and CT scans at the level of the fourth lumbar vertebra of two men (figure 21.8, *b* & *c*).[29] The subject in figure 21.8*c* has considerably more visceral (deep) abdominal fat than subcutaneous abdominal fat.

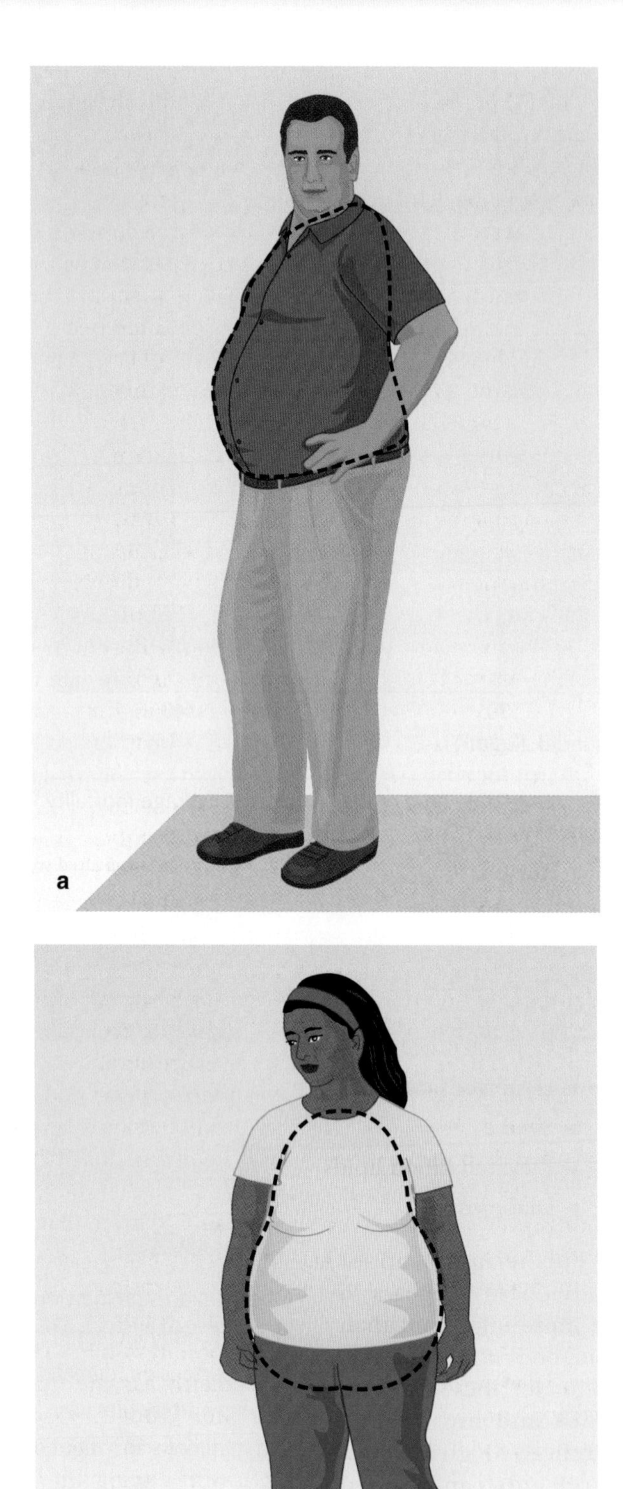

Figure 21.7 *(a)* Upper body (android) obesity; *(b)* lower body (gynoid) obesity.

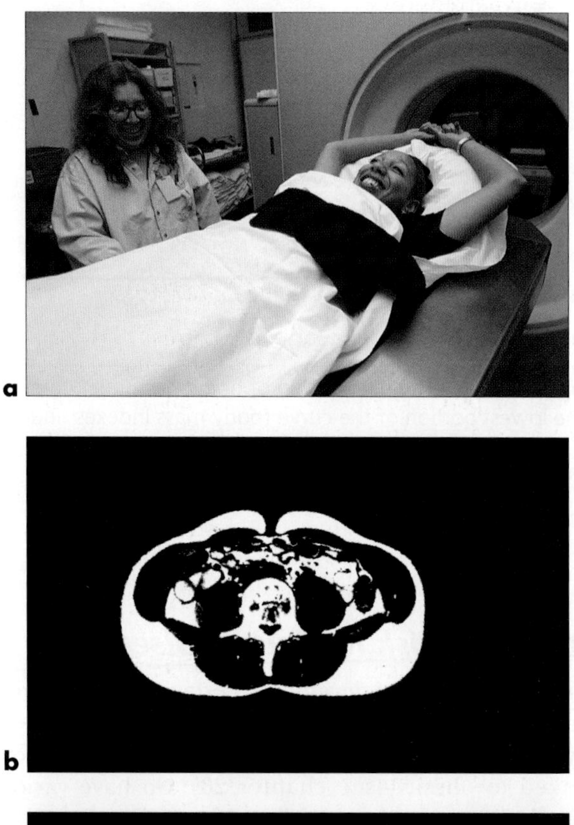

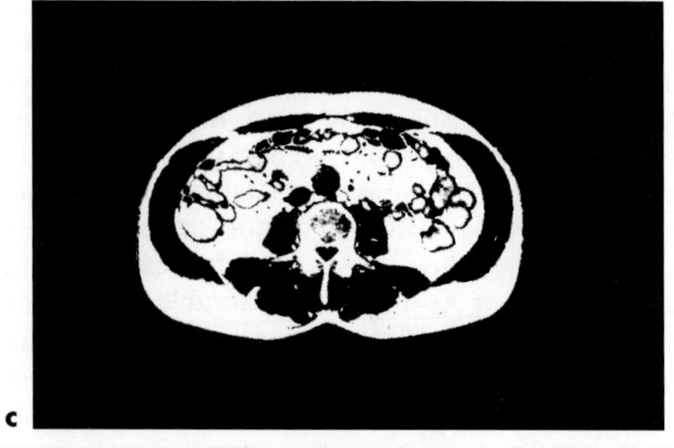

Figure 21.8 *(a)* A subject in the process of undergoing a computed tomography (CT) scan. (*b* and *c*) CT scans at the level of the fourth lumbar vertebra in two people. The subject in *c* has considerably more visceral abdominal fat (light areas) than subcutaneous abdominal fat. The subject in *b* is leaner but also has considerably less of his fat as visceral fat.

Detrimental Effects on Established Diseases

The effects of obesity on existing diseases are not clear. Obesity can contribute to further development of certain diseases and medical conditions, and weight reduction usually is prescribed as an integral part of treatment. Conditions that generally benefit from weight reduction include

- angina pectoris,
- hypertension,
- congestive heart disease,
- myocardial infarction (reduced risk of recurrence),
- varicose veins,
- diabetes, and
- orthopedic problems.

Adverse Psychological Reactions

Emotional or psychological problems might be the cause of obesity in some people. Furthermore, emotional or psychological problems can arise from the condition itself. In our society, obesity carries a social stigma that contributes substantially to the problems of those who have it. Our media typically glamorize only people with extremely lean bodies. Consequently, some obese people might need professional counseling assistance in their efforts to lose weight.

General Treatment of Obesity

In theory, weight control seems to be a simple matter. To maintain weight, the energy consumed by the body in the form of food must equal the total energy expended, which is the sum of the RMR, TEM, and TEA. Ideally, the body normally maintains a balance between caloric intake and caloric expenditure, but when this balance is upset, weight will be lost or gained. Both weight losses and weight gains appear to depend largely on just two factors: dietary intake and physical activity. This is now recognized as an oversimplification, considering the results of the overfeeding study of monozygotic twins discussed earlier, in which considerable variation in weight gains occurred for the same amount of overfeeding.[5] Not everyone responds to the same intervention in the same way. This difference in response must be considered when one is designing treatment programs for individuals attempting to lose weight, and people trying to lose weight must understand this difference so that those who are low responders won't be discouraged. In the past, we have tended to label the low responders as *noncompliers,* but we know now that this is generally not accurate.

Weight loss generally should not exceed 0.45 to 0.9 kg (1-2 lb) per week. Losses greater than this should not be attempted without direct medical supervision. Losing just 0.45 kg (1 lb) of fat a week will result in the loss of 23.4 kg (52 lb) of fat in only a year! Few people become obese that rapidly. Weight loss also should be considered a long-term project. Research and experience have proven that rapid weight losses are usually short-lived, and the lost weight is usually quickly regained because rapid weight losses are generally the result of large losses of body water. The body has built-in safety mechanisms to prevent an imbalance in body fluid levels, so the lost water eventually will be replaced. Thus, a person wishing to lose 9 kg (20 lb) of fat is advised to attempt to attain this goal in a minimum of 2 1/2 to 5 months.

Many special diets have achieved popularity over the years, such as the Drinking Man's Diet, the Beverly Hills Diet, the Cambridge Diet, the California Diet, Dr. Stillman's Diet, The Zone, the South Beach Diet, and the Atkins Diet. Each claims to be the ultimate in effective and comfortable weight loss. Some diets have been developed for use either in the hospital or at home under the supervision of a physician. These are often referred to as very low calorie diets, as they allow only 350 to 500 kcal of food per day. Most of these have been formulated with a certain amount of protein and carbohydrate to minimize the loss of fat-free body mass. Research has shown that many of these are effective, but no single diet has been shown to be more effective than any other. Again, the important factor is the development of a caloric deficit while maintaining a complete, balanced diet that meets the body's nutritional needs. The best diet is the one that meets these criteria and is best suited to individual comfort and personality.

Generally, improper eating habits are at least partially responsible for most weight problems, so no diet should be viewed as a quick fix. A person should learn to make permanent changes in dietary habits, especially reducing the intake of fat and simple sugars. For most people, simply eating a low-fat diet will gradually reduce weight to a desirable level. One potential problem with low-fat diets, however, is that people mistakenly assume that low fat means low calorie, and this often is not the case. For most people, simply reducing total caloric intake by 250 to 500 kcal per day, combined with a selection of low-fat and low simple sugar foods, would be sufficient to accomplish their desired weight loss goals.

Hormones and drugs also have been used to assist patients in weight loss by decreasing their appetite or increasing their RMR. Surgical techniques also are used to treat extreme obesity, but only as a last resort when other treatment procedures have failed and the obesity is life threatening. Intestinal bypass surgery involves surgically bypassing a large segment of the small intestine, thus reducing food absorption. While popular at one time, this procedure is now seldom used because of potential complications. Today, gastric bypass surgery and gastric banding are the most common procedures, both restricting the amount of food that can enter the stomach. While highly effective, these procedures must be considered high risk and should be reserved for those who have extreme obesity or obesity with significant risk factors.

Behavior modification has been proposed as one of the most effective techniques for helping people with weight problems. Major weight losses have been achieved through changes in basic behavior patterns associated with eating. Furthermore, these weight losses appear much more permanent. This approach appeals to most people because the techniques seem to make sense and are often easy to incorporate into a normal daily routine. For example, an individual might not have to consciously reduce the amount of food eaten but simply agree that all eating will be done in one location, which often cuts down on snacking. Or an individual might be allowed to take as much food as desired with the first helping, but no second helpings. Many such simple changes can help regulate eating behavior and result in substantial weight loss.

In review

- The etiology of obesity is not simple; the cause can be any one or a combination of many factors.
- Studies of animals and humans indicate that there is a genetic component to obesity. The disorder also has been linked to hormonal imbalances, emotional trauma, homeostatic imbalances, cultural influences, physical inactivity, and improper diets.
- Overweight and obesity are associated with increased risk of general excess mortality.
- Respiratory problems are quite common among people with obesity. These, in turn, can lead to lethargy and polycythemia.
- Obesity increases the risk of certain chronic degenerative diseases. Upper body (visceral) obesity increases the risk of developing coronary artery disease, hypertension, stroke, elevated blood lipids, diabetes, and the metabolic syndrome. Also, obesity can worsen preexisting health conditions and diseases.
- Emotional or psychological problems may contribute to obesity; and the disorder itself, with the stigma it carries, can be psychologically damaging.
- In the treatment of obesity, it is important to remember that people respond differently to the same intervention. Some have large losses of weight in a relatively short period of time, while others appear to be resistant to weight loss and see only small losses.
- Weight loss generally should not exceed 0.45 to 0.9 kg (1-2 lb) per week. Simple diet modification, reducing intake of fat and simple sugars, is sufficient to help most people lose weight. Behavior modification is also an effective method of weight loss.
- The use of drugs or surgery in the treatment of obesity is generally not recommended unless deemed necessary for the patient's health by a physician.

Role of Physical Activity in Weight Control

Inactivity is a major cause of obesity in the United States. In fact, inactivity may be just as important in the development of obesity as overeating! Thus, increasing physical activity levels must be recognized as an essential component in any program of weight reduction or control.

Changes in Body Composition With Exercise Training

Physical training can alter body composition. Many people have believed that physical activity has little or no influence on changing body composition and that even vigorous exercise burns too few calories to lead to substantial body fat reductions. Yet research has conclusively demonstrated the effectiveness of exercise training in promoting moderate alterations in body composition.

A person who jogs three days a week for 30 min each day at an 11 km/h (7 mph) pace (slightly over 5.4 min/km, or 8.5 min/mi) will expend about 14.5 kcal/min, or 435 kcal, for the 30 min run each day. This results in a total expenditure per week of about 1305 kcal, the equivalent loss of about 0.15 kg (0.33 lb) of adipose tissue (fat plus connective tissue and water) each week just from the exercise period alone. This might lead some people to believe that exercise is a painfully slow way to significantly reduce body fat levels and that there are better and easier ways to lose fat. However, in 52 weeks, providing energy intake remained constant, this person would lose 7.8 kg (17 lb)!

In estimations of an activity's energy cost, typically the average or steady-state rate of energy expenditure for that activity is multiplied by the number of minutes the activity is performed. For example, if the steady-state rate for shoveling snow is 7.5 kcal/min, 1 h of shoveling would require a total of 450 kcal. This would result in an approximate loss of 0.06 kg (0.13 lb) of adipose tissue (450 kcal ÷ 3,500 kcal/0.45 kg of adipose tissue = 0.06 kg, or 450 kcal ÷ 3,500 kcal/lb = 0.13 lb).

But examining the energy expended only during exercise does not give us the full picture. Metabolism remains temporarily elevated after exercise ends. This phenomenon was at one time referred to as the oxygen debt but, as mentioned in chapter 4, is now referred to as the **excess postexercise oxygen consumption (EPOC)**. Returning the metabolic rate back to its preexercise level can require several minutes following light exercise, such as walking; several hours following very heavy exercise, such as playing a football game; and up to 12 to 24 h or even longer for prolonged, exhaustive exercise, such as running a marathon in a hot and humid environment.

> **In focus**
>
> Physical activity is important in both weight maintenance and weight loss. In addition to the calories that are expended during exercise, a substantial expenditure of calories occurs during the postexercise period (EPOC).

The EPOC can require a substantial energy expenditure when considered over the entire recovery period. If, for example, the oxygen consumption following exercise remains elevated by an average of only 0.05 L/min, this will amount to approximately 0.25 kcal/min or 15 kcal/h. If the metabolism remained elevated for 5 h, this would provide an additional expenditure of 75 kcal that would not normally be included in the calculated total energy expenditure for that particular activity. This additional energy expenditure is ignored in most calculations of the energy costs of various activities. The person in this example, by exercising five days per week, would expend 375 kcal, or lose the equivalent of about 0.05 kg (0.1 lb) of fat in one week, or 0.45 kg (1.0 lb) in 10 weeks, from the additional caloric expenditure during the recovery period alone!

Studies have shown significant changes in both weight and body composition with both aerobic and resistance training, which include

- total weight decrease,
- fat mass and relative body fat decrease, and
- either maintained or increased fat-free mass.

Overall, these changes are not large. In a summary of hundreds of individual studies that monitored body composition changes with aerobic training, the expected changes from a typical one-year exercise training program (three times per week, 30-45 min per day, at 55-75% of $\dot{V}O_{2max}$) would be as follows: –3.2 kg (–7.1 lb) total body mass, –5.2 kg (–11.5 lb) fat mass, and +2.0 kg (+4.4 lb) fat-free mass.[36] Furthermore, relative body fat would decrease by nearly 6% (e.g., from 30% body fat to 24% body fat).

> **In focus**
>
> Attempts to lose weight are much more successful when one loses only 0.45 to 0.9 kg (1-2 lb) per week and when dietary restriction is combined with moderate exercise (300-500 kcal per day). This combination minimizes the loss of fat-free mass and maximizes the loss of fat mass.

How Much Activity Is Necessary for Weight Control?

In 2005, a group of scientists from the Mayo Clinic made an important discovery with respect to activity levels and weight control.[21] Ten lean and 10 obese sedentary subjects were instrumented with a very sophisticated system for monitoring even minor changes in body position over a 10-day period. Although both groups were sedentary, the lean subjects spent 350 kcal per day more in postural changes and movement when compared to the obese subjects. As just one example, obese subjects were seated 2 h longer per day than lean subjects. These results point to the importance of simply being active, independent of a formal exercise program.

Although most studies have used aerobic training, a number of studies have used resistance training and have shown impressive decreases in body fat and increases in fat-free mass. The evidence shows that exercise is an important part of any weight loss program. But to maximize losses in body weight and body fat, it is necessary to combine exercise with decreased caloric intake.

Since the 1990s, abdominal visceral fat (figure 21.8 on p. 504) has been identified as a major independent risk factor for cardiovascular diseases and obesity. There is now substantial evidence that physical activity reduces the rate of accumulation of visceral fat and that exercise training actually reduces visceral fat stores.[31] This could be one of the most important health benefits of an active lifestyle!

Mechanisms for Change in Body Weight and Composition

When attempting to explain how exercise causes such changes in body weight and composition, it is necessary to consider both sides of the energy-balance equation. Evaluating energy expenditure requires that we consider each of the three components of energy expenditure: RMR, TEM, and TEA. Evaluating energy intake requires that we also consider the energy that is lost in the feces (energy excreted), which is generally less than 5% of the total caloric intake. Keeping this balance in mind, in the next section we examine some of the possible mechanisms through which exercise might affect body weight and body composition.

The energy-balance equation:

energy intake – energy excreted = RMR + TEM + TEA.

Exercise and Appetite

Some believe that exercise stimulates the appetite to such an extent that food intake is unconsciously increased to at least equal that expended during exercise. In 1954, Jean Mayer, a world-famous nutritionist, reported that animals exercising for periods of from 20 min to 1 h per day had a lower food intake than nonexercising control animals.[23] He concluded from this and other studies that when activity is less than a certain minimal level, food intake does not decrease correspondingly and the animal (or human) begins to accumulate body fat. This led to the theory that a certain minimal level of physical activity is necessary for the body to precisely regulate food intake to balance energy expenditure. A sedentary lifestyle may reduce this regulatory ability, resulting in a positive energy balance and weight gain.

Exercise does, in fact, appear to be a mild appetite suppressant, at least for the first few hours following intense exercise training. Furthermore, studies have shown that the total number of calories consumed per day does not change when a person begins a training program. Although some people interpret this as evidence that exercise does not affect appetite, a more accurate conclusion might be that appetite was affected, in fact suppressed, because caloric intake did not increase in proportion to the additional caloric expenditure from the exercise program. In studies conducted on rats, male rats appear to reduce food intake with exercise training, whereas female rats tend to eat the same or even more than nonexercising control rats.[26] There is no obvious explanation for this sex difference. It is unclear whether this sex difference is present in humans.

The decrease in appetite might occur only with intense levels of exercise, in which the resulting increased catecholamine (epinephrine and norepinephrine) levels might suppress the appetite. The increased body temperature that accompanies either high-intensity activity or almost any activity performed under hot and humid conditions also might suppress appetite. We all know from experience that we desire less food when the weather is hot or when our body temperatures are elevated because of illness. This also might explain why a hard running workout in the heat results in little or no desire to eat, yet a hard swimming workout elicits a relatively strong craving for food. In the pool, provided that the water temperature is well below body core temperature, the heat generated by exercise is lost very effectively, so core temperature typically is not elevated to the same extent.

Regular physical activity may assist in controlling appetite so that caloric intake balances caloric expenditure.

Exercise and Resting Metabolic Rate

The effects of exercise on the components of energy expenditure became a major topic of interest among researchers in the late 1980s and early 1990s. Of obvious interest is how exercise training might affect the RMR, since it represents 60% to 75% of the total calories expended each day. For example, if a 25-year-old man's total daily caloric intake was 2,700 kcal and his RMR accounted for just 60% of that total ($0.60 \times 2700 = 1,620$ kcal RMR), a mere 1% increase in his RMR would require an extra 16 kcal expenditure each day, or 5,840 kcal per year. This small increase in RMR alone would account for the equivalent of an 0.8 kg (1.7 lb) fat loss per year!

The role of physical training in increasing RMR has not been totally resolved. Several cross-sectional studies have shown that highly trained runners have higher RMRs than untrained people of similar age and size. But other studies have not been able to confirm this.[27] Few longitudinal studies have been conducted to determine the change in RMR in untrained people who undergo training for a period of time. Some of these suggest that RMR might increase following training.[7] However, in a study of 40 men and women 17 to 62 years of age (HERITAGE Family Study), a 20-week aerobic training program (three times per week, 35-55 min per day, at 55-75% of $\dot{V}O_{2max}$) failed to increase RMR even though $\dot{V}O_{2max}$ increased by nearly 18%.[39] Because RMR is closely related to the fat-free mass of the body (fat-free tissue is more metabolically active), interest has increased in the use of resistance training to increase fat-free mass in an attempt to increase RMR.[7]

Exercise and the Thermic Effect of a Meal

Several studies have examined the role of individual bouts of exercise and exercise training in increasing the TEM. A single bout of exercise, either before or after a meal, increases the thermic effect of that meal. Less clear is the role of exercise training on the TEM. Some studies have shown increases; others have shown decreases; and yet others have shown no effect at all. As with measuring changes in RMR accompanying exercise training, measurement of the TEM must be timed carefully with the last exercise bout. When measurements are made within 24 h of the last bout, the TEM is typically lower than it is three days afterward.[32]

Exercise and Mobilization of Body Fat

During exercise, fatty acids are freed from their storage sites to be used for energy. Several studies suggest that human growth hormone may be responsible for this increased fatty acid mobilization. Growth hormone levels increase sharply with exercise and remain elevated for up to several hours in the recovery period. Other research has suggested that, with exercise, the adipose tissue is more sensitive to either the sympathetic nervous system or the increasing levels of circulating catecholamines. Either situation would increase lipid mobilization. More recent research suggests that this mobilization occurs in response to a specific fat-mobilizing substance that is highly responsive to elevated levels of activity. Thus, we cannot state with certainty which factors are of greatest importance in mediating this response.

Spot Reduction

Many people, including athletes, believe that exercising a specific area of the body will use the fat in that area, reducing the locally stored fat. Results of several early research studies tended to support this concept of **spot reduction.** But later research suggests that spot reduction is a myth and that exercise, even when localized, draws from almost all of the fat stores of the body, not just from local depots.

One such study used outstanding tennis players, theorizing that they would be ideal subjects for studying spot reduction because they could act as their own controls: Their dominant arms exercise vigorously for several hours every day, whereas their nondominant arms are relatively inactive.[13] Researchers postulated that if spot reduction is a reality, the nondominant (inactive) arm should have substantially more fat than the dominant (active) arm. The players' dominant arms had substantially greater girths attributable to exercise-induced muscle hypertrophy. But the subcutaneous skinfold fat thicknesses in the active and inactive arms showed absolutely no differences.

Another study examined the localized effects of a 27-day intense sit-up training program. Researchers found no difference in the rate at which fat cell diameter changed in the abdomen, the subscapular region, and the gluteal region.[17] This indicates a lack of specific adaptation at the site of the exercise training (the abdomen). Researchers now theorize that fat is mobilized during exercise either mostly from those areas of highest concentration or equally from all areas, thus negating the spot-reduction theory. Decreases in girth can occur with exercise training, but these primarily result from increased muscle tone, not fat loss.

Low-Intensity Aerobics

As we have discussed in earlier chapters, the higher the exercise intensity, the greater the body's reliance on carbohydrate as an energy source. With high-intensity aerobic exercise, carbohydrate can supply up to 90% or more of the body's energy needs. During the late 1980s, various professional exercise groups promoted **low-intensity aerobic exercise** to increase the loss of body fat. These groups theorized that low-intensity aerobic training would allow the body to use more fat as the energy source, hastening the loss of body fat. Indeed, the body uses a higher percentage of fat for energy at lower exercise intensities. However, the total calories expended by the body's use of fat does not necessarily change.

In focus

Low-intensity aerobic activity does not necessarily lead to a greater expenditure of calories from fat. More important, the total caloric expenditure for a given period of time is much less than with high-intensity aerobic activity.

TABLE 21.3 Estimation of Kilocalories Used From Fat and Carbohydrate for a Low- and High-Intensity Aerobic Training Bout of 30 min Duration

Exercise intensity	Average $\dot{V}O_2$ (L/min)	Average RER	% kcal CHO	% kcal fat	kcal for 30 min CHO	kcal for 30 min fat	kcal for 30 min total
Low, 50%	1.50	0.85	50	50	110	110	220
High, 75%	2.25	0.90	67	33	222	110	332

Note. RER = respiratory exchange ratio; CHO = carbohydrate. Subject was a fit but not highly trained 23-year-old woman ($\dot{V}O_{2max}$ = 3.0 L/min).

This is illustrated in table 21.3. In this hypothetical example, a 23-year-old woman with a $\dot{V}O_{2max}$ of 3.0 L/min exercises for 30 min at 50% of her $\dot{V}O_{2max}$ on one day and for 30 min at 75% of her $\dot{V}O_{2max}$ on another. The total calories from fat do not differ between the low- and high-intensity aerobic workouts: In both cases she burns about 110 kcal of fat during 30 min. Most important, however, for the higher-intensity workout she expends about 50% more total calories for the same time period.

Scientists have determined that there is an optimal zone where rates of fat oxidation are at their highest. The Fat_{max} zone, defined as that zone where fat oxidation rates are within 10% of the peak rate, was found to vary from between 55% and 72% of $\dot{V}O_{2max}$.[1] This is illustrated in figure 21.9.

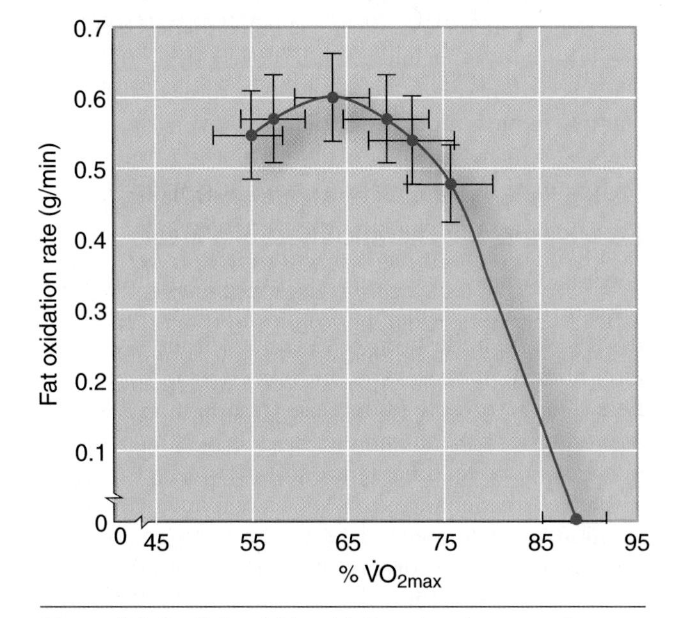

Figure 21.9 Rate of fat oxidation at various exercise intensities, expressed as a percentage of $\dot{V}O_{2max}$.

Reprinted, by permission, from J. Acten, M. Gleeson, and A.E. Jeukendrup, 2002, "Determination of the exercise intensity that elicits maximal fat oxidation," *Medicine and Science in Sports and Exercise* 34: 92-97.

Exercise Gadgets

We seldom get something for nothing. An effortless exercise program would be ideal, of course, but such a program would result in no significant changes in fitness, body composition, or physical dimensions. With the increasing popularity of exercise, many gimmicks and gadgets have appeared on the market. Some of these are legitimate and effective, but unfortunately many are of no practical value for either exercise conditioning or weight loss. Three such devices were evaluated to determine the legitimacy of their claims: the Mark II Bust Developer, the Astro-Trimmer Exercise Belt, and Slim-Skins Vacuum Pants. The first device was claimed to add 2 to 3 in. (5-8 cm) to the bust within three to seven days, and the other two devices were claimed to take inches off of the abdomen, hips, buttocks, and thighs in a matter of minutes. All three devices failed to produce any changes whatsoever when evaluated in highly controlled scientific studies.[37, 38]

Those who are considering weight reduction often cringe at the thought of increasing their physical activity, and who wouldn't prefer immediate results over waiting for a payoff? But reality must be addressed. To gain the benefits from exercise, it is necessary to actually do the work!

Physical Activity and Health Risk Reduction

An important relationship was discovered during the 1990s that suggests another substantial benefit of fitness and activity. For those who are overweight or obese, their overall risk of death from disease is greatly reduced if they are physically active and fit.[2, 35] This is good news for those who seem destined to remain obese or overweight: An active lifestyle and moderate to high fitness levels can greatly reduce their risk of dying from chronic degenerative diseases, such as coronary artery disease and diabetes.

In review

- Inactivity is a major cause of obesity in the United States, perhaps as important as overeating.
- The energy expended by activity includes the steady-state rate of energy expenditure during the activity and also the energy expended after the exercise because the metabolic rate remains elevated for some time after the activity ends, a phenomenon known as excess postexercise oxygen consumption (EPOC).
- Diet alone causes fat loss, but fat-free mass is also lost. With exercise, either alone or with diet, fat is lost, but fat-free mass is either maintained or increased. Possibly the greatest benefit of physical activity and formal exercise is their role in attenuating the accumulation of visceral fat, or in reducing visceral fat stores.
- Simply being active, independent of a formal exercise program, is important in the prevention of obesity.
- Energy intake – energy excreted = RMR + TEM + TEA when one is in energy balance.
- A certain amount of activity appears to be needed for the body to precisely balance energy intake and expenditure.
- Research indicates that exercise can suppress appetite.
- Resting metabolic rate might increase slightly following training, and even a single bout of exercise increases the TEM.
- Exercise increases lipid mobilization from adipose tissue.
- Spot reduction is a myth. Low-intensity aerobic exercise burns no more fat than more vigorous exercise, and more total calories are spent in a more strenuous workout.

DIABETES

Diabetes mellitus, also called simply diabetes, is a disease characterized by high blood glucose levels (**hyperglycemia**) resulting from either an inadequate production of **insulin** by the pancreas, an inability of insulin to facilitate the transport of glucose into the cells, or both. Recall from chapter 2 that insulin is a hormone that reduces the amount of glucose circulating in the blood by facilitating its transport into the cells. We first address the terminology used in defining diabetes and disordered blood sugar control (glycemic control) and then look at the prevalence of diabetes in the United States.

Terminology and Classification

Historically, diabetes mellitus was classified into two major categories: juvenile-onset diabetes (now known as type 1 diabetes) and adult-onset diabetes (now known as type 2 diabetes). This classification was based on the age of onset of diabetes.[22] Unfortunately, there has been an epidemic of type 2 diabetes in children, which largely can be attributed to the increased rates of obesity in children.

Type 1 diabetes is caused by the inability of the pancreas to produce sufficient insulin as a result of failure of the **β-cells** in the pancreas. Thus, this type has also been referred to as **insulin-dependent diabetes mellitus (IDDM).** Type 1 diabetes accounts for only 5% to 10% of all cases of diabetes.

Type 2 diabetes is the result of the ineffectiveness of insulin to facilitate the transport of glucose into the cells and is a result of insulin resistance. It has also been referred to as **non-insulin-dependent diabetes mellitus (NIDDM).** Type 2 diabetes accounts for 90% to 95% of all cases of diabetes. **Insulin resistance** refers to the condition in which a "normal" insulin concentration in the blood produces a less than normal biological response. Insulin's primary function is to facilitate the transport of glucose from the blood into the cell, across the cell membrane. With insulin resistance, the body needs more insulin to transport a given amount of glucose across the cell membrane into the cell. **Insulin sensitivity** is a related term and provides an index of the effectiveness of a given insulin concentration in the blood. As insulin sensitivity increases, insulin resistance decreases.

A third type of diabetes, gestational diabetes, is a form of diabetes that develops in pregnant women and their fetuses in about 4% of all pregnancies. Fortunately, it usually disappears in both mother and baby after delivery. Unfortunately, when gestational diabetes is present, there can be complications during pregnancy.

Another category, **prediabetes,** refers to the condition in those who have impaired fasting glucose, impaired glucose tolerance, or both. Both type 1 and type 2 diabetes are diagnosed on the basis of a plasma glucose level of greater than 125 mg/dL following an 8 h fast. **Impaired fasting glucose** is defined as a plasma glucose level of between 100 and 125 mg/dL, again following an 8 h fast. **Impaired glucose tolerance** is determined by a glucose tolerance test. This involves drinking a solution in which 75 g of anhydrous glucose is dissolved in water. Plasma glucose levels are measured 2 h later. A glucose value of 200 mg/dL or higher is diagnostic of diabetes. Values of 140 to 199 mg/dL represent impaired glucose tolerance, and values below 140 mg/dL are considered normal.

For research studies, an intravenous glucose tolerance test (IVGTT) often is used. A catheter is placed in both arms, with the glucose solution injected into a vein in one arm, and blood samples are withdrawn from a vein in the other arm over the course of 3 h. Samples are taken more frequently during the first 15 to 45 min of the test and less frequently from 60 to 180 min. This allows one to develop a curve of both the glucose and insulin responses to the injected glucose load. The IVGTT is more precise than the oral glucose tolerance test.

There are also symptoms of diabetes that can be used to identify those at risk for diabetes.

These include

- frequent urination,
- excessive thirst,
- unexplained weight loss,
- extreme hunger,
- sudden vision changes,
- tingling or numbness in hands or feet,
- feeling very tired much of the time,
- irritability,
- sores that are slow to heal, and
- more infections than usual.

Prevalence of Diabetes

Approximately 14.6 million Americans have been diagnosed with diabetes. An estimated 6.2 million people likely have diabetes that has not been diagnosed, and an estimated 41 million people are prediabetic. The prevalence of diabetes increased from 4.9% of the U.S. population in 1990 to 7.0% in 2005—a 43% increase. Between 1990 and 1998, the largest increase (76%) occurred in 30- to 39-year-olds. Nearly 21% of people 60 years or older have diabetes. Further, the risk of developing diabetes is 1.7 times higher in Mexican Americans, 1.8 times higher in non-Hispanic blacks, and 2.2 times higher in American Indians compared to non-Hispanic whites. The total prevalence for Asian Americans is unknown. These racial differences in diabetes rates parallel the racial differences in obesity rates discussed earlier in this chapter.

The true prevalence of type 2 diabetes in children has not been established through national epidemiological studies. However, prevalence has been estimated to have increased by as much as 10-fold over the last 20 years. Furthermore, in 10- to 19-year-olds, studies have shown that type 2 diabetes accounted for between 33% and 46% of all diabetes.[22] This is very disturbing, considering that not too long ago we referred to type 2 diabetes as adult-onset diabetes.

Etiology of Diabetes

Heredity appears to play a major role in both type 1 and type 2 diabetes. With type 1 diabetes, the β-cells (insulin-secreting cells) of the pancreas are destroyed. This destruction may be caused by the body's immune system, increased β-cell susceptibility to viruses, or β-cell degeneration.

Type 1 diabetes generally has a sudden onset during childhood or young adulthood. This leads to almost total insulin deficiency, and daily injections of insulin usually are required to control the disease.

In type 2 diabetes, the onset of the disease is more gradual, and the causes are more difficult to establish. Type 2 diabetes often is characterized by any of the following three major metabolic abnormalities: delayed or impaired insulin secretion; impaired insulin action (insulin resistance) in the insulin-responsive tissues of the body, including muscle; or excessive glucose output from the liver.

Obesity plays a major role in the development of type 2 diabetes. With obesity, the β-cells of the pancreas often become less responsive to the stimulation of increased blood glucose concentrations. Furthermore, the target cells throughout the body, including muscle, often undergo a reduction in the number or activation of their insulin receptors, so the insulin in the blood is less effective in transporting glucose into the cell.

Health Problems Associated With Diabetes

Considerable health risks are associated with diabetes. People with this disease have a relatively high mortality rate. Diabetes places a person at increased risk for

- coronary artery disease,
- cerebrovascular disease and stroke;
- hypertension;
- peripheral vascular disease;
- kidney disease;
- eye disorders, including blindness; and
- toxemia during pregnancy.

In focus

Diabetes increases the risk of coronary artery disease, cerebrovascular disease and stroke, hypertension, and peripheral vascular disease. Coronary artery disease, hypertension, obesity, and diabetes may be linked through the common pathway of increased insulin levels in the blood or to target cells' becoming insulin resistant. However, obesity appears to be the trigger.

During the late 1980s, scientists made the important associations among coronary artery disease, hypertension, obesity, and type 2 diabetes. **Hyperinsulinemia** (high blood levels of insulin) and insulin resistance appear to be the important threads linking these disorders, possibly through insulin-mediated sympathetic nervous system stimulation (increased insulin levels cause increased sympathetic nervous system activity).[9] Again, obesity seems to be the trigger setting off this reaction.

General Treatment of Diabetes

The major modes of treatment for type 1 diabetes are insulin administration, diet, and exercise. The dosage of insulin is adjusted to allow normal carbohydrate, fat, and protein metabolism. The type of insulin injected—short acting or intermediate acting—and the time of day at which the injections are administered are also individualized to maintain glycemic control throughout the day.

For type 2 diabetes, the focus has traditionally been on three factors: weight loss, diet, and exercise. However, during the mid- to late-1990s, a number of new drugs were introduced that effectively treat type 2 diabetes. Two types of drugs have been particularly effective: sulfonylureas, to lower blood sugar levels, and biguanides, to reduce hepatic glucose production. Metformin, a biguanide, has been particularly successful in obese patients.

A well-balanced diet generally is prescribed for people with diabetes. In the past, patients were prescribed a low-carbohydrate diet to better control blood sugar levels. However, a low-carbohydrate diet necessitates an increase in dietary fat, which can have a major negative effect on blood lipid levels. Because people with diabetes are already at greater risk for coronary artery disease, this is not desirable. Maintenance of proper blood sugar levels is difficult in patients with obesity, so a reduced-calorie diet is necessary for them to achieve major body fat losses. For many people with type 2 diabetes, weight loss alone can bring blood sugar levels back into the normal range. This can be the most important aspect of the treatment plan for the overweight or obese person with diabetes.

Role of Physical Activity in Diabetes

Indirect scientific evidence has clearly established that a physically active lifestyle reduces the risk of developing type 2 diabetes,[3] but there is little evidence to support this for type 1 diabetes. However, most physicians and scientists agree that physical activity is an important part of the treatment plan for either type of diabetes. Because there is such a disparity between the characteristics and responses of those with type 1 and type 2 diabetes, we discuss each separately.

Type 1 Diabetes

The role of regular exercise and physical training in improving glycemic control (regulation of blood sugar levels) in patients with type 1 diabetes has not been clearly defined and is controversial. The most distinguishing feature differentiating type 1 and type 2 diabetes is that those with type 1 have low blood insulin levels attributable to the inability or reduced ability of the pancreas to produce insulin. Those with type 1 diabetes are prone to **hypoglycemia** (low blood sugar levels) during and immediately after exercise because the liver fails to release glucose at a rate that can keep up with glucose utilization. For these people, exercise can lead to excessive swings in plasma glucose levels that are unacceptable for the management of the disease. The degree of glycemic control during exercise varies tremendously among individuals with type 1 diabetes. As a result, exercise and exercise training can improve glycemic control in some patients, mainly those who are less prone to hypoglycemia, but not in others.[34]

Although glycemic control is generally not improved by exercise in most people with type 1 diabetes, there are other potential benefits of exercise for these patients. Because these patients have a two to three times greater risk for coronary artery disease, exercise may help reduce this risk. Exercise may also help reduce the risk for cerebrovascular and peripheral arterial diseases.

People with uncomplicated type 1 diabetes do not have to restrict physical activity, providing that blood sugar levels are controlled appropriately. A number of athletes who have type 1 diabetes have trained and competed successfully. Monitoring blood sugar levels in an exercising person with type 1 diabetes is important so that diet and insulin dosages can be altered accordingly.

Special attention also should be given to the feet of people with diabetes, as it is common for them to experience peripheral neuropathy (diseased nerves) with some loss of sensation in the feet. Peripheral vascular disease is also more common in patients with diabetes, so the circulation to the extremities, especially the feet, often is impaired significantly. Ulcerations and other lesions on the feet account for more than half of all hospitalizations of diabetic patients.[8] Because weight-bearing exercise places additional stress on the feet, the proper selection of footwear and appropriate preventive foot care are important.

Type 2 Diabetes

Exercise plays a major role in glycemic control for people with type 2 diabetes. Insulin production is generally not of concern in this group, particularly during the early stages of the disease, so the major problem with this form of diabetes is the lack of target cell response to insulin (insulin resistance). Because the cells become resistant to insulin, the hormone cannot perform its function of facilitating glucose transport across the cell membrane, resulting in decreased insulin sensitivity. Muscle contraction has an insulin-like effect.[16] Membrane permeability to glucose increases with muscular contraction, likely attributable to an increase in the number of GLUT-4 glucose transporters associated with the plasma membrane.[15] Thus, acute bouts of exercise decrease insulin resistance and increase insulin sensitivity. This reduces the cells' requirements for insulin, which means that people taking insulin must reduce their dosages. Resistance and aerobic training appear to produce similar effects.[41] This decrease in insulin resistance and increase in insulin sensitivity may primarily be a response to each individual bout of exercise rather than the result of a long-term change associated with training. Some studies have shown that the effect dissipates within 72 h.

In focus

Physical activity has many desirable effects for people with diabetes, particularly those with type 2 diabetes. Glycemic control is improved, primarily in people with type 2 diabetes, possibly attributable to the insulin-like effect of muscle contraction on transporting glucose from the plasma into the cell.

The Diabetes Prevention Program Study

During the mid-1990s, the National Institute of Diabetes and Digestive and Kidney Diseases (NIDDK) of the National Institutes of Health designed and initiated a study to determine whether either a lifestyle intervention (diet and exercise) or an oral diabetes drug (metformin) could prevent or delay the onset of type 2 diabetes in people with impaired glucose tolerance (prediabetics). A total of 3,234 persons 25 years of age and older were randomly assigned to placebo, drug, or lifestyle groups. The goals for the lifestyle modification program were to lose at least 7% of body weight and to participate in physical activity for at least 150 min per week. Both interventions proved very successful, delaying the development of diabetes by 11 years in the lifestyle group and by three years in the metformin group.[20] This landmark study clearly illustrates the power of lifestyle intervention in reducing the risk of a major debilitating disease and its serious consequences.

In review

- Diabetes is a disorder of carbohydrate metabolism characterized by hyperglycemia. It develops from inadequate insulin secretion or utilization.
- Type 1 diabetes involves destruction of the β-cells in the pancreas, and it typically has a sudden, early onset. Type 2 diabetes typically involves impaired insulin secretion or action, excessive liver glucose output, or a combination of these.
- Major modes of treatment for diabetes are drugs (if needed), diet, and exercise.
- In people with type 1 diabetes, glycemic control might or might not be improved with exercise. But these people have a greater risk for coronary artery disease, and exercise certainly can decrease that risk.
- Blood glucose levels must be carefully monitored during exercise, particularly in people with type 1 diabetes, so that diet and insulin dosage can be altered as needed.
- The feet of people with type 1 diabetes deserve special attention, because peripheral neuropathy causes loss of sensation, and impaired peripheral circulation decreases blood flow. These people might not be aware of injuries to their feet, but these injuries can be very serious.
- Type 2 diabetes responds well to exercise. Membrane permeability to glucose improves with exercise, likely associated with an increase in GLUT-4 receptors, which decreases the person's insulin resistance and increases insulin sensitivity.

In Closing

In these last two chapters, we have concluded our look at the role of physical activity in preventing and treating coronary artery disease, hypertension, obesity, and diabetes. We have seen that exercise can decrease individual risk and also can be an integral part of treatment, improving overall health as well as alleviating some symptoms.

With this chapter, we also conclude our journey to an understanding of exercise and sport physiology. We began this book by reviewing how various body systems function during exercise and how they respond to chronic training. We saw how physical activity and performance are affected by the environment, such as extremes of heat, cold, and barometric pressure. We turned our attention to ways in which athletes can, or can attempt to, optimize their performance. Then we considered the unique differences among older and younger participants and male and female participants in sport and exercise. And finally, we examined the role of exercise in the maintenance of health and the development of fitness.

It has been a long journey from cover to cover, but we hope that you will close this book with a new appreciation of physical activity. Perhaps you will leave this book with a new awareness of how your body performs physical activity. Maybe, if you have not yet done so, you will feel compelled to commit to a personal exercise program. And we hope that you now feel some excitement about exercise and sport physiology, realizing that these areas of study affect so many aspects of our lives.

KEY TERMS

β-cells
body mass index (BMI)
diabetes mellitus
excess postexercise oxygen consumption (EPOC)
hyperglycemia
hyperinsulinemia
hypoglycemia
impaired fasting glucose
impaired glucose tolerance
insulin
insulin-dependent diabetes mellitus (IDDM)
insulin resistance
insulin sensitivity
lower body (gynoid) obesity
low-intensity aerobic exercise
non-insulin-dependent diabetes mellitus (NIDDM)

obesity
overweight
prediabetes
resting metabolic rate (RMR)
spot reduction
thermic effect of activity (TEA)
thermic effect of a meal (TEM)
type 1 diabetes
type 2 diabetes
upper body (android) obesity

STUDY QUESTIONS

1. What is the difference between overweight and obesity?
2. What is ideal body weight, and how is it determined?
3. What is body mass index? What is its significance?
4. What is the prevalence of obesity in the United States today? Is there a difference in the prevalence between men and women? Children and adults? Blacks and whites?
5. What are several health-related problems associated with obesity?
6. What is the association among obesity, coronary artery disease, hypertension, insulin resistance, and diabetes?
7. Describe several methods for treating obesity. Which are the most effective?
8. What role does exercise play in the prevention and treatment of obesity?
9. By what mechanisms might exercise effect losses in total weight and fat weight?
10. How effective is spot reduction? Low-intensity aerobic exercise?
11. Describe the two major types of diabetes. How are they caused?
12. What health risks are associated with diabetes?
13. Describe the role of exercise in preventing diabetes and in treating patients with type 1 and type 2 diabetes.

STUDY GUIDE ACTIVITIES

In addition to the activities listed in the chapter opening outline on page 493, two other activities are available in the online study guide, located at

www.HumanKinetics.com/PhysiologyOfSportAndExercise.

The chapter **SUMMARY** reviews the key concepts covered in the chapter, and the end-of-chapter **QUIZ** tests your understanding of the material covered in the chapter.

Glossary

1-repetition maximum (1RM)—The maximal amount of weight that can be lifted just one time.

1RM—*See* 1-repetition maximum.

acclimation—Physiological adaptation to repeated environmental stresses, occurring over a relatively brief period of time (days to weeks). Acclimation often occurs in a laboratory environment.

acclimatization—Physiological adaptation to repeated environmental stress in a natural environment, occurring over months and years of living and exercising in that environment.

ACE—*See* angiotensin converting enzyme.

acetylcholine—A primary neurotransmitter that transmits impulses across the synaptic cleft.

acetyl CoA—*See* acetyl coenzyme A.

acetyl coenzyme A (acetyl CoA)—The compound that forms the common entry point into the Krebs cycle for the oxidation of carbohydrate and fat.

actin—A thin protein filament that acts with myosin filaments to produce muscle action.

action potential—A rapid and substantial depolarization of the membrane of a neuron or muscle cell that is conducted through the cell.

acute altitude (mountain) sickness—Illness characterized by headache, nausea, vomiting, dyspnea, and insomnia. It typically begins 6 to 96 h after one reaches high altitude and lasts several days.

acute exercise—A single bout of exercise.

acute muscle soreness—Soreness or pain felt during and immediately after an exercise bout.

acute overload—an "average" training load, whereby the athlete is stressing the body to the extent necessary to improve both physiological function and performance.

adenosine diphosphate (ADP)—A high-energy phosphate compound from which ATP is formed.

adenosine triphosphatase (ATPase)—An enzyme that splits the last phosphate group off ATP, releasing a large amount of energy and reducing the ATP to ADP and P=$_i$.

adenosine triphosphate (ATP)—A high-energy phosphate compound from which the body derives its energy.

ADH—*See* antidiuretic hormone.

adolescence—The period of life between the end of childhood and the beginning of adulthood. The onset of puberty marks the beginning of adolescence.

ADP—*See* adenosine diphosphate.

adrenaline—A chemical compound that serves as a neurotransmitter throughout the body. Also called epinephrine.

aerobic capacity—*See* maximal oxygen uptake.

aerobic endurance—*See* cardiorespiratory endurance.

aerobic interval training—Repeated, moderate-duration, fast-paced exercise bouts with moderate rest intervals between bouts.

aerobic metabolism—A process occurring in the mitochondria that uses oxygen to produce energy (ATP). Also known as cellular respiration.

aerobic power—Another name for maximal oxygen uptake, or $\dot{V}O_{2max}$.

aerobic training—Training that improves the efficiency of the aerobic energy-producing systems and can improve cardiorespiratory endurance.

afferent division—The sensory division of the peripheral nervous system.

afterload—The pressure against which the heart must pump blood, determined by the peripheral resistance in the large arteries.

air plethysmography—A procedure for assessing body composition by using air displacement to measure body volume, allowing the calculation of body density.

aldosterone—A mineralocorticoid hormone secreted by the adrenal cortex that prevents dehydration by promoting renal absorption of sodium.

alveolar capillary membrane—*See* respiratory membrane.

alveolus—Terminal air sac at the end of the bronchial tree in the lungs, where gas exchange takes place with the capillaries (plural, alveoli).

amenorrhea—The absence (primary amenorrhea) or cessation (secondary amenorrhea) of normal menstrual function.

amino acids—The chief components of proteins that are synthesized by living cells or are obtained by the body through the diet.

α-motor neuron—A neuron innervating extrafusal skeletal muscle fibers.

amphetamine—A central nervous system stimulant proposed to have ergogenic properties.

anabolic steroids—Prescription drugs with the anabolic (growth stimulating) characteristics of testosterone, taken by some athletes to increase body size, muscle mass, and strength.

anabolism—The building up of body tissue; the constructive phase of metabolism.

anaerobic—In the absence of oxygen.

anaerobic glycolysis—The anaerobic breakdown of glucose to lactic acid in the production of energy (ATP).

anaerobic metabolism—The production of energy (ATP) in the absence of oxygen.

anaerobic power—Mean or peak power output in exercise lasting 30 s or less.

anaerobic threshold—The point at which the metabolic demands of exercise can no longer be met by available aerobic sources and at which an increase in anaerobic metabolism occurs, reflected by an increase in blood lactate concentration.

anaerobic training—Training that improves the efficiency of the anaerobic energy-producing systems and can increase muscular strength and tolerance for acid–base imbalances during high-intensity effort.

angiotensin converting enzyme (ACE)—An enzyme that converts angiotensin I to angiotensin II.

anorexia nervosa—A clinical eating disorder characterized by distorted body image, intense fear of fatness or weight gain, amenorrhea, and refusal to maintain more than the minimal normal weight based on age and height.

antidiuretic hormone (ADH)—A hormone secreted by the pituitary gland that regulates fluid and electrolyte balance in the blood by reducing urine production.

arginine vasopressin—*See* antidiuretic hormone.

arterial–mixed venous oxygen difference, or **$(a\text{-}\bar{v})O_2$ difference**—The difference in oxygen content between arterial and mixed venous blood, which reflects the amount of oxygen removed by the whole body.

arterial–venous oxygen difference, or **$(a\text{-}v)O_2$ difference**—The difference in oxygen content between arterial and venous blood at the tissue level.

arteries—Blood vessels that transport blood away from the heart.

arterioles—The smallest arteries that transport blood from larger arteries to the capillaries.

arteriosclerosis—A condition that involves loss of elasticity, thickening, and hardening of the arteries.

atherosclerosis—A form of arteriosclerosis that involves changes in the lining of the arteries and plaque accumulation, leading to progressive narrowing of the arteries.

athlete's heart—A nonpathological enlarged heart, often found in endurance athletes, that results primarily from left ventricular hypertrophy in response to training.

ATP—*See* adenosine triphosphate.

ATPase—*See* adenosine triphosphatase.

ATP-PCr system—The short-term anaerobic energy system that maintains ATP levels. Breakdown of phosphocreatine (PCr) frees P_i, which then combines with ADP to form ATP.

atrioventricular (AV) node—The specialized mass of conducting cells in the heart located at the atrioventricular junction.

atrophy—Loss of size, or mass, of body tissue, such as muscle atrophy with disuse.

autogenic inhibition—Reflex inhibition of a motor neuron in response to excessive tension in the muscle fibers it supplies, as monitored by the Golgi tendon organs.

autoregulation—Local control of blood distribution (through vasodilation) in response to a tissue's changing needs.

AV node—*See* atrioventricular node.

$(a\text{-}v)O_2$ difference—*See* arterial–venous oxygen difference.

$(a\text{-}\bar{v})O_2$ difference—*See* arterial–mixed venous oxygen difference.

axon hillock—A part of the neuron, between the cell body and the axon, that controls traffic down the axon through summation of excitatory and inhibitory postsynaptic potentials.

axon terminal—One of numerous branched endings of an axon. Also known as a terminal fibril.

barometric pressure—The total pressure exerted by the atmosphere at a given altitude.

baroreceptor—Stretch receptor located within the cardiovascular system that senses changes in blood pressure.

basal metabolic rate (BMR)—The lowest rate of body metabolism (energy use) that can sustain life, measured after an overnight sleep in a laboratory under optimal conditions of quiet, rest, and relaxation and after a 12 h fast. *See also* resting metabolic rate.

β-blockers—A class of drugs that block transmission of neural impulses from the sympathetic nervous system, proposed to have ergogenic properties.

BCAA—*See* branched-chain amino acids.

β-cells—Cells in the islets of Langerhans in the pancreas that secrete insulin.

bicarbonate loading—Ingesting bicarbonate to elevate blood pH with hopes of delaying fatigue by increasing the capacity to buffer acids.

bioelectric impedance—A procedure for assessing body composition in which an electrical current is passed through the body. The resistance to current flow through the tissues reflects the relative amount of fat present.

bioenergetics—Term given to the study of metabolic processes that yield or consume energy.

blood doping—Any means by which a person's total volume of red blood cells is increased, typically via transfusion of red blood cells or use of erythropoietin.

blood lipids—Bloodborne fats, such as triglycerides and cholesterol.

BMI—*See* body mass index.

BMR—*See* basal metabolic rate.

body composition—The chemical composition of the body. The model used in this book considers two components: fat-free mass and fat mass.

body density (D_{body})—Body weight divided by body volume.

body mass index (BMI)—A measurement of body overweight or obesity determined by dividing weight (in kilograms) by height (in meters) squared. BMI is highly correlated with body composition.

Borg RPE scale—A numerical scale for rating perceived exertion.

β-oxidation—The first step in fatty acid oxidation, in which fatty acids are broken into separate two-carbon units of acetic acid, each of which is then converted to acetyl CoA.

Boyle's gas law—Law stating that at a constant temperature, the number of gas molecules in a given volume depends on the pressure.

bradycardia—A resting heart rate lower than 60 beats/min.

branched-chain amino acids (BCAA)—Specific amino acids—leucine, isoleucine, and valine—that have been postulated to work in combination with L-tryptophan to delay fatigue, primarily through central nervous system mechanisms.

buffer—A substance that combines with either an acid or a base to maintain a constant acid–base (pH) balance.

bulimia nervosa—A clinical eating disorder characterized by recurrent episodes of binge eating, a feeling of lack of control during these binges, and purging behavior, which may include self-induced vomiting and use of laxatives and diuretics. Sometimes the disorder also includes fasting or excessive exercise behaviors.

CAD—*See* coronary artery disease.

caffeine—A central nervous system stimulant believed by some athletes to have ergogenic properties.

calcitonin—A hormone secreted by the thyroid gland that assists in the control of calcium ion concentrations in the blood.

calorie (cal)—A unit of measure of energy in biological systems, where 1.0 calorie is equal to the amount of heat energy needed to raise the temperature of 1.0 g of water 1 °C, from 15 to 16 °C.

calorimeter—A device for measuring the heat produced by the body (or by specific chemical reactions).

cAMP—*See* cyclic adenosine monophosphate.

capillaries—The smallest vessels transporting blood from the heart to the tissues and the actual sites of exchange between the blood and tissue.

capillary-to-fiber ratio—The number of capillaries per muscle fiber.

carbohydrate—An organic compound formed from carbon, hydrogen, and oxygen; includes starches, sugars, and cellulose.

carbohydrate loading—Increased dietary consumption of carbohydrates, a process used by athletes to increase carbohydrate stores in the body prior to prolonged endurance exercise.

cardiac cycle—The period that includes all events between two consecutive heartbeats.

cardiac hypertrophy—Enlargement of the heart by increases in muscle wall thickness or chamber size or both.

cardiac output ($\dot{Q}$)—The volume of blood pumped out by the heart per minute. $\dot{Q}$ = heart rate × stroke volume.

cardiorespiratory endurance—The ability of the body to sustain prolonged exercise.

cardiovascular deconditioning—A decrease in the cardiovascular system's ability to deliver sufficient oxygen and nutrients.

cardiovascular drift—An increase in heart rate during exercise to compensate for a decrease in stroke volume. This compensation helps maintain a constant cardiac output.

cardiovascular endurance training—*See* aerobic training.

catabolism—The tearing down of body tissue; the destructive phase of metabolism.

catecholamines—Biologically active amines (organic compounds derived from ammonia), such as epinephrine and norepinephrine, that have powerful effects similar to those of the sympathetic nervous system.

central command—Information originating in the brain that is transmitted to the cardiovascular, muscular, or pulmonary systems.

central nervous system (CNS)—System consisting of the brain and spinal cord.

chemoreceptor—A sensory organ capable of reacting to a chemical stimulus.

Cheyne-Stokes breathing—Alternating periods of rapid breathing and slow, shallow breathing including periods in which breathing may actually cease temporarily. A symptom of acute mountain sickness.

childhood—The period of life between the first birthday and the onset of puberty.

chronic adaptation—A physiological change that occurs when the body is exposed to repeated exercise bouts over weeks or months. These changes generally improve the body's efficiency at rest and during exercise.

chronic fatigue syndrome—A syndrome that appears to involve immune system dysfunction. Patients have incapacitating fatigue, sore throat, muscle tenderness or pain, and cognitive dysfunction; the symptoms may vary in severity over time but generally last for months or years.

chronic hypertrophy—An increase in muscle size that results from repeated long-term resistance training.

CK—*See* creatine kinase.

CNS—*See* central nervous system.

cold acclimatization—*See* acclimatization.

concentric contraction—Muscle shortening.

conduction—(1) Transfer of heat through direct molecular contact with a solid object; (2) movement of an electrical impulse, such as through a neuron.

congenital heart disease—A heart defect present at birth that occurs from abnormal prenatal development of the heart or associated blood vessels. Also known as congenital heart defect.

continuous training—Training at a moderate to high intensity without stopping to rest.

contractile velocity (V_o)—The speed of action associated with specific muscle fiber types.

control group—In an experimental design, the nontreated group to which the experimental group is compared.

convection—The transfer of heat or cold via the movement of a gas or liquid across an object, such as the body.

coronary artery disease (CAD)—Progressive narrowing of the coronary arteries.

cortisol—A corticosteroid hormone released from the adrenal cortex that stimulates gluconeogenesis, increases mobilization of free fatty acids, decreases use of glucose, and stimulates catabolism of protein. Also known as hydrocortisone.

creatine—A substance found in skeletal muscles most commonly in the form of PCr. Creatine supplements are often used as ergogenic aids because they are theorized to increase PCr levels, thus enhancing the ATP-PCr energy system by better maintaining muscle ATP levels.

creatine kinase (CK)—The enzyme that facilitates the breakdown of PCr to creatine and P_i.

critical temperature theory—Theory that prolonged exercise in hot environments is limited by attainment of a fixed elevated core temperature.

crossover design—Experimental design in which the control group becomes the experimental group after the first experimental period, and vice versa.

cross-sectional research design—A research design in which a cross section of a population is tested at one specific time and then data from groups within that population are compared.

cross-training—Training for more than one sport at the same time, or training multiple fitness components (such as endurance, strength, and flexibility) within the same period.

cycle ergometer—An exercise device that uses cycling to measure physical work.

cyclic adenosine monophosphate (cAMP)—Intracellular second messenger that mediates hormone action.

DBP—*See* diastolic blood pressure.

dehydration—Loss of body fluids.

delayed-onset muscle soreness (DOMS)—Muscle soreness that develops a day or two after a heavy bout of exercise and that is associated with actual injury within the muscle.

densitometry—The measurement of body density.

dependent variable—The physiological factor that is allowed to vary as another factor (the independent variable) is manipulated. Usually plotted on the *y*-axis.

depolarization—A decrease in the electrical potential across a membrane, as when the inside of a neuron becomes less negative relative to the outside.

detraining—Changes in physiological function in response to a reduction or cessation of regular physical training.

development—Changes that occur in the body starting at conception and continuing through adulthood; differentiation along specialized lines of function, reflecting changes that accompany growth.

diabetes mellitus—A disorder of carbohydrate metabolism characterized by hyperglycemia (high blood sugar levels) and glycosuria (presence of sugar in the urine). The disease develops when there is inadequate production of insulin by the pancreas or inadequate utilization of insulin by the cells.

diastolic blood pressure (DBP)—The lowest arterial pressure, resulting from ventricular diastole (the resting phase).

direct calorimetry—A method that gauges the body's rate and quantity of energy production by direct measurement of the body's heat production.

direct gene activation—The method of action of steroid hormones. They bind to receptors in the cell, and then the hormone–receptor complex enters the nucleus and activates certain genes.

disordered eating—Abnormal eating behavior that ranges from excessive restriction of food intake to pathological behaviors, such as self-induced vomiting and laxative abuse. Disordered eating can lead to clinical eating disorders, such as anorexia nervosa and bulimia nervosa

diuretics—Substances that promote water excretion.

diurnal variation—Fluctuations in physiological responses that occur during a 24 h period.

DOMS—*See* delayed-onset muscle soreness.

dose–response relationship—A relationship between two variables in which one changes predictably as the other increases or decreases.

downregulation—Decreased cellular sensitivity to a hormone, likely the result of a decreased number of cell receptors available to bind with the hormone.

dry heat exchange—Heat transfer by the combined avenues of convection, conduction, and radiation.

dual-energy X-ray absorptiometry (DXA)—A technique used to assess both regional and total-body composition through the use of X-ray absorptiometry.

dynamic contraction—Any muscle action that produces joint movement.

dyspnea—Labored or difficult breathing.

eating disorders—A group of clinical disorders involving eating. *See* anorexia nervosa, bulimia nervosa.

eccentric contraction—Any muscle action in which muscle lengthens.

eccentric training—Training that involves eccentric action.

eccrine sweat glands—Simple sweat glands dispersed over the body surface that respond to increases in core or skin temperature (or both) and facilitate thermoregulation.

ECG—*See* electrocardiogram and exercise electrocardiogram.

EDV—*See* end-diastolic volume.

EF—*See* ejection fraction.

efferent division—The motor division of the peripheral nervous system.

EIAH—*See* exercise-induced arterial hypoxemia.

ejection fraction (EF)—The fraction of blood pumped out of the left ventricle with each contraction, determined by dividing stroke volume by end-diastolic volume and expressed as a percentage.

electrical stimulation training—Stimulation of a muscle via passing an electrical current through it.

electrocardiogram (ECG)—A recording of the heart's electrical activity.

electrocardiograph—A machine used to obtain an electrocardiogram.

electrolyte—A dissolved substance that can conduct an electrical current.

electron transport chain—A series of chemical reactions that convert the hydrogen ion generated by glycolysis and the Krebs cycle into water and produce energy for oxidative phosphorylation.

end branches—Branches coming off the ends of the axons leading to the axon terminals.

end-diastolic volume (EDV)—The volume of blood inside the left ventricle at the end of diastole, just before contraction.

endomysium—A sheath of connective tissue that covers each muscle fiber.

end-systolic volume (ESV)—The volume of blood remaining in the left ventricle at the end of systole, just after contraction.

endurance—The ability to resist fatigue; includes muscular endurance and cardiorespiratory endurance.

energy—The capability of producing force, performing work, or generating heat.

environmental physiology—Study of the effects of the environment (heat, cold, altitude, hyperbaria, etc.) on the function of the body.

ephedrine—a sympathomimetic amine derived from ephedra herbs (also known as ma huang) and is used as a decongestant and as a bronchodilator in the treatment of asthma.

epimysium—The outer connective tissue that surrounds an entire muscle, holding it together.

epinephrine—A catecholamine released from the adrenal medulla that, along with norepinephrine, prepares the body for a fight-or-flight response. It is also a neurotransmitter. *See* catecholamines.

EPOC—*See* excess postexercise oxygen consumption.

EPSP—*See* excitatory postsynaptic potential.

ergogenic—Able to improve work or performance.

ergogenic aid—A substance or phenomenon that can improve work or athletic performance.

ergolytic—Able to impair work or performance.

ergometer—An exercise device that allows the amount and rate of a person's physical work to be controlled (standardized) and measured.

erythropoietin (EPO)—The hormone that stimulates erythrocyte (red blood cell) production.

essential amino acids—The eight or nine amino acids necessary for human growth that the body cannot synthesize and are thus essential parts of our diets.

estrogen—A female sex hormone.

ESV—*See* end-systolic volume.

eumenorrhea—Normal menstrual function.

evaporation—Heat loss through the conversion of water (such as in sweat) to vapor.

excessive training—Training in which volume, intensity, or both are too great or are increased too quickly without proper progression.

excess postexercise oxygen consumption (EPOC)—Elevated oxygen consumption above resting levels after exercise; at one time referred to as oxygen debt.

excitatory postsynaptic potential (EPSP)—A depolarization of the postsynaptic membrane caused by an excitatory impulse.

exercise electrocardiogram (ECG)—A recording of the heart's electrical activity during exercise.

exercise-induced arterial hypoxemia (EIAH)—A decline in arterial PO_2 and arterial oxygen saturation during maximal or near-maximal exercise.

exercise physiology—The study of how body structure and function are altered by exposure to acute and chronic bouts of exercise.

exercise prescription—Individualization of the prescription of exercise duration, frequency, intensity, and mode.

exhaustion—Inability to continue exercise.

expiration—The process by which air is forced out of the lungs through relaxation of the inspiratory muscles and elastic recoil of the lung tissue, which increase the pressure in the thorax.

external respiration—The process of bringing air into the lungs and the resulting exchange of gas between the alveoli and the capillary blood.

extracellular fluid—The 35% to 40% of the water in the body that is outside the cells, including interstitial fluid, blood plasma, lymph, cerebrospinal fluid, and other fluids.

extrinsic neural control—Redistribution of blood at the system or body level through neural mechanisms.

Fartlek training—Developed in the 1930s; the term comes from the Swedish for "speed play." This type of training combines continuous and interval training and stresses both the aerobic and anaerobic energy pathways.

fasciculus—A small bundle of muscle fibers wrapped in a connective tissue sheath within a muscle.

fast-twitch (type II) fiber—A type of muscle fiber with a low oxidative capacity and a high glycolytic capacity; associated with speed or power activities.

fat—A class of organic compounds with limited water solubility that exists in the body in many forms, such as triglycerides, free fatty acids, phospholipids, and steroids.

fat-free mass—The mass (weight) of the body that is not fat, including muscle, bone, skin, and organs.

fatigue—General sensations of tiredness and accompanying decrements in muscular performance.

fat mass—The absolute amount or mass of body fat.

fatty streaks—Early lipid deposits within blood vessels.

female athlete triad—Three interrelated disorders—disordered eating, menstrual dysfunction, and bone mineral disorders—to which some female athletes are prone.

$FEV_{1.0}$—*See* forced expiratory volume in 1 s.

FFA—*See* free fatty acids.

fiber hyperplasia—An increase in the number of muscle fibers.

fiber hypertrophy—An increase in the size of existing individual muscle fibers.

fibromyalgia syndrome—A chronic syndrome that includes muscle pain as its dominant symptom but is also characterized by muscle weakness, migraine-type headaches, and depression.

Fick equation—$\dot{V}O_2 = \dot{Q} \times (a\text{-}\bar{v})O_2$ difference.

Fick's law—Law stating that the net diffusion rate of a gas across a fluid membrane is proportional to the difference in partial pressure, proportional to the area of the membrane, and inversely proportional to the thickness of the membrane.

force—Strength or energy exerted or brought to bear.

forced expiratory volume in 1 s ($FEV_{1.0}$)—The volume of air exhaled in the first second after maximal inhalation.

Frank-Starling mechanism—The mechanism by which an increased amount of blood in the ventricle causes a stronger ventricular contraction to increase the amount of blood ejected.

free fatty acids (FFA)—The components of fat that are used by the body for metabolism.

free radicals—Univalent (unpaired) oxygen intermediates that leak out of the electron transport chain during metabolic processes and may damage tissues.

free weights—Traditional resistance training modality that uses only barbells, dumbbells, and so on to provide resistance.

frostbite—Tissue damage that occurs during cold exposure because circulation to the skin decreases, in an attempt to retain body heat, to the point that the tissue receives insufficient oxygen and nutrients.

gastric emptying—The movement of food mixed with gastric secretions from the stomach into the duodenum.

gender differences—Any differences between females and males. *See also* sex-specific differences.

glucagon—A hormone released by the pancreas that promotes increased breakdown of liver glycogen to glucose (glycogenolysis) and increased gluconeogenesis.

glucocorticoid—a family of steroid hormones produced by the adrenal cortex that help maintain homeostasis through a variety of effects throughout the body

gluconeogenesis—The conversion of protein or fat into glucose.

glucose—Six-carbon sugar that is the primary form of carbohydrate used for metabolism.

glycogen—The form of carbohydrate stored in the body, found predominantly in the muscles and liver.

glycogenesis—The conversion of glucose to glycogen.

glycogen loading—The manipulation of exercise and diet to optimize the body's glycogen storage.

glycogenolysis—The conversion of glycogen to glucose.

glycogen sparing—Increased reliance on fats for energy production during endurance activity, rather than stores of glycogen.

glycolysis—The breakdown of glucose to pyruvic acid.

glycolytic enzymes—Enzymes that are specific to the glycolytic energy system.

glycolytic system—A system that produces energy through glycolysis.

glycosuria—The presence of glucose in the urine.

Golgi tendon organ—A sensory receptor in a muscle tendon that monitors tension.

graded exercise test (GXT)—An exercise test in which the rate of work is increased gradually in 1 to 3 min increments, usually to the point of fatigue or exhaustion.

graded potential—A localized change (depolarization or hyperpolarization) in the membrane potential.

growth—An increase in the size of the body or any of its parts.

growth hormone—An anabolic agent that stimulates fat metabolism and promotes muscle growth and hypertrophy by facilitating amino acid transport into the cells.

GXT—*See* graded exercise test.

habituation—Short-term adaptation to a stress.

HACE—*See* high-altitude cerebral edema.

Haldane transformation—An equation allowing one to calculate the inspired air volume from expired air volume, or expired air volume from inspired air volume.

HAPE—*See* high-altitude pulmonary edema.

HDL—*See* high-density lipoprotein.

HDL-C—*See* high-density lipoprotein cholesterol.

heart failure—A clinical condition in which the myocardium becomes too weak to maintain adequate cardiac output to meet the body's oxygen demands; heart failure usually results from the heart's being damaged or overworked.

heart rate recovery period—The time it takes for heart rate to return to the resting rate following exercise.

heat acclimation—*See* acclimation.

heat cramp—Cramping of the skeletal muscles as a result of excessive dehydration and the associated salt loss.

heat exhaustion—A heat disorder resulting from an inability of the cardiovascular system to meet all the body tissues' needs while also shifting blood to the periphery for cooling, characterized by elevated body temperature, breathlessness, extreme tiredness, dizziness, and rapid pulse.

heatstroke—The most serious heat disorder, resulting from failure of the body's thermoregulatory mechanisms. Heatstroke is characterized by body temperature above 40.5 °C (105 °F), cessation of sweating, and total confusion or unconsciousness and can lead to death.

hematocrit—The percentage of cells or formed elements in the total blood volume. More than 99% of the cells or formed elements are red blood cells.

hematopoiesis—Increased red blood cell concentration by increased production of cells.

hemoconcentration—A relative (not absolute) increase in the cellular content per unit of blood volume, resulting from a reduction in plasma volume.

hemodilution—An increase in blood plasma, resulting in a dilution of the blood's cellular contents.

hemoglobin—The iron-containing pigment in red blood cells that binds oxygen.

hemoglobin saturation—The amount of oxygen bound by each molecule of hemoglobin.

hemorrhagic stroke—Involves bleeding within the brain, which damages nearby brain tissue.

Henry's law—Law stating that gases dissolve in liquids in proportion to their partial pressures, depending also on their solubilities in the specific fluids and on the temperature.

hGH—*See* human growth hormone.

high-altitude cerebral edema (HACE)—A condition of unknown cause in which fluid accumulates in the cranial cavity at altitude; characterized by mental confusion that can progress to coma and death.

high-altitude pulmonary edema (HAPE)—A condition of unknown cause in which fluid accumulates in the lungs at altitude, interfering with ventilation, resulting in shortness of breath and fatigue, and characterized by impaired blood oxygenation, mental confusion, and loss of consciousness.

high-density lipoprotein (HDL)—A cholesterol carrier regarded as a scavenger; theorized to remove cholesterol from the arterial wall and transport it to the liver to be metabolized.

high-density lipoprotein cholesterol (HDL-C)—The cholesterol carried by HDL.

high responders—Those individuals within a population that show clear or exaggerated responses or adaptations to a stimulus.

homeostasis—Maintenance of a constant internal environment.

hormonal agents—A group of hormones proposed to have ergogenic properties.

hormone—A chemical substance produced or released by an endocrine gland and transported by the blood to a specific target tissue.

HR_{max}—*See* maximum heart rate.

human growth hormone (hGH)—A hormone that promotes anabolism and is believed by some athletes to have ergogenic properties.

hydrocortisone—*See* cortisol.

hydrostatic weighing—A method of measuring body volume in which a person is weighed while submerged underwater. The difference between the scale weight on land and the underwater weight (corrected for water density) equals body volume. This value must be further corrected to account for any air trapped in the lungs and other parts of the body.

hyperglycemia—An elevated blood glucose level.

hyperinsulinemia—High levels of insulin in the blood.

hyperplasia—An increase in the number of cells in a tissue or organ. *See also* fiber hyperplasia.

hyperpolarization—An increase in the electrical potential across a membrane.

hypertension—Abnormally high blood pressure. In adults, hypertension is usually defined as a systolic pressure of 140 mmHg or higher or a diastolic pressure of 90 mmHg or higher.

hyperthermia—Elevated body temperature; any temperature above a person's normal resting body temperature.

hypertrophy—Increase in the size or mass of an organ or body tissue. *See also* fiber hypertrophy.

hyperventilation—A breathing rate or tidal volume greater than necessary for normal function.

hypobaric—Referring to an environment, such as that at high altitude, involving low atmospheric pressure.

hypoglycemia—A low blood glucose level.

hyponatremia—A blood sodium concentration below the normal range of 136 to 143 mmol/L.

hypothermia—Low body temperature; any temperature below the given person's normal temperature.

hypoxemia—A decreased oxygen content or concentration within the blood.

hypoxia—A decreased availability of oxygen to the tissues.

hypoxic vasoconstriction—The constriction of blood vessels in response to low levels of oxygen.

IDDM—*See* insulin-dependent diabetes mellitus.

immune function—The body's normal ability to fight infection and illness with antibodies and lymphocytes.

impaired fasting glucose—A plasma glucose level between 110 and 125 mg/dL following an 8 h fast.

impaired glucose tolerance—An abnormal glucose response to an oral glucose load (glucose tolerance test), sometimes seen as a precursor to diabetes.

independent variable—In an experiment, the variable that is manipulated by the experimenter to determine the response of the dependent variable. Usually plotted on the *x*-axis.

indirect calorimetry—A method of estimating energy expenditure by measuring respiratory gases.

infancy—The first year of life.

inhibiting factors—Hormones transmitted from the hypothalamus to the anterior pituitary that inhibit release of some other hormones.

inhibitory postsynaptic potential (IPSP)—A hyperpolarization of the postsynaptic membrane caused by an inhibitory impulse.

inspiration—The active process involving the diaphragm and the external intercostal muscles that expands the thoracic dimensions and thus the lungs. The expansion decreases pressure in the lungs, allowing outside air to rush in.

insulation—Resistance to dry heat exchange.

insulin—A hormone produced by the β-cells in the pancreas that assists glucose entry into cells.

insulin-dependent diabetes mellitus (IDDM)—One of two major categories of diabetes mellitus that is caused by the inability of the pancreas to produce sufficient insulin as a result of failure of the β-cells in the pancreas. This is also known as type 1 diabetes.

insulin resistance—A deficient target cell response to insulin.

insulin sensitivity—An index of the effectiveness of a given insulin concentration on the disposal of glucose.

internal respiration—The exchange of gases between the blood and tissues.

interval training—Repeated, brief, fast-paced exercise bouts with short rest intervals between bouts.

intracellular fluid—The approximately 60% to 65% of total-body water that is contained in the cells.

IPSP—*See* inhibitory postsynaptic potential.

ischemia—A temporary deficiency of blood to a specific area of the body.

ischemic stroke—Brain tissue damage resulting from insufficient oxygen supply to an area of the brain. May be caused by narrowing or blockage of blood vessels supplying the area.

isokinetic training—Resistance training in which the rate of movement is kept constant through the range of motion.

isometric training—Resistance training involving a static action.

Karvonen method—The calculation of training heart rate in which a given percentage of the maximal heart rate reserve is added to the resting heart rate. This method gives an adjusted heart rate that is approximately equivalent to the desired percentage of $\dot{V}O_{2max}$.

kilocalorie (kcal)—The equivalent of 1000 calories. *See* calorie.

Krebs cycle—A series of chemical reactions that involve the complete oxidation of acetyl CoA and produce 2 mol of ATP (energy) along with hydrogen and carbon, which combine with oxygen to form H_2O and CO_2.

lactate—A salt formed from lactic acid.

lactate dehydrogenase (LDH)—A key glycolytic enzyme involved in the conversion of pyruvate to lactate.

lactate threshold—The point during exercise of increasing intensity at which blood lactate begins to accumulate above resting levels, where lactate clearance is no longer able to keep up with lactate production.

L-carnitine—A substance important for fatty acid metabolism because it assists in the transfer of fatty acids from the cytosol (the fluid portion of the cytoplasm, exclusive of organelles) across the inner mitochondrial membrane for β-oxidation.

LDH—*See* lactate dehydrogenase.

LDL—*See* low-density lipoprotein.

LDL-C—*See* low-density lipoprotein cholesterol.

lean body mass—The sum of the body's fat-free mass and essential fat. This is not to be confused with fat-free mass.

lipogenesis—The process of converting protein into fatty acids.

lipolysis—The process of breaking down triglyceride to its basic units to be used for energy.

lipoprotein lipase—The enzyme that breaks down triglycerides to free fatty acids and glycerol, allowing the free fatty acids to enter the cells for use as a fuel or for storage.

lipoproteins—The proteins that carry the blood lipids.

longevity—The length of a person's life.

longitudinal research design—A research design in which subjects are tested initially and then one or more times later to directly measure changes over time resulting from a given intervention.

low-density lipoprotein (LDL)—A cholesterol carrier theorized to be responsible for depositing cholesterol in the arterial wall.

low-density lipoprotein cholesterol (LDL-C)—The cholesterol carried by LDL.

lower body (gynoid) obesity—Obesity that follows the typically female pattern of fat storage, in which fat is stored primarily in the lower body, particularly in the hips, buttocks, and thighs.

low-intensity aerobic exercise—Aerobic exercise performed at low intensity, theoretically to cause the body to burn a higher percentage of fat.

low responders—Those individuals within a population that show little or no response or adaptation to a stimulus.

LSD training—Endurance training involving long, slow distances.

L-tryptophan—An essential amino acid that has been proposed to increase aerobic endurance performance through its effects on the central nervous system. It theoretically acts as an analgesic and delays fatigue.

macrominerals—Those minerals of which the body needs more than 100 mg per day.

MAP—*See* mean arterial pressure.

maturation—The process by which the body takes on the adult form and becomes fully functional. It is often defined by the system or function being considered.

maximal expiratory ventilation ($\dot{V}_{Emax}$)—The highest ventilation that can be achieved during exhaustive exercise.

maximal heart rate reserve—The difference between maximal heart rate and resting heart rate.

maximal oxygen uptake ($\dot{V}O_{2max}$)—The maximal capacity for oxygen consumption by the body during maximal exertion. It is also known as aerobic power, maximal oxygen intake, maximal oxygen consumption, and cardiorespiratory endurance capacity.

maximal voluntary ventilation—The maximal capacity to move air into and out of the lungs, usually measured for 12 s and extrapolated to a per-minute value.

maximum heart rate (HR_{max})—The highest heart rate value attainable during an all-out effort to the point of exhaustion.

mean arterial pressure (MAP)—The average pressure exerted by the blood as it travels through the arteries. It is estimated as follows: $MAP = DBP + [0.333 \times (SBP - DBP)]$.

mechanoreceptors—An end organ that responds to changes in mechanical stress, such as stretch, compression, or distension.

menarche—The onset of menstruation; the first menses.

menses—The menstrual or flow phase of the menstrual cycle.

menstrual cycle—The cycle of uterine changes, averaging 28 days and consisting of the menstrual (flow) phase, the proliferative phase, and the secretory phase.

menstrual dysfunction—Disruption of the normal menstrual cycle; includes oligomenorrhea, primary amenorrhea, and secondary amenorrhea.

metabolic equivalent (MET)—A unit used to estimate the metabolic cost (oxygen consumption) of physical activity. One MET equals the resting metabolic rate of approximately 3.5 ml of $O_2 \cdot kg^{-1} \cdot min^{-1}$.

metabolic syndrome—A term that has been used to link coronary artery disease, hypertension, type 2 diabetes, and upper body obesity to insulin resistance and hyperinsulinemia. This syndrome has also been referred to as syndrome X and the civilization syndrome.

metabolism—All energy-producing and energy-using processes within the body.

microgravity—An environment in which the body experiences a reduced gravitational force.

microminerals (trace elements)—The minerals of which the body needs less than 100 mg per day.

mineralocorticoids—Steroid hormones released from the adrenal cortex that are responsible for electrolyte balance within the body, for example aldosterone.

mitochondrial oxidative enzymes—Oxidative enzymes located in the mitochondria.

MK—*See* myokinase.

mode—Type of exercise.

morphology—The form and structure of the body.

motor division—*See* efferent division.

motor reflex—An involuntary motor response to a given stimulus.

motor unit—The motor nerve and the group of muscle fibers it innervates.

mountain sickness—*See* acute altitude sickness.

muscle buffering capacity—The muscles' ability to tolerate the acid that accumulates in them during anaerobic glycolysis.

muscle fiber—An individual muscle cell.

muscle spindle—A sensory receptor located in the muscle that senses how much the muscle is stretched.

muscular endurance—The ability of a muscle to resist fatigue.

myelination—The process of acquiring a myelin sheath.

myelin sheath—The outer covering of a myelinated nerve fiber, formed by a fatlike substance called myelin.

myocardial infarction—Death of heart tissue that results from insufficient blood supply to part of the myocardium.

myocardium—The muscle of the heart.

myofibril—The contractile element of skeletal muscle.

myoglobin—A compound similar to hemoglobin, but found in muscle tissue, that carries oxygen from the cell membrane to the mitochondria.

myokinase (MK)—A key enzyme in the ATP-PCr energy system.

myosin—One of the proteins that form filaments that produce muscle action.

myosin cross-bridge—The protruding part of a myosin filament. It includes the myosin head, which binds to an active site on an actin filament to produce a power stroke that causes the filaments to slide across each other.

nebulin—A giant protein that coextends with actin and appears to play a regulatory role in mediating actin and myosin interactions.

needs analysis—An assessment of factors that determine the specific training program appropriate for an individual.

negative feedback system—The primary mechanism through which the endocrine system maintains homeostasis. Some body change upsets homeostasis, which triggers release of a hormone to correct the change. Once that correction is accomplished, the hormone is no longer needed, so its secretion decreases.

nerve impulse—The electrical signal conducted along a neuron, which can be transmitted to another neuron or an end organ such as a group of muscle fibers.

neuromuscular junction—The site at which a motor neuron communicates with a muscle fiber.

neuron—A specialized cell in the nervous system responsible for generating and transmitting nerve impulses.

neurotransmitter—A chemical used for communication between a neuron and another cell.

NIDDM—*See* non-insulin-dependent diabetes mellitus.

nonessential amino acids—The 11 or 12 amino acids that the body synthesizes.

non-insulin-dependent diabetes mellitus (NIDDM)—One of two major categories of diabetes mellitus that is caused by the ineffectiveness of insulin to facilitate the transport of glucose into the cells and is a result of insulin resistance. This is also known as type 2 diabetes.

nonresponders—Individuals who show little or no improvement compared with others who undergo the same training program.

nonshivering thermogenesis—The stimulation of metabolism by the sympathetic nervous system to generate more metabolic heat.

nonsteroid hormones—Hormones derived from protein, peptides, or amino acids that cannot easily cross cell membranes.

norepinephrine—A catecholamine released from the adrenal medulla that, along with epinephrine, prepares the body for a fight-or-flight response. It is also a neurotransmitter. *See* catecholamines.

nutritional agents—Nutritional substances proposed to have ergogenic benefits.

obesity—An excessive amount of body fat, generally defined as more than 25% in men and 35% in women; a body mass index of 30 or greater.

oligomenorrhea—Abnormally infrequent or scant menstruation.

oral contraceptives—Drugs used for birth control and other medical purposes, which are believed by some female athletes to have ergogenic properties.

osmolality—The number of solutes (such as electrolytes) dissolved in a fluid divided by the weight of that fluid; usually expressed in units of osmols (or milliosmols) per kg.

ossification—The process of bone formation.

osteopenia—The loss of bone mass with aging.

osteoporosis—Decreased bone mineral content that increases bone porosity.

overload—To train at a level above that at which one normally trains; for example, to impart more physiological strain on muscle during training than the muscle is normally exposed to.

overreaching—A systematic attempt to intentionally overstress the body, allowing the body to adapt even more to the training stimulus above and beyond the adaptation attained during a period of acute overload.

overtraining—The attempt to do more work than can be physically tolerated.

overtraining syndrome—A condition brought on by overtraining and characterized by performance decrements and a general breakdown in physiological function.

overweight—Body weight that exceeds the normal or standard weight for a particular individual based on sex, height, and frame size; a BMI of 25.0 to 29.9.

oxidative capacity of muscle ($\dot{Q}O_2$)—A measure of the muscle's maximal capacity to use oxygen.

oxidative system—The body's most complex energy system, which generates energy by disassembling fuels with the aid of oxygen and has a very high energy yield.

oxygen diffusion capacity—The rate at which oxygen diffuses from one place to another.

oxygen supplementation—The breathing of supplemental oxygen, which is proposed to have ergogenic properties.

oxygen transport system—The components of the cardiovascular and respiratory systems involved in transporting oxygen.

parathyroid hormone (PTH)—The hormone released by the parathyroid gland to regulate plasma calcium concentration and plasma phosphate.

partial pressure—The pressure exerted by an individual gas in a mixture of gases.

partial pressure of oxygen (PO_2)—The pressure exerted by oxygen in a mixture of gases.

pathophysiology—The physiology of a specific disease or disorder.

PCr—*See* phosphocreatine.

PDGF—*See* platelet-derived growth factor.

pericardium—A double-layered outer covering of the heart.

perimysium—The connective tissue sheath surrounding each muscle fasciculus.

peripheral blood flow—Blood flow to the extremities and the skin.

peripheral nervous system (PNS)—That section of the nervous system through which motor nerve impulses are transmitted from the brain and spinal cord to the periphery and sensory nerve impulses are transmitted from the periphery to the brain and spinal cord.

peripheral vascular disease—Diseases of the systemic arteries and veins, especially those to the extremities, that impede adequate blood flow.

peripheral vasoconstriction—*See* vasoconstriction.

PFK—*See* phosphofructokinase.

pharmacological agents—A group of drugs proposed to have ergogenic properties.

phosphate loading—The practice of ingesting sodium phosphate, which has been proposed to have ergogenic properties.

phosphocreatine (PCr)—An energy-rich compound that plays a critical role in providing energy for muscle action by maintaining ATP concentration.

phosphofructokinase (PFK)—A key rate-limiting enzyme of the anaerobic glycolytic energy system.

phosphorylase—A key enzyme of the anaerobic glycolytic energy system.

physical maturity—The point at which the body has attained the adult physical form.

physiological agents—A group of agents normally present in the body that have been proposed to have ergogenic properties.

physiology—The study of the function of organisms.

placebo—An inactive substance, usually provided in a manner identical to that for an active substance, typically to test for real results produced by the test substance versus equally real results of psychological origin.

placebo effect—An effect produced by the subject's expectations after administration of an inactive substance (placebo).

placebo group—The group in an intervention study that receives a placebo rather than a test substance.

plaque—A buildup of lipids, smooth muscle cells, connective tissue, and debris that forms at the site of injury to an artery.

plasmalemma—Plasma membrane, the selectively permeable lipid bilayer coated by proteins that composes the outer layer of a cell.

platelet-derived growth factor (PDGF)—A substance released by blood platelets that promotes the migration of smooth muscle cells from the media of an artery into the intima.

plyometrics—A type of dynamic-action resistance training based on the theory that use of the stretch reflex during jumping will recruit additional motor units.

PNS—*See* peripheral nervous system.

PO_2—*See* partial pressure of oxygen.

POAH—*See* preoptic-anterior hypothalamus.

polycythemia—Increased red blood cells.

power—The rate of performing work; the product of force and velocity. The rate of transformation of metabolic potential energy to work or heat.

power stroke—The tilting of the myosin head, caused by a strong intermolecular attraction between the myosin cross-bridge and the myosin head, that causes the actin and myosin filaments to slide across each other.

prediabetes—a term used to define those who have impaired fasting glucose, impaired glucose tolerance, or both, but are not truly diabetic.

predictive value of an abnormal exercise test—The accuracy with which abnormal test results reflect the presence of a disease.

pregnancy—The state of carrying an embryo or fetus in the body.

preload—The degree to which the myocardium is stretched before it contracts, determined by factors such as central blood volume.

premature ventricular contraction (PVC)—A common cardiac arrhythmia that results in the feeling of skipped or extra beats caused by impulses originating outside the SA node.

preoptic-anterior hypothalamus (POAH)—The area of the midbrain that is the primary controller of thermoregulatory function.

primary amenorrhea—The absence of menarche (the beginning of menstruation) beyond age 18.

primary risk factors—Risk factors that have been conclusively shown to have a strong association with a certain disease. Primary risk factors for coronary artery disease include smoking, hypertension, high blood lipid levels, obesity, and physical inactivity.

principle of hard/easy—The theory that a training program must alternate high-intensity workouts with low-intensity workouts to help the body recover and achieve optimal training adaptation.

principle of individuality—The theory that any training program must consider the specific needs and abilities of the individual for whom it is designed.

principle of orderly recruitment—The theory that motor units generally are activated on the basis of a fixed order of recruitment, in which the motor units within a given muscle appear to be ranked according to the size of the motor neuron.

principle of periodization—The gradual cycling of specificity, intensity, and volume of training to achieve peak levels of fitness for competition.

principle of progressive overload—The theory that, to maximize the benefits of a training program, the training stimulus must be progressively increased as the body adapts to the current stimulus.

principle of reversiblity—The theory that a training program must include a maintenance plan to ensure that the gains from training are not lost.

principle of specificity—The theory that a training program must stress the physiological systems critical for optimal performance in a given sport to achieve desired training adaptations in that sport.

progesterone—A hormone secreted by the ovaries that promotes the luteal phase of the menstrual cycle.

prostaglandins—Substances derived from a fatty acid that act as hormones at the local level.

protein—A class of nitrogen-containing compounds formed by amino acids.

pseudoephedrine—a sympathomimetic amine that is used in over-the-counter medications primarily as a decongestant and in the illicit manufacturing of methamphetamine.

PTH—*See* parathyroid hormone.

puberty—The point at which a person becomes physiologically capable of reproduction.

pulmonary diffusion—The exchange of gases between the lungs and the blood.

pulmonary ventilation—The movement of gases into and out of the lungs.

Purkinje fibers—The terminal branches of the AV bundle that transmit impulses through the ventricles six times faster than through the rest of the cardiac conduction system.

PVC—*See* premature ventricular contraction.

$\dot{Q}$—*See* cardiac output.

$\dot{Q}O_2$—*See* oxidative capacity of muscle.

radiation—The transfer of heat through electromagnetic waves.

rate coding—Refers to the frequency of impulses sent to a muscle. Increased force can be generated through increase in either the number of muscle fibers recruited or the rate at which the impulses are sent. Also called frequency coding.

rating of perceived exertion (RPE)—A person's subjective assessment of how hard he or she is exercising.

rehabilitation programs—Programs designed to reestablish health or fitness following a disability or illness.

relative body fat—The ratio of fat mass to total-body mass, expressed as a percentage.

releasing factors—Hormones transmitted from the hypothalamus to the anterior pituitary that promote release of some other hormones.

renin—An enzyme formed by the kidneys to convert a plasma protein called angiotensinogen into angiotensin II. *See also* renin–angiotensin-aldosterone mechanism.

renin–angiotensin-aldosterone mechanism—The mechanism involved in renal control of blood pressure. The kidneys respond to decreased blood pressure or blood flow by forming renin, which converts angiotensinogen into angiotensin I, which is finally converted to angiotensin II. Angiotensin II constricts arterioles and triggers aldosterone release.

RER—*See* respiratory exchange ratio.

residual volume (RV)—The amount of air that cannot be exhaled from the lungs.

resistance training—Training designed to increase strength, power, and muscular endurance.

respiratory alkalosis—A condition in which increased carbon dioxide clearance allows blood pH to increase.

respiratory centers—Autonomic centers located in the medulla oblongata and the pons that establish breathing rate and depth.

respiratory exchange ratio (RER)—The ratio of carbon dioxide expired to oxygen consumed at the level of the lungs.

respiratory membrane—The membrane separating alveolar air and blood, composed of the alveolar wall, the capillary wall, and their basement membranes.

respiratory pump—Passive movement of blood through the central circulation as a function of pressure changes during breathing.

resting heart rate (RHR)—The heart rate at rest, averaging 60 to 80 beats/min.

resting membrane potential (RMP)—The potential difference between the electrical charges inside a cell and outside the cell, caused by a separation of charges across the membrane.

resting metabolic rate (RMR)—The body's metabolic rate early in the morning following an overnight fast and 8 h of sleep. Determining RMR does not require sleeping overnight in a laboratory or clinical facility. *See also* basal metabolic rate.

retraining—Recovery of conditioning after a period of inactivity.

rhabdomyolysis—Breakdown of muscle fibers resulting in protein buildup in the blood and often in the urine.

rheumatic heart disease—A form of valvular heart disease involving a streptococcal infection that has caused acute rheumatic fever, typically in children between ages 5 and 15.

RHR—*See* resting heart rate.

risk factor—A predisposing factor statistically linked to the development of a disease, such as coronary artery disease.

RMP—*See* resting membrane potential.

RMR—*See* resting metabolic rate.

RPE—*See* rating of perceived exertion.

RV—*See* residual volume.

saltatory conduction—The means of rapid nerve impulse conduction along myelinated neurons.

SA node—*See* sinoatrial node.

sarcolemma—A muscle fiber's cell membrane.

sarcomere—The basic functional unit of a myofibril.

sarcopenia—The loss of muscle mass associated with aging.

sarcoplasm—The gelatin-like cytoplasm in a muscle fiber.

sarcoplasmic reticulum (SR)—A longitudinal system of tubules that is associated with the myofibrils and that stores calcium for muscle action.

satellite cells—Immature cells that can develop into mature cell types, such as myoblasts.

SBP—*See* systolic blood pressure.

SDH—*See* succinate dehydrogenase.

secondary amenorrhea—The cessation of menstruation in a woman with previously normal menstrual function.

second messenger—A substance inside a cell that acts as a messenger after a nonsteroid hormone binds to receptors outside the cell.

sensitivity—A test's ability to correctly identify subjects who fit the criteria being tested, such as coronary artery disease.

sensory division—*See* afferent division.

sensory-motor integration—The process by which the sensory and motor systems communicate and coordinate with each other.

sex-specific differences—True physiological differences between females and males.

shivering—A rapid, involuntary cycle of contraction and relaxation of skeletal muscles that generates heat.

sinoatrial (SA) node—A group of specialized myocardial cells, located in the wall of the right atrium, that control the heart's rate of contraction; the pacemaker of the heart.

size principle—Principle asserting that the size of the motor neuron dictates the order of motor unit recruitment, with small-sized motor neurons being recruited first.

skinfold fat thickness—The most widely applied field technique used to estimate body density, relative body fat, and fat-free mass. It involves measurement with calipers of the skinfold fat at one or more sites.

sliding filament theory—A theory explaining muscle action: A myosin cross-bridge attaches to an actin filament, and then the power stroke drags the two filaments past one another.

slow-twitch (type I) fiber—A type of muscle fiber that has a high oxidative and a low glycolytic capacity, associated with endurance-type activities.

sodium-potassium pump—An enzyme called Na^+-K^+-ATPase, which maintains the resting membrane potential in disequilibrium at –70 mV.

specificity—A test's ability to correctly identify subjects who do not fit the criteria being tested.

specificity of training—The principle that physiological adaptations in response to physical training are highly specific to the nature of the training activity. To maximize benefits, training should be carefully matched to an athlete's specific performance needs.

spirometry—The measurement of lung volumes and capacities.

sport physiology—The application of the concepts of exercise physiology to training athletes and enhancing sport performance.

spot reduction—The practice of exercising a specific area of the body, theoretically to reduce locally stored fat.

sprint training—A form of anaerobic training involving very brief, intense training bouts.

SR—*See* sarcoplasmic reticulum.

ST—*See* slow-twitch fiber.

static-contraction resistance training—Resistance training that emphasizes static muscle action. Also known as isometric resistance training.

static (isometric) muscle contraction—Action in which the muscle contracts without moving, generating force while its length remains static (unchanged). Also known as isometric action.

steady-state heart rate—A heart rate that is maintained constant at submaximal levels of exercise when the rate of work is held constant.

steroid hormones—Hormones with chemical structures similar to cholesterol that are lipid soluble and that diffuse through cell membranes.

strength—The ability of a muscle to exert force—generally the maximal ability.

stroke—A cerebral vascular accident, a condition in which blood supply to some part of the brain is impaired, typically caused by infarction or hemorrhage, so that the tissue is damaged.

stroke volume (SV)—The amount of blood ejected from the left ventricle during contraction; the difference between the end-diastolic volume and the end-systolic volume.

submaximal endurance capacity—The average absolute power output a person can maintain during a fixed period of time on a cycle ergometer, or the average speed or velocity a person can maintain during a fixed period of time. Generally, these tests will last at least 30 min but usually not more than 90 min.

submaximal exercise—All intensities of exercise below maximal exercise intensity.

substrate—Basic fuel source, such as carbohydrates, proteins, and fats.

succinate dehydrogenase (SDH)—A key enzyme of the oxidative enzyme system.

summation—The summing of all individual changes in a neuron's membrane potential.

SV—*See* stroke volume.

swimming flume—A device that uses propeller pumps to circulate water past a swimmer, who attempts to maintain body position by swimming against the current.

synapse—The junction between two neurons.

systolic blood pressure (SBP)—The greatest arterial blood pressure, resulting from systole (the contracting phase of the heart).

T_3—*See* triiodothyronine.

T_4—*See* thyroxine.

tachycardia—A resting heart rate greater than 100 beats/min.

tapering—A reduction in training intensity prior to a major competition to give the body and mind a break from the rigors of intense training.

taper period—A period during which training intensity is reduced, allowing time for tissue damage from intense training to heal and for the body's energy reserves to be fully replenished.

target cells—Cells that possess specific hormone receptors.

TEA—*See* thermic effect of activity.

TEM—*See* thermic effect of a meal.

teratogenic effects—Effects that cause abnormal fetal development.

testosterone—The predominant male sex hormone.

test specificity—Matching the type of ergometer used in testing to the type of activity an athlete usually performs to ensure the most accurate results.

tetanus—Highest tension developed by a muscle in response to stimulation of increasing frequency.

tethered swimming—A method of monitoring a swimmer in which the swimmer is attached to a harness connected to a rope, a series of pulleys, and a pan that contains weights, which allows the swimmer to swim while maintaining a constant position in the pool.

thermal stress—Stress imposed on the body by external temperature.

thermic effect of activity (TEA)—The energy expended in excess of the resting metabolic rate to accomplish a given task or activity.

thermic effect of a meal (TEM)—The energy expended in excess of resting metabolic rate associated with digestion, absorption, transport, metabolism, and storage of ingested food.

thermoreceptors—Sensory receptors that detect changes in body temperature and external temperature and relay this information to the hypothalamus (also called thermoceptors).

thermoregulation—The process by which the thermoregulatory center, located in the hypothalamus, readjusts body temperature in response to small deviations from the set point.

thermoregulatory center—An autonomic nervous center located in the hypothalamus that is responsible for maintaining normal body temperature.

thirst mechanism—A neural mechanism that triggers thirst in response to dehydration.

THR—*See* training heart rate.

threshold—A minimum amount of stimulus needed to elicit a response. Also, the minimum depolarization required to produce an action potential in neurons.

thyrotropin (TSH)—A hormone secreted by the anterior lobe of the pituitary gland that promotes the release of thyroid hormones.

thyroxine (T_4)—A hormone secreted by the thyroid gland that increases the rate of cellular metabolism and the rate and contractility of the heart.

tidal volume—The amount of air inspired or expired during a normal breathing cycle.

titin—A protein that positions the myosin filament to maintain equal spacing between actin filaments.

TLC—*See* total lung capacity.

TLV/RV ratio—The ratio between total lung volume (TLV) and residual volume (RV).

total lung capacity (TLC)—The sum of vital capacity and residual volume.

total peripheral resistance (TPR)—The resistance to the flow of blood through the entire systemic circulation.

trace elements—*See* microminerals.

training effect—Physiological adaptation to repeated bouts of exercise.

training heart rate (THR)—A heart rate goal established by using the heart rate equivalent of a desired percentage of $\dot{V}O_{2max}$. For example, if a training level of 75% $\dot{V}O_{2max}$ is desired, 75% of $\dot{V}O_{2max}$ is calculated, and the heart rate corresponding to this $\dot{V}O_2$ is selected as the THR.

transient hypertrophy—The "pumping up" of muscle that happens during a single exercise bout, resulting mainly from fluid accumulation in the interstitial and intracellular spaces of the muscle.

transverse tubules (T-tubules)—Extensions of the sarcolemma (plasma membrane) that pass laterally through the muscle fiber, allowing nutrients to be transported and nerve impulses to be transmitted rapidly to individual myofibrils.

treadmill—An ergometer in which a motor and pulley system drive a large belt that a person can either walk or run on.

triglycerides—The body's most concentrated energy source and the form in which most fats are stored in the body.

triiodothyronine (T_3)—A hormone released by the thyroid gland that increases the rate of cellular metabolism and the rate and contractility of the heart.

tropomyosin—A tube-shaped protein that twists around actin strands, fitting into the groove between them.

troponin—A complex protein attached at regular intervals to actin strands and tropomyosin.

TSH—*See* thyrotropin.

twitch—the smallest contractile response of a muscle fiber or a motor unit to a single electrical stimulus.

type 1 diabetes—A type of diabetes mellitus that generally has a sudden onset during childhood or young adulthood and leads to almost total insulin deficiency, usually requiring daily insulin injections. Also known as insulin-dependent diabetes mellitus (IDDM) or juvenile-onset diabetes.

type 2 diabetes—A type of diabetes mellitus in which disease onset is more gradual and the causes are more difficult to establish than in type 1 diabetes. Type 2 diabetes is characterized by impaired insulin secretion, impaired insulin action, or excessive glucose output from the liver. Also known as non-insulin-dependent diabetes mellitus (NIDDM).

undertraining—The type of training an athlete would undertake between competitive seasons or during active rest. Generally, physiological adaptations will be minor, and there will be no improvement in performance.

upper body (android) obesity—Obesity that follows the typically male pattern of fat storage, in which fat is stored primarily in the upper body, particularly in the abdomen.

upregulation—An increased cellular sensitivity to a hormone, often caused by increased hormone receptors.

Valsalva maneuver—The process of holding the breath and attempting to compress the contents of the abdominal and thoracic cavities, causing increased intra-abdominal and intrathoracic pressure.

valvular heart disease—A disease involving one or more of the heart valves. Rheumatic heart disease is one example.

variable-resistance training—a technique that allows variation in the resistance applied throughout the range of motion in an attempt to match the ability of the muscle or muscle groups to apply force at any specific point in the range of motion.

vasoconstriction—The constriction or narrowing of blood vessels.

vasodilation—The dilation or widening of blood vessels.

vasopressin—*See* antidiuretic hormone.

VC—*See* vital capacity.

$\dot{V}CO_2$—The volume of CO_2 produced per minute.

$\dot{V}_E$—The volume of air expired per minute.

$\dot{V}_{Emax}$—*See* maximal expiratory ventilation.

$\dot{V}_E/\dot{V}CO_2$—*See* ventilatory equivalent for carbon dioxide.

$\dot{V}_E/\dot{V}O_2$—*See* ventilatory equivalent for oxygen.

veins—Blood vessels that transport blood back to the heart.

ventilatory breakpoint—The point at which ventilation increases disproportionately compared with oxygen consumption.

ventilatory equivalent for carbon dioxide ($\dot{V}_E/\dot{V}CO_2$)—The ratio of the volume of air ventilated ($\dot{V}_E$) to the amount of carbon dioxide produced ($\dot{V}CO_2$).

ventilatory equivalent for oxygen ($\dot{V}_E/\dot{V}O_2$)—The ratio between the volume of air ventilated ($\dot{V}_E$) and the amount of oxygen consumed ($\dot{V}O_2$); indicates breathing economy.

ventilatory threshold—Older name for the ventilatory breakpoint.

ventricular fibrillation—A serious cardiac arrhythmia in which the contraction of the ventricular tissue is uncoordinated, affecting the heart's ability to pump blood. *See also* ventricular tachycardia.

ventricular tachycardia—A serious cardiac arrhythmia consisting of three or more consecutive premature ventricular contractions. *See also* premature ventricular contraction and ventricular fibrillation.

venules—Small vessels that transport blood from the capillaries to the veins and then back to the heart.

very low density lipoprotein (VLDL)—A lipoprotein carrier of cholesterol.

very low density lipoprotein cholesterol (VLDL-C)—The cholesterol carried by VLDL.

vital capacity (VC)—The maximal volume of air expelled from the lungs after maximal inhalation.

vitamin—One of a group of unrelated organic compounds that perform specific functions to promote growth and to maintain health. Vitamins act primarily as catalysts in chemical reactions.

VLDL—*See* very low density lipoprotein.

VLDL-C—*See* very low density lipoprotein cholesterol.

$\dot{V}O_2$—The volume of oxygen consumed per minute.

$\dot{V}O_2$ drift—A slow increase in $\dot{V}O_2$ during prolonged submaximal exercise at a constant power output.

$\dot{V}O_{2max}$—*See* maximal oxygen uptake.

wet-bulb globe temperature (WBGT)—A measurement of temperature that simultaneously accounts for conduction, convection, evaporation, and radiation, providing a single temperature reading to estimate the cooling capacity of the surrounding environment. The apparatus for measuring WBGT consists of a dry bulb, a wet bulb, and a black globe.

windchill—A chill factor created by the increase in the rate of heat loss via convection and conduction caused by wind.

work—Force expressed through distance, or a displacement, independent of time.

References and Selected Readings

Introduction

References

1. Åstrand, P.-O. (1991). Influence of Scandinavian scientists in exercise physiology. *Scandinavian Journal of Medicine and Science in Sports,* **1,** 3-9.
2. Åstrand, P.-O., & Rhyming, I. (1954). A nomogram for calculation of aerobic capacity (physical fitness) from pulse rate during submaximal work. *Journal of Applied Physiology,* **7,** 218-221.
3. Bainbridge, F.A. (1931). *The physiology of muscular exercise.* London: Longmans, Green.
4. Brown, R.C., & Kenyon, G.S. (1968). *Classical studies on physical activity.* Englewood Cliffs, NJ: Prentice Hall.
5. Buskirk, E.R. (1996). Early history of exercise physiology in the United States: Part I. A contemporary historical perspective. In J.D. Messengale & R.A. Swanson (Eds.), *History of exercise and sport science* (pp. 55-74). Champaign, IL: Human Kinetics.
6. Buskirk, E.R., & Taylor, H.L. (1957). Maximal oxygen uptake and its relation to body composition, with special reference to chronic physical activity and obesity. *Journal of Applied Physiology,* **11,** 72-78.
7. Cooper, K.H. (1968). *Aerobics.* New York: Evans.
8. Dill, D.B. (1938). *Life, heat, and altitude.* Cambridge, MA: Harvard University Press.
9. Dill, D.B. (1968). Historical review of exercise physiology science. In R. Warren & R.E. Johnson (Eds.), *Science and medicine of exercise and sports* (2nd ed., pp. 42-48). New York: Harper.
10. Dill, D.B. (1985). *The hot life of man and beast.* Springfield, IL: Charles C Thomas.
11. Fletcher, W.M., & Hopkins, F.G. (1907). Lactic acid in amphibian muscle. *Journal of Physiology,* **35,** 247-254.
12. Flint, A., Jr. (1871). On the physiological effects of severe and protracted muscular exercise; with special reference to the influence of exercise upon the excretion of nitrogen. *New York Medical Journal,* **13,** 609-697.
13. Foster, M. (1970). *Lectures on the history of physiology.* New York: Dover.
14. Horvath, S.M., & Horvath, E.C. (1973). *The Harvard Fatigue Laboratory: Its history and contributors.* Englewood Cliffs, NJ: Prentice Hall.
15. LaGrange, F. (1889). *Physiology of bodily exercise.* London: Kegan Paul International.
16. McArdle, W.D., Katch, F.I., & Katch, V.L. (2001). *Exercise physiology: Energy, nutrition, and human performance* (5th ed.). Baltimore: Williams & Wilkins.
17. Reilly, T., & Brooks, G.A. (1990). Selective persistence of circadian rhythms in physiological responses to exercise. *Chronobiology International,* **7,** 59-67.
18. Robinson, S. (1938). Experimental studies of physical fitness in relation to age. *Arbeitsphysiologie,* **10,** 251-327.
19. Séguin, A., & Lavoisier, A. (1793). Premier mémoire sur la respiration des animaux. *Histoire et Mémoires de l'Academie Royale des Sciences,* **92,** 566-584.
20. Taylor, H.L., Buskirk, E.R., & Henschel, A. (1955). Maximal oxygen intake as an objective measure of cardiorespiratory performance. *Journal of Applied Physiology,* **8,** 73-80.
21. Tipton, C.M. (2003). *Exercise physiology: People and ideas.* New York: Oxford University Press.
22. Zuntz, N., & Schumberg, N.A.E.F. (1901). *Studien Zur Physiologie des Marches* (p. 211). Berlin: A. Hirschwald.

Selected Readings

Beecher, C.E. (1858). *Physiology and calisthenics.* New York: Harper & Brothers.

Bergstrom, J. (1962). Muscle electrolytes in man. *Scandinavian Journal of Clinical Investigation,* **14**(Suppl.), 1-110.

Berryman, J.W. (1995). *Out of many, one: A history of the American College of Sports Medicine.* Champaign, IL: Human Kinetics.

Consolazio, C.F., Johnson, R.E., & Pecora, L.J. (1963). *Physiological measurements of metabolic functions in man.* New York: McGraw-Hill.

Hill, A.V. (1927). *Muscular movement in man: The factors governing speed and recovery from fatigue.* London: McGraw-Hill.

Hill, A.V. (1970). *First and last experiments in muscle mechanics.* Cambridge, UK: Cambridge University Press.

McCurdy, J.H., & Larson, L. (1939). *The physiology of exercise.* Philadelphia: Lea & Febiger.

Park, R.J. (1995). History of research on physical activity and health: Selected topics, 1867 to the 1950s. *Quest,* **47,** 274-287.

Sargent, D.A. (1921). The physical test of a man. *American Physical Education Review,* **26,** 188-194.

Trine, M.R., & Morgan, W.P. (1995). Influence of time of day on psychological responses to exercise. *Sports Medicine,* **20,** 328-337.

Chapter 1

References

1. Brooks, G.A., Fahey, T.D., & Baldwin, K.M. (2005). *Exercise physiology: Human bioenergetics and its applications* (4th ed.). New York: McGraw-Hill.
2. Close, R. (1967). Properties of motor units in fast and slow skeletal muscles of the rat. *Journal of Physiology* (London), **193,** 45-55.
3. Costill, D.L., Daniels, J., Evans, W., Fink, W., Krahenbuhl, G., & Saltin, B. (1976). Skeletal muscle enzymes and fiber composition in male and female track athletes. *Journal of Applied Physiology,* **40,** 149-154.
4. Costill, D.L., Fink, W.J., Flynn, M., & Kirwan, J. (1987). Muscle fiber composition and enzyme activities in elite female distance runners. *International Journal of Sports Medicine,* **8,** 103-106.
5. Costill, D.L., Fink, W.J., & Pollock, M.L. (1976). Muscle fiber composition and enzyme activities of elite distance runners. *Medicine and Science in Sports,* **8,** 96-100.
6. MacIntosh, B.R., Gardiner, P.F., & McComas, A.J. (2006). *Skeletal muscle form and function* (2nd ed.). Champaign, IL: Human Kinetics.

Selected Readings

Andersen, J.L., Schjerling, P., & Saltin, B. (2000). Muscle, genes and athletic performance. *Scientific American,* **283,** 48-55.

Burke, R.E., & Edgerton, V.R. (1975). Motor unit properties and selective involvement in movement. *Exercise and Sport Sciences Reviews,* **3,** 31-81.

Enoka, R.M. (2002). *Neuromechanics of human movement* (3rd ed.). Champaign, IL: Human Kinetics.

Essen-Gustavsson, B., & Borges, O. (1986). Histochemical and metabolic characteristics of human skeletal muscle in relation to age. *Acta Physiologica Scandinavica,* **126,** 107-114.

Gardiner, P.F. (2001). *Neuromuscular aspects of physical activity.* Champaign, IL: Human Kinetics.

Gordon, T., & Pattullo, M.C. (1993). Plasticity of muscle fiber and motor unit types. *Exercise and Sport Sciences Reviews,* **21,** 331-362.

Heckman, C.J., & Sandercock, T.G. (1996). From motor unit to whole muscle properties during locomotor movements. *Exercise and Sport Sciences Reviews,* **24,** 109.

Lindstedt, S.L., LaStayo, P.C., & Reich, T.E. (2001). When active muscles lengthen: Properties and consequences of eccentric contractions. *News in the Physiological Sciences,* **16,** 256-261.

Roy, R.R., Baldwin, K.M., & Edgerton, V.R. (1991). The plasticity of skeletal muscle: Effects of neuromuscular activity. *Exercise and Sport Sciences Reviews,* **19,** 269-312.

Staron, R.S. (1997). Human skeletal muscle fiber types: Delineation, development and distribution. *Canadian Journal of Applied Physiology,* **22,** 307-327.

Chapter 2

Selected Readings

Borer, K.T. (2003). *Exercise endocrinology.* Champaign, IL: Human Kinetics.

Galbo, H. (1983). *Hormonal and metabolic adaptation to exercise.* New York: Thieme-Stratton.

Gastin, P.B. (2001). Energy system interaction and relative contribution during maximal exercise. *Sports Medicine,* **31,** 725-741.

Hargreaves, M. (Ed.). (1995). *Exercise metabolism.* Champaign, IL: Human Kinetics.

Holloszy, J.O., & Hansen, P.A. (1996). Regulation of glucose transport into skeletal muscle. *Reviews of Physiology, Biochemistry and Pharmacology,* **129,** 99-193.

Houston, M.E. (1995). *Biochemistry primer for exercise science.* Champaign, IL: Human Kinetics.

Hultman, E. (1995). Fuel selection, muscle fibre. *Proceedings of the Nutrition Society,* **54,** 107-121.

Katz, A., & Sahlin, K. (1990). Role of oxygen in regulation of glycolysis and lactate production in human skeletal muscle. *Exercise and Sport Sciences Reviews,* **18,** 1-28.

Mooren, F.C., & Volker, K. (Eds.). (2005). *Molecular and cellular exercise physiology.* Champaign, IL: Human Kinetics.

Richter, E.A., & Sutton, J.R. (1994). Hormonal adaptation to physical activity. In C. Bouchard, R. Shephard, & T. Stephens (Eds.), *Physical activity, fitness, and health: International proceedings and consensus statement* (pp. 331-342). Champaign, IL: Human Kinetics.

Chapter 3

References

1. Edstrom, L., & Grimby, L. (1986). Effect of exercise on the motor unit. *Muscle and Nerve,* **9,** 104-126.
2. Guyton, A.C., & Hall, J.E. (2000). *Textbook of medical physiology* (10th ed.). Philadelphia: Saunders.
3. Marieb, E.N. (1995). *Human anatomy and physiology* (3rd ed.). New York: Benjamin/Cummings.
4. Petajan, J.H., Gappmaier, E., White, A.T., Spencer, M.K., Mino, L., & Hicks, R.W. (1996). Impact of aerobic training on fitness and quality of life in multiple sclerosis. *Annals of Neurology,* **39,** 432-441.
5. Pette, D., & Vrbova, G. (1985). Neural control of phenotypic expression in mammalian muscle fibers. *Muscle and Nerve,* **8,** 676-689.

Selected Readings

Bawa, P. (2002). Neural control of motor output: Can training change it? *Exercise and Sport Sciences Reviews,* **30,** 59-63.

Binder, M.D., Heckman, C.J., & Powers, R.K. (1996). The physiological control of motoneuron activity. In L.B. Rowell & J.T. Shepherd (Eds.), *Handbook of physiology:*

Section 12. Exercise: Regulation and integration of multiple systems (pp. 3-53). New York: Oxford University Press.

Cope, T.C., & Pinter, M.J. (1995). The size principle: Still working after all these years. *News in the Physiological Sciences,* **10,** 280-286.

Edgerton, V.R., Bodine-Fowler, S., Roy, R.R., Ishihara, A., & Hodgson, J.A. (1996). Neuromuscular adaptation. In L.B. Rowell & J.T. Shepherd (Eds.), *Handbook of physiology: Section 12. Exercise: Regulation and integration of multiple systems* (pp. 54-88). New York: Oxford University Press.

Enoka, R.M. (2002). *Neuromechanics of human movement* (3rd ed.). Champaign, IL: Human Kinetics.

Gandevia, S.C. (1996). Kinesthesia: Roles for afferent signals and motor commands. In L.B. Rowell & J.T. Shepherd (Eds.), *Handbook of physiology: Section 12. Exercise: Regulation and integration of multiple systems* (pp. 128-172). New York: Oxford University Press.

Henneman, E., & Mendell, L.M. (1981). Functional organization of motoneuron pool and its inputs. In V.B. Brooks (Ed.), *Handbook of physiology: Section 1. The nervous system: Motor control, part I* (Vol. 2, pp. 423-507). Bethesda, MD: American Physiological Society.

Prochazka, A. (1996). Proprioceptive feedback and movement regulation. In L.B. Rowell & J.T. Shepherd (Eds.), *Handbook of physiology: Section 12. Exercise: Regulation and integration of multiple systems* (pp. 89-127). New York: Oxford University Press.

Roy, R.R., Baldwin, K.M., & Edgerton, V.R. (1991). The plasticity of skeletal muscle: Effects of neuromuscular activity. *Exercise and Sport Sciences Reviews,* **19,** 269-312.

Seals, D.R., & Victor, R.G. (1991). Regulation of muscle sympathetic nerve activity during exercise in humans. *Exercise and Sport Sciences Reviews,* **19,** 313-349.

Chapter 4

References

1. Bar-Or, O. (1987). The Wingate Anaerobic Test: An update on methodology, reliability and validity. *Sports Medicine,* **4,** 381-394.
2. Barstow, T.J., Jones, A.M., Nguyen, P.H., & Casaburi, R. (1996). Influence of muscle fiber type and pedal frequency on oxygen uptake kinetics of heavy exercise. *Journal of Applied Physiology,* **81,** 1642-1650.
3. Costill, D.L. (1986). *Inside running: Basics of sports physiology.* Indianapolis: Benchmark Press.
4. Gaesser, G.A., & Poole, D.C. (1996). The slow component of oxygen uptake kinetics in humans. *Exercise and Sport Sciences Reviews,* **24,** 35-70.
5. Galloway, S.D.R., & Maughan, R.J. (1997). Effects of ambient temperature on the capacity to perform prolonged cycle exercise in man. *Medicine and Science in Sports and Exercise,* **29,** 1240-1249.
6. Hawley, J.A. (1997). Carbohydrate loading and exercise performance: An update. *Sports Medicine,* **24,** 73-81.
7. Hill, D.W. (1993). The critical power concept: A review. *Sports Medicine,* **16,** 237-254.
8. Medbø, J.I., Mohn, A.C., Tabata, I., Bahr, R., Vaage, O., & Sejersted, O.M. (1988). Anaerobic capacity determined by maximal accumulated O_2 deficit. *Journal of Applied Physiology,* **64,** 50-60.
9. Westerblad, H., Allen, D.G., & Lännergren, J. (2002). Muscle fatigue: Lactic acid or inorganic phosphate the major cause? *News in the Physiological Sciences,* **17,** 17-21.
10. Zuntz, N., & Hagemann, O. (1898). *Untersuchungen uber den Stroffwechsel des Pferdes bei Ruhe und Arbeit.* Berlin: Parey.

Selected Readings

Bangsbo, J. (1998). Quantification of anaerobic energy production during intense exercise. *Medicine and Science in Sports and Exercise,* **30,** 47-52.

Bergstrom, J. (1967). Local changes of ATP and phosphocreatine in human muscle tissue in connection with exercise. In *Physiology of muscular exercise* (Monograph No. 15, pp. 191-196). New York: American Heart Association.

Costill, D.L., Coyle, E., Dalsky, G., Evans, W., Fink, W., & Hoopes, D. (1977). Effects of elevated plasma FFA and insulin on muscle glycogen usage during exercise. *Journal of Applied Physiology,* **43,** 695-699.

Fitts, R.H. (1994). Cellular mechanisms of muscle fatigue. *Physiological Reviews,* **74,** 49-84.

Gladden, L.B. (2000). Muscle as a consumer of lactate. *Medicine and Science in Sports and Exercise,* **32,** 764-771.

Hargreaves, M., & Spriet, L. (2006). Exercise metabolism (2nd ed.). Champaign, IL: Human Kinetics.

Holloszy, J.O., & Hansen, P.A. (1996). Regulation of glucose transport into skeletal muscle. *Reviews of Physiology, Biochemistry and Pharmacology,* **129,** 99-193.

Holloszy, J.O., & Kohrt, W.M. (1996). Regulation of carbohydrate and fat metabolism during and after exercise. *Annual Review of Nutrition,* **16,** 121-138.

MacIntosh, B.R., & Rassier, D.E. (2002). What is fatigue? *Canadian Journal of Applied Physiology,* **27,** 42-55.

Richardson, R.S. (1998). Oxygen transport: Air to muscle cell. *Medicine and Science in Sports and Exercise,* **30,** 53-59.

Romijn, J.A., Coyle, E.F., Sidossis, L.S., Zhang, X.-J., & Wolfe, R.R. (1995). Relationship between fatty acid delivery and fatty acid oxidation during strenuous exercise. *Journal of Applied Physiology,* **79,** 1939-1945.

Chapter 5

Selected Readings

Buckwalter, J.B., & Clifford, P.S. (2001). The paradox of sympathetic vasoconstriction in exercising skeletal muscle. *Exercise and Sport Sciences Reviews,* **29,** 159-163.

Green, D.J., O'Driscoll, G., Blanksby, B.A., & Taylor, R.R. (1996). Control of skeletal muscle blood flow during dynamic exercise: Contribution of endothelium-derived nitric oxide. *Sports Medicine,* **21,** 119-146.

Rowell, L.B. (1986). *Human circulation: Regulation during physical stress.* New York: Oxford University Press.

Rowell, L.B. (1993). *Human cardiovascular control.* New York: Oxford University Press.

Rowell, L.B. (2004). Ideas about control of skeletal and cardiac muscle blood flow (1876-2003): Cycles of revision and new vision. *Journal of Applied Physiology,* **97,** 384-392.

Saltin, B., Bouschel, R., Secher, N., & Mitchell, J. (Eds.). (2000). *Exercise and circulation in health and disease.* Champaign, IL: Human Kinetics.

Silverthorn, D.U. (2007). *Human physiology: An integrated approach* (4th ed.). Upper Saddle River, NJ: Benjamin Cummings.

Chapter 6

Selected Reading

Dempsey, J.A. (2006). Challenges for future research in exercise physiology as applied to the respiratory system. *Exercise and Sport Sciences Reviews,* **34,** 92-98.

Chapter 7

References

1. Hermansen, L. (1981). Effect of metabolic changes on force generation in skeletal muscle during maximal exercise. In R. Porter & J. Whelan (Eds.), *Human muscle fatigue: Physiological mechanisms* (pp. 75-88). London: Pitman Medical.
2. McKirnan, M.D., Gray, C.G., & White, F.C. (1991). Effects of feeding on muscle blood flow during prolonged exercise in miniature swine. *Journal of Applied Physiology,* **70,** 1097-1104.
3. Poliner, L.R., Dehmer, G.J., Lewis, S.E., Parkey, R.W., Blomqvist, C.G., & Willerson, J.T. (1980). Left ventricular performance in normal subjects: A comparison of the responses to exercise in the upright and supine position. *Circulation,* **62,** 528-534.
4. Powers, S.K., Martin, D., & Dodd, S. (1993). Exercise-induced hypoxaemia in elite endurance athletes: Incidence, causes and impact on $\dot{V}O_2$max. *Sports Medicine,* **16,** 14-22.
5. Tanaka, H., Monahan, D.K., & Seals, D.R. (2001). Age-predicted maximal heart rate revisited. *Journal of the American College of Cardiology,* **37,** 153-156.
6. Turkevich, D., Micco, A., & Reeves, J.T. (1988). Noninvasive measurement of the decrease in left ventricular filling time during maximal exercise in normal subjects. *American Journal of Cardiology,* **62,** 650-652.
7. Wasserman, K., & McIlroy, M.B. (1964). Detecting the threshold of anaerobic metabolism in cardiac patients during exercise. *American Journal of Cardiology,* **14,** 844-852.
8. Zhou, B., Conlee, R.K., Jensen, R., Fellingham, G.W., George, J.D., & Fisher, A.G. (2001). Stroke volume does not plateau during graded exercise in elite male distance runners. *Medicine and Science in Sports and Exercise,* **33,** 1849-1854.

Selected Readings

Buckwalter, J.B., & Clifford, P.S. (2001). The paradox of sympathetic vasoconstriction in exercising skeletal muscle. *Exercise and Sport Sciences Reviews,* **29,** 159-163.

Dempsey, J.A. (2006). Challenges for future research in exercise physiology as applied to the respiratory system. *Exercise and Sport Sciences Reviews,* **34,** 92-98.

Forster, H.V. (2000). Exercise hyperpnea: Where do we go from here? *Exercise and Sport Sciences Reviews,* **28,** 133-137.

Hughson, R.L., & Tschakovsky, M.E. (1999). Cardiovascular dynamics at the onset of exercise. *Medicine and Science in Sports and Exercise,* **31,** 1005-1010.

Laughlin, M.H., & Armstrong, R.B. (1985). Muscle blood flow during locomotor exercise. *Exercise and Sport Sciences Reviews,* **13,** 95-136.

Raven, P.B., Potts, J.T., & Shi, X. (1997). Baroreflex regulation of blood pressure during dynamic exercise. *Exercise and Sport Sciences Reviews,* **25,** 368-389.

Rowell, L.B. (1986). *Human circulation: Regulation during physical stress.* New York: Oxford University Press.

Saltin, B., Bouschel, R., Secher, N., & Mitchell, J. (Eds.). (2000). *Exercise and circulation in health and disease.* Champaign, IL: Human Kinetics.

Saltin, B., & Rowell, L.B. (1980). Functional adaptations to physical activity and inactivity. *Federation Proceedings,* **39,** 1506-1513.

Senay, L.C., Jr., & Pivarnik, J.M. (1985). Fluid shifts during exercise. *Exercise and Sport Sciences Reviews,* **13,** 335-387.

Chapter 8

References

1. American College of Sports Medicine. (2002). ACSM position stand: Progression models in resistance training for healthy adults. *Medicine and Science in Sports and Exercise,* **34,** 364-380.
2. Baechle, T.R., & Earle, R.W. (Eds.). (2000). *Essentials of strength training and conditioning* (2nd ed.). Champaign, IL: Human Kinetics.
3. Fleck, S.J., & Kraemer, W.J. (2004). *Designing resistance training programs* (3rd ed.). Champaign, IL: Human Kinetics.
4. Fox, E.L., & Mathews, D.K. (1974). Interval training conditioning for sports and general fitness. Philadelphia: Saunders.
5. Kraemer, W.J., & Ratamess, N.A. (2004). Fundamentals of resistance training: Progression and exercise prescription. *Medicine and Science in Sports and Exercise,* **36,** 674-688.
6. Stone, M.H., O'Bryant, H.S., & Garhammer, J. (1981). A hypothetical model for strength training. *Journal of Sports Medicine and Physical Fitness,* **21**(4): 336, 342-351.

Selected Readings

Bird, S.P., Tarpenning, K.M., & Marino, F.E. (2005). Designing resistance training programmes to enhance muscular fitness: A review of the acute programme variables. *Sports Medicine,* **35,** 841-851.

Chu, D.A. (1998). *Jumping into plyometrics* (2nd ed.). Champaign, IL: Human Kinetics.

Crewther, B., Cronin, J., & Keogh, J. (2005). Possible stimuli for strength and power adaptation. *Sports Medicine,* **35,** 967-989.

Komi, P.V. (Ed.). (1992). *Strength and power in sport.* Boston: Blackwell Scientific.

Rhea, M.R., Alvar, B.A., Burkett, L.N., & Ball, S.D. (2003). A meta-analysis to determine the dose response for strength development. *Medicine and Science in Sports and Exercise,* **35,** 456-464.

Chapter 9

References

1. Aagaard, P., Simonsen, E.B., Andersen, J.L., Magnusson, S.P., Halkjaer-Kristensen, J., & Dyhre-Poulsen, P. (2000). Neural inhibition during maximal eccentric and concentric quadriceps contraction: Effects of resistance training. *Journal of Applied Physiology,* **89,** 2249-2257.
2. American College of Sports Medicine. (2002). ACSM position stand: Progression models in resistance training for healthy adults. *Medicine and Science in Sports and Exercise,* **34,** 364-380.
3. Andersen, J.L., Schjerling, P., & Saltin, B. (2000). Muscle, genes and athletic performance. *Scientific American,* **283,** 48-55.
4. Appell, H.-J. (1990). Muscular atrophy following immobilisation: A review. *Sports Medicine,* **10,** 42-58.
5. Armstrong, R.B. (1984). Mechanisms of exercise-induced delayed-onset muscular soreness: A brief review. *Medicine and Science in Sports and Exercise,* **16,** 529-538.
6. Armstrong, R.B., Warren, G.L., & Warren, J.A. (1991). Mechanisms of exercise-induced muscle fibre injury. *Sports Medicine,* **12,** 184-207.
7. Duchateau, J., & Enoka, R.M. (2002). Neural adaptations with chronic activity patterns in able-bodied humans. *American Journal of Physical Medicine and Rehabilitation,* **81**(11 Suppl.), 517-527.
8. Ebbeling, C.B., & Clarkson, P.M. (1989). Exercise-induced muscle damage and adaptation. *Sports Medicine,* **7,** 207-234.
9. Enoka, R.M. (1988). Muscle strength and its development: New perspectives. *Sports Medicine,* **6,** 146-168.
10. Enoka, R.M. (1997). Neural adaptations with chronic physical activity. *Journal of Biomechanics,* **30,** 447-455.
11. Gonyea, W.J. (1980). Role of exercise in inducing increases in skeletal muscle fiber number. *Journal of Applied Physiology,* **48,** 421-426.
12. Gonyea, W.J., Sale, D.G., Gonyea, F.B., & Mikesky, A. (1986). Exercise induced increases in muscle fiber number. *European Journal of Applied Physiology,* **55,** 137-141.
13. Graves, J.E., Pollock, M.L., Leggett, S.H., Braith, R.W., Carpenter, D.M., & Bishop, L.E. (1988). Effect of reduced training frequency on muscular strength. *International Journal of Sports Medicine,* **9,** 316-319.
14. Green, H.J., Klug, G.A., Reichmann, H., Seedorf, U., Wiehrer, W., & Pette, D. (1984). Exercise-induced fibre type transitions with regard to myosin, parvalbumin, and sarcoplasmic reticulum in muscles of the rat. *Pflugers Archiv: European Journal of Physiology,* **400,** 432-438.
15. Hagerman, F.C., Hikida, R.S., Staron, R.S., Sherman, W.M., & Costill, D.L. (1984). Muscle damage in marathon runners. *Physician and Sportsmedicine,* **12,** 39-48.
16. Hakkinen, K., Alen, M., & Komi, P.V. (1985). Changes in isometric force and relaxation-time, electromyographic and muscle fibre characteristics of human skeletal muscle during strength training and detraining. *Acta Physiologica Scandinavica,* **125,** 573-585.
17. Hawke, T.J., & Garry, D.J. (2001). Myogenic satellite cells: Physiology to molecular biology. *Journal of Applied Physiology,* **91,** 534-551.
18. Kraemer, W.J. (2000). Physiological adaptations to anaerobic and aerobic endurance training programs. In T.R. Baechle & R.W. Earle (Eds.), *Essentials of strength training and conditioning* (2nd ed., p. 150). Champaign, IL: Human Kinetics.
19. McCall, G.E., Byrnes, W.C., Dickinson, A., Pattany, P.M., & Fleck, S.J. (1996). Muscle fiber hypertrophy, hyperplasia, and capillary density in college men after resistance training. *Journal of Applied Physiology,* **81,** 2004-2012.
20. Schwane, J.A., Johnson, S.R., Vandenakker, C.B., & Armstrong, R.B. (1983). Delayed-onset muscular soreness and plasma CPK and LDH activities after downhill running. *Medicine and Science in Sports and Exercise,* **15,** 51-56.
21. Schwane, J.A., Watrous, B.G., Johnson, S.R., & Armstrong, R.B. (1983). Is lactic acid related to delayed-onset muscle soreness? *Physician and Sportsmedicine,* **11**(3), 124-131.
22. Schwellnus, M.P. (1999). Skeletal muscle cramps during exercise. *Physician and Sportsmedicine,* **27**(12), 109-115.
23. Shepstone, T.N., Tang, J.E., Dallaire, S., Schuenke, M.D., Staron, R.S., & Phillips, S.M. (2005). Short-term high- vs. low-velocity isokinetic lengthening training results in greater hypertrophy of the elbow flexors in young men. *Journal of Applied Physiology,* **98,** 1768-1776.
24. Sjöström, M., Lexell, J., Eriksson, A., & Taylor, C.C. (1991). Evidence of fibre hyperplasia in human skeletal muscles from healthy young men? A left-right comparison of the fibre number in whole anterior tibialis muscles. *European Journal of Applied Physiology,* **62,** 301-304.
25. Staron, R.S., Karapondo, D.L., Kraemer, W.J., Fry, A.C., Gordon, S.E., Falkel, J.E., Hagerman, F.C., & Hikida, R.S. (1994). Skeletal muscle adaptations during early phase of heavy resistance training in men and women. *Journal of Applied Physiology,* **76,** 1247-1255.

26. Staron, R.S., Leonardi, M.J., Karapondo, D.L., Malicky, E.S., Falkel, J.E., Hagerman, F.C., & Hikida, R.S. (1991). Strength and skeletal muscle adaptations in heavy-resistance-trained women after detraining and retraining. *Journal of Applied Physiology,* **70,** 631-640.
27. Staron, R.S., Malicky, E.S., Leonardi, M.J., Falkel, J.E., Hagerman, F.C., & Dudley, G.A. (1990). Muscle hypertrophy and fast fiber type conversions in heavy-resistance-trained women. *European Journal of Applied Physiology,* **60,** 71-79.
28. Talag, T.S. (1973). Residual muscular soreness as influenced by concentric, eccentric and static contractions. *Research Quarterly,* **44,** 458-469.
29. Tidball, J.G. (1995). Inflammatory cell response to acute muscle injury. *Medicine and Science in Sports and Exercise,* **27,** 1022-1032.
30. Warren, G.L., Ingalls, C.P., Lowe, D.A., & Armstrong, R.B. (2001). Excitation-contraction uncoupling: Major role in contraction-induced muscle injury. *Exercise and Sport Sciences Reviews,* **29,** 82-87.

Selected Readings

Aagaard, P. (2003). Training-induced changes in neural function. *Exercise and Sport Sciences Reviews,* **31,** 61-67.

Armstrong, R.B. (1986). Muscle damage and endurance events. *Sports Medicine,* **3,** 370-381.

Carroll, T.J., Riek, S., & Carson, R.G. (2001). Neural adaptations to resistance training. *Sports Medicine,* **31,** 829-840.

Fry, A.C. (2004). The role of resistance exercise intensity on muscle fibre adaptations. *Sports Medicine,* **34,** 663-679.

Gabriel, D.A., Kamen, G., & Frost, G. (2006). Neural adaptations to resistive exercise: Mechanisms and recommendations for training procedures. *Sports Medicine,* **36,** 133-149.

Hawke, T.J. (2005). Muscle stem cells and exercise training. *Exercise and Sport Sciences Reviews,* **33,** 63-68.

Kendall, B., & Eston, R. (2002). Exercise-induced muscle damage and the potential protective role of estrogen. *Sports Medicine,* **32,** 103-123.

Kendall, T.L., Black, C.D., Elder, C.P., Gorgey, A., & Dudley, G.A. (2006). Determining the extent of neural activation during maximal effort. *Medicine and Science in Sports and Exercise,* **38,** 1470-1475.

Kraemer, W.H., & Ratamess, N.A. (2005). Hormonal responses and adaptations to resistance exercise and training. *Sports Medicine,* **35,** 339-361.

Proske, U., & Allen, T.J. (2005). Damage to skeletal muscle from eccentric exercise. *Exercise and Sport Sciences Reviews,* **33,** 98-104.

Chapter 10

References

1. Armstrong, R.B., & Laughlin, M.H. (1984). Exercise blood flow patterns within and among rat muscles after training. *American Journal of Physiology,* **246,** H59-H68.
2. Bouchard, C. (1990). Discussion: Heredity, fitness, and health. In C. Bouchard, R.J. Shephard, T. Stephens, J.R. Sutton, & B.D. McPherson (Eds.), *Exercise, fitness, and health* (pp. 147-153). Champaign, IL: Human Kinetics.
3. Bouchard, C., An, P., Rice, T., Skinner, J.S., Wilmore, J.H., Gagnon, J., Pérusse, L., Leon, A.S., & Rao, D.C. (1999). Familial aggregation of $\dot{V}O_2max$ response to exercise training: Results from the HERITAGE Family Study. *Journal of Applied Physiology,* **87,** 1003-1008.
4. Bouchard, C., Dionne, F.T., Simoneau, J.-A., & Boulay, M.R. (1992). Genetics of aerobic and anaerobic performances. *Exercise and Sport Sciences Reviews,* **20,** 27-58.
5. Bouchard, C., Lesage, R., Lortie, G., Simoneau, J.A., Hamel, P., Boulay, M.R., Pérusse, L., Theriault, G., & Leblanc, C. (1986). Aerobic performance in brothers, dizygotic and monozygotic twins. *Medicine and Science in Sports and Exercise,* **18,** 639-646.
6. Costill, D.L., Coyle, E.F., Fink, W.F., Lesmes, G.R., & Witzmann, F.A. (1979). Adaptations in skeletal muscle following strength training. *Journal of Applied Physiology: Respiratory Environmental Exercise Physiology,* **46,** 96-99.
7. Costill, D.L., Fink, W.J., Ivy, J.L., Getchell, L.H., & Witzmann, F.A. (1979). Lipid metabolism in skeletal muscle of endurance-trained males and females. *Journal of Applied Physiology,* **28,** 251-255.
8. Dempsey, J.A. (1986). Is the lung built for exercise? *Medicine and Science in Sports and Exercise,* **18,** 143-155.
9. Ehsani, A.A., Ogawa, T., Miller, T.R., Spina, R.J., & Jilka, S.M. (1991). Exercise training improves left ventricular systolic function in older men. *Circulation,* **83,** 96-103.
10. Ekblom, B., Goldbarg, A.M., & Gullbring, B. (1972). Response to exercise after blood loss and reinfusion. *Journal of Applied Physiology,* **33,** 175-180.
11. Fagard, R.H. (1996). Athlete's heart: A meta-analysis of the echocardiographic experience. *International Journal of Sports Medicine,* **17,** S140-S144.
12. Hagberg, J.M., Ehsani, A.A., Goldring, D., Hernandez, A., Sinacore, D.R., & Holloszy, J.O. (1984). Effect of weight training on blood pressure and hemodynamics in hypertensive adolescents. *Journal of Pediatrics,* **104,** 147-151.
13. Hermansen, L., & Wachtlova, M. (1971). Capillary density of skeletal muscle in well-trained and untrained men. *Journal of Applied Physiology,* **30,** 860-863.
14. Holloszy, J.O., Oscai, L.B., Mole, P.A., & Don, I.J. (1971). Biochemical adaptations to endurance exercise in skeletal muscle. In B. Pernow & B. Saltin (Eds.), *Muscle metabolism during exercise* (pp. 51-61). New York: Plenum Press.
15. Jacobs, I., Esbjörnsson, M., Sylvén, C., Holm, I., & Jansson, E. (1987). Sprint training effects on muscle myoglobin, enzymes, fiber types, and blood lactate. *Medicine and Science in Sports and Exercise,* **19,** 368-374.
16. Jansson, E., Esbjörnsson, M., Holm, I., & Jacobs, I. (1990). Increase in the proportion of fast-twitch muscle fibres by sprint training in males. *Acta Physiologica Scandinavica,* **140,** 359-363.

17. Klissouras, V. (1971). Adaptability of genetic variation. *Journal of Applied Physiology,* **31,** 338-344.

18. MacDougall, J.D., Hicks, A.L., MacDonald, J.R., McKelvie, R.S., Green, H.J., & Smith, K.M. (1998). Muscle performance and enzymatic adaptations to sprint interval training. *Journal of Applied Physiology,* **84,** 2138-2142.

19. Martino, M., Gledhill, N., & Jamnik, V. (2002). High $\dot{V}O_2$max with no history of training is primarily due to high blood volume. *Medicine and Science in Sports and Exercise,* **34,** 966-971.

20. McCarthy, J.P., Pozniak, M.A., & Agre, J.C. (2002). Neuromuscular adaptations to concurrent strength and endurance training. *Medicine and Science in Sports and Exercise,* **34,** 511-519.

21. McGuire, D.K., Levine, B.D., Williamson, J.W., Snell, P.G., Blomqvist, C.G., Saltin, B., & Mitchell, J.H. (2001). A 30-year follow-up of the Dallas Bedrest and Training Study: II. Effect of age on cardiovascular adaptation to exercise training. *Circulation,* **104,** 1358-1366.

22. Milliken, M.C., Stray-Gundersen, J., Peshock, R.M., Katz, J., & Mitchell, J.H. (1988). Left ventricular mass as determined by magnetic resonance imaging in male endurance athletes. *American Journal of Cardiology,* **62,** 301-305.

23. Pirnay, F., Dujardin, J., Deroanne, R., & Petit, J.M. (1971). Muscular exercise during intoxication by carbon monoxide. *Journal of Applied Physiology,* **31,** 573-575.

24. Prud'homme, D., Bouchard, C., LeBlanc, C., Landrey, F., & Fontaine, E. (1984). Sensitivity of maximal aerobic power to training is genotype-dependent. *Medicine and Science in Sports and Exercise,* **16,** 489-493.

25. Rico-Sanz, J., Rankinen, T., Joanisse, D.R., Leon, A.S., Skinner, J.S., Wilmore, J.H., Rao, D.C., & Bouchard, C. (2003). Familial resemblance for muscle phenotypes in The Heritage Family Study. *Medicine and Science in Sports and Exercise,* **35**(8): 1360-1366.

26. Saltin, B., Nazar, K., Costill, D.L., Stein, E., Jansson, E., Essen, B., & Gollnick, P.D. (1976). The nature of the training response: Peripheral and central adaptations to one-legged exercise. *Acta Physiologica Scandinavica,* **96,** 289-305.

27. Saltin, B., & Rowell, L.B. (1980). Functional adaptations to physical activity and inactivity. *Federation Proceedings,* **39,** 1506-1513.

28. Sawka, M.N., Convertino, V.A., Eichner, E.R., Schnieder, S.M., & Young, A.J. (2000). Blood volume: Importance and adaptations to exercise training, environmental stresses, and trauma/sickness. *Medicine and Science in Sports and Exercise,* **32,** 332-348.

29. Strømme, S.B., Ingjer, F., & Meen, H.D. (1977). Assessment of maximal aerobic power in specifically trained athletes. *Journal of Applied Physiology,* **42,** 833-837.

30. Wilmore, J.H., Stanforth, P.R., Gagnon, J., Rice, T., Mandel, S., Leon, A.S., Rao, D.C., Skinner, J.S., & Bouchard, C. (2001). Cardiac output and stroke volume changes with endurance training: The HERITAGE Family Study. *Medicine and Science in Sports and Exercise,* **33,** 99-106.

31. Wilmore, J.H., Stanforth, P.R., Hudspeth, L.A., Gagnon, J., Daw, E.W., Leon, A.S., Rao, D.C., Skinner, J.S., & Bouchard, C. (1998). Alterations in resting metabolic rate as a consequence of 20 wk of endurance training: The HERITAGE Family Study. *American Journal of Clinical Nutrition,* **68,** 66-71.

Selected Readings

Convertino, V.A. (1991). Blood volume: Its adaptation to endurance training. *Medicine and Science in Sports and Exercise,* **23,** 1338-1348.

Joyner, M.J., & Shastry, S. (2000). Vascular endothelial growth factor and capillary density in exercise training. *Exercise and Sport Sciences Reviews,* **28,** 97-98.

Lash, J.M. (1998). Training-induced alterations in contractile function and excitation-contraction coupling in vascular smooth muscle. *Medicine and Science in Sports and Exercise,* **30,** 60-66.

MacRae, H.S.-H., Dennis, S.C., Bosch, A.N., & Noakes, T.D. (1992). Effects of training on lactate production and removal during progressive exercise in humans. *Journal of Applied Physiology,* **72,** 1649-1656.

Mier, C.M., Turner, M.J., Ehsani, A.A., & Spina, R.J. (1997). Cardiovascular adaptations to 10 days of cycle exercise. *Journal of Applied Physiology,* **83,** 1900-1906.

Moore, R.L., & Korzick, D.H. (1995). Cellular adaptations of the myocardium to chronic exercise. *Progress in Cardiovascular Diseases,* **37,** 371-396.

Perrault, H., & Turcotte, R.A. (1994). Exercise-induced cardiac hypertrophy: Fact or fallacy? *Sports Medicine,* **17,** 288-308.

Rowell, L.B. (1986). *Human circulation regulation during physical stress.* New York: Oxford University Press.

Saltin, B. (1990). Cardiovascular and pulmonary adaptation to physical activity. In C. Bouchard, R.J. Shephard, T. Stephens, J.R. Sutton, & B.D. McPherson (Eds.), *Exercise, fitness, and health* (pp. 187-203). Champaign, IL: Human Kinetics.

Sexton, W.L., Korthuis, R.J., & Laughlin, M.H. (1988). High-intensity exercise training increases vascular transport capacity of rat hindquarters. *American Journal of Physiology,* **254,** H274-H278.

Chapter 11

References

1. American College of Sports Medicine. (2007). Exertional heat illness during training and competition. *Medicine and Science in Sports and Exercise,* **39**(3), 556-572.

2. American College of Sports Medicine. (2006). Prevention of cold injuries during exercise. *Medicine and Science in Sports and Exercise,* **38**(11), 2012-2029.

3. Febbraio, M.A. (2000). Does muscle function and metabolism affect exercise performance in the heat? *Exercise and Sport Sciences Reviews,* **28,** 171-176.

4. Fink, W., Costill, D.L., Van Handel, P., & Getchell, L. (1975). Leg muscle metabolism during exercise in the heat and cold. *European Journal of Applied Physiology,* **34,** 183-190.

5. Gisolfi, C.V., & Wenger, C.B. (1984). Temperature regulation during exercise: Old concepts, new ideas. *Exercise and Sport Sciences Reviews,* **12,** 339-372.

6. King, D.S., Costill, D.L., Fink, W.J., Hargreaves, M., & Fielding, R.A. (1985). Muscle metabolism during exercise in the heat in unacclimatized and acclimatized humans. *Journal of Applied Physiology,* **59,** 1350-1354.

7. Rowell, L.B. (1974). Human cardiovascular adjustments to heat stress. *Physiological Reviews,* **54,** 75-159.

8. Webb, P. (1951). Air temperature in respiratory tracts of resting subjects in the cold. *Journal of Applied Physiology,* **4,** 378-382.

9. Young, A.J. (1996). Homeostatic responses to prolonged cold exposure: Human cold acclimation. In M.J. Fregley & C.M. Blatteis (Eds.), *Handbook of physiology: Section 4. Environmental physiology* (pp. 419-438). New York: Oxford University Press.

Selected Readings

Armstrong, L.E. (2000). *Performing in extreme environments.* Champaign, IL: Human Kinetics.

Cheuvront, S.N., & Haymes, E.M. (2001). Thermoregulation and marathon running. *Sports Medicine,* **31,** 743-762.

Doubt, T.J. (1991). Physiology of exercise in the cold. *Sports Medicine,* **11,** 367-381.

Gonzalez-Alonso, J., Calbet, J.A.L., & Nielsen, B. (1999). Metabolic and thermodynamic responses to dehydration-induced reductions in muscle blood flow in exercising humans. *Journal of Physiology,* **520,** 577-589.

Inter-Association Task Force on Exertional Heat Illnesses. (2003). Inter-association task force on exertional heat illnesses consensus statement. *NATA News* 6.03: 24-29.

Kenney, W.L. (1997). Thermoregulation at rest and during exercise in healthy older adults. *Exercise and Sport Sciences Reviews,* **25,** 41-76.

Kenney, W.L., & Munce, T.A. (2003). Aging and human thermoregulation. *Journal of Applied Physiology,* **95,** 2598-2603.

Nadel, E.R. (Ed.). (1977). *Problems with temperature regulation during exercise.* New York: Academic Press.

Pandolf, K.B., Sawka, M.N., & Gonzalez, R.R. (Eds.). (1988). *Human performance physiology and environmental medicine at terrestrial extremes.* Indianapolis: Benchmark Press.

Young, A.J. (1990). Energy substrate utilization during exercise in extreme environments. *Exercise and Sport Sciences Reviews,* **18,** 65-117.

Chapter 12

References

1. Brooks, G.A., Wolfel, E.E., & Groves, B.M. (1992). Muscle accounts for glucose disposal but not blood lactate appearance during exercise after acclimatization to 4,300 m. *Journal of Applied Physiology,* **72,** 2435-2445.

2. Brosnan, M.J., Martin, D.T., Hahn, A.G., Gore, C.J., & Hawley, J.A. (2000). Impaired interval exercise responses in elite female cyclists at moderate simulated altitude. *Journal of Applied Physiology,* **89,** 1819-1824.

3. Buskirk, E.R., Kollias, J., Piconreatigue, E., Akers, R., Prokop, E., & Baker, P. (1967). Physiology and performance of track athletes at various altitudes in the United States and Peru. In R.F. Goddard (Ed.), *The effects of altitude on physical performance* (pp. 65-71). Chicago: Athletic Institute.

4. Daniels, J., & Oldridge, N. (1970). Effects of alternate exposure to altitude and sea level on world-class middle-distance runners. *Medicine and Science in Sports,* **2,** 107-112.

5. Forster, P.J.G. (1985). Effect of different ascent profiles on performance at 4200 m elevation. *Aviation, Space, and Environmental Medicine,* **56,** 785-794.

6. Levine, B.D., & Stray-Gundersen, J. (1997). "Living high–training low": Effect of moderate-altitude acclimatization with low-altitude training on performance. *Journal of Applied Physiology,* **83,** 102-112.

7. Norton, E.G. (1925). *The fight for Everest: 1924.* London: Arnold.

8. Pugh, L.C.G.E., Gill, M., Lahiri, J., Milledge, J., Ward, M., & West, J. (1964). Muscular exercise at great altitudes. *Journal of Applied Physiology,* **19,** 431-440.

9. Stray-Gundersen, J., Chapman, R.F., & Levine, B.D. (2001). "Living high–training low" altitude training improves sea level performance in male and female elite runners. *Journal of Applied Physiology,* **91,** 1113-1120.

10. Sutton, J., & Lazarus, L. (1973). Mountain sickness in the Australian Alps. *Medical Journal of Australia,* **1,** 545-546.

11. Sutton, J.R., Reeves, J.T., Wagner, P.D., Groves, B.M., Cymerman, A., Malconian, M.K., Rock, P.B., Young, P.M., Walter, S.D., & Houston, C.S. (1988). Operation Everest II: Oxygen transport during exercise at extreme simulated altitude. *Journal of Applied Physiology,* **64,** 1309-1321.

12. Ward, M.P., Milledge, J.S., & West, J.B. (1989). *High altitude medicine and physiology.* Philadelphia: University of Pennsylvania Press.

13. West, J.B., Boyer, S.J., Graber, D.J., Hackett, P.M., Maret, K.M., Milledge, J.S., Peters, R.M., Pizzo, C.J., Samata, M., Sarnquist, F.N., Schoene, R.B., & Winslow, R.M. (1983). Maximal exercise at extreme altitudes on Mount Everest. *Journal of Applied Physiology,* **55,** 688-698.

14. West, J.B., Peters, R.M., Aksnes, G., Maret, K.H., Milledge, J.S., & Schoene, R.B. (1986). Nocturnal periodic breathing at altitudes of 6300 and 8050 m. *Journal of Applied Physiology,* **61,** 280-287.

Selected Readings

Cerretelli, P., & Hoppeler, H. (1996). Morphologic and metabolic response to chronic hypoxia: The muscle system. In M.J. Fregly & C.M. Blatteis (Eds.), *Handbook of physiology:*

Section 4. Environmental physiology (Vol. 2, pp. 1155-1181). New York: Oxford University Press.

Coote, J.H. (1995). Medicine and mechanisms in altitude sickness. *Sports Medicine,* **20,** 148-159.

Fulco, C.S., & Cymerman, A. (1990). Human performance and acute hypoxia. In K. Pandolf, M. Sawka, & R. Gonzalez (Eds.), *Human performance physiology and environmental medicine at terrestrial extremes* (pp. 467-495). Indianapolis: Benchmark Press.

Grover, R.F., Weil, J.V., & Reeves, J.T. (1986). Cardiovascular adaptation to exercise at high altitude. *Exercise and Sport Sciences Reviews,* **14,** 269-302.

Heath, D., & Williams, D.R. (1989). *High-altitude medicine and pathology.* London: Butterworths.

Margaria, R. (Ed.). (1967). *Exercise at altitude.* Amsterdam: Excerpta Medica Foundation.

Pigman, E.C. (1991). Acute mountain sickness: Effects and implications for exercise at intermediate altitude. *Sports Medicine,* **12,** 71-79.

West, J.B., Boyer, S.J., Graber, D.J., Hackett, P.H., Maret, K.H., Milledge, J.S., Peters, R.M., Jr., Pizzo, C.J., Samaja, M., Sarnquist, F.H., Schoene, R.B., & Winslow, R.M. (1983). Maximal exercise at extreme altitudes on Mount Everest. *Journal of Applied Physiology,* **55,** 688-698.

Westerterp, K.R. (2001). Energy and water balance at high altitude. *News in the Physiological Sciences,* **16,** 134-137.

Young, A.J., & Young, P.M. (1990). Human acclimatization to high terrestrial altitude. In K. Pandolf, M. Sawka, & R. Gonzalez (Eds.), *Human performance physiology and environmental medicine at terrestrial extremes* (pp. 497-543). Indianapolis: Benchmark Press.

Chapter 13

References

1. Armstrong, L.E., & VanHeest, J.L. (2002). The unknown mechanism of the overtraining syndrome. *Sports Medicine,* **32,** 185-209.
2. Costill, D.L. (1998). Training adaptations for optimal performance. Paper presented at the VIII International Symposium on Biomechanics and Medicine of Swimming, June 28, University of Jyväskylä, Finland.
3. Costill, D.L., Fink, W.J., Hargreaves, M., King, D.S., Thomas, R., & Fielding, R. (1985). Metabolic characteristics of skeletal muscle during detraining from competitive swimming. *Medicine and Science in Sports and Exercise,* **17,** 339-343.
4. Costill, D.L., King, D.S., Thomas, R., & Hargreaves, M. (1985). Effects of reduced training on muscular power in swimmers. *Physician and Sportsmedicine,* **13**(2), 94-101.
5. Costill, D.L., Maglischo, E., & Richardson, A. (1991). *Handbook of sports medicine: Swimming.* London: Blackwell.
6. Costill, D.L., Thomas, R., Robergs, R.A., Pascoe, D.D., Lambert, C.P., Barr, S.I., & Fink, W.J. (1991). Adaptations to swimming training: Influence of training volume. *Medicine and Science in Sports and Exercise,* **23,** 371-377.
7. Coyle, E.F., Martin, W.H., III, Sinacore, D.R., Joyner, M.J., Hagberg, J.M., & Holloszy, J.O. (1984). Time course of loss of adaptations after stopping prolonged intense endurance training. *Journal of Applied Physiology,* **57,** 1857-1864.
8. Fitts, R.H., Costill, D.L., & Gardetto, P.R. (1989). Effect of swim-exercise training on human muscle fiber function. *Journal of Applied Physiology,* **66,** 465-475.
9. Fleck, S.J., & Kraemer, W.J. (2004). *Designing resistance training programs* (3rd ed.). Champaign, IL: Human Kinetics.
10. Fry, R.W., Morton, A.R., & Keast, D. (1991). Overtraining in athletes: An update. *Sports Medicine,* **12,** 32-65.
11. Hickson, R.C., Foster, C., Pollock, M.L., Galassi, T.M., & Rich, S. (1985). Reduced training intensities and loss of aerobic power, endurance, and cardiac growth. *Journal of Applied Physiology,* **58,** 492-499.
12. Houmard, J.A., Costill, D.L., Mitchell, J.B., Park, S.H., Hickner, R.C., & Roemmish, J.N. (1990). Reduced training maintains performance in distance runners. *International Journal of Sports Medicine,* **11,** 46-51.
13. Houmard, J.A., Scott, B.K., Justice, C.L., & Chenier, T.C. (1994). The effects of taper on performance in distance runners. *Medicine and Science in Sports and Exercise,* **26,** 624-631.
14. Kraemer, W.J., & Ratamess, N.A. (2003). Endocrine responses and adaptations to strength and power training. In P.V. Komi (Ed.), *Strength and power in sport* (pp. 379-380). Oxford: Blackwell Scientific.
15. Krivickas, L.S. (2006). Recurrent rhabdomyolysis in a collegiate athlete: A case report. *Medicine and Science in Sports and Exercise,* **38,** 407-410.
16. Lemmer, J.T., Hurlbut, D.E., Martel, G.F., Tracy, B.L., Ivey, F.M., Metter, E.J., Fozard, J.L., Fleg, J.L., & Hurley, B.F. (2000). Age and gender responses to strength training and detraining. *Medicine and Science in Sports and Exercise,* **32,** 1505-1512.
17. Mujika, I., & Padilla, S. (2003). Scientific bases for precompetition tapering strategies. *Medicine and Science in Sports and Exercise,* **35,** 1182-1187.
18. Nieman, D.C. (1994). Exercise, infection, and immunity. *International Journal of Sports Medicine,* **15,** S131-S141.
19. Nieman, D.C. (1997). Immune response to heavy exertion. *Journal of Applied Physiology,* **82,** 1385-1394.
20. O'Toole, M.L. (1998). Overreaching and overtraining in endurance athletes. In R.B. Kreider, A.C. Fry, & M.L. O'Toole (Eds.), *Overtraining in sport* (pp. 10, 13). Champaign, IL: Human Kinetics.
21. Saltin, B., Blomqvist, G., Mitchell, J.H., Johnson, R.L., Jr., Wildenthal, K., & Chapman, C.B. (1968). Response to submaximal and maximal exercise after bed rest and training. *Circulation,* **38**(Suppl. 7).
22. Selye, H. (1956). *The stress of life.* New York: McGraw-Hill.

23. Shepherd, R.J. (2001). Chronic fatigue syndrome: An update. *Sports Medicine,* **31,** 167-194.

24. Smith, L.L. (2000). Cytokine hypothesis of overtraining: A physiological adaptation to excessive stress? *Medicine and Science in Sports and Exercise,* **32,** 317-331.

25. Springer, B.L., & Clarkson, P.M. (2003). Two cases of exertional rhabdomyolysis precipitated by personal trainers. *Medicine and Science in Sports and Exercise,* **35,** 1499-1502.

26. Trappe, T., Trappe, S., Lee, G., Widrick, J., Fitts, R., & Costill, D. (2006). Cardiorespiratory responses to physical work during and following 17 days of bed rest and spaceflight. *Journal of Applied Physiology,* **100,** 951-957.

27. Watenpaugh, D.E., & Hargens, A.R. (1996). The cardiovascular system in microgravity. In M.J. Fregly & C.M. Blatteis (Eds.), *Handbook of physiology: Environmental physiology* (Vol. 1, pp. 631-674). New York: Oxford University Press.

Selected Readings

Clarkson, P.M., & Eichner, E.R. (2006). Invited commentary—exertional rhabdomyolysis: Does elevated blood creatine kinase foretell renal failure? *Current Sports Medicine Reports,* **5,** 57-60.

Fry, A.C., & Kraemer, W.J. (1997). Resistance exercise overtraining and overreaching: Neuroendocrine responses. *Sports Medicine,* **23,** 106-129.

Halson, S.L., & Jeukendrup, A.E. (2004). Does overtraining exist? An analysis of overreaching and overtraining research. *Sports Medicine,* **34,** 967-981.

Kreider, R.B., Fry, A.C., & O'Toole, M.L. (Eds.). (1998). *Overtraining in sport.* Champaign, IL: Human Kinetics.

Mujika, I., & Padilla, S. (2001). Cardiorespiratory and metabolic characteristics of detraining in humans. *Medicine and Science in Sports and Exercise,* **33,** 413-421.

Mujika, I., & Padilla, S. (2001). Muscular characteristics of detraining in humans. *Medicine and Science in Sports and Exercise,* **33,** 1297-1303.

Mujika, I., Padilla, S., Pyne, D., & Busso, T. (2004). Physiological changes associated with the pre-event taper in athletes. *Sports Medicine,* **34,** 891-927.

Nieman, D.C., & Pedersen, B.K. (1999). Exercise and immune function. *Sports Medicine,* **27,** 73-80.

Trinity, J.D., Pahnke, M.D., Reese, E.C., & Coyle, E.F. (2006). Maximal mechanical power during a taper in elite swimmers. *Medicine and Science in Sports and Exercise,* **38,** 1643-1649.

Urhausen, A., & Kindermann, W. (2002). Diagnosis of overtraining: What tools do we have? *Sports Medicine,* **32,** 95-102.

Chapter 14

References

1. American College of Sports Medicine, American Dietetic Association, and Dietitians of Canada. (2000). Nutrition and athletic performance. Joint position statement. *Medicine and Science in Sports and Exercise,* **32,** 2130-2145.

2. Armstrong, L.E., Costill, D.L., & Fink, W.J. (1985). Influence of diuretic-induced dehydration on competitive running performance. *Medicine and Science in Sports and Exercise,* **17,** 456-461.

3. Åstrand, P.-O. (1967). Diet and athletic performance. *Federation Proceedings,* **26,** 1772-1777.

4. Barr, S.I., Costill, D.L., & Fink, W.J. (1991). Fluid replacement during prolonged exercise: Effects of water, saline or no fluid. *Medicine and Science in Sports and Exercise,* **23,** 811-817.

5. Beaton, L.J., Allan, D.A., Tarnopolsky, M.A., Tiidus, P.M., & Phillips, S.M. (2002). Contraction-induced muscle damage is unaffected by vitamin E supplementation. *Medicine and Science in Sports and Exercise,* **34,** 798-805.

6. Cheuvront, S.N. (1999). The Zone diet and athletic performance. *Sports Medicine,* **27,** 213-228.

7. Coombes, J.S., & Hamilton, K.L. (2000). The effectiveness of commercially available sports drinks. *Sports Medicine,* **29,** 181-209.

8. Costill, D.L., Bowers, R., Branam, G., & Sparks, K. (1971). Muscle glycogen utilization during prolonged exercise on successive days. *Journal of Applied Physiology,* **31,** 834-838.

9. Costill, D.L., Coyle, E., Dalsky, G., Evans, W., Fink, W., & Hoopes, D. (1977). Effects of elevated plasma FFA and insulin on muscle glycogen usage during exercise. *Journal of Applied Physiology,* **43**(4), 695-699.

10. Dougherty, K.A., Baker, L.B., Chow, M., & Kenney, W.L. (2006). Two percent dehydration impairs and six percent carbohydrate drink improves boys basketball skills. *Medicine and Science in Sports and Exercise,* **38,** 1650-1658.

11. Fairchild, T.J., Fletcher, S., Steele, P., Goodman, C., Dawson, B., & Fournier, P.A. (2002). Rapid carbohydrate loading after a short bout of near maximal-intensity exercise. *Medicine and Science in Sports and Exercise,* **34,** 980-986.

12. Foster-Powell, K., Holt, S.H.A., & Brand-Miller, J.C. (2002). International table of glycemic index and glycemic load values: 2002. *American Journal of Clinical Nutrition,* **76,** 5-56.

13. Friedmann, B., Weller, E., Mairbäurl, H., & Bärtsch, P. (2001). Effects of iron repletion on blood volume and performance capacity in young athletes. *Medicine and Science in Sports and Exercise,* **33,** 741-746.

14. Frizzell, R.T., Lang, G.H., Lowance, D.C., & Lathan, S.R. (1986). Hyponatremia and ultramarathon running. *Journal of the American Medical Association,* **255,** 772-774.

15. Gollnick, P.D., Piehl, K., & Saltin, B. (1974). Selective glycogen depletion pattern in human muscle fibres after exercise of varying intensity and at varying pedaling rates. *Journal of Physiology,* **241,** 45-57.

16. Hartmann, A., Nieb, A.M., Grünert-Fuchs, M., Poch, B., & Speit, G. (1995). Vitamin E prevents exercise-induced DNA damage. *Mutation Research,* **346,** 195-202.

17. Hew-Butler, T., Almond, C., Ayus, J.C., Dugas, J., Meeuwisse, W., Noakes, T., Reid, S., Siegel, A., Speedy, D., Stuempfle, K., Verbalis, J., & Weschler, L. (2005). Consensus statement of the 1st International Exercise-Associated Hyponatremia Consensus Development Conference, Cape Town, South Africa 2005. *Clinical Journal of Sports Medicine,* **15,** 206-211.

18. Ivy, J.L. (2004). Regulation of muscle glycogen repletion, muscle protein synthesis and repair following exercise. *Journal of Sports Science and Medicine,* **3,** 131-138.

19. Ivy, J.L., Goforth, H.W., Damon, B.M., McCauley, T.R., Parsons, E.C., & Price, T.B. (2002). Early postexercise muscle glycogen recovery is enhanced with a carbohydrate-protein supplement. *Journal of Applied Physiology,* **93,** 1337-1344.

20. Ivy, J.L., Katz, A.L., Cutler, C.L., Sherman, W.M., & Coyle, E.F. (1988). Muscle glycogen synthesis after exercise: Effect of time of carbohydrate ingestion. *Journal of Applied Physiology,* **64,** 1480-1485.

21. Ivy, J.L., Lee, M.C., Brozinick, J.T., Jr., & Reed, M.J. (1988). Muscle glycogen storage after different amounts of carbohydrate ingestion. *Journal of Applied Physiology,* **65,** 2018-2023.

22. Jeukendrup, A., & Gleeson, M. (2004). *Sport nutrition: An introduction to energy production and performance.* Champaign, IL: Human Kinetics.

23. Montain, S.J., Sawka, M.N., & Wenger, C.B. (2001). Hyponatremia associated with exercise: Risk factors and pathogenesis. *Exercise and Sport Sciences Reviews,* **29,** 113-117.

24. Schabort, E.J., Bosch, A.N., Weltan, S.M., & Noakes, T.D. (1999). The effect of a preexercise meal on time to fatigue during prolonged cycling exercise. *Medicine and Science in Sports and Exercise,* **31,** 464-471.

25. Sears, B. (1995). *The Zone.* New York: HarperCollins.

26. Sears, B. (2000). The Zone diet and athletic performance [letter]. *Sports Medicine,* **29,** 289-291.

27. Sherman, W.M., Costill, D.L., Fink, W.J., & Miller, J.M. (1981). Effect of exercise diet manipulation on muscle glycogen and its subsequent utilization during performance. *International Journal of Sports Medicine,* **2,** 114-118.

28. Wilmore, J.H., Brown, C.H., & Davis, J.A. (1977). Body physique and composition of the female distance runner. *Annals of the New York Academy of Sciences,* **301,** 764-776.

29. Wilmore, J.H., Morton, A.R., Gilbey, H.J., & Wood, R.J. (1998). Role of taste preference on fluid intake during and after 90 min of running at 60% of $\dot{V}O_2$max in the heat. *Medicine and Science in Sports and Exercise,* **30,** 587-595.

30. Wolfe, R.R. (2006). Skeletal muscle protein metabolism and resistance exercise. *Journal of Nutrition,* **136,** 525S-528S.

Selected Readings

American College of Sports Medicine. (2007). Exercise and fluid replacement: Position stand. *Medicine and Science in Sports and Exercise,* **39,** 377-390.

Berardi, J.M., Price, T.B., Noreen, E.E., & Lemon, P.W.R. (2006). Postexercise muscle glycogen recovery enhanced with a carbohydrate-protein supplement. *Medicine and Science in Sports and Exercise,* **38,** 1106-1113.

Hawley, J.A., Brouns, F., & Jeukendrup, A. (1998). Strategies to enhance fat utilisation during exercise. *Sports Medicine,* **25,** 241-257.

Hermansen, L., Hultman, E., & Saltin, B. (1967). Muscle glycogen during prolonged severe exercise. *Acta Physiologica Scandinavica,* **71,** 129-139.

Heymsfield, S.B., Lohman, T.G., Wang, Z-M., & Going, S.B. (Eds.). (2005). *Human body composition* (2nd ed.). Champaign, IL: Human Kinetics.

Heyward, V.H., & Wagner, D.R. (2004). *Applied body composition assessment* (2nd ed.). Champaign, IL: Human Kinetics.

Manore, M., & Thompson, J. (2000). *Sport nutrition for health and performance.* Champaign, IL: Human Kinetics.

Maughan, R.J., & Murray, R. (2001). *Sports drinks: Basic science and practical aspects.* Boca Raton: CRC Press.

Shirreffs, S.M., & Maughan, R.J. (2000). Rehydration and recovery of fluid balance after exercise. *Exercise and Sport Sciences Reviews,* **28,** 27-32.

Chapter 15

References

1. Alvois, L., Robinson, N., Saudan, D., Baume, N., Mangin, P., & Saugy, M. (2006). Central nervous system stimulants and sport practice. *British Journal of Sports Medicine,* **40**(Suppl. I), i16-i20.

2. American College of Sports Medicine Consensus Statement. (2000). The physiological and health effects of oral creatine supplementation. *Medicine and Science in Sports and Exercise,* **32,** 706-717.

3. American College of Sports Medicine Position Stand. (1996). The use of blood doping as an ergogenic aid. *Medicine and Science in Sports and Exercise,* **28**(6), i-xii.

4. Ariel, G., & Saville, W. (1972). Anabolic steroids: The physiological effects of placebos. *Medicine and Science in Sports and Exercise,* **4,** 124-126.

5. Bannister, R.G., & Cunningham, D.J.C. (1954). The effects on respiration and performance during exercise of adding oxygen to the inspired air. *Journal of Physiology,* **125,** 118-137.

6. Berning, J.M., Adams, K.J., & Stamford, B.A. (2004). Anabolic steroid usage in athletics: Facts, fiction, and public relations. *Journal of Strength and Conditioning Research,* **18,** 908-917.

7. Bhasin, S., Storer, T.W., Berman, N., Callegari, C., Clevenger, B., Phillips, J., Bunnell, T.J., Tricker, R., Shirazi, A., & Casaburi, R. (1996). The effects of supraphysiologic

doses of testosterone on muscle size and strength in normal men. *New England Journal of Medicine,* **335,** 1-7.

8. Birkeland, K.I., Stray-Gundersen, J., Hemmersbach, P., Hallén, J., Haug, E., & Bahr, R. (2000). Effect of rhEPO administration on serum levels of sTfR and cycling performance. *Medicine and Science in Sports and Exercise,* **32,** 1238-1243.
9. Broeder, C.E., Quindry, J., Brittingham, K., Panton, L., Thomson, J., Appakondu, S., Bruel, K., Byrd, R., Douglas, J., Earnest, C., Mitchell, C., Olson, M., Roy, T., & Yarlagadda, C. (2000). The Andro Project: Physiological and hormonal influences of androstenedione supplementation in men 35 to 65 years old participating in a high-intensity resistance training program. *Archives of Internal Medicine,* **160,** 3093-3104.
10. Bronson, F.H., & Matherne, C.M. (1997). Exposure to anabolic-androgenic steroids shortens life span of male mice. *Medicine and Science in Sports and Exercise,* **29,** 615-619.
11. Brown, G.A., Vukovich, M.D., Sharp, R.L., Reifenrath, T.A., Parsons, K.A., & King, D.S. (1999). Effect of oral DHEA on serum testosterone and adaptations to resistance training in young men. *Journal of Applied Physiology,* **87,** 2274-2283.
12. Buick, F.J., Gledhill, N., Froese, A.B., Spriet, L., & Meyers, E.C. (1980). Effect of induced erythrocythemia on aerobic work capacity. *Journal of Applied Physiology,* **48,** 636-642.
13. Calfee, R., & Fadale, P. (2006). Popular ergogenic drugs and supplements in young athletes. *Pediatrics,* **117,** e577-e589.
14. Costill, D.L., Dalsky, G.P., & Fink, W.J. (1978). Effects of caffeine ingestion on metabolism and exercise performance. *Medicine and Science in Sports,* **10,** 155-158.
15. Costill, D.L., Verstappen, F., Kuipers, H., Janssen, E., & Fink, W. (1984). Acid-base balance during repeated bouts of exercise: Influence of HCO-3. *International Journal of Sports Medicine,* **5,** 228-231.
16. Davis, J.M. (1995). Carbohydrates, branched-chain amino acids, and endurance: The central fatigue hypothesis. *International Journal of Sport Nutrition,* **5,** S29-S38.
17. Eichner, E.R. (1989). Ergolytic drugs. *Sports Science Exchange,* **2**(15), 1-4.
18. Ekblom, B., & Berglund, B. (1991). Effect of erythropoietin administration on maximal aerobic power. *Scandinavian Journal of Medicine and Science in Sports,* **1,** 88-93.
19. Ekblom, B., Goldbarg, A.N., & Gullbring, B. (1972). Response to exercise after blood loss and reinfusion. *Journal of Applied Physiology,* **33,** 175-180.
20. Evans, N.A. (2004). Current concepts in anabolic-androgenic steroids. *American Journal of Sports Medicine,* **32,** 534-542.
21. Forbes, G.B. (1985). The effect of anabolic steroids on lean body mass: The dose response curve. *Metabolism,* **34,** 571-573.
22. Gledhill, N. (1985). The influence of altered blood volume and oxygen transport capacity on aerobic performance. *Exercise and Sport Sciences Reviews,* **13,** 75-93.
23. Goforth, H.W., Jr., Campbell, N.L., Hodgdon, J.A., & Sucec, A.A. (1982). Hematologic parameters of trained distance runners following induced erythrocythemia [abstract]. *Medicine and Science in Sports and Exercise,* **14,** 174.
24. Graham, T.E. (2001). Caffeine and exercise: Metabolism, endurance and performance. *Sports Medicine,* **31,** 785-807.
25. Hartgens, F., & Kuipers, H. (2004). Effects of androgenic-anabolic steroids in athletes. *Sports Medicine,* **34,** 513-554.
26. Heinonen, O.J. (1996). Carnitine and physical exercise. *Sports Medicine,* **22,** 109-132.
27. Hervey, G.R., Knibbs, A.V., Burkinshaw, L., Morgan, D.B., Jones, P.R.M., Chettle, D.R., & Vartsky, D. (1981). Effects of methandienone on the performance and body composition of men undergoing athletic training. *Clinical Science,* **60,** 457-461.
28. Ivy, J.L., Costill, D.L., Fink, W.J., & Lower, R.W. (1979). Influence of caffeine and carbohydrate feedings on endurance performance. *Medicine and Science in Sports and Exercise,* **11,** 6-11.
29. Juhn, M.S. (2003). Popular sports supplements and ergogenic aids. *Sports Medicine,* **33,** 921-939.
30. King, D.S., Sharp, R.L., Vukovich, M.D., Brown, G.A., Reifenrath, T.A., Uhl, N.L., & Parsons, K.A. (1999). Effect of oral androstenedione on serum testosterone and adaptations of resistance training in young men: A randomized controlled trial. *Journal of the American Medical Association,* **281,** 2020-2028.
31. Linderman, J., & Fahey, T.D. (1991). Sodium bicarbonate ingestion and exercise performance: An update. *Sports Medicine,* **11,** 71-77.
32. Magkos, F., & Kavouras, S.A. (2004). Caffeine and ephedrine: Physiological, metabolic and performance-enhancing effects. *Sports Medicine,* **34,** 871-889.
33. Maughan, R.J., King, D.S., & Lea, T. (2004). Dietary supplements. *Journal of Sports Sciences,* **22,** 95-113.
34. Nissen, S.L., & Sharp, R.L. (2003). Effect of dietary supplements on lean mass and strength gains with resistance exercise: A meta-analysis. *Journal of Applied Physiology,* **94,** 651-659.
35. Pärssinen, M., & Seppälä, T. (2002). Steroid use and long-term health risks in former athletes. *Sports Medicine,* **32,** 83-94.
36. Roth, D.A., & Brooks, G.A. (1990). Lactate transport is mediated by a membrane-bound carrier in rat skeletal muscle sarcolemmal vesicles. *Archives of Biochemistry and Biophysics,* **279,** 377-385.
37. Rudman, D., Feller, A.G., Nagraj, H.S., Gergans, G.A., Lalitha, P.Y., Goldberg, A.F., Schlenker, R.A., Cohn, L., Rudman, I.W., & Mattson, D.E. (1990). Effects of human

growth hormone in men over 60 years old. *New England Journal of Medicine,* **323,** 1-6.

38. Smith-Rockwell, M., Nickols-Richardson, S.M., & Thye, F.W. (2001). Nutrition knowledge, opinions, and practices of coaches and athletic trainers at a division 1 university. *International Journal of Sports Nutrition and Exercise Metabolism,* **11,** 174-185.
39. Spriet, L.L. (1991). Blood doping and oxygen transport. In D.R. Lamb & M.H. Williams (Eds.), *Ergogenics: Enhancement of performance in exercise and sport* (pp. 213-242). Dubuque, IA: Brown & Benchmark.
40. Spriet, L.L., & Gibala, M.J. (2004). Nutritional strategies to influence adaptations to training. *Journal of Sports Sciences,* **22,** 127-141.
41. Tamaki, T., Uchiyama, S., Uchiyama, Y., Akatsuka, A., Roy, R.R., & Edgerton, V.R. (2001). Anabolic steroids increase exercise tolerance. *American Journal of Physiology: Endocrinology and Metabolism,* **280,** E973-E981.
42. Tokish, J.M., Kocher, M.S., & Hawkins, R.J. (2004). Ergogenic aids: A review of basic science, performance, side effects, and status in sports. *American Journal of Sports Medicine,* **32,** 1543-1553.
43. van Hall, G., Raaymakers, J.S.H., Saris, W.H.M., & Wagenmakers, A.J.M. (1995). Ingestion of branched-chain amino acids and tryptophan during sustained exercise in man: Failure to affect performance. *Journal of Physiology,* **486,** 789-794.
44. Villareal, D.T., & Holloszy, J.O. (2006). DHEA enhances effects of weight training on muscle mass and strength. *American Journal of Physiology: Endocrinology and Metabolism.* **291,** E1003-1008.
45. Williams, M.H. (Ed.). (1983). *Ergogenic aids in sport.* Champaign, IL: Human Kinetics.
46. Williams, M.H., Wesseldine, S., Somma, T., & Schuster, R. (1981). The effect of induced erythrocythemia upon 5-mile treadmill run time. *Medicine and Science in Sports and Exercise,* **13,** 169-175.
47. Winter, F.D., Snell, P.G., & Stray-Gundersen, J. (1989). Effects of 100% oxygen on performance of professional soccer players. *Journal of the American Medical Association,* **262,** 227-229.
48. Yarasheski, K.E. (1994). Growth hormone effects on metabolism, body composition, muscle mass, and strength. *Exercise and Sport Sciences Reviews,* **22,** 285-312.
49. Yesalis, C.E. (Ed.). (2000). *Anabolic steroids in sport and exercise* (2nd ed.). Champaign, IL: Human Kinetics.

Selected Readings

Bahrke, M.S., & Yesalis, C.E. (2002). *Performance-enhancing substances in sport and exercise.* Champaign, IL: Human Kinetics.

Brown, G.A., Vukovich, M., & King, D.S. (2006). Testosterone prohormone supplements. *Medicine and Science in Sports and Exercise,* **38,** 1451-1461.

Reents, S. (2000). *Sport and exercise pharmacology.* Champaign, IL: Human Kinetics.

Wadler, G.I., & Hainline, B. (1989). *Drugs and the athlete.* Philadelphia: Davis.

Williams, M.H. (1989). *Beyond training.* Champaign, IL: Leisure Press.

Williams, M.H., Kreider, R.B., & Branch, J.D. (1999). *Creatine: The power supplement.* Champaign, IL: Human Kinetics.

Yesalis, C.E. (2000). *Anabolic steroids in sport and exercise* (2nd ed.). Champaign, IL: Human Kinetics.

Chapter 16

References

1. Bar-Or, O. (1983). *Pediatric sports medicine for the practitioner: From physiologic principles to clinical applications.* New York: Springer-Verlag.
2. Bar-Or, O. (1989). Temperature regulation during exercise in children and adolescents. In C.V. Gisolfi & D.R. Lamb (Eds.), *Perspectives in exercise science and sports medicine: Youth, exercise and sport* (pp. 335-362). Carmel, IN: Benchmark Press.
3. Bass, S.L. (2000). The prepubertal years: A uniquely opportune stage of growth when the skeleton is most responsive to exercise? *Sports Medicine,* **30,** 73-78.
4. Beneke, R., Hütler, M., Jung, M., & Leithäuser, R.M. (2005). Modeling the blood lactate kinetics at maximal short-term exercise conditions in children, adolescents, and adults. *Journal of Applied Physiology,* **99,** 499-504.
5. Clarke, H.H. (1971). *Physical and motor tests in the Medford boys' growth study.* Englewood Cliffs, NJ: Prentice Hall.
6. Cureton, K.J., Sloniger, M.A., Black, D.M., McCormack, W.P., & Rowe, D.A. (1997). Metabolic determinants of the age-related improvement in one-mile run/walk performance in youth. *Medicine and Science in Sports and Exercise,* **29,** 259-267.
7. Daniels, J., Oldridge, N., Nagle, F., & White, B. (1978). Differences and changes in $\dot{V}O_2$ among young runners 10 to 18 years of age. *Medicine and Science in Sports and Exercise,* **10,** 200-203.
8. Eriksson, B.O. (1972). Physical training, oxygen supply and muscle metabolism in 11-13-year-old boys. *Acta Physiologica Scandinavica* (Suppl. 384), 1-48.
9. Falk, B., & Eliakim, A. (2003). Resistance training, skeletal muscle and growth. *Pediatric Endocrinology Reviews,* **1,** 120-127.
10. Fleck, S.J., & Kraemer, W.J. (2004). *Designing resistance training programs* (3rd ed.). Champaign, IL: Human Kinetics.
11. Froberg, K., & Lammert, O. (1996). Development of muscle strength during childhood. In O. Bar-Or (Ed.), *The child and adolescent athlete* (p. 28). London: Blackwell.
12. Kraemer, W.J., & Fleck, S.J. (2005). *Strength training for young athletes* (2nd ed.). Champaign, IL: Human Kinetics.

13. Krahenbuhl, G.S., Morgan, D.W., & Pangrazi, R.P. (1989). Longitudinal changes in distance-running performance of young males. *International Journal of Sports Medicine,* **10,** 92-96.

14. Mahon, A.D., & Vaccaro, P. (1989). Ventilatory threshold and $\dot{V}O_2$max changes in children following endurance training. *Medicine and Science in Sports and Exercise,* **21,** 425-431.

15. Malina, R.M. (1989). Growth and maturation: Normal variation and effect of training. In C.V. Gisolfi & D.R. Lamb (Eds.), *Perspectives in exercise science and sports medicine: Youth, exercise and sport* (pp. 223-265). Carmel, IN: Benchmark Press.

16. Malina, R.M., Bouchard, C., & Bar-Or, O. (2004). *Growth, maturation, and physical activity* (2nd ed.). Champaign, IL: Human Kinetics.

17. Ramsay, J.A., Blimkie, C.J.R., Smith, K., Garner, S., MacDougall, J.D., & Sale, D.G. (1990). Strength training effects in prepubescent boys. *Medicine and Science in Sports and Exercise,* **22,** 605-614.

18. Robinson, S. (1938). Experimental studies of physical fitness in relation to age. *Arbeitsphysiologie,* **10,** 251-323.

19. Rogers, D.M., Olson, B.L., & Wilmore, J.H. (1995). Scaling for the $\dot{V}O_2$-to-body size relationship among children and adults. *Journal of Applied Physiology,* **79,** 958-967.

20. Rowland, T.W. (1985). Aerobic response to endurance training in prepubescent children: A critical analysis. *Medicine and Science in Sports and Exercise,* **17,** 493-497.

21. Rowland, T.W. (1989). Oxygen uptake and endurance fitness in children: A developmental perspective. *Pediatric Exercise Science,* **1,** 313-328.

22. Rowland, T.W. (1991). "Normalizing" maximal oxygen uptake, or the search for the holy grail (per kg). *Pediatric Exercise Science,* **3,** 95-102.

23. Rowland, T.W. (2005). *Children's exercise physiology* (2nd ed.). Champaign, IL: Human Kinetics.

24. Santos, A.M.C., Welsman, J.R., De Ste Croix, M.B.A., & Armstrong, N. (2002). Age- and sex-related differences in optimal peak power. *Pediatric Exercise Science,* **14,** 202-212.

25. Sjödin, B., & Svedenhag, J. (1992). Oxygen uptake during running as related to body mass in circumpubertal boys: A longitudinal study. *European Journal of Applied Physiology,* **65,** 150-157.

26. Turley, K.R., & Wilmore, J.H. (1997). Cardiovascular responses to treadmill and cycle ergometer exercise in children and adults. *Journal of Applied Physiology,* **83,** 948-957.

Selected Readings

Armstrong, L.E., & Maresh, C.M. (1995). Exercise-heat tolerance of children and adolescents. *Pediatric Exercise Science,* **7,** 239-252.

Armstrong, N., & Welsman, J.R. (2000). Development of aerobic fitness during childhood and adolescence. *Pediatric Exercise Science,* **12,** 128-149.

Bailey, D.A., Faulkner, R.A., & McKay, H.A. (1996). Growth, physical activity, and bone mineral acquisition. *Exercise and Sport Sciences Reviews,* **24,** 233-266.

Bar-Or, O. (Ed.). (1996). *The child and adolescent athlete.* London: Blackwell.

Beunen, G., & Thomis, M. (2000). Muscular strength development in children and adolescents. *Pediatric Exercise Science,* **12,** 174-197.

Falk, B. (1998). Effects of thermal stress during rest and exercise in the paediatric population. *Sports Medicine,* **25,** 221-240.

Janssen, I., Heymsfield, S.B., Wang, Z., & Ross, R. (2000). Skeletal muscle mass and distribution in 468 men and women aged 18-88 yr. *Journal of Applied Physiology,* **89,** 81-88.

Krahenbuhl, G.S., Skinner, J.S., & Kohrt, W.M. (1985). Developmental aspects of maximal aerobic power in children. *Exercise and Sport Sciences Reviews,* **13,** 503-538.

Turley, K.R. (1997). Cardiovascular responses to exercise in children. *Sports Medicine,* **24**(4), 241-257.

Van Praagh, E. (2000). Development of anaerobic function during childhood and adolescence. *Pediatric Exercise Science,* **12,** 150-173.

Williams, C.A. (1997). Children's and adolescents' anaerobic performance during cycle ergometry. *Sports Medicine,* **24**(4), 227-240.

Chapter 17

References

1. Buskirk, E.R., & Hodgson, J.L. (1987). Age and aerobic power: The rate of change in men and women. *Federation Proceedings,* **46,** 1824-1829.

2. Connelly, D.M., Rice, C.L., Roos, M.R., & Vandervoort, A.A. (1999). Motor unit firing rates and contractile properties in tibialis anterior of young and old men. *Journal of Applied Physiology,* **87,** 843-852.

3. Costill, D.L. (1986). *Inside running: Basics of sports physiology.* Indianapolis: Benchmark Press.

4. Dill, D.B., Robinson, S., & Ross, J.C. (1967). A longitudinal study of 16 champion runners. *Journal of Sports Medicine and Physical Fitness,* **7,** 4-27.

5. Doherty, T.J., Vandervoort, A.A., Taylor, A.W., & Brown, W.F. (1993). Effects of motor unit losses on strength in older men and women. *Journal of Applied Physiology,* **74,** 868-874.

6. Fitzgerald, M.D., Tanaka, H., Tran, Z.V., & Seals, D.R. (1997). Age-related declines in maximal aerobic capacity in regularly exercising vs. sedentary women: A meta-analysis. *Journal of Applied Physiology,* **83,** 160-165.

7. Frontera, W.R., Meredith, C.N., O'Reilly, K.P., Knuttgen, W.G., & Evans, W.J. (1988). Strength conditioning in older men: Skeletal muscle hypertrophy and improved function. *Journal of Applied Physiology,* **64,** 1038-1044.

8. Goodrick, C.L. (1980). Effects of long-term voluntary wheel exercise on male and female Wistar rats: 1. Longevity, body weight and metabolic rate. *Gerontology,* **26,** 22-33.

9. Hagerman, F.C., Walsh, S.J., Staron, R.S., Hikida, R.S., Gilders, R.M., Murray, T.F., Toma, K., & Ragg, K.E. (2000). Effects of high-intensity resistance training on untrained older men. I. Strength, cardiovascular, and metabolic responses. *Journals of Gerontology Series A: Biological Sciences and Medical Sciences,* **55,** B336-B346.

10. Häkkinen, K., Kraemer, W.J., Pakarinen, A., Triplett-McBride, T., McBride, J.M., Häkkinen, A., Alen, M., McGuigan, M.R., Bronks, R., & Newton, R.U. (2002). Effects of heavy resistance/power training on maximal strength, muscle morphology, and hormonal response patterns in 60-75-year-old men and women. *Canadian Journal of Applied Physiology,* **27,** 213-231.

11. Häkkinen, K., Pakarinen, A., Kraemer, W.J., Häkkinen, A., Valkeinen, H., & Alen, M. (2001). Selective muscle hypertrophy, changes in EMG and force, and serum hormones during strength training in older women. *Journal of Applied Physiology,* **91,** 569-580.

12. Hameed, M., Harridge, S.D.R., & Goldspink, G. (2002). Sarcopenia and hypertrophy: A role for insulin-like growth factor-1 and aged muscle? *Exercise and Sport Sciences Reviews,* **30,** 15-19.

13. Hawkins, S.A., Marcell, T.J., Jaque, S.V., & Wiswell, R.A. (2001). A longitudinal assessment of change in $\dot{V}O_2$max and maximal heart rate in master athletes. *Medicine and Science in Sports and Exercise,* **33,** 1744-1750.

14. Hikida, R.S., Staron, R.S., Hagerman, F.C., Walsh, S., Kaiser, E., Shell, S., & Hervey, S. (2000). Effects of high-intensity resistance training on untrained older men. II. Muscle fiber characteristics and nucleo-cytoplasmic relationships. *Journals of Gerontology Series A: Biological Sciences and Medical Sciences,* **55,** B347-B354.

15. Holloszy, J.O. (1997). Mortality rate and longevity of food-restricted exercising male rats: A reevaluation. *Journal of Applied Physiology,* **82,** 399-403.

16. Jackson, A.S., Beard, E.F., Wier, L.T., Ross, R.M., Stuteville, J.E., & Blair, S.N. (1995). Changes in aerobic power of men, ages 25-70 yr. *Medicine and Science in Sports and Exercise,* **27,** 113-120.

17. Jackson, A.S., Wier, L.T., Ayers, G.W., Beard, E.F., Stuteville, J.E., & Blair, S.N. (1996). Changes in aerobic power of women, ages 20-64 yr. *Medicine and Science in Sports and Exercise,* **28,** 884-891.

18. Janssen, I., Heymsfield, S.B., Wang, Z., & Ross, R. (2000). Skeletal muscle mass and distribution in 468 men and women aged 18-88 yr. *Journal of Applied Physiology,* **89,** 81-88.

19. Johnson, M.A., Polgar, J., Weihtmann, D., & Appleton, D. (1973). Data on the distribution of fiber types in thirty-six human muscles: An autopsy study. *Journal of Neurological Science,* **1,** 111-129.

20. Kenney, W.L. (1997). Thermoregulation at rest and during exercise in healthy older adults. *Exercise and Sport Sciences Reviews,* **25,** 41-77.

21. Kohrt, W.M., Malley, M.T., Coggan, A.R., Spina, R.J., Ogawa, T., Ehsani, A.A., Bourey, R.E., Martin, W.H., III, & Holloszy, J.O. (1991). Effects of gender, age, and fitness level on response of $\dot{V}O_2$max to training in 60-71 yr olds. *Journal of Applied Physiology,* **71,** 2004-2011.

22. Kohrt, W.M., Malley, M.T., Dalsky, G.P., & Holloszy, J.O. (1992). Body composition of healthy sedentary and trained, young and older men and women. *Medicine and Science in Sports and Exercise,* **24,** 832-837.

23. Lexell, J., Downham, D.Y., Larson, Y., Bruhn, E., & Morsing, B. (1995). Heavy-resistance training in older Scandinavian men and women: Short- and long-term effects on arm and leg muscles. *Scandinavian Journal of Medicine and Science in Sports,* **5,** 329-341.

24. Lexell, J., Taylor, C.C., & Sjostrom, M. (1988). What is the cause of the aging atrophy? Total number, size, and proportion of different fiber types studied in whole vastus lateralis muscle from 15- to 83-year-old men. *Journal of Neurological Science,* **84,** 275-294.

25. Maharam, L.G., Bauman, P.A., Kalman, D., Skolnik, H., & Perle, S.M. (1999). Masters athletes: Factors affecting performance. *Sports Medicine,* **28,** 273-285.

26. Meredith, C.N., Frontera, W.R., Fisher, E.C., Hughes, V.A., Herland, J.C., Edwards, J., & Evans, W.J. (1989). Peripheral effects of endurance training in young and old subjects. *Journal of Applied Physiology,* **66,** 2844-2849.

27. Meusel, H. (1984). Health and well-being for older adults through physical exercises and sport—outline of the Giessen model. In B. McPherson (Ed.), *Sport and aging* (pp. 107-115). Champaign, IL: Human Kinetics.

28. Paterson, D.H., Cunningham, D.A., Koval, J.J., & St. Croix, C.M. (1999). Aerobic fitness in a population of independently living men and women aged 55-86 years. *Medicine and Science in Sports and Exercise,* **31,** 1813-1820.

29. Proctor, D.N., Shen, P.H., Dietz, N.M., Eickhoff, T.J., Lawler, L.A., Ebersold, E.J., Loeffler, D.L., & Joyner, M.J. (1998). Reduced leg blood flow during dynamic exercise in older endurance-trained men. *Journal of Applied Physiology,* **85,** 68-75.

30. Robinson, S. (1938). Experimental studies of physical fitness in relation to age. *Arbeitsphysiologie,* **10,** 251-323.

31. Saltin, B. (1986). The aging endurance athlete. In J.R. Sutton & R.M. Brock (Eds.), *Sports medicine for the mature athlete* (pp. 59-80). Indianapolis: Benchmark Press.

32. Shephard, R.J. (1997). *Aging, physical activity, and health.* Champaign, IL: Human Kinetics.

33. Spirduso, W.W. (1995). *Physical dimensions of aging.* Champaign, IL: Human Kinetics.

34. Tanaka, H., Monahan, K.D., & Seals, D.R. (2001). Age-predicted maximal heart rate revisited. *Journal of the American College of Cardiology,* **37,** 153-156.

35. Tanaka, H., & Seals, D. (1997). Age and gender interactions in physiological functional capacity: Insight from swimming performance. *Journal of Applied Physiology,* **82,** 846-851.

36. Trappe, S.W., Costill, D.L., Fink, W.J., & Pearson, D.R. (1995). Skeletal muscle characteristics among distance runners: A 20-yr follow-up study. *Journal of Applied Physiology,* **78,** 823-829.

37. Trappe, S.W., Costill, D.L., Goodpaster, B.H., & Pearson, D.R. (1996). Calf muscle strength in former elite distance runners. *Scandinavian Journal of Medicine and Science in Sports,* **6,** 205-210.

38. Trappe, S.W., Costill, D.L., Vukovich, M.D., Jones, J., & Melham, T. (1996). Aging among elite distance runners: A 22-yr longitudinal study. *Journal of Applied Physiology,* **80,** 285-290.

39. Wiswell, R.A., Jaque, S.V., Marcell, T.J., Hawkins, S.A., Tarpenning, K.M., Constantino, N., & Hyslop, D.M. (2000). Maximal aerobic power, lactate threshold, and running performance in master athletes. *Medicine and Science in Sports and Exercise,* **32,** 1165-1170.

Selected Readings

Daley, M.J., & Spinks, W.L. (2000). Exercise, mobility and aging. *Sports Medicine,* **29,** 1-12.

Deschenes, M.R. (2004). Effects of aging on muscle fibre type and size. *Sports Medicine,* **34,** 809-824.

Hawkins, S.A., & Wiswell, R.A. (2003). Rate and mechanism of maximal oxygen consumption decline with aging: Implications for exercise training. *Sports Medicine,* **33,** 877-888.

Hunter, G.R., McCarthy, J.P., & Bamman, M.M. (2004). Effects of resistance training on older adults. *Sports Medicine,* **34,** 329-348.

Kenney, W.L., Holowatz, L.A., & DeGroot, D.W. (2004). Extremes of heat tolerance: Life at the precipice of thermoregulatory failure. *Journal of Thermal Biology,* **29,** 479-485.

Kenney, W.L., & Munce, T.A. (2003). Aging and human temperature regulation. *Journal of Applied Physiology,* **95,** 2598-2603.

Lindle, R.S., Metter, E.J., Lynch, N.A., Fleg, J.L., Fozard, J.L., Tobin, J., Roy, T.A., & Hurley, B.F. (1997). Age and gender comparisons of muscle strength in 654 women and men aged 20-93 yr. *Journal of Applied Physiology,* **83,** 1581-1587.

Porter, M.M., Vandervoort, A.A., & Lexell, J. (1995). Aging of human muscle: Structure, function and adaptability. *Scandinavian Journal of Medicine and Science in Sports,* **5,** 129-142.

Proctor, D.N., & Joyner, M.J. (1997). Skeletal muscle mass and the reduction of $\dot{V}O_2$max in trained older subjects. *Journal of Applied Physiology,* **82,** 1411-1415.

Seals, D.R. (2003). Habitual exercise and the age-associated decline in large artery compliance. *Exercise and Sport Sciences Reviews,* **31,** 68-72.

Chapter 18

References

1. American College of Obstetricians and Gynecologists Committee Opinion. (2002). Exercise during pregnancy and the postpartum period. *Obstetrics and Gynecology,* **99,** 171-173.

2. American Psychiatric Association. (1994). *Diagnostic and statistical manual of mental disorders* (4th ed.). Washington, DC: American Psychiatric Association.

3. Åstrand, I. (1960). Aerobic work capacity in men and women with special reference to age. *Acta Physiologica Scandinavica,* **49**(Suppl. 169).

4. Åstrand, P.-O., Rodahl, K., Dahl, H.A., & Strømme, S.B. (2003). *Textbook of work physiology: Physiological bases of exercise* (4th ed.). Champaign, IL: Human Kinetics.

5. Bjorntorp, P. (1986). Fat cells and obesity. In K.D. Brownell & J.P. Foreyt (Eds.), *Handbook of eating disorders* (pp. 88-98). New York: Basic Books.

6. Cann, C.E., Martin, M.C., Genant, H.K., & Jaffe, R.B. (1984). Decreased spinal mineral content in amenorrheic women. *Journal of the American Medical Association,* **251,** 626-629.

7. Charkoudian, N., & Joyner, M.J. (2004). Physiologic considerations for exercise performance in women. *Clinics in Chest Medicine,* **25,** 247-255.

8. Costill, D.L., Fink, W.J., Flynn, M., & Kirwan, J. (1987). Muscle fiber composition and enzyme activities in elite female distance runners. *International Journal of Sports Medicine,* **8**(Suppl. 2), 103-106.

9. Costill, D.L., & Winrow, E. (1970). Maximal oxygen intake among marathon runners. *Archives of Physical Medicine and Rehabilitation,* **51,** 317-320.

10. Cureton, K., Bishop, P., Hutchinson, P., Newland, H., Vickery, S., & Zwiren, L. (1986). Sex differences in maximal oxygen uptake: Effect of equating haemoglobin concentration. *European Journal of Applied Physiology,* **54,** 656-660.

11. Cureton, K.J., & Sparling, P.B. (1980). Distance running performance and metabolic responses to running in men and women with excess weight experimentally equated. *Medicine and Science in Sports and Exercise,* **12,** 288-294.

12. Davis, J.A., Wilson, L.D., Caiozzo, V.J., Storer, T.W., & Pham, P.H. (2006). Maximal oxygen uptake at the same fat-free mass is greater in men than women. *Clinical Physiology and Functional Imaging,* **26,** 61-66.

13. Drinkwater, B.L. (1973). Physiological responses of women to exercise. *Exercise and Sport Sciences Reviews,* **1,** 125-153.

14. Drinkwater, B.L., Bruemner, B., & Chesnut, C.H., III. (1990). Menstrual history as a determinant of current bone density in young athletes. *Journal of the American Medical Association,* **263,** 545-548.

15. Drinkwater, B.L., Nilson, K., Chesnut, C.H., III, Bremner, W.J., Shainholtz, S., & Southworth, M.B. (1984). Bone

mineral content of amenorrheic and eumenorrheic athletes. *New England Journal of Medicine,* **311,** 277-281.

16. Drinkwater, B.L., Nilson, K., Ott, S., & Chesnut, C.H., III. (1986). Bone mineral density after resumption of menses in amenorrheic athletes. *Journal of the American Medical Association,* **256,** 380-382.
17. Fagard, R.H., Thijs, L.B., & Amery, A.K. (1995). The effect of gender on aerobic power and exercise hemodynamics in hypertensive adults. *Medicine and Science in Sports and Exercise,* **27,** 29-34.
18. Fink, W.J., Costill, D.L., & Pollock, M.L. (1977). Submaximal and maximal working capacity of elite distance runners: Part II. Muscle fiber composition and enzyme activities. *Annals of the New York Academy of Sciences,* **301,** 323-327.
19. Frisch, R.E. (1983). Fatness and reproduction: Delayed menarche and amenorrhea of ballet dancers and college athletes. In P.E. Garfinkel, P.L. Darby, & D.M. Garner (Eds.), *Anorexia nervosa: Recent developments in research* (pp. 343-363). New York: Liss.
20. Fu, Q., & Levine, B.D. (2005). Cardiovascular response to exercise in women. *Medicine and Science in Sports and Exercise,* **37,** 1433-1435.
21. Gadpaille, W.J., Sanborn, C.F., & Wagner, W.W. (1987). Athletic amenorrhea, major affective disorders, and eating disorders. *American Journal of Psychiatry,* **144,** 939-942.
22. Hermansen, L., & Andersen, K.L. (1965). Aerobic work capacity in young Norwegian men and women. *Journal of Applied Physiology,* **20,** 425-431.
23. Hicks, A.L., Kent-Braun, J., & Ditor, D.S. (2001). Sex differences in human skeletal muscle fatigue. *Exercise and Sport Sciences Reviews,* **29,** 109-112.
24. Janssen, I., Heymsfield, S.B., Wang, Z., & Ross, R. (2000). Skeletal muscle mass and distribution in 468 men and women aged 18-88 yr. *Journal of Applied Physiology,* **89,** 81-88.
25. Loucks, A.B., & Thuma, J.R. (2003). Luteinizing hormone pulsatility is disrupted at a threshold of energy availability in regularly menstruating women. *Journal of Clinical Endocrinology and Metabolism,* **88,** 297-311.
26. Malina, R.M. (1983). Menarche in athletes: A synthesis and hypothesis. *Annals of Human Biology,* **10,** 1-24.
27. Mier, C.M., Domenick, M.A., Turner, N.S., & Wilmore, J.H. (1996). Changes in stroke volume and maximal aerobic capacity with increased blood volume in men and women. *Journal of Applied Physiology,* **80,** 1180-1186.
28. Mier, C.M., Domenick, M.A., & Wilmore, J.H. (1997). Changes in stroke volume with β-blockade before and after 10 days of exercise training in men and women. *Journal of Applied Physiology,* **83,** 1660-1665.
29. Mittendorfer, B., & Rennie, M.J. (2006). Swings and roundabouts for muscle gain and loss: Differences between sexes? *Journal of Applied Physiology,* **100,** 375-376.
30. Otis, C.L., Drinkwater, B., Johnson, M., Loucks, A., & Wilmore, J. (1997). The female athlete triad. *Medicine and Science in Sports and Exercise,* **29**(5), i-ix.
31. Pate, R.R., Sparling, P.B., Wilson, G.E., Cureton, K.J., & Miller, B.J. (1987). Cardiorespiratory and metabolic responses to submaximal and maximal exercise in elite women distance runners. *International Journal of Sports Medicine,* **8**(Suppl. 2), 91-95.
32. Pivarnik, J.M. (1994). Maternal exercise during pregnancy. *Sports Medicine,* **18,** 215-217.
33. Pollock, M.L. (1977). Submaximal and maximal working capacity of elite distance runners: Part I. Cardiorespiratory aspects. *Annals of the New York Academy of Sciences,* **301,** 310-322.
34. Porter, M.M., Stuart, S., Boij, M., & Lexell, J. (2002). Capillary supply of the tibialis anterior muscle in young, healthy, and moderately active men and women. *Journal of Applied Physiology,* **92,** 1451-1457.
35. Proctor, D.N., Beck, K.C., Shen, P.H., Eickhoff, T.J., Halliwill, J.R., & Joyner, M.J. (1998). Influence of age and gender on cardiac output-$\dot{V}O_2$ relationships during submaximal cycle ergometry. *Journal of Applied Physiology,* **84,** 599-605.
36. Redman, L.M., & Loucks, A.B. (2005). Menstrual disorders in athletes. *Sports Medicine,* **35,** 747-755.
37. Robinson, S. (1938). Experimental studies of physical fitness in relation to age. *Arbeitsphysiologie,* **10,** 251-323.
38. Saltin, B., & Åstrand, P.-O. (1967). Maximal oxygen uptake in athletes. *Journal of Applied Physiology,* **23,** 353-358.
39. Saltin, B., Henriksson, J., Nygaard, E., & Andersen, P. (1977). Fiber types and metabolic potentials of skeletal muscles in sedentary man and endurance runners. *Annals of the New York Academy of Sciences,* **301,** 3-29.
40. Saltin, B., & Rowell, L.B. (1980). Functional adaptations to physical activity and inactivity. *Federation Proceedings,* **39,** 1506-1513.
41. Schantz, P., Randall-Fox, E., Hutchison, W., Tyden, A., & Åstrand, P.-O. (1983). Muscle fibre type distribution, muscle cross-sectional area and maximal voluntary strength in humans. *Acta Physiologica Scandinavica,* **117,** 219-226.
42. Stager, J.M., Wigglesworth, J.K., & Hatler, L.K. (1990). Interpreting the relationship between age of menarche and prepubertal training. *Medicine and Science in Sports and Exercise,* **22,** 54-58.
43. Turley, K.R., & Wilmore, J.H. (1997). Cardiovascular responses to submaximal exercise in 7- to 9-yr old boys and girls. *Medicine and Science in Sports and Exercise,* **29,** 824-832.
44. Wilmore, J.H. (1991). Eating and weight disorders in the female athlete. *International Journal of Sport Nutrition,* **1,** 104-117.
45. Wilmore, J.H., & Brown, C.H. (1974). Physiological profiles of women distance runners. *Medicine and Science in Sports,* **6,** 178-181.
46. Wilmore, J.H., Stanforth, P.R., Gagnon, J., Rice, T., Mandel, S., Leon, A.S., Rao, D.C., Skinner, J.S., & Bouchard, C. (2001). Cardiac output and stroke volume changes with endurance training: The HERITAGE

Family Study. *Medicine and Science in Sports and Exercise,* **33,** 99-106.

47. Wilmore, J.H., Wambsgans, K.C., Brenner, M., Broeder, C.E., Paijmans, I., Volpe, J.A., & Wilmore, K.M. (1992). Is there energy conservation in amenorrheic compared to eumenorrheic distance runners? *Journal of Applied Physiology,* **72,** 15-22.
48. Wolfe, L.A., Brenner, I.K.M., & Mottola, M.F. (1994). Maternal exercise, fetal well-being and pregnancy outcome. *Exercise and Sport Sciences Reviews,* **22,** 145-194.
49. Wolfe, L.A., Hall, P., Webb, K.A., Goodman, L., Monga, M., & McGrath, M.J. (1989). Prescription of aerobic exercise during pregnancy. *Sports Medicine,* **8,** 273-301.
50. Wolfe, L.A., Ohtake, P.J., Mottola, M.F., & McGrath, M.J. (1989). Physiological interactions between pregnancy and aerobic exercise. *Exercise and Sport Sciences Reviews,* **17,** 295-351.

Selected Readings

American College of Sports Medicine. (2004). ACSM position stand on physical activity and bone health. *Medicine and Science in Sports and Exercise,* **36,** 1985-1996.

Borer, K.T. (2005). Physical activity in the prevention and amelioration of osteoporosis in women: Interaction of mechanical, hormonal and dietary factors. *Sports Medicine,* **35,** 779-830.

Drinkwater, B.L. (Ed.). (2000). *Women in sport.* Oxford, UK: Blackwell Science.

Harber, V.J. (2000). Menstrual dysfunction in athletes: An energetic challenge. *Exercise and Sport Sciences Reviews,* **28,** 19-23.

Loucks, A.B. (2001). Physical health of the female athlete: Observations, effects, and causes of reproductive disorders. *Canadian Journal of Applied Physiology,* **26,** S176-S185.

Loucks, A.B., Stachenfeld, N.S., & DiPietro, L. (2006). The female athlete triad: Do female athletes need to take special care to avoid low energy availability? *Medicine and Science in Sports and Exercise,* **38,** 1694-1700.

Manore, M.M. (2002). Dietary recommendations and athletic menstrual dysfunction. *Sports Medicine,* **32,** 887-901.

Shephard, R.J. (2000). Exercise and training in women, Part I: Influence of gender on exercise and training responses; and Part II: Influence of menstrual cycle and pregnancy on exercise responses. *Canadian Journal of Applied Physiology,* **25,** 19-34, 35-54.

Sparling, P.B. (Ed.). (1987). A comprehensive profile of elite women distance runners. *International Journal of Sports Medicine,* **8**(Suppl. 2), 71-136.

Winters-Stone, K.M., & Snow, C.M. (2005). Osteoporosis. In J.S. Skinner (Ed.), *Exercise testing and exercise prescription for special cases: Theoretical basis and clinical application* (pp. 171-187). Philadelphia: Lippincott, Williams & Wilkins.

Wolfe, L.A. (2005). Pregnancy. In J.S. Skinner (Ed.), *Exercise testing and exercise prescription for special cases: Theoretical basis and clinical application* (pp. 377-391). Philadelphia: Lippincott Williams & Wilkins.

Chapter 19

References

1. American College of Sports Medicine. (1998). The recommended quantity and quality of exercise for developing and maintaining cardiorespiratory and muscular fitness, and flexibility in healthy adults. *Medicine and Science in Sports and Exercise,* **30,** 975-991.
2. American College of Sports Medicine. (2006). *Guidelines for exercise testing and prescription* (7th ed.). Philadelphia: Lippincott Williams & Wilkins.
3. Booth, F.W., Chakravarthy, M.V., Gordon, S.E., & Spangenburg, E.E. (2002). Waging war on physical inactivity: Using modern molecular ammunition against an ancient enemy. *Journal of Applied Physiology,* **93,** 3-30.
4. Booth, F.W., Gordon, S.E., Carlson, C.J., & Hamilton, M.T. (2000). Waging war on modern chronic disease: Primary prevention through exercise biology. *Journal of Applied Physiology,* **88,** 774-787.
5. Borg, G.A.V. (1998). *Borg's perceived exertion and pain scales.* Champaign, IL: Human Kinetics.
6. Byrne, N.M., Hills, A.P., Hunter, G.R., Weinsier, R.L., & Schutz, Y. (2005). Metabolic equivalent: One size does not fit all. *Journal of Applied Physiology,* **99,** 1112-1119.
7. Cooper, K.H. (1968). *Aerobics.* New York: Evans.
8. Davis, J.A., & Convertino, V.A. (1975). A comparison of heart rate methods for predicting endurance training intensity. *Medicine and Science in Sports,* **7,** 295-298.
9. Fletcher, G.F., Balady, G.J., Amsterdam, E.A., Chaitman, B., Eckel, R., Fleg, J., Froelicher, V.F., Leon, A.S., Piña, I.L., Rodney, R., Simons-Morton, D.G., Williams, M.A., & Bazzarre, T. (2001). Exercise standards for testing and training: A statement for healthcare professionals from the American Heart Association. *Circulation,* **104,** 1694-1740.
10. Fletcher, G.F., Blair, S.N., Blumenthal, J., Caspersen, C., Chaitman, B., Epstein, S., Falls, H., Froelicher, E.S.S., Froelicher, V.F., & Piña, I.L. (1992). Statement on exercise: Benefits and recommendations for physical activity programs for all Americans. *Circulation,* **86,** 340-344.
11. Karvonen, M.J., Kentala, E., & Mustala, O. (1957). The effects of training heart rate: A longitudinal study. *Annales Medicinae Experimentalis et Biologiae Fenniae,* **35,** 307-315.
12. Kirshenbaum, J., & Sullivan, R. (1983). Hold on there, America. *Sports Illustrated,* **58**(5), 60-74.
13. Lauer, M., Sivarajan Froelicher, E., Williams, M., & Kligfield, P. (2005). Exercise testing in asymptomatic adults. *Circulation,* **112,** 771-776.
14. Manini, T.M., Everhart, J.E., Patel, K.V., Schoeller, D.A., Colbert, L.H., Visser, M., Tylavsky, F., Bauer, D.C., Goodpaster, B.H., & Harris, T.B. (2006). Daily activity energy expenditure and mortality among older adults. *Journal of the American Medical Association,* **296,** 171-179.
15. National Institutes of Health, Consensus Development Panel on Physical Activity and Cardiovascular Health.

(1996). Physical activity and cardiovascular health. *Journal of the American Medical Association,* **276,** 241-246.

16. Pate, R.R., Pratt, M., Blair, S.N., Haskell, W.L., Macera, C.A., Bouchard, C., Buchner, D., Ettinger, W., Heath, G.W., King, A.C., Kriska, A., Leon, A.S., Marcus, B.H., Morris, J., Paffenbarger, R.S., Patrick, K., Pollock, M.L., Rippe, J.M., Sallis, J., & Wilmore, J.H. (1995). Physical activity and public health: A recommendation from the Centers for Disease Control and Prevention and the American College of Sports Medicine. *Journal of the American Medical Association,* **273,** 402-407.
17. Persinger, R., Foster, C., Gibson, M., Fater, D.C.W., & Porcari, J.P. (2004). Consistency of the talk test for exercise prescription. *Medicine and Science in Sports and Exercise,* **36,** 1632-1636.
18. Pollock, M.L., Franklin, B.A., Balady, G.J., Chaitman, B.L., Fleg, J.L., Fletcher, B., Limacher, M., Piña, I.L., Stein, R.A., Williams, M., & Bazzarre, T. (2000). Resistance exercise in individuals with and without cardiovascular disease: Benefits, rationale, safety and prescription. *Circulation,* **101,** 828-833.
19. Swain, D.P., & Leutholtz, B.C. (1997). Heart rate reserve is equivalent to $\%\dot{V}O_2$ reserve, not to $\%\dot{V}O_2max$. *Medicine and Science in Sports and Exercise,* **29,** 410-414.
20. U.S. Department of Health and Human Services. (1996). *Physical activity and health: A report of the Surgeon General.* Atlanta: U.S. Department of Health and Human Services, Centers for Disease Control and Prevention, National Center for Chronic Disease Prevention and Health Promotion.
21. U.S. Department of Health and Human Services. (2000, November). *Healthy people 2010: Understanding and improving health* (2nd ed.). Washington, DC: U.S. Government Printing Office.

Selected Readings

Bouchard, C., Blair, S.N., & Haskell, W.L. (2007). Physical activity and health. Champaign, IL: Human Kinetics.

Graves, J.E., & Franklin, B.A. (Eds.). (2001). *Resistance training for health and rehabilitation.* Champaign, IL: Human Kinetics.

Haskell, W.L. (1994). Health consequences of physical activity: Understanding and challenges regarding dose-response. *Medicine and Science in Sports and Exercise,* **26,** 649-660.

Heyward, V.H. (2002). *Advanced fitness assessment and exercise prescription* (4th ed.). Champaign, IL: Human Kinetics.

Mazzeo, R.S., & Tanaka, H. (2001). Exercise prescription for the elderly: Current recommendations. *Sports Medicine,* **31,** 809-818.

Meyers, J.N. (1996). *Essentials of cardiopulmonary exercise testing.* Champaign, IL: Human Kinetics.

Nieman, D.C. (1998). *The exercise-health connection.* Champaign, IL: Human Kinetics.

Roberts, S.O., Robergs, R.A., & Hanson, P. (1997). *Clinical exercise testing and prescription: Theory and application.* Boca Raton, FL: CRC Press.

Sanders, L.F., & Duncan, G.E. (2006). Population-based reference standards for cardiovascular fitness among U.S. adults: NHANES 1999-2000 and 2001-2002. *Medicine and Science in Sports and Exercise,* **38,** 701-707.

Skinner, J.S. (Ed.). (2005). *Exercise testing and exercise prescription for special cases: Theoretical basis and clinical application* (3rd ed.). Philadelphia: Lea & Febiger.

Chapter 20

References

1. Albert, C.M., Mittleman, M.A., Chae, C.U., Lee, I.-M., Hennekens, C.H., & Manson, J.E. (2000). Triggering of sudden death from cardiac causes by vigorous exertion. *New England Journal of Medicine,* **343,** 1355-1361.
2. American College of Sports Medicine Position Stand. (1994). Exercise for patients with coronary artery disease. *Medicine and Science in Sports and Exercise,* **26**(3), i-v.
3. American College of Sports Medicine Position Stand. (2004). Exercise and hypertension. *Medicine and Science in Sports and Exercise,* **36,** 533-553.
4. American Heart Association. (2006). *Heart and stroke facts.* Dallas: American Heart Association.
5. American Heart Association. (2006). Heart disease and stroke statistics—2006 update. *Circulation,* **113,** e85-e151.
6. Berenson, G.S., Srinivasan, S.R., Bao, W., Newman, W.P., Tracy, R.E., & Wattigney, W.A. (1998). Association between multiple cardiovascular risk factors and atherosclerosis in children and young adults. The Bogalusa Heart Study. *New England Journal of Medicine,* **338,** 1650-1656.
7. Blair, S.N., Goodyear, N.N., Gibbons, L.W., & Cooper, K.H. (1984). Physical fitness and incidence of hypertension in healthy normotensive men and women. *Journal of the American Medical Association,* **252,** 487-490.
8. Blair, S.N., & Jackson, A.S. (2001). Guest editorial: Physical fitness and activity as separate heart disease risk factors: A meta-analysis. *Medicine and Science in Sports and Exercise,* **33,** 762-764.
9. Blair, S.N., Kohl, H.W., Paffenbarger, R.S., Clark, D.G., Cooper, K.H., & Gibbons, L.W. (1989). Physical fitness and all-cause mortality: A prospective study of healthy men and women. *Journal of the American Medical Association,* **262,** 2395-2401.
10. Braganza, D.M., & Bennett, M.R. (2001). New insights into atherosclerotic plaque rupture. *Postgraduate Medical Journal,* **77,** 94-98.
11. Carnethon, M.R., Gulati, M., & Greenland, P. (2005). Prevalence and cardiovascular disease correlates of low cardiorespiratory fitness in adolescents and adults. *Journal of the American Medical Association,* **294,** 2981-2988.
12. Caspersen, C.J. (1987). Physical inactivity and coronary heart disease. *Physician and Sportsmedicine,* **15**(11), 43-44.
13. Conroy, M.B., Cook, N.R., Manson, J.E., Buring, J.E., & Lee, I-M. (2005). Past physical activity, current physical

activity, and risk of coronary heart disease. *Medicine and Science in Sports and Exercise,* **37,** 1251-1256.

14. Cooper, K.H., Pollock, M.L., Martin, R.P., White, S.R., Linnerud, A.C., & Jackson, A. (1976). Physical fitness levels vs. selected coronary risk factors: A cross-sectional study. *Journal of the American Medical Association,* **236,** 166-169.
15. Corti, R., Hutter, R., Badimon, J.J., & Fuster, V. (2004). Evolving concepts in the triad of atherosclerosis, inflammation and thrombosis. *Journal of Thrombosis and Thrombolysis,* **17,** 35-44.
16. Dunn, A.L., & Dishman, R.K. (1991). Exercise and the neurobiology of depression. *Exercise and Sport Sciences Reviews,* **19,** 41-98.
17. Durstine, J.L., Grandjean, P.W., Cox, C.A., & Thompson, P.D. (2002). Lipids, lipoproteins, and exercise. *Journal of Cardiopulmonary Rehabilitation,* **22,** 385-398.
18. Ehsani, A.A. (1987). Cardiovascular adaptations to endurance exercise training in ischemic heart disease. *Exercise and Sport Sciences Reviews,* **15,** 53-66.
19. Enos, W.F., Holmes, R.H., & Beyer, J. (1953). Coronary disease among United States soldiers killed in action in Korea. *Journal of the American Medical Association,* **152,** 1090-1093.
20. Franklin, B.A., & Kahn, J.K. (1996). Delayed progression or regression of coronary atherosclerosis with intensive risk factor modification: Effects of diet, drugs, and exercise. *Sports Medicine,* **22,** 306-320.
21. Gibbons, L.W., Blair, S.N., Cooper, K.H., & Smith, M. (1983). Association between coronary heart disease risk factors and physical fitness in healthy adult women. *Circulation,* **67,** 977-983.
22. Haskell, W.L., Sims, C., Myll, J., Bortz, W.M., St. Goar, F.G., & Alderman, E.L. (1993). Coronary artery size and dilating capacity in ultradistance runners. *Circulation,* **87,** 1076-1082.
23. Hu, F.B., Stampfer, M.J., Colditz, G.A., Ascherio, A., Rexrode, K.M., Willett, W.C., & Manson, J.E. (2000). Physical activity and risk of stroke in women. *Journal of the American Medical Association,* **283,** 2961-2967.
24. Joint National Committee on Prevention, Detection, Evaluation, and Treatment of High Blood Pressure. (2003). The seventh report of the Joint National Committee on Prevention, Detection, Evaluation, and Treatment of High Blood Pressure. *Journal of the American Medical Association,* **289,** 2560-2572.
25. Kannel, W.B., & Dawber, T.R. (1972). Atherosclerosis as a pediatric problem. *Journal of Pediatrics,* **80,** 544-554.
26. Knez, W.L., Coombes, J.S., & Jenkins, D.G. (2006). Ultra-endurance exercise and oxidative damage: Implications for cardiovascular health. *Sports Medicine,* **36,** 429-441.
27. Kramsch, D.M., Aspen, A.J., Abramowitz, B.M., Kreimendahl, T., & Hood, W.B. (1981). Reduction of coronary atherosclerosis by moderate conditioning exercise in monkeys on an atherogenic diet. *New England Journal of Medicine,* **305,** 1483-1489.
28. LaPorte, R.E., Adams, L.L., Savage, D.D., Brenes, G., Dearwater, S., & Cook, T. (1984). The spectrum of physical activity, cardiovascular disease and health: An epidemiologic perspective. *American Journal of Epidemiology,* **120,** 507-517.
29. Lee, C.D., & Blair, S.N. (2002). Cardiorespiratory fitness and stroke mortality in men. *Medicine and Science in Sports and Exercise,* **34,** 592-595.
30. Lee, I.-M., & Paffenbarger, R.S., Jr. (1996). Do physical activity and physical fitness avert premature mortality? *Exercise and Sport Sciences Reviews,* **24,** 135-171.
31. Leon, A.S., & Connett, J. (1991). Physical activity and 10.5 year mortality in the Multiple Risk Factor Intervention Trial (MRFIT). *International Journal of Epidemiology,* **20,** 690-697.
32. Leon, A.S., Connett, J., Jacobs, D.R., & Rauramaa, R. (1987). Leisure-time physical activity levels and risk of coronary heart disease and death. *Journal of the American Medical Association,* **258,** 2388-2395.
33. McNamara, J.J., Molot, M.A., Stremple, J.F., & Cutting, R.T. (1971). Coronary artery disease in combat casualties in Vietnam. *Journal of the American Medical Association,* **216,** 1185-1187.
34. Montoye, H.J., Metzner, H.L., Keller, J.B., Johnson, B.C., & Epstein, F.H. (1972). Habitual physical activity and blood pressure. *Medicine and Science in Sports and Exercise,* **4,** 175-181.
35. Morgan, W.P. (1994). Physical activity, fitness and depression. In C. Bouchard, R.J. Shephard, & T. Stephens (Eds.), *Physical activity, fitness, and health* (pp. 851-867). Champaign, IL: Human Kinetics.
36. Morris, J.N., Adam, C., Chave, S.P.W., Sirey, C., Epstein, L., & Sheehan, D.J. (1973). Vigorous exercise in leisure-time and the incidence of coronary heart disease. *Lancet,* **1,** 333-339.
37. Morris, J.N., Heady, J.A., Raffle, P.A.B., Roberts, C.G., & Parks, J.W. (1953). Coronary heart-disease and physical activity of work. *Lancet,* **265,** 1053-1057, 1111-1120.
38. Morris, J.N., Pollard, R., Everitt, M.G., Chave, S.P.W., & Semmence, A.M. (1980). Vigorous exercise in leisure-time: Protection against coronary heart disease. *Lancet,* **2,** 1207-1210.
39. Paffenbarger, R.S., Hyde, R.T., Wing, A.L., & Hsieh, C.-C. (1986). Physical activity, all-cause mortality, and longevity of college alumni. *New England Journal of Medicine,* **314,** 605-613.
40. Petersen, A.M.W., & Pedersen, B.K. (2005). The anti-inflammatory effect of exercise. *Journal of Applied Physiology,* **98,** 1154-1162.
41. Petruzzello, S.J., Landers, D.M., Hatfield, B.D., Kubitz, K.A., & Salazar, W. (1991). A meta-analysis on the anxiety-reducing effects of acute and chronic exercise: Outcomes and mechanisms. *Sports Medicine,* **11,** 143-182.
42. Powell, K.E., Thompson, P.D., Caspersen, C.J., & Kendrick, J.S. (1987). Physical activity and the incidence of

coronary heart disease. *Annual Reviews in Public Health,* **8,** 253-287.

43. Ross, R. (1986). The pathogenesis of atherosclerosis—an update. *New England Journal of Medicine,* **314,** 488-500.
44. Ross, R. (1999). Atherosclerosis—an inflammatory disease. *New England Journal of Medicine,* **340,** 115-126.
45. Siscovick, D.S., Weiss, N.S., Fletcher, R.H., & Lasky, T. (1984). The incidence of primary cardiac arrest during vigorous exercise. *New England Journal of Medicine,* **311,** 874-877.
46. Tanasescu, M., Leitzmann, M.F., Rimm, E.B., Willett, W.C., Stampfer, M.J., & Hu, F.B. (2002). Exercise type and intensity in relation to coronary heart disease in men. *Journal of the American Medical Association,* **288,** 1994-2000.
47. Taylor, R.S., Brown, A., Ebrahim, S., Jolliffe, J., Noorani, H., Rees, K., Skidmore, B., Stone, J.A., Thompson, D.R., & Oldridge, N. (2004). Exercise-based rehabilitation for patients with coronary heart disease: Systematic review and meta-analysis of randomized controlled trials. *American Journal of Medicine,* **116,** 682-692.
48. Thompson, P.D. (1982). Cardiovascular hazards of physical activity. *Exercise and Sport Sciences Reviews,* **10,** 208-235.
49. Walther, C., Gielen, S., & Hambrecht, R. (2004). The effect of exercise training on endothelial function in cardiovascular disease in humans. *Exercise and Sport Sciences Reviews,* **32,** 129-134.
50. Whang, W., Manson, J.E., Hu, F.B., Chae, C.U., Rexrode, K.M., Willett, W.C., Stampfer, M.J., & Albert, C.M. (2006). Physical exertion, exercise, and sudden cardiac death in women. *Journal of the American Medical Association,* **295,** 1399-1403.
51. Williams, P.T. (2001). Physical fitness and activity as separate heart disease risk factors: A meta-analysis. *Medicine and Science in Sports and Exercise,* **33,** 754-761.
52. Wilmore, J.H., Constable, S.H., Stanforth, P.R., Tsao, W.Y., Rotkis, T.C., Paicius, R.M., Mattern, C.M., & Ewy, G.A. (1982). Prevalence of coronary heart disease risk factors in 13- to 15-year-old boys. *Journal of Cardiac Rehabilitation,* **2,** 223-233.
53. Wilmore, J.H., & McNamara, J.J. (1974). Prevalence of coronary heart disease risk factors in boys 8 to 12 years of age. *Journal of Pediatrics,* **84,** 527-533.

Selected Readings

American College of Sports Medicine. (2006). *ACSM's guidelines for exercise testing and prescription* (7th ed.). Baltimore: Lippincott Williams & Wilkins.

American Heart Association. (1996). Cardiovascular preparticipation screening of competitive athletes: Exercise, sudden death. *Medicine and Science in Sports and Exercise,* **28,** 1445-1452.

American Heart Association. (2003). Exercise and physical activity in the prevention and treatment of atherosclerotic cardiovascular disease. *Circulation,* **107,** 3109-3116.

American Heart Association. (2005). Cardiac rehabilitation and secondary prevention of coronary heart disease. *Circulation,* **111,** 369-376.

Caspersen, C.J. (1989). Physical activity epidemiology: Concepts, methods, and applications to exercise science. *Exercise and Sport Sciences Reviews,* **17,** 423-473.

Franklin, B.A., & Gordon, N.F. (2005). *Contemporary diagnosis and management in cardiovascular exercise.* Newton, PA: Handbooks in Health Care.

Franklin, B.A., & Shephard, R.J. (2000). Avoiding repeat cardiac events. *Physician and Sportsmedicine,* **28**(9), 31-58.

Harris, S.S., Caspersen, C.J., DeFriese, G.H., & Estes, E.H., Jr. (1989). Physical activity counseling for healthy adults as a primary preventive intervention in the clinical setting. *Journal of the American Medical Association,* **261,** 3590-3598.

Leon, A.S. (Ed.). (1997). *Physical activity and cardiovascular health: A national consensus.* Champaign, IL: Human Kinetics.

Wannamethee, S.G., & Shaper, A.G. (2001). Physical activity in the prevention of cardiovascular disease: An epidemiological perspective. *Sports Medicine,* **31,** 101-114.

Chapter 21

References

1. Achten, J., Gleeson, M., & Jeukendrup, A.E. (2002). Determination of the exercise intensity that elicits maximal fat oxidation. *Medicine and Science in Sports and Exercise,* **34,** 92-97.
2. Barlow, C.E., Kohl, H.W., III, Gibbons, L.W., & Blair, S.N. (1995). Physical fitness, mortality and obesity. *International Journal of Obesity,* **19**(Suppl. 4), 41-44.
3. Bassuk, S.S., & Manson, J.E. (2005). Epidemiological evidence for the role of physical activity in reducing risk of type 2 diabetes and cardiovascular disease. *Journal of Applied Physiology,* **99,** 1193-1204.
4. Bouchard, C. (1991). Heredity and the path to overweight and obesity. *Medicine and Science in Sports and Exercise,* **23,** 285-291.
5. Bouchard, C., Tremblay, A., Després, J.-P., Nadeau, A., Lupien, P.J., Theriault, G., Dussault, J., Moorjani, S., Pinault, S., & Fournier, G. (1990). The response to long-term overfeeding in identical twins. *New England Journal of Medicine,* **322,** 1477-1482.
6. Bray, G.A. (1985). Obesity: Definition, diagnosis and disadvantages. *Medical Journal of Australia,* **142,** S2-S8.
7. Broeder, C.E., Burrhus, K.A., Svanevik, L.S., & Wilmore, J.H. (1992). The effects of either high intensity resistance or endurance training on resting metabolic rate. *American Journal of Clinical Nutrition,* **55,** 802-810.
8. Chisholm, D.J. (1992). Diabetes mellitus. In J. Bloomfield, P.A. Fricker, & K.D. Fitch (Eds.), *Textbook of science*

and medicine in sport (pp. 555-561). Boston: Blackwell Scientific.

9. Daly, P.A., & Landsberg, L. (1991). Hypertension in obesity and NIDDM: Role of insulin and sympathetic nervous system. *Diabetes Care,* **14,** 240-248.
10. Flegal, K.M., Carroll, M.D., Kuczmarski, R.J., & Johnson, C.L. (1998). Overweight and obesity in the United States: Prevalence and trends, 1960-1994. *International Journal of Obesity,* **22,** 39-47.
11. Flegal, K.M., Carroll, M.D., Ogden, C.L., & Johnson, C.L. (2002). Prevalence and trends in obesity among US adults, 1999-2000. *Journal of the American Medical Association,* **288,** 1723-1727.
12. Ford, E.S., Giles, W.H., & Dietz, W.H. (2002). Prevalence of the metabolic syndrome among US adults. *Journal of the American Medical Association,* **287,** 356-359.
13. Gwinup, G., Chelvam, R., & Steinberg, T. (1971). Thickness of subcutaneous fat and activity of underlying muscles. *Annals of Internal Medicine,* **74,** 408-411.
14. Harnack, L.J., Jeffery, R.W., & Boutelle, K.N. (2000). Temporal trends in energy intake in the United States: An ecologic perspective. *American Journal of Clinical Nutrition,* **71,** 1478-1484.
15. Holloszy, J.O. (2005). Exercise-induced increase in muscle insulin sensitivity. *Journal of Applied Physiology,* **99,** 338-343.
16. Ivy, J.L. (1987). The insulin-like effect of muscle contraction. *Exercise and Sport Sciences Reviews,* **15,** 29-51.
17. Katch, F.I., Clarkson, P.M., Kroll, W., McBride, T., & Wilcox, A. (1984). Effects of sit up exercise training on adipose cell size and adiposity. *Research Quarterly for Exercise and Sport,* **55,** 242-247.
18. Keesey, R.E. (1986). A set-point theory of obesity. In K.D. Brownell & J.P. Foreyt (Eds.), *Handbook of eating disorders: Physiology, psychology, and treatment of obesity, anorexia, and bulimia* (pp. 63-87). New York: Basic Books.
19. Keys, A., Brozek, J., Henschel, A., Mickelsen, O., & Taylor, H.L. (1950). *The biology of human starvation.* Minneapolis: University of Minnesota Press.
20. Knowler, W.C., Barrett-Connor, E., Fowler, S.E., Hamman, R.F., Lachin, J.M., Walker, E.A., & Nathan, D.M. (2002). Reduction in the incidence of type 2 diabetes with lifestyle intervention or metformin. *New England Journal of Medicine,* **346,** 393-403.
21. Levine, J.A., Lanningham-Foster, L.M., McCrady, S.K., Krizan, A.C., Olson, L.R., Kane, P.H., Jensen, M.D., & Clark, M.M. (2005). Interindividual variation in posture allocation: Possible role in human obesity. *Science,* **307,** 584-586.
22. Ludwig, D.S., & Ebbeling, C.B. (2001). Type 2 diabetes mellitus in children. *Journal of the American Medical Association,* **286,** 1426-1430.
23. Mayer, J., Marshall, N.B., Vitale, J.J., Christensen, J.H., Mashayekhi, M.B., & Stare, F.J. (1954). Exercise, food intake, and body weight in normal rats and genetically obese adult mice. *American Journal of Physiology,* **177,** 544-548.
24. National Institutes of Health. (2000). *The practical guide: Identification, evaluation, and treatment of overweight and obesity in adults* (NIH Publication No. 00-4084). Washington, DC: U.S. Department of Health and Human Services.
25. Ogden, C.L., Carroll, M.D., Curtin, L.R., McDowell, M.A., Tabak, C.J., & Flegal, K.M. (2006). Prevalence of overweight and obesity in the United States, 1999-2004. *Journal of the American Medical Association,* **295,** 1549-1555.
26. Oscai, L.B. (1973). The role of exercise in weight control. *Exercise and Sport Sciences Reviews,* **1,** 103-123.
27. Poehlman, E.T. (1989). A review: Exercise and its influence on resting energy metabolism in man. *Medicine and Science in Sports and Exercise,* **21,** 515-525.
28. Schulz, L.O., Bennett, P.H., Ravussin, E., Kidd, J.R., Kidd, K.K., Esparza, J., & Valencia, M.E. (2006). Effects of traditional and western environments on prevalence of type 2 diabetes in Pima Indians in Mexico and the U.S. *Diabetes Care,* **29,** 1866-1871.
29. Seidell, J.C., Deurenberg, P., & Hautvast, J.G.A.J. (1987). Obesity and fat distribution in relation to health—current insights and recommendations. *World Review of Nutrition and Dietetics,* **50,** 57-91.
30. Sims, E.A.H. (1976). Experimental obesity, dietary-induced thermogenesis and their clinical implications. *Clinics in Endocrinology and Metabolism,* **5,** 377-395.
31. Slentz, C.A., Aiken, L.B., Houmard, J.A., Bales, C.W., Johnson, J.L., Tanner, C.J., Duscha, B.D., & Kraus, W.E. (2005). Inactivity, exercise and visceral fat. STRRIDE: A randomized, controlled study of exercise intensity and amount. *Journal of Applied Physiology,* **99,** 1613-1618.
32. Tremblay, A., Nadeau, A., Fournier, G., & Bouchard, C. (1988). Effect of a three-day interruption of exercise-training on resting metabolic rate and glucose-induced thermogenesis in trained individuals. *International Journal of Obesity,* **12,** 163-168.
33. Troiano, R.P., Flegal, K.M., Kuczmarski, R.J., Campbell, S.M., & Johnson, C.L. (1995). Overweight prevalence and trends for children and adolescents: The National Health and Nutrition Examination Surveys, 1963 to 1991. *Archives of Pediatric Adolescent Medicine,* **149,** 1085-1091.
34. Vitug, A., Schneider, S.H., & Ruderman, N.B. (1988). Exercise and type I diabetes mellitus. *Exercise and Sport Sciences Reviews,* **16,** 285-304.
35. Welk, G.J., & Blair, S.N. (2000). Physical activity protects against the health risks of obesity. *Research Digest: President's Council on Physical Fitness and Sports,* **3**(12), 1-6.
36. Wilmore, J.H. (1996). Increasing physical activity: Alterations in body mass and composition. *American Journal of Clinical Nutrition,* **63,** 456S-460S.
37. Wilmore, J.H., Atwater, A.E., Maxwell, B.D., Wilmore, D.L., Constable, S.H., & Buono, M.J. (1985). Alterations in body size and composition consequent to Astro-Trimmer and Slim-Skins training programs. *Research Quarterly for Exercise and Sport,* **56,** 90-92.

38. Wilmore, J.H., Atwater, A.E., Maxwell, B.D., Wilmore, D.L., Constable, S.H., & Buono, M.J. (1985). Alterations in breast morphology consequent to a 21-day bust developer program. *Medicine and Science in Sports and Exercise,* **17,** 106-112.

39. Wilmore, J.H., Stanforth, P.R., Hudspeth, L.A., Gagnon, J., Daw, E.W., Leon, A.S., Rao, D.C., Skinner, J.S., & Bouchard, C. (1998). Alterations in resting metabolic rate as a consequence of 20-wk of endurance training: The HERITAGE Family Study. *American Journal of Clinical Nutrition,* **68,** 66-71.

40. World Health Organization. (1998). *Obesity: Preventing and managing the global epidemic. Report of a WHO consultation on obesity.* Geneva: WHO.

41. Yaspelkis, B.B. (2006). Resistance training improves insulin signaling and action in skeletal muscle. *Exercise and Sport Sciences Reviews,* **34,** 42-46.

Selected Readings

American College of Sports Medicine Position Stand. (2000). Exercise and type 2 diabetes. *Medicine and Science in Sports and Exercise,* **32,** 1345-1360.

American College of Sports Medicine Position Stand. (2001). Appropriate intervention strategies for weight loss and prevention of weight regain in adults. *Medicine and Science in Sports and Exercise,* **33,** 2145-2156.

Grundy, S.M., Blackburn, G., Higgins, M., Lauer, R., Perri, M.G., & Ryan, D. (1999). Physical activity in the prevention and treatment of obesity and its comorbidities: Evidence report of independent panel to assess the role of physical activity in the treatment of obesity and its comorbidities. *Medicine and Science in Sports,* **31,** 1493-1500.

Hill, J.O., & Wyatt, H.R. (2005). Role of physical activity in preventing and treating obesity. *Journal of Applied Physiology,* **99,** 765-770.

Jakicic, J.M., & Otto, A.D. (2006). Treatment and prevention of obesity: What is the role of exercise? *Nutrition Reviews,* **64,** S57-S61.

LaMonte, M.J., & Blair, S.N. (2006). Physical activity, cardiorespiratory fitness, and adiposity: Contributions to disease risk. *Current Opinion in Clinical Nutrition and Metabolic Care,* **9,** 540-546.

LaMonte, M.J., Blair, S.N., & Church, T.S. (2005). Physical activity and diabetes prevention. *Journal of Applied Physiology,* **99,** 1205-1213.

Ross, R., Freeman, J.A., & Janssen, I. (2000). Exercise alone is an effective strategy for reducing obesity and related comorbidities. *Exercise and Sport Sciences Reviews,* **28,** 165-170.

Sigal, R.J., Wasserman, D.H., Kenny G.P., & Castaneda-Sceppa, C. (2004). Physical activity/exercise and type 2 diabetes. *Diabetes Care,* **27,** 2518-2539.

Stiegler, P., & Cunliffe, A. (2006). The role of diet and exercise for the maintenance of fat-free mass and resting metabolic rate during weight loss. *Sports Medicine,* **36,** 239-262.

Volek, J.S., VanHeest, J.L., & Forsythe, C.E. (2005). Diet and exercise for weight loss. *Sports Medicine,* **35,** 1-9.

Index

Note: The italicized *f* and *t* following page numbers refer to figures and tables, respectively.

R

About the Authors

Jack H. Wilmore, PhD, is the Margie Gurley Seay Centennial professor emeritus of the department of kinesiology and health education at the University of Texas at Austin. He retired in 2003 from Texas A&M University as a distinguished professor in the department of health and kinesiology. From 1985 to 1997, Wilmore was the chair of the department of kinesiology and health education at the University of Texas at Austin. During that time he was also a Margie Gurley Seay Endowed Centennial professor. Prior to that, he served on the faculties at the University of Arizona, the University of California, and Ithaca College. Wilmore earned his PhD in physical education from the University of Oregon in 1966.

Wilmore has published 53 chapters, more than 320 peer-reviewed research papers, and 15 books on exercise physiology. He is one of five principal investigators for the HERITAGE Family Study, a large multicenter clinical trial investigating the possible genetic basis for the variability in the responses of physiological measures and risk factors for cardiovascular disease and type 2 diabetes to endurance exercise training. Wilmore's research interests include determining the role of exercise in the prevention and control of both obesity and coronary heart disease. He is also interested in determining the mechanisms accounting for alterations in physiological function with training and detraining and factors limiting the performance of elite athletes.

A former president of the American College of Sports Medicine, Wilmore was the recipient of the American College of Sports Medicine's Honor Award in 2006. In addition to serving as chair for many ACSM organizational committees, Wilmore served on the United States Olympic Committee's Sports Medicine Council and chaired their Research Committee. He is currently a member of the American Physiological Society and a fellow and former president of the American Academy of Kinesiology and Physical Education.

Wilmore has served as a consultant for several professional sports teams, the California Highway Patrol, the President's Council on Physical Fitness and Sport, NASA, and the U.S. Air Force. He has served on editorial boards for journals such as *Medicine and Science in Sports and Exercise, International Journal of Obesity, Sports Medicine, Journal of Pediatric Exercise Science, Journal of Sports Nutrition, Physician and Sportsmedicine,* and *Clinical Exercise Physiology.*

In his free time Wilmore enjoys Bible study, running, walking, and playing with his grandchildren. He and his wife, Dottie, have three daughters (Wendy, Kristi, and Melissa) and six grandchildren.

David L. Costill, PhD, is the emeritus John and Janice Fisher chair in exercise science at Ball State University in Muncie, Indiana. He established the Ball State University Human Performance Laboratory in 1966 and served as its director for over 32 years.

Costill has written and coauthored more than 400 publications over the course of his career, including books, peer-reviewed articles, and lay publications. He served as the editor in chief of the *International Journal of Sports Medicine* for 12 years. Between 1971 and 1998, he averaged 25 U.S. and international lecture trips each year. He was president of the ACSM from 1976 to 1977, a member of its board of trustees for 12 years, and a recipient of ACSM Citation and Honor Awards. Many of his former students are now leaders in the field of exercise physiology.

Costill received his PhD in physical education and physiology from Ohio State University in 1965. He and his wife, Judy, have two daughters, Jill and Holly. In his leisure time, Costill is a private pilot, experimental airplane builder, competitive masters swimmer, and runner.

W. Larry Kenney, PhD, is a professor of physiology and kinesiology at Pennsylvania State University in University Park, Pennsylvania. Working at Penn State's Noll Laboratory, Kenney is currently researching the effects of aging and elevated cholesterol on the control of blood flow in human skin. He is also studying the effects of heat and dehydration on the skill performance of athletes and the effects of heat and cold on health and well-being as well as exercise and sport performance.

Kenney served as president of the American College of Sports Medicine from 2003 to 2004 and is currently the chair of the Gatorade Sports Science Institute in Barrington, Illinois. He is a fellow of the American College of Sports Medicine, a fellow of the American Academy of Kinesiology and Physical Education, and a member of the American Physiological Society.

For his service to the university and his field, Kenney has been awarded Penn State University's Faculty Scholar Medal, the Evan G. and Helen G. Pattishall Distinguished Research Career Award, and the Pauline Schmitt Russell Distinguished Research Career Award.

Kenney is a member of the editorial and advisory boards for several journals, including *Medicine and Science in Sports and Exercise, Current Sports Medicine Reports* (inaugural board member), and *Exercise and Sport Sciences Reviews.* He has also served on the editorial and advisory boards of the *Journal of Applied Physiology, Human Performance, Fitness Management,* and *ACSM's Health & Fitness Journal* (inaugural board member).

Kenney received his PhD in physiology from Penn State University in 1983. He and his wife, Patti, have three children: Matt, Alex, and Lauren. In his free time he enjoys golfing, running, and coaching youth sports.